Theory of MACHINES

(IN SI UNITS)

S S RATTAN

Department of Mechanical Engineering
Regional Engineering College
Kurukshetra

Tata McGraw-Hill Publishing Company Limited

NEW DELHI

McGraw-Hill Offices

New Delhi New York St Louis San Francisco Auckland Bogotá Guatemala Hamburg Lisbon London Madrid Mexico Milan Montreal Panama Paris San Juan São Paulo Singapore Sydney Tokyo Toronto

To the memory of
My Father

Tata McGraw-Hill

23rd reprint 2004
RADYCDRIRZBQB

ISBN 0-07-460320-5

Published by the Tata McGraw-Hill Publishing Company Limited, 7 West Patel Nagar, New Delhi 110 008 and printed at Adarsh Printers, Naveen Shahdara, Delhi 110 032.

The McGraw-Hill Companies

PREFACE

Mechanisms and machines have considerable fascination for most students of mechanical engineering since the theoretical principles involved have immediate application to practical problems. The main objective of writing this book has been to present the concepts in a logical, innovative and lucid manner. The basic theory presented in the book has been evolved out of simple and readily understood principles. A sincere effort has been made to maintain physical concepts in the various derivations. The use of simple mathematical methods instead of more elegant but less obvious methods has been preferred so that those with limited mathematical skill can easily understand the exposition.

An effort has been made to give a balanced presentation of the graphical and algebraic approaches. The method chosen is that which gives a sufficiently accurate solution with the least labour. Graphical techniques have been used frequently along with the simple algebraic methods as they usually economise on effort and time. The graphical differentiation technique given in the chapter on acceleration analysis would be of considerable help to the students in their practical classes.

Computers are increasingly playing a significant role in the modern world of advanced technology. Some computer-aided problem-solving techniques have been given in the chapters on computer-aided analysis and synthesis of mechanisms.

A number of worked examples have been included to reinforce the concepts. The International System of Units (SI) has been adopted throughout the book. A bibliography at the end has been added for the convenience of students interested in additional information.

An author is always helped and influenced by many during the writing of his book. Dr N P Mehta, Dr O N Kaul and Dr B S Gill of Regional Engineering College, Kurukshetra, are only a few of many whose help I would like to acknowledge.

It is but natural that some errors creep into a work of such volume. I would appreciate if any errors and shortcomings are brought to my knowledge. Suggestions for the improvement of the book would be also welcome.

Finally, I would like to make an affectionate acknowledgement to my wife Neena, and my children Ravneet and Jasmeet, for putting up with it all so cheerfully.

S S Rattan

CONTENTS

SYMBOLS AND ABBREVIATIONS

(Bold face letters indicate vector quantities. The same letters in italic indicate their magnitudes)

a	area, addendum
ab	velocity of B relative to A
A	area, amplitude
b	width
c	damping coefficient
C, **C**	couple
C	centre distance
d	diameter, pitch diameter, length
D	diameter, operator
e	eccentricity, maximum fluctuation of energy, piston offset
E	Young's modulus, effort, energy
f, **f**	acceleration
f	frequency, force, stress
F, **F**	force
F	degree of freedom, friction
g	acceleration due to gravity
G	gear ratio, modulus of rigidity, centre of mass
h	length, height, lift
I	mass moment of inertia, instantaneous centre
J	polar moment of inertia
k	radius of gyration
K	coefficient of fluctuation of speed, ratio
l	length
L	length, number of loops, lead
m	mass, module
M	mass
n	number, ratio, speed
N	rotational speed (rpm), number of links
p	pitch, pressure, circular pitch
P	power, pressure, diametral pitch, planet gear, number of pairs
q	torsional stiffness
r, **r**	radius
r	crank length, length

R, $\mathbf{R}$	reaction
R	radius
s	displacement, stiffness, length
S	slip
t	time, thickness, number of teeth
T	time period, number of teeth
T, $\mathbf{T}$	torque, tension
u	displacement
v, $\mathbf{v}$	velocity
w	width
w, $\mathbf{w}$	weight
x	displacement
$\dot{x}$	velocity (derivative of x with respect to time)
$\ddot{x}$	acceleration
x, y, z	cartesian coordinates
X	displacement
y	deflection, amplitude
θ	angle, angle of lap, angle between shafts
$\dot{\theta}$	angular velocity
$\ddot{\theta}$	angular acceleration
α	angle, angle of approach
α, $\boldsymbol{\alpha}$	angular acceleration
β	angle, angle of recess
γ	angle, pitch angle
ψ	angle, helix angle
δ	angle, angle of dwell, angle of action, increment of a quantity, logarithmic decrement
μ	coefficient of friction
ω, ω	angular velocity (rad/s)
ρ	mass density
λ	angle, lead angle of worm
ε	transmissibility
ζ	damping factor
φ	angle, pressure angle, angle of friction
π	3.1416
η	efficiency
Δ	displacement
Σ	sum of quantities

Subscripts

1, 2 etc. number of degrees of freedom

a	ascent approach, addendum, axial arm
ab	A relative to B
A	annular gear
b	binary link, base circle
c	cam, connecting rod, centrifugal, centripetal, crossed belt,

	countermass, critical
d	descent, damped
e	engine
f	flank, friction
g	gear, gyroscope, gravity
h	horizontal
i	inner, input, inertia
m	mean
min	minimum
max	maximum
n	nose, normal, natural
o	initial value, output, outer, open belt, other links, without friction
p	primary, pinion, piston, crankpin, pitch line
P	planet
r	roller, recess, rack, rod, radial, resultant
s	sleeve, spring, secondary
ss	steady state
S	sum
t	tensile, ternary link, tangential
v	vertical
w	wheel
x	x component
y	y component
z	z component

Superscripts

c	centripetal
l	leading
t	trailing, tangential, total
s	sliding
cr	coriolis

Abbreviations

CF	complimentary function
IC	internal combustion
IDC	inner-dead centre
KE	kinetic energy
MF	magnification factor
MOI	moment of inertia
ODC	outer-dead centre
PE	potential energy
SE	strain energy
SHM	simple-harmonic motion
TF	transfer function
VR	velocity ratio

1

MECHANISMS AND MACHINES

1.1 INTRODUCTION

The study of mechanisms involves their analysis as well as synthesis. Analysis is the study of motions and forces concerning different parts of an existing mechanism while synthesis involves the design of the different parts. In a mechanism, the various parts are so proportioned and related that the motion of one imparts requisite motions to the others, and the parts are able to withstand the forces impressed upon them. But the study of the relative motions of the parts is independent of their strength and absolute dimensions.

In a reciprocating engine, the displacement of the piston depends upon the lengths of the connecting rod and the crank (Fig.1.1). It is independent of the bearing strength of the parts or whether they are able to withstand the forces or not. Thus for the study of motions, it is immaterial if a machine part is made of mild steel, cast iron or wood. Also, it is not necessary to know the actual shape and area of cross-section of the part.

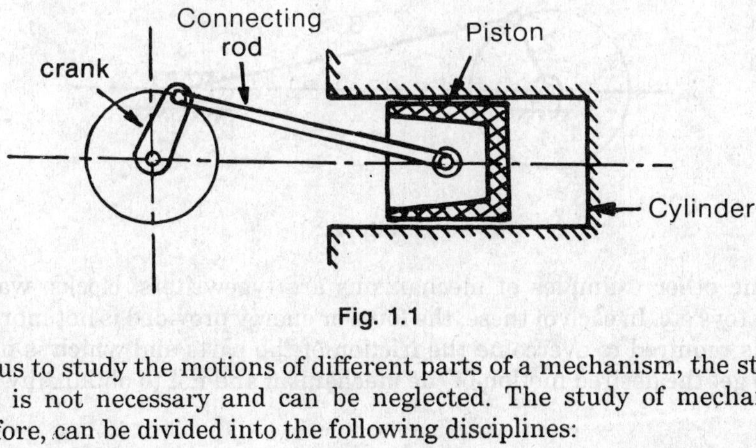

Fig. 1.1

Thus to study the motions of different parts of a mechanism, the study of forces is not necessary and can be neglected. The study of mechanisms, therefore, can be divided into the following disciplines:

Kinematics

It deals with the relative motions of different parts of a mechanism without taking into consideration the forces producing the motions. Thus it is the study, from a geometric point of view, to know the displacement, velocity and acceleration of a part of a mechanism.

Dynamics

It involves the calculations of forces impressed upon different parts of a mechanism. The forces can be either static or dynamic.

1.2 MECHANISM AND MACHINE

If a number of bodies (usually rigid) are assembled in such a way that the motion of one causes constrained and predictable motions to the others, it is known as a mechanism. Thus, mechanism transmits and modifies a motion.

A machine is a mechanism or a combination of mechanisms which, apart from imparting definite motions to the parts, also transmits and modifies the available mechanical energy into some kind of desired work. It is neither a source of energy nor a producer of work but helps in proper utilisation of the same. The motive power has to be derived from external sources.

A slider-crank mechanism (Fig.1.2) converts the reciprocating motion of a slider into a rotary motion of the crank or vice-versa. However, when it is used as an automobile engine by adding valve mechanism etc. it becomes a machine which converts the available energy (force on the piston) into the desired energy (torque of the crank-shaft). The torque is used to move a vehicle. Reciprocating pumps, reciprocating compressors and steam engines are other examples of machines derived from the slider-crank mechanism.

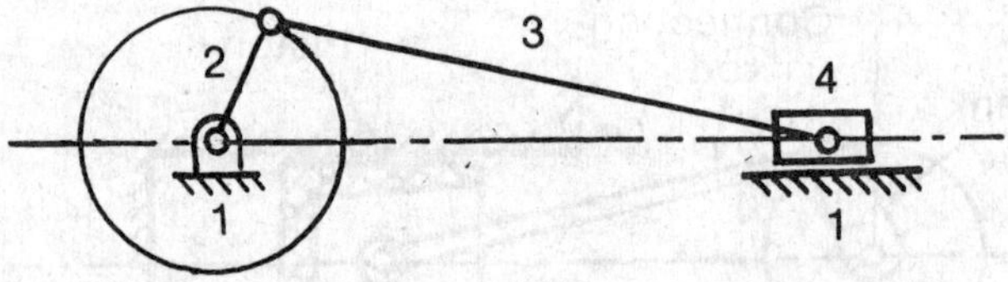

Fig. 1.2

Some other examples of mechanisms are typewriters, clocks, watches, spring toys etc. In each of these, the force or energy provided is not more than what is required to overcome the friction of the parts and which is utilised just to get the desired motion of the mechanism and not to obtain any useful work.

1.3 RIGID AND RESISTANT BODIES

A body is said to be rigid if under the action of forces, it does not suffer any distortion or the distance between any two points on it remains constant.

Resistant bodies are those which are rigid for the purposes they have to serve. Apart from rigid bodies, there are some semi-rigid bodies which are normally flexible, but under certain loading conditions act as rigid bodies for the limited purpose and thus are resistant bodies. A belt is rigid when subjected to tensile forces. Therefore, the belt in a belt-drive acts as a resistant body. Similarly, fluids can also act as resistant bodies when compressed as in case of a hydraulic press. For some purposes, springs are also resistant bodies.

These days resistant bodies are usually referred to as rigid bodies.

1.4 LINK

A mechanism is made of a number of resistant bodies out of which some may have motions relative to the others. A resistant body or a group of resistant bodies with rigid connections preventing their relative movement is known as a *link*. A link may also be defined as a member or a combination of members of a mechanism, connecting other members and having motion relative to them. Thus a link may consist of one or more resistant bodies. A slider-crank mechanism consists of four links: frame and guides, crank, connecting-rod and slider. However, the frame may consist of bearings for the crankshaft. The crank link may have crankshaft and flywheel also, forming one link having no relative motion of these.

A link is also known as *kinematic link* or *element*.

Links can be classified into binary, ternary, quaternary, etc. depending upon its ends on which revolute or turning pairs (Sec. 1.5) can be placed. The links shown in Fig.1.3 are rigid links and there is no relative motion between the joints within the link.

Fig. 1.3

1.5 KINEMATIC PAIR

A kinematic pair or simply a pair is a joint of two links having relative motion between them. In a slider-crank mechanism (Fig.1.2), link 2 rotates relative to link 1 and constitutes a revolute or turning pair. Similarly, links 2, 3 and 3, 4 constitute turning pairs. Link 4 (slider) reciprocates relative to link 1 and is a sliding pair.

Types of Kinematic Pairs

Kinematic pairs can be classified according to

– nature of contact
– nature of mechanical constraint
– nature of relative motion.

Kinematic Pairs according to Nature of Contact

(a) Lower Pair A pair of links having surface or area contact between the members is known as a lower pair. The contact surfaces of the two links are similar.

Examples Nut turning on a screw, shaft rotating in a bearing, all pairs of a slider-crank mechanism, universal joint, etc.

(b) Higher Pair When a pair has a point or line contact between the links, it is known as a higher pair. The contact surfaces of the two links are dissimilar.

Examples Wheel rolling on a surface, cam and follower pair, tooth gears, ball and roller bearings, etc.

Kinematic Pairs according to Nature of Mechanical Constraint

(a) Closed Pair When the elements of a pair are held together mechanically, it is known as a closed pair. The two elements are geometrically identical; one is solid and full and the other is hollow or open. The latter not only envelops the former but also encloses it. The contact between the two can be broken only by destruction of at least one of the members.

All the lower pairs and some of the higher pairs are closed pairs. A cam and follower pair (higher pair) shown in Fig. 1.4a and a screw pair (lower pair) belong to the closed pair category.

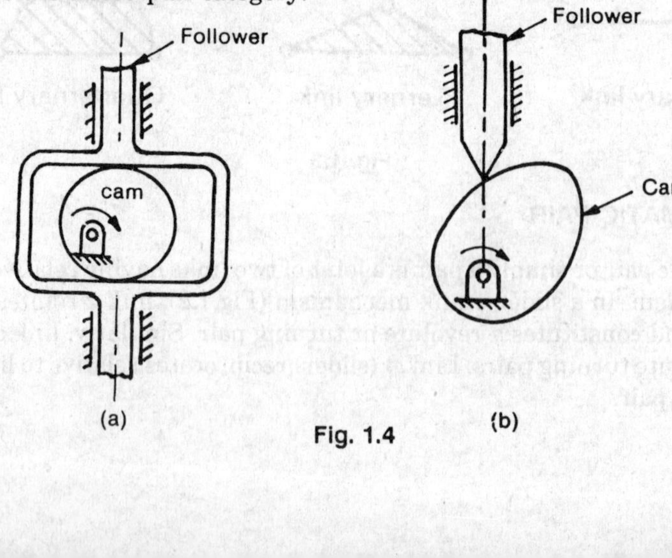

(a) (b)

Fig. 1.4

(b) Unclosed Pair When two links of a pair are in contact either due to force of gravity or some spring action, they constitute an unclosed pair. In this, the links are not held together mechanically, e.g. cam and follower pair of Fig.1.4b.

Kinematic Pairs according to Nature of Relative Motion

(a) Sliding Pair If two links have a sliding motion relative to each other, they form a sliding pair.

A rectangular rod in a rectangular hole in a prism is a sliding pair (Fig.1.5a).

(b) Turning Pair When one link has a turning or revolving motion relative to the other, they constitute a turning or revolving pair (Fig.1.5b).

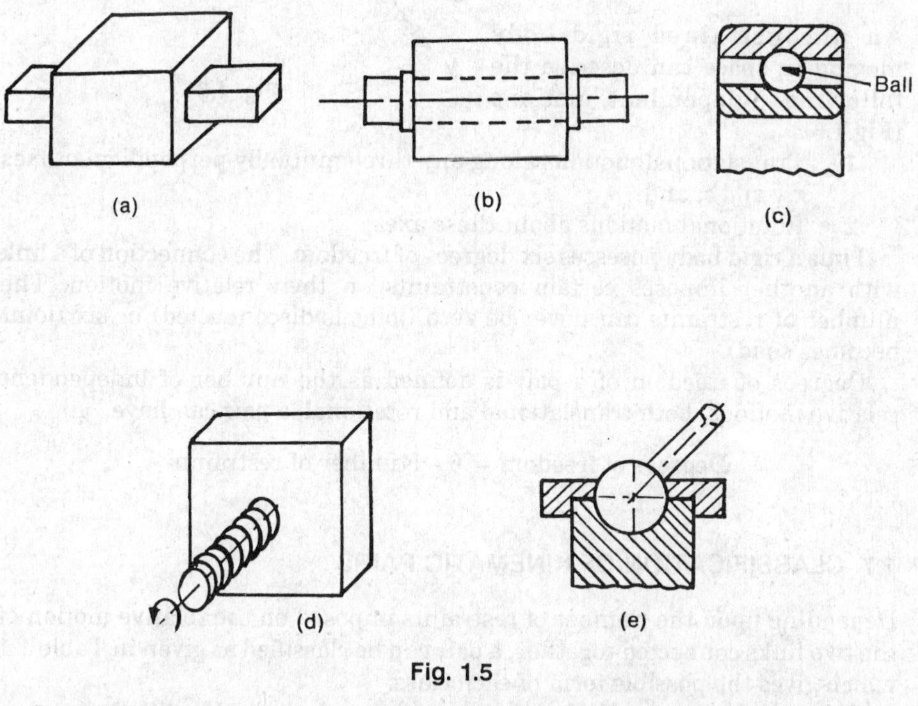

Fig. 1.5

In a slider-crank mechanism, all pairs except the slider and guide pair are turning pairs. A circular shaft revolving inside a bearing is a turning pair.

(c) Rolling Pair When the links of a pair have a rolling motion relative to each other, they form a rolling pair, e.g. a rolling wheel on a flat surface, ball and roller bearings, etc. In a ball bearing (Fig.1.5c), the ball and the shaft constitute one rolling pair whereas the ball and the bearing is the second rolling pair.

(d) Screw Pair (Helical Pair) If two mating links have a turning as well as sliding motion between them, they form a screw pair. This is achieved by cutting matching threads on the two links.

The lead screw and the nut of a lathe is a screw pair (Fig.1.5 d)

(e) Spherical Pair When one link in the form of a sphere turns inside a fixed link, it is a spherical pair.

The ball and socket joint is a spherical pair (Fig.1.5 e).

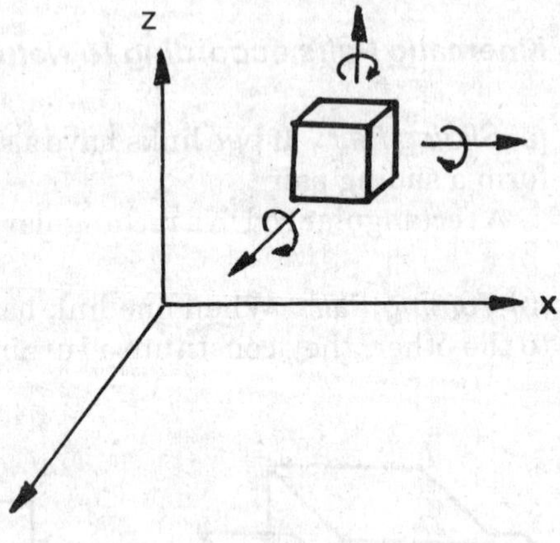

Fig. 1.6

1.6 DEGREES OF FREEDOM

An unconstrained rigid body moving in space can describe the following independent motions (Fig.1.6):

1. Translational motions along any three mutually perpendicular axes x,y and z, and
2. Rotational motions about these axes.

Thus a rigid body posesses six degrees of freedom. The connection of a link with another imposes certain constraints on their relative motion. The number of restraints can never be zero (joint is disconnected) or six (joint becomes solid).

Degrees of freedom of a pair is defined as the number of independent relative motions, both translational and rotational, a pair can have.

$$\text{Degrees of freedom} = 6 - \text{Number of restraints}$$

1.7 CLASSIFICATION OF KINEMATIC PAIRS

Depending upon the number of restraints imposed on the relative motion of the two links connected together, a pair can be classified as given in Table 1.1 which gives the possible form of each class.

Different forms of each class have also been shown in Fig.1.7. Remember that a particular relative motion between two links of a pair must be independent of the other relative motions that the pair can have. A screw and nut pair permits translational and rotational motions. However, as the two motions cannot be accomplished independently, a screw and nut pair is a kinematic pair of the fifth class and not the fourth class.

Table 1.1

Class	Number of restraints	Form	Restraint on translatory motion	Restraint on rotary motion	Kinematic pair	Fig.1.7
I	1	1st	1	0	Sphere—plane	a
II	2	1st	2	0	Sphere—cylinder	b
		2nd	1	1	Cylinder—plane	c
III	3	1st	3	0	Spheric	d
		2nd	2	1	Sphere—slotted cylinder	e
		3rd	1	2	Prism—plane	f
IV	4	1st	3	1	Slotted—spheric	g
		2nd	2	2	Cylinder	h
V	5	1st	3	2	Cylinder (collared)	i
		2nd	2	3	Prismatic	j

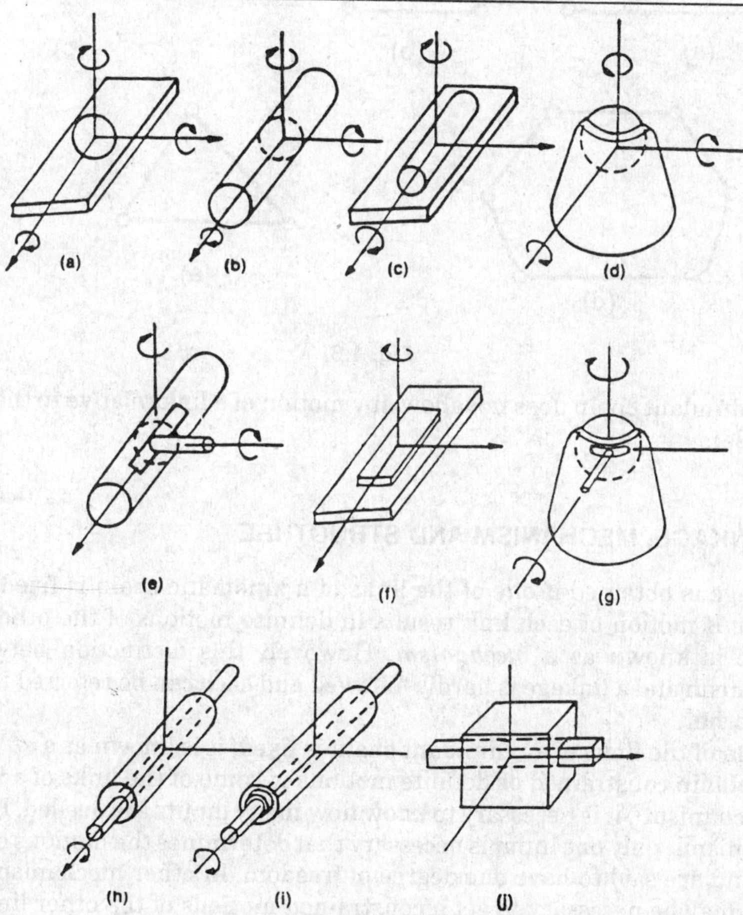

Fig. 1.7

1.8 KINEMATIC CHAIN

A kinematic chain is an assembly of links in which the relative motions of the links is possible and the motion of each relative to the others is definite (Fig.1.8 a, b, and c).

In case, the motion of a link results in indefinite motions of other links, it is a non-kinematic chain (Fig.1.8d). However, some authors prefer to call all chains having relative motions of the links as kinematic chains.

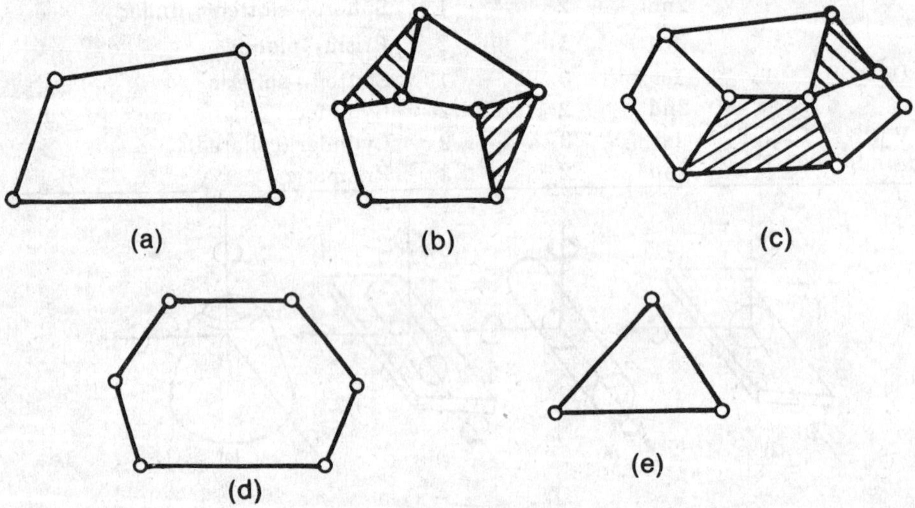

Fig. 1.8

A redundant chain does not allow any motion of a link relative to the other (Fig.1.8e).

1.9 LINKAGE, MECHANISM AND STRUCTURE

A linkage is obtained if one of the links of a kinematic chain is fixed to the ground. If motion of each link results in definite motions of the others, the linkage is known as a *mechanism*. However, this distinction between a mechanism and a linkage is hardly followed and each can be referred in place of the other.

If one of the links of a redundant chain is fixed, it is known as a *structure*.

To obtain constrained or definite motions of some of the links of a linkage (or mechanism), it is necessary to know how many inputs are needed. In some mechanisms, only one input is necessary that determines the motions of other links and are said to have one degree of freedom. In other mechanisms, two inputs may be necessary to get a constrained motions of the other links and are said to have two degrees of freedom and so on.

The degree of freedom of a structure is zero or less. A structure with negative degrees of freedom is known as a *superstructure*.

1.10 CLASSIFICATION OF MECHANISMS

A mechanism may consist of a number of pairs belonging to different classes having different number of restraints. It is also possible that some of the restraints imposed on the individual links are common or general to all the links of the mechanism. According to the number of these general or common restraints, a mechanism may be classified into different orders. A zero order mechanism will have no such general restraint. Of course, some of the pairs may have individual restraints. A first order mechanism has one general restraint, a second order mechanism has two general restraints and so on up to fifth order. A sixth order mechanism cannot exist since all the links become stationary and no movement is possible.

Degrees of freedom of a mechanism in space can be determined as follows: Let

N = total number of links in a mechanism

F = degrees of freedom

P_1 = number of pairs having one degree of freedom

P_2 = number of pairs having two degrees of freedom and so on.

In a mechanism, one link is fixed.

Therefore,

Number of movable links = $N - 1$

Number of degrees of freedom of $(N - 1)$ movable links = $6(N - 1)$

Each pair having one degree of freedom imposes 5 restraints on the mechanism reducing its degrees of freedom by $5P_1$.

Each pair having two degrees of freedom will impose 4 restraints reducing the degrees of freedom of the mechanism by $4P_2$

Similarly, other pairs having 3, 4 and 5 degrees of freedom reduce the degrees of freedom of the mechanism.

$$F = 6\,(N - 1) - 5P_1 - 4P_2 - 3P_3 - 2P_4 - 1P_5 \tag{1.1}$$

The above criterion is hardly needed to find the degrees of freedom, as space mechanisms, especially of the zero order are not practical. Most of the mechanisms are two-dimensional such as a four-link or a slider-crank mechanism in which displacement is possible along two axes (one restraint) and rotation about only one axis (two restraints). Thus there are three general restraints.

Therefore, for plane mechanisms, the following relation may be used to find the degrees of freedom,

$$F = 3\,(N - 1) - 2P_1 - 1P_2 \tag{1.2}$$

This is known as Gruebler's criterion for degrees of freedom of plane mechanisms in which each movable link possesses three degrees of freedom.

Each pair with one degree of freedom imposes two further restraints on the mechanisms, thus reducing its degrees of freedom. Similarly, each pair with two degrees of freedom reduces the degrees of freedom of the mechanism at the rate of one restraint each.

For linkages with turning pairs only, $P_2 = 0$

$$F = 3\,(N - 1) - 2P_1 \tag{1.3}$$

Most of the linkages are expected to have one degree of freedom so that with one input to any of the links, a constrained motion of the others is obtained.

Then,

$$1 = 3\,(N - 1) - 2P_1$$

or

$$2P_1 = 3\,N - 4 \tag{1.4}$$

As P_1 and N are to be whole numbers, the relation can be satisfied only if N is even. For possible linkages made of binary links only,

$N = 4,$	$P_1 = 4$	No excess turning pair
$N = 6,$	$P_1 = 7$	One excess turning pair
$N = 8,$	$P_1 = 10$	Two excess turning pairs

and so on.

Thus with the increase in the number of links, the number of excess turning pairs goes on increasing. To get the required number of turning pairs from the required number of binary links is not possible. Therefore, the excess or the additional pairs or joints can be obtained only from the links having more than two ends, i.e. ternary or quaternary links, etc.

For a six-link chain, some of the possible types are shown in Figs.1.9 a, b and c. However, the chain of Fig.1.9 c is not a six-link chain but a four-link chain as links 1, 2 and 3 are, in fact, one link only with no relative motion of these links.

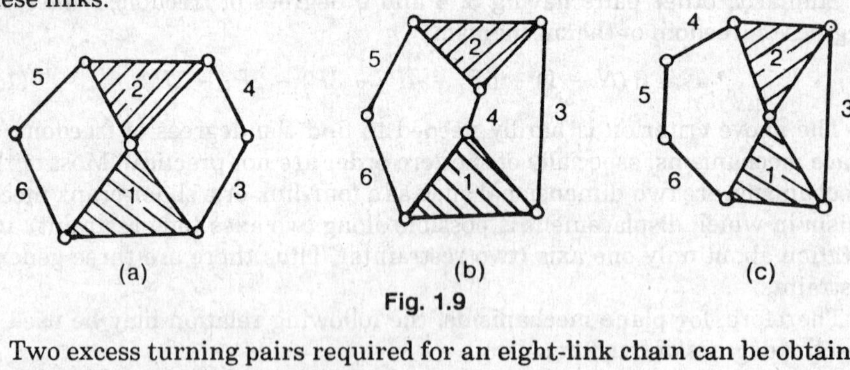

Fig. 1.9

Two excess turning pairs required for an eight-link chain can be obtained by using (apart from binary links):

four ternary links (Figs.1.10 a and b)
two quaternary links (Fig.1.10 c)
one quaternary and two ternary links (Fig.1.10 d).

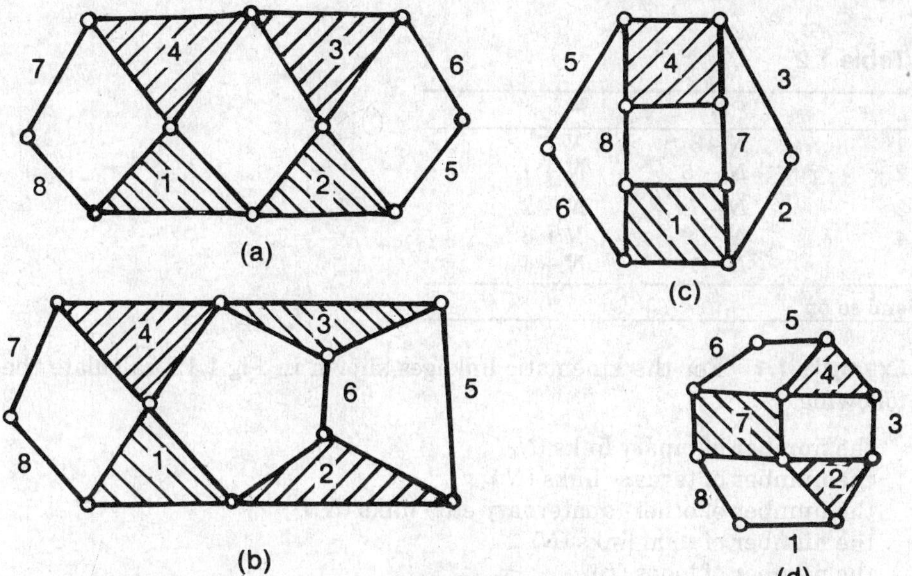

Fig. 1.10

Now, consider the kinematic chain shown in Fig.1.11. It has 8 links, but only three ternary links. However, the links 6, 7 and 8 constitute a double pair so that the total number of pairs is again 10. The degree of freedom of such a linkage will be

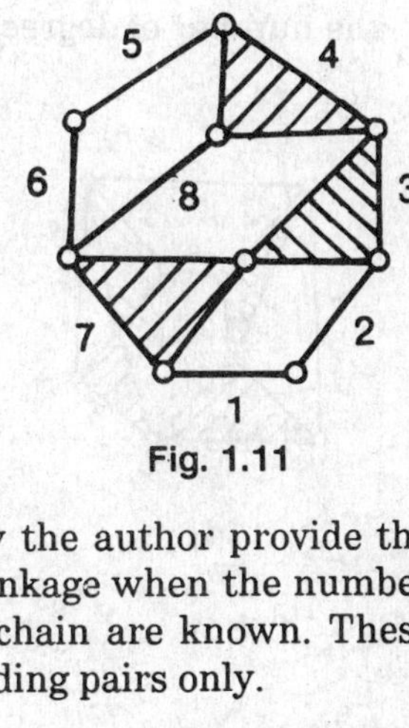

Fig. 1.11

$$F = 3\,(8 - 1) - 2 \times 10$$

$$= 1$$

This shows that the number of ternary or quaternary links in a chain can be reduced by providing double joints also.

The following empirical relations formulated by the author provide the degrees of freedom and the number of joints in a linkage when the number of links and the number of loops in a kinematic chain are known. These relations are valid for linkages with turning and sliding pairs only.

$$F = N - (2L + 1) \tag{1.5}$$

$$P_1 = N + (L - 1) \tag{1.6}$$

where

L = Number of loops in a linkage.

Thus, for different number of loops in a linkage, the degrees of freedom and the number of pairs are as shown in Table 1.2.

For example, if in a linkage, there are 4 loops and 11 links, its degrees of freedom will be 2 and number of joints 15. Similarly, if a linkage has 3 loops, it will require 8 links to have one degree of freedom, 9 links to have 2 degrees of freedom, 7 links to have –1 degree of freedom, etc.

Table 1.2

L	F	P_1
1	$N-3$	N
2	$N-5$	$N+1$
3	$N-7$	$N+2$
4	$N-9$	$N+3$
5	$N-11$	$N+4$
and so on		

Example 1.1 For the kinematic linkages shown in Fig.1.12, calculate the following:

the number of binary links (N_b)
the number of ternary links (N_t)
the number of other (quaternary etc.) links (N_o)
the number of total links (N)
the number of loops (L)
the number of joints or pairs (P_1)
the number of degrees of freedom (F)

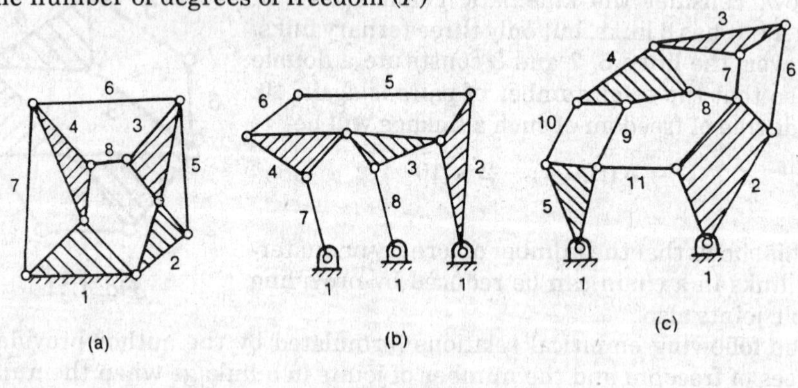

Fig. 1.12

Solution:

(a) $N_b = 4$; $N_t = 4$; $N_o = 0$; $N = 8$; $L = 4$
$P_1 = 11$ by counting

or

$P_1 = (N + L - 1) = 11$

$F = 3(N-1) - 2P_1$ or $F = N - (2L + 1)$
$= 3(8-1) - 2 \times 11$ or $= 8 - (2 \times 4 + 1)$
$= -1$ or $= -1$

The linkage has negative degree of freedom and thus is a superstructure.

(b) $N_b = 4$; $N_t = 4$; $N_o = 0$; $N = 8$; $L = 3$
$P_1 = 10$ (by counting) or $P_1 = (N + L - 1) = 10$
$F = N - (2L + 1) = 8 - (2 \times 3 + 1) = 1$

or $F = 3(N - 1) - 2P_1 = 3(8 - 1) - 2 \times 10 = 1$

that is, the linkage has a constrained motion when one of the seven moving links is driven by an external source.

(c) $N_b = 7$; $N_t = 2$; $N_Q = 2$; $N = 11$

$L = 5$; $P_1 = 15$

$F = N - (2L + 1) = 11 - (2 \times 5 + 1) = 0$

Therefore, the linkage is a structure.

Example 1.2 State whether the linkages shown in Fig.1.13 are mechanisms with one degree of freedom. If not, make suitable changes. The number of links should not be varied by more than 1.

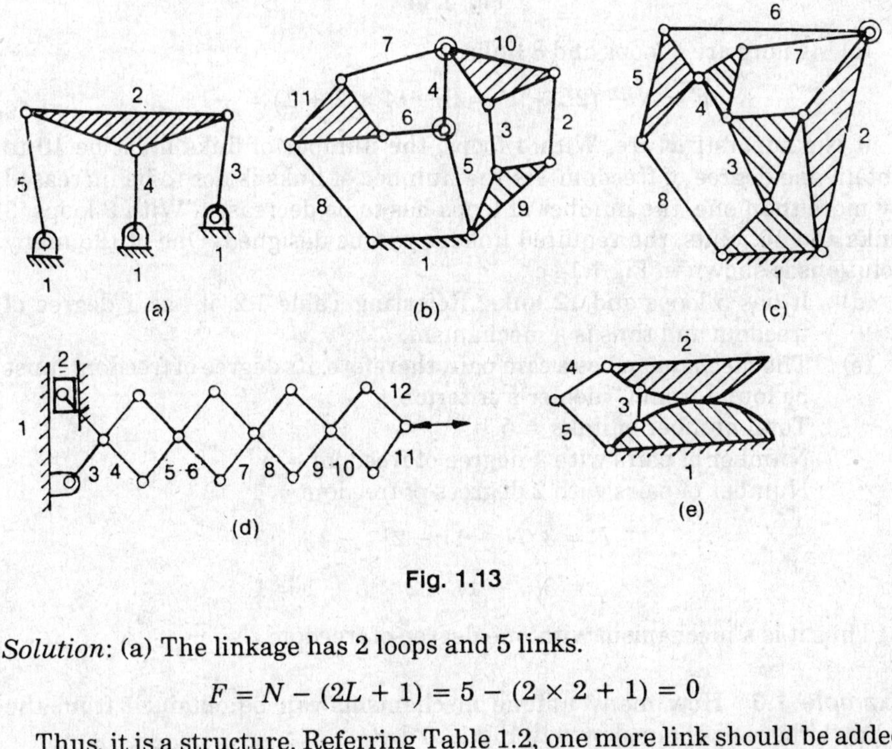

Fig. 1.13

Solution: (a) The linkage has 2 loops and 5 links.

$$F = N - (2L + 1) = 5 - (2 \times 2 + 1) = 0$$

Thus, it is a structure. Referring Table 1.2, one more link should be added to the linkage to make it a mechanism of $F = 1$. One of the possible solutions has been shown in Fig.1.14 a.

(b) The linkage has 4 loops and 11 links. Referring Table 1.2, it has 2 degrees of freedom. With 4 loops and 1 degree of freedom, the number of links should be 10 and the number of joints 13. Three excess joints can be formed by

6 ternary links, or 4 ternary links and 1 quaternary link, or 2 ternary links, and 2 quaternary links, or 3 quaternary links, or a combination of ternary and quaternary links with double joints.

Figure 1.14 b shows one of the possible solutions.

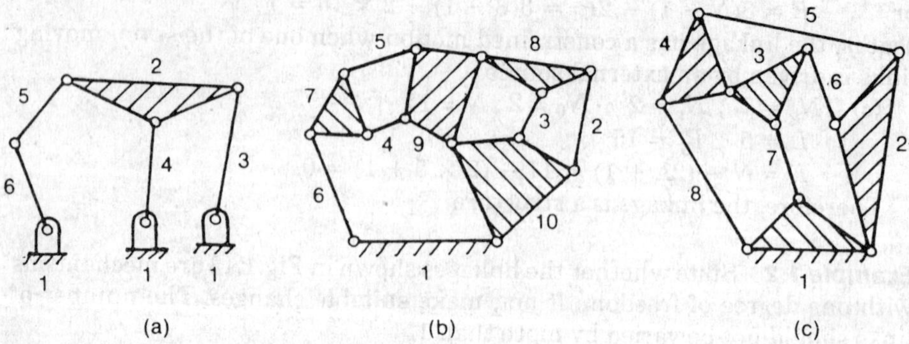

Fig. 1.14

(c) There are 4 loops and 8 links.

$$F = N - (2L + 1) = 8 - (4 \times 2 + 1) = -1$$

It is a superstructure. With 4 loops, the number of links must be 10 to obtain one degree of freedom. As the number of links is not to be increased by more than one, the number of loops has to be decreased. With 3 loops, 8 links and 10 joints, the required linkage can be designed. One of the many solutions is shown in Fig.1.14 c.

(d) It has 5 loops and 12 links. Referring Table 1.2, it has 1 degree of freedom and thus is a mechanism.

(e) The mechanism has a cam pair, therefore, its degree of freedom must be found from Gruebler's criterion.
Total number of links = 5
Number of pairs with 1 degree of freedom = 5
Number of pairs with 2 degrees of freedom = 1

$$F = 3(N - 1) - 2P_1 - P_2$$

$$= 3(5 - 1) - 2 \times 5 - 1 = 1.$$

Thus it is a mechanism with one degree of freedom.

Example 1.3 How many unique mechanisms can be obtained from the 8-link kinematic chain shown in Fig.1.10 d.

Solution: The kinematic chain has 8 links in all. A unique mechanism is obtained by fixing one of the links to the ground each time and retaining only one out of the symmetric mechanisms thus obtained.

The given kinematic chain is symmetric about links 3 or 7. Thus identical inversions (mechanisms) are obtained if the links 2, 1, 8 or 4, 5, 6 are fixed. In addition, two more unique inversions are obtained by fixing links 3 and 7 in turn. Therefore, in all five unique mechanisms can be obtained from the 8-link kinematic chain as shown in Fig.1.15.

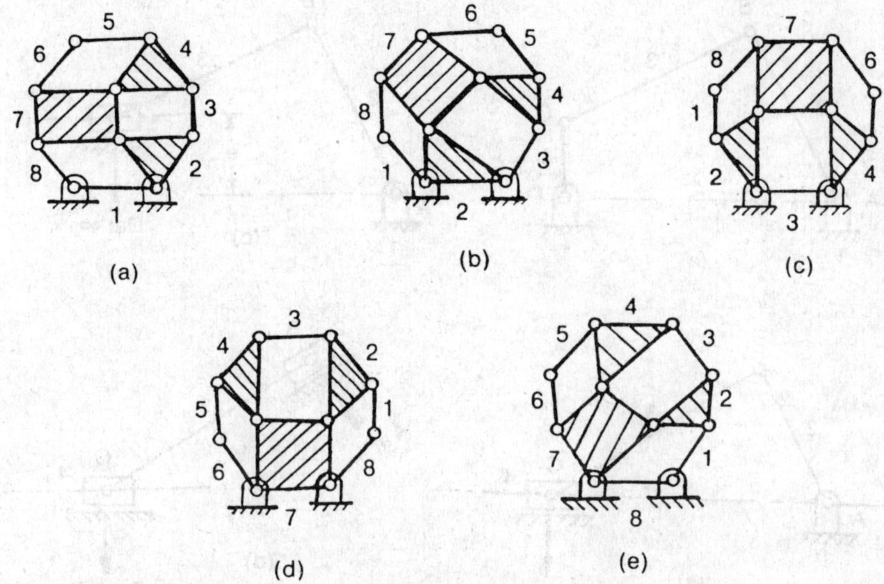

Fig. 1.15

Example 1.4 A linkage has 11 links and 4 loops. Calculate its degree of freedom and the number of ternary and quaternary links it will have if it has only single turning pairs.

Solution:

$$F = N - (2L + 1) = 11 - (2 \times 4 + 1) = 2$$

$$P_1 = N + (L - 1) = 11 + (4 - 1) = 14$$

The linkage has 3 excess joints and if all the joints are single turning pairs, the excess joints can be provided either or by

6 ternary links
4 ternary links and one quaternary link
2 ternary links and two quaternary links
3 quaternary links.

1.11 EQUIVALENT MECHANISMS

It is possible to replace turning pairs of plane mechanisms by other type of pairs having one or two degrees of freedom, such as sliding pairs or cam pairs. This can be done according to some set rules so that the new mechanisms also have the same degrees of freedom and are kinematically similar.

1. Sliding Pairs in Place of Turning Pairs

Figure 1.16a shows a four-link mechanism. Let the length of link 4 be

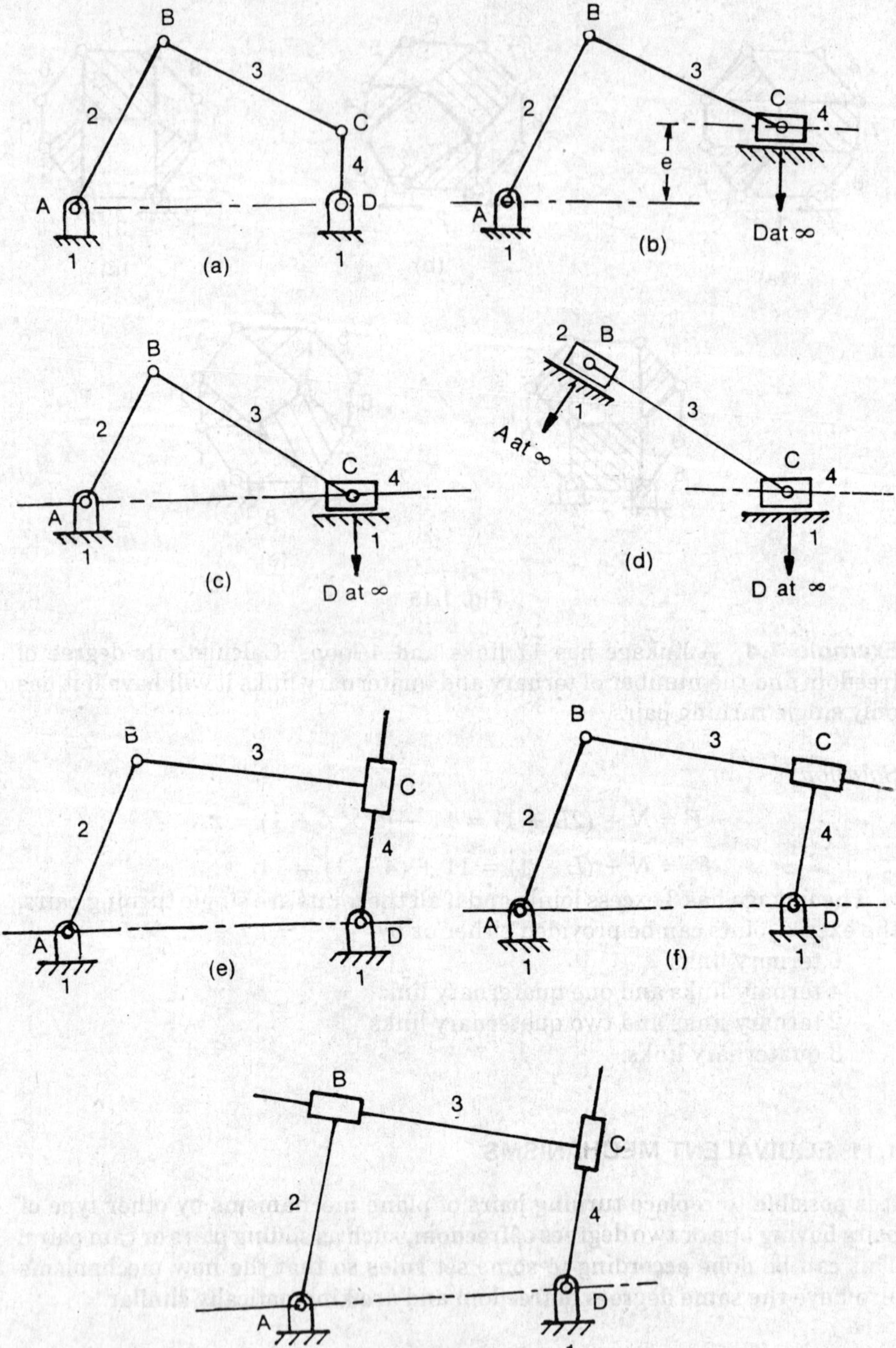
B
3
2
C
4
A
D
1
1
(a)
B
3
C
4
2
e
A
1
D at ∞
(b)
B
2
3
C
4
A
1
1
D at ∞
(c)
2 B
1
A at ∞
3
C
4
1
D at ∞
(d)
B
3
C
2
4
A
D
1
1
(e)
B
3
C
2
4
A
D
1
1
(f)
B
3
C
2
4
A
D
1
1
(g)

Fig. 1.16

increased to infinity so that D lies at infinity. Now, with the rotation of link 2, C will have a linear motion perpendicular to the axis of link 4. The same motion of C can be obtained if link 4 is replaced by a slider, and guides are provided for its motion as shown in Fig.1.16 b. In this case, the axis of the slider does not pass through A and there is an eccentricity. Figure 1.16 c shows a slider-crank mechanism with no eccentricity.

Note that the axis of the sliding pair must be in the plane of the linkage or parallel to it.

Similarly, the turning pair at A can also be replaced by a sliding pair by providing a slider with guides at B (Fig.1.16 d).

In case, the axes of the two sliding pairs are in one line or parallel, the two sliders along with link 3 act as one link with no relative motion among these links. Then the arrangement ceases to be a linkage. Thus in order to replace two turning pairs in a linkage with sliding pairs, the axis of the sliding pairs must intersect.

In the same way, the turning pairs at B and C can be replaced by sliding pairs by fixing a slider to any of the two links forming the pair (Figs. 16 e and f). Figure 1.16 g shows both of the turning pairs at B and C replaced by sliding pairs.

2. Spring in Place of Turning Pairs

The action of a spring is to elongate or to shorten as it becomes in tension or in compression. A similar variation in length is accomplished by two binary links joined by a turning pair. In Fig.1.17 a, the length AB varies as OB is moved away or towards point A. Figure 1.17 b shows a 6-link mechanism in which links 4 and 5 have been shown replaced by a spring.

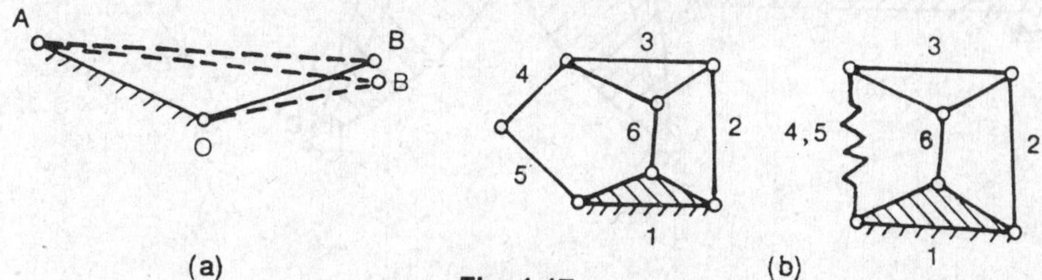

Fig. 1.17

Remember that spring is not a rigid link but is simulating the action of two binary links joined by a turning pair. Therefore, to find the degree of freedom of such a mechanism, the spring has to be replaced by the binary links.

3. Cam Pair in Place of Turning Pair

A cam pair has two degrees of freedom. For linkages with one degree of freedom, application of Gruebler's equation yields,

$$F = 3(N - 1) - 2P_1 - 1P_2$$

or

$$1 = 3N - 3 - 2P_1 - 1 \times 1$$

or
$$P_1 = \frac{3N - 5}{2}$$

This shows that to have one cam pair in a mechanism with one degree of freedom, the number of links and turning pairs should be as below:

$$N = 3, \quad P_1 = 2$$
$$N = 5, \quad P_1 = 5$$
$$N = 7, \quad P_1 = 8$$
$$N = 9, \quad P_1 = 11 \text{ and so on.}$$

A comparison of this with linkages having turning pairs only (Table 1.2) indicates that a cam pair can be replaced by one binary link with two turning pairs at each end.

Figure 1.18 a shows link *CD* (of a four-link mechanism) with two turning pairs at its ends replaced by a cam pair. The centres of curvatures at the point of contact *X* of the two cams lie at *D* and *C*. Figures 1.18 b and c show link *BC* with turning pairs at *B* and *C* replaced by a cam pair. The centres of curvatures at the point of contact *X* lie at *B* and *C* respectively. Figure 1.18 d shows equivalent mechanism for disc cam with reciprocating curved-face

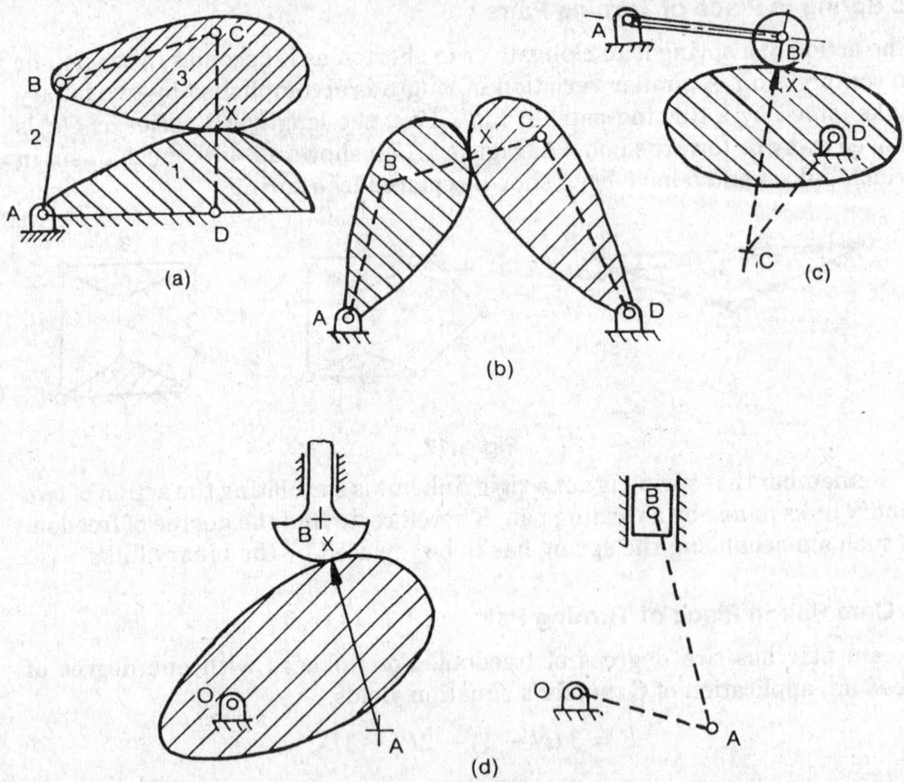

Fig. 1.18

follower. The centres of curvature of the cam and the follower at the instant lie at *A* and *B* respectively.

Example 1.5 Draw the equivalent kinematic chains with turning pairs for the chains shown in Fig.1.19.

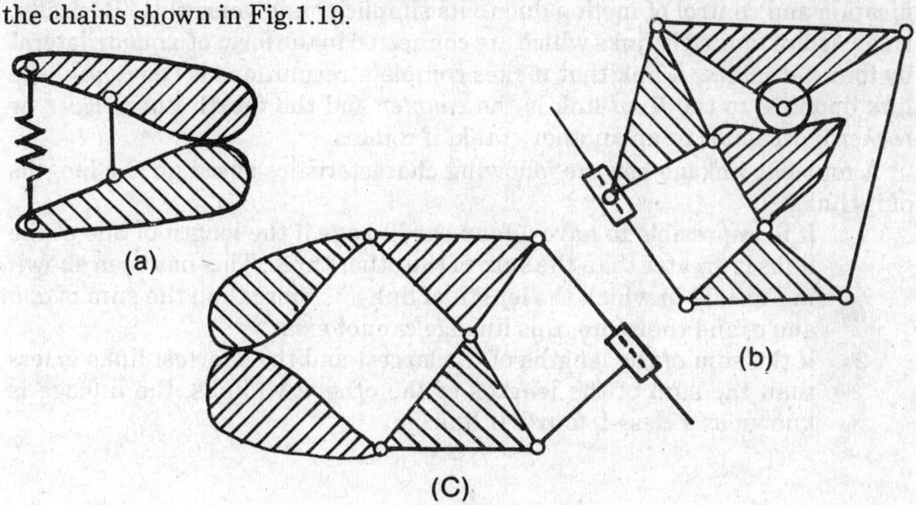

Fig. 1.19

Solution:

(a) A spring is equivalent to two binary links connected by a turning pair. A cam follower is equivalent of one binary link with turning pairs at each end. The equivalent chain with turning pairs is shown in Fig.1.20 a.

(b) A slider pair can be replaced by one link with a turning pair at the other end. A cam pair with a roller follower can be replaced by a binary link with turning pairs at each end similar to the case of a curved-face follower of Fig.1.18 d. The equivalent chain is shown in Fig.1.20 b.

(c) The equivalent chain has been shown in Fig.1.20 c.

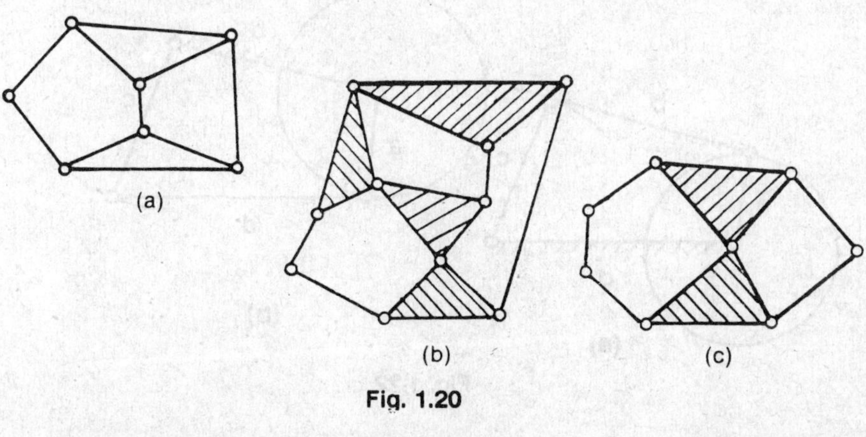

Fig. 1.20

1.12 FOUR–LINK (BAR) MECHANISM

A four–link mechanism or linkage is the most fundamental of the plane kinematic linkages. It is a much preferred mechanical device for the mechanisation and control of motion due to its simplicity and versatility. Basically, it consists of four rigid links which are connected in the form of a quadrilateral by four pin-joints. A link that makes complete revolutions is the *crank*, the link opposite to the *fixed* link is the *coupler* and the fourth link a *lever* or *rocker* if oscillates or an another crank, if rotates.

A four-bar linkage has the following characteristics based on the lengths of its links:

1. It is impossible to have a four-bar linkage if the length of one of the links is greater than the sum of the other three. This has been shown in Fig.1.21 in which the length of link *d* is more than the sum of *a*, *b* and *c*, and therefore, this linkage cannot exist.
2. If the sum of the lengths of the largest and the shortest links is less than the sum of the lengths of the other two links, the linkage is known as a class-I, four-bar linkage.

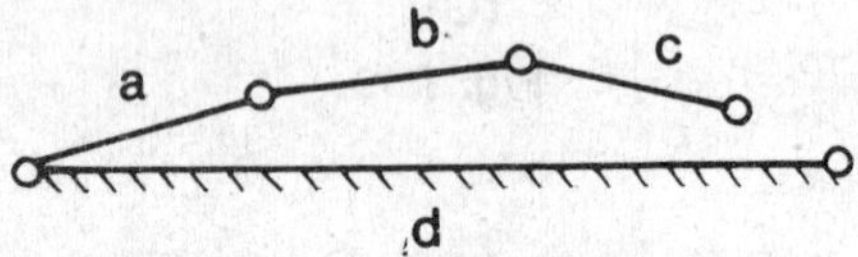

Fig. 1.21

In such a case, if the shortest link is made a crank (which can revolve) and any of the adjacent links is fixed, the link opposite to the crank oscillates (rocks). In Fig.1.22 a, *a* is the crank, *d* is fixed and *c* oscillates whereas in Fig.1.22 b, *b* is fixed and *c* oscillates. The mechanism is known as a crank-rocker or a crank-lever mechanism or a rotary-oscillating converter.

If different links of the four-link mechanism obtained above, are fixed,

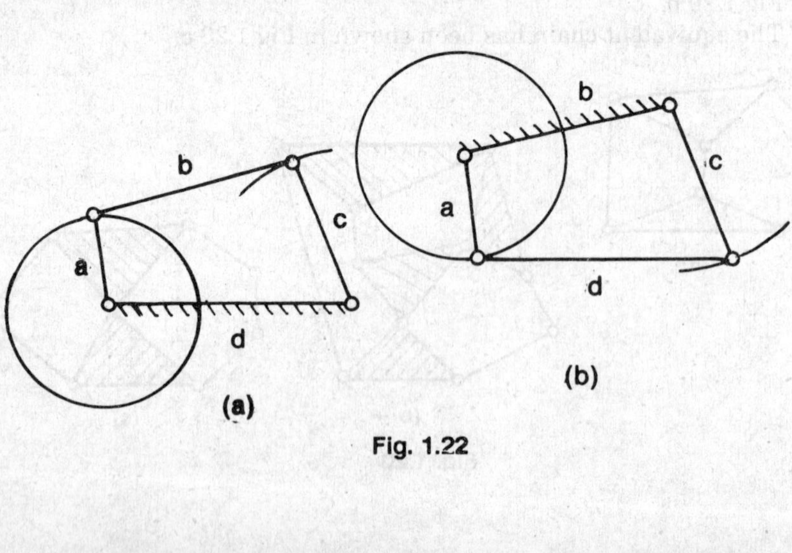

Fig. 1.22

different mechanisms (known as inversions) are obtained. However, by fixing the links adjacent to the shortest link results in the same type of mechanism as shown above.

Shortest Link Fixed If the shortest link *a* is fixed, the adjacent links *b* and *d* would make complete revolutions (Fig.1.23 a). The mechanism thus obtained is known as crank-crank or double-crank or drag-crank mechanism or rotary-rotary converter.

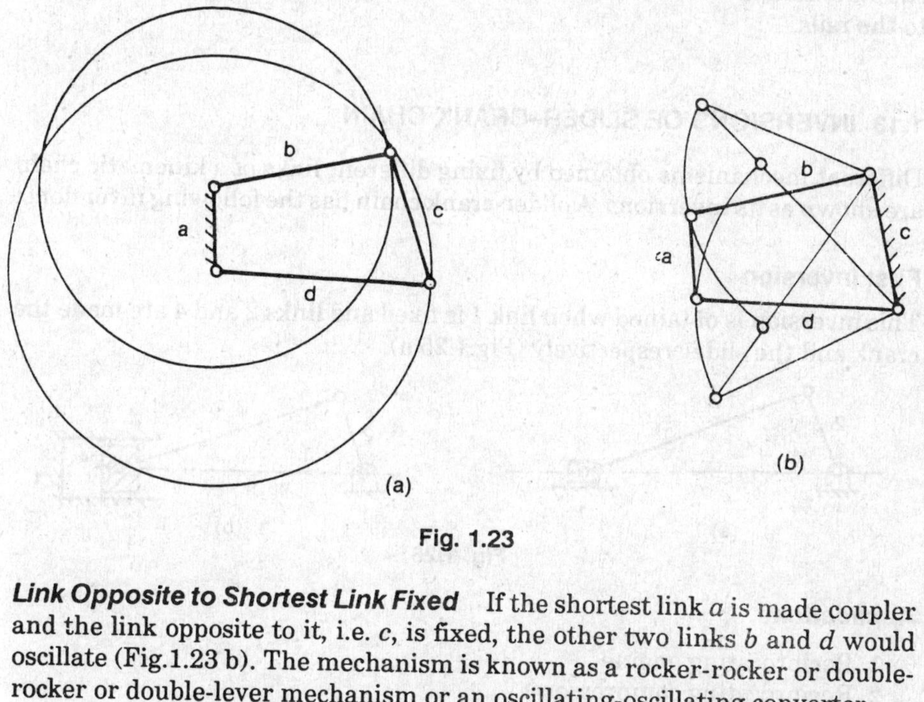

Fig. 1.23

Link Opposite to Shortest Link Fixed If the shortest link *a* is made coupler and the link opposite to it, i.e. *c*, is fixed, the other two links *b* and *d* would oscillate (Fig.1.23 b). The mechanism is known as a rocker-rocker or double-rocker or double-lever mechanism or an oscillating-oscillating converter.

3. When the sum of the lengths of the largest and the shortest links is more than the sum of lengths of the other two links, the linkage is known as a class-II, four-bar linkage.

In such a linkage, fixing of any of the links always results in a rocker-rocker mechanism. In other words, the mechanism and its inversions give the same type of motion (of a double-rocker mechanism).

Parallel-Crank Four-Bar Linkage If in a four-bar linkage, two opposite links are parallel and equal in length, then any of the links can be made fixed. The two links adjacent to the fixed link will always act as two cranks. The four links form a parallelogram in all the positions of the cranks, provided the cranks rotate in the same sense as shown in Fig.1.24.

The use of such a mechanism is made in the coupled wheels of locomotive in which the rotary motion of one wheel is transmitted to the other wheel. For kinematic analysis, link *d* is treated as fixed and the relative motions of

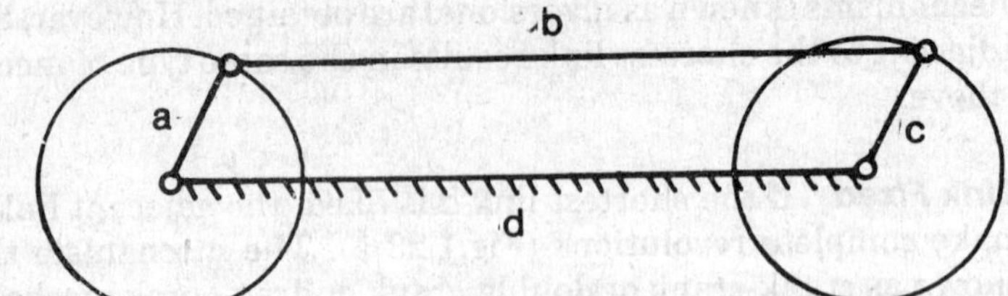

Fig. 1.24

the other links are found. However, in fact, *d* has a translatory motion parallel to the rails.

1.13 INVERSIONS OF SLIDER–CRANK CHAIN

Different mechanisms obtained by fixing different links of a kinematic chain are known as its inversions. A slider-crank chain has the following inversions:

First Inversion

This inversion is obtained when link 1 is fixed and links 2 and 4 are made the crank and the slider respectively (Fig.1.25 a).

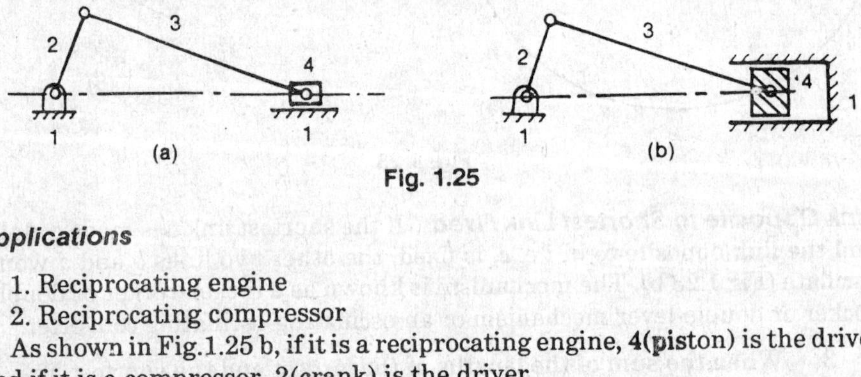

Fig. 1.25

Applications

1. Reciprocating engine
2. Reciprocating compressor

As shown in Fig.1.25 b, if it is a reciprocating engine, 4(piston) is the driver and if it is a compressor, 2(crank) is the driver.

Second Inversion

(a) Fixing of link 2 of a slider-crank chain results in the second inversion.

The slider-crank mechanism of Fig.1.25 a can also be drawn as shown in Fig.1.26 a. Further, when its link 2 is fixed instead of link 1, link 3 along with the slider at its end *B* becomes a crank. This makes link 1 to rotate about *O* along with the slider which also reciprocates on it (Fig.1.26 b).

Applications:

1. Whitworth quick-return mechanism
2. Rotary engine.

Whitworth Quick-Return Mechanism It is a mechanism used in workshops to cut metals. The forward stroke takes a little longer and cuts the metal whereas the return stroke is idle and takes a shorter period.

Slider 4 rotates in a circle about A and slides on link 1 (Fig.1.26c). C is a point on link 1 where link 5 is pivoted. The other end of link 5 is pivoted to the tool, the forward stroke of which cuts the metal. The axis of motion of slider 6 (tool) passes through O and is perpendicular to OA, the fixed link. The crank 3 rotates in the counter-clockwise direction.

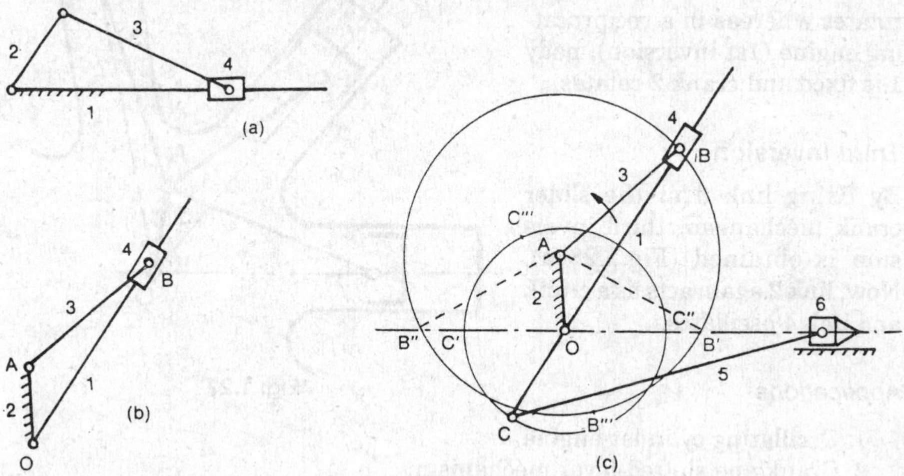

Fig. 1.26

Initially, let the slider 4 be at B' so that C be at C'. Cutting tool 6 will be in the extreme left position. With the movement of the crank, the slider traverses the path $B'BB''$ whereas point C moves through $C'CC''$. Cutting tool 6 will have the forward stroke. Finally, slider B assumes the position B'' and cutting tool 6 is in the extreme right position. The time taken for the forward stroke of slider 6 is proportional to the obtuse angle $B''AB'$ at A.

Similarly, slider 4 completes the rest of the circle through path $B''B'''B'$ and C passes through $C''C'''C'$. There is backward stroke of tool 6. The time taken in this is proportional to the acute angle $B''AB'$ at A.

Let

$$\theta = \text{obtuse angle } B'AB'' \text{ at } A.$$

$$\beta = \text{acute angle } B'AB'' \text{ at } A.$$

Then,

$$\frac{\text{Time of cutting}}{\text{Time of return}} = \frac{\theta}{\beta}$$

Rotary Engine Referring Fig.1.26 b, it can be observed that with the rotation of link 3, link 1 rotates about O and the slider 4 reciprocates on it.

This also implies that if the slider is made to reciprocate on link 1, crank 3 will rotate about A and the link 1 about O.

In a rotary engine, the slider is replaced by a piston and link 1 by a cylinder pivoted at O. Moreover, instead of one cylinder, seven or nine cylinders symmetrically placed at regular intervals in the same plane or in parallel planes, are used. All the cylinders rotate about the same fixed centre and form a balanced system. The fixed link 2 is also common to all cylinders (Fig. 1.27).

Thus in a rotary engine, crank 2 is fixed and body 1 rotates whereas in a reciprocating engine (1st inversion), body 1 is fixed and crank 2 rotates.

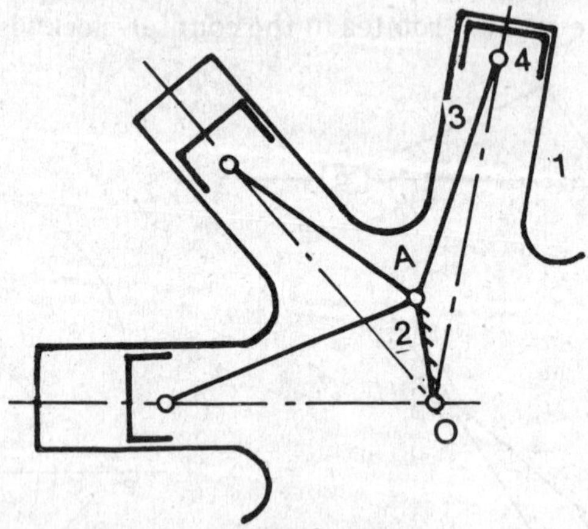

Fig. 1.27

Third Inversion

By fixing link 3 of the slider crank mechanism, third inversion is obtained (Fig.1.28 a). Now, link 2 again acts as a crank and link 4 oscillates.

Applications

1. Oscillating cylinder engine
2. Crank and slotted-lever mechanism.

Oscillating Cylinder Engine As shown in Fig.1.28 b, link 4 is made in the form of a cylinder and a piston is fixed to the end of link 1. The piston reciprocates inside the cylinder pivoted to the fixed link 3. The arrangement is known as oscillating cylinder engine, in which as the piston reciprocates in the oscillating cylinder, the crank rotates.

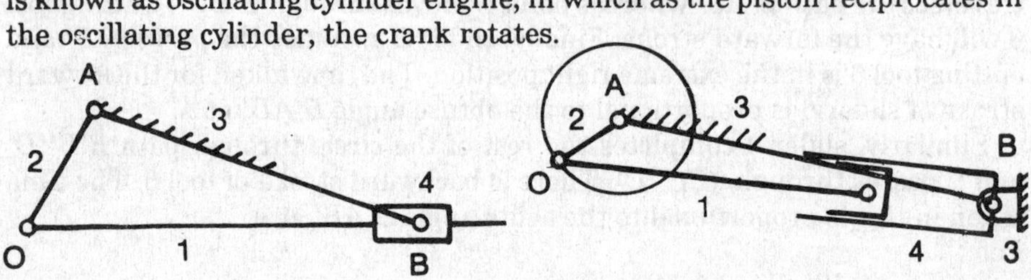

Fig. 1.28

(a) (b)

Crank and Slotted-Lever Mechanism If the cylinder of an oscillating cylinder engine is made in the form of a guide and the piston in the form of a slider, the arrangement as shown in Fig. 1.29 a is obtained. As the crank rotates about A, guide 4 oscillates about B. At a point C on the guide, link 5 is pivoted, the other end of which is connected to the cutting tool through a pivoted joint.

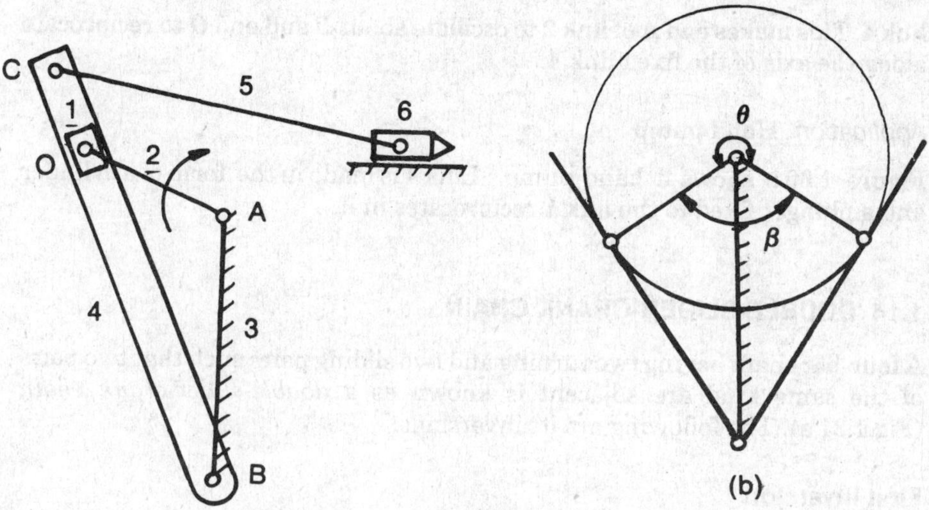

(a) (b)

Fig. 1.29

Figure 1.29 b shows the extreme positions of the oscillating guide 4. The time of the forward stroke is proportional to the angle θ whereas for the return stroke, it is proportional to β provided the crank rotates clockwise.

Comparing a crank and slotted-lever quick-return mechanism with a Whitworth quick-return mechanism, the following observations are made:

1. Crank 3 of the Whitworth mechanism is longer than its fixed link 2 whereas the crank 2 of the slotted lever mechanism is shorter than its fixed link 3.
2. Coupler link 1 of the Whitworth mechanism makes complete rotations about its pivoted joint *O* with the fixed link. However, coupler link 4 of the slotted-lever mechanism oscillates about its pivot *B*.
3. The coupler link holding the tool can be pivoted to the main coupler link at any convenient point *C* in both cases. However, for the same displacement of the tool, it is more convenient if the point *C* is taken on the extension of the main coupler link (towards the pivot with the fixed link) in case of the Whitworth mechanism and beyond the extreme position of the slider in the slotted lever mechanism.

Fourth Inversion If link 4 of the slider-crank mechanism is fixed, the fourth inversion is obtained (Fig. 1.30 a). Link 3 can oscillate about the fixed pivot *B* on

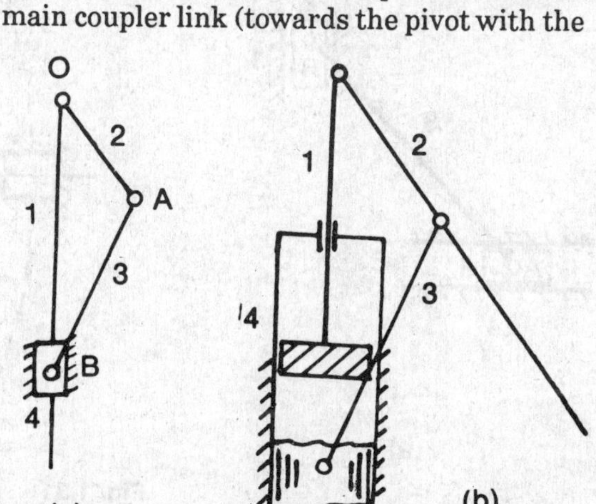

(a) (b)

Fig. 1.30

link 4. This makes end A of link 2 to oscillate about B and end O to reciprocate along the axis of the fixed link 4.

Application Hand pump

Figure 1.30 b shows a hand-pump. Link 4 is made in the form of a cylinder and a plunger fixed to the link 1 reciprocates in it.

1.14 DOUBLE-SLIDER-CRANK CHAIN

A four-bar chain having two turning and two sliding pairs such that two pairs of the same kind are adjacent is known as a *double-slider-crank chain* (Fig.1.31 a). The following are its inversions:

First Inversion

This inversion is obtained when link 1 is fixed and the two adjacent pairs 23 and 34 are turning pairs and the other two pairs 12 and 14 are sliding pairs.

Application Elliptical trammel

Elliptical Trammel Figure 1.31b shows an elliptical trammel in which the fixed link 1 is in the form of guides for sliders 2 and 4. With the movement of

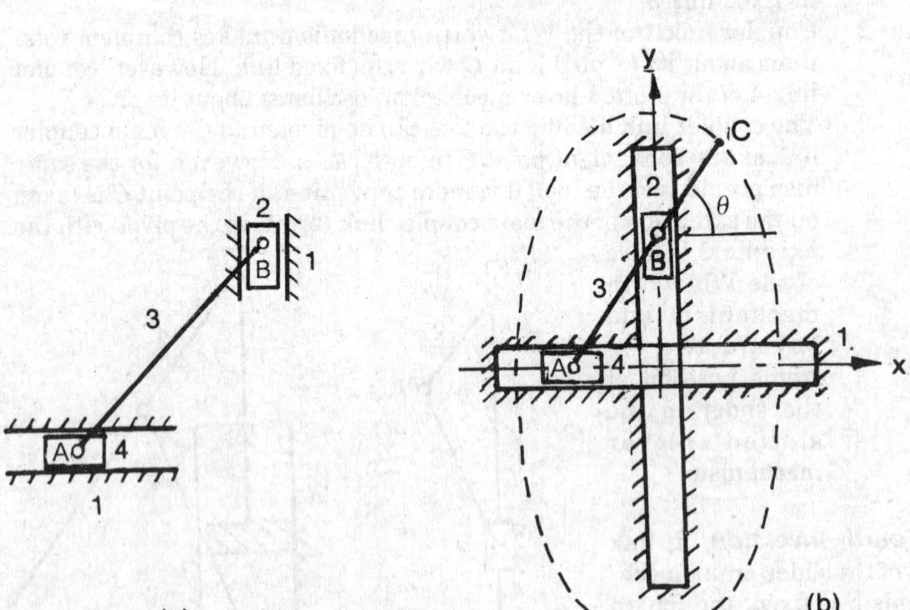

Fig. 1.31

the sliders, any point C on the link 3, except the mid-point of AB will trace an ellipse on a fixed plate. The mid-point of AB will trace a circle.

Let at any instant, link 3 make angle θ with the x-axis. Considering the displacements of the sliders from the centre of the trammel,

$$x = BC\cos\theta \text{ and } y = AC\sin\theta$$

$$\therefore \qquad \frac{x}{BC} = \cos\theta \text{ and } \frac{y}{AC} = \sin\theta$$

Squaring and adding,

$$\frac{x^2}{(BC)^2} + \frac{y^2}{(AC)^2} = \cos^2\theta + \sin^2\theta = 1$$

This is the equation of an ellipse. Therefore, the path traced by C is an ellipse with the semi-major and semi-minor axes being equal to AC and BC respectively.

When C is the mid-point of AB; $AC = BC$,

and
$$\frac{x^2}{(AC)^2} + \frac{y^2}{(AC)^2} = 1$$

$$x^2 + y^2 = (AC)^2$$

which is the equation of a circle with AC $(=BC)$ as the radius of the circle.

Second Inversion

If any of the slide-blocks of the first inversion is fixed, the second inversion of the double-slider-crank chain is obtained. When link 4 is fixed, end B of crank 3 rotates about A and link 1 reciprocates in the horizontal direction.

Application Scotch yoke

Scotch Yoke A scotch yoke mechanism (Fig.1.32) is used to convert the rotary motion into a sliding motion. As crank 3 rotates, the horizontal portion of link 1 slides or reciprocates in the fixed link 4.

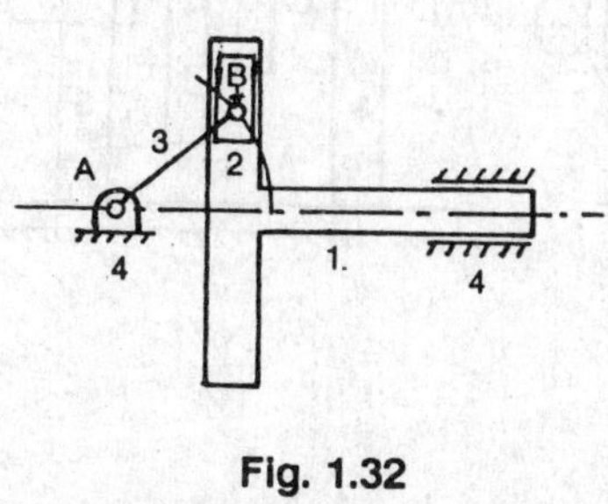

Fig. 1.32

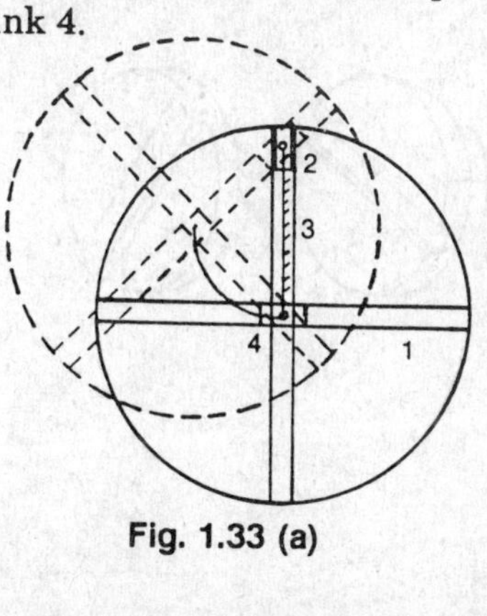

Fig. 1.33 (a)

Third Inversion

This inversion is obtained when link 3 of the first inversion is fixed and link 1 is free to move.

The rotation of link 1 has been shown in Fig.1.33 a in which the full lines show the initial position. With rotation of the link 4 through 45° in the clockwise direction, the links 1 and 2 rotate through the same angle whereas the mid-point of link 1 rotates through 90° in a circle with length of link 3 as diameter. Thus the angular velocity of the mid-point of link 1 is twice that of links 2 and 4.

The sliding velocity of link 1 relative to link 4 will be maximum when the mid-point of link 1 is at the axis of link 4. In this position, the sliding velocity is equal to the tangential velocity of the mid-point of link 1.

i.e. maximum sliding velocity = tangential velocity of mid-point of link 1
= angular velocity of mid-point of link 1 × Radius
= (2 × angular velocity of link 4) × (distance between axes of links 2 and 4)/2

The sliding velocity of link 1 relative to link 4 is zero when the mid-point of 1 is on the axis of link 2.

If the rotating links 2 and 4 of the mechanism are replaced by two shafts, one can act as the driver and the other as the driven shaft with their axes at the pivots of links 2 and 4.

Figure 1.33 b shows an actual Oldham's coupling which is used to connect two parallel shafts when the distance between their axes is small. The two shafts have flanges at the ends and are supported in the fixed bearings representing link 3. In the flange 2, a slot is cut in which tongue *X* of link 1 is fitted and has a sliding motion. Link 1 is made circular and has another tongue *Y* at right angles to the first and which fits in the recess of the flange of shaft 4. Thus the intermediate link 1 slides in the two slots in the two flanges while having the rotary motion.

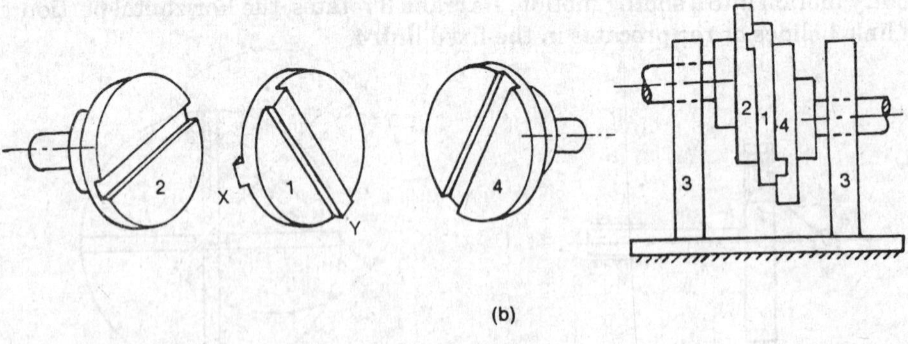

(b)

Fig. 1.33 (b)

As mentioned earlier, the mid-point of the intermediate piece describes a circle with distance between the axes of the shafts as diameter. The maximum sliding velocity of each tongue in the slot will be the peripheral velocity of the mid-point of the intermediate disc along the circular path.

Maximum sliding velocity = peripheral velocity along the circular path

= angular velocity of shaft × distance between shafts.

Example 1.6 The distance between two parallel shafts is 18 mm and they are connected by an Oldham's coupling. The driving shaft revolves at 160 rpm. What will be the maximum speed of sliding of the tongue of the intermediate piece along its groove?

Solution:

$$\omega = \frac{2\pi \times 160}{60} = 16.75 \text{ rad/s}$$

$$\text{Maximum velocity of sliding} = \omega \times d$$

$$= 16.75 \times 0.018$$

$$= 0.302 \text{ m/s.}$$

PROBLEMS

1.1 For the kinematic linkages shown in Fig. 1.34, find the number of binary links (N_b), ternary links (N_t), other links (N_o), total links N, loops L, joints of pairs (P_1), and degree of freedom (F).

(*a*) $N_b = 3; N_t = 4; N_0 = 0; N = 7; L = 3; P_1 = 9; F = 0$
(*b*) $N_b = 7; N_t = 5; N_0 = 0; N = 12; L = 4; P_1 = 15; F = 3$
(*c*) $N_b = 8; N_t = 2; N_0 = 1; N = 11; L = 5; P_1 = 15; F = 0$

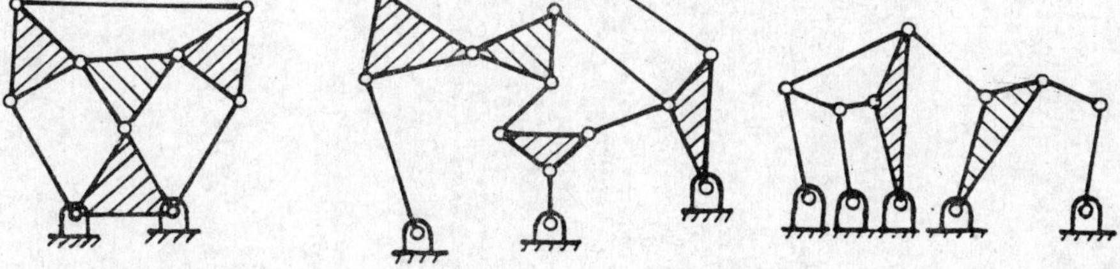

(a) (b) (c)

Fig. 1.34

1.2 Show that the linkages shown in Fig. 1.35 are structures. Suggest some changes to make them mechanisms having one degree of freedom. Number of links should not be changed by more than ± 1.

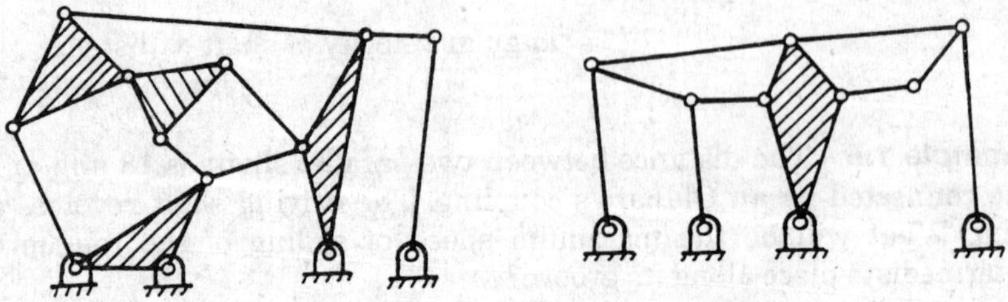

(a) (b)

Fig. 1.35

1.3 A linkage has 14 links and the number of loops is 5. Calculate its
(i) degrees of freedom
(ii) number of joints
(iii) maximum number of ternary links that can be had.
Assume that all the pairs are turning pairs.

(3; 18; 8)

2

VELOCITY ANALYSIS

2.1 INTRODUCTION

For a proper study of the motions of the different parts of a machine one needs to know their velocities and accelerations at different moments. To facilitate such study, a machine or a mechanism is represented by a skeleton or a line diagram, commonly known as a *configuration diagram*.

Velocities and accelerations in machines can be determined either analytically or graphically. With the invention of calculators and computers, it has become convenient to make use of analytical methods. A graphical analysis is more direct and is accurate to an acceptable degree and thus cannot be neglected. This chapter is devoted to the study of graphical methods of velocity analysis. The analytical approach will be discussed in Chapter 4.

2.2 ABSOLUTE AND RELATIVE MOTIONS

Strictly speaking, all motions are relative since an arbitrary set of axes or planes is required to define a motion. Usually, the earth is taken to be a fixed reference plane and all motions relative to it are termed as absolute motions.

If a train moves in a particular direction, the motion of the train is referred to as the absolute motion of the train or motion of the train relative to the earth. Now, suppose a man moves inside the train. Then, the motion of the man will be described in two different ways with different meanings:

1. Motion of the man relative to the train. It is equivalent to the motion of the man assuming the train to be stationary.
2. Motion of the man or absolute motion of the man or motion of the man relative to the earth = Motion of man relative to the train + Motion of train relative to the earth.

2.3 VECTORS

Problems involving relative motions are conveniently solved by the use of vectors. A vector is a line which represents a vector quantity such as force, velocity, acceleration etc.

Characteristics of a Vector

1. Length of the vector **ab** (Fig. 2.1) drawn to a convenient scale, represents the magnitude of the quantity (written as *ab*).
   Direction of the line is parallel to the direction in which the quantity acts.
   The initial end **a** of the line is the tail and the final end **b**, the head. An arrowhead on the line indicates the direction-sense of the quantity which is always from the tail to the head i.e. **a** to **b**.
   If the sense is as shown in Fig. 2.1 a, the vector is read as **ab** and if the sense is opposite (Fig. 2.1 b), the vector is read as **ba**. This implies that **ab** = – **ba**.

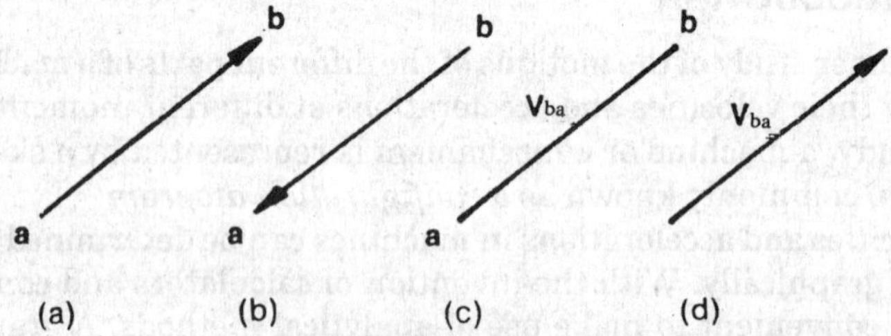

Fig. 2.1

2. Vector **ab** may also represent a vector quantity of a body *B* relative a body *A* such as velocity of *B* relative to *A*.
   If body *A* is fixed, **ab** represents the absolute velocity of *B*. If both the bodies *A* and *B* are in motion, the velocity of *B* relative to *A* means the velocity of *B* assuming the body *A* to be fixed for the moment.
   The vector **ab** can also be shown as $\mathbf{v}_{ba}$ (Fig. 2.1c), meaning the velocity of *B* relative to *A* provided *a* and *b* are indicated at the ends or an arrowhead is put on the vector (Fig. 2.1d).
3. Vector **ab** may also represent a vector quantity of a point *B* relative to a point *A* in the same body.
   If a vector $\mathbf{v}_{ba}$ or **ab** represents the velocity of *B* relative to *A*, the same vector in the opposite sense represents the velocity of *A* relative to *B* and will be read as $\mathbf{v}_{ab}$ or **ba**.

2.4 ADDITION AND SUBTRACTION OF VECTORS

Let

$\mathbf{v}_{ao}$ = velocity of *A* relative to *O*
$\mathbf{v}_{ba}$ = velocity of *B* relative to *A*
$\mathbf{v}_{bo}$ = velocity of *B* relative to *O*

The law of vector addition states that the velocity of *B* relative to *O* is equal to the vectorial sum of the velocity of B relative to A and the velocity of A relative to O.

Velocity of B relative to O = velocity of B relative to A + velocity of A relative to O, (2.1)

i.e.
$$\mathbf{v}_{bo} = \mathbf{v}_{ba} + \mathbf{v}_{ao}$$
$$= \mathbf{v}_{ao} + \mathbf{v}_{ba}$$

or

$$\mathbf{ob} = \mathbf{oa} + \mathbf{ab}$$

Take the vector **oa** and place the vector **ab** at the end of the vector **oa. ob** is given by the closing side of the two vectors (Fig. 2.2)

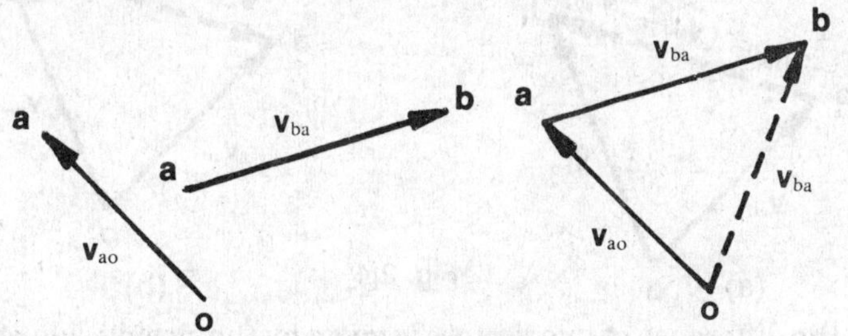

Fig. 2.2

Note that the arrows of the two vectors to be added are in the same order and that of the resultant in the opposite order.

Any number of vectors can be added as follows:

1. Take the first vector.
2. At the end of the first vector, place the beginning of the second vector.
3. At the end of the second vector, place the beginning of the third vector and so on.
4. Joining of the beginning of the first vector and the end of the last vector represents the sum of the vectors. Figure 2.3 shows the addition of four vectors.

$$\mathbf{v}_{do} = \mathbf{v}_{dc} + \mathbf{v}_{cb} + \mathbf{v}_{ba} + \mathbf{v}_{ao} \quad (2.2)$$
$$= \mathbf{v}_{ao} + \mathbf{v}_{ba} + \mathbf{v}_{cb} + \mathbf{v}_{dc}$$
$$\mathbf{od} = \mathbf{oa} + \mathbf{ab} + \mathbf{bc} + \mathbf{cd}$$

Equation 2.1 may be written as,

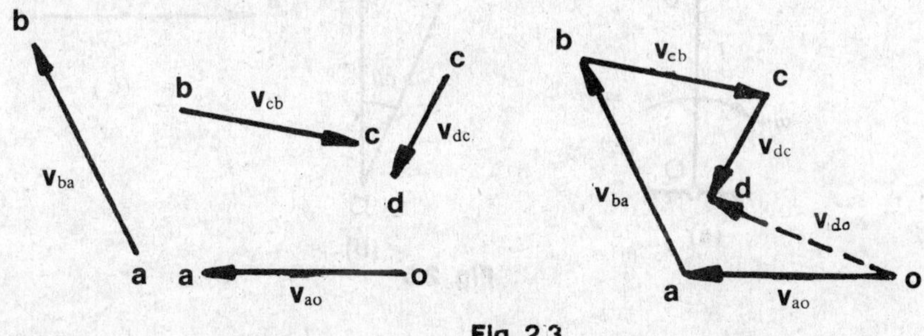

Fig. 2.3

Vel. of B rel. to A = Vel. of B rel. to O − Vel. of A rel. to O (2.3)

$$\mathbf{v}_{ba} = \mathbf{v}_{bo} - \mathbf{v}_{ao}$$

or $$\mathbf{ab} = \mathbf{ob} - \mathbf{oa}$$

This shows that in Fig. 2.2, **ab** also represents the subtraction of **oa** from **ob** (Fig. 2.4a).

Also $$\mathbf{v}_{ab} = -\mathbf{v}_{ba} = \mathbf{v}_{ao} - \mathbf{v}_{bo}$$

or $$\mathbf{ba} = \mathbf{oa} - \mathbf{ob}$$

This has been shown in Fig. 2.4 b.

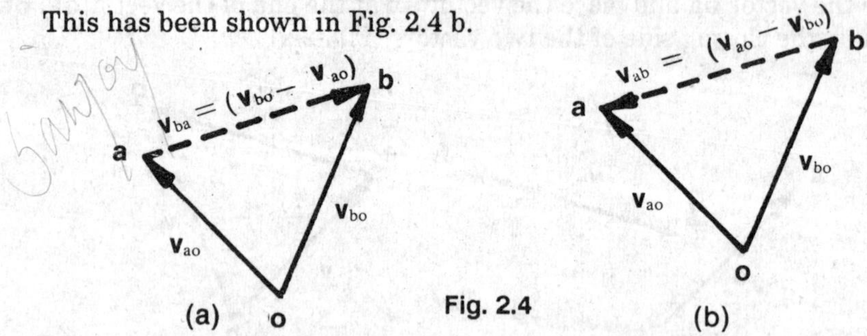

Fig. 2.4

Thus, the difference of two vectors is given by the closing side of a triangle, the other two sides of which are formed by placing the two vectors tail to tail the sense being towards the vector quantity from which the other is subtracted.

2.5 MOTION OF A LINK

Let a rigid link OA, of length r, rotate about a fixed point O with a uniform angular velocity ω rad/s in the counter-clockwise direction (Fig. 2.5 a). OA turns through a small angle $\delta\theta$ in a small interval of time δt. Then A will travel along the arc AA' as shown in Fig. 2.5 b.

$$\text{Velocity of } A \text{ relative to } O = \frac{Arc\ AA'}{\delta t}$$

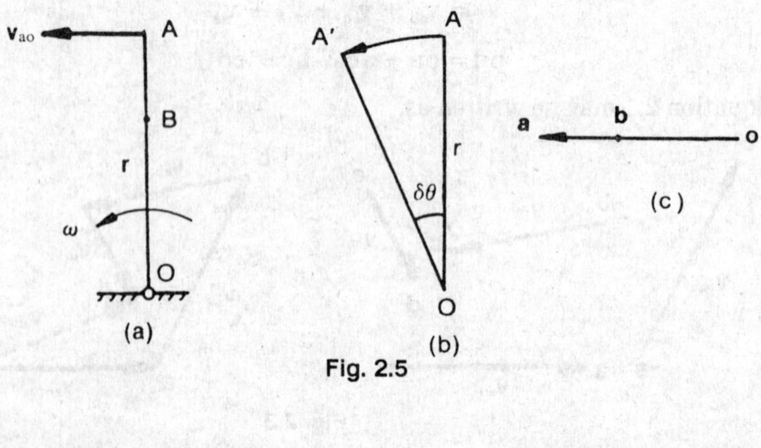

Fig. 2.5

or

$$\mathbf{v}_{ao} = \frac{r\delta\theta}{\delta t}$$

In the limits, when $\delta t \to 0$

$$\mathbf{v}_{ao} = r\frac{d\theta}{dt} \tag{2.4}$$

$$= r\,\omega$$

The direction of $\mathbf{v}_{ao}$ is along the displacement of A. Also, as δt approaches zero ($\delta t \to 0$), AA' will be perpendicular to OA. Thus, velocity of A is ωr and is perpendicular to OA. This can be represented by a vector **oa** (Fig. 2.5 c). The fact that the direction of velocity vector is perpendicular to the link also emerges from the fact that A can neither approach nor recede from O and thus, the only possible motion of A relative to O is in a direction perpendicular to OA.

Consider a point B on the link OA.

Velocity of $B = \omega\ OB$ perpendicular to OB.

If **ob** represents the velocity of B, it can be observed that

$$\frac{\mathbf{ob}}{\mathbf{oa}} = \frac{\omega\ OB}{\omega\ OA} = \frac{OB}{OA} \tag{2.5}$$

i.e. **b** divides the velocity vector in the same ratio as B divides the link.

Remember, the velocity vector $\mathbf{v}_{ao}$ (Fig. 2.5c) represents the velocity of A at a particular instant. At other instants, when the link OA assumes other positions, the velocity vectors will have their directions changed accordingly.

Also, the magnitude of the instantaneous linear velocity of a point on a rotating body is proportional to its distance from the axis of rotation.

2.6 FOUR-LINK MECHANISM

Figure 2.6a shows a four-link mechanism (quadric-cycle mechanism) $ABCD$ in which AD is the fixed link and BC is the coupler. AB is the driver rotating at an angular speed of ω rad/s in the clockwise direction. In case the link AB has also an angualr acceleration, ω wil be the angular velocity at the instant under consideration. It is required to find the absolute velocity of C (or velocity of C relative to A).

Writing the velocity vector equation,

Vel. of C rel. to A = Vel. of C rel. to B + vel. of B rel. to A

$$\mathbf{v}_{ca} = \mathbf{v}_{cb} + \mathbf{v}_{ba} \tag{2.6}$$

The velocity of any point relative to any other point on a fixed link is always zero. Thus all the points on a fixed link are represented by one point in the velocity diagram. In Fig. 2.6 a, the points A and D, both lie on the fixed link AD. Therefore, the velocity of C relative to A is the same as velocity of C relative to D.

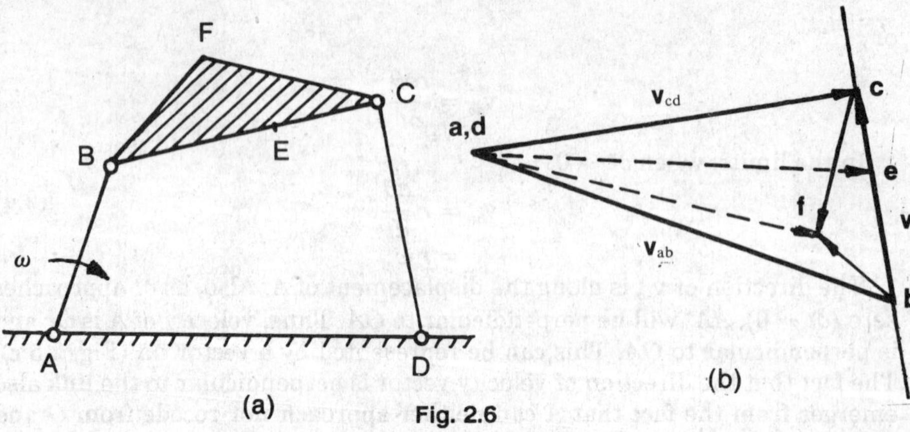

Fig. 2.6

Equation (2.5) may be written as,

$$\mathbf{v}_{cd} = \mathbf{v}_{ba} + \mathbf{v}_{cb}$$

or

$$\mathbf{dc} = \mathbf{ab} + \mathbf{bc}$$

where $\mathbf{v}_{ba}$ or **ab** $= \omega\ AB$; ⊥ to AB

$\mathbf{v}_{cb}$ or **bc** is unknown in magnitude ; ⊥ to BC

$\mathbf{v}_{cd}$ or **dc** is unknown in magnitude ; ⊥ to DC

The velocity diagram is constructed as follows:

1. Take the first vector **ab** as it is completely known.
2. To add vector **bc** to **ab**, draw a line ⊥ BC through **b**, of any length. Since the direction-sense of **bc** is unknown, it can lie on either side of **b**. A convenient length of the line can be taken on both sides of **b**.
3. Through **d**, draw a line ⊥ DC to locate the vector **dc**. The intersection of this line with the line of vector **bc** locates the point **c**.
4. Mark arrowheads on the vectors **bc** and **dc** to give the proper sense. Then **dc** is the magnitude and also represents the direction of the velocity of C relative to A (or D). It is also the absolute velocity of point C (A and D being fixed points).

Intermediate Point

The velocity of an intermediate point on any of the links can be found easily by dividing the corresponding velocity vector in the same ratio as the point divides the link. For point E on the link BC,

$$\frac{\mathbf{be}}{\mathbf{bc}} = \frac{BE}{BC}$$

ae represents the absolute velocity of E.

Offset Point

Write the vector equation for point F,

$$\mathbf{v}_{fb} + \mathbf{v}_{ba} = \mathbf{v}_{fc} + \mathbf{v}_{cd}$$

or

$$\mathbf{v}_{ba} + \mathbf{v}_{fb} = \mathbf{v}_{cd} + \mathbf{v}_{fc}$$

or

$$\mathbf{ab} + \mathbf{bf} = \mathbf{dc} + \mathbf{cf}$$

The vectors $\mathbf{v}_{ba}$ and $\mathbf{v}_{cd}$ are already there on the velocity diagram.

$\mathbf{v}_{fb}$ is $\perp BF$, draw a line $\perp BF$ through b;

$\mathbf{v}_{fc}$ is $\perp CF$, draw a line $\perp CF$ through **c**.

The intersection of the two lines locates the point **f**.

af or **df** indicates the velocity of F relative to A (or D) or the absolute velocity of F.

Also note that the triangle **bfc** is similar to triangle BFC and the order of the letters is the same (clockwise).

2.7 ANGULAR VELOCITY OF LINKS

1. Angular Velocity of BC

(a) Velocity of C relative to B, v_{cb} = **bc** (Fig. 2.6)
Point C relative to B moves in the direction-sense given by $\mathbf{v}_{cb}$. Thus C moves in the counter-clockwise direction about B.

$$v_{cb} = \omega_{cb} \times BC = \omega_{cb} \times CB$$

$$\therefore \quad \omega_{cb} = \frac{v_{cb}}{CB}$$

(b) Velocity of B relative to C, v_{bc} = **cb**
B relative to C moves in a direction-sense given by $\mathbf{v}_{bc}$ (opposite to **bc**) i.e. B moves in the counter-clockwise direction about C with magnitude ω_{bc}, given by $\frac{v_{bc}}{BC}$.

It can be seen that the magnitude of $\omega_{cb} = \omega_{bc}$ as $v_{cb} = v_{bc}$ and the direction of rotation is the same. Therefore, angular velocity of a link about one extremity is the same as the angular velocity about the other. In general, the angular velocity of link BC is ω_{bc} $(= \omega_{cb})$ in the counter-clockwise direction.

2. Angular Velocity of CD

Velocity of C relative to D,

$$v_{cd} = \mathbf{dc}$$

It is seen that C relative to D moves in a direction-sense given by $\mathbf{v}_{cd}$ or C moves in the clockwise direction about D.

$$\omega_{cd} = \frac{v_{cd}}{CD}$$

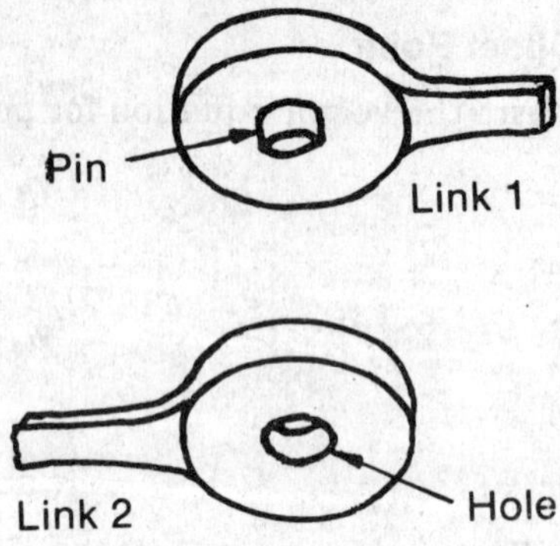

Fig. 2.7

2.8 VELOCITY OF RUBBING

Figure 2.7 shows two ends of the two links of a turning pair. A pin is fixed to one of the links whereas a hole is provided in the other to fit the pin. When joined, the surface of the hole of one link will rub on the surface of pin of the other link. The velocity of rubbing of the two surfaces will depend upon the angular velocity of a link relative to the other.

Pin at A (Fig. 2.6 a)

The pin at A joins links AD and AB. AD being fixed, the velocity of rubbing will depend upon the angular velocity of AB only.

Let r_a = radius of the pin at A.

Then velocity of rubbing = $r_a\,\omega$

Pin at D

Let r_d = radius of the pin at D.

Velocity of rubbing = $r_d\,\omega_{cd}$

Pin at B

$\omega_{ba} = \omega_{ab} = \omega$ clockwise

$\omega_{bc} = \omega_{cb} = \dfrac{v_{bc}}{BC}$ counter-clockwise

Since the directions of the two angular velocities of links AB and BC are in the opposite directions, the angular velocity of one link relative to the other is the sum of the two velocities.

Let r_b = radius of the pin at B

Velocity of rubbing = $r_b\,(\omega_{ab} + \omega_{bc})$

Pin at C

$\omega_{bc} = \omega_{cb}$ counter-clockwise

$\omega_{dc} = \omega_{cd}$ clockwise

Let r_c = radius of the pin at C.

Velocity of rubbing = $r_d\,(\omega_{bc} + \omega_{dc})$

In case it is found that the angular velocities of the two links joined together are in the same direction, the velocity of rubbing will be the difference of the angular velocities multiplied by the radius of the pin.

2.9 SLIDER-CRANK MECHANISM

Figure 2.8 a shows a slider-crank mechanism in which OA is the crank moving with uniform angular velocity ω rad/s in the clockwise direction. At point B, a slider moves on the fixed guide G. AB is the coupler joining A and B. It is required to find the velocity of the slider at B.

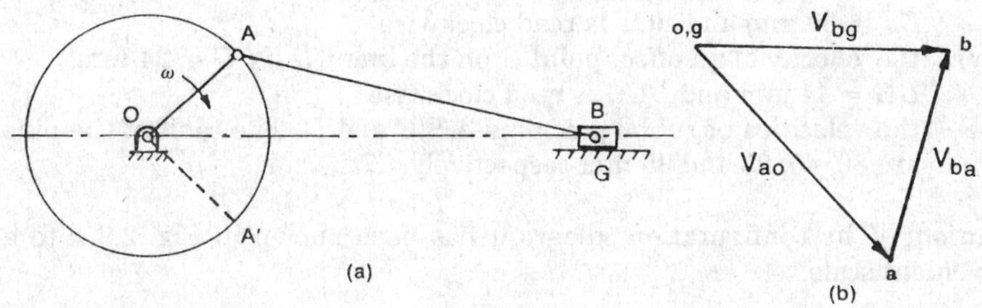

Fig. 2.8

Writing the velocity vector equation,

$$\text{Vel. of } B \text{ rel. to } O = \text{Vel. of } B \text{ rel to } A + \text{Vel. of } A \text{ rel. to } O$$

$$\mathbf{v}_{bo} = \mathbf{v}_{ba} + \mathbf{v}_{ao}$$

$$\mathbf{v}_{bg} = \mathbf{v}_{ao} + \mathbf{v}_{ba}$$

or $$\mathbf{gb} = \mathbf{oa} + \mathbf{ab}$$

$\mathbf{v}_{bo}$ is replaced by $\mathbf{v}_{bg}$ as O and G are two points on a fixed link with zero relative velocity between them.

Take the vector $\mathbf{v}_{ao}$ which is completely known.

$\mathbf{v}_{ao} = \omega\, OA$; $\perp$ to OA

$\mathbf{v}_{ba}$ is $\perp AB$, draw a line $\perp AB$ through **a**;

Through **g** (or **a**), draw a line parallel to the motion of B (to locate the vector $\mathbf{v}_{bg}$).

The intersection of the two lines locates the point **b**.

gb (or **ob**) indicates the velocity of the slider B relative to the guide G. This is also the absolute velocity of the slider (G is fixed). The slider moves towards right as indicated by **gb**. When the crank assumes the position OA' while rotating, it will be found that the vector **gb** lies on the left of **g** indicating that B moves towards left.

For the given configuration, the coupler AB has angular velocity in the counter-clockwise direction, the magnitude being $\frac{v_{ba} \text{ (or } ab)}{BA \text{ (or } AB)}$

Example 2.1 **In a four-link mechanism, the dimensions of the links are as under:**

AB = 50 mm, BC = 66 mm, CD = 56 mm and AD = 100 mm

At the instant when $\angle DAB$ = 60°, the link AB has an angular velocity of 10.5 rad/s in the counter-clockwise direction. Determine

(i) the velocity of point C

(ii) the velocity of point E on the link BC when BE = 40 mm

(iii) the angular velocities of the links BC and CD

(iv) the velocity of an offset point F on the link BC if BF = 45 mm, CF = 30 mm and BCF is read clockwise

(v) the velocity of an offset point G on the link CD if CG = 24 mm, DG = 44 mm and DCG is read clockwise

(vi) the velocities of rubbing at pins A, B, C and D. The radii of the pins are 30, 40, 25 and 35 mm respectively.

Solution: The configuration diagram has been shown in Fig. 2.9 a to a convenient scale.

Writing the vector equation,

Vel. of C rel. to A = Vel. of C rel. to B + Vel. of B rel. to A

$$\mathbf{v}_{ca} = \mathbf{v}_{cb} + \mathbf{v}_{ba}$$

or

$$\mathbf{v}_{cd} = \mathbf{v}_{ba} + \mathbf{v}_{cb}$$

or

$$\mathbf{dc} = \mathbf{ab} + \mathbf{bc}$$

We have,

$$v_{ba} = \omega_{ba} \times BA = 10.5 \times 0.05 = 0.525 \text{ m/s}$$

Take the vector $\mathbf{v}_{ba}$ to a convenient scale in the proper direction and sense (Fig. 2.9 b).

$\mathbf{v}_{cb}$ is $\perp BC$, draw a line $\perp BC$ through **b**;

$\mathbf{v}_{cd}$ is $\perp DC$, draw a line $\perp DC$ through **d**.

The intersection of the two lines locates point **c**.

(i) v_c = **ac** (or **dc**) = 0.39 m/s

(ii) Locate point **e** on **bc** such that $\frac{\mathbf{be}}{\mathbf{bc}} = \frac{BE}{BC}$

bc = 0.34 m/s from the velocity diagram.

$$\mathbf{be} = 0.34 \times \frac{0.040}{0.066} = 0.206 \text{ m/s}$$

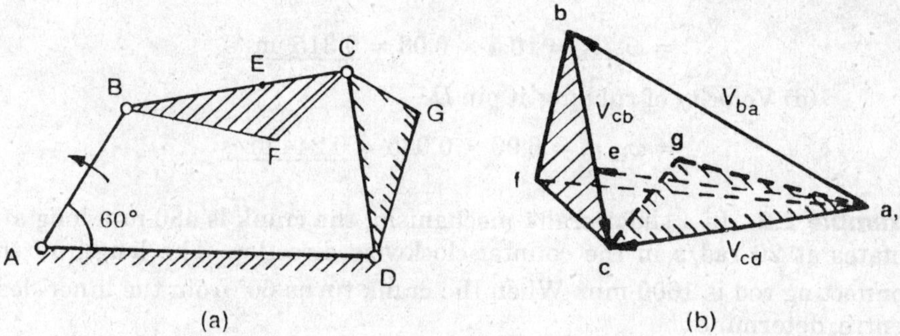

Fig. 2.9

Therefore, v_e = **ae** (or **de**) = 0.415 m/s

(iii) $\omega_{cb} = \dfrac{v_{cb}}{CB} = \dfrac{0.340}{0.066} = \underline{5.15\text{ rad/s}}$ clockwise

$\omega_{cd} = \dfrac{v_{cd}}{CD} = \dfrac{0.390}{0.056} = \underline{6.96\text{ rad/s counter-clockwise}}$

(iv) $\mathbf{v}_{fb}$ is $\perp BF$, draw a line $\perp BF$ through **b**;
$\mathbf{v}_{fc}$ is $\perp CF$, draw a line $\perp CF$ through **c**;
The intersection locates the point **f**.
v_f (i.e. v_{fa} or v_{fd}) = **af** = 0.495 m/s.

(v) $\mathbf{v}_{gd}$ is $\perp DG$, draw **dg** $\perp DG$ through **d**;
$\mathbf{v}_{gc}$ is $\perp CG$, draw **cg** $\perp CG$ through **c**.
The intersection locates the point **g**.

$$v_g = \mathbf{dg} = \underline{0.305\ \text{m/s}}$$

However, the velocity of G could be found directly since G is a point on the link DC which rotates about a fixed point D and the velocity of C is already known.

$$\frac{v_g}{v_c} = \frac{DG}{DC}$$

or

$$v_g = 0.390 \times \frac{0.044}{0.056} = \underline{0.306\ \text{m/s}}$$

(iv) (a) ω_{ba} (or ω_{ab}) is counter-clockwise and ω_{cb} (or ω_{bc}) is clockwise,
Velocity of rubbing at pin $B = (\omega_{ab} + \omega_{cb})r_b$
$= (10.5 + 5.15) \times 0.040$
$= \underline{0.626\ \text{m/s.}}$

(b) ω_{dc} is counter-clockwise and ω_{bc} is clockwise,
Velocity of rubbing at pin $C = (\omega_{dc} + \omega_{bc})\, r_c$
$= (6.96 + 5.15) \times 0.025$
$= \underline{0.303\ \text{m/s}}$

(c) Velocity of rubbing at pin A

$$= \omega_{ba}\, r_a = 10.5 \times 0.03 = \underline{0.315 \text{ m/s}}$$

(d) Velocity of rubbing at pin D

$$= \omega_{cd}\, r_d = 6.96 \times 0.035 = \underline{0.244 \text{ m/s}}$$

Example 2.2 In a slider-crank mechanism, the crank is 480 mm long and rotates at 20 rad/s in the counter-clockwise direction. The length of the connecting rod is 1600 mm. When the crank turns 60° from the inner-dead centre, determine

(i) the velocity of the slider
(ii) the velocity of a point E located at a distance 450 mm on the connecting rod extended
(iii) the position and velocity of a point F on the connecting rod having the least absolute velocity
(iv) the angular velocity of the connecting rod
(v) the velocities of rubbing at the pins of the crankshaft, crank and the cross-head having diameters 80, 60 and 100 mm respectively.

Solution:

Figure 2.10 a shows the configuration diagram to a convenient scale.

$$v_{ao} = \omega_{ao} \times OA = 20 \times 0.48 = 9.6 \text{ m/s}$$

The vector equation is, $\mathbf{v}_{bo} = \mathbf{v}_{ba} + \mathbf{v}_{ao}$

or

$$\mathbf{v}_{bg} = \mathbf{v}_{ba} + \mathbf{v}_{ao}$$

or

$$\mathbf{v}_{bg} = \mathbf{v}_{ao} + \mathbf{v}_{ba}$$

or

$$\mathbf{gb} = \mathbf{oa} + \mathbf{ab}$$

Take the vector $\mathbf{v}_{ao}$ to a convenient scale in the proper direction and sense (Fig. 2.10 b).

$\mathbf{v}_{ba}$ is $\perp AB$, draw a line $\perp AB$ through **a**;

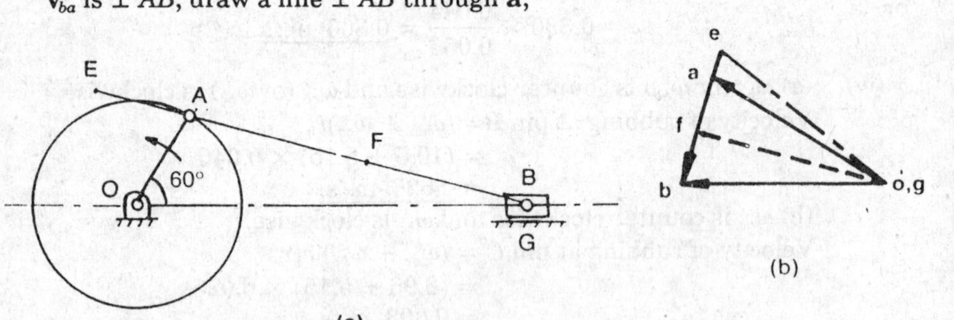

Fig. 2.10

The slider B has a linear motion relative to the guide G. Draw a line parallel to the direction of motion of the slider through **g** (or **a**). Thus the point **b** is located.

(i) Velocity of the slider, v_b = **ob** = 9.7 m/s

(ii) Locate point **e** on **ba** extended such that $\frac{\mathbf{ae}}{\mathbf{ba}} = \frac{AE}{BA}$

ba = 5.25 m/s on measuring from the diagram.

$$\mathbf{ae} = 5.25 \times \frac{0.45}{1.60} = 1.48 \text{ m/s}$$

$$v_e = \mathbf{oe} = \underline{10.2 \text{ m/s}}$$

(iii) To locate a point F on the connecting rod which has the least velocity relative to the crankshaft or has the least absolute velocity, draw **of** ⊥ **ab** through **o**

Locate point F on AB such that $\frac{AF}{AB} = \frac{\mathbf{af}}{\mathbf{ab}}$

or

$$AF = 1.60 \times \frac{2.76}{5.25} = 0.84 \text{ m}$$

$$v_f = \mathbf{of} = \underline{9.45 \text{ m/s}}$$

(iv) $\omega_{ba} = \frac{v_{ba}}{AB} = \frac{5.25}{1.60} = 3.28$ rad/s clockwise

(v) (a) velocity of rubbing at the pin of the crankshaft (at A)

$$= \omega_{ao}\, r_o = 9.6 \times 0.04 = \underline{0.384 \text{ m/s}} \quad \left(r_o = \frac{80}{2} = 40 \text{ mm}\right)$$

(b) ω_{oa} is counter-clockwise and ω_{ba} is clockwise.

Velocity of rubbing at the crank pin $A = (\omega_{oa} + \omega_{ba})\, r_a$

$$= (9.6 + 3.28) \times 0.03$$

$$= \underline{0.386 \text{ m/s}}$$

(c) At the cross-head, the slider does not rotate and only the connecting rod has the angular motion.

Velocity of rubbing at the cross-head pin at B

$$= \omega_{ab}\, r_b = 3.28 \times 0.05 = \underline{0.164 \text{ m/s}}$$

Example 2.3 An engine crankshaft drives a reciprocating pump through a mechanism shown in Fig. 2.11 a. The crank rotates in the clockwise direction at 160 rpm. The diameter of the pump piston at F is 200 mm. Dimensions of the various links are

OA = 170 mm (crank) CD = 170 mm

AB = 660 mm DE = 830 mm

BC = 510 mm

For the position of the crank shown in the diagram, determine

(i) the velocity of the crosshead E

(ii) the velocity of rubbing at the pins A, B, C and D, the diameters being 40, 30, 30 and 50 mm respectively.

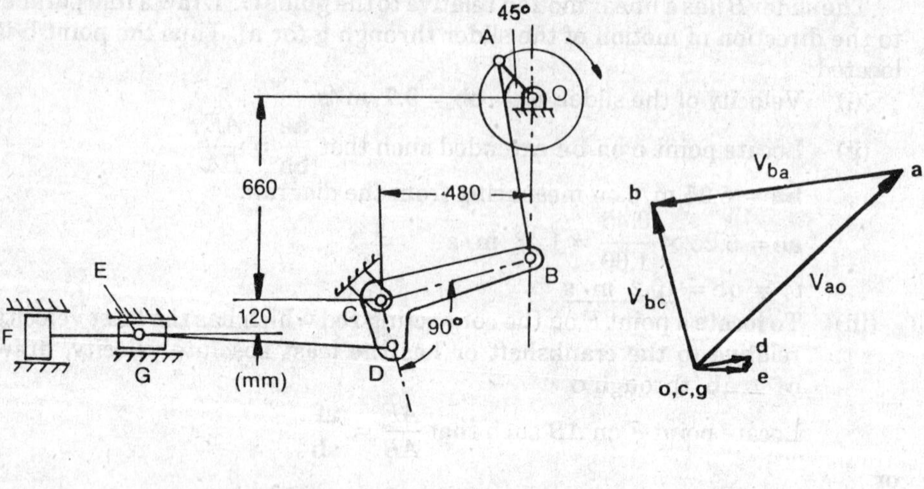

(a) (b)

Fig. 2.11

(iii) the torque required at the shaft O to overcome a pressure of 300 kN/m^2 at the pump piston at F.

Solution:

$$\omega_{ao} = \frac{2\pi N}{60} = \frac{2\pi \times 160}{60} = 16.76 \text{ rad/s}$$

$$v_{ao} = \omega_{ao} \times OA = 16.76 \times 0.17 = 2.85 \text{ m/s}$$

Writing the vector equation for the mechanism $OABC$,

$$\mathbf{v}_{bo} = \mathbf{v}_{ba} + \mathbf{v}_{ao}$$

or

$$\mathbf{v}_{bc} = \mathbf{v}_{ao} + \mathbf{v}_{ba}$$

or

$$\mathbf{cb} = \mathbf{oa} + \mathbf{ab}$$

Take the vector $\mathbf{v}_{ao}$ to a convenient scale (Fig. 2.11b)
$\mathbf{v}_{ba}$ is $\perp AB$, draw a line $\perp AB$ through **a**;
$\mathbf{v}_{bc}$ is $\perp BC$, draw a line $\perp BC$ through **c**.
The intersection of the two lines locates the point **b**. Velocity of B relative to C is upwards for the given configuration. Therefore, the link BCD moves counter-clockwise about pivot C.

$$\frac{v_{dc}}{v_{bc}} = \frac{DC}{BC} \quad \text{or} \quad v_{dc} = 1.71 \times \frac{0.17}{0.51} = 0.57 \text{ m/s} \quad \text{(perpendicular to } DC\text{)}$$

Writing the vector equation for the mechanism CDE,

$$\mathbf{v}_{ec} = \mathbf{v}_{ed} + \mathbf{v}_{dc}$$

or

$$\mathbf{v}_{eg} = \mathbf{v}_{dc} + \mathbf{v}_{ed}$$

or

$$\mathbf{ge} = \mathbf{cd} + \mathbf{de}$$

Take $\mathbf{v}_{dc}$ in the proper direction and sense from **c**;
$\mathbf{v}_{ed}$ is $\perp DE$, draw a line $\perp DE$ through **d**.
For $\mathbf{v}_{eg}$, draw a line through **g**, parallel to the direction of motion of the slider E in the guide G.
This way the point **e** is located.

(i) The velocity of the crosshead,

$$v_e = \mathbf{oe} = \underline{0.54 \text{ m/s}}$$

(ii) (a) ω_{oa} and ω_{ba}, both are clockwise.

$$\omega_{ba} = \frac{\mathbf{ab}}{AB} = \frac{2.49}{0.66} = 3.77 \text{ rad/s}$$

Velocity of rubbing at pin $A = (\omega_{oa} - \omega_{ba})\, r_a$
$= (16.76 - 3.77) \times \dfrac{0.04}{2}$
$= \underline{0.26 \text{ m/s}}$

(b) ω_{ab} is clockwise and ω_{cb} is counter-clockwise.

$$\omega_{cb} = \frac{v_{cb}}{CB} = \frac{1.71}{0.51} = 3.35 \text{ rad/s}$$

Velocity of rubbing at $B = (\omega_{ab} + \omega_{cb})\, r_b$
$= (3.77 + 3.35) \times 0.015$ . . . . $(\omega_{ab} = \omega_{ba})$
$= \underline{0.107 \text{ m/s}}$

(c) Velocity of rubbing at $C = \omega_{bc} \cdot r_c = 3.35 \times \dfrac{0.03}{2} = \underline{0.05 \text{ m/s}}$

(d) ω_{cd} and ω_{ed}, both are counter-clockwise

$$\omega_{cd} = \omega_{bc} = 3.35 \text{ rad/s ... } (BCD \text{ is one link})$$

$$\omega_{ed} = \frac{v_{ed}}{ED} = \frac{0.15}{0.83} = 0.18 \text{ rad/s}$$

Velocity of rubbing at $D = (\omega_{cd} - \omega_{ed})\, r_d$
$= (3.35 - 0.18) \times \dfrac{0.05}{2}$
$= 0.079 \text{ m/s}$

(iii) Work input = work output
$T\omega = Fv$

where

T = torque on the crankshaft
ω = angular velocity of the crank

F = force on the piston
v = velocity of the piston = $v_f = v_e$
Thus, neglecting losses,

$$T = \frac{Fv}{\omega} = \frac{\pi}{4}(0.02)^2 \times 300 \times 10^3 \times \frac{0.54}{16.76}$$

$$= \underline{303.66 \text{ N.m}}$$

Example 2.4 Figure 2.12 a shows a mechanism in which OA = 300 mm, AB = 600 mm, AC = 1200 mm and BD = 1200 mm. OD is horizontal for the given configuration. If OA rotates at 200 rpm in the clockwise direction, find:

(i) the linear velocities of C and D, and
(ii) the angular velocities of links AC and BD.

Solution:

$$\omega_a = \frac{2\pi \times 200}{60} = 20.94 \text{ rad/s}$$

$$v_a = \omega_a OA = 20.94 \times 0.03 = 6.28 \text{ m/s}$$

Writing the vector equation for the mechanism OAC,

$$\mathbf{v}_{co} = \mathbf{v}_{ca} + \mathbf{v}_{ao}$$

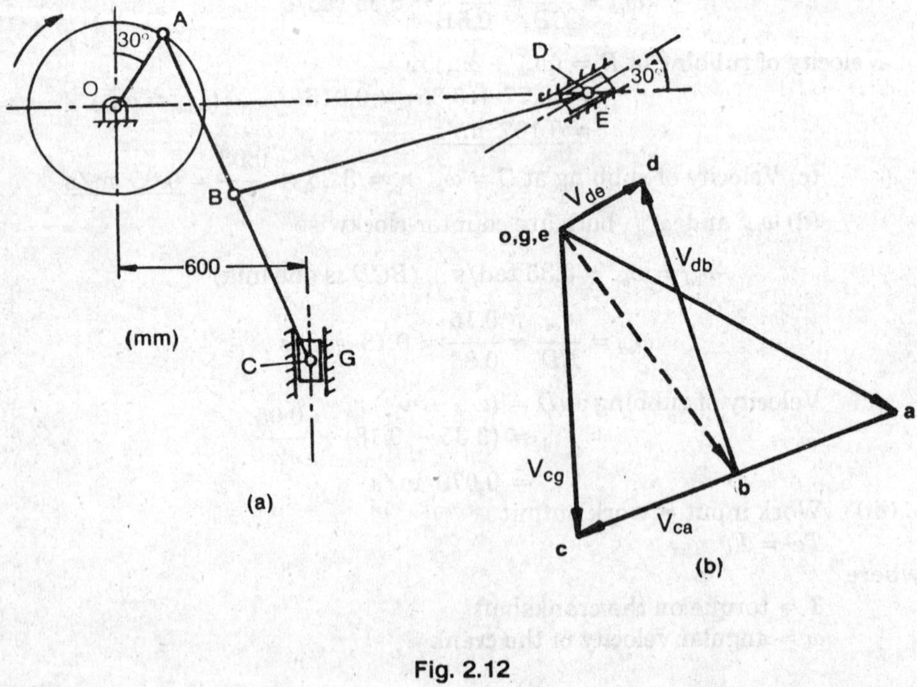

Fig. 2.12

or

$$\mathbf{v}_{cg} = \mathbf{v}_{ao} + \mathbf{v}_{ca}$$

or

$$\mathbf{gc} = \mathbf{oa} + \mathbf{ac}$$

Take the vector $\mathbf{v}_{ao}$ to a convenient scale (Fig. 2.12).
$\mathbf{v}_{ca}$ is $\perp$ AC, draw a line $\perp$ AC through **a**;
$\mathbf{v}_{cg}$ is vertical, draw a vertical line through **g** (or **o**).

The intersection of the two lines locates the point **c**. Locate point **b** on **ac** as usual. Join **ob** which gives $\mathbf{v}_{bo}$. Writing the vector equation for the mechanism *OABD*,

$$\mathbf{v}_{do} = \mathbf{v}_{db} + \mathbf{v}_{bo}$$

or

$$\mathbf{v}_{de} = \mathbf{v}_{bo} + \mathbf{v}_{db}$$

or

$$\mathbf{ed} = \mathbf{ob} + \mathbf{bd}$$

$\mathbf{v}_{db}$ is $\perp$ BD, draw a line $\perp$ BD through **b**.

For $\mathbf{v}_{de}$, draw a line through **e**, parallel to the line of stroke of the piston in the guide E.

The intersection locates the point **d**.

$v_c = \mathbf{oc} = \underline{5.2 \text{ m/s}}$

$v_d = \mathbf{od} = \underline{1.55 \text{ m/s}}$

$$\omega_{ac} = \omega_{ca} = \frac{v_{ca}}{AC} = \frac{5.7}{1.20} = \underline{4.75 \text{ rad/s}} \text{ clockwise}$$

$$\omega_{bd} = \omega_{db} = \frac{v_{db}}{BD} = \frac{5.17}{1.20} = \underline{4.31 \text{ rad/s}} \text{ counter-clockwise.}$$

Example 2.5 For the position of the mechanism shown in Fig. 2.13 a, determine the velocity of the slider B if the velocity of the slider A is known.

Solution: The velocity vector equation is

$$\mathbf{v}_{bo} = \mathbf{v}_{ba} + \mathbf{v}_{ao}$$

or

$$\mathbf{v}_{bg} = \mathbf{v}_{ao} + \mathbf{v}_{ba}$$

or

$$\mathbf{gb} = \mathbf{oa} + \mathbf{ab}$$

Take $\mathbf{v}_{ao}$ to a convenient scale (Fig. 2.13 b).
$\mathbf{v}_{ba}$ is $\perp$ AB, draw a line $\perp$ AB through **a**.

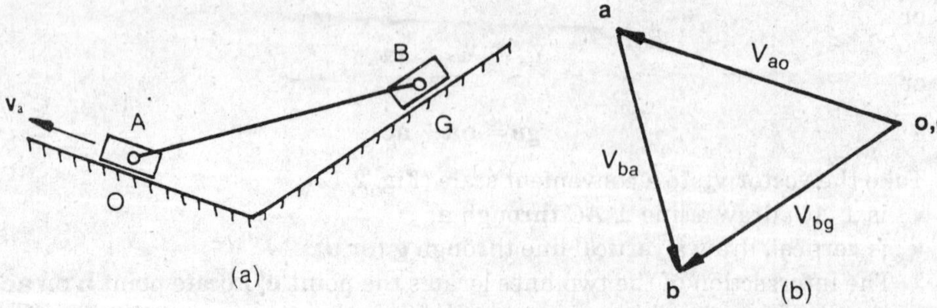

Fig. 2.13

For $\mathbf{v}_{bg}$, draw a line through **g** parallel to the line of stroke of the slider B on the guide G.

The intersection of the two lines locates the point **b**.

Velocity of B = **ob** = $\mathbf{v}_{bo}$

Example 2.6 The lengths of various links of a mechanism shown in Fig. 2.14 a are as follows:

OA = 150 mm	CD = 125 mm
AC = 600 mm	BD = 500 mm
CQ = QD = 145 mm	OQ = 625 mm

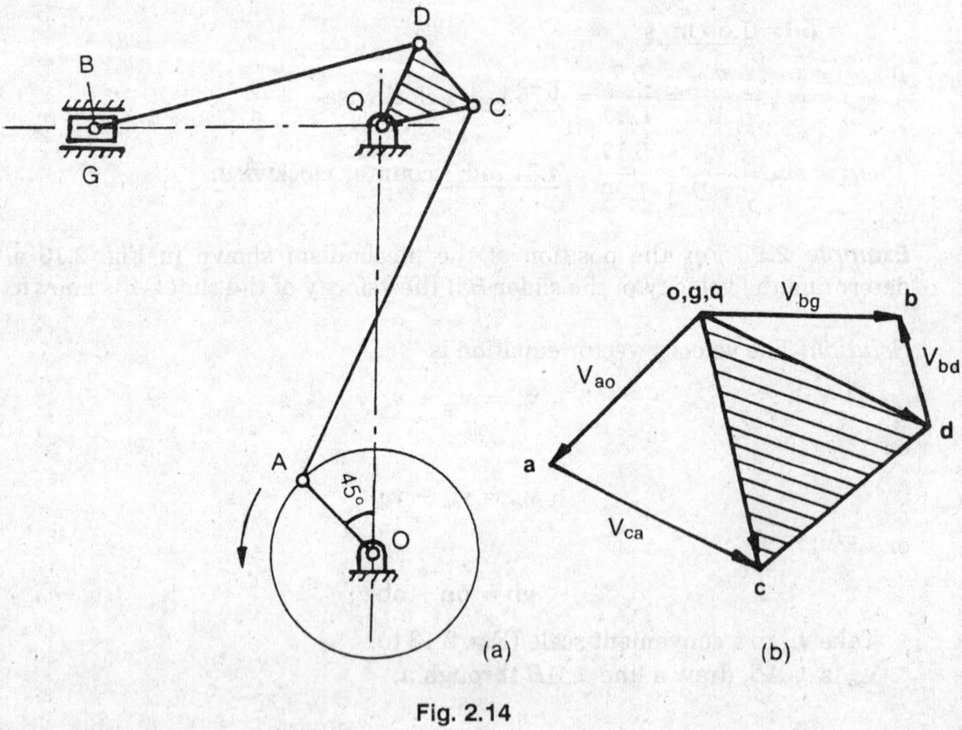

Fig. 2.14

The crank OA rotates at **60 rpm** in the counter-clockwise direction. Determine the velocity of the slider B and the angular velocity of the link BD when the crank has turned an angle of 45° with the vertical.

Solution:

$$v_a = \frac{2\pi N}{60} \times OA = \frac{2\pi \times 60}{60} \times 0.15 = 0.94 \text{ m/s}$$

Take the vector $\mathbf{v}_a$ to a convenient scale (Fig. 2.14 b) and complete the velocity diagram for the mechanism $OACQ$.

Now CQD is one link. Make Δ **cqd** similar to Δ CQD such that **cqd** reads clockwise as CQD is clockwise. This locates the point **d**. Complete the velocity diagram for the mechanism QDB.

$$v_b = \mathbf{ob} = 0.9 \text{ m/s}$$

$$\omega_{bd} = \frac{v_{bd}}{BD} = \frac{0.49}{0.50} = \underline{0.98 \text{ rad/s}} \text{ clockwise}$$

Example 2.7 Figure 2.15 a shows an Andrew variable stroke engine mechanism. The lengths of the cranks OA and QB are 90 mm and 45 mm respectively. The diameters of wheels with centres O and Q are 250 mm and 120 mm respectively. Other lengths are shown in the diagram in mm. There is a rolling contact between the two wheels. If OA rotates at 100 rpm, determine

(i) the angular velocity of the slider D
(ii) the angular velocities of links BC and CD
(iii) the torque at QB when force required at D is 3 kN.

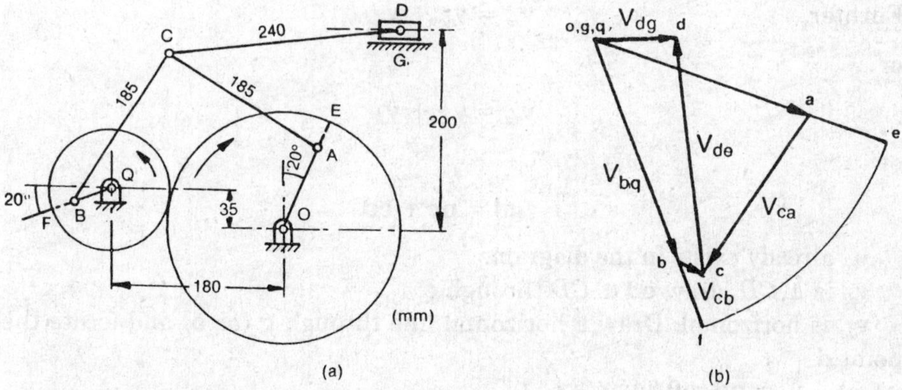

Fig. 2.15

Solution:

$$v_a = \frac{2\pi N}{60} = \frac{2\pi \times 100}{60} \times 0.09 = 0.943 \text{ m/s}$$

$$v_e = v_a \frac{OE}{OA} = 0.943 \times \frac{0.125}{0.09} = 1.309 \text{ m/s}$$

$$v_f = v_e = 1.309 \text{ m/s}$$

$$v_b = v_f \frac{QB}{QF} = 1.309 \times \frac{0.045}{0.06} = 0.982 \text{ m/s}$$

Or obtain $\mathbf{v}_b$ graphically as follows:

Take vector $\mathbf{v}_a$ to a convenient scale (Fig. 2.15 b). Produce **oa** to **e** such that **oe**/**oa** = OE/OA. Rotate **oe** to **of** so that **of** is perpendicular to QF. Mark point **b** on **qf** such that **qb**/**qf** = QB/QF.

Now,

$$\mathbf{v}_{co} = \mathbf{v}_{cq}$$

or

$$\mathbf{v}_{ca} + \mathbf{v}_{ao} = \mathbf{v}_{cb} + \mathbf{v}_{bq}$$

or

$$\mathbf{v}_{ao} + \mathbf{v}_{ca} = \mathbf{v}_{bq} + \mathbf{v}_{cb}$$

or

$$\mathbf{oa} + \mathbf{ac} = \mathbf{qb} + \mathbf{bc}$$

$\mathbf{v}_{ao}$ and $\mathbf{v}_{bq}$ are already there in the velocity diagram.

$\mathbf{v}_{ca}$ is $\perp AC$, draw a line $\perp AC$ through **a**;

$\mathbf{v}_{cb}$ is $\perp BC$, draw a line $\perp BC$ through **b**.

Thus point **c** is located.

Furhter,

$$\mathbf{v}_{do} = \mathbf{v}_{dc} + \mathbf{v}_{co}$$

or

$$\mathbf{v}_{dg} = \mathbf{v}_{co} + \mathbf{v}_{dc}$$

or

$$\mathbf{gd} = \mathbf{oc} + \mathbf{cd}$$

$\mathbf{v}_{co}$ already exists in the diagram.

$\mathbf{v}_{dc}$ is $\perp CD$, draw **cd** $\perp CD$ through **c**;

$\mathbf{v}_{dg}$ is horizontal. Draw a horizontal line through **g** (or **o**) and locate the point **d**.

(i) $\mathbf{v}_d = \mathbf{od} = \underline{0.34 \text{ m/s}}$

(ii) $\omega_{bc} = \frac{v_{bc}}{BC} = \frac{0.12}{0.185} = \underline{0.649 \text{ rad/s}}$ clockwise

$$\omega_{cd} = \frac{v_{dc}}{DC} = \frac{1.0}{0.24} = \underline{4.17 \text{ rad/s}} \text{ counter-clockwise}$$

(iii) $T\omega = F_d v_d$

$F_d = 3000$ N

$v_d =$ **od**(=**gd**) $= 0.34$ m/s

$$T = \frac{3000 \times 0.34}{(2\pi \times 100)/60} = 97.4 \text{ N.m}$$

Example 2.8 The mechanism of a stone-crusher is shown in Fig. 2.16 a along with various dimensions of links in mm. If crank *OA* rotates at a uniform velocity of 120 rpm, determine the velocity of the point *K* (jaw) when the crank *OA* is inclined at an angle of 30° to the horizontal. What will be the torque required at the crank *OA* to overcome a horizontal force of 40 kN at *K*?

Solution:

$$\omega_{ao} = \frac{2\pi \times 120}{60} = 12.6 \text{ rad/s}$$

$$v_{ao} = \omega_{ao} \times OA = 12.6 \times 0.1 = 1.26 \text{ m/s}$$

Write the vector equation for the mechansim *QABQ* and complete the velocity diagram as usual (Fig. 2.16 b). Make Δ **bac** similar to Δ*BAC* (both are read clockwise).

Write the vector equation for the mechanism *OACDM* and complete the velocity diagram. Make Δ**dmk** similar to Δ *DMK* (both are read clockwise).

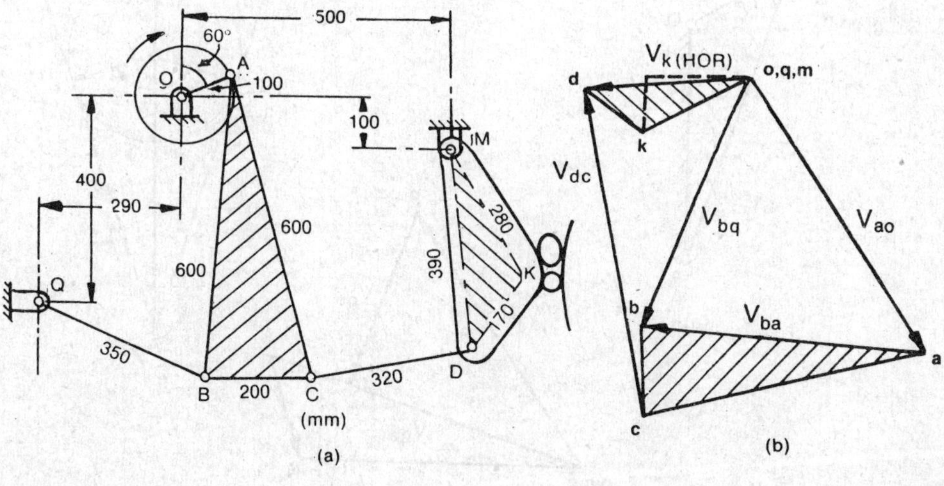

Fig. 2.16

$$v_k = \mathbf{ok} = 0.45 \text{ m/s}$$

$$v_k \text{ (horizontal)} = 0.39 \text{ m/s}$$

$$T\omega_{ao} = F_k\, v_k \text{ (horizontal)}$$

$$T = \frac{40\,000 \times 0.39}{12.6} = \underline{1242 \text{ N.m}}$$

2.10 CRANK AND SLOTTED LEVER MECHANISM

Figure 2.17a shows a crank and slotted-lever mechanism which is a form of quick-return motion used for slotting and shaping machines. OP is the crank rotating at an angular velocity of ω rad/s in the clockwise direction about centre O. At the end P of the crank, a slider S is pivoted which can move on an oscillating link AR.

A cutting tool is fixed to the ram S which is connected to R by the link RS. It is required to find the velocity of S at the given instant.

Let Q be a point on link AR immediately below P. As the crank rotates, there is relative movement of the points P and Q along AR.

Writing the vector equation for the mechanism OPA,

Vel. of Q rel. to O = Vel. of Q rel. to P + Vel. of P rel. to O

$$\mathbf{v}_{qo} = \mathbf{v}_{qp} + \mathbf{v}_{po}$$

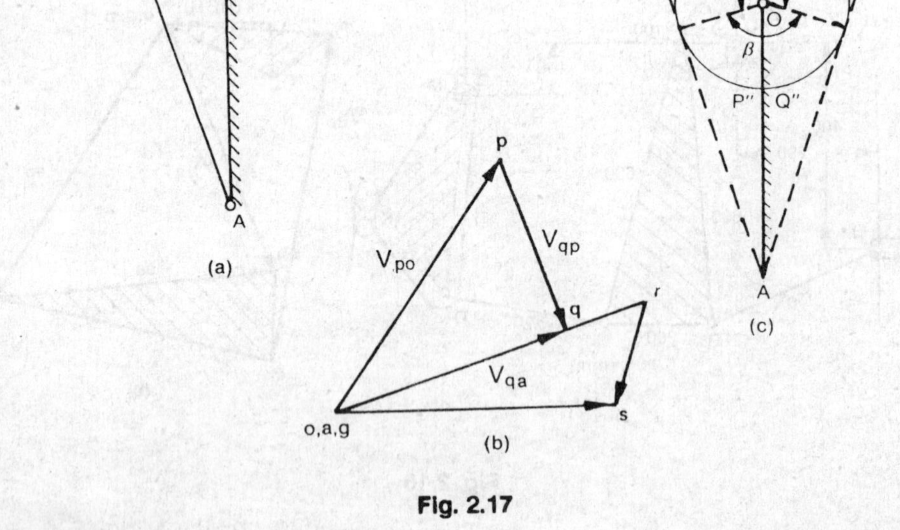

Fig. 2.17

or

$$\mathbf{v}_{qa} = \mathbf{v}_{po} + \mathbf{v}_{qp}$$

or

$$\mathbf{aq} = \mathbf{op} + \mathbf{pq}$$

In this equation,

$\mathbf{v}_{po}$ or $\mathbf{op} = \omega\, OP$; $\perp$ to OP

$\mathbf{v}_{qp}$ or $\mathbf{pq}$ is unknown in magnitude ; $||$ to AR

$\mathbf{v}_{qa}$ or $\mathbf{aq}$ is unknown in magnitude ; $\perp$ to AR

Take the vector $\mathbf{v}_{po}$ which is fully known (Fig. 2.17 b).

$\mathbf{v}_{qp}$ is $||$ AR, draw a line $||$ to AR through **p**;

$\mathbf{v}_{qa}$ is $\perp$ AR, draw a line $\perp$ AR through **a** (or **o**).

The intersection locates the point **q**.

The vector equation for the above could also have been written as:

Vel. of P rel. to A = Vel. of P rel. to Q + Vel. of Q rel. to A

$$\mathbf{v}_{pa} = \mathbf{v}_{pq} + \mathbf{v}_{qa}$$

or

$$\mathbf{v}_{po} = \mathbf{v}_{qa} + \mathbf{v}_{pq}$$

or

$$\mathbf{op} = \mathbf{aq} + \mathbf{qp}$$

Take the vector $\mathbf{v}_{po}$ which is completely known.

$\mathbf{v}_{qa}$ is $\perp$ AR, draw a line $\perp$ AR through **a**;

$\mathbf{v}_{pq}$ is $||$ AR, draw a line $||$ AR through **p**.

The intersection locates the point **q**. Observe that the velocity diagrams obtained in the two cases are the same except that the direction of $\mathbf{v}_{pq}$ is the reverse of that of $\mathbf{v}_{qp}$.

As the vectors **oq** and **qp** are perpendicular to each other, the vector $\mathbf{v}_{po}$ may be assumed to have two components, one perpendicular to AR and the other parallel to AR.

The component of velocity along AR, i.e. **qp** indicates the relative velocity between Q and P or the velocity of sliding of the block on link AR.

Now, the velocity of R is perpendicular to AR. As the velocity of Q perpendicular to AR is known, the point **r** will lie on vector **aq** produced such that $\mathbf{ar}/\mathbf{aq} = AR/AQ$

To find the velocity of ram S, write the velocity vector equation,

$$\mathbf{v}_{so} = \mathbf{v}_{sr} + \mathbf{v}_{ro}$$

or

$$\mathbf{v}_{sg} = \mathbf{v}_{ro} + \mathbf{v}_{sr}$$

or

$$\mathbf{gs} = \mathbf{or} + \mathbf{rs}$$

$\mathbf{v}_{ro}$ is already there in the diagram. Draw a line through **r** perpendicular to *RS* for the vector $\mathbf{v}_{sr}$ and a line through **g**, parallel to the line of motion of the slider *S* on the guide *G*, for the vector $\mathbf{v}_{sg}$. In this way the point **s** is located.

The velocity of the ram *S* = **os** (or **gs**) towards right for the given position of the crank.

Also $\omega_{rs} = \dfrac{v_{rs}}{RS}$ clockwise

Usually, the coupler *RS* is long and its obliquity is neglected.

Then **or** ≈ **os**

Referring Fig. 2.17 c,

$$\frac{\text{Time of cutting}}{\text{Time of return}} = \frac{\theta}{\beta}$$

When the crank assumes the position *OP′* during the cutting stroke, the component of velocity along *AR* (i.e. **pq**) is zero and **oq** is maximum (=**op**)

Let

r = length of crank (=*OP*)

l = length of slotted lever (= *AR*)

c = distance between fixed centres (=*AO*)

ω = angular velocity of the crank.

Thus, during the cutting stroke,

$$v_{s\,\max} = \omega \times OP' \times \frac{AR}{AQ} = \omega r \times \frac{l}{c + r}$$

This is by neglecting the obliquity of the link *RS*, i.e. assuming the velocity of *S* equal to that of *R*.

Similarly, during the return stroke,

$$v_{s\,\max} = \omega \times OP'' \times \frac{AR}{AQ''} = \omega r \times \frac{l}{c - r}$$

$$\frac{v_{s\,\max}\,(\text{cutting})}{v_{s\,\max}(\text{return})} = \frac{\omega r \dfrac{l}{c + r}}{\omega r \dfrac{l}{c - r}} = \frac{c - r}{c + r}$$

Example 2.9 For the inverted slider-crank mechanism shown in Fig. 2.18 a, find the angular velocity of the link *QR* and the sliding velocity of the block on the link *QR*. The crank *OA* is 300 mm long and rotates at 20 rad/s in the clockwise direction. *OQ* is 650 mm and ∠*QOA* = 40°.

Solution:

$v_{ao} = 20 \times 0.30 = 6$ m/s

The velocity vector equation can be written as usual.

$\mathbf{v}_{aq} = \mathbf{v}_{ab} + \mathbf{v}_{bq}$ or $\mathbf{v}_{bo} = \mathbf{v}_{ba} + \mathbf{v}_{ao}$

$\mathbf{v}_{ao} = \mathbf{v}_{bq} + \mathbf{v}_{ab}$ $\quad\mathbf{v}_{bq} = \mathbf{v}_{ao} + \mathbf{v}_{ba}$

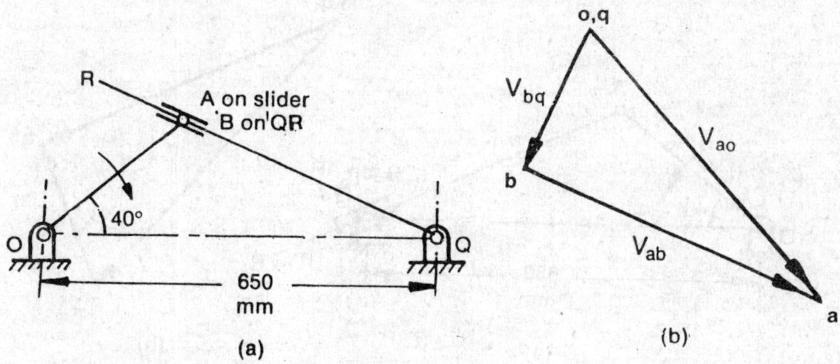

Fig. 2.18

$$\mathbf{oa} = \mathbf{qb} + \mathbf{ba} \qquad \mathbf{qb} = \mathbf{oa} + \mathbf{ab}$$

$\mathbf{v}_{oa}$ is fully known and after taking this vector, draw lines for $\mathbf{v}_{bq}$ and $\mathbf{v}_{ab}$ (or $\mathbf{v}_{ba}$) and locate the point **b**. Obviously, the direction-sense of $\mathbf{v}_{ab}$ is opposite to that of $\mathbf{v}_{ba}$. Figure 2.18 b shows the solution of the first equation.

$$\omega_{qr} = \omega_{qb} = \frac{\mathbf{v}_{qb} \text{ or } \mathbf{v}_{bq}}{BQ}$$

$$= \frac{2.55}{0.46} \qquad (BQ = 0.46 \text{ m on measuring})$$

$$= \underline{5.54 \text{rad/s}} \text{ counter-clockwise}$$

Sliding velocity of block = v_{ba} or **ab** = $\underline{5.45 \text{ m/s}}$

Example 2.10 For the position of the mechanism shown in Fig.2.19a, calculate the angular velolcity of the link *AR*. *OA* is 300 mm long and rotates at 20 rad/s in the clockwise direction. OQ = 650 mm and $\angle QOA = 40°$.

Solution:

$$v_{ao} = 20 \times 0.3 = 6 \text{ m/s}$$

Writing the vector equation,

$$\mathbf{v}_{bo} = \mathbf{v}_{ba} + \mathbf{v}_{ao} \quad \text{or} \quad \mathbf{v}_{aq} = \mathbf{v}_{ab} + \mathbf{v}_{bq}$$

Solving the first one,

$$\mathbf{v}_{bq} = \mathbf{v}_{ao} + \mathbf{v}_{ba}$$

or

$$\mathbf{qb} = \mathbf{oa} + \mathbf{ab}$$

Take $\mathbf{v}_{ao}$ to a convenient scale (Fig. 2.19 b).

$\mathbf{v}_{ba}$ is $\perp AB$, draw a line $\perp AB$ through **a**;

$\mathbf{v}_{bq}$ is along *AB*, draw a line || to *AB* through **q**.

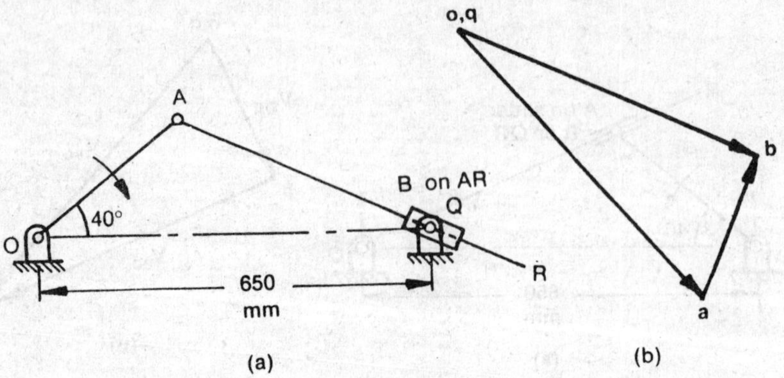

Fig. 2.19

The intersection locates the point **b**.

$$\omega_{ar} = \omega_{ab} = \frac{v_{ab} \text{ or } v_{ba}}{AB} = \frac{2.55}{0.46}$$

$$= \underline{5.54 \text{ rad/s}} \text{ counter-clockwise}$$

Example 2.11 In the pump mechansim shown in Fig.2.20a, OA = 320 mm AC = 680 mm and OQ = 650 mm. For the given configuration, determine

(i) the angular velocity of the cylinder
(ii) the sliding velocity of the plunger
(iii) the absolute velocity of the plunger.

The crank OA rotaes at 20 rad/s clockwise.

Solution:

$$v_{ao} = 0.32 \times 20 = 6.4 \text{ m/s}$$

Method I Produce AC to R. Line AC passes through the pivot Q. Let B be a point on AR beneath Q.

Writing the vector equation,

$$\mathbf{v}_{bo} = \mathbf{v}_{ba} + \mathbf{v}_{ao} \quad \text{or} \quad \mathbf{v}_{aq} = \mathbf{v}_{ab} + \mathbf{v}_{bq}$$

Solving any of these equations leads to same velocity diagram except for the direction of $\mathbf{v}_{ba}$ and $\mathbf{v}_{ab}$.

Taking the latter equation,

$$\mathbf{v}_{aq} = \mathbf{v}_{ab} + \mathbf{v}_{bq}$$

or

$$\mathbf{v}_{ao} = \mathbf{v}_{bq} + \mathbf{v}_{ab}$$

or

$$\mathbf{oa} = \mathbf{qb} + \mathbf{ba}$$

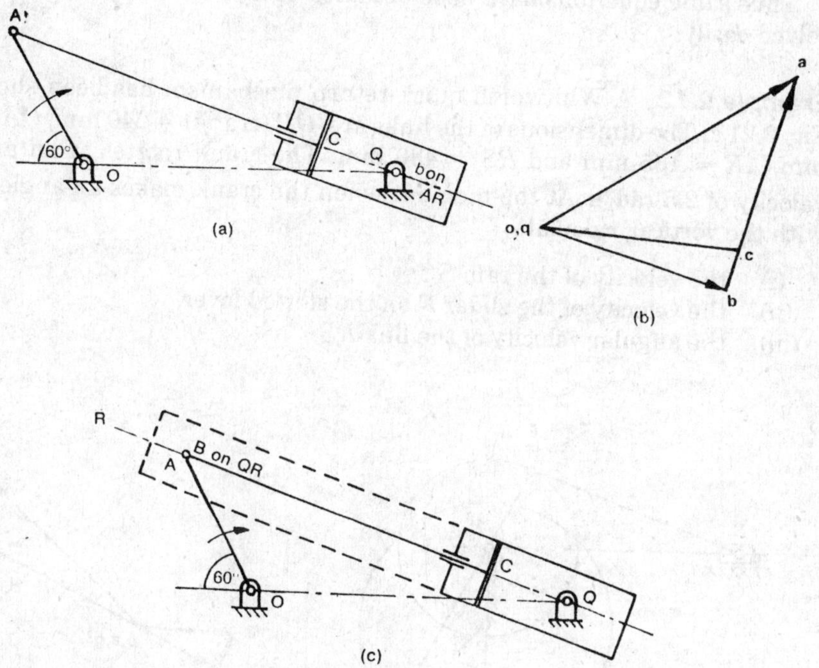

Fig. 2.20

Complete the velocity triangle as usual (Fig. 2.20 b)

Locate point **c** on **ab** such that $\dfrac{\mathbf{ac}}{\mathbf{ab}} = \dfrac{AC}{AB}$

(i) Angular velocity of cylinder = Angular velocity of AR or AB

$$= \frac{v_{ab}}{AB}$$

$$= \frac{4.77}{0.85} = \underline{5.61 \text{ rad/s}} \text{ clockwise}$$

(ii) Sliding velocity of plunger = Velocity of B relative to Q
$= \mathbf{qb} = \underline{4.1 \text{ m/s}}$

(iii) Absolute velocity of plunger = **oc** or **qc** = $\underline{4.22 \text{ m/s}}$

Method II Link AC is integrated with the plunger and thus A can be considered to be a point on it. Assume the cylinder to be of such a length that a point B is located on it just beneath the point A (Fig. 2.20 c).

Writing the vector equation,

$$\mathbf{v}_{bo} = \mathbf{v}_{ba} + \mathbf{v}_{ao} \qquad \text{or} \qquad \mathbf{v}_{aq} + \mathbf{v}_{ab} + \mathbf{v}_{bq}$$

$$\mathbf{v}_{bq} = \mathbf{v}_{ao} + \mathbf{v}_{ba} \qquad \mathbf{v}_{ao} = \mathbf{v}_{bq} + \mathbf{v}_{ab}$$

$$\mathbf{qb} = \mathbf{oa} + \mathbf{ab} \qquad \mathbf{oa} = \mathbf{qb} + \mathbf{ba}$$

Thus same equations have been obtained as in method-I and thus can be solved easily.

Example 2.12 A Whitworth quick-return mechanism has been shown in Fig. 2.21 a. The dimensions of the links are: OP (crank) = 240 mm, OA = 150 mm, AR = 165 mm and RS = 430 mm. The crank rotates at an angular velocity of 2.5 rad/s. At the moment when the crank makes an angle of 45° with the vertical, calculate

(i) the velocity of the ram S
(ii) the velocity of the slider P on the slotted lever
(iii) the angular velocity of the link RS.

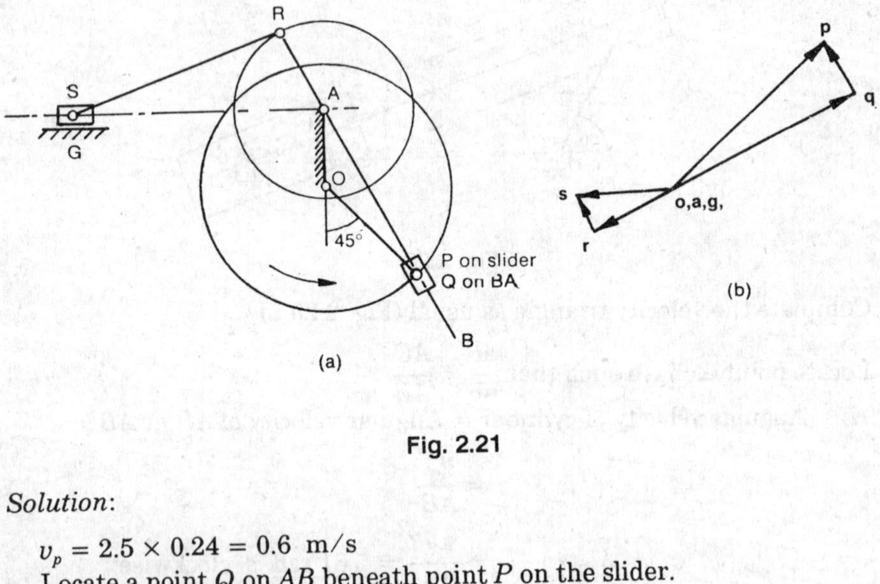

Fig. 2.21

Solution:

$v_p = 2.5 \times 0.24 = 0.6$ m/s
Locate a point Q on AB beneath point P on the slider.
Solve any of the following velocity vector equations,

$$\mathbf{v}_{pa} = \mathbf{v}_{pq} + \mathbf{v}_{qa} \qquad \text{or} \qquad \mathbf{v}_{qo} = \mathbf{v}_{qp} + \mathbf{v}_{po}$$

Produce **qa** to **r** such that $\dfrac{\mathbf{ar}}{\mathbf{qa}} = \dfrac{AR}{QA}$

Now, $\mathbf{v}_{sa} = \mathbf{v}_{sr} + \mathbf{v}_{ra}$
Complete the velocity diagram as indicated by this equation.

(i) $\mathbf{v}_s = \mathbf{gs} = \underline{0.276 \text{ m/s}}$
(ii) $\mathbf{v}_{pq} = \mathbf{qp} = \underline{0.177 \text{ m/s}}$
(iii) $\omega_{rs} = \dfrac{v_{rs} \text{ or } v_{sr}}{RS} = \dfrac{0.12}{0.43} = \underline{0.279 \text{ rad/s}}$ clockwise

2.11 INSTANTANEOUS CENTRE (I-CENTRE)

Let there be a plane body p having a non-linear motion relative to another plane body q. At any instant, the linear velocities of two points A and B on the body p are $\mathbf{v}_a$ and $\mathbf{v}_b$ respectively in the directions as shown in Fig. 2.22.

If a line is drawn perpendicular to the direction of $\mathbf{v}_a$ at A, the body can be imagined to rotate about some point on this line. Similarly, centre of rotation of the body also lies on a line perpendicular to the direction of $\mathbf{v}_b$ at B. If the intersection of the two lines is at I, the body p will be rotating about I at the instant. This point I is known as the *instantaneous centre* for the body p. The position of the instantaneous centre changes with the motion of the body. In our further discussions, this centre will be called the I-centre.

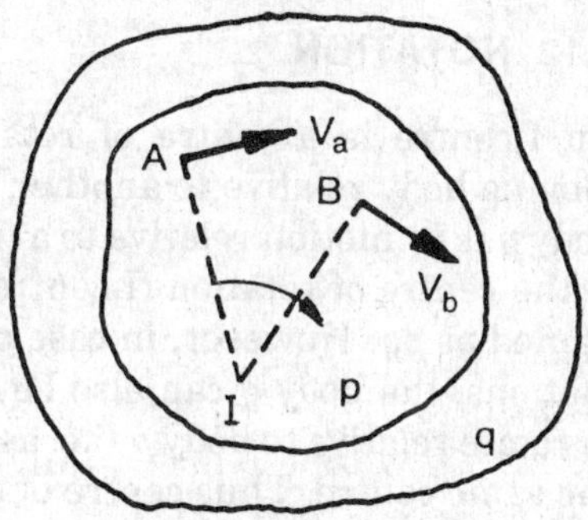

Fig. 2.22

In case the perpendiculars to $\mathbf{v}_a$ and $\mathbf{v}_b$ at A and B respectively meet outside the body p, the I-centre will lie outside the body p (Fig. 2.23 a). If the directions of $\mathbf{v}_a$ and $\mathbf{v}_b$ are parallel and the perpendiculars at A and B meet at infinity, the I-centre of the body lies at infinity. This is the case when the body has a linear motion (Fig. 2.23 b).

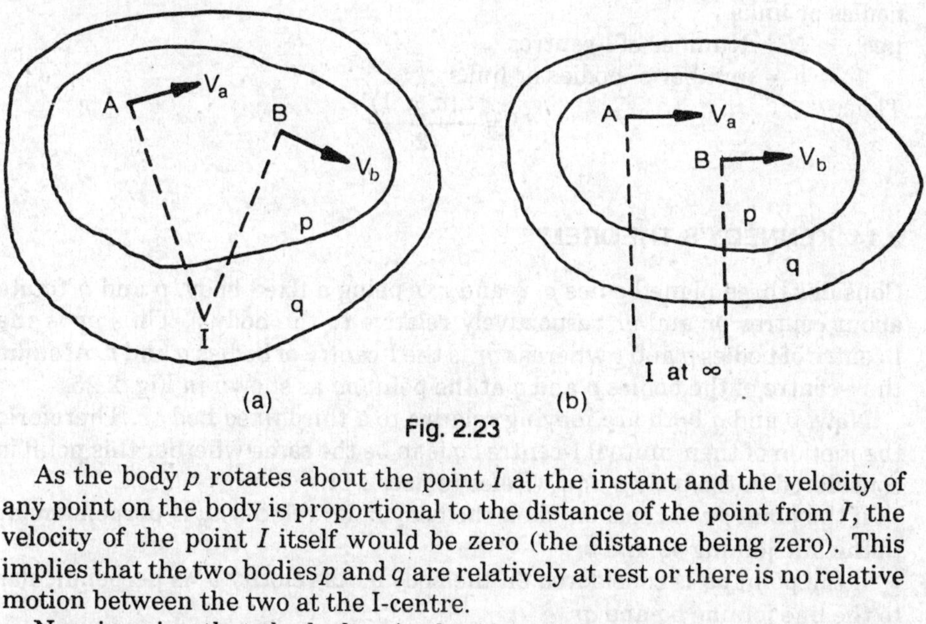

Fig. 2.23

As the body p rotates about the point I at the instant and the velocity of any point on the body is proportional to the distance of the point from I, the velocity of the point I itself would be zero (the distance being zero). This implies that the two bodies p and q are relatively at rest or there is no relative motion between the two at the I-centre.

Now imagine that the body q is also in motion relative to a third body r

(Fig. 2.24). Then the motion of the point I relative to the third body would be the same whether this point is considered on the body p or q.

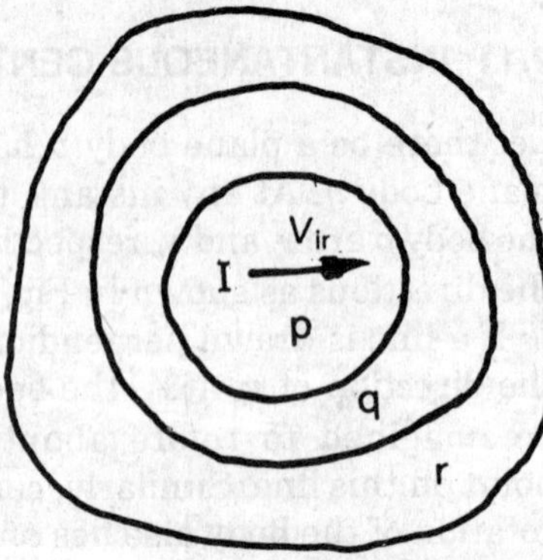

Fig. 2.24

2.12 NOTATION

An I-centre is a centre of rotation of a moving body relative to another body. If a body p is in motion relative to a fixed body q, the centre of rotation (I-centre) may be named as pq. However, in case of relative motions, the body q can also be imagined to rotate relative to body p (i.e. as if the body p is fixed for the moment) about the same centre. Thus centre of rotation or I-centre can be named qp also.

This shows that the I-centre of two bodies p and q in relative motion can be named either pq or qp meaning the same thing. In general, the I-centres will be named in the ascending order of the alphabets or digits, i.e. 13,35, pq, eg, etc.

2.13 NUMBER OF I-CENTRES

For two bodies having relative motion between them, there is an I-centre. Thus in a mechanism, number of I-centres will be equal to possible pairs of bodies or links.

Let N = Number of I-centres

n = number of bodies or links

Then,

$$N = \frac{n(n-1)}{2}$$

2.14 KENNEDY'S THEOREM

Consider three plane bodies p, q and r; r being a fixed body. p and q rotate about centres pr and qr respectively relative to the body r. Thus pr is the I-centre of bodies p and r whereas qr is the I-centre of bodies q and r. Assume the I-centre of the bodies p and q at the point pq as shown in Fig. 2.25.

Now, p and q both are moving relative to a third fixed body r. Therefore, the motion of their mutual I-centre pq is to be the same whether this point is considered in the body p or q. (Refer to Sec. 2.11.)

If the point pq is considered on the body p, its velocity v_p is perpendicular to the line joining pq and pr.

If the point pq is considered on the body q, its velocity v_q is perpendicular to the line joining pq and qr.

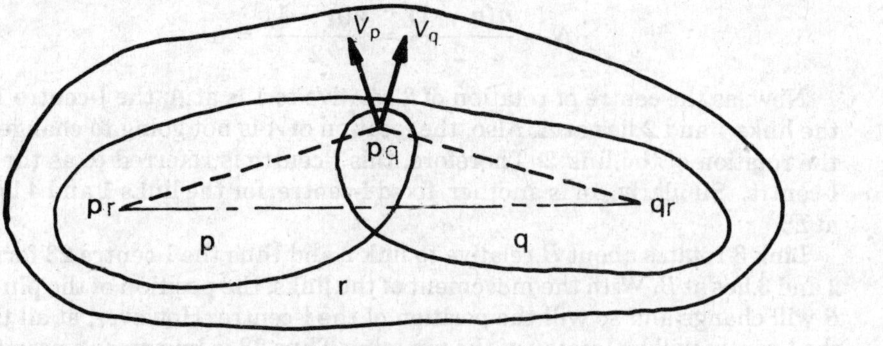

Fig. 2.25

It is found that the two velocities of the I-centre *pq* are in different directions which is impossible. Therefore, the I-centre of the bodies *p* and *q* cannot be at the assumed position of *pq*. The velocities v_p and v_q of the I-centre will be same only if this centre lies on the line joining *pr* and *qr*.

In words, if three plane bodies have relative motion among themselves, their I-centres must lie on a straight line. This is known as Kennedy's theorem.

2.15 LOCATING I-CENTRES

The procedure to locate I-centres of a mechanism is being illustrated with the help of following example of a four-link mechanism.

Figure 2.26 shows a four-link mechanism *ABCD*, the links of which have been named as 1, 2, 3 and 4. The number of I-centres is given by

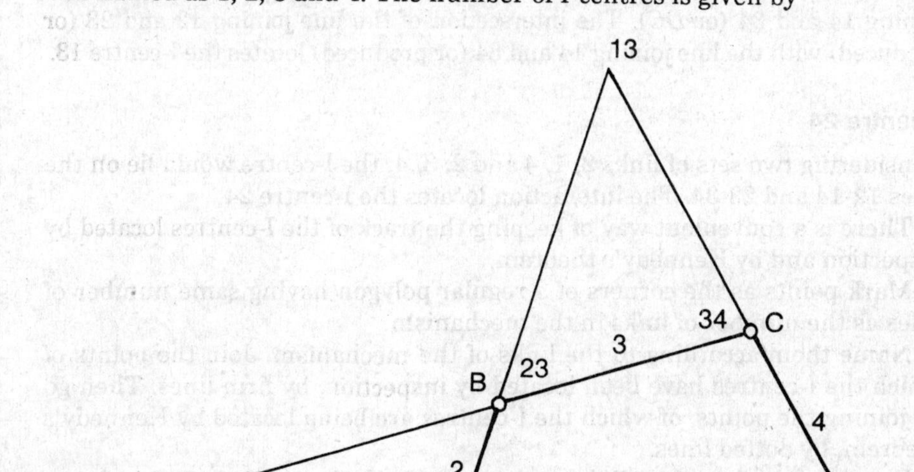

Fig. 2.26

$$N = \frac{n(n-1)}{2} = \frac{n(n-1)}{2} = 6$$

Now, as the centre of rotation of 2 relative to 1 is at *A*, the I-centre 12 for the links 1 and 2 lies at *A*. Also, the location of *A* is not going to change with the rotation of the link 2. Therefore, this I-centre is referred to as the *fixed* I-centre. Similarly, 14 is another fixed I-centre for the links 1 and 4 located at *D*.

Link 3 rotates about *B* relative to link 2 and thus the I-centre 23 for links 2 and 3 lies at *B*. With the movement of the links, the position of the pin-joint *B* will change and so will the position of the I-centre. However, at all times, the I-centre will be located at the pin joint. Thus 23 is known as a *permanent* but not a fixed I-centre. Similarly, 34 is another permanent but not a fixed I-centre for the links 3 and 4.

The above I-centres have been located by inspection only. The other two I-centres 13 and 24 which are neither fixed nor permanent can be located easily by applying Kennedy's theorem as explained below:

I-centre 13

First consider three links 1, 2 and 3. One more link 2 has been added to links 1 and 3 with the condition that the I-centres 12 and 23 are already known and the third I-centre 13 is to be located.

Now, as the three links 1, 2, and 3 have relative motions among themselves, their I-centres lie on a straight line. Thus I-centre 13 lies on the line joining 12 and 23 (or line *AB*).

Similarly, consider the links 1, 4, and 3. Their I-centres are 14, 34 and 13. Out of these, 14 and 34 are already known. Therefore, I-centre 13 lies on the line joining 14 and 34 (or *DC*). The intersection of the line joining 12 and 23 (or produced) with the line joining 14 and 34 (or produced) locates the I-centre 13.

I-centre 24

Considering two sets of links 2, 1, 4 and 2, 3, 4; the I-centre would lie on the lines 12-14 and 23-34. The interaction locates the I-centre 24.

There is a convenient way of keeping the track of the I-centres located by inspection and by Kennedy's theorem.

Mark points as the corners of a regular polygon having same number of sides as the number of links in the mechanism.

Name them according to the links of the mechanism. Join the points of which the I-centres have been located by inspection, by firm lines. Then go on joining the points, of which the I-centres are being located by Kennedy's theorem, by dotted lines.

For example, for a four link mechanism, mark the points 1, 2, 3 and 4 as shown in Fig. 2.27 a. Join 12, 23, 34 and 14 (or 41) by firm lines after locating these I-centres by inspection (Fig. 2.27 b). In Fig. 2.27 c these centres have been encircled for the record.

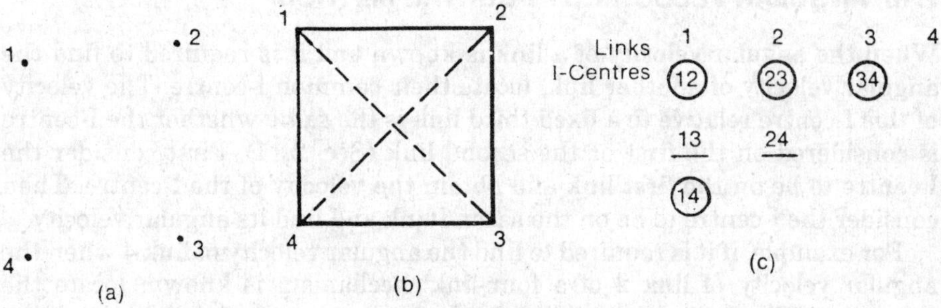

Fig. 2.27

To find the I-centre 13, join 1 to 3 by a dotted line (Fig. 2.27 b). The construction shows that the I-centre lies on the line joining I-centres 12 and 23, and the line joining 14 and 34 (or 43). Locate the I-centre actually on the intersection of the two lines in the configuration diagram of the mechanism. In Fig. 2.27 c, 13 is underlined to note that the I-centre has been located by Kennedy's theorem. Similarly, find the I-centre 24 by joining 2 and 4 and locate the point on the intersection of the lines 12-14 and 23-34.

Rules to Locate I-centres by Inspection

1. In a pivoted joint, the centre of the pivot is the I-centre for the two links of the pivot (Fig. 2.28 a).
2. In a sliding motion, the I-centre lies at infinity in a direction perpendicular to the path of motion of the slider. This is because the sliding motion is equivalent to a rotary motion of the links with radius of curvature as infinity (Fig. 2.28 b).
3. In a pure rolling contact of the two links, the I-centre lies at the point of contact at the given instant (Fig. 2.28 c). It is because the two points of contact on the two bodies have the same linear velocity and thus there is no relative motion of the two at the point of contact which is the I-centre (Refer Sec. 2.11).

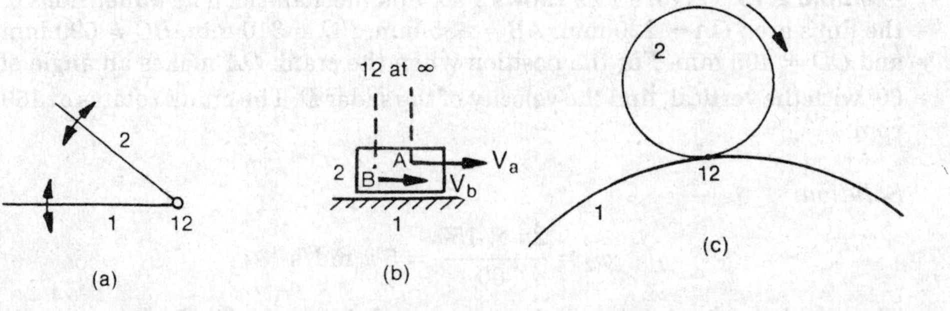

Fig. 2.28

2.16 ANGULAR VELOCITY BY I-CENTRE METHOD

When the angular velocity of a link is known and it is required to find the angular velocity of another link, locate their common I-centre. The velocity of this I-centre relative to a fixed third link is the same whether the I-centre is considered on the first or the second link (Sec. 2.11). First consider the I-centre to be on the first link and obtain the velocity of the I-centre. Then consider the I-centre to be on the second link and find its angular velocity.

For example, if it is required to find the angular velocity of link 4 when the angular velocity of link 2 of a four-link mechanism is known, locate the I-centre 24 (Fig. 2.26). Imagine link 2 to be in the form of a flat disk containing point 24 and revolving about 12 or *A*. Then

$$v_{24} = \omega_2\,(24 - 12)$$

Now, imagine link 4 to be large enough to contain point 24 and revolving about 14 or *D*. Then

$$v_{24} = \omega_4\,(24 - 12)$$

or

$$\omega_4 = \frac{v_{24}}{24 - 14} = \omega_2\left(\frac{24 - 12}{24 - 14}\right)$$

or

$$\frac{\omega_4}{\omega_2} = \frac{24 - 12}{24 - 14}$$

In words, the angular velocities of two links vary inversely as the distances from their common I-centre to their respective centres of rotation relative to the frame.

In the above case, the points 12 and 14 lie on the same side of 24 on the line 24-14 and the direction of rotation of the two links (2 and 4) is the same, i.e. clockwise or counter-clockwise. Had they been on the opposite sides of the common I-centre, the direction would have been opposite.

Example 2.13 Figure 2.29 shows a six-link mechanism. The dimensions of the links are : *OA* = 220 mm, *AB* = 485 mm, *BQ* = 310 mm, *BC* = 590 mm and *CD* = 400 mm. For the position when the crank *OA* makes an angle of 60° with the vertical, find the velocity of the slider *D*. The crank rotates at 150 rpm.

Solution:

$$\omega_2 = \frac{2\pi \times 150}{60} = 5\,\pi \text{ rad/s}$$

The velocity of a point *A* on link 2 is known. It is required to find the velocity of a point on link 6. Thus locate the I-centre 26 as follows:

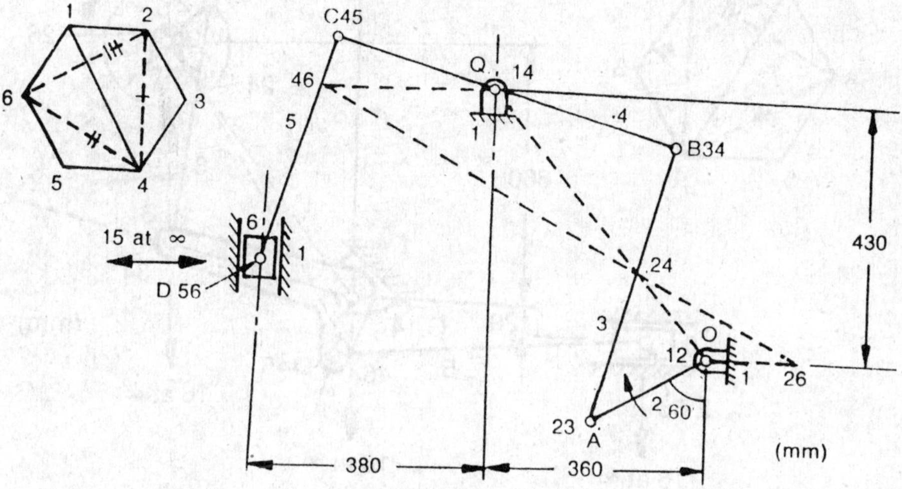

Fig. 2.29

(a) Locate I-centres 12, 23, 34, 45, 56, 16 and 14 by inspection.

(b) Locate 24 which lies on the intersection of 21-14 and 23-34.

(c) Locate 46 which lies on the intersection of 45-56 and 14-16 (16 is at ∞).

(d) Locate 26 which is the intersection of 24-46 and 21-16.

Imagine link 2 to be in the form of a flat disc containing point 26 and revolving about O with an angular velocity of 5π rad/s.

Then,

$$v_{26} = \omega_2 \times (12 - 26) = 5\pi \times 0.145 = 2.28 \text{ m/s.}$$

The velocity of point 26 is in the vertically downward direction if OA rotates clockwise.

Imagine link 6 (slider) to be large enough to contain point 26. Now, the slider can have motion in the vertical direction only and the velocity of a point 26 on it is known; it implies that all the points on the slider move with the same velocity.

Thus velocity of the slider, $v_d = v_{26} = \underline{2.28 \text{ m/s}}$

Example 2.14 Solve Example 2.3 by the instantaneous centre method.

Solution Draw the configuration to a suitable scale as shown in Fig. 2.30.

(a) To find the velocity of E or the link 6, is required to locate the I-centre 26 as the velocity of a point A on link 2 is known.

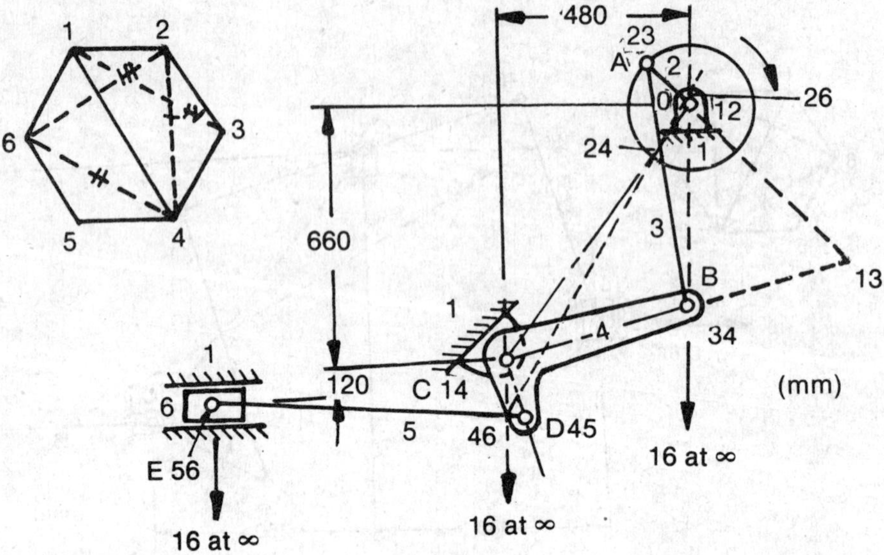

Fig. 2.30

After locating I-centres by inspection, locate I-centres 24, 46 and 26 by Kennedy's theorem.

First consider 26 to be on the crank 2.

$$v_{26} = \omega\ (12 - 26) = 16.76 \times 0.032 = 0.536 \text{ m/s (horizontal)}$$

When the point 26 is considered on the link 6, all points on it will have the same velocity as the point 26.

Velocity of the crosshead = $\underline{0.536 \text{ m/s}}$

(b) (i) To find the velocity of rubbing at A (or 23), ω_2 and ω_3 are required. Locate I-centre 13. Then

$$\omega_3\ (23 - 13) = \omega_2\ (23 - 12)$$

$$\therefore \quad \omega_3 = 16.76 \times \frac{0.17}{0.756}$$

$$= \underline{3.77 \text{ rad/s}}$$

ω_3 is clockwise as 13 and 12 lie on the same side of 23.

Velocity of rubbing at $A = (\omega_3 - \omega_2)\ r_a$

$$= (16.76 - 3.77) \times \frac{0.04}{2}$$

$$= \underline{0.26 \text{ m/s}}$$

(ii) For velocity of rubbing at B, ω_3 and ω_4 are required. ω_3 was calculated above.

$$\omega_4\ (34 - 14) = \omega_3\ (34 - 13)$$

$$\omega_4 = 3.77 \times \frac{0.45}{0.51} = 3.33 \text{ rad/s}$$

ω_4 is counter-clockwise as 14 and 13 lie on the opposite sides of 34 and ω_3 is clockwise.

Thus velocity of rubbing at B can be calculated.

(iii) As ω_4 is known, the velocity of rubbing at C can be known.

Similarly, locate the I-centre 15 and obtain ω_5 from the relation, $\omega_5 = \omega_4 \left(\frac{45 - 14}{45 - 15}\right)$ and determine the velocity of rubbing at D.

Torque is determined in the same way as in Example 2.3.

Example 2.15 The configuration diagram of a wrapping machine is given in Fig. 2.31. Determine the velocity of the point P on the bell-crank lever 6 if crank 2 rotates at ω_2 rad/s.

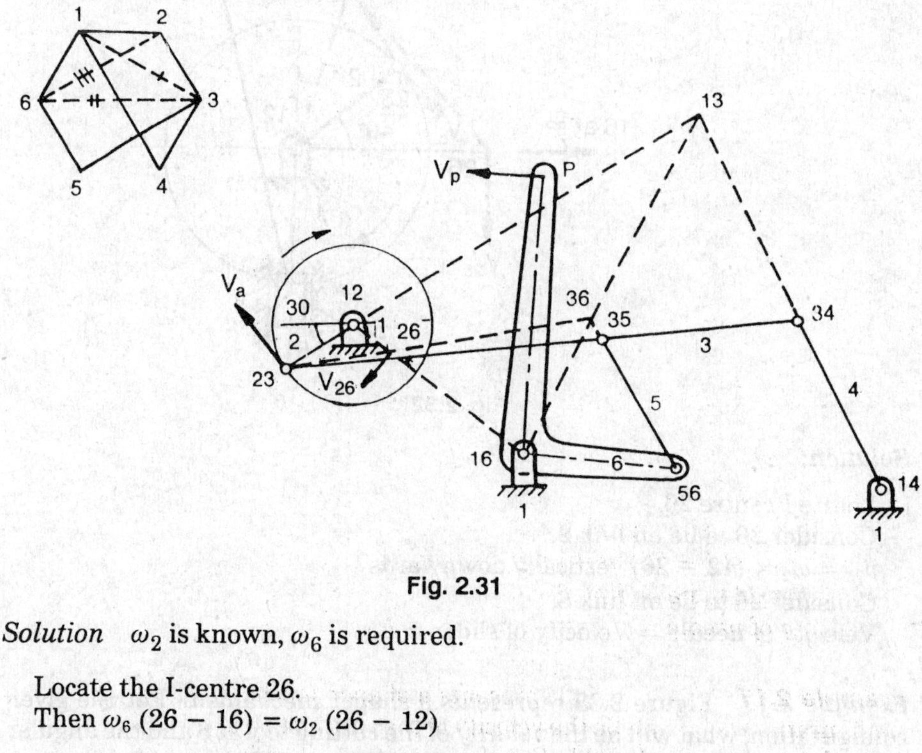

Fig. 2.31

Solution ω_2 is known, ω_6 is required.

Locate the I-centre 26.

Then ω_6 (26 – 16) = ω_2 (26 – 12)

$$\omega_6 = \omega_2 \times \frac{26 - 12}{26 - 16}$$

It is counter-clockwise as 16 and 12 lie on the opposite sides of 26 and ω_2 is clockwise.

Thus $$v_p = \omega_6 \times (16 - P)$$

$$= \omega_2 \times \frac{26 - 12}{26 - 16} \times (16 - P)$$

Example 2.16 Figure 2.32 shows the mechanism of a sewing machine needle box. For the given configuration, find the velocity of the needle fixed to the slider 6 when the crank 2 rotates at ω_2 rad/s.

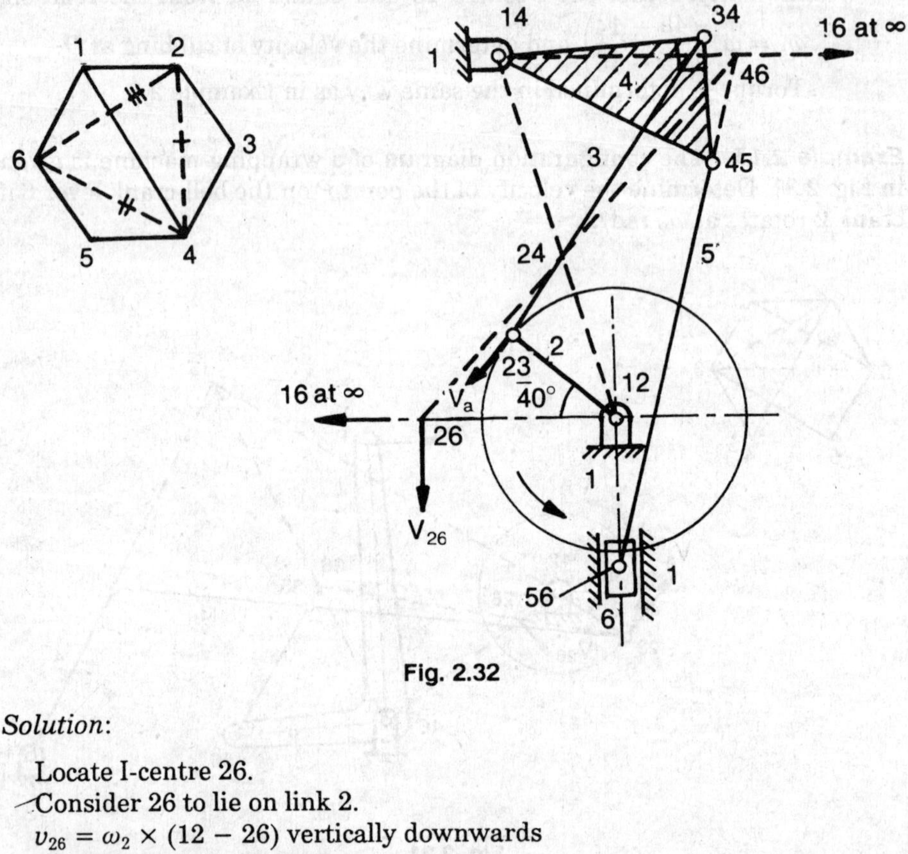

Fig. 2.32

Solution:

Locate I-centre 26.

Consider 26 to lie on link 2.

$v_{26} = \omega_2 \times (12 - 26)$ vertically downwards

Consider 26 to lie on link 6,

Velocity of needle = Velocity of slider = v_{26}

Example 2.17 Figure 2.33 represents a shaper mechanism. For the given configuration, what will be the velocity of the cutting tool at 6 and the angular velocities of the links 4 and 5. Crank 2 rotates at ω_2 rad/s.

Solution:

Locate the I-centre 26.

(i) $v_6 = v_{26} = \omega_2 \times (12 - 26)$

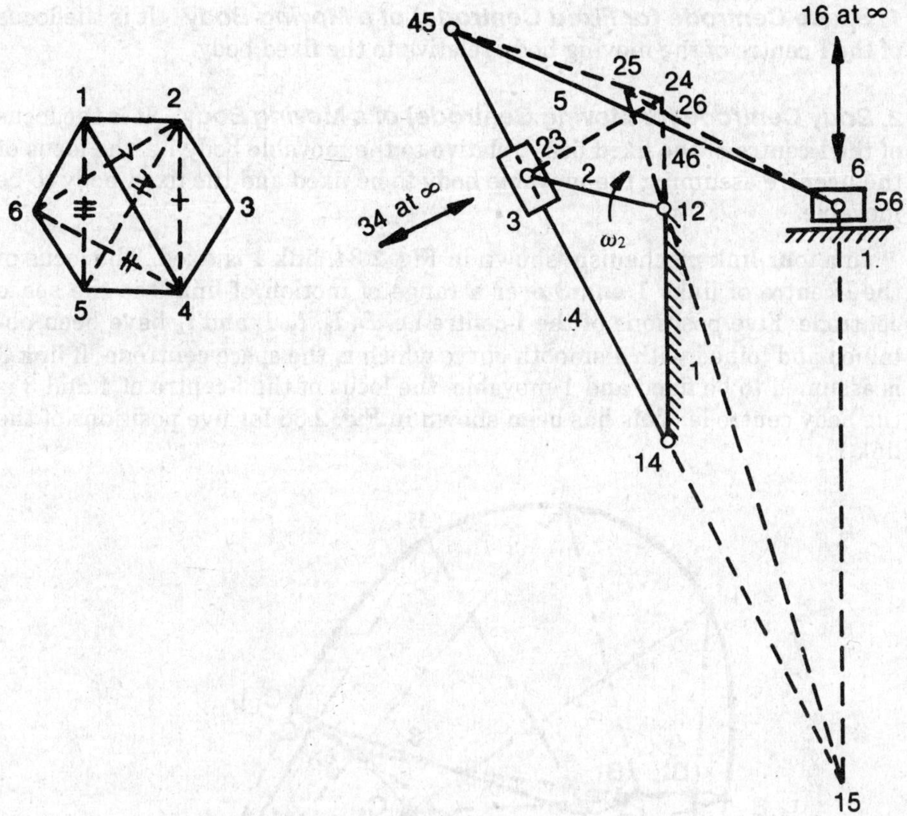

Fig. 2.33

(ii) $\omega_4 = \omega_2 \left(\dfrac{24 - 12}{24 - 14}\right)$ clockwise

Similarly,

$$\omega_5 = \omega_2 \left(\frac{25 - 12}{25 - 15}\right) \text{ clockwise}$$

or

$$\omega_5 = \omega_4 \left(\frac{45 - 14}{45 - 15}\right) \text{ clockwise}$$

2.17 CENTRODE

A centrode is the locus of the I-centre of a plane body relative to another plane body for the range of motion specified or during a finite period of time.

There are two types of centrodes:

1. Space Centrode (or Fixed Centrode) of a Moving Body It is the locus of the I-centre of the moving body relative to the fixed body.

2. Body Centrode (or Moving Centrode) of a Moving Body It is the locus of the I-centre of the fixed body relative to the movable body i.e. the locus of the I-centre assuming the movable body to be fixed and the fixed body to be movable.

In a four-link mechanism shown in Fig. 2.34, link 1 is fixed. The locus of the I-centre of links 1 and 3 over a range of motion of link 3 is the space centrode. Five positions of the I-centre i.e. I_1, I_2, I_3, I_4 and I_5 have been obtained and joined with a smooth curve which is the space centrode. If link 3 is assumed to be fixed and 1 movable, the locus of the I-centre of 1 and 3 is the body centrode. This has been shown in Fig. 2.35 for five positions of the link 1.

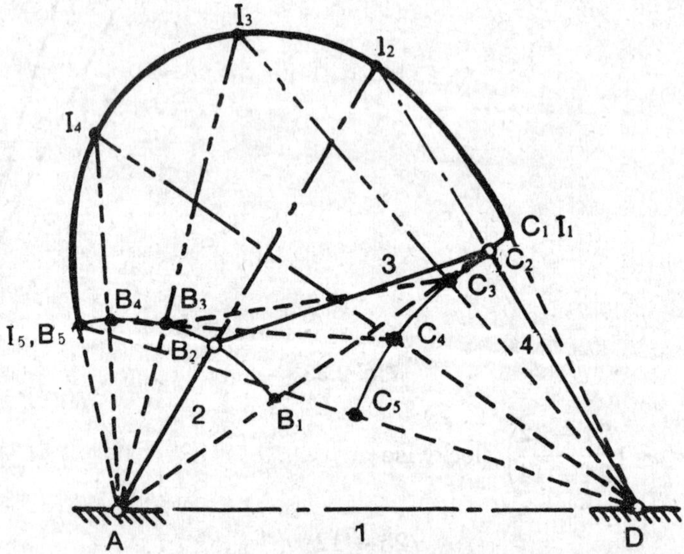

Fig. 2.34

Comparing Figs. 2.34 and 2.35, observe that the first position $AB_1C_1(I_1)D$ of Fig. 2.34 is exactly similar to the first position $A_1BC(I_1')D_1$ of Fig. 2.35. The second positions of the two figures are also similar. Similarly, the third position $AB_3I_3C_3D$ of Fig. 2.34 is exactly similar to the third position $A_3BI_3'CD_3$ of Fig. 2.35 and so on. Thus Δ s $B_2C_2I_2$, $B_3C_3I_3$ and $B_4C_4I_4$ are similar to Δ s BCI_2', BCI_3' and BCI_4' respectively. This implies that the positions of the I-centre of Fig. 2.35 can be obtained directly by constructing on B_2C_2 Δ s, similar to $B_2C_2I_2$ (already exists), $B_3C_3I_3$ and $B_4C_4I_4$. I_1' lies on C_2 and I_5' on B_2.

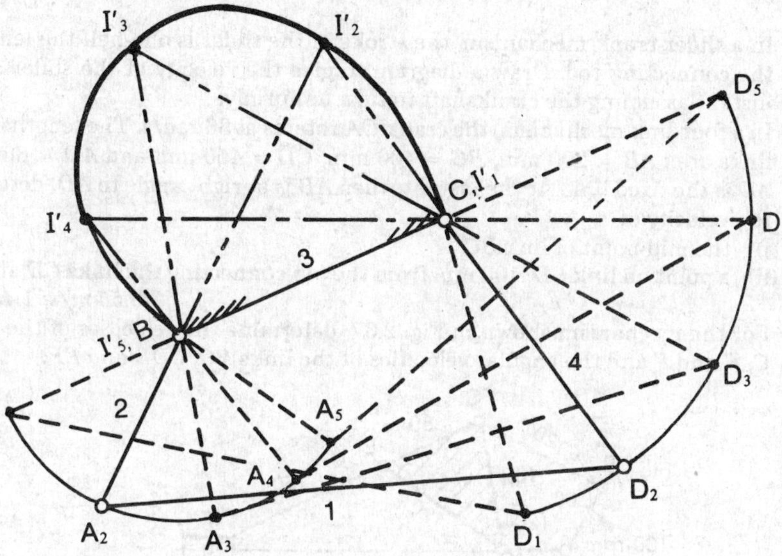

Fig. 2.35

In Fig. 2.36, space and body centrodes of link 3 relative to 1 have been obtained in the same diagram considering four positions of link 3.

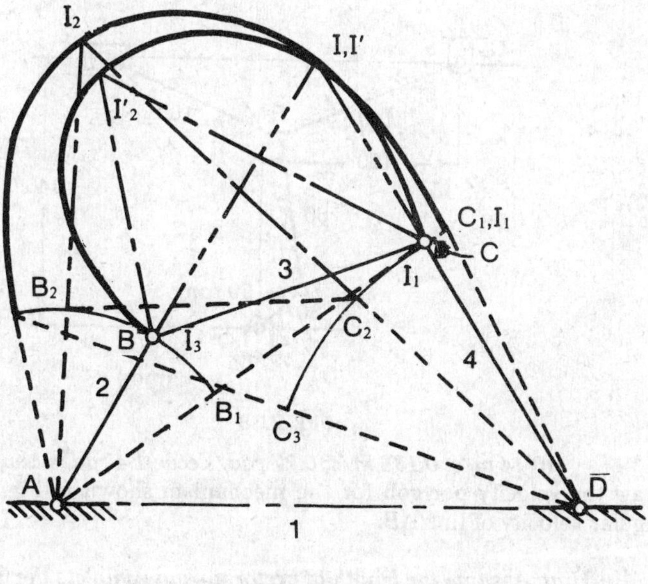

Fig. 2.36

PROBLEMS

2.1 In a slider-crank mechanism, the stroke of the slider is one-half the length of the connecting rod. Draw a diagram to give the velocity of the slider at any instant assuming the crankshaft to turn uniformly.

2.2 In a four-link mechanism, the crank *OA* rotates at 36 rad/s. The lengths of the links are : *AB* = 200 mm, *BC* = 400 mm, *CD* = 450 mm and *AD* = 600 mm. *AD* is the fixed link. At the instant when *AB* is at right angle to *AD*, determine the velocity of

(i) the mid-point of link *BC*

(ii) a point on link *CD*, 100 mm from the pin connecting the links *CD* and *AD*.

(6.55 m/s; 1.45 m/s)

2.3 For the mechanism shown in Fig. 2.37, determine the velocities of the points *C*, *E* and *F* and the angular velocities of the links *BC*, *CDE* and *EF*.

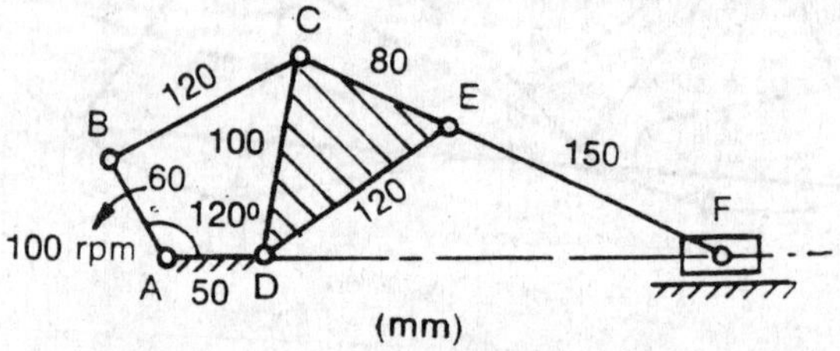

Fig. 2.37

(0.82 m/s; 0.88 m/s; 0.85 m/s.; 4.5 rad/s ccw; 8.15 rad/s ccw; 5.4 rad/s cw)

2.4 A toggle mechanism is shown in Fig. 2.38 along with the dimensions of the links in mm. Find the velocities of the points *B* and *C* and the angular velocities of links *AB*, *BQ* and *BC*. The crank rotates at 50 rpm in the clockwise direction.

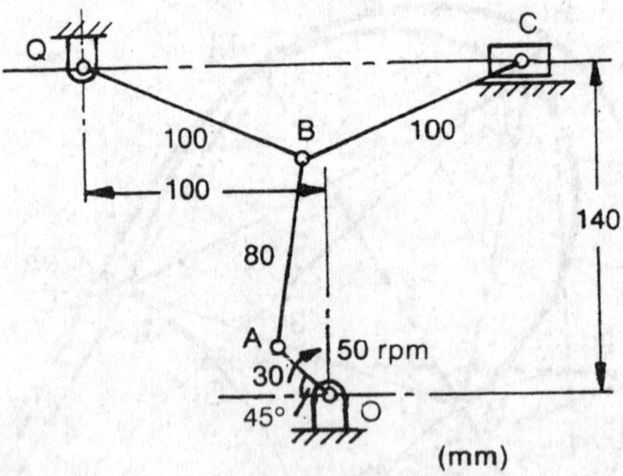

Fig. 2.38

(0.14 m/s; 0.138 m/s; 0.77 rad/s ccw; 1.4 rad/s ccw; 1.38 rad/s cw)

2.5 Draw the velocity polygon for the mechanism shown in Fig. 2.39. Find the angular velocity of link *AB*.

(4.33 rad/s cw)

(Hint: Assume the length of vector $\mathbf{v}_{ef}$ *and complete the velocity polygon. Determine the velocity scale.)*

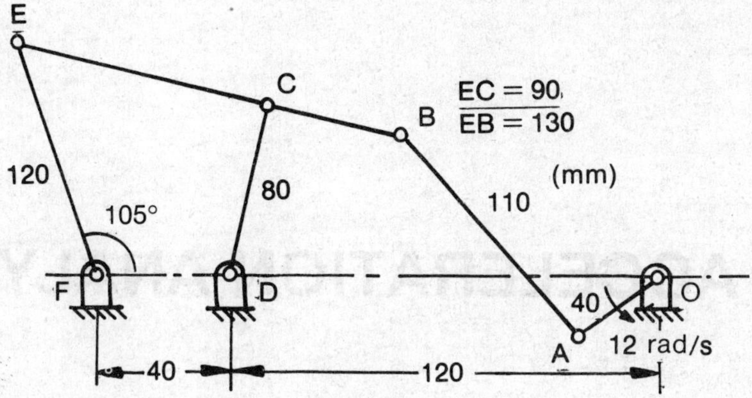

Fig. 2.39

2.6 For the mechanism shown in Fig. 2.40, determine the angular velocity of link *AB*.

(57.6 rad/s ccw)

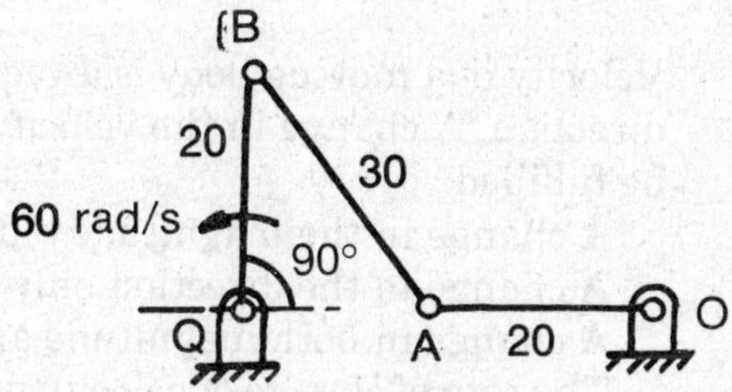

Fig. 2.40

2.7 In the mechanism shown in Fig. 2.41, *O* and *A* are fixed. *CD* = 200 mm, *OA* = 60 mm, *AC* = 50 mm and *OB* (crank) = 150 mm. ∠*OAD* = 90°. Determine the velocity of slider *D* for counter-clockwise rotation of *OB* at 80 rpm.

2.8 The crank *OP* of a crank and slotted lever mechanism (Fig. 2.17) rotates at 100 rpm in the counter-clockwise direction. Various lengths of the links are *OP* = 90 mm, *OA* = 300 mm, *AR* = 480 mm and *RS* = 330 mm. The slider moves along an axis perpendicular to *AO* and is 120 mm from *O*. Determine the velocity of the slider when the angle *AOP* is 135° and also the maximum velocity of the slider.

(0.954 m/s; 2.16 m/s)

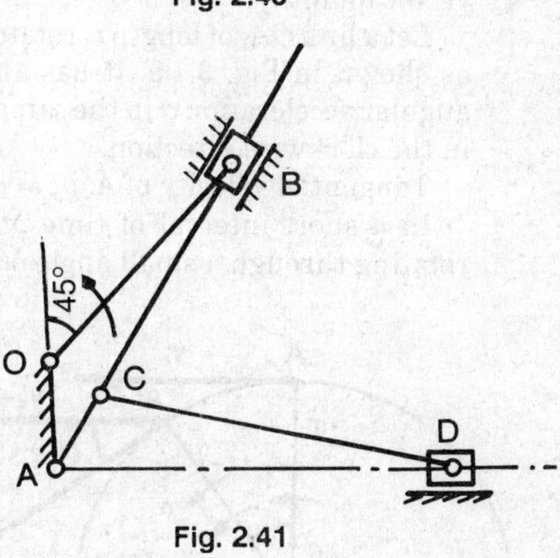

Fig. 2.41

2.9 Solve problem 2.4 using the I-centre method.

2.10 Locate all the I-centres for the mechanism shown in Fig. 2.38 and solve the problem 2.4 using the I-centre method.

2.11 Solve Example 2.4 (Fig. 2.12) using the I-centre method.

2.12 Solve Example 2.8 (Fig. 2.16) using the I-centre method.

3

ACCELERATION ANALYSIS

3.1 ACCELERATION

Velocity of a moving body is a vector quantity having the magnitude and the direction. A change in the velocity require any of the following conditions to be fulfilled:

A change in the magnitude only

A change in the direction only

A change in both magnitude and direction.

The rate of change of velocity with respect to time is known as *acceleration* and acts in the direction of the change in velocity. Thus acceleration is also a vector quantity.

Let a link OA, of length r, rotate in a circular path in the clockwise direction as shown in Fig. 3.1 a. It has an instantaneous angular velocity ω and an angular acceleration α in the same direction, i.e. the angular velocity increases in the clockwise direction.

Tangential velocity of A, $v_a = \omega r$

In a short interval of time δt, let OA assume the new position OA' by rotating through a small angle $\delta\theta$.

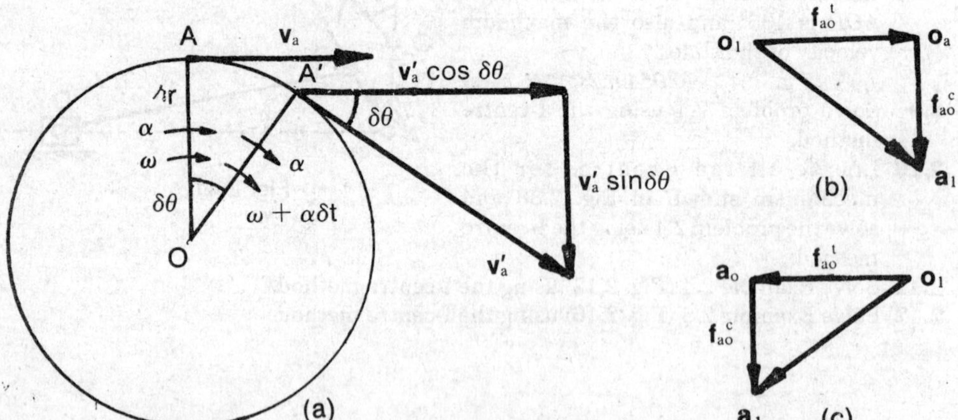

Fig. 3.1

Angular velocity of OA', $\omega'_a = \omega + \alpha\,\delta t$
Tangential velocity of A', $v'_a = (\omega + \alpha\,\delta t)\, r$
the tangential velocity of A' may be considered to have two components; one perpendicular to OA and the other parallel to OA.

Change of Velocity Perpendicular to OA

Velocity of $A \perp$ to $OA = v_a$
Velocity of $A' \perp$ to $OA = v'_a \cos\delta\theta$
$\therefore$ Change of velocity $= v'_a \cos\delta\theta - v_a$

$$\text{Acceleration of } A \perp \text{ to } OA = \frac{(\omega + \alpha\cdot\delta t)\, r\cos\delta\theta - \omega\, r}{\delta t}$$

In the limit, as $\delta t \to \theta$, $\cos\delta\theta \to 1$
$\therefore$ Acceleration of $A \perp$ to $OA = \alpha\, r$

$$= \left(\frac{d\,\omega}{dt}\right) r \quad \ldots \qquad \left(\alpha = \frac{d\,\omega}{dt}\right)$$

$$= \frac{dv}{dt}$$

This represents the rate of change of velocity in the tangential direction of the motion of A relative to O and thus is known as the *tangential acceleration* of A relative to O ($\mathbf{f}^t_{ao}$).

Change of Velocity Parallel to OA

Velocity of A parallel to $OA = 0$
Velocity of A' parallel to $OA = v'_a \sin\delta\,\theta$
$\therefore$ change of velocity $= v'_a \sin\delta\,\theta - 0$

$$\text{Acceleration of } A \text{ parallel to } A = \frac{(\omega + \alpha\,\delta t)\, r\sin\delta\,\theta}{\delta\, t}$$

In the limit, as $\delta\, t \to 0$, $\sin\delta\theta \to \delta\theta$

$$\text{Acceleration of } A \text{ parallel to } OA = \omega r\frac{d\theta}{dt}$$

$$= \omega\, r\,\omega \quad \ldots \qquad \left(\omega = \frac{d\theta}{dt}\right)$$

$$= \omega^2\, r.$$

$$= \frac{v}{r} \quad \ldots \qquad (v = \omega\, r)$$

This represents the rate of change of velocity along OA, the direction being from A towards O or towards the centre of rotation. This acceleration is known as the *centripetal* or the *radial acceleration* of A relative to $O(\mathbf{f}^c_{ao})$.

Figure 3.1b shows the centripetal and the tangential components of the acceleration acting on A. Note that

1. When $\alpha = 0$, i.e. *OA* rotates with uniform angular velocity, $\mathbf{f}_{ao}^{t} = 0$ and thus $\mathbf{f}_{ao}^{c}$ represents the total acceleration.
2. When $\omega = 0$, i.e. *A* has a linear motion, $\mathbf{f}_{ao}^{c} = 0$ and thus the tangential acceleration is the total acceleration.
3. When α is negative or the link *OA* decelerates, tangential acceleration will be negative or its direction will be as shown in Fig. 3.1 c.

Total acceleration vectors are denoted by small letters with a subscript '1' attached. The meeting point of its components may be denoted by any of the small letters used for the total acceleration with a subscript of the other.

For example, components of the total acceleration $\mathbf{o}_1\,\mathbf{a}_1$ can be written in either of the two ways:

1. $\mathbf{o}_1\,\mathbf{o}_a$ and $\mathbf{o}_a\,\mathbf{a}_1$ as in Fig. 3.1 b,
2. $\mathbf{o}_1\,\mathbf{a}_o$ and $\mathbf{a}_o\,\mathbf{a}_1$ as in Fig. 3.1 c.

3.2 FOUR-LINK MECHANISM

The configuration and the velocity diagrams of a four-link mechanism discussed in Sec. 2.6 have been reproduced in Figs. 3.2 a and b. Let α = Angular acceleration of *AB*, assumed positive, i.e. the speed increases in the clockwise direction.

For the construction of the acceleration diagram, a vector equation for the same can be formed similar to the one applied to the velocity diagram.

Acc. of *C* rel. to *A* = Acc. of *C* rel. to *B* + Acc. of *B* rel. to *A*

$$\mathbf{f}_{ca} = \mathbf{f}_{cb} + \mathbf{f}_{ba}$$

or

$$\mathbf{f}_{cd} = \mathbf{f}_{ba} + \mathbf{f}_{cb}$$

or

$$\mathbf{d}_1\,\mathbf{c}_1 = \mathbf{a}_1\,\mathbf{b}_1 + \mathbf{b}_1\,\mathbf{c}_1$$

Each acceleration may have a centripetal and a tangential component. Thus the equation can be expanded as below:

$$\mathbf{f}_{cd}^{c} + \mathbf{f}_{cd}^{t} = \mathbf{f}_{ba}^{c} + \mathbf{f}_{ba}^{t} + \mathbf{f}_{cb}^{c} + \mathbf{f}_{cb}^{t}$$

or

$$\mathbf{d}_1\,\mathbf{c}_d + \mathbf{c}_d\,\mathbf{c}_1 = \mathbf{a}_1\,\mathbf{b}_a + \mathbf{b}_a\,\mathbf{b}_1 + \mathbf{b}_1\,\mathbf{c}_b + \mathbf{c}_b\,\mathbf{c}_1$$

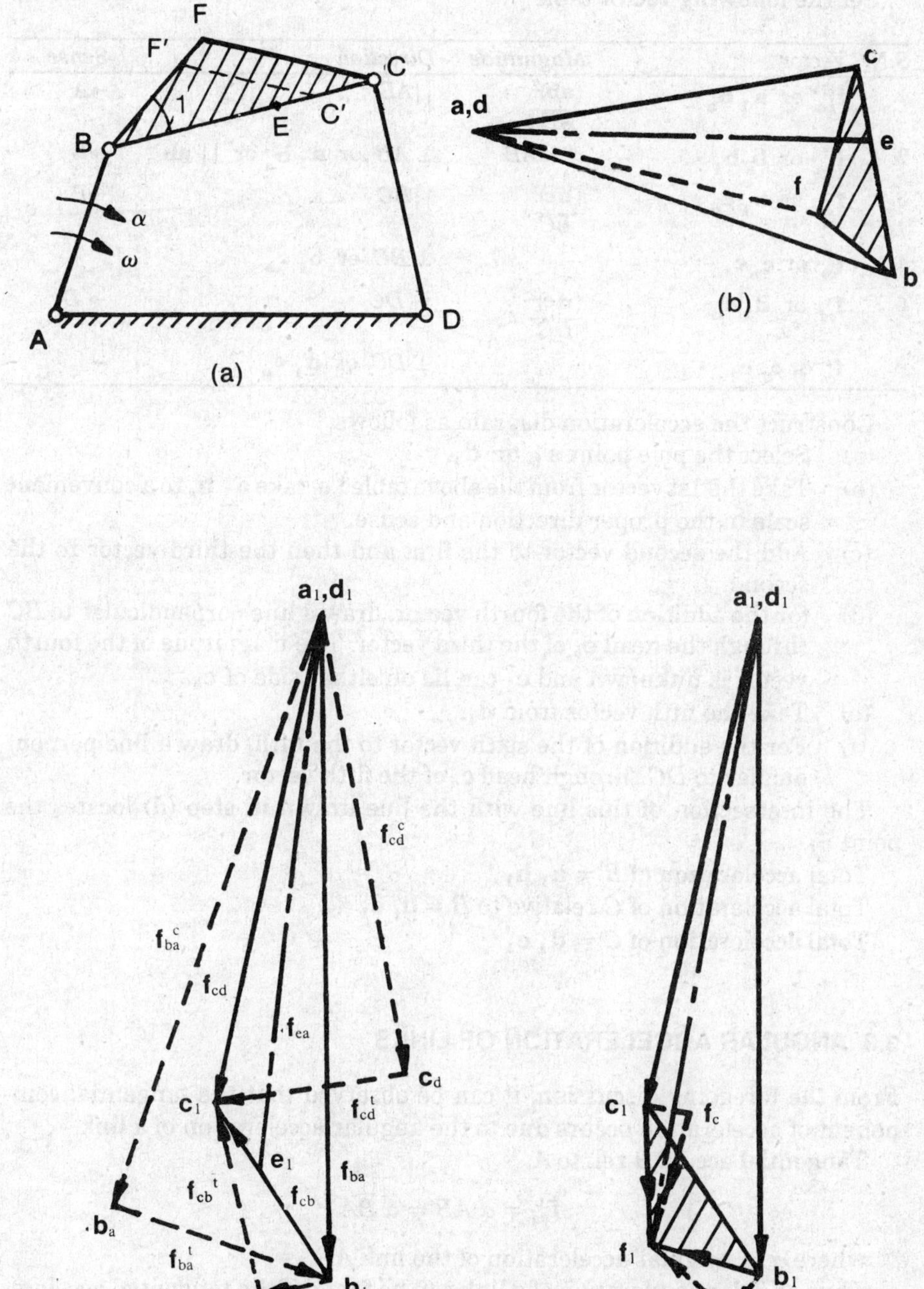
F
F′
C
1
E
C′
B
α
ω
A
D
(a)
a,d
c
e
f
b
(b)
a₁,d₁
f_cd^c
f_ba^c
f_cd
f_ea
c_d
c₁
f_cd^t
e₁
f_ba
f_cb^t
f_cb
b_a
f_ba^t
b₁
c_b
f_cb^c
(c)
a₁,d₁
c₁
f_c
f₁
b₁
f_b
(d)

Fig. 3.2

Set the following vector table:

S.N.	Vector	Magnitude	Direction	Sense
1.	$\mathbf{f}^c_{ba}$ or $\mathbf{a}_1\,\mathbf{b}_a$	$\frac{(\mathbf{ab})^2}{AB}$	$\parallel AB$	$\rightarrow A$
2.	$\mathbf{f}^t_{ba}$ or $\mathbf{b}_a\,\mathbf{b}_1$	$\alpha \times AB$	$\perp AB$ or $\mathbf{a}_1\,\mathbf{b}_a$ or $\parallel \mathbf{ab}$	$\rightarrow \mathbf{b}$
3.	$\mathbf{f}^c_{cb}$ or $\mathbf{b}_1\,\mathbf{c}_b$	$\frac{(\mathbf{bc})^2}{BC}$	$\parallel BC$	$\rightarrow B$
4.	$\mathbf{f}^t_{cb}$ or $\mathbf{c}_b\,\mathbf{c}_1$	–	$\perp BC$ or $\mathbf{b}_1\,\mathbf{c}_b$	–
5.	$\mathbf{f}^c_{cd}$ or $\mathbf{d}_1\,\mathbf{c}_d$	$\frac{(\mathbf{dc})^2}{DC}$	$\parallel DC$	$\rightarrow D$
6.	$\mathbf{f}^t_{cd}$ or $\mathbf{c}_d\,\mathbf{c}_1$	–	$\perp DC$ or $\mathbf{d}_1\,\mathbf{c}_d$	–

Construct the acceleration diagram as follows:

(a) Select the pole point $\mathbf{a}_1$ or $\mathbf{d}_1$.

(b) Take the 1st vector from the above table i.e. take $\mathbf{a}_1\,\mathbf{b}_a$ to a convenient scale in the proper direction and sense.

(c) Add the second vector to the first and then the third vector to the second.

(d) for the addition of the fourth vector, draw a line perpendicular to BC through the head $\mathbf{c}_b$ of the third vector. The magnitude of the fourth vector is unknown and $\mathbf{c}_1$ can lie on either side of $\mathbf{c}_b$.

(e) Take the fifth vector from $\mathbf{d}_1$.

(f) For the addition of the sixth vector to the fifth, draw a line perpendicular to DC through head $\mathbf{c}_d$ of the fifth vector.

The intersection of this line with the line drawn in step (d) locates the point $\mathbf{c}_1$.

Total acceleration of $B = \mathbf{a}_1\,\mathbf{b}_1$

Total acceleration of C relative to $B = \mathbf{b}_1\,\mathbf{c}_1$

Total acceleration of $C = \mathbf{d}_1\,\mathbf{c}_1$

3.3 ANGULAR ACCELERATION OF LINKS

From the foregoing discussion, it can be observed that the tangential component of acceleration occurs due to the angular acceleration of a link.

Tangential acc. of B rel. to A,

$$\mathbf{f}^t_{ba} = \alpha\, AB = \alpha\, BA$$

where α = Angular acceleration of the link AB

Thus angular acceleration of a link can be found if the tangential acceleration is known.

Referring to Fig. 3.2,

Tangential acc. of C rel. to B,

$$\mathbf{f}^t_{cb} = \mathbf{c_b\, c_1}$$

i.e. *C* relative to *B* moves in a direction $\mathbf{c_b}$ to $\mathbf{c_1}$ or *C* moves in a counter-clockwise direction about *B*.

As $f^t_{cb} = \alpha_{cb}\, CB$

$\therefore\ \alpha_{cb} = f^t_{cb}/CB$

$$\text{Tangential acc. of } B \text{ rel. to } C, f^t_{bc} = \mathbf{c_1\, c_b}$$

i.e., *B* moves in a counter-clockwise direction about *C* with magnitude, $\alpha_{bc} = f^t_{bc}/BC$ which is the same as α_{cb}.

Thus angular acceleration of a link about one extremity is the same in magnitude and direction as the angular acceleration about the other.

$$\text{Tangential acc. of } C \text{ rel. to } D, \mathbf{f}^t_{cd} = \mathbf{c_d\, c_1}$$

i.e., *C* relative to *D* moves in a direction from $\mathbf{c_d}$ to $\mathbf{c_1}$ or *C* moves in the counter-clockwise direction about *D*.

$$\alpha_{cd} = \frac{f^t_{cd}}{CD} = \frac{\mathbf{c_d\, c_1}}{CD}$$

3.4 ACCELERATION OF INTERMEDIATE AND OFFSET POINTS

Intermediate Point

The acceleration of intermediate points on the links can be obtained by dividing the acceleration vectors in the same ratio as the points divide the links. For point E on the link BC (Fig. 3.2),

$$\frac{BE}{BC} = \frac{\mathbf{b_1\, e_1}}{\mathbf{b_1\, c_1}}$$

$\mathbf{a_1\, e_1}$ gives the total acceleration of point E.

Offset Points

The acceleration of an offset point on a link, such as *F* on *BC* (Fig. 3.2), can be determined by applying any of the following methods:

1. Writing the vector equation,

$$\begin{aligned} \mathbf{f}_{fb} + \mathbf{f}_{ba} &= \mathbf{f}_{fc} + \mathbf{f}_{cd} \\ \text{or} \quad \mathbf{f}_{ba} + \mathbf{f}_{fb} &= \mathbf{f}_{cd} + \mathbf{f}_{fc} \\ \text{or} \quad \mathbf{f}_{ba} + \mathbf{f}^c_{fb} + \mathbf{f}^t_{fb} &= \mathbf{f}_{cd} + \mathbf{f}^c_{fc} + \mathbf{f}^c_{fc} \\ \text{or} \quad \mathbf{a_1\, b_1} + \mathbf{b_1\, f_b} + \mathbf{f_b\, f_1} &= \mathbf{d_1\, c_1} + \mathbf{c_1\, f_c} + \mathbf{f_c\, f_1} \end{aligned}$$

The equation can be easily solved graphically as shown in Figure 3.2d. $\mathbf{a_1\, f_1}$ represents the acceleration of *F* relative to *A* or *D*.

2. **Writing the vector equation,**

$$\mathbf{f}_{fa} = \mathbf{f}_{fb} + \mathbf{f}_{ba}$$
$$= \mathbf{f}_{ba} + \mathbf{f}_{fb}$$
$$= \mathbf{f}_{ba} + \mathbf{f}^c_{fb} + \mathbf{f}^t_{fb}$$

or $$\mathbf{a_1 f_1} = \mathbf{a_1 b_1} + \mathbf{b_1 f_b} + \mathbf{f_b f_1}$$

$\mathbf{f}_{ba}$ already exists on the acceleration diagram.

$\mathbf{f}^c_{fb} = \dfrac{(bf)^2}{BF}$, || to BF, direction towards B.

$$\mathbf{f}^t_{fb} = \alpha_{fb} \times fb = \alpha_{cb} \times FB$$
$$= \frac{\mathbf{f}^t_{cb}}{CB} \times FB \perp \text{ to } FB; \text{ direction } \mathbf{b} \text{ to } \mathbf{f}$$

$\alpha_{fb} = \alpha_{cb}$, because angular acceleration of all the points on the link BCF about the point B is the same (counter-clockwise).

In this way $\mathbf{f}_{fa}$ can be found.

3. It can be proved that the triangle $\mathbf{b_1\, c_1\, f_1}$ will be similar to the triangle BCF in Figs. 3.2d and a.

Let ω' = Angular velocity of the link BCF
α = angular acceleration of the link BCF

Referring to Figs. 3.2a and 3.3,

$$\frac{\mathbf{b_1 f_b}}{\mathbf{b_1 c_b}} = \frac{\omega'^2\, BF}{\omega'^2\, BC} = \frac{BF}{BC} = \frac{\alpha\, BF}{\alpha\, BC} = \frac{\mathbf{f_b f_1}}{\mathbf{c_b c_1}}$$

$\mathbf{b_1\, f_b\, f_1}$ and $\mathbf{b_1\, c_b\, c_1}$ are two right-angled triangles in which the ratio of the two corresponding sides is the same as proved above. Therefore. the two triangles are similar.

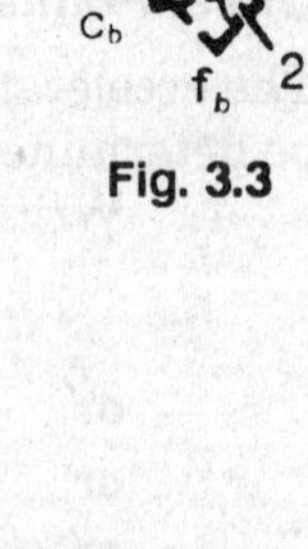

Fig. 3.3

$$\frac{\mathbf{b_1 f_1}}{\mathbf{b_1 c_1}} = \frac{BF}{BC} = k$$

Also $\angle \mathbf{f_b\, b_1\, f_1} = \angle \mathbf{c_b\, b_1\, c_1}$

or $\angle \mathbf{f_b\, b_1\, f_1} - \angle \mathbf{c_b\, b_1\, f_1} = \angle\, \mathbf{c_b\, b_1\, c_1} - \angle \mathbf{c_b\, b_1\, f_1}$

or $\angle 3 = \angle 2$

$= \angle 1$ $\quad (\because \mathbf{b_1\, f_b} \parallel BF,\ \mathbf{b_1\, c_b} \parallel BC)$

Now in Δs $\mathbf{b_1\, f_1\, c_1}$ and BFC,

$$\angle 3 = \angle 1$$

and $$\frac{\mathbf{b_1 f_1}}{\mathbf{b_1 c_1}} = \frac{BF}{BC} = k$$

the two triangles are similar.

Thus to find the acceleration of an offset point on a link, a triangle, similar to the one formed in the configuration diagram, can be made on the acceleration image of the link in such a way that the sequence of letters is the same, i.e. $\mathbf{b_1\, f_1\, c_1}$ is clockwise, so should be BFC.

An easier method of making triangle $\mathbf{b_1 f_1 c_1}$ similar to BFC is by marking BC' on BC equal to $\mathbf{b_1 c_1}$ and drawing a line parallel to CF, meeting BF in F'. $BC'F'$ is the exact size of the triangle to be made on $\mathbf{b_1 c_1}$.

Take $\mathbf{b_1 f_1} = BF'$ and $\mathbf{c_1 f_1} = C'F'$.

Thus the point $\mathbf{f_1}$ is obtained.

3.5 SLIDER-CRANK MECHANISM

The configuration and the velocity diagrams of a slider-crank mechanism discussed in Sec. 2.9 have been reproduced in Figs. 3.4 a and b.

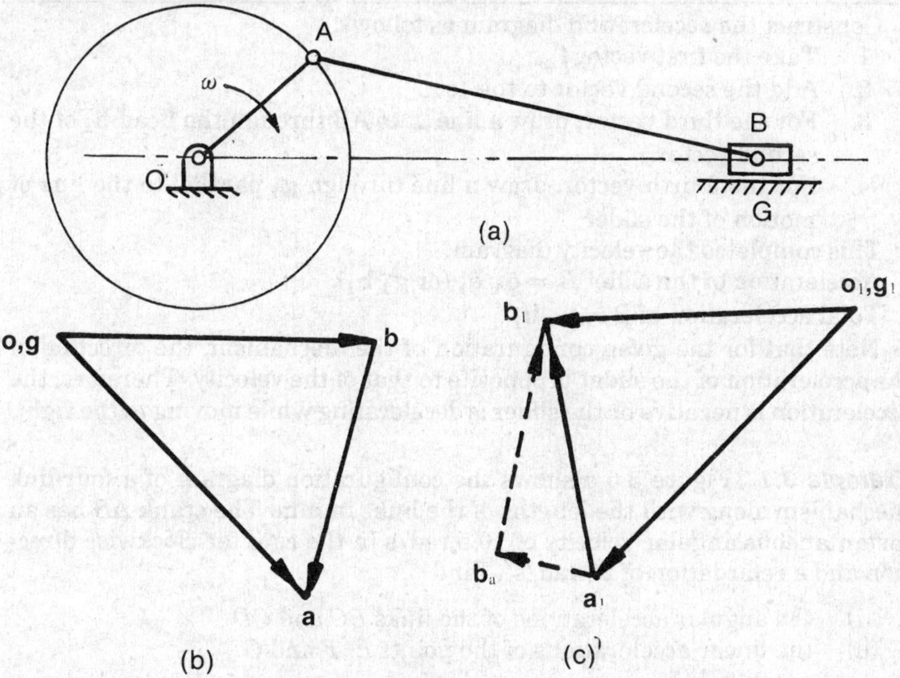

Fig. 3.4

Writing the acceleration equation,

Acc. of B rel. to O = Acc. of B rel. to A + Acc. of A rel. to O

$$\mathbf{f}_{bo} = \mathbf{f}_{ba} + \mathbf{f}_{ao}$$

$$\mathbf{f}_{bg} = \mathbf{f}_{ao} + \mathbf{f}_{ba}$$

$$= \mathbf{f}_{ao} + \mathbf{f}^c_{ba} + \mathbf{f}^t_{ba}$$

$$\mathbf{g_1 b_1} = \mathbf{o_1 a_1} + \mathbf{a_1 b_a} + \mathbf{b_a b_1}$$

The crank OA rotates at a uniform velocity, therefore, the acceleration of A relative to O has only the centripetal component. Similarly, the slider moves in a linear direction and thus has no centripetal component.

Setting the vector table:

S.N.	Vector	Magnitude	Direction	Sense
1.	$\mathbf{f}_{ao}$ or $\mathbf{o}_1\mathbf{a}_1$	$\dfrac{(\mathbf{oa})^2}{OA}$	$\parallel OA$	$\rightarrow O$
2.	$\mathbf{f}^c_{ba}$ or $\mathbf{a}_1\mathbf{b}_a$	$\dfrac{(\mathbf{ab})^2}{AB}$	$\parallel AB$	$\rightarrow A$
3.	$\mathbf{f}^t_{ba}$ or $\mathbf{b}_a\mathbf{b}_1$	–	$\perp AB$	–
4.	$\mathbf{f}_{bg}$ or $\mathbf{g}_1\mathbf{b}_1$	—	$\parallel$ to line of motion of B	—

Construct the acceleration diagram as follows:

1. Take the first vector $\mathbf{f}_{ao}$.
2. Add the second vector to the 1st.
3. For the third vector, draw a line $\perp$ to AB through the head $\mathbf{b}_a$ of the second vector.
4. For the fourth vector, draw a line through $\mathbf{g}_1$ parallel to the line of motion of the slider.

This completes the velocity diagram.

Acceleration of the slider $B = \mathbf{o}_1\mathbf{b}_1$ (or $\mathbf{g}_1\mathbf{b}_1$)

Total acceleration of $B = \mathbf{a}_1\mathbf{b}_1$

Note that for the given configuration of the mechanism, the direction of the acceleration of the slider is opposite to that of the velocity. Therefore, the acceleration is negative or the slider is decelerating while moving to the right.

Example 3.1 Figure 3.5 a shows the configuration diagram of a four-link mechanism along with the lengths of the links in mm. The crank AB has an instantaneous angular velocity of 10.5 rad/s in the counter-clockwise direction and a retardation of 26 rad/s^2. Find

(i) the angular accelerations of the links BC and CD
(ii) the linear accelerations of the points E, F and G.

Solution:

$$v_b = 10.5 \times 0.05 = 0.525 \text{ m/s}$$

Complete the velocity diagram as shown in Fig. 3.5 b.

Writing the vector equation,

$$\text{Acc. of } C \text{ rel. to } A = \text{Acc. of } C \text{ rel. to } B + \text{Acc. of } B \text{ rel. to } A$$

$$\mathbf{f}_{ca} = \mathbf{f}_{cb} + \mathbf{f}_{ba}$$

or
$$\mathbf{f}_{cd} = \mathbf{f}_{ba} + \mathbf{f}_{cb}$$

or
$$\mathbf{d}_1\mathbf{c}_1 = \mathbf{a}_1\mathbf{b}_1 + \mathbf{b}_1\mathbf{c}_1$$

Each vector has a centripetal and a tangential component,

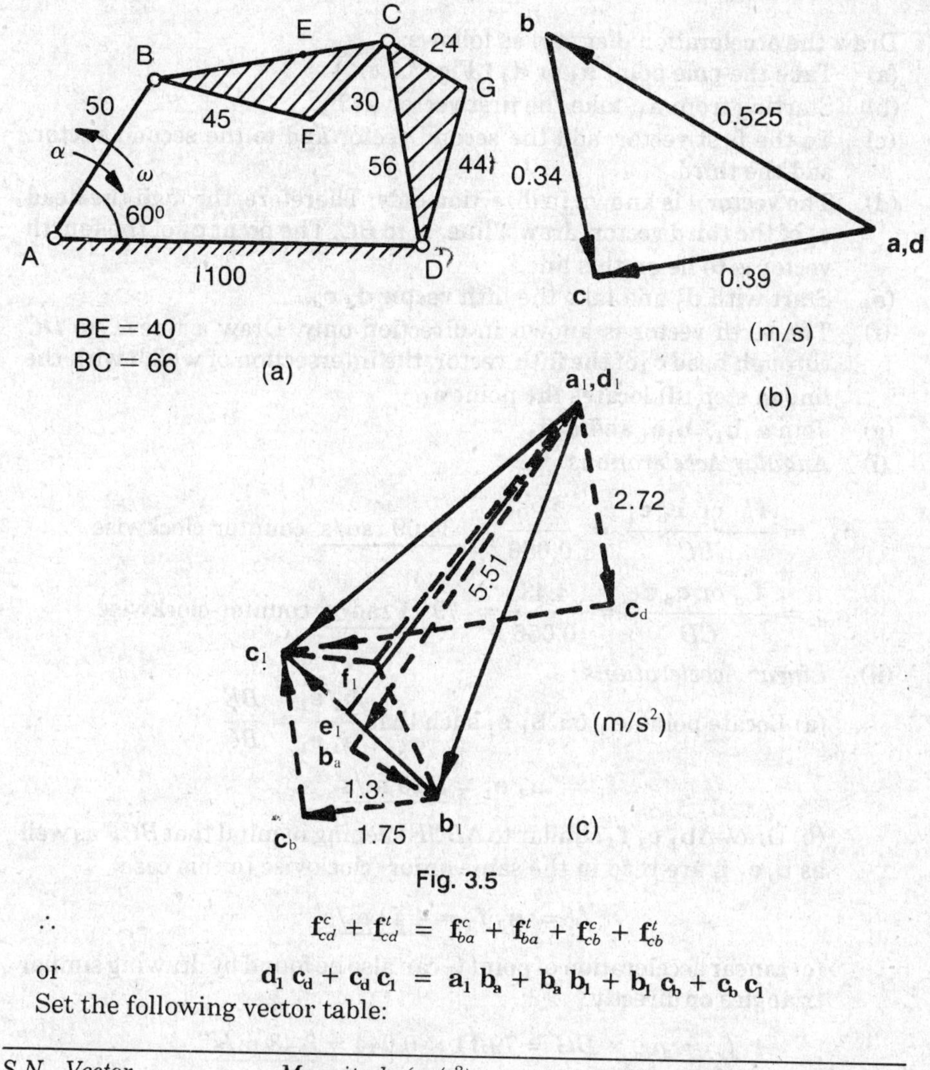

Fig. 3.5

$$\therefore \qquad f_{cd}^{c} + f_{cd}^{t} = f_{ba}^{c} + f_{ba}^{t} + f_{cb}^{c} + f_{cb}^{t}$$

or

$$\mathbf{d_1 c_d} + \mathbf{c_d c_1} = \mathbf{a_1 b_a} + \mathbf{b_a b_1} + \mathbf{b_1 c_b} + \mathbf{c_b c_1}$$

Set the following vector table:

S.N.	*Vector*	*Magnitude (m/s²)*	*Direction*	*Sense*
1.	f_{ba}^{c} or $\mathbf{a_1 b_a}$	$\dfrac{(\mathbf{ab})^2}{AB} = \dfrac{(0.525)^2}{0.05} = 5.51$	$\parallel AB$	$\to A$
2.	f_{ba}^{t} or $\mathbf{b_a b_1}$	$\alpha \times AB = 26 \times 0.05 = 1.3$	$\perp AB$ or $\parallel \mathbf{ab}$	$\to \mathbf{a}$
3.	f_{cb}^{c} or $\mathbf{b_1 c_b}$	$\dfrac{(\mathbf{bc})^2}{BC} = \dfrac{(0.34)^2}{0.066} = 1.75$	$\parallel BC$	$\to B$
4.	f_{cb}^{t} or $\mathbf{c_b c_1}$	—	$\perp BC$	—
5.	f_{cd}^{c} or $\mathbf{d_1 c_d}$	$\dfrac{(\mathbf{dc})^2}{DC} = \dfrac{(0.39)^2}{0.56} = 2.72$	$\parallel DC$	$\to D$
6.	f_{cd}^{t} or $\mathbf{c_d c_1}$	—	$\perp DC$	—

Draw the acceleration diagram as follows:

(a) Take the pole point $\mathbf{a}_1$ or $\mathbf{d}_1$ (Fig. 3.5 c).

(b) Starting from $\mathbf{a}_1$, take the first vector $\mathbf{a}_1\,\mathbf{b}_a$.

(c) To the first vector, add the second vector and to the second vector, add the third.

(d) The vector 4 is known in direction only. Therefore, through the head $\mathbf{c}_b$ of the third vector, draw a line, ⊥ to *BC*. The point $\mathbf{c}_1$ of the fourth vector is to lie on this line.

(e) Start with $\mathbf{d}_1$ and take the fifth vector $\mathbf{d}_1\,\mathbf{c}_d$.

(f) The sixth vector is known in direction only. Draw a line ⊥ to *DC* through head $\mathbf{c}_d$ of the fifth vector, the intersection of which with the line in step (d) locates the point $\mathbf{c}_1$.

(g) Join $\mathbf{a}_1\mathbf{b}_1$, $\mathbf{b}_1\mathbf{c}_1$ and $\mathbf{c}_1\mathbf{a}_1$

(i) *Angular Accelerations*

$$\alpha_{bc} = \frac{\mathbf{f}^t_{cb} \text{ or } \mathbf{c}_b\,\mathbf{c}_1}{BC} = \frac{2.25}{0.066} = \underline{34.09 \text{ rad/s}^2} \text{ counter-clockwise}$$

$$\alpha_{cd} = \frac{\mathbf{f}^t_{cd} \text{ or } \mathbf{c}_d\,\mathbf{c}_1}{CD} = \frac{4.43}{0.056} = \underline{79.11 \text{ rad/s}^2} \text{ counter-clockwise}$$

(ii) *Linear Accelerations*

(a) Locate point $\mathbf{e}_1$ on $\mathbf{b}_1\,\mathbf{c}_1$ such that $\dfrac{\mathbf{b}_1\,\mathbf{e}_1}{\mathbf{b}_1\,\mathbf{c}_1} = \dfrac{BE}{BC}$

$$\mathbf{f}_e = \mathbf{a}_1\,\mathbf{e}_1 = \underline{5.15 \text{ m/s}^2}.$$

(b) Draw $\Delta\mathbf{b}_1\,\mathbf{c}_1\,\mathbf{f}_1$ similar to ΔBCF keeping in mind that *BCF* as well as $\mathbf{b}_1\,\mathbf{c}_1\,\mathbf{f}_1$ are read in the same order (clockwise in this case)

$$\therefore\ f_f = \mathbf{a}_1\,\mathbf{f}_1 = \underline{4.42 \text{ m/s}^2}$$

(c) Linear acceleration of point G can also be found by drawing similar triangles or directly.

$$f_g = \alpha_{cd} \times DG = 79.11 \times 0.044 = \underline{3.48 \text{ m/s}^2}$$

Example 3.2 For the configuration of a slider-crank mechanism shown in Fig. 3.6 a, calculate

(i) The acceleration of the slider at *B*

(ii) The acceleration of point *E*

(iii) The acceleration of link *AB*.

OA rotates at 20 rad/s counter-clockwise.

Solution:

$$v_a = 20 \times 0.48 = 9.6 \text{ m/s}$$

Complete the velocity diagram as shown in Fig. 3.6 b.

Writing the vector equation,

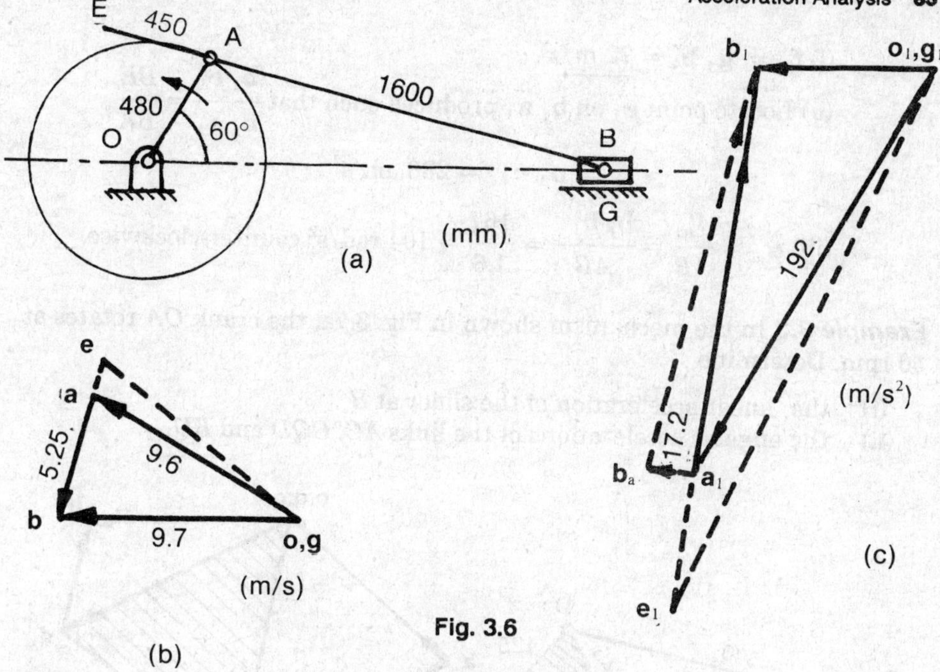

Fig. 3.6

$$\mathbf{f}_{bo} = \mathbf{f}_{ba} + \mathbf{f}_{ao}$$

or

$$\mathbf{f}_{bg} = \mathbf{f}_{ao} + \mathbf{f}_{ba}$$
$$= \mathbf{f}^c_{ao} + \mathbf{f}^c_{ba} + \mathbf{f}^t_{ba}$$

or

$$\mathbf{g}_1\mathbf{b}_1 = \mathbf{o}_1\mathbf{a}_1 + \mathbf{a}_1\mathbf{b}_a + \mathbf{b}_a\mathbf{b}_1$$

Set the vector table as given below:

S.N.	*Vector*	*Magnitude (m/s²)*	*Direction*	*Sense*
1.	$\mathbf{f}^c_{ao}$ or $\mathbf{o}_1\mathbf{a}_1$	$\frac{(\mathbf{oa})^2}{OA} = \frac{(9.6)^2}{0.48} = 192$	$\parallel OA$	$\rightarrow O$
2.	$\mathbf{f}^c_{ba}$ or $\mathbf{a}_1\mathbf{b}_a$	$\frac{(\mathbf{ab})^2}{AB} = \frac{(5.52)^2}{1.60} = 17.2$	$\parallel AB$	$\rightarrow A$
3.	$\mathbf{f}^t_{ba}$ or $\mathbf{b}_a\mathbf{b}_1$	—	$\perp AB$	—
4.	$\mathbf{f}_{bg}$ or $\mathbf{g}_1\mathbf{b}_1$	—	$\parallel$ to slider motion	—

The acceleration diagram is drawn as follows:

(a) Take the pole point $\mathbf{o}_1$ or $\mathbf{g}_1$ (Fig. 3.6 c).

(b) Take the first vector $\mathbf{o}_1\mathbf{a}_1$ and to it add the second vector.

(c) For the third vector, draw a line $\perp$ to AB through the head $\mathbf{b}_a$ of the second vector.

(d) For the fourth vector, draw a line $\parallel$ to the line of motion of the slider through $\mathbf{g}_1$. The intersection of this line with the line drawn in step (d) locates point $\mathbf{b}_1$.

(e) Join $\mathbf{a}_1\mathbf{b}_1$.

(i) $f_b = \mathbf{g_1 b_1} = 72 \text{ m/s}^2$

(ii) Locate point $\mathbf{e_1}$ on $\mathbf{b_1 a_1}$ produced such that $\dfrac{\mathbf{b_1 e_1}}{\mathbf{b_1 a_1}} = \dfrac{BE}{BA}$

$$f_e = \mathbf{o_1 e_1} = 236 \text{ m/s}^2$$

(iii) $\alpha_{ab} = \dfrac{f^t_{ba}}{AB} = \dfrac{\mathbf{b_a b_1}}{AB} = \dfrac{167}{1.6} = 104 \text{ rad/s}^2$ counter-clockwise.

Example 3.3 In the mechanism shown in Fig. 3.7a, the crank *OA* rotates at 60 rpm. Determine

(i) the linear acceleration of the slider at *B*

(ii) the angular accelerations of the links *AC*, *CQD* and *BD*.

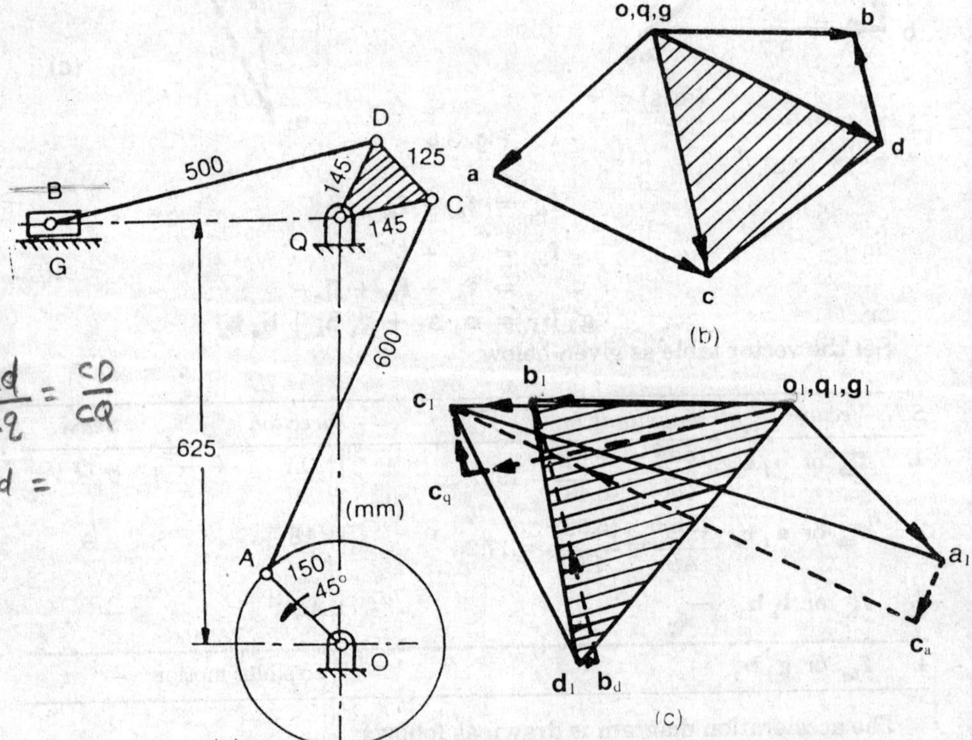

Fig. 3.7

Solution:

$$v_a = \frac{2\pi \times 60}{60} \times 0.15 = 0.94 \text{ m/s}$$

Complete the velocity diagram as shown in Fig. 3.7b.

It is a six-link mechanism. First consider the four-link mechanism *OACQ* and write the vector equation

$$\mathbf{f}_{co} = \mathbf{f}_{ca} + \mathbf{f}_{ao}$$

or $$\mathbf{f}_{cq} = \mathbf{f}_{ao} + \mathbf{f}_{ca}$$

or $$\mathbf{q}_1\,\mathbf{c}_1 = \mathbf{o}_1\,\mathbf{a}_1 + \mathbf{a}_1\,\mathbf{c}_1$$

Links *AC* and *CQ* each can have centripetal and tangential components.

$$\mathbf{f}^c_{cq} + \mathbf{f}^t_{cq} = \mathbf{f}_{ao} + \mathbf{f}^t_{ca} + \mathbf{f}^t_{ca}$$

or $$\mathbf{q}_1\,\mathbf{c}_q + \mathbf{c}_q\,\mathbf{c}_1 = \mathbf{o}_1\,\mathbf{a}_1 + \mathbf{a}_1\,\mathbf{c}_a + \mathbf{c}_a\,\mathbf{c}_1$$

Set the following vector table:

S.N.	*Vector*	*Magnitude (m/s²)*	*Direction*	*Sense*
1.	$\mathbf{f}_{ao}$ or $\mathbf{o}_1\,\mathbf{a}_1$	$\frac{(\mathbf{oa})^2}{OA} = \frac{(0.94)^2}{0.15} = 5.92$	$\parallel OA$	$\to O$
2.	$\mathbf{f}^c_{ca}$ or $\mathbf{a}_1\,\mathbf{c}_a$	$\frac{(\mathbf{ac})^2}{AC} = \frac{(1.035)^2}{0.60} = 1.79$	$\parallel AC$	$\to A$
3.	$\mathbf{f}^t_{ca}$ or $\mathbf{c}_a\,\mathbf{c}_1$	—	$\perp AC$	—
4.	$\mathbf{f}^c_{cq}$ or $\mathbf{q}_1\,\mathbf{c}_q$	$\frac{(\mathbf{qc})^2}{QC} = \frac{(1.14)^2}{0.145} = 8.96$	$\parallel QC$	$\to Q$
5.	$\mathbf{f}^t_{cq}$ or $\mathbf{c}_q\,\mathbf{c}_1$	—	$\perp QC$	—

Complete the acceleration vector diagram $\mathbf{o}_1\,\mathbf{a}_1\,\mathbf{c}_1\,\mathbf{q}_1$ as usual (Fig. 3.7c).

Draw $\Delta\mathbf{c}_1\,\mathbf{q}_1\,\mathbf{d}_1$ similar to ΔCQD such that both are read in the same sense, i.e. clockwise.

Write the vector equation for the slider-crank mechanism *QDB*,

$$\mathbf{f}_{bq} = \mathbf{f}_{bd} + \mathbf{f}_{dq}$$

or $$\mathbf{f}_{bg} = \mathbf{f}_{dq} + \mathbf{f}_{bd}$$

or $$\mathbf{g}_1\,\mathbf{b}_1 = \mathbf{q}_1\,\mathbf{d}_1 + \mathbf{d}_1\,\mathbf{b}_1$$

From this equation $\mathbf{q}_1\,\mathbf{d}_1$ is already drawn in the diagram and $\mathbf{g}_1\,\mathbf{b}_1$ is a linear acceleration component.

$$\mathbf{f}_{bg} = \mathbf{f}_{dq} + \mathbf{f}^c_{bd} + \mathbf{f}^t_{bd}$$

or $$\mathbf{g}_1\,\mathbf{b}_1 = \mathbf{q}_1\,\mathbf{d}_1 + \mathbf{d}_1\,\mathbf{b}_d + \mathbf{b}_d\,\mathbf{b}_1$$

Set the following vector table:

S.N.	*Vector*	*Magnitude (m/s²)*	*Direction*	*Sense*
1.	$\mathbf{f}_{dq}$ or $\mathbf{q}_1\,\mathbf{d}_1$	Already drawn	—	—
2.	$\mathbf{f}^c_{bd}$ or $\mathbf{d}_1\,\mathbf{b}_d$	$\frac{(\mathbf{db})^2}{DB} = \frac{(0.495)^2}{0.50} = 0.49$	$\parallel DB$	$\to D$
3.	$\mathbf{f}^t_{bd}$ or $\mathbf{b}_d\,\mathbf{b}_1$	—	$\perp DB$	—
4.	$\mathbf{f}_{bg}$ or $\mathbf{g}_1\,\mathbf{b}_1$	—	$\parallel$ to slider motion	—

Complete the acceleration vector diagram $\mathbf{q}_1\,\mathbf{d}_1\,\mathbf{b}_1\,\mathbf{g}_1$.

(i) $f_g = \mathbf{g_1 b_1} = 7\ \text{m/s}^2$ towards left

(ii) $$\alpha_{ac} = \frac{\mathbf{f}^t_{ca} \text{ or } \mathbf{c_a c_1}}{AC} = \frac{13.8}{0.6} = \underline{23\ \text{rad/s}^2} \text{ counter-clockwise}$$

$$\alpha_{cqd} = \frac{\mathbf{f}^t_{cq} \text{ or } \mathbf{c_q c_1}}{QC} = \frac{2.0}{0.145} = \underline{13.8\ \text{rad/s}^2} \text{ counter-clockwise}$$

$$\alpha_{bd} = \frac{\mathbf{f}^t_{bd} \text{ or } \mathbf{b_d b_1}}{BD} = \frac{7.2}{0.5} = \underline{14.4\ \text{rad/s}^2} \text{ clockwise.}$$

Example 3.4 An Andrew variable-stroke engine mechanism is shown in Fig. 3.8 a. The crank *OA* rotates at 100 rpm. Find

(i) The linear acceleration of the slider at *D*
(ii) The angular acceleration of the links *AC*, *BC* and *CD*.

Solution:

$$v_a = \frac{2\pi \times 100}{60} \times 0.09 = 0.94\ \text{m/s}$$

Complete the velocity diagram as shown in Fig. 3.8 b. The procedure is explained in Example 2.7.

Write the acceleration vector equation noting that the cranks *OA* and *QB* rotate at different uniform speeds.

For the linkage *OACBQ*,

$$\mathbf{f}_{ca} + \mathbf{f}_{ao} = \mathbf{f}_{cb} + \mathbf{f}_{bq}$$

$$\mathbf{f}_{ao} + \mathbf{f}_{ca} = \mathbf{f}_{bq} + \mathbf{f}_{cb}$$

$$\mathbf{o_1 a_1} + \mathbf{a_1 c_1} = \mathbf{q_1 b_1} + \mathbf{b_1 c_1}$$

Links *AC* and *BC* each has two components,

$$\mathbf{f}_{ao} + \mathbf{f}^c_{ca} + \mathbf{f}^t_{ca} = \mathbf{f}_{bq} + \mathbf{f}^c_{cb} + \mathbf{f}^t_{cb}$$

or $$\mathbf{o_1 a_1} + \mathbf{a_1 c_a} + \mathbf{c_a c_1} = \mathbf{q_1 b_1} + \mathbf{b_1 c_b} + \mathbf{c_b c_1}$$

Set the following vector table:

S.N.	*Vector*	*Magnitude (m/s²)*	*Direction*	*Sense*
1.	$\mathbf{f}_{ao}$ or $\mathbf{o_1 a_1}$	$\frac{(\mathbf{oa})^2}{OA} = \frac{(0.94)^2}{0.09} = 9.87$	‖ *OA*	→ *O*
2.	$\mathbf{f}^c_{ca}$ or $\mathbf{a_1 c_a}$	$\frac{(\mathbf{ac})^2}{AC} = \frac{(0.81)^2}{0.185} = 3.55$	‖ *AC*	→ *A*
3.	$\mathbf{f}^t_{ca}$ or $\mathbf{c_a c_1}$	—	⊥ *AC*	—
4.	$\mathbf{f}_{bq}$ or $\mathbf{q_1 b_1}$	$\frac{(\mathbf{qb})^2}{QB} = \frac{(1.0)^2}{0.045} = 22.2$	‖ *QB*	→ *Q*
5.	$\mathbf{f}^c_{cb}$ or $\mathbf{b_1 c_1}$	$\frac{(\mathbf{bc})^2}{BC} = \frac{(0.12)^2}{0.185} = 0.078$	‖ *BC*	→ *B*
6.	$\mathbf{f}^t_{cb}$ or $\mathbf{c_b c_a}$	—	⊥ *BC*	—

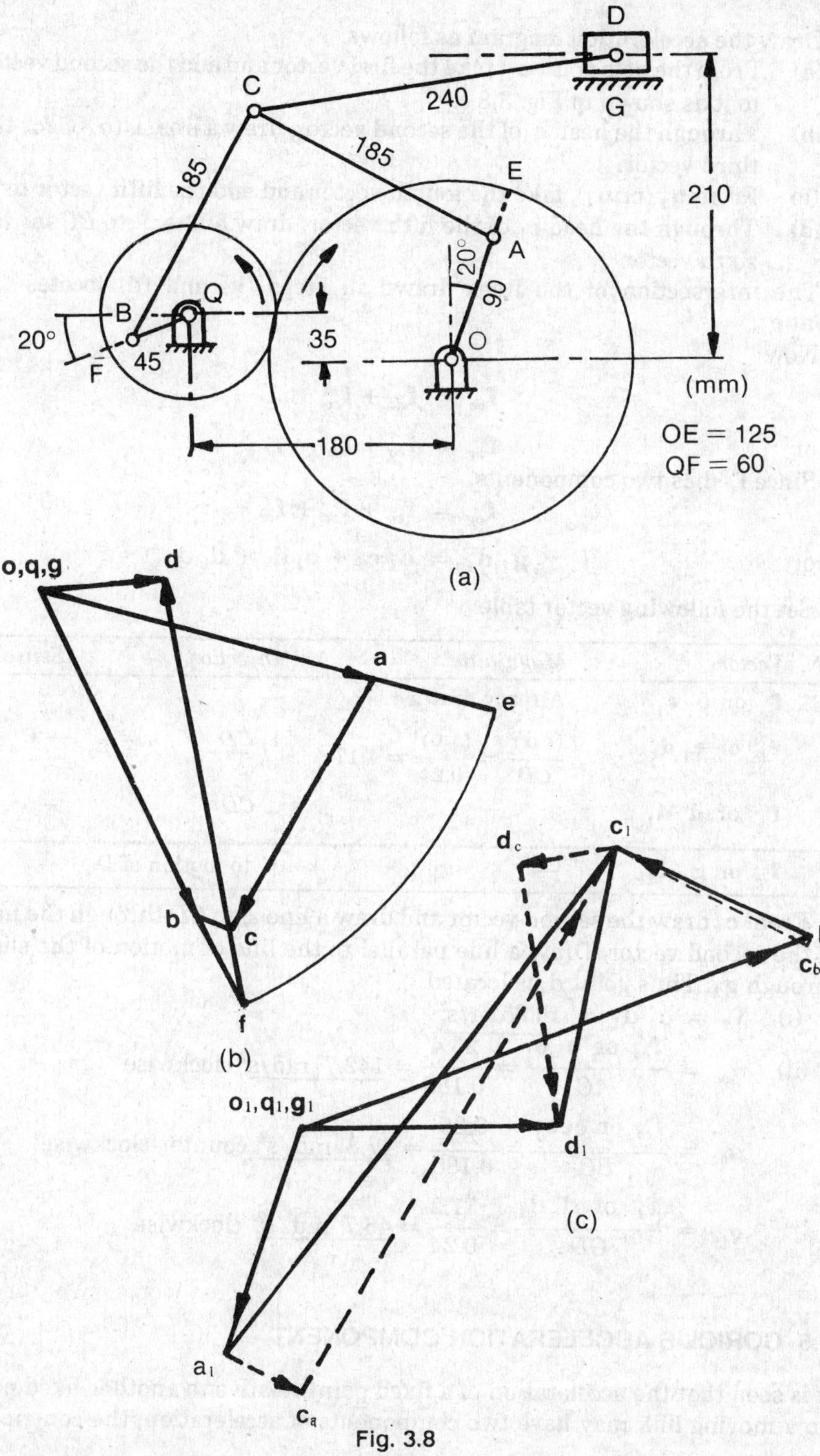
D
C
240
G
185
185
E
210
A
20°
90
Q
B
20°
35
O
F
45
(mm)
180
OE = 125
QF = 60
(a)
o,q,g
d
a
e
b
c
f
(b)
c1
dc
b1
cb
o1,q1,g1
d1
(c)
a1
ca

Fig. 3.8

Draw the acceleration diagram as follows:

(a) From the pole point $\mathbf{o}_1$, take the first vector and add the second vector to it as shown in Fig. 3.8 c.

(b) Through the head $\mathbf{c}_a$ of the second vector, draw a line ⊥ to *AC* for the third vector.

(c) From $\mathbf{q}_1$ (or $\mathbf{o}_1$), take the fourth vector and add the fifth vector to it.

(d) Through the head $\mathbf{c}_b$ of the fifth vector, draw a line ⊥ to *BC* for the sixth vector.

The intersection of the lines drawn in steps (b) and (d) locates the point $\mathbf{c}_1$.

Now

$$\mathbf{f}_{do} = \mathbf{f}_{dc} + \mathbf{f}_{co}$$

or

$$\mathbf{f}_{dg} = \mathbf{f}_{co} + \mathbf{f}_{dc}$$

Since $\mathbf{f}_{dc}$ has two components,

$$\mathbf{f}_{dg} = \mathbf{f}_{co} + \mathbf{f}^c_{dc} + \mathbf{f}^t_{dc}$$

or

$$\mathbf{g}_1\mathbf{d}_1 = \mathbf{o}_1\mathbf{c}_1 + \mathbf{c}_1\mathbf{d}_c + \mathbf{d}_c\mathbf{d}_1$$

Set the following vector table:

S.N.	*Vector*	*Magnitude*	*Direction*	*Sense*
1.	$\mathbf{f}_{co}$ or $\mathbf{o}_1\mathbf{c}_1$	Already drawn	—	—
2.	$\mathbf{f}^c_{dc}$ or $\mathbf{c}_1\mathbf{d}_c$	$\frac{(\mathbf{cd})^2}{CD} = \frac{(1.0)^2}{0.24} = 4.17$	‖ *CD*	→ *C*
3.	$\mathbf{f}^t_{dc}$ or $\mathbf{d}_c\mathbf{d}_1$	—	⊥ *CD*	—
4.	$\mathbf{f}_{dg}$ or $\mathbf{g}_1\mathbf{d}_1$	—	‖ to motion of *D*	—

From $\mathbf{c}_1$ draw the second vector and draw a line ⊥ to *CD* through the head of the second vector. Draw a line parallel to the line of motion of the slider through $\mathbf{g}_1$. Thus point $\mathbf{d}_1$ is located.

(i) $f_d = \mathbf{o}_1\mathbf{d}_1 = \underline{10.65 \text{ m/s}^2}$

(ii) $\alpha_{ac} = \dfrac{\mathbf{f}^t_{ca} \text{ or } \mathbf{c}_a\mathbf{c}_1}{AC} = \dfrac{26.4}{0.185} = \underline{142.7 \text{ rad/s}^2}$ clockwise

$\alpha_{bc} = \dfrac{\mathbf{f}^t_{cb} \text{ or } \mathbf{c}_b\mathbf{c}_1}{BC} = \dfrac{8.85}{0.185} = \underline{47.8 \text{ rad/s}^2}$ counter-clockwise

$\alpha_{cd} = \dfrac{\mathbf{f}^t_{dc} \text{ or } \mathbf{d}_c\mathbf{d}_1}{CD} = \dfrac{11.2}{0.24} = \underline{46.7 \text{ rad/s}^2}$ clockwise

3.6 CORIOLIS ACCELERATION COMPONENT

It is seen that the acceleration of a fixed point relative to another fixed point on a moving link may have two components of acceleration; the centripetal and the tangential. Now, a case will be taken where the acceleration of a moving point relative to a fixed point on a moving link will be considered.

Let a link AR rotate about a fixed point A on it (Fig. 3.9). P is a point on a slider on the link.

At any given instant,

Let ω = Angular velocity of the link

α = Angular acceleration of the link

$\mathbf{v}$ = Linear velocity of the slider of the link

$\mathbf{f}$ = Linear acceleration of the slider on the link

r = Radial distance of point P on the slider.

In a short interval of time δt, let $\delta\theta$ be the angular displacement of the link and δr, the radial displacement of the slider in the outward direction.

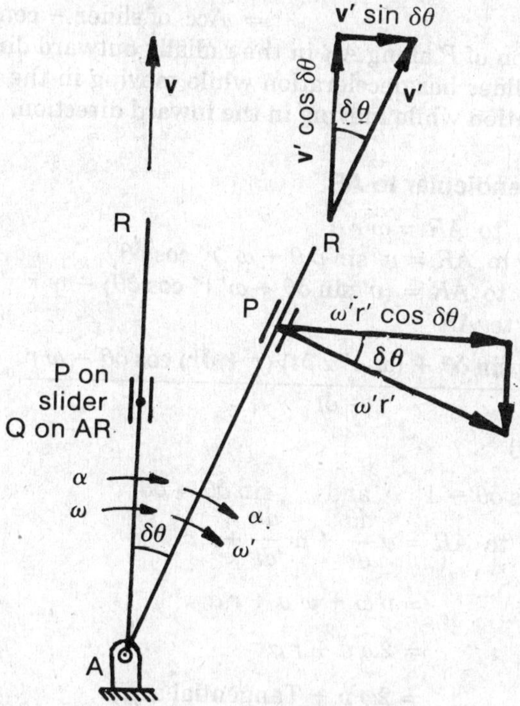

Fig. 3.9

After the short interval of time δt, let

$\omega' = \omega + \alpha r$ = Angular velocity of the link

$v' = v + f \cdot \delta t$ = Linear velocity of the slider on the link

$r' = r + \delta r$ = Radial distance of the slider

Acceleration of P Parallel to AR

Initial veclocity of P along $AR = v = v_{pq}$

Final velocity of P along $AR = v' \cos\delta\theta - \omega' r' \sin\delta\theta$

Change of velocity along $AR = (v' \cos\delta\theta - \omega' r' \sin\delta\theta) - v$

Acceleration of P along AR

$$= \frac{(v + f\,\delta t)\cos\delta\theta - (\omega + \alpha\,\delta t)(r + \delta r)\sin\delta\theta - v}{\delta t}$$

In the limit, as $\delta t \to 0$
$\cos\delta\theta \to 1$ and $\sin\delta\theta \to \theta$

$$\text{Acceleration of } P \text{ along } AR = f - \omega r \frac{d\theta}{dt}$$
$$= f - \omega r \omega$$
$$= f - \omega^2 r$$
$$= \text{Acc. of slider} - \text{cent. acc.}$$

This is the acceleration of P along AR in the radially outward direction. f will be negative if the slider has deceleration while moving in the outward direction or has acceleration while moving in the inward direction.

Acceleration of P Perpendicular to AR

Initial velocity of $P \perp$ to $AR = \omega r$
Final velocity of $P \perp$ to $AR = v' \sin\delta\theta + \omega' r' \cos\delta\theta$
Change of velocity $\perp$ to $AR = (v' \sin\delta\theta + \omega' r' \cos\delta\theta) - \omega r$
Acceleration of $P \perp$ to AR

$$= \frac{(v + f\,\delta t)\sin\delta\theta + (\omega + \alpha\,\delta t)(r + \delta r)\cos\delta\theta - \omega r}{\delta t}$$

In the limit, as $\delta t \to 0$

$$\cos\delta\theta \to 1 \quad \text{and} \quad \sin\delta\theta \to \delta\theta$$

$$\text{Acceleration of } P \perp \text{ to } AR = v\frac{d\theta}{dt} + \omega\frac{dr}{dt} + r\alpha$$
$$= v\omega + \omega v + r\alpha$$
$$= 2\omega v + r\alpha$$
$$= 2\omega v + \text{Tangential acc.}$$

This is the acceleration of P perpendicular to AR. The component $2\,\boldsymbol{\omega}\mathbf{v}$ is known as the *Coriolis acceleration component*. It is positive if both $\boldsymbol{\omega}$ and $\mathbf{v}$ are either positive or negative.

Thus Coriolis component is positive if
the link AR rotates clockwise and the slider moves radially outwards
the link rotates counter-clockwise and the slider moves radially inwards.
Otherwise, the Coriolis component will be negative.

These observations can be summarized into the following rule:

The direction of the Coriolis acceleration component is obtained by rotating the radial velocity vector $\mathbf{v}$ through 90° in the direction of rotation of the link.

Let Q be a point on the link AR immediately beneath the point P at the instant. Then

$$\text{Acc. of } P \text{ rel. to } A = \text{Acc. of } P \text{ rel. to } Q + \text{Acc. of } Q \text{ rel. to } A$$

or

$$\text{Acc. of } P \parallel \text{ to } AR + \text{Acc. of } P \perp \text{ to } AR = \text{Acc. of } P \text{ rel. to } Q + \text{Acc. } Q \text{ rel. to } A$$

or
$$(f - \omega^2 r) + (2\,\omega\, v + r\,\alpha) = f_{pq} + (r\,\alpha - \omega^2 r)$$

or
$$f_{pq} = f + 2\,\omega\, v$$

$$= \text{linear acc. of slider on the link} + \text{Coriolis acc. component } (\perp \text{ to link})$$

$$= f_{pq}^{s} + f_{pq}^{cr}$$

3.7 CRANK AND SLOTTED LEVER MECHANISM

The configuration and the velocity diagrams of a slotted lever mechanism have been shown in Figs. 3.10 a and b respectively. The crank *OP* rotates at uniform angular velocity of ω rad/s clockwise.

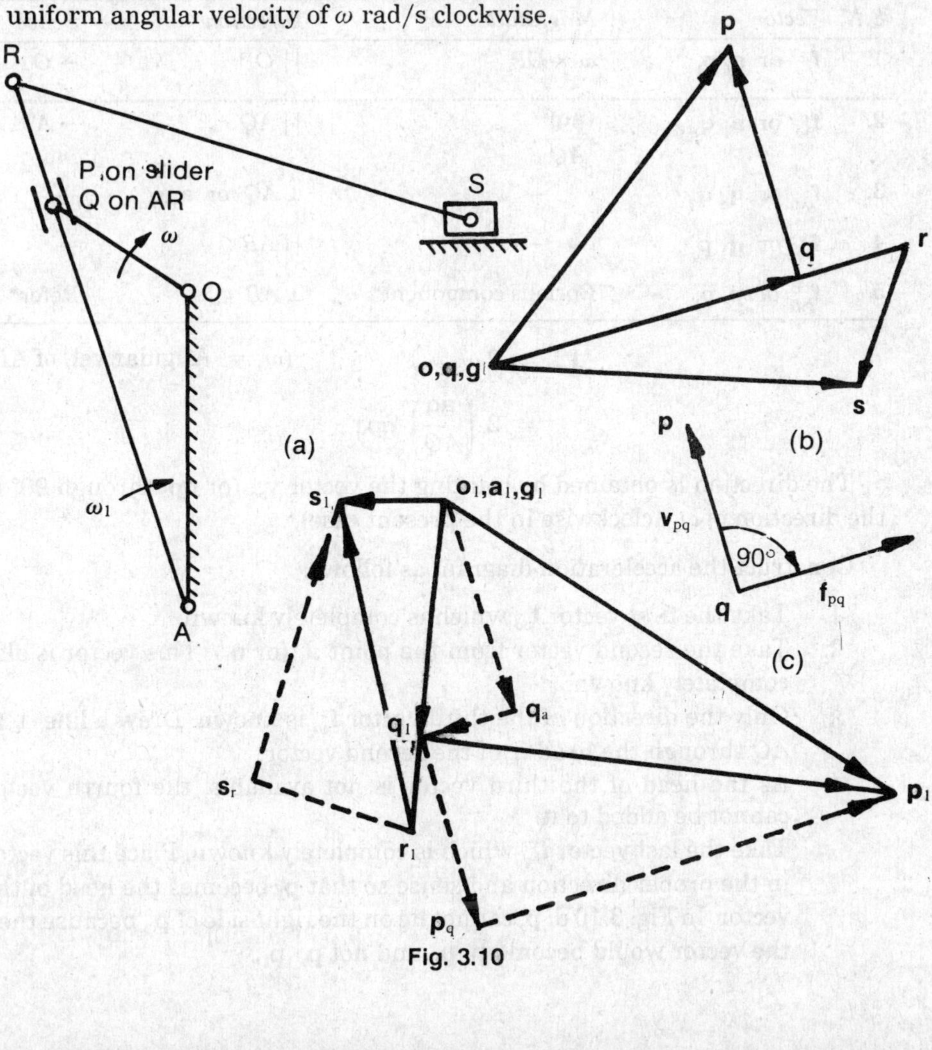

Fig. 3.10

Writing the vector equation,

$$\mathbf{f}_{pa} = \mathbf{f}_{pq} + \mathbf{f}_{qa} \text{ or } \mathbf{f}_{qo} = \mathbf{f}_{qp} + \mathbf{f}_{po}$$

Any of these equations can be solved graphically. Both will lead to the same acceleration diagram except for the direction of the vectors $\mathbf{f}_{pq}$ and $\mathbf{f}_{qp}$.

Considering the first equation,

$$\mathbf{f}_{pa} = \mathbf{f}_{pq} + \mathbf{f}_{qa}$$

or

$$\mathbf{f}_{po} = \mathbf{f}_{qa} + \mathbf{f}_{pq}$$

$$= \mathbf{f}^{c}_{qa} + \mathbf{f}^{t}_{qa} + \mathbf{f}^{s}_{pq} + \mathbf{f}^{cr}_{pq}$$

or

$$\mathbf{o}_1\,\mathbf{p}_1 = \mathbf{a}_1\,\mathbf{q}_a + \mathbf{q}_a\,\mathbf{q}_1 + \mathbf{q}_1\,\mathbf{p}_q + \mathbf{p}_q\,\mathbf{p}_1$$

Set the following vector table:

S.N.	Vector	Magnitude	Direction	Sense
1.	$\mathbf{f}_{po}$ or $\mathbf{o}_1\,\mathbf{p}_1$	$\omega \times OP$	$\parallel OP$	$\to O$
2.	$\mathbf{f}^{c}_{qa}$ or $\mathbf{a}_1\,\mathbf{q}_a$	$\dfrac{(\mathbf{aq})^2}{AQ}$	$\parallel AQ$	$\to A$
3.	$\mathbf{f}^{t}_{qa}$ or $\mathbf{q}_a\,\mathbf{q}_1$	—	$\perp AQ$ or $\mathbf{a}_1\,\mathbf{q}_a$	—
4.	$\mathbf{f}^{s}_{pq}$ or $\mathbf{q}_1\,\mathbf{p}_q$	—	$\parallel AR$	—
5.	$\mathbf{f}^{cr}_{pq}$ or $\mathbf{p}_q\,\mathbf{p}_1$	Coriolis component*	$\perp AR$	Refer*

$$^*\mathbf{f}^{cr}_{pq} = 2\,\omega_1\,v_{pq} \qquad (\omega_1 = \text{Angular vel. of } AR)$$

$$= 2\left(\frac{\mathbf{aq}}{AQ}\right)\mathbf{qp};$$

The direction is obtained by rotating the vector $\mathbf{v}_{pq}$ (or $\mathbf{qp}$) through 90° in the direction of ω_1 (clockwise in the present case).

Construct the acceleration diagram as follows:

1. Take the first vector $\mathbf{f}_{po}$ which is completely known.
2. Take the second vector from the point $\mathbf{a}_1$ (or $\mathbf{o}_1$). This vector is also completely known.
3. Only the direction of the third vector $\mathbf{f}^{t}_{qa}$ is known. Draw a line ⊥ to AQ through the head $\mathbf{q}_a$ of the second vector.
4. As the head of the third vector is not available, the fourth vector cannot be added to it.

   Take the last vector $\mathbf{f}^{cr}_{pq}$ which is completely known. Place this vector in the proper direction and sense so that $\mathbf{p}_1$ becomes the head of the vector. In Fig. 3.10 d, $\mathbf{p}_q$ cannot lie on the right side of $\mathbf{p}_3$ because then the vector would become $\mathbf{p}_1\,\mathbf{p}_q$ and not $\mathbf{p}_q\,\mathbf{p}_1$.

5. For the fourth vector; draw a line parallel to AR through the point $\mathbf{p}_q$ of the fifth vector.
The intersection of this line with the line drawn in step 3 locates the point $\mathbf{q}_1$;
Total acc. of P rel. to Q, $\mathbf{f}_{pq} = \mathbf{q}_1\mathbf{p}_1$
Total acc. of Q rel. to A, $\mathbf{f}_{qa} = \mathbf{a}_1\mathbf{q}_1$
The acceleration of R relative to A is given on $\mathbf{a}_1\mathbf{q}_1$ produced such that

$$\frac{\mathbf{a}_1\mathbf{r}_1}{\mathbf{a}_1\mathbf{q}_1} = \frac{AR}{AQ}$$

Note that in the present case, the sliding acceleration $\mathbf{a}_1\mathbf{q}_a$ is in the opposite direction to the sliding velocity **qp**. This signifies that the slider is decelerating.

Also,

$$\mathbf{f}_{sa} = \mathbf{f}_{sr} + \mathbf{f}_{ra}$$

$$\mathbf{f}_{sg} = \mathbf{f}_{ra} + \mathbf{f}_{sr}$$

$$= \mathbf{f}_{ra} + \mathbf{f}^c_{sr} + \mathbf{f}^t_{sr}$$

$$\mathbf{g}_1\mathbf{s}_1 = \mathbf{a}_1\mathbf{r}_1 + \mathbf{r}_1\mathbf{s}_r + \mathbf{s}_r\mathbf{s}_1$$

The equation can be solved as usual.
Total acc. of S rel. to R, $\mathbf{f}_{sr} = \mathbf{r}_1\mathbf{s}_1$
Acceleration of $S = \mathbf{g}_1\mathbf{s}_1$ or $\mathbf{a}_1\mathbf{s}_1$ or $\mathbf{o}_1\mathbf{s}_1$
The direction of $\mathbf{g}_1\mathbf{s}_1$ is opposite to the direction of motion of the slider S indicating that the slider is decelerating.

Example 3.5 For the Whitworth quick-return mechanism of Example 2.12, determine the following when the crank OP rotates at an angular velocity of 2.5 rad/s and also has an angular deceleration of 20 rad/s^2.

(i) The acceleration of the slider S
(ii) The angular acceleration of links AR and RS.

Solution:

The configuration and the velocity diagrams have been reproduced in Figs 3.11a and b.

Writing the acceleration vector equation,

$$\mathbf{f}_{qo} = \mathbf{f}_{qp} + \mathbf{f}_{po} \quad \text{or} \quad \mathbf{f}_{pa} = \mathbf{f}_{pq} + \mathbf{f}_{qa}$$

Both the equations lead to the same acceleration diagram except that the direction sense of the acceleration vectors between P and Q are reversed. Taking the first one,

$$\mathbf{f}_{qo} = \mathbf{f}_{qp} + \mathbf{f}_{po}$$

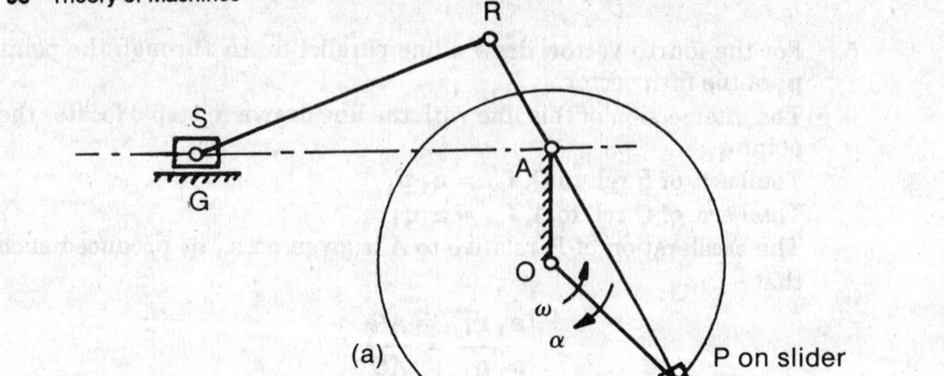

Fig. 3.11

$$\mathbf{f}_{qa} = \mathbf{f}_{po} + \mathbf{f}_{qp}$$

or $$\mathbf{a_1 q_1} = \mathbf{o_1 p_1} + \mathbf{p_1 q_1}$$

Each has two components,

$$\mathbf{f}^c_{qa} + \mathbf{f}^t_{qa} = \mathbf{f}^c_{po} + \mathbf{f}^{cr}_{qp} + \mathbf{f}^s_{qp}$$

or $$\mathbf{a_1 q}_a + \mathbf{q}_a \mathbf{q_1} = \mathbf{o_1 p}_o + \mathbf{p}_o \mathbf{p_1} + \mathbf{p_1 q}_p + \mathbf{q}_p \mathbf{q_1}$$

Set the following vector table:

S.N.	*Vector*	*Magnitude (m/s²)*	*Direction*	*Sense*
1.	$\mathbf{f}^c_{po}$ or $\mathbf{o_1 p_o}$	$\dfrac{(\mathbf{op})^2}{OP} = \dfrac{(0.6)^2}{0.24} = 1.5$	$\parallel OP$	$\to O$
2.	$\mathbf{f}^t_{po}$ or $\mathbf{p_o p_1}$	$\alpha_{op} \times OP = 20 \times 0.24 = 0.48$	$\perp OP$ or $\parallel \mathbf{op}$	$\to \mathbf{o}$
3.	$\mathbf{f}^{cr}_{qp}$ or $\mathbf{p_1 q_p}$	$2\,\omega_{ba}\,\mathbf{v_{qp}} = 0.38$*	$\perp AQ$	Refer*
4.	$\mathbf{f}^s_{qp}$ or $\mathbf{q_p q_1}$	—	$\parallel AQ$	—
5.	$\mathbf{f}^c_{qa}$ or $\mathbf{a_1 q_a}$	$\dfrac{(\mathbf{aq})^2}{AQ} = \dfrac{(0.585)^2}{0.365} = 0.93$	$\parallel AQ$	$\to A$
6.	$\mathbf{f}^t_{qa}$ or $\mathbf{q_a q_1}$	—	$\perp AQ$	—

$$*f^{cr}_{qp} = 2\,\omega_{ba}\, v_{qp} = 2\frac{v_{qa}}{QA} v_{qp} \qquad (\omega_{ba} = \omega_{qa})$$

$$= 2 \times \frac{0.585}{0.365} \times 0.118 \qquad (v_{qa} = \mathbf{aq} = 0.585)$$

$$= 0.38 \text{ m/s}^2$$

Direction of $\mathbf{f}^{cr}_{qp}$ is obtained by rotating $\mathbf{v}_{qp}$ through 90° in the direction of angular movement of link *QA* or *BA* (counter-clockwise in this case). Draw the acceleration diagram as follows:

1. From the pole point $\mathbf{o}_1$, take the first vector and add to it the second vector.
2. Add the third vector to the second vector. For the fourth vector, draw a line to *AQ*, through the head $\mathbf{q}_p$ of the third vector.
3. From the pole point $\mathbf{a}_1$ or $\mathbf{o}_1$, take the fifth vector and for the sixth vector, draw a line to *AQ* through the head $\mathbf{q}_a$ of the fifth vector. This way the point $\mathbf{q}_1$ is located.
4. Join $\mathbf{q}_1$ and $\mathbf{a}_1$ and extend to $\mathbf{r}_1$ such that

$$\frac{\mathbf{a}_1\,\mathbf{r}_1}{\mathbf{a}_1\,\mathbf{q}_1} = \frac{AR}{AQ}$$

Writing the vector equation,

$$\mathbf{f}_{so} = \mathbf{f}_{sr} + \mathbf{f}_{ro}$$

or

$$\mathbf{f}_{sg} = \mathbf{f}_{ro} + \mathbf{f}_{sr}$$

$$= \mathbf{f}_{ro} + \mathbf{f}^{c}_{sr} + \mathbf{f}^{t}_{sr}$$

or

$$\mathbf{g}_1\,\mathbf{s}_1 = \mathbf{o}_1\,\mathbf{r}_1 + \mathbf{r}_1\,\mathbf{s}_r + \mathbf{s}_r\,\mathbf{s}_1$$

$\mathbf{f}_{ro}$ is already available on the acceleration diagram. $\mathbf{f}_{sg}$ is horizontal.

$$\mathbf{f}^{c}_{sr} = \frac{(\mathbf{r\,s})^2}{RS} = \frac{(0.12)^2}{0.43} = 0.033 \text{ m/s}^2$$

Complete the vector diagram as usual.

$$\mathbf{f}_s = \mathbf{o}_1\,\mathbf{s}_1 = \underline{0.39 \text{ m/s}^2}$$

$$\alpha_{ar} = \alpha_{qa} = \frac{\mathbf{f}^{t}_{qa} \text{ or } \mathbf{q}_a\,\mathbf{q}_1}{QA} = \frac{0.57}{0.365} = \underline{1.56 \text{ rad/s}^2} \text{ clockwise}$$

$$\alpha_{rs} = \frac{\mathbf{f}^{t}_{rs} \text{ or } \mathbf{s}_1\,\mathbf{s}_r}{RS} = \frac{0.24}{0.43} = \underline{0.558 \text{ rad/s}^2} \text{ clockwise.}$$

Example 3.6 One cylinder of a rotary engine is shown in the configuration diagram shown in Fig. 3.12 a. *OA* is the fixed crank, 200 mm long. *OP* is the connecting rod and is 520 mm long. The line of stroke is along *AR* and at the instant is inclined at 30° to the vertical. The body of the engine consisting of cylinders rotate at a uniform speed of 400 rpm about the fixed centre *A*. Determine

(i) Acceleration of piston (slider) inside the cylinder
(ii) Angular acceleration of the connecting rod.

Solution:

Let Q be a point on AR beneath the point P.

$$v_{qa} = \frac{2\pi N}{60} \times QA = \frac{2\pi \times 400}{60} \times 0.68 = 28.5 \text{ m/s}$$

The velocity vector equation is

$$\mathbf{v}_{pa} = \mathbf{v}_{pq} + \mathbf{v}_{qa} \quad \text{or} \quad \mathbf{v}_{qo} = \mathbf{v}_{qp} + \mathbf{v}_{po}$$

Taking the first one,

$$\mathbf{v}_{po} = \mathbf{v}_{qa} + \mathbf{v}_{qp}$$

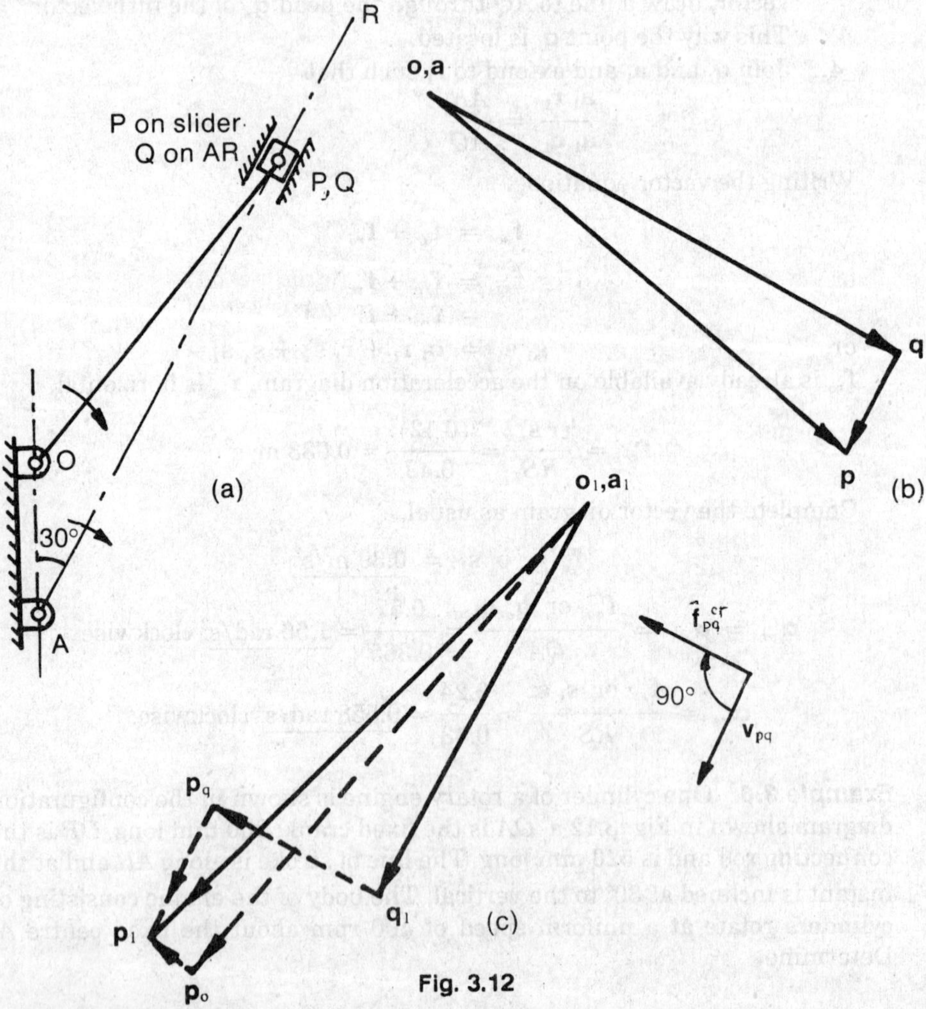

Fig. 3.12

or

$$\mathbf{op} = \mathbf{aq} + \mathbf{qp}$$

Take vector $\mathbf{v}_{qa}$ to a convenient scale (Fig. 3.12b).
$\mathbf{v}_{pq}$ is || to AR, draw a line || to AR through $\mathbf{q}$.
$\mathbf{v}_{po}$ is ⊥ to OP, draw $\mathbf{op}$, a line ⊥ to OP through $\mathbf{o}$.
The intersection locates the point $\mathbf{p}$.
Similarly, writing the acceleration vector equation,

$$\mathbf{f}_{pa} = \mathbf{f}_{pq} + \mathbf{f}_{qa}$$

or

$$\mathbf{f}_{po} = \mathbf{f}_{qa} + \mathbf{f}_{pq}$$

or

$$\mathbf{o}_1\mathbf{p}_1 = \mathbf{a}_1\mathbf{q}_1 + \mathbf{q}_1\mathbf{p}_1$$

Expanding,

$$\mathbf{f}^c_{po} + \mathbf{f}^t_{po} = \mathbf{f}_{qa} + \mathbf{f}^{cr}_{pq} + \mathbf{f}^s_{pq}$$

or

$$\mathbf{o}_1\mathbf{p}_o + \mathbf{p}_o\mathbf{p}_1 = \mathbf{a}_1\mathbf{q}_1 + \mathbf{q}_1\mathbf{p}_q + \mathbf{p}_q\mathbf{p}_1$$

Set the vector table:

S.N.	Vector	Magnitude (m/s²)	Direction	Sense
1.	$\mathbf{f}_{qa}$ or $\mathbf{a}_1\mathbf{q}_1$	$\frac{(\mathbf{aq})^2}{AQ} = \frac{(28.5)^2}{0.68} = 1194$	$\parallel AQ$	→ A
2.	$\mathbf{f}^{cr}_{pq}$ or $\mathbf{q}_1\mathbf{p}_q$	486*	⊥ AQ	—
3.	$\mathbf{f}^s_{pq}$ or $\mathbf{p}_q\mathbf{p}_1$	—	$\parallel AQ$	—
4.	$\mathbf{f}^c_{po}$ $\mathbf{o}_1\mathbf{p}_o$	$\frac{(\mathbf{op})^2}{OP} = \frac{(29.3)^2}{0.52} = 1651$	$\parallel OP$	→ $\mathbf{o}$
5.	$\mathbf{f}^t_{po}$ or $\mathbf{p}_o\mathbf{p}_1$	—	⊥ OP	—

$$*\mathbf{f}^{cr}_{pq} = 2\,\omega_{ra}\,\mathbf{v}_{pq}$$

$$= 2\frac{v_{qa}}{QA}\,\mathbf{qp} \qquad (\omega_{ra} = \omega_{qa})$$

$$= 2 \times \frac{28.5}{0.68} \times 5.8$$

$$= 486 \text{ m/s}^2$$

Direction of $\mathbf{f}^{cr}_{pq}$ is obtained by rotating $\mathbf{v}_{pq}$ through 90° in the direction of ω_{qa} (clockwise). Draw the vector diagram as follows:

1. Take the first vector from the pole point $\mathbf{a}_1$ or $\mathbf{o}_1$ (Fig. 3.12c).
2. Add the second vector to the first vector.
3. Through the head of the second vector, draw a line parallel to AQ for the third vector.
4. Take the fourth vector from the pole point $\mathbf{o}_1$.
5. Through the head of the fourth vector, draw a line perpendicular to OP for the fifth vector.

The intersection of the lines drawn in steps (3) and (5) locates the point $\mathbf{p}_1$.

(i) Acceleration of the slider inside the cylinder,

$$\mathbf{f}^{s}_{pq} = \mathbf{p_q\,p_1} = \underline{390 \text{ m/s}^2}$$

(ii) Angular acceleration of the connecting rod,

$$\alpha_{op} = \frac{f^{t}_{po} \text{ or } \mathbf{p_o p_1}}{OP} = \frac{150}{0.52} = \underline{288.5 \text{ rad/s}^2 \text{ counter-clockwise.}}$$

Example 3.7 For the pump mechanism of Example 2.11, determine

(i) Linear (sliding) acceleration of slider C relative to cylinder walls
(ii) Angular acceleration of the piston rod.

Solution: Configuration and velocity diagrams have been reproduced in Figs. 3.13 a and b.

The problem can be solved either of the two methods discussed for velocity diagram in Example 2.11. Using the latter configuration.

Writing the acceleration vector equation,

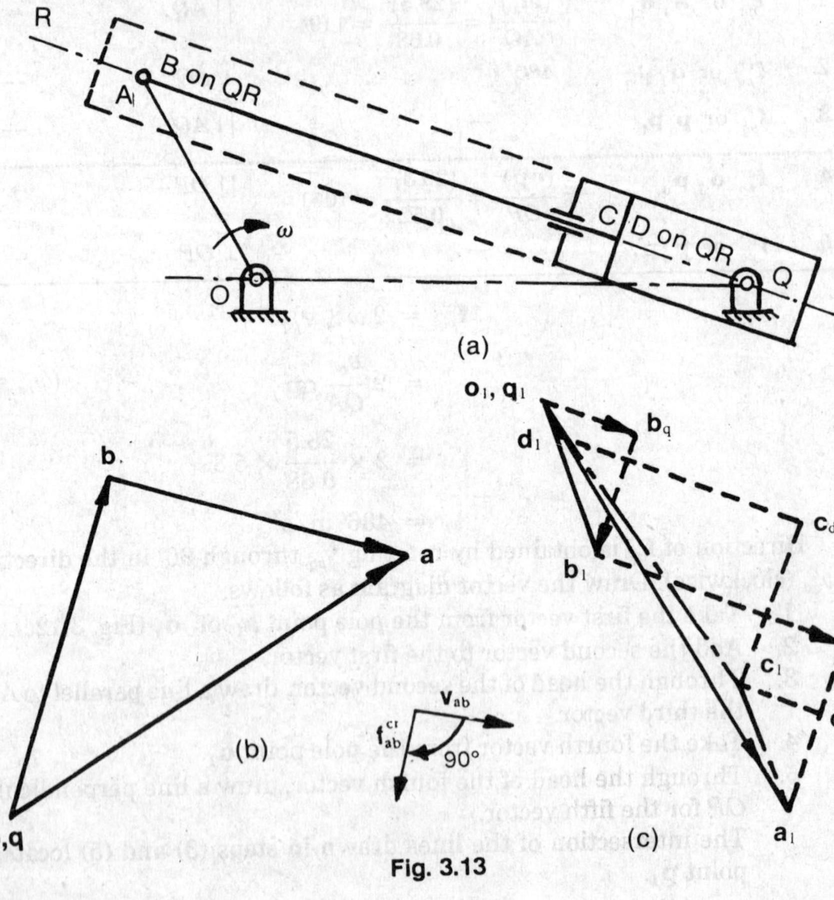

Fig. 3.13

$$\mathbf{f}_{aq} = \mathbf{f}_{ab} + \mathbf{f}_{bq}$$

or
$$\mathbf{f}_{ao} = \mathbf{f}_{bq} + \mathbf{f}_{ab}$$
$$= \mathbf{f}_{bq}^{c} + \mathbf{f}_{bq}^{t} + \mathbf{f}_{ab}^{s} + \mathbf{f}_{ab}^{cr}$$

or
$$\mathbf{o}_1\,\mathbf{a}_1 = \mathbf{q}_1\,\mathbf{b}_\mathbf{q} + \mathbf{b}_\mathbf{q}\,\mathbf{b}_1 + \mathbf{b}_1\,\mathbf{a}_\mathbf{b} + \mathbf{a}_\mathbf{b}\,\mathbf{a}_1$$

Set the vector table.

S.N.	*Vector*	*Magnitude (m/s^2)*	*Direction*	*Sense*
1.	$\mathbf{f}_{ao}$ or $\mathbf{o}_1\,\mathbf{a}_1$	$\dfrac{(\mathbf{oa})^2}{OA} = \dfrac{(6.4)^2}{0.32} = 128$	$\parallel OA$	$\to O$
2.	$\mathbf{f}_{bq}^{c}$ or $\mathbf{q}_1\,\mathbf{b}_\mathbf{q}$	$\dfrac{(\mathbf{bq})^2}{BQ} = \dfrac{(4.77)^2}{0.85} = 26.8$	$\parallel QB$	$\to Q$
3.	$\mathbf{f}_{bq}^{t}$ or $\mathbf{b}_\mathbf{q}\,\mathbf{b}_1$	—	$\perp QB$	—
4.	$\mathbf{f}_{ab}^{s}$ or $\mathbf{b}_1\,\mathbf{a}_\mathbf{b}$	—	$\parallel QB$	—
5	$\mathbf{f}_{ab}^{cr}$ or $\mathbf{a}_\mathbf{b}\,\mathbf{a}_1$	47.1*	$\perp QB$	—

$$*\mathbf{f}_{ab}^{cr} = 2\,\omega_{rq}\,v_{ab} = 2\,\frac{v_{bq}}{BQ}\,ba \qquad (\omega_{rq} = \omega_{bq})$$
$$= 2 \times \frac{4.77}{0.85} \times 4.2 = 47.1 \text{ m/s}^2$$

The direction of $\mathbf{f}_{ab}^{cr}$ is obtained by rotating $\mathbf{v}_{ba}$ through 90° in the direction of ω_{bq} (clockwise).

Draw the acceleration diagram as given below:

1. Take the first vector.
2. From point $\mathbf{q}_1$ (pole point), take the second vector and through the head of it, draw a line perpendicular to QB for the 3rd vector.
3. Take the fifth vector from the point $\mathbf{a}_1$ such that the vector is in the proper direction and sense.
4. For the fourth vector, draw a line parallel to QB through the tail $\mathbf{a}_\mathbf{b}$ of the fifth vector.

The intersection of the lines drawn in steps (2) and (4) locates point $\mathbf{b}_1$.

Let D be a point on QB beneath point C.

Acc. of C rel. to cylinder walls $= f_{cd}^{s} = f_{ab}^{s} = \underline{72 \text{ m/s}^2}$

Writing the vector equation,

$$\mathbf{f}_{cq} = \mathbf{f}_{cd} + \mathbf{f}_{dq}$$
$$= \mathbf{f}_{dq} + \mathbf{f}_{cd}$$
$$= \mathbf{f}_{dq} + \mathbf{f}_{cd}^{s} + \mathbf{f}_{cd}^{cr}$$

or
$$\mathbf{q}_1\,\mathbf{c}_1 = \mathbf{q}_1\,\mathbf{d}_1 + \mathbf{d}_1\,\mathbf{c}_d + \mathbf{c}_d\,\mathbf{c}_1$$

$\mathbf{f}_{cd}^{s} = \mathbf{f}_{ab}^{s} = \underline{72 \text{ m/s}^2}$ parallel to QB

$\mathbf{f}_{cd}^{cr} = 2\,\boldsymbol{\omega}_{rq}\,\mathbf{v}_{cd} = 2\,\boldsymbol{\omega}_{rq}\,\mathbf{v}_{ab} = \mathbf{f}_{ab}^{cr}$

Locate point $\mathbf{d}_1$ on $\mathbf{q}_1\,\mathbf{b}_1$ such that $\dfrac{\mathbf{q}_1\,\mathbf{d}_1}{\mathbf{q}_1\,\mathbf{b}_1} = \dfrac{QD}{QB}$

To the vector $\mathbf{f}_{dq}$, add the vector $\mathbf{f}^s_{cd}$ and then $\mathbf{f}^{cr}_{cd}$.
Thus point $\mathbf{c}_1$ is located.
Join $\mathbf{a}_1$ to $\mathbf{c}_1$.
$\mathbf{a}_1\,\mathbf{c}_1$ represents total acceleration of C relative to A. This has two components.

$$\mathbf{f}^t_{ca} \perp \text{ to } CA = \mathbf{a}_1\,\mathbf{c}_a$$

and $$\mathbf{f}^c_{ca} \;||\; \text{to } CA \;(= \mathbf{c}_a\,\mathbf{c}_1)$$

$$\alpha_{ca} = \frac{\mathbf{f}^t_{ca}}{CA} = \frac{14}{0.68} = \underline{20.59 \text{ rad/s}^2} \text{ counter-clockwise}$$

3.8 KLEIN'S CONSTRUCTION

Slider Crank Mechanism

In Klein's construction, the velocity and the acceleration diagrams are made on the configuration diagram itself. The line that represents the crank in the configuration diagram also represents the velocity and the acceleration of its moving end in the velocity and the acceleration diagrams respectively.

In Fig. 3.14, OAB represents the configuration of a slider-crank mechanism. Its velocity and acceleration diagrams are as shown in Figs. 3.4b and c.

Length of the crank = $OA = r$

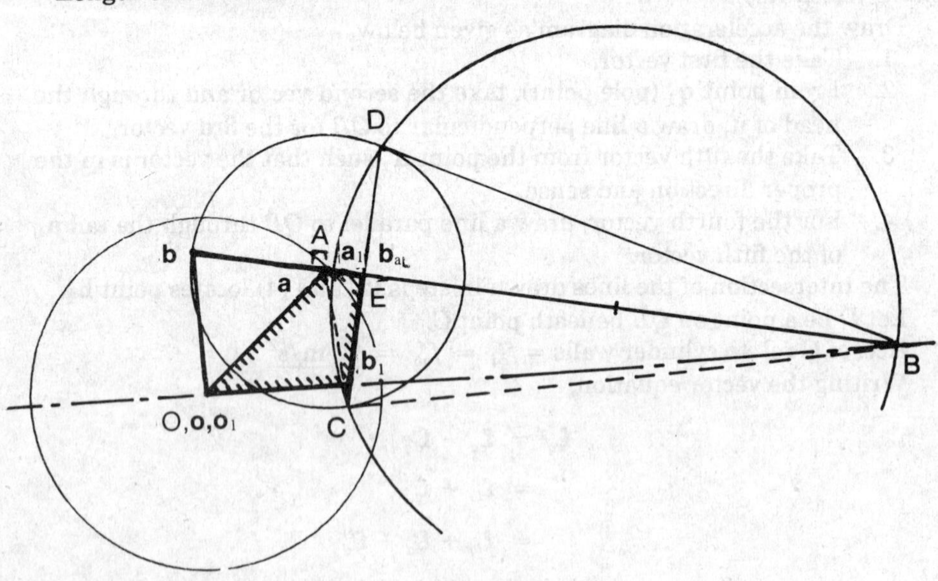

Fig. 3.14

Velocity Diagram

For velocity diagram, let r represents v_{ao}, to some scale. Then for the velocity diagram, length $oa = \omega r = OA$.

From this, the scale for the velocity diagram is known.

Produce BA and draw a line perpendicular to OB through O. The intersection of the two lines locates point **b**. The figure **oab** is the velocity diagram which is similar to the velocity diagram of Fig. 3.4 b rotated through 90° in a direction opposite to that of the crank.

Acceleration Diagram

For acceleration diagram, let r represents f_{ao}.

$\therefore \quad \mathbf{o}_1\,\mathbf{a}_1 = \omega^2 r = OA$

This provides the scale for the acceleration diagram.

Make the following construction:

1. Draw a circle with **ab** as the radius and **a** as the center.
2. Draw another circle with AB as diameter.
3. Join the points of intersections C and D of the two circles. Let it meet OB at $\mathbf{b}_1$ and AB at E.

Then $\mathbf{o}_1\,\mathbf{a}_1\,\mathbf{b}_a\,\mathbf{b}_1$ is the required acceleration diagram which is similar to the acceleration diagram of Fig. 3.4 c rotated through 180°.

The proof is as follows:

Join AC and BC.

AEC and ABC are two right-angled triangles in which angle CAB is common. Therefore, the triangles are similar.

$$\frac{AE}{AC} = \frac{AC}{AB}$$

or

$$AE = \frac{(AC)^2}{AB}$$

or

$$\mathbf{a}_1\,\mathbf{b}_a = \frac{(\mathbf{ab})^2}{AB} = f_{ba}^c$$

Thus this acceleration diagram has all the sides parallel to that of acceleration diagram of Fig. 3.4 c and also has two sides $\mathbf{o}_1\,\mathbf{a}_1$ and $\mathbf{a}_1\,\mathbf{b}_a$ representing the corresponding magnitudes of the acceleration. Thus the two diagrams are similar.

3.9 VELOCITY AND ACCELERATION FROM DISPLACEMENT-TIME CURVE

Sometimes displacement-time data for a moving point in a mechanism are available and it is required to find the velocities and accelerations at various instants. This can be done easily by graphical differentiation that uses the following rules:

1. Velocity is the derivative of displacement with respect to time and is proportional to the slope of the tangent to the displacement-time curve for any instant.
2. Acceleration is the derivative of velocity with respect to time and is proportional to the slope of the tangent to the velocity-time curve for any instant.

Figure 3.15 a shows a displacement-time curve of a point in a mechanism. At point C, LN is the tangent where L and N points are chosen arbitrarily.

Then

$$v_c = \frac{k_s\, NM}{k_t\, LM} \quad \text{(i)}$$

where v_c = velocity at C

k_s = displacement scale

k_t = time scale

and NM, LM = actual drawing distances

To plot the velocity-time curve, select a convenient point P (known as pole point) as shown in Fig. 3.15 b. Draw a line PT parallel to LN. Then it can be said that TO is the magnitude of the velocity at point C to some scale which can be found as follows.

Let k_v = velocity scale

Then

$$v_c = k_v\, TO \quad \text{(ii)}$$

from (i) & (ii).

$$k_v\, TO = \frac{k_s\, NM}{k_t\, LM}$$

$$= \frac{k_s\, TO}{k_t\, PO}$$

or

$$k_v = \frac{k_s}{k_t} \frac{1}{PO}$$

Thus scale of k_v is known.

Select more points on the displacement-time curve and draw tangents to the curve. From the pole point P draw lines parallel to these tangents meeting the y-axis. Project points obtained on this axis to the corresponding ordinates. Using a french curve complete the velocity-time curve.

In the same way, the acceleration-time curve can be drawn taking another pole point P_1 for that (Fig. 3.15 c). The acceleration scale k_f will be given by

$$k_f = \frac{k_v}{k_t} \frac{1}{P_1O_1}$$

Note that:

1. The derivative is positive if a curve rises and is negative if falls.

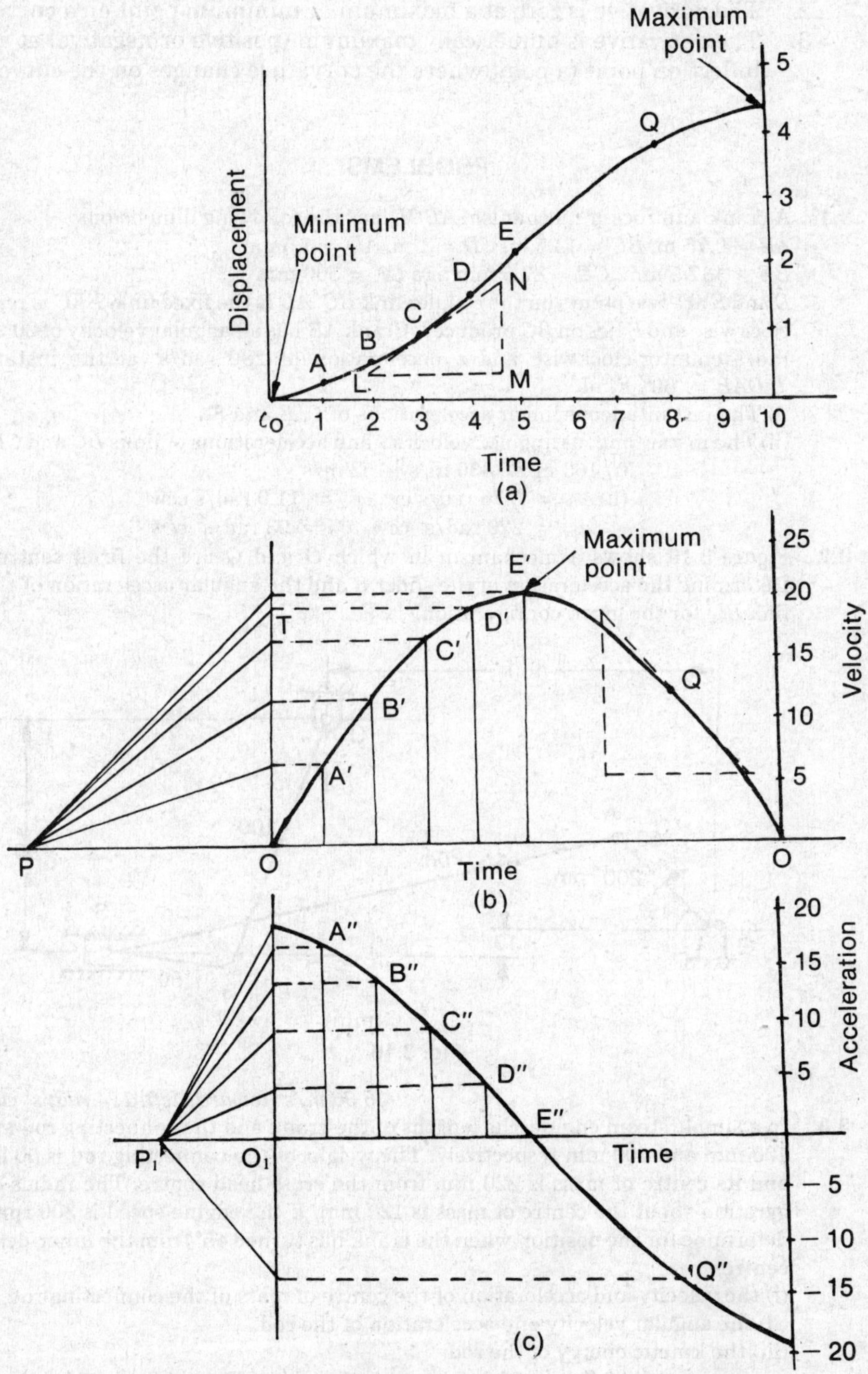

Maximum
point
Minimum
point
Displacement
Q
E
D
N
C
B
A
L
M
t_0
1
2
3
4
5
6
7
8
9
10
5
4
3
2
1
Time
(a)
Maximum
point
E′
D′
T
C′
B′
A′
Q
Velocity
25
20
15
10
5
P
O
O
Time
(b)
A″
B″
C″
D″
E″
P_1
O_1
Time
Q″
Acceleration
20
15
10
5
– 5
– 10
– 15
– 20
(c)

Fig. 3.15

2. The derivative is zero at a maximum or minimum point on a curve.
3. The derivative is numerically maximum (positive or negative) at an inflection point (a point where the curvature changes on the curve).

PROBLEMS

3.1 A crank and rocker mechanism *ABCD* has the following dimensions:
AB = 0.75 m, BC = 1.25 m, CD = 1 m, AD = 1.5 m.
BE = 437.5 mm, CE = 87.5 mm and CF = 500 mm
E and *F* are two points on the coupler link *BC*. *AD* is the fixed link. *BEC* is read clockwise and *F* lies on *BC* produced. Crank *AB* has an angular velocity of 20.94 rad/s counter-clockwise and a deceleration of 280 rad/s^2 at the instant $\angle DAB$ = 60°. Find
(i) The instantaneous linear accelerations of *C*, *E* and *F*
(ii) The instantaneous angular velocities and accelerations of links *BC* and *CD*.

(i) 163 m/s^2, 330 m/s^2, 152 m/s^2
(ii) ω_{bc} = 5.76 rad/s cw, ω_{cd} = 11.6 rad/s ccw
α_{bc} = 229 rad/s^2 ccw, α_{cd} = 93 rad/s^2 ccw

3.2 Figure 3.16 shows a mechanism in which *O* and *Q* are the fixed centres. Determine the acceleration of the slider *S* and the angular acceleration of the link *BQ* for the given configuration.

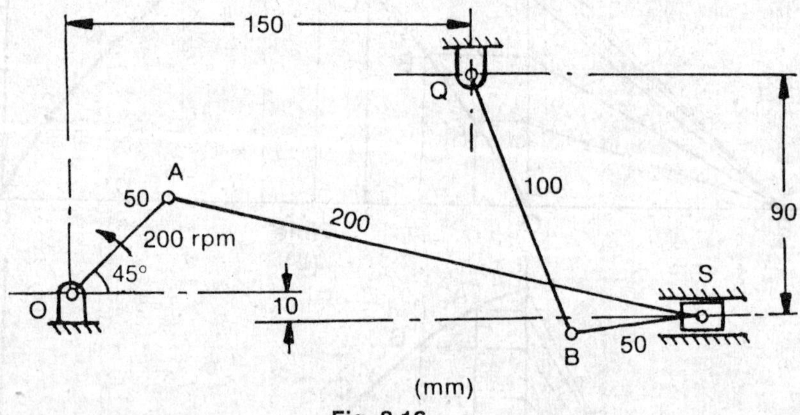

Fig. 3.16

(36.96 m/s^2 towards left; 114 rad/s^2 cw)

3.3 In a simple steam engine, the lengths of the crank and the connecting rod are 100 mm and 400 mm respectively. The weight of the connecting rod is 50 kg and its centre of mass is 220 mm from the cross-head centre. The radius of gyration about the centre of mass is 120 mm. If the engine speed is 300 rpm, determine for the position when the crank has turned 45° from the inner-dead centre,
(i) the velocity and acceleration of the centre of mass of the connecting rod
(ii) the angular velocity and acceleration of the rod.
(iii) the kinetic energy of the rod.

[(i) 2.7 m/s, 86.3 m/s^2 (ii) 5.65 rad/s, 183 rad/s^2 (iii) 188.4 N.m]

3.4 From the data of a reciprocating pump given in Example 2.3, find the linear acceleration of the cross-head E and the angular accelerations of the links BCD and DE.

3.5 In a crank and slotted-lever quick-return mechanism (Fig. 3.10a), the distance between the fixed centres O and A is 250 mm. Other lengths are: OP = 100 mm, AR = 400 mm, RS = 150 mm and $\angle AOP$ = 120°. Uniform speed of the crank is 60 rpm clockwise. Line of stroke of the ram is perpendicular to OA and is 450 mm from A. Calculate the velocity and the acceleration of the ram S.

(0.95 m/s; 3.08 m/s^2)

3.6 For the inverted slider-crank mechanism of Example 2.9, determine the angular acceleration of the link QR.

3.7 In the pump mechanism shown in Fig. 3.13a, the crank OA is 50 mm long and the piston rod AC 150 mm long. The lengths OQ and CQ are 250 mm and 80 mm respectively. The crank rotates at 300 rpm in the clockwise direction. Determine

(a) the velocity of the piston relative to walls
(b) the angular velocities of rod AC and the cylinder
(c) the sliding acceleration of the piston relative to cylinder walls
(d) the velocity of piston (absolute)
(e) the angular acceleration of the piston rod BC.

[(a) 1.49 m/s (b) 2.17 rad/s ccw of both, rod AC and cylinder (c) 16.75 m/s^2 (d) 1.5 m/s (e) 180 rad/s^2 ccw]

4

COMPUTER-AIDED ANALYSIS OF MECHANISMS

4.1 INTRODUCTION

The analyses of the velocity and the acceleration, given in chapters 2 and 3, depend upon the graphical approach and are suitable for finding out the velocity and the acceleration of the links of a mechanism in one or two positions of the crank. To draw velocity and acceleration diagrams again and again for different positions of the crank is not very convenient. In this chapter, analytical expressions for the displacement, velocity and acceleration in terms of the general parameters have been derived. A desk-calculator or a digital computer facilitates the calculation work.

4.2 FOUR-LINK MECHANISM

Displacement Analysis

A four-link mechanism shown in Fig. 4.1 is in equilibrium. *a, b, c* and *d* represent the magnitudes of the links *AB, BC, CD* and *DA* respectively. θ, β and φ are the angles of *AB, BC,* and *DC* respectively with the *x*-axis (taken along *AD*). *AD* is the fixed link.

Considering the links as vectors and writing the vector displacement relationships.

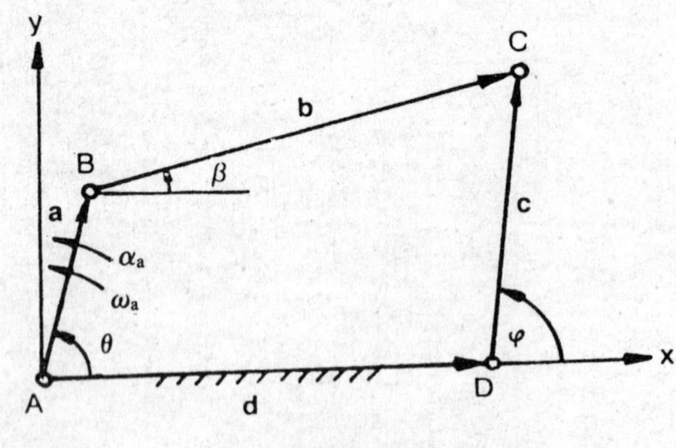

Fig. 4.1

Displacement along x-axis

$$d = -c \cos \varphi + b \cos \beta + a \cos \theta \tag{4.1}$$

or

$$b \cos \beta = c \cos \varphi - a \cos \theta + d$$

or

$$\begin{aligned}(b \cos \beta)^2 &= (c \cos \varphi - a \cos \theta + d)^2 \\ &= c^2 \cos^2 \varphi + a^2 \cos^2 \theta + d^2 \\ &\quad - 2ac \cos \theta \cos \varphi - 2ad \cos \theta + 2cd \cos \varphi \end{aligned} \tag{4.2}$$

Displacement along y-axis

$$0 = -c \sin \varphi + b \sin \beta + a \sin \theta \tag{4.3}$$

or

$$b \sin \beta = c \sin \varphi - a \sin \theta$$

or

$$\begin{aligned}(b \sin \beta)^2 &= (c \sin \varphi - a \sin \theta)^2 \\ &= c^2 \sin^2 \varphi + a^2 \sin^2 \theta - 2ac \sin \theta \sin \varphi \end{aligned} \tag{4.4}$$

Adding Eqs (4.2) and (4.4),

$$b^2 = c^2 + a^2 + d^2 - 2ac \cos \theta \cos \varphi - 2ad \cos \theta + 2cd \cos \varphi - 2ac \sin \theta \sin \varphi$$

Put

$$a^2 - b^2 + c^2 + d^2 = 2k$$

Then,

$$2cd \cos \varphi - 2ac \cos \theta \cos \varphi - 2ac \sin \theta \sin \varphi - 2ad \cos \theta + 2k = 0$$

or

$$cd \cos \varphi - ac \cos \theta \cos \varphi - ac \sin \theta \sin \varphi - ad \cos \theta + k = 0 \tag{4.5}$$

From trigonometric identities,

$$\sin \varphi = \frac{2 \tan \left(\dfrac{\varphi}{2}\right)}{1 + \tan^2 \left(\dfrac{\varphi}{2}\right)}$$

$$\cos \varphi = \frac{1 - \tan^2 \left(\dfrac{\varphi}{2}\right)}{1 + \tan^2 \left(\dfrac{\varphi}{2}\right)}$$

$$cd\left[\frac{1-\tan^2(\varphi/2)}{1+\tan^2(\varphi/2)}\right]-ac\cos\theta\left[\frac{1-\tan^2(\varphi/2)}{1+\tan^2(\varphi/2)}\right]$$

$$-ac\sin\theta\left[\frac{2\tan(\varphi/2)}{1+\tan^2(\varphi/2)}\right]-ad\cos\theta+k=0$$

Multiplying throughout by $\left[1+\tan^2\left(\dfrac{\varphi}{2}\right)\right]$

$$cd-cd\tan^2\left(\frac{\varphi}{2}\right)-ac\cos\theta+ac\cos\theta\tan^2\left(\frac{\varphi}{2}\right)-2ac\sin\theta\tan\left(\frac{\varphi}{2}\right)$$

$$-ad\cos\theta-ad\cos\theta\tan^2\left(\frac{\varphi}{2}\right)+k+k\tan^2\left(\frac{\varphi}{2}\right)=0$$

$$[k-a(d-c)\cos\theta-cd]\tan^2\left(\frac{\varphi}{2}\right)+[-2ac\sin\theta]\tan\left(\frac{\varphi}{2}\right)$$

$$+[k-a(d+c)\cos\theta+cd]=0$$

or

$$A\tan^2\left(\frac{\varphi}{2}\right)+B\tan\left(\frac{\varphi}{2}\right)+C=0 \tag{4.6}$$

where

$$A=k-a(d-c)\cos\theta-cd$$
$$B=-2ac\sin\theta$$
$$C=k-a(d+c)\cos\theta+cd$$

Equation (4.6) is a quadratic in $\tan\left(\dfrac{\varphi}{2}\right)$. Its two roots are,

$$\tan\left(\frac{\varphi}{2}\right)=\frac{-B\pm\sqrt{B^2-4AC}}{2A}$$

or

$$\varphi=2\tan^{-1}\left[\frac{-B\pm\sqrt{B^2-4AC}}{2A}\right] \tag{4.7}$$

Thus the position of the output link, given by angle φ can be calculated if the magnitudes of the links and the position of the input link are known, i.e. a, b, c, d and θ are known.

A relation between the coupler link position β and the input link position θ can also be found as below:

Equations (4.1) and (4.3) can be written as,

$$c\cos\varphi=a\cos\theta+b\cos\beta-d \tag{4.8}$$

$$c\sin\varphi=a\sin\theta+b\sin\beta \tag{4.9}$$

Squaring and adding the two equations,

$$c^2 = a^2 + b^2 + d^2 + 2ab\cos\theta\cos\beta - 2bd\cos\beta - 2ad\cos\theta + 2ab\sin\theta\sin\beta$$

Put $\quad a^2 + b^2 - c^2 + d^2 = 2k'$

$$-2bd\cos\beta + 2ab\cos\theta\cos\beta + 2ab\sin\theta\sin\beta - 2ad\cos\theta + 2k' = 0$$

$$-bd\cos\beta + ab\cos\theta\cos\beta + ab\sin\theta\sin\beta - ad\cos\theta + k' = 0 \quad (4.10)$$

The Eqn. (4.10) is identical to Eqn. (4.6a) and can be obtained from the same by substituting β for φ, $-b$ for c and k' for k.

Thus the solution of Eqn. (4.10) will be,

$$\beta = 2\tan^{-1}\left[\frac{-E \pm \sqrt{E^2 - 4DF}}{2D}\right] \quad (4.11)$$

Where $D = k' - a(d + b)\cos\theta + bd$

$E = 2ab\sin\theta$

$F = k' - a(d - b)\cos\theta - bd$

β can also be found directly from relation (4.3) after calculating φ.

Velocity Analysis

Let ω_a, ω_b and ω_c be the angular velocities of the links AB, BC and CD respectively.

Eqns. (4.1) and (4.3) can be written as,

$$a\cos\omega_a t + b\cos\omega_b t - c\cos\omega_c t - d = 0 \quad (4.12)$$

$$a\sin\omega_a t + b\sin\omega_b t - c\sin\omega_c t = 0 \quad (4.13)$$

Differentiating Eqns. (4.12) and (4.13) with respect to time,

$$-a\,\omega_a\sin\theta - b\,\omega_b\sin\beta + c\,\omega_c\sin\varphi = 0 \quad (4.14)$$

$$a\,\omega_a\cos\theta + b\,\omega_b\cos\beta - c\,\omega_c\cos\varphi = 0 \quad (4.15)$$

Multiply (4.14) by $\cos\beta$ and (4.15) by $\sin\beta$ and add,

$$a\,\omega_a(\sin\beta\cos\theta - \cos\beta\sin\theta) + c\,\omega_c(\sin\varphi\cos\beta - \cos\varphi\sin\beta) = 0$$

or

$$a\,\omega_a\sin(\beta - \theta) + c\,\omega_c\sin(\varphi - \beta) = 0$$

or

$$\omega_c = -\frac{a\,\omega_a\sin(\beta - \theta)}{c\sin(\varphi - \beta)} \quad (4.16)$$

Multiply Eqn. (4.14) by $\cos\varphi$ and Eqn. (4.15) by $\sin\varphi$ and add,

$$a\,\omega_a(\sin\varphi\cos\theta - \cos\varphi\sin\theta) + b\,\omega_b(\sin\varphi\cos\beta - \cos\varphi\sin\beta) = 0$$

or

$$a\,\omega_a\sin(\varphi - \theta) + b\,\omega_b\sin(\varphi - \beta) = 0$$

or

$$\omega_b = -\frac{a\,\omega_a \sin(\varphi - \theta)}{b \sin(\varphi - \beta)} \tag{4.17}$$

Since $a, b, c, \theta, \beta, \varphi$ and ω_a are already known, ω_c and ω_b can be calculated from Eqns. (4.17) and (4.18) respectively.

Acceleration Analysis

Let α_a, α_b and α_c be the angular accelerations of the links a, b and c respectively.

Rewriting Eqs. (4.14) and (4.15) as below,

$$a\omega_a \sin \omega_a t + b\omega_b \sin \omega_b t - c\omega_c \sin \omega_c t = 0 \tag{4.18}$$

$$a\omega_a \cos \omega_a t + b\omega_b \cos \omega_b t - c\omega_c \cos \omega_c t = 0 \tag{4.19}$$

Differentiating Eqs. (4.19) and (4.20) with respect to time,

$$-a\alpha_a \sin\theta - a\omega_a^2 \cos\theta - b\,\alpha_b \sin\beta - b\omega_b^2 \cos\beta + c\alpha_c \sin\varphi + c\omega_c^2 \cos\varphi = 0 \tag{4.20}$$

$$a\alpha_a \cos\theta - a\omega_a^2 \sin\theta + b\alpha_b \cos\beta - b\omega_b^2 \sin\beta - c\alpha_c \cos\varphi + c\omega_c^2 \sin\varphi = 0 \tag{4.21}$$

Where

$$\alpha_a = \frac{d\omega_a}{dt},\ \alpha_b = \frac{d\omega_b}{dt} \text{ and } \alpha_c = \frac{d\omega_c}{dt}.$$

Multiply Eq. (4.20) by $\cos\varphi$ and Eq. (4.21) by $\sin\varphi$ and add,

$$a\alpha_a (\sin\varphi\cos\theta - \cos\varphi\sin\theta) - a\omega_a^2(\cos\theta\cos\varphi + \sin\theta\sin\varphi) + b\alpha_b (\cos\beta\sin\varphi - \cos\beta\sin\varphi) - b\omega_b^2 (\cos\beta\cos\varphi + \sin\beta\sin\varphi) + c\omega_c^2 = 0$$

or

$$a\alpha_a \sin(\varphi-\theta) - a\omega_a^2 \cos(\varphi-\theta) + b\alpha_b \sin(\varphi-\beta) - b\omega_b^2 \cos(\varphi-\beta) + c\omega_c^2 = 0$$

or

$$\alpha_b = -\frac{a\alpha_a \sin(\varphi-\theta) - a\omega_a^2 \cos(\varphi-\theta) - b\omega_b^2 \cos(\varphi-\beta) + c\omega_c^2}{b \sin(\varphi - \beta)} \tag{4.22}$$

Multiply Eq. (4.20) by $\cos\beta$ and Eq. (4.21) by $\sin\beta$ and add,

$$a\alpha_a (\sin\beta\cos\theta - \cos\beta\sin\theta) - a\omega_a^2 (\sin\beta\sin\theta + \cos\beta\cos\theta) - b\omega_b^2 + c\alpha_c (\sin\varphi\cos\beta - \cos\varphi\sin\beta) + a\omega_c^2 (\sin\beta\sin\varphi + \cos\beta\cos\varphi) = 0$$

or

$$a\alpha_a \sin(\beta - \theta) - a\omega_a^2 \cos(\beta - \theta) - b\omega_b^2 + c\alpha_c \sin(\varphi - \beta) + c\omega_c^2 \cos(\varphi - \beta) = 0$$

or

$$\alpha_c = -\frac{a\alpha_a \sin(\beta-\theta) - a\omega_a^2 \cos(\beta-\theta) - b\omega_b^2 + c\omega_c^2 \cos(\varphi-\beta)}{c \sin(\varphi - \beta)} \quad (4.23)$$

Figure 4.2 shows a programme in Fortran for solving such a problem. The various input variables are

```
C          PROGRAMME TO FIND THE VELOCITY AND ACCELERATION
C          IN A FOUR-LINK MECHANISM
           DIMENSION PH(2),PHI(2),PP(2),BET(2),BT(2),VELC(2),
        1  VELB(2),ACCC(2),ACCB(2),C1(2),C2(2),C3(2),C4(2),
        2  B1(2),B2(2),B3(2),B4(2)
           READ(*,*)A,B,C,D,VELA,ACCA,THETA
           PI=4.*ATAN(1.0)
           THET=0
           IHT=180/THETA
           DTHET=PI/IHT
           DO 10 J=1,2*IHT
           THET=(J-1)*DTHET
           AK=((A**2-B**2+C**2+D**2)*0.5)
           TH=THET*180/PI
           AA=AK-A*(D-C)*COS(THET)-(C*D)
           BB=-2.0*A*C*SIN(THET)
           CC=AK-A*(D+C)*COS(THET)+(C*D)
           AB=BB**2-4*AA*CC
           IF(AB.LT.0)GO TO 10
           PHH=SQRT(AB)
           PH(1)=-BB+PHH
           PH(2)=-BB-PHH
           DO 5 I=1,2
           PHI(I)=ATAN(PH(I)*.5/AA)*2
           PP(I)=PHI(I)*180/PI
           BET(I)=ASIN((C*SIN(PHI(I))-A*SIN(THET))/B)
           BT(I)=BET(I)*180/PI
           VELC(I)=(A*VELA*SIN(BET(I)-THET))/(C*SIN(BET(I)-PHI(I)))
           VELB(I)=(A*VELA*SIN(PHI(I)-THET))/(B*SIN(BET(I)-PHI(I)))
           C1(I)=A*ACCA*SIN(BET(I)-THET)
           C2(I)=A*VELA**2*COS(BET(I)-THET)+B*VELB(I)**2
           C3(I)=C*VELC(I)**2*COS(PHI(I)-BET(I))
           C4(I)=C*SIN(BET(I)-PHI(I))
           ACCC(I)=(C1(I)-C2(I)+C3(I))/C4(I)
           B1(I)=A*ACCA*SIN(PHI(I)-THET)
           B2(I)=A*VELA**2*COS(PHI(I)-THET)
           B3(I)=B*VELB(I)**2*COS(PHI(I)-BET(I))-C*VELC(I)**2
           B4(I)=B*SIN(BET(I)-PHI(I))
5          ACCB(I)=(B1(I)-B2(I)-B3(I))/B4(I)
           IF(J.NE.1)GO TO 9
           WRITE(*,7)
7          FORMAT(4X,'THET',4X,'PHI',4X,'BETA',4X,'VELC',4X,'VELB',
        1  4X,'ACCC',4X,'ACCB')
9          WRITE(*,6)TH,PP(1),BT(1),VELC(1),VELB(1),ACCC(1),ACCB(1)
6          FORMAT(8F8.2)
           WRITE(*,2)PP(2),BT(2),VELC(2),VELB(2),ACCC(2),ACCB(2)
2          FORMAT(8X,8F8.2)
10         CONTINUE
           STOP
           END
```

Fig. 4.2

A, B, C, D	Magnitudes of links *AB*, *BC*, *CD* and *DA* respectively (mm)
VELA	Angular velocity of the input link *AB* (m/s)
ACCA	Angular acceleration of the input link (m/s^2) (acceleration is taken positive, deceleration negative)
THETA	The interval of the input angle i.e. the results are to be taken with a difference of 10°, 20° or 30° etc. starting from zero.

The output variables are:

THET	Angular displacement of the input link *AB* (degrees)
PHI	Angular displacement of the output link *DC* (degrees)
BETA	Angular displacement of the coupler link *BC* (degrees)
VELC	Angular velocity of the output link *DC* (rad/s)
VELB	Angular velocity of the coupler link *BC* (rad/s)
ACCC	Angular acceleration of the output link (rad/s^2)
ACCB	Angular acceleration of the coupler link (rad/s^2)

The results are obtained at an interval of THETA degrees of the input link starting from zero, in sets of two possible solutions for each position of the input link. In case, the input link AB is not a crank, the results are obtained for the possible positions only. Also, in case the angular acceleration of *AB* is given, it is considered at all positions at the same velocity, which may not be practical.

Example 4.1 In a four-link mechanism, the dimensions of the links are as under:

AB = 50 mm, *BC* = 66 mm, *CD* = 56 mm and *AD* = 100 mm.

AD is the fixed link. Considering an angular velocity of 10.5 rad/s in the counter-clockwise direction and retardation of 26 rad/s^2 at all positions of the input link *AB*, determine the following at an interval of 30° of the link *AB*:

(i) Angular displacements of the links *CD* and *BC*.
(ii) Angular velocities and the angular accelerations of the same links.

Solution: The results obtained using the programme of Fig. 4.2 are given in Fig. 4.3.

```
50 66 56 100 10.5 -26 30
```

THET	PHI	BETA	VELC	VELB	ACCC	ACCB
.00	-103.21	-55.69	-10.50	-10.50	-124.46	77.77
	103.21	55.69	-10.50	-10.50	176.46	-25.77
30.00	-135.97	-75.58	-10.39	-2.22	110.77	201.19
	88.38	27.99	.38	-7.79	194.43	104.01
60.00	-160.35	-70.29	-7.15	5.15	50.04	94.32
	100.35	10.29	7.15	-5.15	77.26	32.98
90.00	179.17	-48.18	-8.50	10.81	-133.10	162.60
	127.70	-4.95	12.70	-6.61	175.62	-120.08

```
270.00 -127.70     4.95    12.70    -6.61 -238.51   152.83
       -179.17    48.18    -8.50    10.81  175.19 -216.15
300.00 -100.35   -10.29     7.15    -5.15 -112.68   -7.48
        160.35    70.29    -7.15     5.15  -14.63 -119.83
330.00  -88.38   -27.99      .38    -7.79 -196.30  -65.43
        135.97    75.58   -10.39    -2.22  -59.33 -190.21
Stop - Program terminated.
```

Fig. 4.3

[Compare these values of α_b and α_c at 60° with the values obtained graphically in Example 3.1(i)]

4.3 SLIDER-CRANK MECHANISM

Figure 4.4 shows a slider-crank mechanism in which the stroke-line of the slider does not pass through the axis of rotation of the crank. β in clockwise direction from x-axis is taken as negative.

Let e = eccentricity (distance CD).

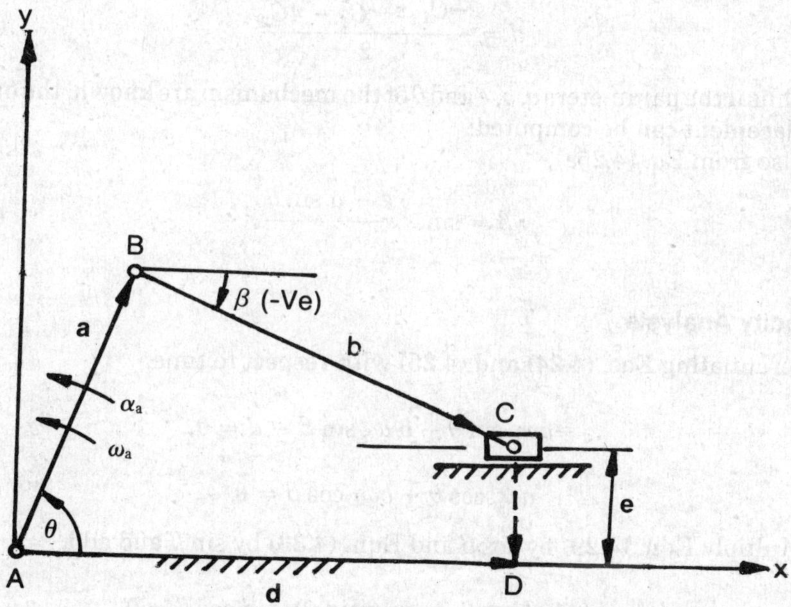

Fig. 4.4

Displacement along x-axis,

$$a \cos \theta + b \cos (-\beta) = d \tag{4.24}$$

or

$$b \cos \beta = d - a \cos \theta \tag{4.24a}$$

Displacement along y-axis,

$$a \sin \theta = b \sin (-\beta) + e \tag{4.25}$$

or

$$b \sin \beta = e - a \sin \theta \tag{4.25a}$$

Squaring Eqs. (4.24a) and (4.25a) and adding,

$$b^2 = a^2 \cos^2 \theta + d^2 - 2ad \cos \theta + a^2 \sin^2 \theta + e^2 - 2ae \sin \theta$$

$$= a^2 + e^2 + d^2 - 2ae \sin \theta - 2ad \cos \theta$$

or

$$d^2 - (2a \cos \theta)d + a^2 - b^2 + e^2 - 2ae \sin \theta = 0$$

or

$$d^2 + C_1 d + C_2 = 0 \tag{4.26}$$

where $C_1 = -2a \cos \theta$

$C_2 = a^2 - b^2 + e^2 - 2ae \sin \theta$

Equation (4.26) is a quadric in d. Its two roots are,

$$d = \frac{-C_1 \pm \sqrt{C_1^2 - 4C_2}}{2} \tag{4.27}$$

Thus if the parameters a, b, e and θ of the mechanism are known, the output displacement can be computed.

Also from Eq. (4.25a),

$$\beta = \sin^{-1} \frac{e - a \sin \theta}{b} \tag{4.28}$$

Velocity Analysis

Differentiating Eqs. (4.24) and (4.25) with respect to time,

$$-a\omega_a \sin \theta - b\, \omega_b \sin \beta - \dot{d} = 0 \tag{4.29}$$

$$a\omega_a \cos \theta + b\omega_b \cos \beta = 0 \tag{4.30}$$

Multiply Eqn. (4.29) by $\cos \beta$ and Eqn. (4.30) by $\sin \beta$ and add,

$$a\omega_a (\sin \beta \cos \theta - \cos \beta \sin \theta) - \dot{d} \cos \beta = 0$$

$$\dot{d} = \frac{a\omega_a \sin(\beta - \theta)}{\cos\beta} \tag{4.31}$$

From Eqn. (4.30),

$$\omega_b = \frac{a\,\omega_a \cos\theta}{b\cos\beta} \tag{4.32}$$

Acceleration Analysis

Differentiating Eqs. (4.29) and (4.30) with respect to time,

$$-[a\alpha_a \sin\theta + a\omega_a^2 \cos\theta\,] - [b\alpha_b \sin\beta + b\omega_b^2 \sin\beta\,] - \ddot{d} = 0 \tag{4.33}$$

$$[a\alpha_a \cos\theta - a\omega_a^2 \sin\theta] + [b\alpha_b \cos\beta - b\,\omega_b^2 \sin\beta] = 0 \tag{4.34}$$

Multiply Eqn. (4.33) by $\cos\beta$ and Eqn. (4.34) by $\sin\beta$ and add,

$$a\alpha_a \sin(\beta - \theta) - a\omega_a^2 \cos(\beta - \theta) - b\omega_b^2 - \ddot{d}\cos\beta = 0$$

or

$$\ddot{d} = \frac{a\alpha_a \sin(\beta - \theta) - a\omega_a^2 \cos(\beta - \theta) - b\omega_b^2}{\cos\beta} \tag{4.35}$$

From Eqn. (4.34)

$$\alpha_b = -\frac{a\alpha_a \cos\theta - a\omega_a^2 \sin\theta - b\omega_b^2 \sin\beta}{b\cos\beta} \tag{4.36}$$

α_b provides the angular acceleration of the coupler link whereas $\ddot{d}$ gives the linear acceleration of the slider.

Figure 4.5 shows a programme to solve this type of problem. The input variables are

A, B, E	The magnitudes a, b and e respectively (mm)
VELA	Angular velocity of the input link AB (m/s)
ACCA	Angular acceleration of the input link (m/s^2)
THETA	The interval of the input angle (degrees)

The output variables are:

THET	Angular displacement of the input link (degrees)
BET	Angular displacement of link AB (rad/s)
VELS	Linear velocity of the slider (m/s)
VELB	Angular velocity of link AB (rad/s)
ACCS	Linear acceleration of the slider (m/s^2)
ACCB	Angular acceleration of link AB (rad/s^2)

```
C      PROGRAMME TO FIND THE VELOCITY AND ACCELERATION
C      IN A SLIDER-CRANK MECHANISM
       READ(*,*)A,B,E,VELA,ACCA,THETA
       PI=4*ATAN(1.)
       THET=0
       IHE=180/THETA
       DTHET=PI/IHE
       DO 10 I=1;2*IHE
       THET=(I-1)*DTHET
       BET=ASIN((E-A*SIN(THET))/B)
       VELS=-A*VELA*SIN(THET-BET)/(COS(BET)*1000)
       VELB=-A*VELA*COS(THET)/B*COS(BET)
       ACC1=A*ACCA*SIN(BET-THET)-B*VELB**2
       ACC2=A*VELA**2*COS(BET-THET)
       ACCS=(ACC1-ACC2)/(COS(BET)*1000)
       ACC3=A*ACCA*COS(THET)-A*VELA**2*SIN(THET)
       ACC4=B*VELB**2*SIN(BET)
       ACCB=-(ACC3-ACC4)/(B*COS(BET))
       IF(I.EQ.1)WRITE(*,8)
8      FORMAT(3X,'THET',5X,'BET',4X,'VELS',4X,'VELB',4X,
     1 'ACCS',4X,'ACCB')
10     WRITE(*,4)THET*180/PI,BET*180/PI,VELS,VELB,ACCS,ACCB
4      FORMAT(6F8.2)
       STOP
       END
```

Fig. 4.5

It is to be noted that the values are computed assuming the same velocity and acceleration at all positions of the input link *OA*.

Example 4.2 In a slider-crank mechanism, the lengths of the crank and the connecting rod are 480 mm and 1600 mm respectively. It has an eccentricity of 100 mm. Assuming a velocity of 20 rad/s and an acceleration of 10 rad/s^2 at all positions of the link *OA*, calculate the following:

(i) The velocity and the acceleration of the slider
(ii) The angular velocity and the angular acceleration of the connecting rod.

Solution: The input and the output have been shown in Fig. 4.6. The results have been obtained at an interval of 30° of the input link *OA*.

```
480 1600 100 20 10 30
  THET        BET       VELS      VELB      ACCS      ACCB
   .00       3.58        .60     -5.99   -249.19      -.76
 30.00      -5.02      -5.53     -5.18   -203.64     55.27
 60.00     -11.38      -9.28     -2.94    -81.29    102.74
 90.00     -13.74      -9.60       .00     42.14    123.53
120.00     -11.38      -7.35      2.94    111.67    105.80
150.00      -5.02      -4.07      5.18    129.64     60.49
180.00       3.58       -.60      5.99    134.21      5.25
210.00      12.27       2.99      5.08    146.44    -53.14
240.00      18.80       6.68      2.84    142.32   -105.45
270.00      21.25       9.60       .00     79.48   -128.76
300.00      18.80       9.95     -2.84    -48.04   -108.62
330.00      12.27       6.61     -5.08   -184.31    -58.45
Stop - Program terminated.
```

Fig. 4.6

4.4 COUPLER CURVES

A coupler curve is the locus of a point on the coupler link. A four-link mechanism *ABCD* with a coupler point *E* (offset) is shown in Fig. 4.7. Let the *x*-axis be along the fixed link AD.

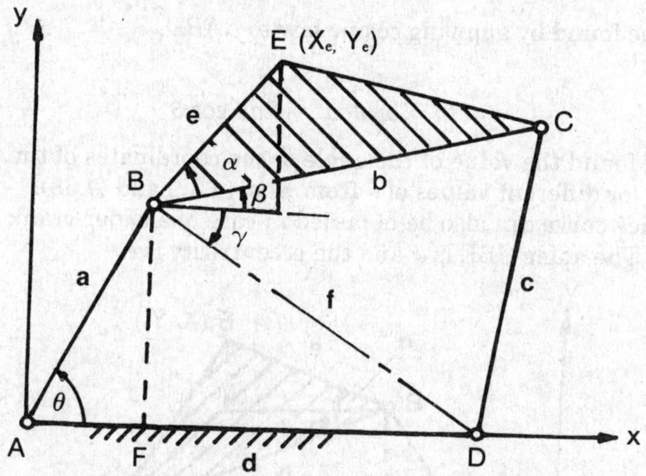

Fig. 4.7

Let $BE = e$ and $\angle CBE = \alpha$

Angles β and γ are defined as shown in the diagram.

Let X_e and Y_e be the coordinates of the point *E*.

Then,

$$X_e = a\cos\theta + e\cos(\alpha + \beta) \tag{4.37}$$

$$Y_e = a \sin\theta + e \sin(\alpha + \beta) \tag{4.38}$$

In these equations a, e, θ and α are known. To know the coordinates X_e and Y_e, it is necessary to express β in terms of known parameters, i.e. a, b, c, d, e, θ and α.

In $\Delta\, BDC$, applying Cosine law,

$$\cos(\beta + \gamma) = \frac{b^2 + f^2 - c^2}{2bf}$$

or

$$\beta + \gamma = \cos^{-1}\left[\frac{b^2 + f^2 - c^2}{2bf}\right]$$

or

$$\beta = \cos^{-1}\left[\frac{b^2 + f^2 - c^2}{2bf}\right] - \gamma \tag{4.39}$$

where $\tan\gamma = \dfrac{BF}{FD} = \dfrac{BF}{AD - AF} = \dfrac{a \sin\theta}{d - a\cos\theta}$

or

$$\gamma = \tan^{-1}\left[\frac{a \sin\theta}{d - a\cos\theta}\right]$$

f^2 can be found by applying cosine law to ΔABD, i.e.,

$$f^2 = a^2 + d^2 - 2ad\cos\theta$$

Having found the value of the angle β, the coordinates of the point E can be known for different values of θ from Eqs. (4.37) and (4.38).

A coupler curve can also be obtained in case of a slider-crank mechanism (Fig. 4.8). The angle CBE is α and the eccentricity is e.

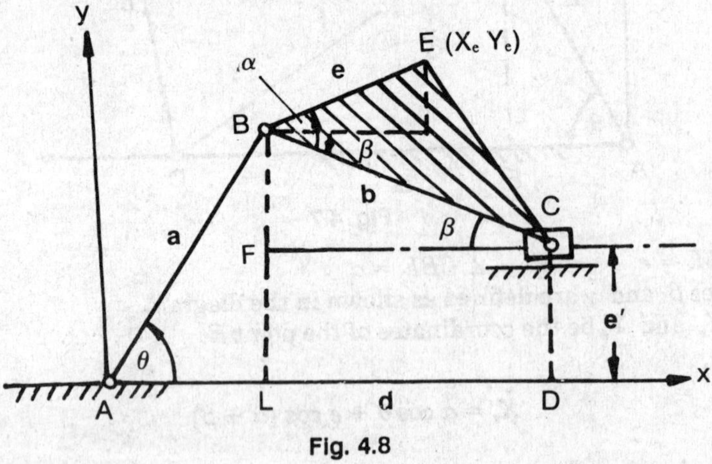

Fig. 4.8

Draw $BL \perp AD$ and $CF \perp BL$

$$X_e = a\cos\theta + e\cos(\alpha - \beta) \tag{4.40}$$

$$Y_e = a\sin\theta + e\sin(\alpha - \beta) \tag{4.41}$$

β (negative) can be expressed in terms of known parameters as below:

$$\sin\beta = \frac{BF}{BC} = \frac{BL - FL}{BC} = \frac{a\sin\theta - e'}{b}$$

$$\beta = \sin^{-1}\left[\frac{a\sin\theta - e'}{b}\right] \tag{4.42}$$

Figure 4.9 shows a programme to find the coordinates of the coupler point for both the above cases. The input variables are:

```
C     PROGRAMME TO FIND THE COORDINATES OF A COUPLER
C     POINT OF A FOUR-LINK MECHANISM OR A SLIDER-
C     CRANK MECHANISM
      READ(*,*)A,B,C,D,E,ALPH,CASE
      PI=4*ATAN(1.0)
      THET=0
5     IF(CASE.EQ.2)GO TO 9
      GAMM=ATAN(A*SIN(THET)/(D-A*COS(THET)))
      SQ=A**2+D**2-2*A*D*COS(THET)
      IF(SQ.LT.0)GO TO 8
      F=SQRT(SQ)
      BET=ACOS((B**2+F**2-C**2)/(2*B*F))-GAMM
      IF(CASE.EQ.1)GO TO 11
9     BET=ASIN((C-A*SIN(THET))/B)
11    XE=A*COS(THET)+E*COS(ALPH*PI/180+BET)
      YE=A*SIN(THET)+E*SIN(ALPH*PI/180+BET)
      IF(THET.EQ.0)WRITE(*,7)
7     FORMAT(4X,'THET',5X,'XE',5X,'YE')
      WRITE(*,3)THET*180/PI,XE,YE
3     FORMAT(3F8.2)
8     THET=THET+PI/6
      TH=2.*PI-PI/6.
      IF(THET.GT.TH)GO TO 10
      GO TO 5
10    STOP
      END
```

Fig. 4.9

A, B, E The magnitudes *a*, *b* and *e* respectively (mm)
CASE 1, in case of a four-link mechanism
2, in case of a slider-crank mechanism
C The magnitude *c* (case 1) or eccentricity *e′* (case 2)
D The magnitude *d* (case 1) or 0 (case 2)
ALPH The angle α (degrees)

The output variables are:

THET Angular displacement of the link *AB* (degrees)
XE *X*-coordinates of the point *E*
YE *Y*-coordinates of the point *E*

Example 4.3 Draw a coupler curve of the coupler point E of a four-link mechanism having the following data:

AB = 50 mm, BC = 66 mm, CD = 90 mm, AD = 100 mm, BE = 30 mm $\angle\ CBE$ = 40° (refer to Fig. 4.7)

Solution: The input and the output have been shown in Fig. 4.10 using the programme of Fig. 4.9 for the given data. The required coupler curve has been shown in Fig. 4.11.

50 66 90 100 30 40 1

THET	XE	YE
.00	26.73	18.93
30.00	35.27	53.90
60.00	29.79	72.92
90.00	11.70	77.62
120.00	-9.51	68.99
150.00	-26.10	49.58
180.00	-34.41	25.63
210.00	-35.43	3.95
240.00	-28.72	-13.53
270.00	-15.08	-24.06
300.00	1.75	-24.35
330.00	16.54	-11.44

Stop - Program terminated.

Fig. 4.10

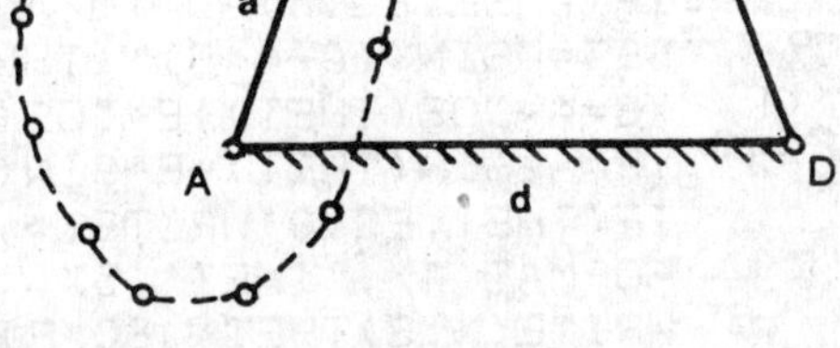

Fig. 4.11

PROBLEMS

4.1 Derive expressions for the displacement, velocity and acceleration analyses of an inverted slider-crank mechanism.

4.2 In a four-link mechanism (Fig. 4.1), the dimensions of the links are AB = 30 mm, BC = 80 mm, CD = 40 mm and AD = 75 mm. If OA rotates at a constant angular velocity of 30 rad/s in the clockwise direction, calculate the angular velocities and the angular accelerations of links BC and CD for values of θ at an interval of 30°.

4.3 In a slider-crank mechanism (Fig. 4.4), the crank length AB is 50 mm and the connecting rod is 160 mm long. e = 15 mm. For the angle $\theta = 45°$, angular velocity of AB = 8 rad/s with an angular acceleration of 12 rad/s (both clockwise), find the linear velocity and the acceleration of the slider and the angular velocity and the angular acceleration of the connecting rod.

4.4 Derive expressions to find the angular displacement, angular velocity and the angular acceleration of the link EF of a six-link mechanism shown in Fig. 4.12. AB is the input link having an angular velocity of ω rad/s in the counter-clockwise direction.

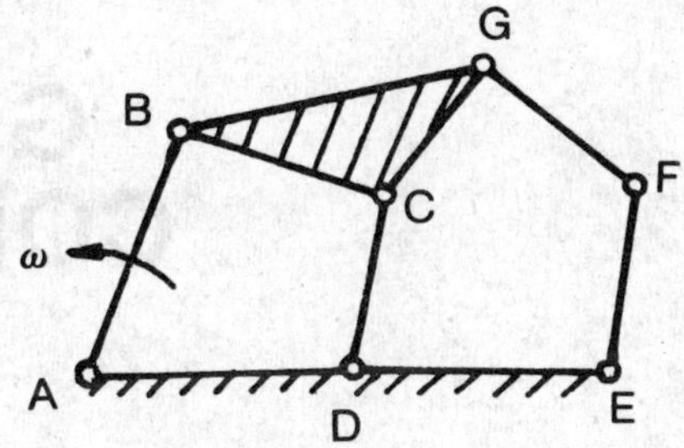

Fig. 4.12

4.5 Derive expressions for the coupler curves of an inverted slider-crank mechanism.

4.6 For the data of Example 4.3, take some more coupler points by taking different values of BE and $\angle\alpha$ and draw coupler curves for the same. Make a cardboard model of the mechanism and obtain the coupler curves by rotating the crank through 360°.

5

GRAPHICAL AND COMPUTER-AIDED SYNTHESIS OF MECHANISMS

PART A GRAPHICAL METHODS

5.1 POLE

Consider a four-link mechanism $ABCD$ in two positions AB_1C_1D and

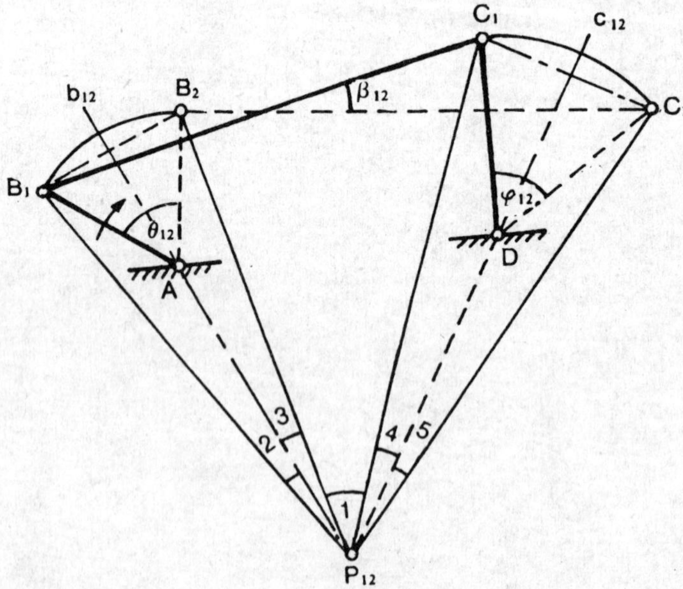

Fig. 5.1

AB_2C_2D. The coupler link BC has moved from the position B_1C_1 to B_2C_2. The input link AB and the output link DC have moved through angles θ_{12} and φ_{12} respectively in the clockwise direction (Fig. 5.1).

A pole P_{12} of the coupler link BC is its centre of rotation with respect to the fixed link for the motion of the coupler from B_1C_1 to B_2C_2. Each point on the link BC describes a circular arc with centre at the pole P_{12}. Thus, a line joining the two positions of a point on the link is a chord of the circle and the mid-normal (perpendicular bisector) of the chord passes through the centre of rotation P_{12}. B and C are also two points on the link BC. B moves from B_1 to B_2 while C from C_1 to C_2. Therefore, B_1B_2 and C_1C_2 are the chords of the two circles and their mid-normals b_{12} and c_{12} also pass through or intersect at the centre of their rotation i.e. at P_{12}.

Properties of Pole Point

1. As $AB_1 = AB_2$, the mid-normal b_{12} of B_1B_2 passes through the fixed pivot A. Similarly, the mid-normal of C_1C_2 passes through pivot D.
2. The coupler link BC is rotated about P_{12} from the position B_1C_1 *to* B_2C_2,
   $\therefore\ \Delta\ B_1P_{12}C_1 \equiv \Delta\ B_2P_{12}C_2$
   $\therefore\ \angle 2 + \angle 3 + \angle 1 = \angle 1 + \angle 4 + \angle 5$
   i.e. Angle subtended by B_1C_1 at P_{12} = Angle subtended by B_2C_2 at P_{12} or the angle subtended by BC at P_{12} in two positions is the same.

(3) From (2), $\angle 2 + \angle 3 + \angle 1 = \angle 1 + \angle 4 + \angle 5$
or $\angle 2 + \angle 3 = \angle 4 + \angle 5$
i.e. B_1B_2 and C_1C_2 subtend equal angles at P_{12}.

(4) P_{12} lies on the mid-normal of B_1B_2,
$\therefore\ \angle 2 = \angle 3$
Similarly, $\angle 4 = \angle 5$

(5) $\angle 2 + \angle 3 = \angle 4 + \angle 5$
But $\angle 2 = \angle 3$ and $\angle 4 = \angle 5$,
$\therefore\ \angle 2 = \angle 4$
and $\angle 3 = \angle 5$
i.e. the input and the output links subtend equal angles at P_{12} in their corresponding positions.

(6) $\angle 2 + \angle 3 + \angle 1 = \angle 1 + \angle 4 + \angle 5 = \angle 1 + \angle 4 + \angle 3 \qquad (\angle 3 = \angle 5)$
i.e. the angle subtended by the coupler link is equal to that subtended by the fixed pivots A and D.

(7) The triangle $B_1P_{12}C_1$ moves as one link about P_{12} to the position $B_2P_{12}C_2$,
Angular displacement of coupler B_1C_1

$$= \text{Angular displacement of } P_{12}C_1$$

i.e. $\angle\beta_{12} = \angle 4 + \angle 5$

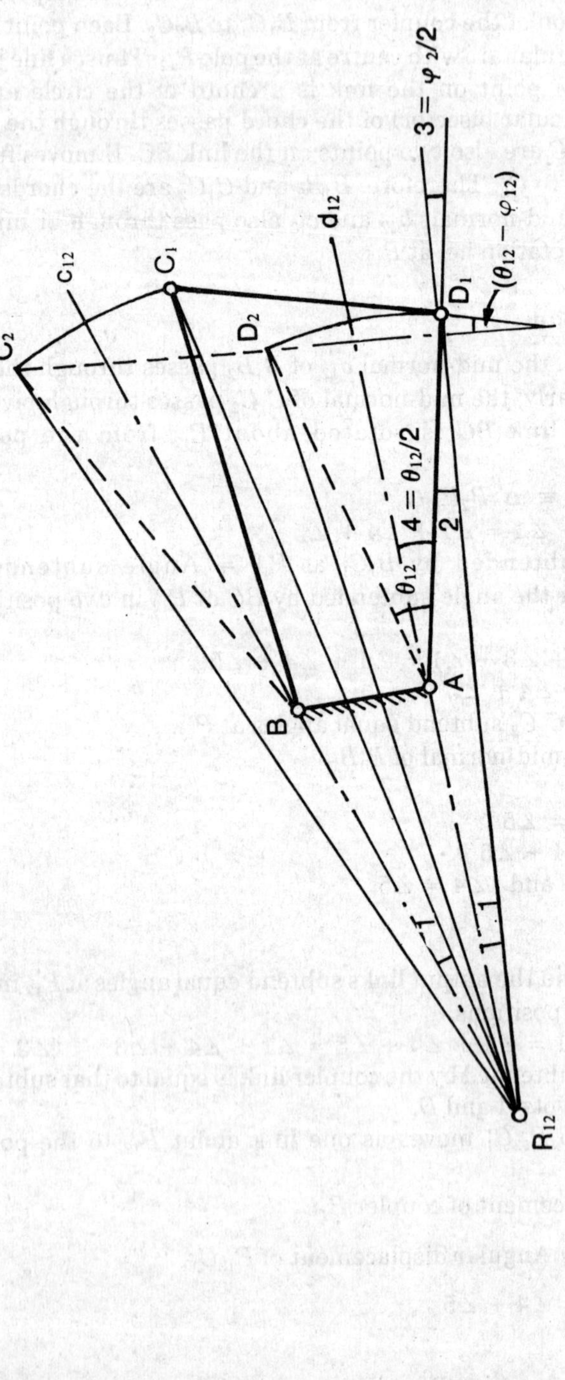

Fig. 5.2

5.2 RELATIVE POLE

A pole of a moving link is the centre of its rotation relative to a fixed link. However, if the rotation of the link is considered relative to another moving link, the pole is known as the relative pole. The relative pole can be found by fixing the link of reference and observing the motion of the other link in the reverse direction.

For the four-link mechanism of Fig. 5.2, the pole of *BC* relative to *AB* is at the pivot *B*. The pole of *DC* relative to *AB* can be found as follows:

Let θ_{12} = angle of rotation of *AB* (clockwise)

φ_{12} = angle of rotation of *DC* (clockwise)

Make the following constructions:

1. Assume *A* and *B* as the fixed pivots and rotate AD_1 about *A* through angle θ_{12} in the counter-clockwise direction (opposite to the direction of rotation of *AB*). Let D_2 be the new position of *D* after rotation. Thus *AB* and *AD* assume the same relative positions after the rotation of *AD* (*AB* fixed) as would have been after the rotation of *AB* with *AD* fixed.
2. Locate the point C_2 by drawing arcs with centres *B* and D_2 and radii equal to BC_1 and D_1C_1 respectively. Then ABC_2D_2 is known as the inversion of ABC_1D_1.
3. Draw mid-normals of D_1D_2 and C_1C_2 which pass through *A* and *B* and intersect at R_{12} which is the required relative pole.

Now $(\varphi_{12} - \theta_{12})$ = Angle of rotation of the output link *DC* relative to the input link *AB*.

This angle will be negative if $DC > AB$ and is positive if $DC < AB$.

Angular displacement of $R_{12}D_1$ = angular displacement of D_1C_1

(Refer Sec. 5.1(vii))

$$\angle D_1R_{12}D_2 = -(\angle\varphi_{12} - \angle\theta_{12}) \quad \text{(assuming } DC > AB\text{)}$$

or

$$2\angle 1 = -(\angle\varphi_{12} - \angle\theta_{12})$$

or

$$\angle 1 = -\frac{1}{2}(\angle\varphi_{12} - \angle\theta_{12})$$

In $\Delta AR_{12}D_1$,

$$\angle 4 = \angle 1 + \angle 2$$

or

$$\frac{1}{2}\angle\theta_{12} = -\frac{1}{2}(\angle\varphi_{12} - \angle\theta_{12}) + \angle 3 \quad (\angle 2 = \angle 3)$$

$$= -\frac{1}{2}\angle\varphi_{12} + \frac{1}{2}\angle\theta_{12} + \angle 3$$

$$\angle 3 = \frac{1}{2}\angle\varphi_{12}$$

The conclusion, just arrived, provides a method to locate the pole of the output link *DC* relative to the input link *AB*.

Procedure

1. Join A and D, the centres of the fixed pivots.
2. Rotate AD about A through an angle $\theta_{12}/2$ in a direction opposite to that of AB.
3. Again rotate AD about D through an angle $\varphi_{12}/2$ in a direction opposite to that of DC. The point of intersection of the two positions of AD after rotation about A and D, is the required relative pole R_{12}.

The angles subtended by D_1D_2 and C_1C_2 at R_{12} are the same,

i.e. $$\angle D_1R_{12}D_2 = \angle C_1R_{12}C_2$$

or $$2\,\angle D_1R_{12}A = 2\,\angle C_1R_{12}B$$

or $$\angle D_1R_{12}A = \angle C_1R_{12}B$$

Thus, it is also concluded that the angle subtended by the fixed pivots (A and D) at the relative pole is equal to the angle subtended by the coupler BC (Refer Sec. 5.1 also).

Now, consider the slider-crank mechanism of Fig. 5.3. In this, if C reciprocates through a horizontal distance s, its centre of rotation will lie at infinity on a vertical line where the point D can also be assumed to lie. Then AD will also be a vertical line through A. Rotate this line about A through $\theta_{12}/2$ in the counter-clockwise direction as usual. Rotating AD about D through $\varphi_{12}/2$ would mean a vertical line towards left of A, at a distance of $s/2$. The intersection of the two lines locates R_{12}.

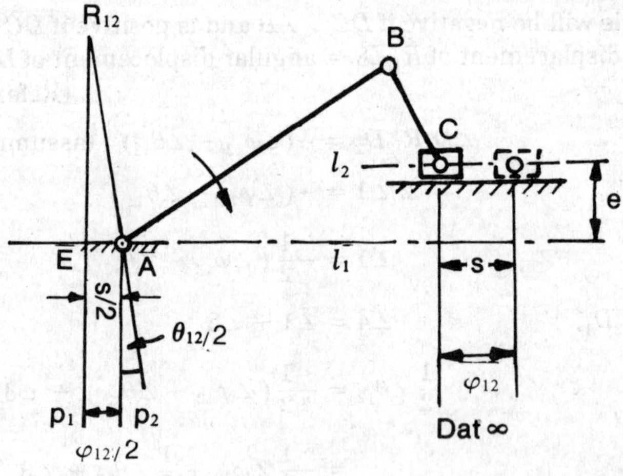

Fig. 5.3

Thus the procedure to locate the relative pole of a slider-crank mechanism will be as under:

1. Draw two parallel lines l_1 and l_2 at a distance e apart (if there is an eccentricity).

2. Select a line segment AE of length $s/2$ on the line l_1 such that E is measured in a direction opposite to the motion of the slider.
3. At A and E, draw perpendicular lines p_1 and p_2 respectively to the line l_1.
4. Make the angle $\theta_{12}/2$ at the point A with the line p_1 in a direction opposite to the rotation of the input link.

The intersection of this line with the line p_2 locates the relative pole R_{12}.

5.3 DESIGN OF MECHANISMS BY RELATIVE POLE METHOD

(a) Four-link Mechanism

Let for a four-link mechanism, the positions of the pivots A and D along with the angular displacements θ_{12} (angle between θ_1 and θ_2) and φ_{12} (angle between φ_1 and φ_2) of the driver and the driven links respectively, be known.

To design the mechanism (Fig. 5.4), first locate the relative pole R_{12} by the procedure given in Sec. 5.2.

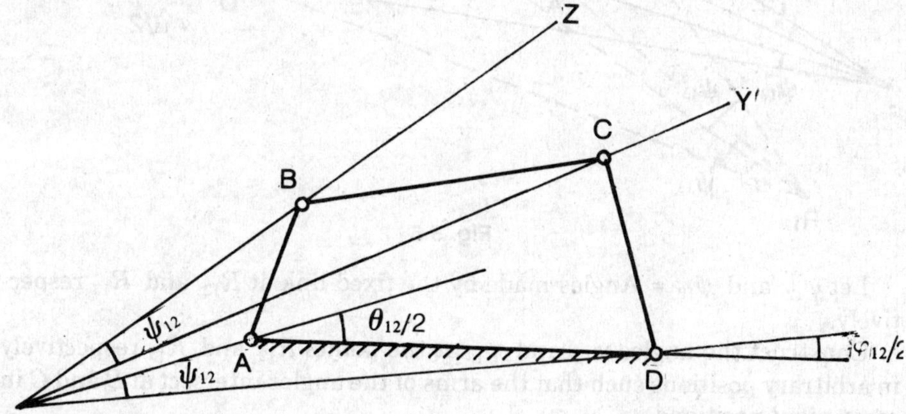

Fig. 5.4

Now, angle subtended by the coupler BC at R_{12}

= Angle subtended by the fixed pivots A and D at R_{12}

$$= 1/2 \angle\theta_{12} - 1/2 \angle\varphi_{12} \qquad \text{(assuming } DC > AB\text{)}$$

$$= \angle\psi_{12}$$

Adopt any of the following alternatives to design the required mechanism:

1. At point R_{12}, construct an angle ψ_{12} at an arbitrary position. Join any two points on the two arms of the angle to obtain the coupler link BC of the mechanism. Join AB and DC to have the driver and the driven links respectively.

2. Locate the point C arbitrary so that DC is the output link. Construct an angle $CR_{12}Z = \psi_{12}$. Take any point B on $R_{12}Z$. Join AB and BC.
3. Instead of locating point C as above, locate the point B arbitrary so that AB is the input link. Construct an angle $BR_{12}Y = \psi_{12}$. Take any point C on $R_{12}Y$. Join BC and DC.

Then $ABCD$ is the required four-link mechanism.

If instead of two positions, three positions of the input link ($\theta_1, \theta_2, \theta_3$) and the output link ($\varphi_1, \varphi_2, \varphi_3$) are known, find R_{12} and R_{13} as shown in Fig. 5.5.

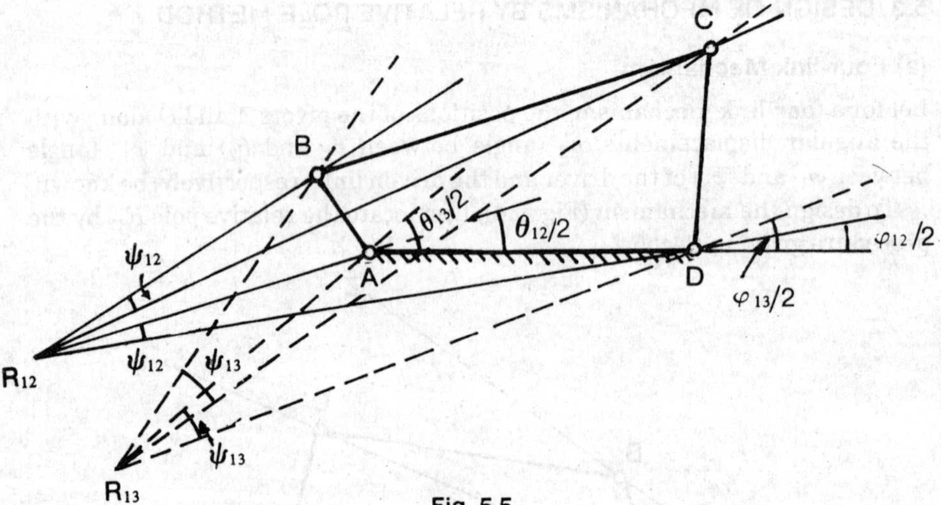

Fig. 5.5

Let ψ_{12} and ψ_{13} = Angles made by the fixed link at R_{12} and R_{13} respectively.

Construct the angles ψ_{12} and ψ_{13} at the points R_{12} and R_{13} respectively in arbitrary positions such that the arms of the angles intersect at B and C in convenient positions.

(b) Slider-crank Mechanism

For a slider-crank mechanism,

Let θ_{12} = angular displacement of input link ($\angle$between θ_1 and θ_2)

s_{12} = linear displacement of the slider

e = eccentricity

Locate the relative pole R_{12} as shown in Fig. 5.6.

Adopt the following procedure to design the required mechanism:

At the point R_{12}, construct an angle equal to $\theta_{12}/2$ ($\because \varphi_{12} \approx 0, \therefore \psi \approx \theta_{12}/2$) in an arbitrary (but convenient) position. The intersection of an arm of this angle with the line l_2 provides the position of the slider. Select an arbitrary point B on the other arm of the angle so that ABC is the required slider-crank mechanism.

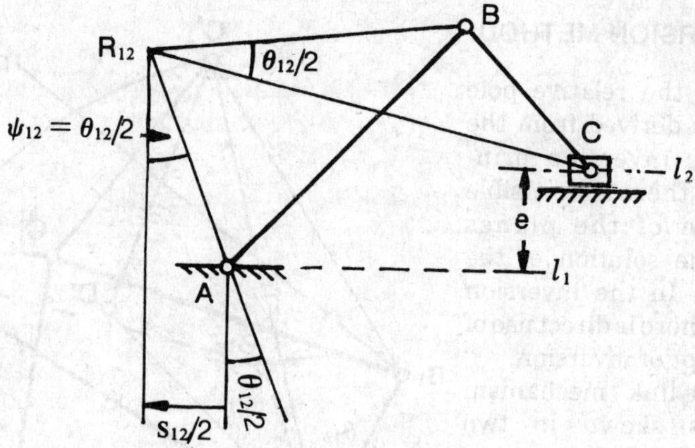

Fig. 5.6

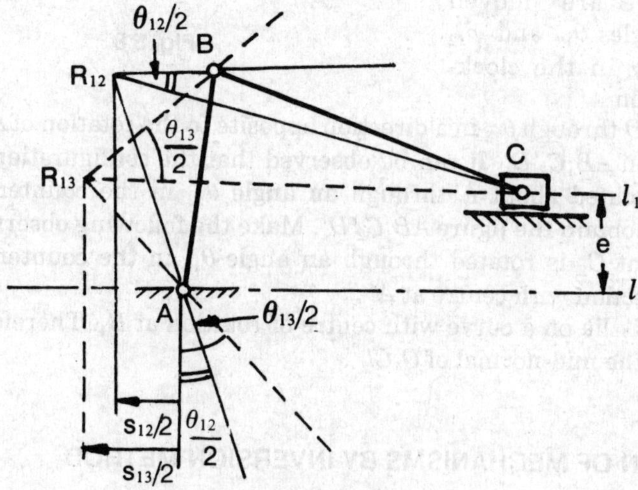

Fig. 5.7

If instead of two positions, three positions of the input link (θ_1, θ_2, θ_3) and the slider (s_1, s_2, s_3) are known, find R_{12} and R_{13} as shown in Fig. 5.7.

Now, $\dfrac{\theta_{12}}{2}$ = angle made by the fixed link at R_{12}

$\dfrac{\theta_{13}}{2}$ = angle made by the fixed link at R_{13}

Therefore, construct angle $\theta_{12}/2$ at R_{12} in an arbitrary position locating the point C. Draw the angle $\theta_{13}/2$ at R_{13} with an arm along $R_{13}C$. Intersection of the two arms (not through C) of the two angles locates the point B.

5.4 INVERSION METHOD

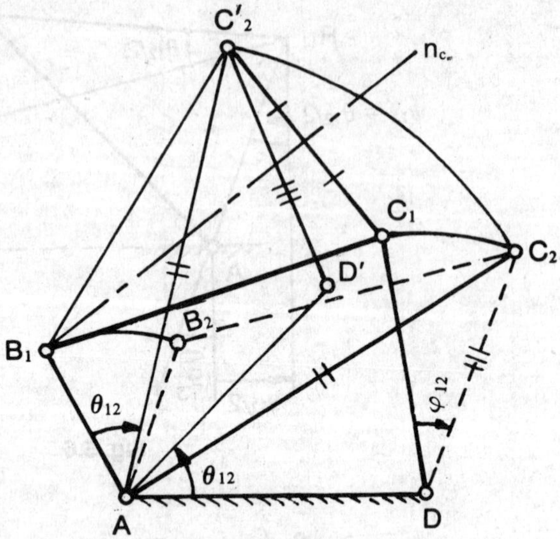

Fig. 5.8

Basically, the relative pole method is derived from the kinematic inversion principle. But there is no visible inversion of the planes during the solution of the problems. In the inversion method, there is direct use of the concept of inversion.

A four-link mechanism *ABCD* is shown in two positions AB_1C_1D and AB_2C_2D in Fig. 5.8. The input and the output links *AB* and *DC* are moved through angles θ_{12} and φ_{12} respectively in the clockwise direction.

Rotate *AD* through θ_{12} in a direction opposite to the rotation of *AB* and get the inversion $AB_1C_2'D'$. It can be observed that the configuration AB_2C_2D has been rotated about A through an angle θ_{12} in the counter-clockwise direction to obtain the figure $AB_1C_2'D'$. Make the following observations:

1. Point C_2 is rotated through an angle θ_{12} in the counter-clockwise direction with centre at *A*.
2. $C_1C'_2$ lie on a curve with centre of rotation at B_1. Therefore, B_1 lies on the mid-normal of $C_1C'_2$.

5.5 DESIGN OF MECHANISMS BY INVERSION METHOD

(a) Four-link Mechanism

Let the distance between the fixed pivots *A* and *D*, and the angles θ_{12} and φ_{12} be known. To design the mechanism, proceed as follows:

1. Draw a line segment *AD* of length equal to the distance between the fixed pivots (Fig. 5.9).
2. At point *D*, construct an angle $C_1DC_2 = \varphi_{12}$ (clockwise) at an arbitrary position, selecting a suitable output crank length $DC_1 = DC_2$.
3. Rotate point C_2 in the counter-clockwise direction through an angle θ_{12} with A as centre and obtain the point C_2'.
4. Join $C_1C'_2$ and draw its mid-normal. Select a suitable point B_1 on it.

AB_1C_1D is the required four-link mechanism in which B_1C_1 is the coupler.

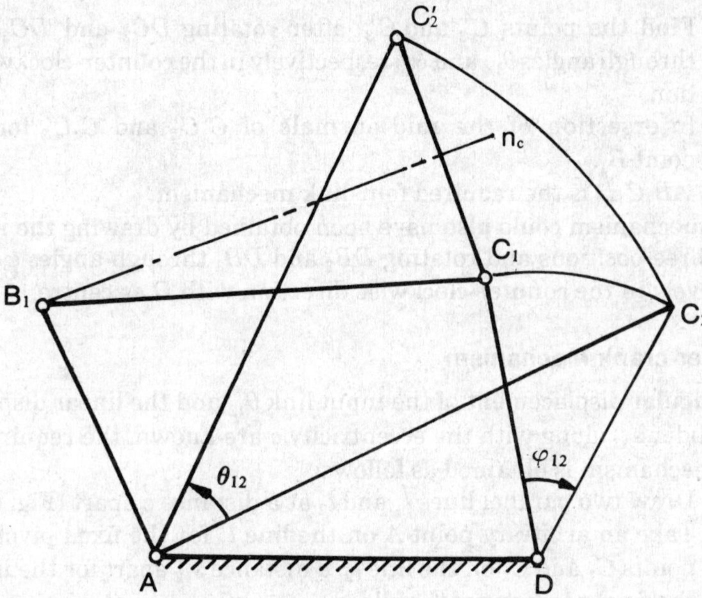

Fig. 5.9

If three positions of the input link ($\theta_1, \theta_2, \theta_3$) and three positions of the output link ($\varphi_1, \varphi_2, \varphi_3$) are known, proceed as given below:

1. Draw the output link DC in three positions DC_1, DC_2 and DC_3 as its angular positions are known (Fig. 5.10).

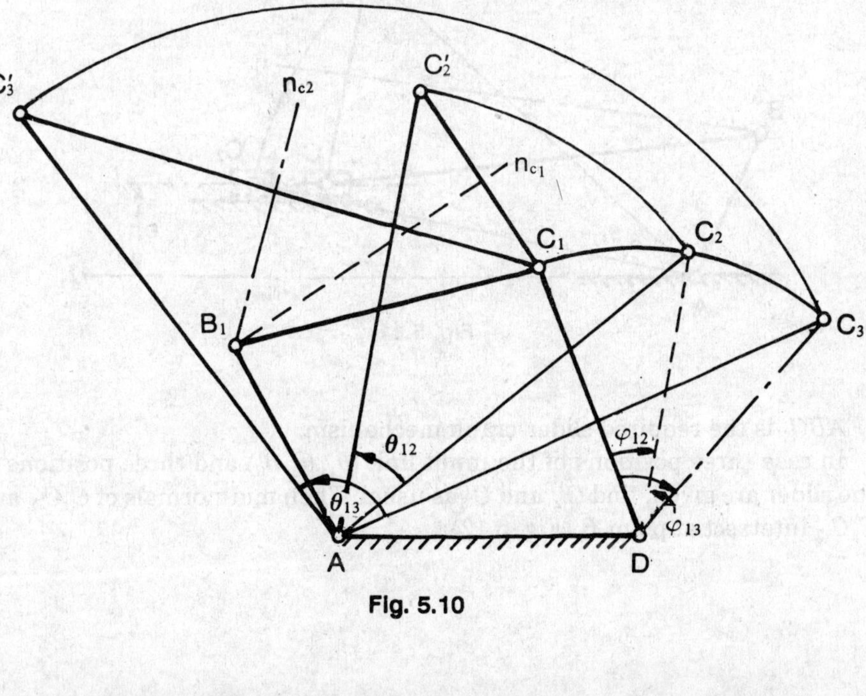

Fig. 5.10

2. Find the points C'_2 and C'_3 after rotating DC_2 and DC_3 about A through angles θ_{12} and θ_{13} respectively in the counter-clockwise direction.
3. Intersection of the mid-normals of $C_1C'_2$ and $C_1C'_3$ locates the point B_1.

Then AB_1C_1D is the required four-link mechanism.

The mechanism could also have been obtained by drawing the input link AB in three positions and rotating DB_2 and DB_3 through angles φ_{12} and φ_{13} respectively in the counter-clockwise direction with D as centre.

(b) Slider-crank Mechanism

If the angular displacement of the input link θ_{12} and the linear displacement of the slider s_{12} along with the eccentricity e are known, the required slider-crank mechanism is obtained as follows:

1. Draw two parallel lines l_1 and l_2 at a distance e apart (Fig. 5.11).
2. Take an arbitrary point A on the line l_1 for the fixed pivot and two points C_1 and C_2 on the line l_2, a distance s_{12} apart for the initial and the final positions of the slider.
3. Rotate the point C_2 about A through an angle θ_{12} in the counter-clockwise direction to obtain the point C_2'.
4. Join $C_1C'_2$ and draw its mid-normal n_c. Take an arbitrary but convenient point B on it.

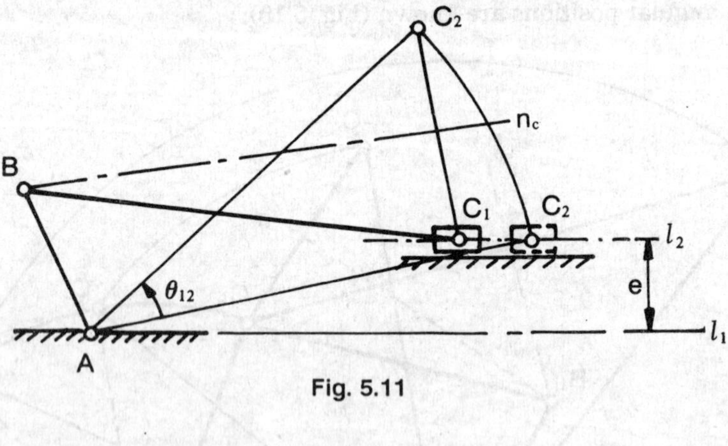

Fig. 5.11

ABC_1 is the required slider-crank mechanism.

In case three positions of the input link (θ_1, θ_2, θ_3) and three positions of the slider are given, find C'_2 and C'_3 as usual. Then mid-normals of $C_1C'_2$ and $C_1C'_3$ intersect at point B (Fig. 5.12).

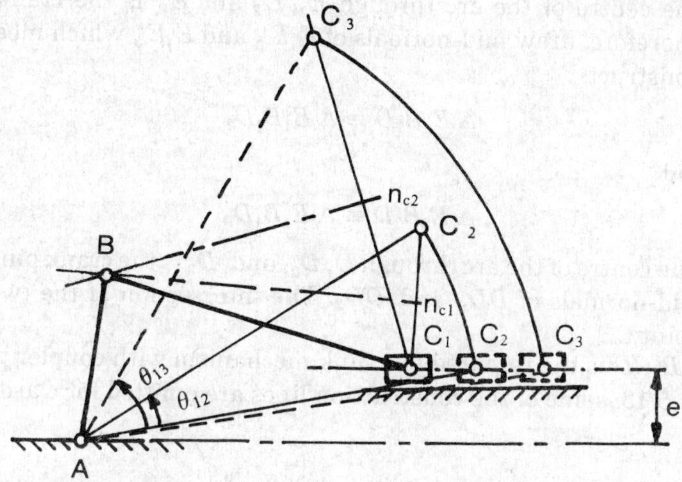

Fig. 5.12

(c) Four-link Mechanism with Three Positions of Input Link and of Coupler Point

Three positions of the input link (θ_1, θ_2, θ_3) and three positions of the coupler point (E_1, E_2, E_3) are known. To design such a mechanism, select a suitable location of the fixed link AD and proceed as follows (Fig. 5.13):

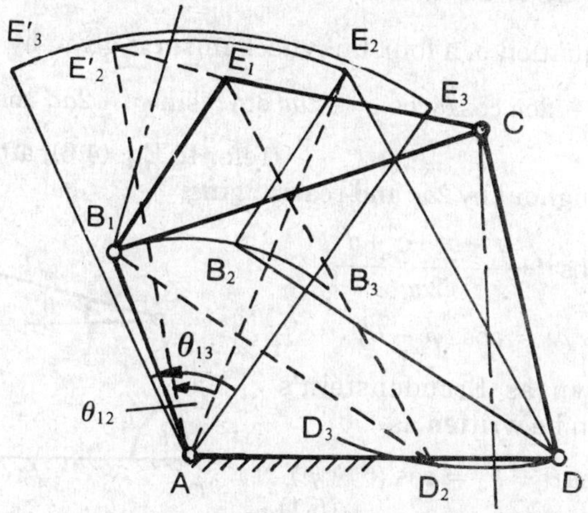

Fig. 5.13

1. Draw the input link in three positions AB_1, AB_2 and AB_3 and locate the three positions of coupler point E_1, E_2 and E_3.
2. Rotate E_2 through an angle θ_{12} in the counter-clockwise direction with A as centre and obtain the point E'_2. Similarly, rotate E_3 through an angle θ_{13} in the counter-clockwise direction with centre A and obtain the point E'_3.

3. The centre of the arc through E_1, E'_2 and E'_3 is the crank pin B_1. Therefore, draw mid-normals of $E_1E'_2$ and $E_1E'_3$ which meet at B_1.
4. Construct

$$\Delta\, E_2B_2D \equiv \Delta\, E_1B_1D_2$$

and

$$\Delta\, E_3B_3D \equiv \Delta\, E_1B_1D_3$$

5. The centre of the arc through D, D_2 and D_3 is the crank pin C. Draw mid-normals of DD_2 and DD_3. The intersection of the two locates point C.

Thus AB_1CD is the required four-link mechanism with coupler point E_1. (In Fig. 5.13, some of the construction lines are omitted for clarity sake.)

PART B COMPUTER-AIDED SYNTHESIS OF MECHANISMS

5.6 TO COORDINATE ANGULAR DISPLACEMENTS OF INPUT AND OUTPUT LINKS

Displacement equation of a four-link mechanism is given by (Fig. 5.14),

$$b^2 = c^2 + a^2 + d^2 - 2ac\cos\theta\cos\varphi - 2ac\sin\theta\sin\varphi - 2ad\cos\theta + 2cd\cos\varphi$$

(refer to Eq. (4.6), after rearranging

Dividing throughout by $2ac$ and rearranging,

$$\frac{d}{a}\cos\varphi - \frac{d}{c}\cos\theta + \frac{a^2 - b^2 + c^2 + d^2}{2ac}$$

$$= \cos(\theta - \varphi) = \cos(\varphi - \theta)$$

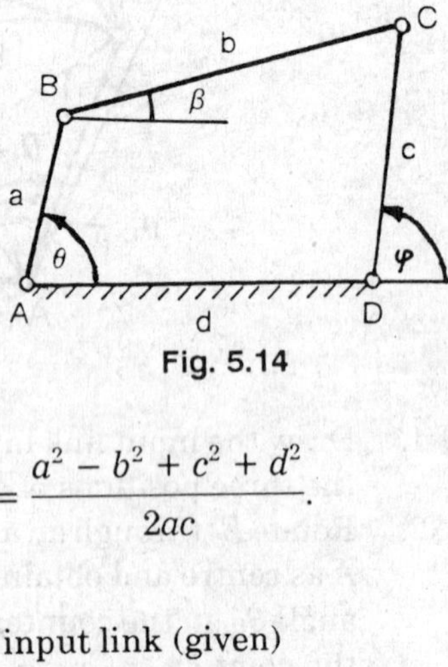

Fig. 5.14

This is known as Freudenstein's equation and can be written as,

$$k_1\cos\varphi + k_2\cos\theta + k_3 = \cos(\theta - \varphi) \tag{5.1}$$

where

$$k_1 = \frac{d}{a};\quad k_2 = -\frac{d}{c};\quad \text{and}\quad k_3 = \frac{a^2 - b^2 + c^2 + d^2}{2ac}.$$

Let

$$\theta_1, \theta_2, \theta_3 = \text{Three positions of input link (given)}$$

and

$$\varphi_1, \varphi_2, \varphi_3 = \text{Three positions of output link (given)}$$

It is required to find the values of a, b, c and d to form a four-link mechanism giving the prescribed motions of the input and the output links.

Equation (5.1) can be written as,

$$k_1 \cos \varphi_1 + k_2 \cos \theta_1 + k_3 = \cos (\theta_1 - \varphi_1)$$

$$k_1 \cos \varphi_2 + k_2 \cos \theta_2 + k_3 = \cos (\theta_2 - \varphi_2)$$

$$k_1 \cos \varphi_3 + k_2 \cos \theta_3 + k_3 = \cos (\theta_3 - \varphi_3)$$

k_1, k_2 and k_3 can be evaluated by Gaussian elimination method (Example 5.3) or by the Cramer's rule.

Using Cramer's rule,

$$\Delta = \begin{vmatrix} \cos \varphi_1 & \cos \theta_1 & 1 \\ \cos \varphi_2 & \cos \theta_2 & 1 \\ \cos \varphi_3 & \cos \theta_3 & 1 \end{vmatrix}$$

$$\Delta_1 = \begin{vmatrix} \cos (\theta_1 - \varphi_1) & \cos \theta_1 & 1 \\ \cos (\theta_2 - \varphi_2) & \cos \theta_2 & 1 \\ \cos (\theta_1 - \varphi_3) & \cos \theta_3 & 1 \end{vmatrix}$$

$$\Delta_2 = \begin{vmatrix} \cos \varphi_1 & \cos (\theta_1 - \varphi_1) & 1 \\ \cos \varphi_2 & \cos (\theta_2 - \varphi_2) & 1 \\ \cos \varphi_3 & \cos (\theta_3 - \varphi_3) & 1 \end{vmatrix}$$

$$\Delta_3 = \begin{vmatrix} \cos \varphi_1 & \cos \theta_1 & \cos (\theta_1 - \varphi_1) \\ \cos \varphi_2 & \cos \theta_2 & \cos (\theta_2 - \varphi_2) \\ \cos \varphi_3 & \cos \theta_3 & \cos (\theta_3 - \varphi_3) \end{vmatrix}$$

k_1, k_2 and k_3 are given by,

$$k_1 = \frac{\Delta_1}{\Delta}; \; k_2 = \frac{\Delta_2}{\Delta}; \; k_3 = \frac{\Delta_3}{\Delta}$$

Knowing k_1, k_2 and k_3, the values of a, b, c and d can be computed from the relations

$$k_1 = \frac{d}{a}, k_2 = -\frac{d}{c} \text{ and } k_3 = \frac{a^2 - b^2 + c^2 + d^2}{2ac}$$

Value of either a or d can be assumed to be unity to get the proportionate values of other parameters.

Figure 5.15 shows a program in Fortran for solving such a problem. The input variables are

```
C         PROGRAMME TO COORDINATE ANGULAR DISPLACEMENTS OF
C         THE INPUT AND OUTPUT LINKS IN THREE POSITIONS
          READ(*,*)TH1,TH2,TH3,PH1,PH2,PH3
          RAD=4*ATAN(1.0)/180
          P1=COS(PH1*RAD)
          P2=COS(PH2*RAD)
          P3=COS(PH3*RAD)
          T1=COS(TH1*RAD)
          T2=COS(TH2*RAD)
          T3=COS(TH3*RAD)
          A1=COS((TH1-PH1)*RAD)
          A2=COS((TH2-PH2)*RAD)
          A3=COS((TH3-PH3)*RAD)
          DEL=P1*(T2-T3)+T1*(P3-P2)+(P2*T3-P3*T2)
          DEL1=A1*(T2-T3)+T1*(A3-A2)+(A2*T3-A3*T2)
          DEL2=P1*(A2-A3)+A1*(P3-P2)+(P2*A3-P3*A2)
          DEL3=P1*(T2*A3-T3*A2)+T1*(A2*P3-A3*P2)+A1*(P2*T3-P3*T2)
          A=DEL/DEL1
          C=-DEL/DEL2
          B=SQRT(A**2+C**2+1.-2.*A*C*DEL3/DEL)
          WRITE(*,3)A,B,C,1.
3         FORMAT(6X,'A',7X,'B',7X,'C',7X,'D'/4F8.2)
          STOP
          END
```

Fig. 5.15

TH1, TH2, TH3	Angular displacements of the input link (degrees)
PH1, PH2, PH3	Angular displacements of the output link (degrees)

The output variables are

A, B, C, D	Ratio of magnitudes of the links *AB*, *BC*, *CD* and *AD* respectively

Least-square Technique

The above synthesis technique is used to synthesize a mechanism where three finitely separated positions of the input and the output links are known. It is observed that a four-link mechanism can be designed precisely up to five positions of the input and the output links, provided θ and φ are measured from some arbitrary reference. In such cases, the synthesis equations become non-linear and have to be solved using other means than the Cramer's rule.

It is not possible to design a mechanism for more than five positions of the input and the output links. However, it is possible to design a mechanism which gives least deviation from the specified positions and provides the average performance. To achieve this, an approximated solution of the problem is sought which gives the least error. A method known as the *least-square technique* is useful in synthesizing such a mechanism.

Considering Freudenstein's equation,

$$k_1 \cos \varphi_i + k_2 \cos \theta_i + k_3 - \cos (\theta_i - \varphi_i) = 0$$

Summing up for n values of θ and φ and defining,

$$S = \sum_{i=1}^{n} [k_1 \cos \varphi_i + k_2 \cos \theta_i + k_3 - \cos (\theta_i - \varphi_i)]^2$$

Conditions for this to be minimum are,

$$\frac{\partial S}{\partial k_1} = 0, \quad \frac{\partial S}{\partial k_2} = 0 \text{ and } \frac{\partial S}{\partial k_3} = 0$$

i.e

$$\sum_{i=1}^{n} 2\left[k_1 \cos \varphi_i + k_2 \cos \theta_i + k_3 - \cos(\theta_i - \varphi_i)\right] \cos \varphi_i = 0$$

or

$$k_1 \Sigma \cos^2 \varphi_i + k_2 \Sigma \cos \theta_i \cos \varphi_i + k_3 \Sigma \cos \varphi_i = \Sigma \cos(\theta_i - \varphi_i)\cos \varphi_i \quad (5.2)$$

Similarly,

$$k_1 \Sigma \cos \varphi_i \cos \theta_i + k_2 \Sigma \cos^2 \theta_i + k_3 \Sigma \cos \theta_i = \Sigma \cos(\theta_i - \varphi_i) \cos \theta_i \quad (5.3)$$

and

$$k_1 \Sigma \cos \varphi_i + k_2 \Sigma \cos \theta_i + k_3 \Sigma 1 = \Sigma \cos(\theta_i - \varphi_i) \quad (5.4)$$

These are three simultaneous linear, non-homogeneous equations in three unknowns k_1, k_2 and k_3. These can be solved by using Cramer's rule or other means.

Figure 5.16 shows a programme to find the ratio of different links using the least-square technique. The input variables are

I	Number of specified positions
TH(J)	Angular displacements of the input link *AB* for $J = 1$ to I (degrees)

```
C      PROGRAMME TO COORDINATE ANGULAR DISPLACEMENTS OF THE
C      INPUT AND OUTPUT LINKS IN MORE THAN THREE POSITIONS
       DIMENSION TH(15),PH(15)
       READ(*,*)I
       READ(*,*)(TH(J),J=1,I),(PH(J),J=1,I)
       RAD=4*ATAN(1.0)/180
       DO 10 K=1,I
       P1=P1+(COS(PH(K)*RAD))**2
       P2=P2+(COS(TH(K)*RAD))*(COS(PH(K)*RAD))
       P3=P3+(COS(PH(K)*RAD))
       T1=P2
       T2=T2+(COS(TH(K)*RAD))**2
       T3=T3+(COS(TH(K)*RAD))
       B1=P3
       B2=T3
       B3=I
       TT=COS((TH(K)-PH(K))*RAD)
       A1=A1+TT*COS(PH(K)*RAD)
       A2=A2+TT*COS(TH(K)*RAD)
10     A3=A3+TT
       DEL=P1*(T2*B3-T3*B2)+T1*(B2*P3-B3*P2)+B1*(P2*T3-P3*T2)
       DEL1=A1*(T2*B3-T3*B2)+T1*(B2*A3-B3*A2)+B1*(A2*T3-A3*T2)
       DEL2=P1*(A2*B3-A3*B2)+A1*(B2*P3-B3*P2)+B1*(P2*A3-P3*A2)
       DEL3=P1*(T2*A3-T3*A2)+T1*(A2*P3-A3*P2)+A1*(P2*T3-P3*T2)
       A=DEL/DEL1
       C=-DEL/DEL2
       B=SQRT(A**2+C**2+1.-2.*A*C*DEL3/DEL)
       WRITE(*,3)A,B,C,1.
3      FORMAT(6X,'A',7X,'B',7X,'C',7X,'D'/4F8.2)
       STOP
       END
```

Fig. 5.16

PH(J) Angular displacements of the output link *DC* for $J = 1$ to I (degrees)

The output variables are

A, B, C, D Ratio of magnitudes of the links *AB*, *BC*, *CD* and *AD* respectively

Example 5.1 Design a four-link mechanism to coordinate three positions of the input and the output links as follows:

$$\theta_1 = 20° \qquad \varphi_1 = 35°$$
$$\theta_2 = 35° \qquad \varphi_2 = 45°$$
$$\theta_3 = 50° \qquad \varphi_3 = 60°$$

Solution: The input and the output have been shown in Fig. 5.17.

```
20 35 50 35 45 60
       A          B          C          D
     1.56        .66       1.33       1.00
Stop - Program terminated.
```

Fig. 5.17

The mechanism is shown in Fig. 5.18.

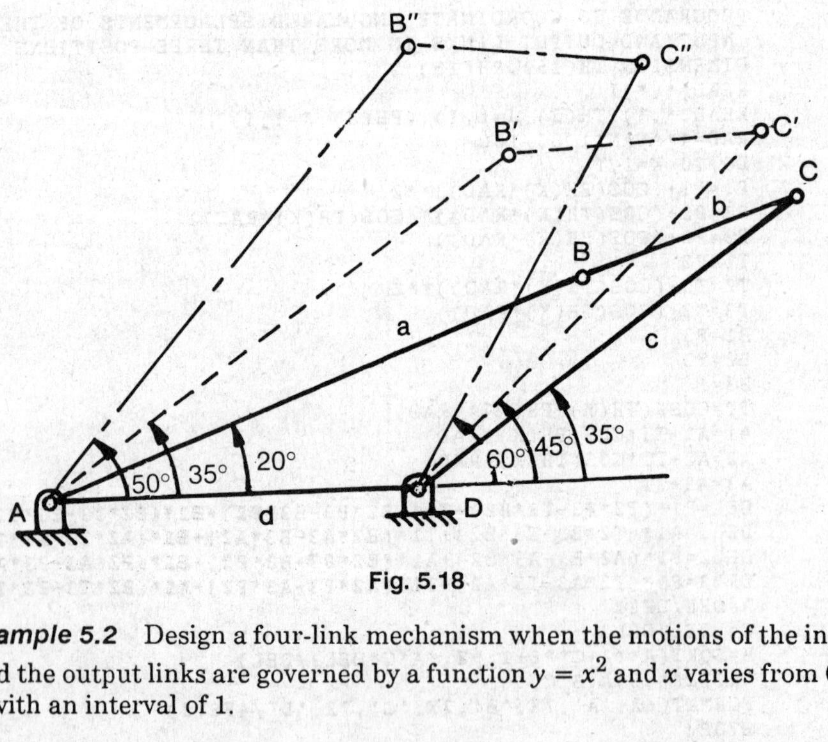

Fig. 5.18

Example 5.2 Design a four-link mechanism when the motions of the input and the output links are governed by a function $y = x^2$ and x varies from 0 to 2 with an interval of 1.

Solution: The angular displacement of the input link is governed by x whereas that of the output link by y.

Let θ vary from 50° to 150° (i.e. through 100°)
φ vary from 80° to 160° (i.e. through 80°).
(The values have been chosen arbitrarily)
In the present case,
x is 0, 1 and 2, i.e. a variation of (2 – 0) = 2
The corresponding values of y are 0, 1^2 and 2^2 or 0, 1 and 4 i.e. a variation of (4 – 0) = 4.
Thus when θ is 50°, φ = 80°

$$\text{when } \theta \text{ is } \left(50° + 100° \times \frac{1}{2}\right) = 100°, \ \varphi \text{ is } \left(80° + 80° \times \frac{1}{4}\right) = 100°$$

$$\text{when } \theta \text{ is } \left(50° + 100° \times \frac{2}{2}\right) = 150°, \ \varphi \text{ is } \left(80° + 80° \times \frac{4}{4}\right) = 160°$$

This can be written in a tabular form also:

Position	x	y	θ	φ
1	0	0	50°	80°
2	1	1	100°	100°
3	2	4	150°	160°

Thus we have the following equations,

$$k_1 \cos 80° + k_2 \cos 50° + k_3 = \cos 30°$$

$$k_2 \cos 100° + k_2 \cos 100° + k_3 = \cos 0°$$

$$k_3 \cos 160° + k_2 \cos 150° + k_3 = \cos 10°$$

Using Cramer's rule,

$$\Delta = -0.3850$$

$$\Delta_1 = -0.1052 \qquad k_1 = 0.273 = \frac{d}{a}$$

$$\Delta_2 = 0.1079 \qquad k_2 = -0.280 = -\frac{d}{c}$$

$$\Delta_3 = -0.3844 \qquad k_3 = 0.988 = \frac{a^2 - b^2 + c^2 + d^2}{2ac}$$

which gives

a = 3.66 units
b = 1.02 units
c = 3.57 units
and d = 1 unit

Figure 5.19 shows the required mechanism.

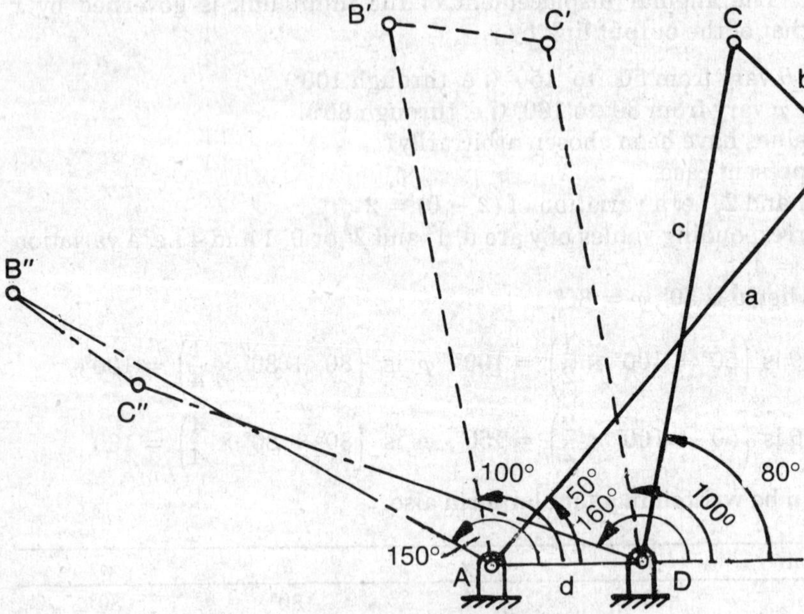

Fig. 5.19

Example 5.3 Design a four-link mechanism to coordinate the motions of the input and the output links governed by a function $y = \log x$ for $0 < x \leq 8$. Take $\delta x = 1$. The range for θ is from 15° to 120° whereas for φ, it is from 20° to 150°.

Solution: The angular positions of the input and the output links are tabulated below:

x	y	θ	φ
1	0	15°	20°
2	0.69	30°	*63°
3	1.10	45°	89°
4	1.39	60°	107°
5	1.61	75°	121°
6	1.79	90°	132°
7	1.95	105°	142°
8	2.08	120°	150°

$$*20° + (150° - 20°) \times \frac{0.69}{2.08} = 63°$$

It is required to design the mechanism so that the input and the output links pass through eight specified positions. It is not possible to design such

```
8
15 30 45 60 75 90  105  120
20 63 89  107  121  132  142  150
      A            B          C          D
    2.42         .91        2.37       1.00
Stop - Program terminated.
```

Fig. 5.20

a mechanism since at the most, the motion through five positions may be coordinated. However, using the least-square technique, a mechanism may be devised which gives the least deviation from the specified positions.

The input and the output have been shown in Fig. 5.20 using the programme given in Fig. 5.16.

Figure 5.21 shows the required mechanism which will give the least deviation from the specified positions.

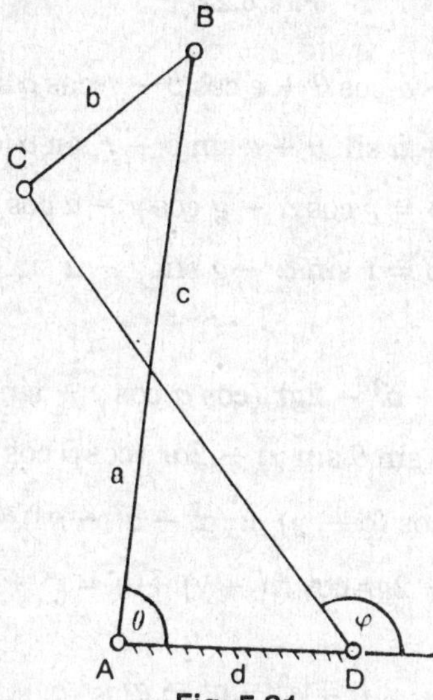

Fig. 5.21

5.7 TO COORDINATE ANGULAR DISPLACEMENTS OF INPUT LINK AND POSITIONS OF A COUPLER POINT

A four-link mechanism ABCD with a coupler point E is shown in Fig. 5.22. Three positions of the input link (θ_1, θ_2, θ_3) and three positions of the coupler point E given by three values of r and α, i.e. r_1, r_2, r_3 and $\alpha_1, \alpha_2, \alpha_3$ are known. It is required to find the dimensions of a, c, e and f along with the locations of the pivots A and D given by g, γ and h, ψ respectively.

For the loop OABE,

$$\mathbf{s}_{eo} = \mathbf{s}_{eb} + \mathbf{s}_{ba} + \mathbf{s}_{ao}$$

or

$$\mathbf{s}_{ao} + \mathbf{s}_{ba} + \mathbf{s}_{eb} - \mathbf{s}_{eo} = 0$$

Therefore, along the x- and y-axis, we have

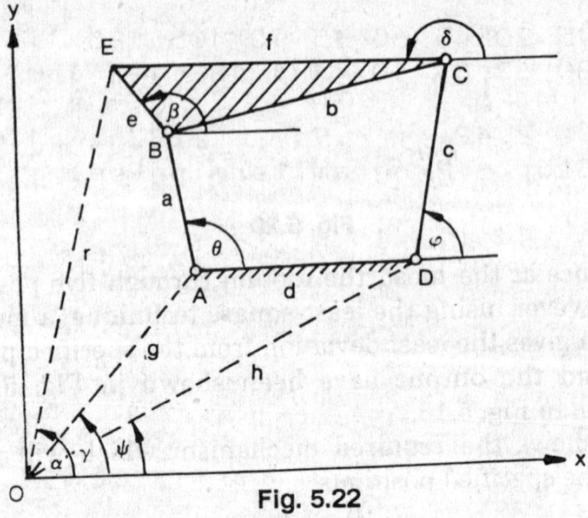

Fig. 5.22

$$g \cos \gamma + a \cos \theta + e \cos \beta - r \cos \alpha = 0 \tag{5.5}$$

and
$$g \sin \gamma + a \sin \theta + e \sin \beta - r \sin \alpha = 0 \tag{5.6}$$

or
$$e \cos \beta = r \cos \alpha - g \cos \gamma - a \cos \theta$$

and
$$e \sin \beta = r \sin \alpha - g \sin \gamma - a \sin \theta$$

Squaring and adding,

$$e^2 = r^2 + g^2 + a^2 - 2gr (\cos \alpha \cos \gamma + \sin \alpha \sin \gamma)$$
$$+ 2ag (\cos \theta \cos \gamma + \sin \theta \sin \gamma) - 2ar (\cos \theta \cos \alpha + \sin \theta \sin \alpha)$$

or $2ar \cos (\theta - \alpha) + 2gr \cos (\alpha - \gamma) + (e^2 - a^2 - g^2) = r^2 + 2ag \cos (\theta - \gamma)$

or
$$2ar \cos (\theta - \alpha) + 2gr \cos (\alpha - \gamma) + k = r^2 + 2ag \cos (\theta - \gamma) \tag{5.7}$$

where

$$k = e^2 - a^2 - g^2 \tag{5.8}$$

Inserting the values of r_1, r_2, r_3; $\alpha_1, \alpha_2, \alpha_3$ and $\theta_1, \theta_2, \theta_3$, we obtain three equations. The unknowns are a, g, e and γ. Thus, for three equations, there are four unknowns and therefore, equations cannot be solved. However, the value of one of the unknowns can be assumed. Assuming the value of γ, we are left with three unknowns a, g, e and there are three equations to solve them.

Even now, the equations cannot be solved as such, as these are non-linear equations. However, by making the following substitutions, these can be solved easily.

Let

$$\left.\begin{aligned} a &= l_a + \lambda m_a \\ g &= l_g + \lambda m_g \\ k &= l_k + \lambda m_k \end{aligned}\right\} \tag{5.9}$$

where $$\begin{aligned}\lambda &= ag \\ &= (l_a + \lambda m_a)(l_g + \lambda m_g) \\ &= l_a l_g + \lambda l_a m_g + \lambda l_g m_a + \lambda^2 m_a m_g\end{aligned}$$

or $$m_a m_g \lambda^2 + (l_a m_g + l_g m_a - 1)\lambda + l_a l_g = 0$$

or $$A\lambda^2 + B\lambda + C = 0$$

or

$$\lambda = \frac{-B \pm \sqrt{B^2 - 4AC}}{2A} \tag{5.10}$$

where

$$A = m_a m_g$$

$$B = l_a m_g + l_g m_a - 1$$

$$C = l_a l_g$$

Thus Eq. (5.7) becomes,

$$2(l_a+\lambda m_a)r \cos(\theta-\alpha) + 2(l_g+\lambda m_g)r \cos(\alpha-\gamma)+l_k+\lambda m_k = r^2+2\,\lambda \cos(\theta-\gamma)$$

Separating the components into two groups; one with and the other without λ.

$$l_a[2r\cos(\theta-\alpha)] + l_g\,[2r\cos(\alpha-\gamma)] + l_k = r^2 \tag{5.11}$$

$$m_a[2r\cos(\theta-\alpha)] + m_g\,[2r\cos(\alpha-\gamma)] + m_k = 2\cos(\theta-\gamma) \tag{5.12}$$

From Eq. (5.11), three equations can be written as,

$$l_a\,[2r_1\cos(\theta_1-\alpha_1)] + l_g\,[2r_1\cos(\alpha_1-\gamma)] + l_k = r_1^2 \tag{5.13}$$

$$l_a\,[2r_2\cos(\theta_2-\alpha_2)] + l_g\,[2r_2\cos(\alpha_2-\gamma)] + l_k = r_2^2 \tag{5.14}$$

$$l_a\,[2r_3\cos(\theta_3-\alpha_3)] + l_g\,[2r_3\cos(\alpha_3-\gamma)] + l_k = r_3^2 \tag{5.15}$$

These are three linear equations in l_a, l_g and l_k and can be solved applying Cramer's rule or by other means.

Similarly, m_a, m_g and m_k can also be found.

As l_a, l_g, l_k and m_a, m_g, m_k have been found, a, g and k can be calculated from the relations of Eq. (5.9).

Also, $$e = \sqrt{k + a^2 + g^2}$$ [from Eq. (5.8)]

From Eq. (5.5), three values of β can be found,

$$e \cos \beta = r \cos \alpha - g \cos \gamma - a \cos \theta$$

$$\beta_1 = \cos^{-1}\left[\frac{r_1 \cos \alpha_1 - g \cos \gamma - a \cos \theta_1}{e}\right] \tag{5.16}$$

Similarly, β_2 and β_3 can be found.

Thus, we have obtained the values of a, e, g, γ and β. The whole procedure can be repeated for the loop ODCE. The following equations are formed,

$$h \cos \psi + c \cos \varphi + f \cos \delta - r \cos \alpha = 0 \tag{5.17}$$

$$h \sin \psi + c \sin \varphi + f \sin \delta - r \sin \alpha = 0 \tag{5.18}$$

These equations are similar to Eqs. (5.5) and (5.6).

Assuming

$$f = l_f + \lambda' m_f$$

$$h = l_h + \lambda' m_h$$

$$p = l_p + \lambda' m_p$$

two sets of equations similar to Eqs. (5.11) and (5.12) are obtained by eliminating φ as given below:

$$l_f [2r \cos (\delta - \alpha)] + l_h [2r \cos (\alpha - \psi)] + l_p = r^2 \tag{5.19}$$

$$m_f [2r \cos (\delta - \alpha)] + m_h [2r \cos (\alpha - \psi)] + m_p = 2 \cos (\delta - \psi) \tag{5.20}$$

In these equations α and r are known. ψ can be assumed. Also assuming δ_1, the values of δ_2 and δ_3 can be found as follows:

The angular displacements of the coupler link BCE is the same at the points B and C,

$$\therefore \quad \delta_2 - \delta_1 = \beta_2 - \beta_1$$

or

$$\delta_2 = \delta_1 + (\beta_2 - \beta_1) \tag{5.21}$$

Similarly,

$$\delta_3 = \delta_1 + (\beta_3 - \beta_1) \tag{5.22}$$

Solving the Eqs. (5.19) and (5.20), the values of f, h and c can be known.

As the points A, B, E, C and D are located, the dimensions a, b, c, d, e and f can be obtained.

Figure 5.23 shows a programme for the solution of such a problem.

```
C     TO COORDINATE THREE POSITIONS OF THE INPUT LINK
C     AND THE COUPLER POINT
      READ(*,*)TB1,TB2,TB3,R1,R2,R3,AL1,AL2,AL3,GAMM,
     1 SI,DEL1,CASE
      RAD=4.*ATAN(1.)/180
      T11=TB1*RAD
      T22=TB2*RAD
      T33=TB3*RAD
      A11=AL1*RAD
      A22=AL2*RAD
      A33=AL3*RAD
      GG=GAMM*RAD
      SS=SI*RAD
      D11=DEL1*RAD
      IT=0
10    IT=IT+1
      P1=2*R1*COS(T11-A11)
      P2=2*R2*COS(T22-A22)
      P3=2*R3*COS(T33-A33)
      T1=2*R1*COS(A11-GG)
      T2=2*R2*COS(A22-GG)
      T3=2*R3*COS(A33-GG)
      C1=R1**2
      C2=R2**2
      C3=R3**2
      CALL DLL(P1,P2,P3,T1,T2,T3,C1,C2,C3,AK1,AK2,AK3)
      ALA=AK1
      ALG=AK2
      ALK=AK3
      C1=2*COS(T11-GG)
      C2=2*COS(T22-GG)
      C3=2*COS(T33-GG)
      CALL DLL(P1,P2,P3,T1,T2,T3,C1,C2,C3,AK1,AK2,AK3)
      AMA=AK1
      AMG=AK2
      AMK=AK3
      AA=AMA*AMG
      BB=ALA*AMG+ALG*AMA-1
      CC=ALA*ALG
      SQ=BB**2-4*AA*CC
      IF(SQ.LT.0)GO TO 7
      ALL=SQRT(SQ)
      AL1=(-BB-ALL)/(2*AA)
      AL2=(-BB+ALL)/(2*AA)
      A1=ALA+AL1*AMA
      G1=ALG+AL1*AMG
      A2=ALA+AL2*AMA
      G2=ALG+AL2*AMG
      E1=SQRT(ALK+AL1*AMK+A1**2+G1**2)
      E2=SQRT(ALK+AL2*AMK+A2**2+G2**2)
      IF(IT.EQ.1)GO TO 13
      IF(IT.NE.3)GO TO 29
      IF(CASE.NE.2)GO TO 31
      WRITE(*,6)G21,E21,A21,G1,E1,A1
      WRITE(*,6)G21,E21,A21,G2,E2,A2
      GO TO 27
31    WRITE(*,6)G21,A21,E21,G1,E1,A1
      WRITE(*,6)G21,A21,E21,G2,E2,A2
      GO TO 27
```

```
29    IF(CASE.EQ.2)WRITE(*,8)G12,E12,A12,G1,E1,A1
      IF(CASE.EQ.1)WRITE(*,8)G12,A12,E12,G1,E1,A1
8     FORMAT(5X,'G',7X,'A',7X,'E',7X,'H',7X,'C',7X,'F'/6F8.2)
      IF(CASE.EQ.1)WRITE(*,6)G12,A12,E12,G2,E2,A2
      IF(CASE.EQ.2)WRITE(*,6)G12,E12,A12,G2,E2,A2
6     FORMAT(6F8.2)
13    IF(IT.EQ.2)GO TO 11
      G12=G1
      A12=A1
      E12=E1
      G21=G2
      A21=A2
      E21=E2
      IF(CASE.EQ.2)GO TO 41
      BET1=ACOS((R1*COS(A11)-G1*COS(GG)-A1*COS(T11))/E1)
      BET2=ACOS((R2*COS(A22)-G1*COS(GG)-A1*COS(T22))/E1)
      BET3=ACOS((R3*COS(A33)-G1*COS(GG)-A1*COS(T33))/E1)
      D22=D11+BET2-BET1
      D33=D11+BET3-BET1
      IF (CASE.EQ.1)GO TO 33
41    D22=D11+T22-T11
      D33=D11+T33-T11
33    A3=A2
      G3=G2
      E3=E2
      P11=T11
      P22=T22
      P33=T33
      T11=D11
      T22=D22
      T33=D33
      GS=GG
      GG=SS
7     IF(IT.EQ.1)GO TO 10
      IF(CASE.EQ.2)GO TO 35
11    BET1=ACOS((R1*COS(A11)-G3*COS(GS)-A3*COS(P11))/E3)
      BET2=ACOS((R2*COS(A22)-G3*COS(GS)-A3*COS(P22))/E3)
      BET3=ACOS((R3*COS(A33)-G3*COS(GS)-A3*COS(P33))/E3)
      D22=D11+BET2-BET1
      D33=D11+BET3-BET1
      IF(CASE.EQ.1)GO TO 37
35    D22=D11+T22-T11
      D33=D11+T33-T11
37    T11=D11
      T22=D22
      T33=D33
      IF(IT.EQ.2)GO TO 10
27    STOP
      END
      SUBROUTINE DLL(P1,P2,P3,T1,T2,T3,C1,C2,C3,AK1,AK2,AK3)
      DEL=P1*(T2-T3)+T1*(P3-P2)+(P2*T3-P3*T2)
      DEL1=C1*(T2-T3)+T1*(C3-C2)+(C2*T3-C3*T2)
      DEL2=P1*(C2-C3)+C1*(P3-P2)+(P2*C3-P3*C2)
      DEL3=P1*(T2*C3-T3*C2)+T1*(C2*P3-C3*P2)+C1*(P2*T3-P3*T2)
      AK1=DEL1/DEL
      AK2=DEL2/DEL
      AK3=DEL3/DEL
      RETURN
      END
```

Fig. 5.23

The input variables are:

TB1, TB2, TB3	Angular displacements of the input link *AB* (degrees)
R1, R2, R3	Radial distances of the coupler point from origin (mm)
AL1, AL2, AL3	Angular displacements of the coupler point (degrees)
Gamm	Assumed value of angle γ (degrees)
SI	Assumed value of angle ψ (degrees)
DEL1	Assumed value of angle δ_1 (degrees)
Case	1

The output variables are:

G, A, E, H, C, F	Dimensions *g, a, e, h, c* and *f* respectively (mm)

If more than three positions of the input link along with the same numbe of positions of the coupler point are known, the mechanism can be synthesizec using the least-square technique. The deviations of the coupler point E from the prescribed positions will be minimum in such a design. Thus,

$$S_1 = \Sigma \left[l_a \{2r \cos(\theta - \alpha)\} + l_g \{2r \cos(\alpha - \gamma)\} + l_k - r^2\right]^2$$

$$S_2 = \Sigma \left[m_a \{2r \cos(\theta - \alpha)\} + m_g \{2r \cos(\alpha - \gamma)\} + m_k - 2\cos(\theta - \alpha)\right]^2$$

For minimum deviations,

$$\frac{\delta S_1}{\delta l_a} = 0 \qquad \frac{\delta S_1}{\delta l_g} = 0 \qquad \frac{\delta S_1}{\delta l_k} = 0$$

and

$$\frac{\delta S_2}{\delta l_a} = 0 \qquad \frac{\delta S_2}{\delta l_g} = 0 \qquad \frac{\delta S_2}{\delta l_k} = 0$$

when $\dfrac{\delta S_1}{\delta l_a} = 0,$

$$\sum_1^n 2[l_a 2r \cos(\theta - \alpha) + l_g \cdot 2r \cos(\alpha - \gamma) + l_k - r^2] \cdot 2r \cos(\theta - \alpha) = 0$$

or

$$l_a \sum_1^n [2r \cos(\theta - \alpha)]^2 + l_g \sum_1^n [2r \cos(\alpha - \gamma)][2r \cos(\theta - \alpha)]$$
$$+ l_k \sum_1^n [2r \cos(\theta - \alpha)] = \Sigma[2r \cos(\theta - \alpha)]r^2$$

Similarly for $\dfrac{\delta S_1}{\delta l_g} = 0$ and $\dfrac{\delta S_1}{\delta l_k} = 0,$

$$l_a\sum_1^n [2r\cos(\theta-\alpha)][2r\cos(\alpha-\gamma)] + l_g\sum_1^n [2r\cos(\alpha-\gamma)]^2 + l_k\sum_1^n [2r\cos(\alpha-\gamma)] = \sum_1^n [2r\cos(\alpha-\gamma)]r^2$$

$$l_a\sum_1^n [2r\cos(\theta-\alpha)] + l_g\sum_1^n [2r\cos(\alpha-\gamma)] + l_k = \sum_1^n r^2$$

Inserting n values of r, α, θ (given) and one value of γ (assumed), l_a, l_g and l_k can be calculated by using the Cramer's rule etc.

Similarly, using the conditions $\frac{\delta S_2}{\delta l_a} = 0$, $\frac{\delta S_2}{\delta l_g} = 0$ and $\frac{\delta S_2}{\delta l_k} = 0$, the values of m_a, m_g and m_k can be found. The rest of the procedure is as given earlier for three values of α, θ and r.

Example 5.4 Design a four-link mechanism to coordinate three positions of the input link with three positions of the coupler point, the data for which is given below:

$\theta_1 = 110°$	$r_1 = 80$ mm	$\alpha_1 = 65°$
$\theta_2 = 77°$	$r_2 = 90$ mm	$\alpha_2 = 56°$
$\theta_3 = 50°$	$r_3 = 96$ mm	$\alpha_3 = 48°$

Solution: The procedure is as follows:

(a) Assume values of γ, ψ and δ

Let $\gamma = 20°$

$\psi = 10°$

and $\delta_1 = 150°$

(b) Solve the simultaneous Eqs. (5.13), (5.14) and (5.15) and obtain

$l_a = 8.14$ $l_g = 38.57$ $l_k = 1115.3$

(c) Write three equations from Eq. (5.12) and obtain

$m_a = 0.00958$ $m_g = 0.0173$ $m_k = -3.045$

(d) Find two values of λ using Eqn. (5.10).

$\lambda_1 = 945$ $\lambda_2 = 2001$

(e) Using $\lambda = 945$, obtain a, g and k from relations given in Eq. (5.9). Also find e from the relation of Eq. (5.8).

$a = 17.2$ $g = 55$ $e = 39.4$

(f) Using three relations for β [Eq. (5.16)], find β_1, β_2 β_3.

$\beta_1 = 107.7°$ $\beta_2 = 97.5°$ $\beta_3 = 87.7°$

Thus all the parameters for the loop *OABEO* are known.

(g) Obtain δ_2 and δ_3 from relations of Eqs. (5.21) and (5.22).

$\delta_2 = 139.8°$ $\delta_3 = 130.1°$

The same procedure is adopted for the loop *ODCEO*.

(h) Solving Eqs. (5.19) and (5.20), the following values are obtained:

$l_f = 71.15$ $\quad l_h = 62.16$ $\quad l_p = 1688$
$m_f = 0.0262$ $\quad m_h = 0.00342$ $\quad m_p = -2.21$

(i) Write three equations similar to Eq. (5.9) in λ' and obtain
$\lambda'_1 = -9507$ $\quad \lambda'_2 = 5197$
When $\lambda' = -9507$, $f = -320.3$; $h = 29.7$ and $c = 355.2$
When $\lambda' = 5197$ $f = 65.1$; $h = 79.9$ and $c = 28.5$

(j) When $\lambda = 2001$, another set of a, g and e and two more sets of f, h and c are obtained.

Figure 5.24 shows the input and the four sets of values obtained by using the programme of Fig. 5.23.

```
110 77 50 80 90 96 65 56 48 20  10 150  1
     G          A        E         H        C         F
 54.95      17.20     39.40     29.69    355.19  -320.26
 54.95      17.20     39.40     79.91     28.50    65.04
 73.25      27.31     33.68     23.31     96.77   -47.12
 73.25      27.31     33.68    103.50     20.07    71.12
Stop - Program terminated.
```

Fig. 5.24

Figure 5.25 shows the solution obtained from the second set.

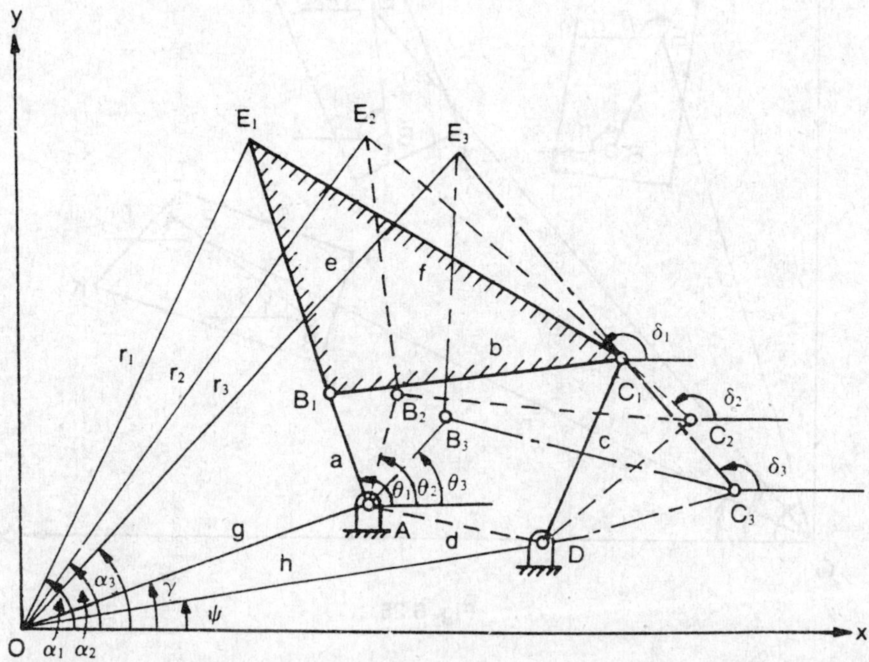

Fig. 5.25

5.8 TO GUIDE A RIGID BODY THROUGH THREE FINITELY SEPARATED POSITIONS

A rigid body in its three finitely separated positions is shown in Fig. 5.26. In order to synthesize a four-link mechanism to guide the body through these positions, locate a convenient point E on the body in its three positions and define the same by the radial distances from the assumed origin and its angular positions i.e. by $r_1, \alpha_1; r_2, \alpha_2$ and r_3, α_3. Also note down the angles β'_1, β'_2 and β'_3 of any line on the body as shown in the figure. Assuming that the body is fixed to the coupler link, it can be observed that if any point is taken on the line, the angle with the x-axis would be the same as at E. So, the problem reduces to that of locating points A and D (refer Fig. 5.22) and finding a point B on the line such that the angular displacements at B are β_1, β_2 and β_3 equal to β'_1, β'_2 and β'_3 respectively.

The analysis is similar to that given in Section 5.7 except that now θ is to be eliminated from the Eqs. (5.5) and (5.6) instead of β. The equations formed are exactly the same if θ is replaced by β in Eqs. (5.11) and (5.12). Also as β is directly known, there is no need of using Eq. 5.16.

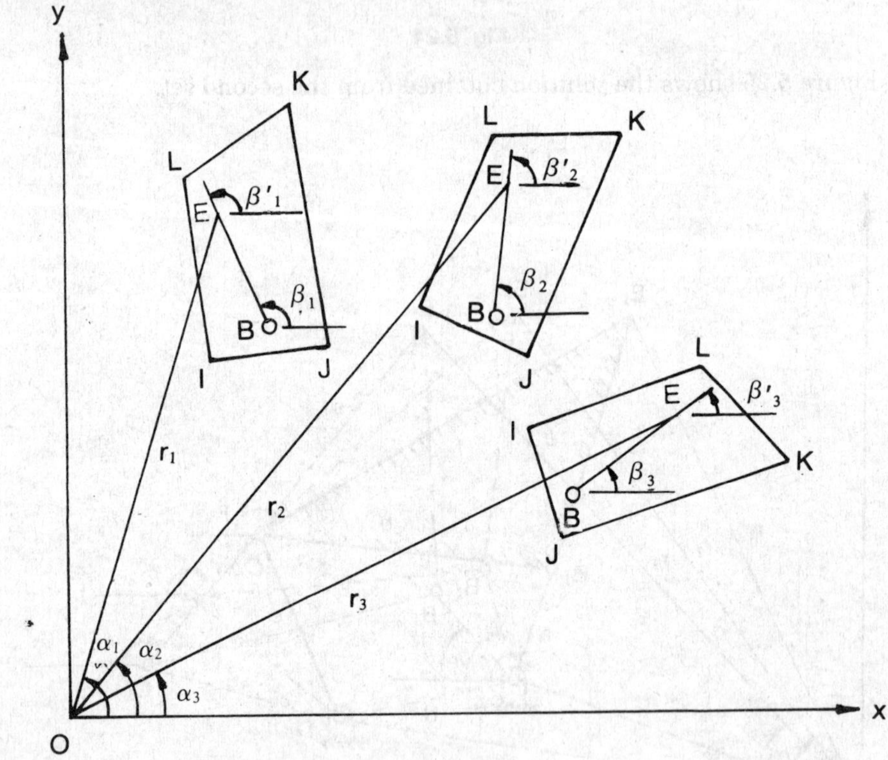

Fig. 5.26

The programme given in Fig. 5.23 also solves this type of problem. The input variables are

TB1, TB2, TB3	Angles β_1, β_2 and β_3 respectively (degrees)
Case	2

The rest of the variables of the input as well as of the output are as before.

Example 5.5 Design a four-link mechanism to guide a rigid body through three finitely separated positions given by

$\beta_1 = 105°$	$r_1 = 80$ mm	$\alpha_1 = 65°$
$\beta_2 = 95°$	$r_2 = 90$ mm	$\alpha_2 = 56°$
$\beta_3 = 50°$	$r_3 = 96$ mm	$\alpha_3 = 48°$

Solution: Assume the values of γ, ψ and δ_1.

Let $\gamma = 20°$ $\psi = 10°$ and $\delta_1 = 150°$

Figure 5.27 shows the input and the four sets of values of the output obtained by using the programme of Fig. 5.23.

```
105 95 85 80 90 96 65 56 48 20 10 150 2
    G          A          E          H          C          F
  27.04      49.76     102.24      32.33     240.67    -201.61
  27.04      49.76     102.24      81.82      27.25      66.79
  53.75      16.14      40.60      32.33     240.67    -201.61
  53.75      16.14      40.60      81.82      27.25      66.79
Stop - Program terminated.
```

Fig. 5.27

PROBLEMS

5.1 Design a four-link mechanism to coordinate three positions of the input and the output links given by

$\theta_1 = 25°$ $\varphi_1 = 30°$

$\theta_2 = 35°$ $\varphi_2 = 40°$

$\theta_3 = 50°$ $\varphi_3 = 60°$

5.2 Design a four-link mechanism when the motions of the input and the output links are governed by the function $y = 2x^2$. x varies from 2 to 4 with an interval of 1.

5.3 Design a four-link mechanism to coordinate the motions of the input and the output links governed by a function $y = 2 \log x$ for $2 < x < 12$. Take $\Delta x = 1$. Assume suitable ranges for θ and φ.

5.4 Design a four-link mechansim to guide a rigid body through three positions of the input link with three position of the coupler point, the data for which is given below:

$\theta_1 = 40°$	$r_1 = 90$ mm	$\alpha_1 = 78°$
$\theta_2 = 55°$	$r_2 = 40$ mm	$\alpha_2 = 90°$
$\theta_3 = 70°$	$r_3 = 75$ mm	$\alpha_3 = 95°$

5.5 Design a four-link mechanism, the coupler point of which traces a coupler curve that is approximated by ten positions given by the following data

$\theta_1 = 160°$	$r_1 = 57$ mm	$\alpha_1 = 70°$
$\theta_2 = 130°$	$r_2 = 76$ mm	$\alpha_2 = 65°$
$\theta_3 = 98°$	$r_3 = 88$ mm	$\alpha_3 = 55°$
$\theta_4 = 73°$	$r_4 = 98$ mm	$\alpha_4 = 45°$
$\theta_5 = 32°$	$r_5 = 92$ mm	$\alpha_5 = 30°$
$\theta_6 = -15°$	$r_6 = 89$ mm	$\alpha_6 = 20°$
$\theta_7 = -25°$	$r_7 = 82$ mm	$\alpha_7 = 19°$
$\theta_8 = -70°$	$r_8 = 53$ mm	$\alpha_8 = 25°$
$\theta_9 = -125°$	$r_9 = 38$ mm	$\alpha_9 = 50°$
$\theta_{10} = -165°$	$r_{10} = 42$ mm	$\alpha_{10} = 70°$

6

LOWER PAIRS

6.1 INTRODUCTION

Lower pairs usually comprise turning (pivoted) and sliding pairs. Mechanisms with pivoted links are widely used in machines and the required movements of links are produced by using them in a variety of forms and methods. In this chapter, some of the more common mechanisms will be studied. Pantographs are used to copy the curves on reduced or enlarged scales. Some pivoted-link mechanisms are used to guide reciprocating parts either exactly or approximately in straight paths to eliminate the friction of the straight guides of the sliding pairs. However, these days, sliders are also being used to get linear motions.

An exact straight-line mechanism guides a reciprocating part in an exact straight line. On the other hand, an approximate straight line mechanism is designed in such a way that the middle and the two extreme positions of the guided point are in a straight line and the intermediate positions deviate as little as possible from the line.

Although this chapter will be restricted to the more elementary aspects of the analysis of mechanisms, the possibilities of their use in the mechanisms and the machines of daily use can easily be glimpsed. Moreover, systematic design techniques are being developed, to permit the use of these mechanisms in devising means for accurate control of the processes and the machines, being needed in modern technology.

6.2 PANTOGRAPH

A pantograph is a four-bar linkage used to produce paths exactly similar to the ones traced out by a point on the linkage. The paths so produced are, usually, on an enlarged or a reduced scale.

The four links of a pantograph are arranged in such a way that a parallelogram *ABCD* is formed. Thus $AB = DC$ and $BC = AD$. If some point O in one of the links is made fixed and three other points P, Q and R on the other three links are located in such a way that $OPQR$ is a straight line, it can be shown that the points P, Q and R always move parallel and similar to each other over any path, straight or curved. Their motions will be proportional to their distances from the fixed point.

Let O, P, Q and R lie on links CD, DA, AB and BC respectively. $ABCD$ is the initial assumed position as shown in Fig. 6.1.

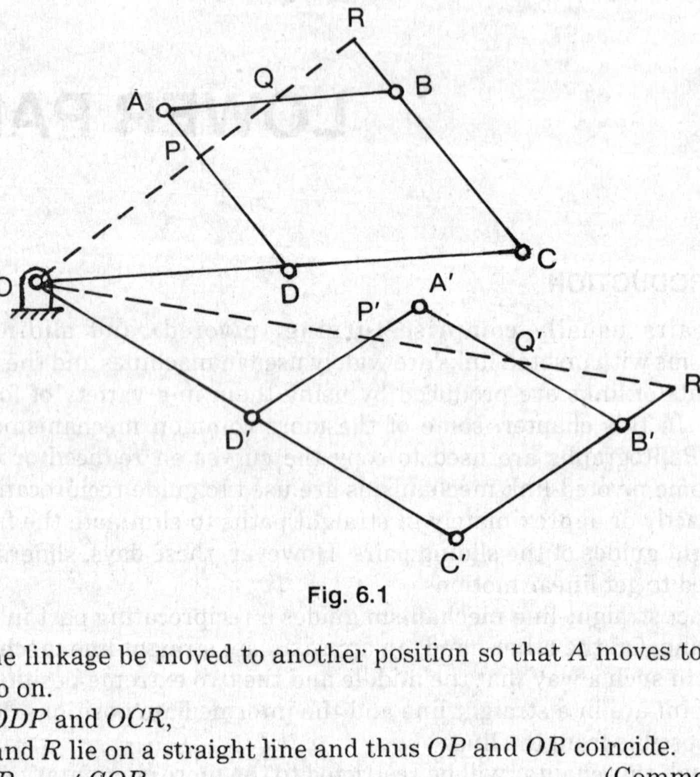

Fig. 6.1

Let the linkage be moved to another position so that A moves to A', B to B' and so on.

In $\Delta\ ODP$ and OCR,

O, P and R lie on a straight line and thus OP and OR coincide.

$\angle DOP = \angle COR$ (Common angle)

$\angle ODP = \angle OCR$ ($\because DP \parallel CR$)

Therefore, Δs are similar and

$$\frac{OD}{OC} = \frac{OP}{OR} = \frac{DP}{CR} \qquad \text{(i)}$$

Now, $A'B' = AB = DC = D'C'$

and $B'C' = BC = AD = A'D'$

Therefore, $A'B'C'D'$ is again a parallelogram.

In $\Delta s\ OD'P'$ and $OC'R'$,

$$\frac{OD'}{OC'} = \frac{OD}{OC} = \frac{DP}{CR} \qquad \text{[from (i)]}$$

$$= \frac{D'P'}{C'R'}$$

and

$$\angle OD'P' = \angle OC'R' \qquad (D'P' \,||\, C'B' \text{ as } A'B'C'D' \text{ is a } ||\text{gm})$$

Thus Δs are similar.

$$\therefore \qquad \angle D'OP' = \angle C'OR'$$

or O, P' and R' lie on a straight line.

Now

$$\frac{OP}{OR} = \frac{OD}{OC} \qquad \text{[from (i)]}$$

$$= \frac{OD'}{OC'}$$

$$= \frac{OP'}{OR'} \qquad (\because \Delta s\ OD'P' \text{ and } OC'R' \text{ are similar})$$

This shows that as the linkage is moved, the ratio of the distances of P and R from the fixed point remains same or the two points are displaced proportional to their distances from the fixed point. This will be true for all the positions of the links. Thus P and R will trace exactly similar paths.

Similarly, it can also be proved that P and Q trace similar paths. Thus P, Q and R trace similar paths when the linkage is given motion.

6.3 STRAIGHT LINE MECHANISMS

1. Paucellier Mechanism

A Paucellier mechanism consists of eight links (Fig. 6.2) such that,

$$OA = OQ; \qquad AB = AC$$

and

$$BP = PC = CQ = QB$$

OA is the fixed link and OQ is a rotating link. It can be proved that as the

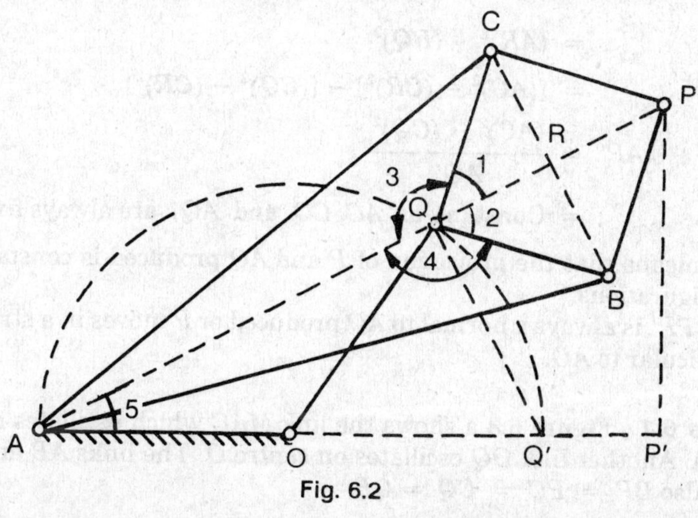

Fig. 6.2

link OQ moves around O, P moves in a straight line perpendicular to OA. All the joints are pin-jointed.

Since $BPCQ$ is a rhombus,

QP always bisect the angle BQC.

i.e.,

$$\angle 1 = \angle 2 \qquad \text{(i)}$$

in all the positions

Also in Δs AQC and AQB,

AQ is common,

$AC = AB$

$QC = QB$

$\therefore$ Δs are congruent in all positions.

or

$$\angle 3 = \angle 4 \qquad \text{(ii)}$$

Adding (i) and (ii),

$\angle 1 + \angle 3 = \angle 2 + \angle 4$

But $\angle 1 + \angle 2 + \angle 3 + \angle 4 = 360°$

$\therefore$ $\angle 1 + \angle 3 = \angle 2 + \angle 4 = 180°$

or A, Q and P lie on a straight line.

Let PP' be the perpendicular on AO produced.

Δs AQQ' and APP' are similar because $\angle 5$ is common and $\angle AQQ' = \angle AP'P = 90°$

$$\therefore \qquad \frac{AQ}{AP'} = \frac{AQ'}{AP}$$

or $\quad AQ' \cdot AP' = (AQ)(AP)$

$$= (AR - RQ)(AR + RP)$$

$$= (AR - RQ)(AR + RQ) \qquad \ldots (RP = RQ)$$

$$= (AR)^2 - (RQ)^2$$

$$= [(AC)^2 - (CR)^2] - [(CQ)^2 - (CR)^2]$$

or $\quad AP' = \dfrac{(AC)^2 - (CQ)^2}{AQ'}$

= Constant, as AC, CQ and AQ' are always fixed.

This means that the projection of P and AQ produced is constant for all the configurations.

Thus PP' is always a normal to AO produced or P moves in a straight line perpendicular to AO.

Example 6.1 Figure 6.3 a shows the link MAC which oscillates on a fixed centre A. Another link OQ oscillates on centre O. The links AB and AC are equal. Also $BP = PC = CQ = QB$.

Locate the position of O such that,

(i) P moves in a straight line

(ii) P moves in a circle with centre A

(iii) P moves in a circle with centre at OA produced

(iv) P moves in a circle with centre O and Q moves in a straight line (Modify the lengths of links if necessary).

Solution:

(i) As in the Paucellier mechanism, O is located by drawing a straight line through A and perpendicular to the motion of P such that $AO = OQ$ (Fig. 6.3a).

(ii) If O is made to coincide with A, AQ would be equal to OQ. Thus Q and P will be fixed on AP. Q will rotate about A and thus P will also rotate in a circle about A with AP as the radius (Fig. 6.3b).

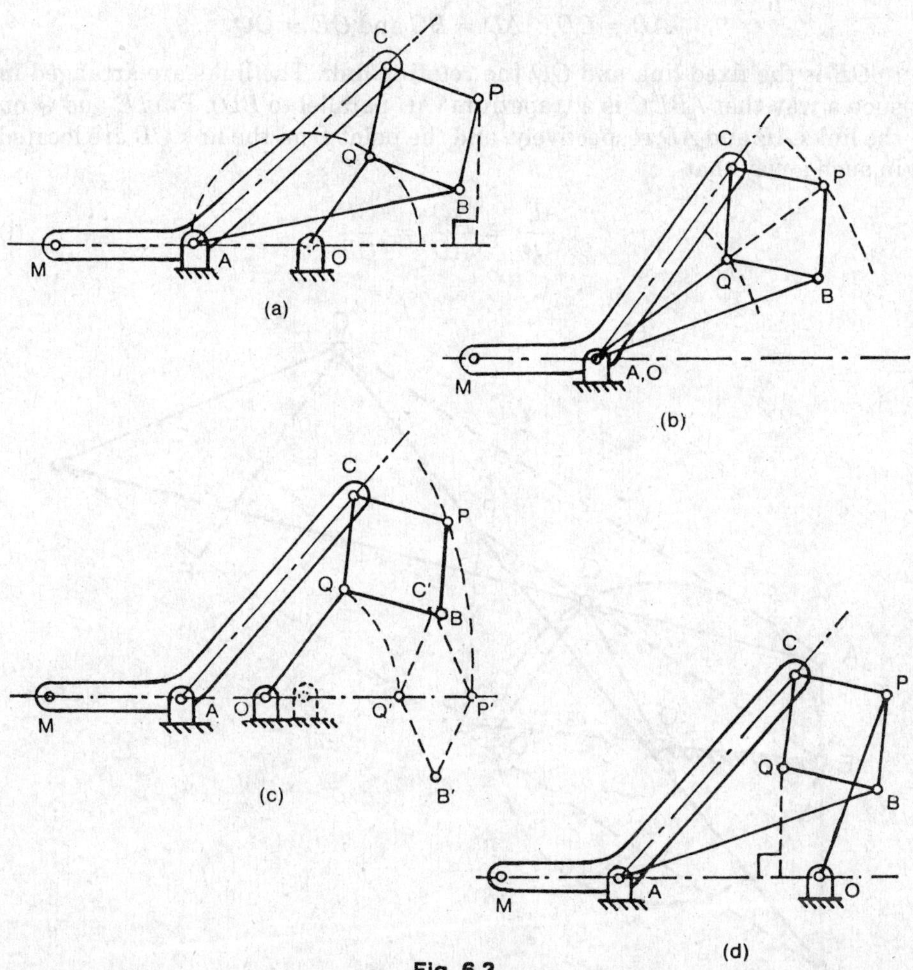

Fig. 6.3

(iii) From the above two cases, it can be observed that in (i) P moves in a circle with the centre at infinity on OA produced and in (ii) P moves in circle with the centre at A. Thus if P is to move in a circle with the centre in-between A and infinity on OA produced, O must lie in-between O and A or in other words OQ should be greater than OA (Fig. 6.3c).

(iv) The mechanism will be similar to the Paucellier mechanism. P is to be joined with O by a link so that P moves in a circle about O and $OA = OP$. The lengths can be modified in two ways (Fig. 6.3d),
(a) OA is increased and OA and OP are made equal.
(b) Lengths AB and AC are reduced in such a way that $OA = OP$.

2. Hart Mechanism

A Hart mechanism consists of six links as shown in Fig. 6.4 such that,

$$AB = CD; \quad AD = BC \text{ and } OE = OQ$$

OE is the fixed link and OQ the rotating link. The links are arranged in such a way that $ABDC$ is a trapezium (AC parallel to BD). Pins E and Q on the links AB and AD respectively, and the point P on the link CB are located in such a way that

$$\frac{AE}{AB} = \frac{AQ}{AD} = \frac{CP}{CB} \qquad \text{(i)}$$

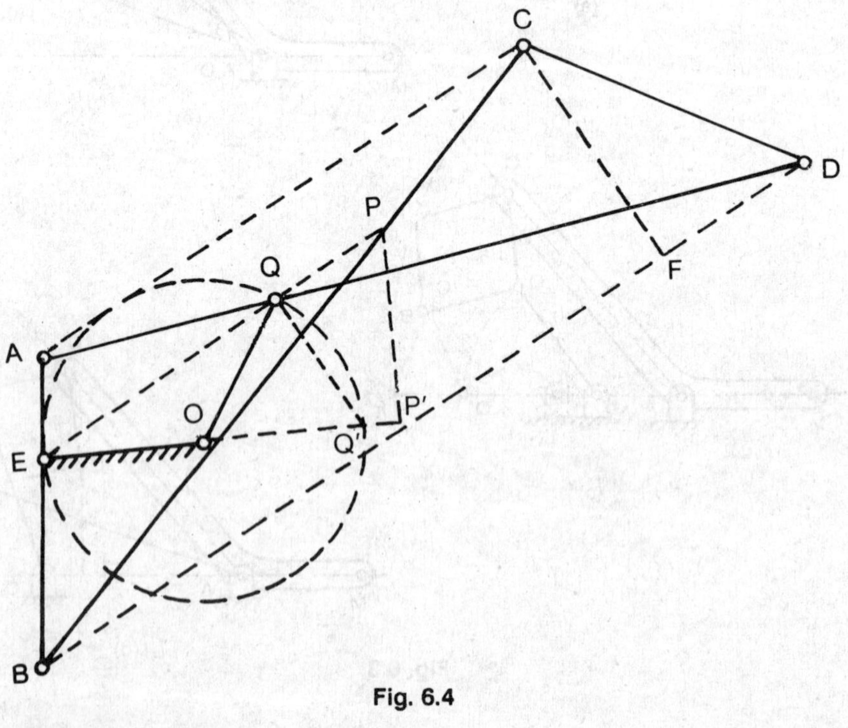

Fig. 6.4

It can be shown that as OQ rotates about O, P moves in a line perpendicular to EO produced.

In ΔABD, $\dfrac{AE}{AB} = \dfrac{AQ}{AD}$ (Given)

Therefore, EQ is parallel to BD and thus parallel to AC.

In ΔABC, $\dfrac{AE}{AB} = \dfrac{CP}{CB}$ (Given)

Therefore, EP is parallel to AC and thus parallel to BD.

Now, EQ and EP are both parallel to AC and BD and have a point E in common; therefore, EQP is a straight line.

Δs AEQ and ABD are similar ($\because EQ \,||\, BD$).

$$\therefore \quad \frac{EQ}{BD} = \frac{AE}{AB} \quad \text{or} \quad EQ = BD \times \frac{AE}{AB} \qquad \text{(ii)}$$

Δs BEP and BAC are similar ($\because EP \,||\, AC$).

$$\therefore \quad \frac{EP}{AC} = \frac{BE}{BA} \quad \text{or} \quad EP = AC \times \frac{BE}{AB} \qquad \text{(iii)}$$

Δs EQQ' and $EP'P$ are similar, because $\angle QEQ'$ or $\angle PEP'$ is common and $\angle EQQ' = \angle EP'P = 90°$.

$$\therefore \quad \frac{EQ}{EP'} = \frac{EQ'}{EP}$$

or

$$EQ' \times EP' = EQ \times EP$$

$$= \left(BD \times \frac{AE}{AB}\right)\left(AC \times \frac{BE}{AB}\right) \quad \text{[from (ii) and (iii)]}$$

or

$$EP' = \frac{AE \times BE}{(EQ')(AB)^2}[(BD)(AC)]$$

$$= \frac{AE \times BE}{(EQ')(AB)^2}[(BF + FD)(BF - FD)]$$

$$= \frac{AE \times BE}{(EQ')(AB)^2}[(BF)^2 - (FD)^2]$$

$$= \frac{AE \times BE}{(EQ')(AB)^2}\left[\{(BC)^2 - (CF)^2\} - \{(CD)^2 - (CF)^2\}\right]$$

$$= \frac{AE \times BE}{(EQ')(AB)^2}\left[(BC)^2 - (CD)^2\right]$$

$=$ constant, as all the parameters are fixed.

Thus EP' is always constant. Therefore, the projection of P on EO produced is always the same point or P moves in a straight line perpendicular to EO.

Example 6.2 A circle with EQ' as diameter has a point Q on its circumference. P is a point on EQ produced such that if Q turns about E, $EQ \cdot EP$ is constant. Prove that the point P moves in a straight line perpendicular to EQ'.

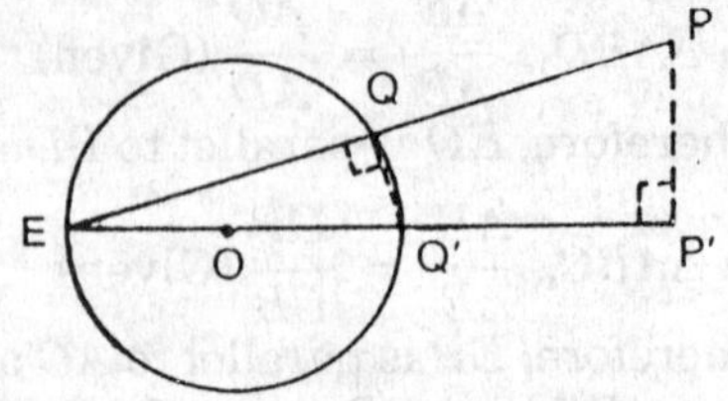

Fig. 6.5

Solution: Let PP' be perpendicular to EQ' produced (Fig. 6.5).

For any position of Q on the circumference of the circle with diameter EQ', Δs EQQ' and $EP'P$ are similar ($\angle QEQ'$ is common and $\angle EQQ' = \angle EP'P = 90°$).

$$\therefore \quad \frac{EQ}{EQ'} = \frac{EP'}{EP}$$

or
$$EQ' \cdot EP' = EQ \cdot EP$$

or
$$EP' = \frac{EQ \cdot EP}{EQ'}$$

= constant, as EQ' is fixed and $EQ \cdot EP$ = Constant

Thus EP' will be constant for all positions of Q. Therefore, the location of P' is fixed which means that P moves in a straight line perpendicular to EQ'.

3. Scott-Russel Mechanism

A Scott-Russel mechanism consists of three movable links; OQ, PS and slider S which moves along OS. OQ is the crank (Fig. 6.6). The links are connected in such a way that

$$QO = QP = QS$$

It can be proved that P moves in a straight line perpendicular to OS as the slider S moves along OS.

As $QO = QP = QS$, a circle can be drawn passing through O, P and S with PS as the diameter and Q as the centre.

Now, O lies on the circumference of the circle, $\angle POS$ is a right angle. This is true for all the positions of S and is possible only if P moves in a straight line perpendicular to OS at O.

Note that in such a mechanism, the path of P is through the joint O which is not desired. This can be avoided if the links are proportioned in a way that QS is the mean proportional between OQ and QP, i.e.

$$\frac{OQ}{QS} = \frac{QS}{QP}$$

However, in this case P will approximately traverse a straight line perpen-

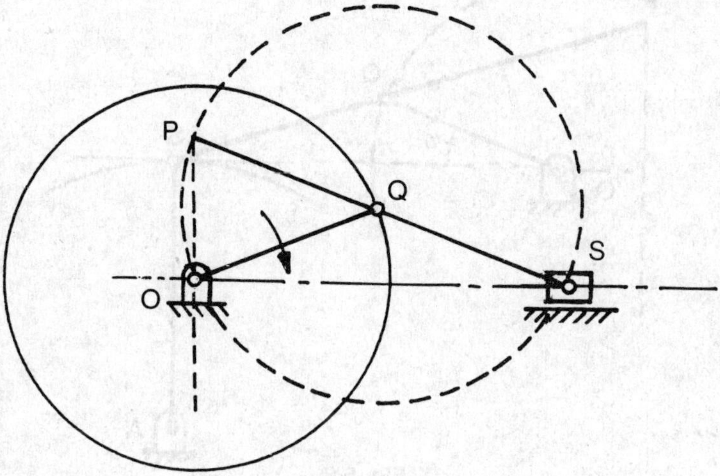

Fig. 6.6

dicular to OS and that also for small movements of S or for small values of angle θ (Fig. 6.7). A mathematical proof of this, being not simple, is omitted here. However, by drawing the mechanism in a number of positions, the fact can be verified.

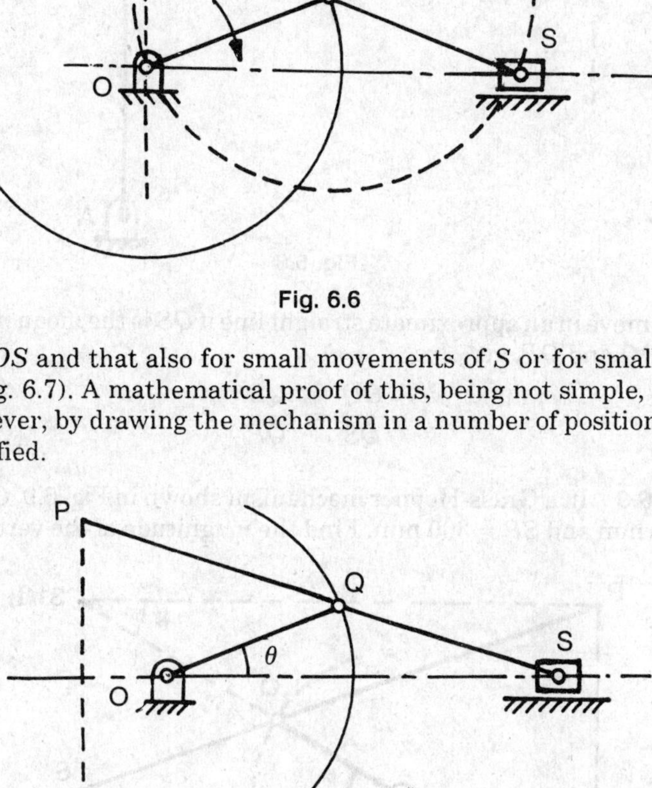

Fig. 6.7

Usually, this is known as the *modified Scott-Russel mechanism.*

4. Grass-Hopper Mechanism

This mechanism is a derivation of the modified Scott-Russel mechanism in which the sliding pair at S is replaced by a turning pair. This is achieved by replacing the slider with a link AS perpendicular to OS in the mean position. AS is pin-jointed at A (Fig. 6.8).

If the length AS is large enough, S moves in an approximated straight line perpendicular to AS (or in line with OS) for small angular movements. P

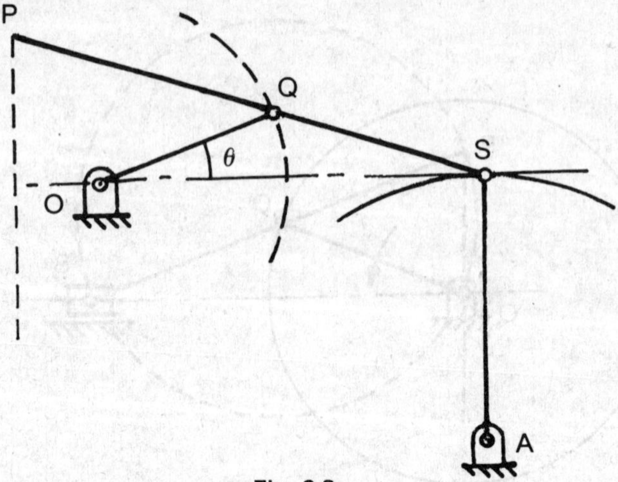

Fig. 6.8

again will move in an approximate straight line if QS is the mean proportional between OQ and QP, i.e.

$$\frac{OQ}{QS} = \frac{QS}{QP}$$

Example 6.3 In a Grass-Hopper mechanism shown in Fig. 6.9, OQ = 8 mm, SQ = 120 mm and SP = 300 mm. Find the magnitude of the vertical force at

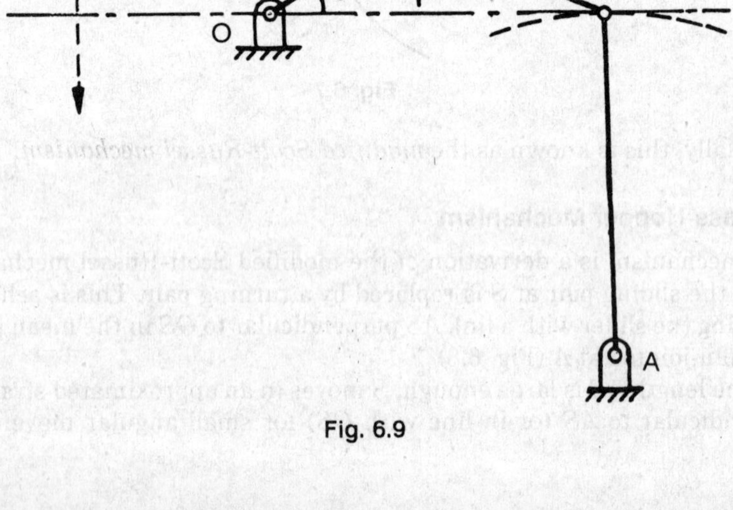

Fig. 6.9

P necessary to resist a torque of 100 N.m applied to the link OQ when it makes angles of 0°, 10° and 20° with the horizontal.

Solution: $OQ = 80$ mm

$$QS = 120 \text{ mm}$$
$$QP = 300 - 120 = 180 \text{ mm}$$
$$\frac{OQ}{QS} = \frac{QS}{QP}, \text{ i.e. } \frac{80}{120} = \frac{120}{180}$$

As the condition for the dimensions of the Grass-Hopper mechanism is satisfied, P moves in an approximate straight line for small angles of OQ with the horizontal.

Now $$F_p \times v_p = T_q \times \omega_q$$

$$F_p = \frac{T_q\,\omega_q}{v_p} = \frac{T_q}{v_p}\frac{v_q}{OQ} \quad \text{(i)}$$

Locate the I-centre (instantaneous centre) of link SP. It is at 31 as the directions of motions of points P and Q on it are known.

$$\frac{v_q}{v_p} = \frac{IQ}{IP} = \frac{OQ}{OS} \quad (\because \Delta s\ IQP \text{ and } OQS \text{ are similar})$$

(i) becomes, $F_p = \dfrac{T_q}{OS}$

When $\varphi = 0°$, $OS = 80 + 120 = 200$ mm

$$F_p = \frac{100}{0.2} = 500\,N$$

When $\varphi = 10°$,

$$OS = 80 \cos 10° + \sqrt{(120)^2 - (80 \sin 10°)^2}$$
$$= 198 \text{ mm}$$
$$F_p = \frac{100}{0.198} = \underline{505.05 \text{ N}}$$

When $\varphi = 20°$,

$$OS = 80 \cos 20° + \sqrt{(120)^2 - (80 \sin 20°)^2}$$
$$= 192 \text{ mm}$$
$$F_p = \frac{100}{0.192} = \underline{520.8 \text{ N}}$$

As the angle φ increases, P moves in only approximate straight line and thus the calculations for F_p are not exact.

5. Watt Mechanism

It is a very simple mechanism. It has four links OQ, OA, QB and AB. OQ is a fixed link. Links OA and QB can oscillate about centres O and Q respectively. It is seen that if P is a point on the link AB such that $PA/PB = QB/OA$, then for small oscillations of OA and QB, P will trace an approximate straight line. This has been shown in Fig. 6.10 for three positions.

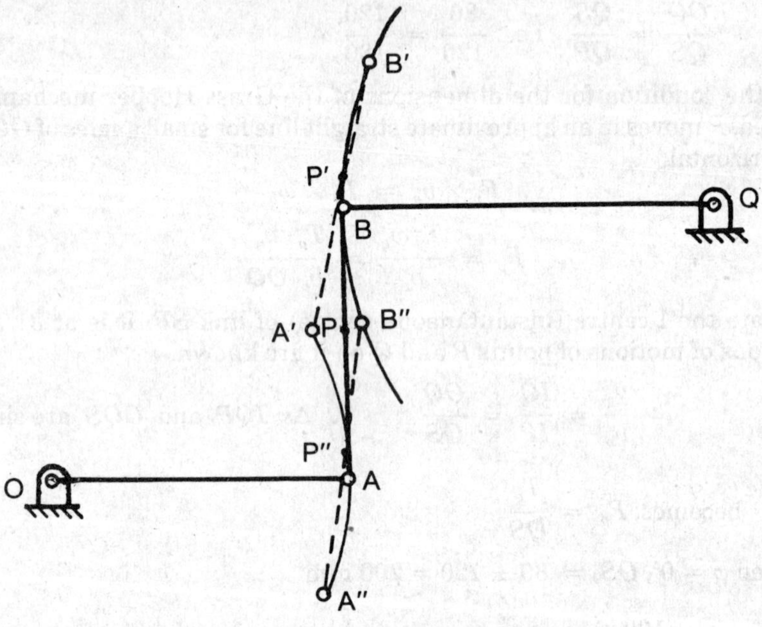

Fig. 6.10

In earlier times, the mechanism was used by Watt to guide the piston, as it was difficult to machine plane surfaces.

6. Tchebicheff Mechanism

It consists of four links OA, QB, AB and OQ (fixed) as shown in Fig. 6.11. The links OA and QB are equal and crossed. P, the mid-point of AB, is the tracing point. The proportions of the links are taken in such a way that P, A and B lie on vertical lines when on extreme positions i.e. when directly above O or Q.

Let $AB = 1$ unit

$OA = QB = x$ units

and $OQ = y$ units

When AB is on the extreme left position, A and B assume the positions A' and B', respectively.

In $\Delta OQB'$,

$$(QB')^2 - (OQ)^2 = (OB')^2$$

$$(QB')^2 - (OQ)^2 = (OB')^2$$

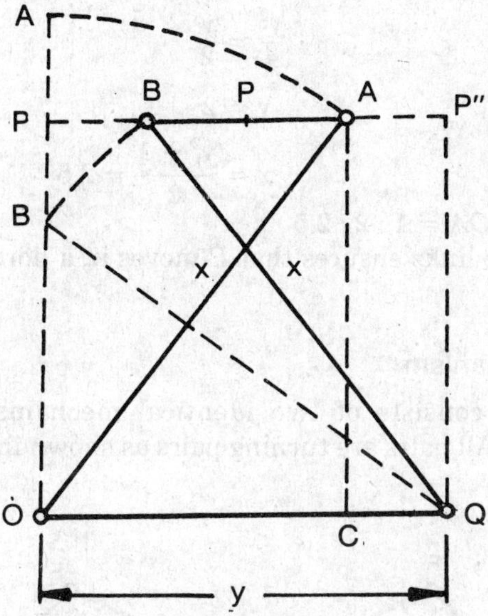

Fig. 6.11

$$x^2 - xy^2 = (OA' - A'B')^2 = (x - 1)^2 = x^2 - 2x + 1$$

or
$$2x - 1 = y^2$$

$$x = \frac{y^2 + 1}{2} \qquad \text{(i)}$$

In ΔOAC,

$$(OA)^2 - (AC)^2 = (OC)^2$$

$$(OA)^2 - (OP')^2 = (OQ - CQ)^2$$

$$(OA)^2 - (OA' - A'P')^2 = \left[OQ - (PP'' - AP)\right]^2$$

$$x^2 - \left(x - \frac{1}{2}\right)^2 = \left[y - \left(\frac{y}{2} - \frac{1}{2}\right)\right]^2$$

$$= \left(\frac{y}{2} + \frac{1}{2}\right)^2$$

or
$$x^2 - (x^2 + \frac{1}{4} - x) = \frac{y^2}{4} + \frac{1}{4} + \frac{y}{2}$$

$$x = \frac{y^2}{4} + \frac{y}{2} + \frac{1}{2} \qquad \text{(ii)}$$

From Eqs (i) and (ii),

$$\frac{y^2}{2} + \frac{1}{2} = \frac{y^2}{4} + \frac{y}{2} + \frac{1}{2}$$

$$\frac{y^2}{4} = \frac{y}{2}$$

or $$y = 2$$

and $$x = \frac{y^2 + 1}{2} = 2.5$$

Thus $AB : OQ : OA = 1 : 2 : 2.5$

This ratio of the links ensures that P moves in a horizontal straight line parallel to OQ.

7. Kempe's Mechanism

This mechanism consists of two identical mechanisms $ABCDEF$ and $A'\,B'\,C'\,D'\,E'\,F'$. All pairs are turning pairs as shown in Fig. 6.12.

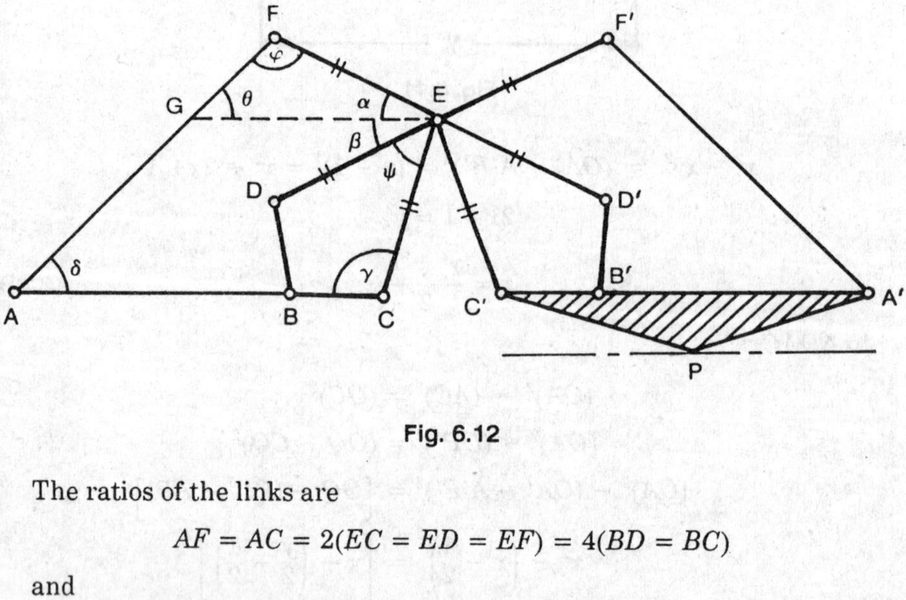

Fig. 6.12

The ratios of the links are

$$AF = AC = 2(EC = ED = EF) = 4(BD = BC)$$

and

$$A'F' = A'C' = 2(EC' = ED' = EF') = 4(B'D' = B'C')$$

Links DEF' and $D'EF$ are rigid links having turning pairs at E and at the ends.

It can be proved that if ABC is a fixed horizontal link, $A'B'C'$ also remains horizontal (in line with ABC) and thus any point on the link such as P will move in a horizontal straight line.

Quadrilaterals $ACEF$ and $EDBC$ are similar because,

$$\frac{AC}{ED} = \frac{CE}{DB} = \frac{EF}{BC} = \frac{FA}{CE} = 2$$

and

$$\angle\varphi = \angle\gamma \quad \text{(angles between corresponding sides)}$$

(∴ in quadrilateral $ACEF$, $AC = AF$ and $EC = EF$)

∴ $$\angle\delta = \angle\psi \quad \text{(corresponding angles of two quadrilaterals)}$$

Draw $EG \parallel CA$

In $\Delta\, EFG$,

$$\angle\alpha + \angle\varphi + \angle\theta = \pi$$

or

$$\angle\alpha = \pi - \angle\varphi - \angle\theta$$

$$= \pi - \angle\varphi - \angle\delta \qquad (\because EG \parallel CA)$$

$$= \pi - \angle\gamma - \angle\psi$$

But as CE cuts EG and CA, two parallel lines,

$$\angle\gamma + \angle\psi + \angle\beta = \pi$$

or

$$\angle\gamma = \pi - \angle\psi - \angle\beta$$

∴

$$\angle\alpha = \pi - (\pi - \angle\psi - \angle\beta) - \angle\psi$$

$$= \angle\beta$$

Thus for all configurations, $\angle\alpha = \angle\beta$, i.e. inclination of ED and EF is same to EG or CA and the two identical parts of the mechanism always remain symmetrical.

Hence, if ABC is a fixed horizontal link, $A'B'C'$ also remains horizontal in line with ABC and any point P on it traces a horizontal path.

8. Parallel Linkages

If the opposite links of a four-link mechanism are made equal, the linkage will always form a parallelogram. The following types of parallel linkages are used universely:

1. Parallel Ruler As shown in Fig. 6.13, in a parallel ruler, all the horizontal links have the same length, i.e.

$$AB = CD = EF = GH = IJ$$

The lengths of the opposite links of each parallelogram should also be equal i.e.

$$AC = BD,\ CE = DF,\ EG = FH \text{ and } GI = HJ$$

It can be seen that any number of parallelograms can be used to form this ruler. The dimensions of the mechanism ensures that IJ moves parallel to AB.

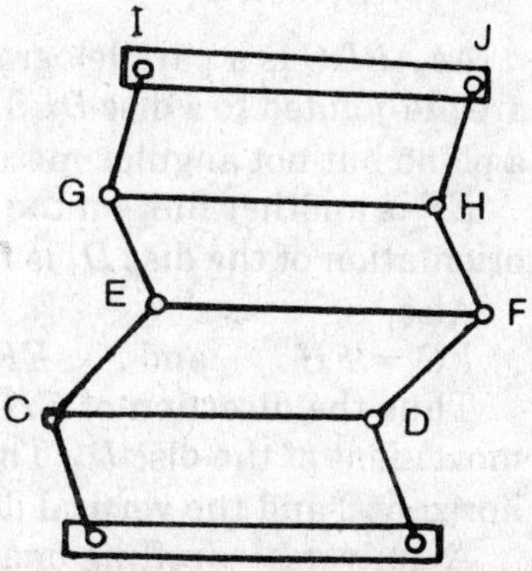

Fig. 6.13

2. Lazy Tongs In this mechanism (Fig. 6.14), O is pin-jointed and is a fixed point. The point A slides in the vertical guides while all other points are pin-jointed. All the links are of equal length. As A moves vertically, P will move in a horizontal line. The use of such a mechanism can be made in supporting a bulb (of a table lamp) or telephone etc.

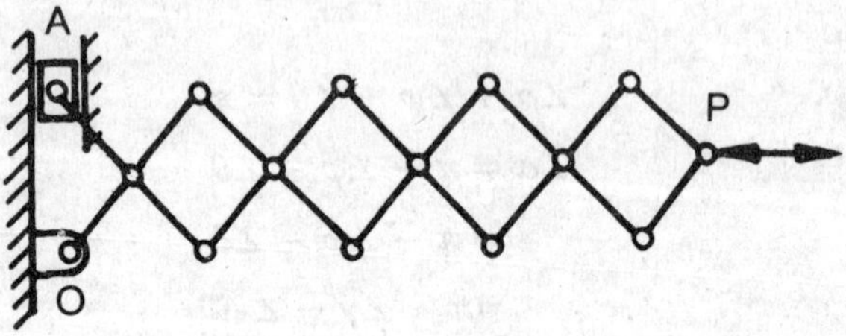

Fig. 6.14

3. Universal Drafting Machine In such a mechanism (Fig. 6.15), two parallelograms of the links are formed.

$AB = CD$ and $AC = BD$

The link AB is fixed.

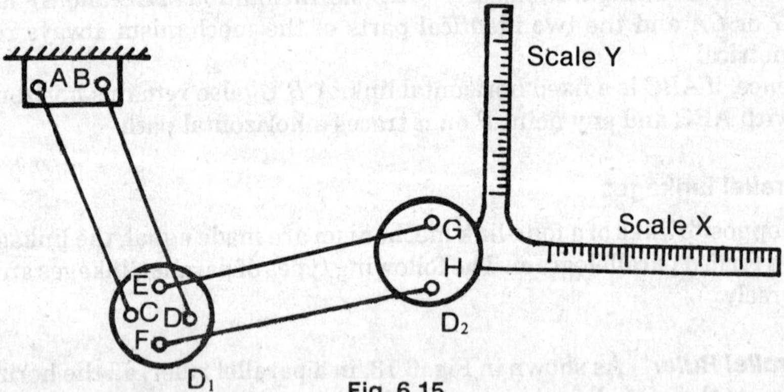

Fig. 6.15

As $ABDC$ is a parallelogram, CD always remains parallel to AB. C and D are pin-jointed to a disc D_1. Thus the disc D_1 can have translatory motions in a plane but not angular motions.

EF is another link on the disc D_1 pin-jointed at the ends E and F. As the orientation of the disc D_1 is fixed, the direction of EF is also fixed.

Also,

$EG = FH$ and $EF = GH$

Thus the direction of GH is always parallel to EF or there is no angular movement of the disc D_2. Therefore, scales X and Y will always be along the horizontal and the vertical directions.

A universal drafting machine is extensively used as a substitute for T-square and set-squares.

6.4 ENGINE INDICATORS

An indicator of a reciprocating engine is an instrument that keeps the graphical record of pressure inside the cylinder during the piston stroke.

An indicator consists of an indicator cylinder with a piston. The indicator cylinder is connected to the engine cylinder. Thus varying pressure of the gas or steam is communicated to the indicator piston, the displacement of which is constrained by a spring to get a direct measure of the gas or the steam pressure. The displacement is recorded by a pencil on paper, wrapped on a drum, to a suitable scale with the help of a straight line mechanism.

The following are the usual types of indicators:

1. Simplex Indicator

This indicator employs the mechanism of a pantograph. As shown in Fig. 6.16, O is the fixed pivot whereas $ABCD$ is a parallelogram formed by the four links. R is a point on the link CB produced to trace the path of A or P (movement of piston). Also refer to Fig. 6.1.

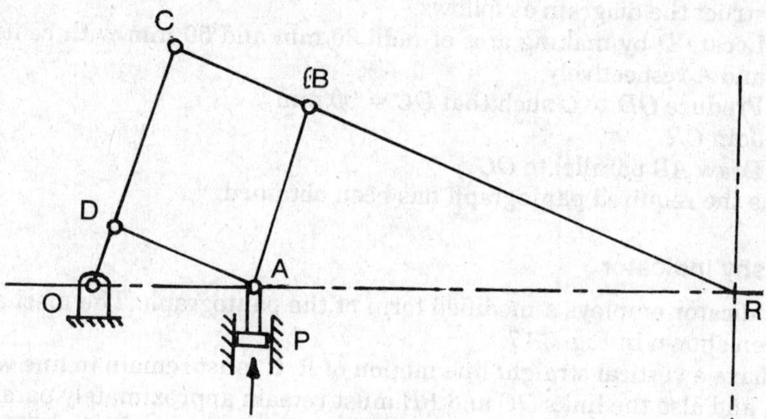

Fig. 6.16

P moves in a vertical straight line within the guides or the indicator cylinder. Its movement is controlled by the steam or the gas pressure to be measured. Thus R also moves in a vertical straight line recording the variation of pressure with the help of a pencil recorder.

Example 6.4 Design a pantograph for an indicator to be used to obtain the indicator diagram of an engine. The distance between the fixed point and the tracing point is 180 mm. The indicator diagram should be three times the gas pressure inside the cylinder of the engine.

Solution:

Refer to Fig. 6.16,

$$OR = 180 \text{ mm} \quad \text{and} \quad \frac{OR}{OA} = 3 \quad \text{(given)}$$

or
$$\frac{180}{OA} = 3$$

or
$$OA = 60 \text{ mm}$$

The relationship of the different arms of a simplex indicator is as follows:

$$\frac{OR}{OA} = \frac{OC}{OD} = \frac{CR}{CB} = 3$$

Choose convenient dimensions of OD and DA. Let these be 30 mm and 50 mm respectively.

Thus as $ABCD$ is to be a parallelogram and the above relation is to be fulfilled, the other dimensions will be

$$OC = 30 \times 3 = 90 \text{ mm}$$

$$CR = 50 \times 3 = 150 \text{ mm}$$

Construct the diagram as follows:

1. Locate D by making arcs of radii 30 mm and 50 mm with centres O and A respectively.
2. Produce OD to C such that OC = 90 mm.
3. Join CR.
4. Draw AB parallel to OC.

Thus the required pantograph has been obtained.

2. Crosby Indicator

This indicator employs a modified form of the pantograph. The mechanism has been shown in Fig. 6.17.

To have a vertical straight line motion of R, it must remain in line with O and P, and also the links OC and PB must remain approximately parallel.

As P lies on the link 3 and R on 5, locate the I-centres 31 and 51. If the directions of velocities of any two points on a link are known, the I-centre can be located easily which is the intersection of the perpendiculars to the directions of velocities at the two points.

First locate 31 as the directions of velocities of P and E on the link 3 are known.

The direction of velocity of P is vertical.

Therefore 31 lies on a horizontal line through P.

The direction of velocity of E is $\perp$ to QE.

Therefore 31 lies of QE (or QE produced).

The intersection of QE produced with the horizontal line through P locates the point 31.

Thus the link 3 has its centre of rotation at 31 (link 1 is fixed) and the velocity of any point on the link is proportional to its distance from 31, the direction being perpendicular to a line joining the point with the I-centre.

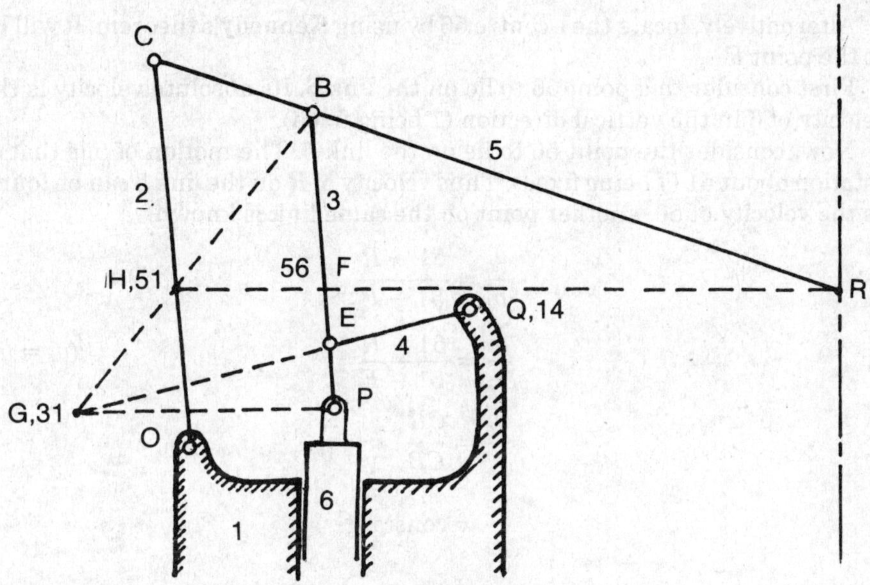

Fig. 6.17

To locate 51, the directions of velocities of *B* and *C* are known.

The direction of velocity of *B* is ⊥ to 31 - B.

Therefore 51 lies on 31-B.

The direction of velocity of *C* is ⊥ to *OC*.

Therefore 51 lies on *OC*.

Thus 51 can be located.

Now, the link 5 has its centre of rotation at 51. The direction of velocity of point *R* on this link will be perpendicular to 51-*R*. To have a vertical motion of *R*, it must lie on a horizontal line through 51.

The ratio of the velocities of *R* and *P* is given by,

$$\frac{v_r}{v_p} = \frac{v_r}{v_b}\frac{v_b}{v_p} \qquad (B \text{ is common to 3 and 5})$$

$$= \frac{51 - R}{51 - B}\,\frac{31 - B}{31 - P}$$

$$= \frac{51 - R}{51 - B}\,\frac{51 - B}{51 - F} \qquad (\because \Delta\text{s } BPG \text{ and } BFH \text{ are similar})$$

$$= \frac{51 - R}{51 - F}$$

$$= \frac{CR}{CB} \qquad (\because \Delta\text{s } CRH \text{ and } BRF \text{ are similar})$$

$$= \text{constant}$$

This shows that the velocity or the displacement of *R* will be proportional to that of *P*.

Alternatively, locate the I-centre 56 by using Kennedy's theorem. It will be at the point F.

First consider this point 56 to lie on the link 6. Its absolute velocity is the velocity of 6 in the vertical direction (1 being fixed).

Now, consider the point 56 to lie on the link 5. The motion of 5 is that of rotation about 51 (1 being fixed). Thus velocity of R on the link 5 can be found as the velocity of 56, another point on the same link is known.

$$\frac{v_r}{v_f} = \frac{51 - R}{51 - F}$$

or
$$\frac{v_r}{v_p} = \frac{51 - R}{51 - F} \qquad (v_f = v_p)$$

$$\frac{v_r}{v_p} = \frac{CR}{CB}$$

$$\frac{v_r}{v_p} = \text{constant}$$

3. Thomson Indicator

A Thomson indicator employs a Grass-hopper mechanism $OCEQ$. R is the tracing point which lies on CE produced as shown in Fig. 6.18.

The best position of the tracing point R is obtained as discussed below.

Locate the I-centres 31 and 51 as in case of a Crosby indicator. The directions of velocities of two points C and E on the link 5 are known; therefore, first locate the I-centre 51.

The direction of velocity of C is ⊥ to OC.

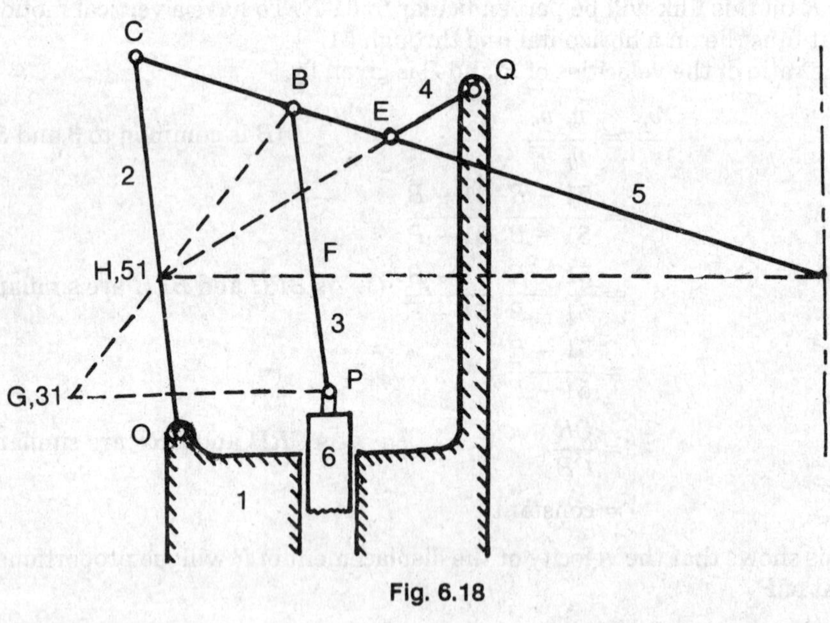

Fig. 6.18

Therefore 51 lies on OC.

The direction of velocity of E is $\perp$ to QE,

Therefore 51 lies on QE (or QE produced).

Thus 51 can be located.

Now, the directions of velocities of two points B and P on the link 3 are known.

The direction of velocity of B is $\perp$ to 51-B, (B is on the link 5 also)

Therefore 31 lies on the line 51B.

The direction of velocity of P is vertical,

Therefore 31 lies on a horizontal line through P.

Thus 31 can be located.

As R is to move in a vertical direction, it must lie on a horizontal line through the I-centre of the link 5 on which the pointer lies.

Similar to the case of a Crosby indicator, the velocity ratio is given by,

$$\frac{v_r}{v_p} = \frac{CR}{CB} = \text{Constant}$$

Therefore, the velocity or the displacement of R is proportional to that of P.

It is to be remembered that since OC and PB does not remain parallel for all positions, R moves in an approximate vertical line. However, the variations are negligible.

4. Dobbie McInnes Indicator This indicator is similar to a Thomson indicator, the difference being that the link 3 is pivoted to a point in link 4 instead of a point on link 5. Thus the motion of the indicator piston is imparted to the link 4.

The indicator is shown in Fig.6.19.

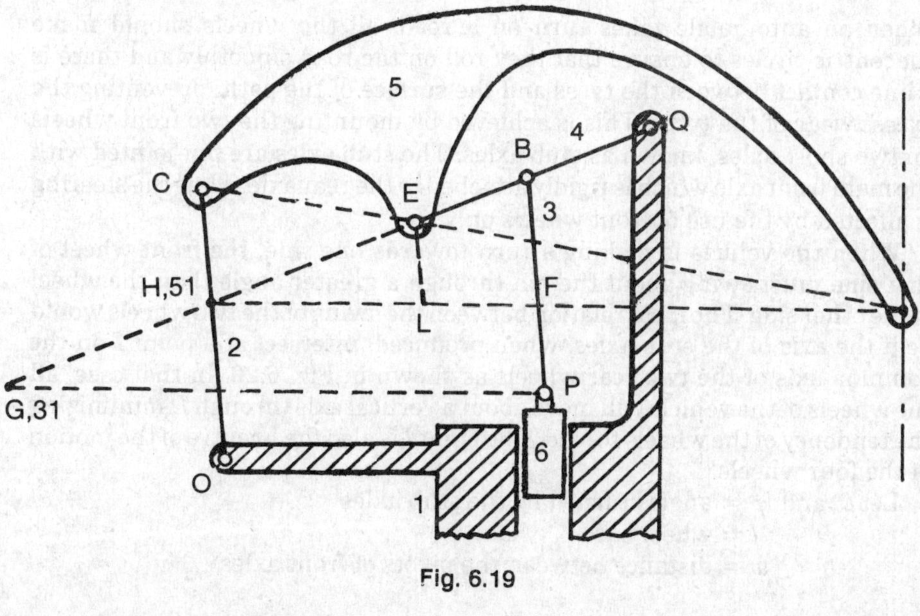

Fig. 6.19

Locate the I-centre 31 as before.

Locate R by finding the intersection of CE and a horizontal line through 51.

Locate I-centre 31 as usual.

Now

$$\frac{v_r}{v_p} = \frac{v_r}{v_e} \times \frac{v_e}{v_b} \times \frac{v_b}{v_p}$$

$$= \frac{51-R}{51-E}\,\frac{QE}{QB}\,\frac{31-B}{31-P}$$

$$= \frac{51-R}{51-E}\,\frac{QE}{QB}\,\frac{51-B}{51-F} \quad (\because \Delta s\ BPG \text{ and } BFH \text{ are similar})$$

$$= \frac{51-R}{51-E}\,\frac{QE}{QB}\,\frac{51-E}{51-L} \quad (\because \Delta s\ BFH \text{ and } ELH \text{ are similar})$$

$$= \frac{51-R}{51-L}\,\frac{QE}{QB}$$

$$= \frac{CR}{CE}\,\frac{QE}{QB}$$

$$= \text{constant}$$

This expression also gives approximately the ratio of the displacement of R to that of P.

6.5 AUTOMOBILE STEERING GEARS

When an automobile takes turn on a road, all the wheels should make concentric circles to ensure that they roll on the road smoothly and there is a line contact between the tyres and the surface of the path, preventing the excess wear of the tyres. This is achieved by mounting the two front wheels on two short axles, known as stub axles. The stub axles are pin-jointed with the main front axle which is rigidly attached to the rear axle. Thus the steering is affected by the use of front wheels only.

When the vehicle is making a turn towards one side, the front wheel of that side must swing about the pin through a greater angle than the wheel of the other side. The ideal relation between the swing of the two wheels would be if the axis of the stub axles, when produced, intersect at a point I on the common axis of the two rear wheels as shown in Fig. 6.20. In that case, all the wheels of the vehicle will move about a vertical axis through I, minimizing the tendency of the wheels to skid. The point I is also the I-centre of the motion of the four wheels.

Let θ and φ = angles turned by the stub axles

l = wheel base

w = distance between the pivots of front axles

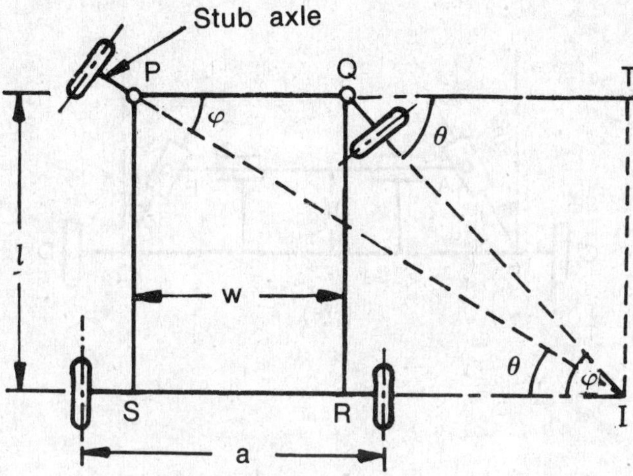

Fig. 6.20

Then,

$$\cot \varphi = \frac{PT}{TI} \quad \text{and} \quad \cot \theta = \frac{QT}{TI}$$

$$\cot \varphi - \cot \theta = \frac{PT - QT}{TI} = \frac{PQ}{TI} = \frac{w}{l} \tag{6.1}$$

This is known as the fundamental equation of correct gearing. Mechanisms that fulfil this fundamental equation are known as steering gears.

6.6 TYPES OF STEERING GEARS

There are two main types of steering gears:

1. the Davis steering Gear, and
2. the Ackermann steering gear

A *Davis steering gear* has sliding pairs which means more friction and easy wearing. The gear fulfils the fundamental equation of gearing in all the positions. However, due to easy wearing it becomes inaccurate after some time.

An *Ackermann steering gear* has only turning pairs and thus is preferred. Its drawback is that it fulfils the fundamental equation of correct gearing at the middle and the two extreme position and not in all positions.

Davis Steering Gear

A Davis steering gear shown in Fig. 6.21a consists of two arms *PK* and *QL* fixed to the stub axles *PC* and *QD* to form two similar bell-crank levers *CPK* and *DQL* pivoted at *P* and *Q* respectively. A cross link *AB*, constrained to slide parallel to *PQ*, is pin-jointed at its ends to two sliders. The sliders S_1 and S_2 are free to slide on the links *PK* and *QL* respectively.

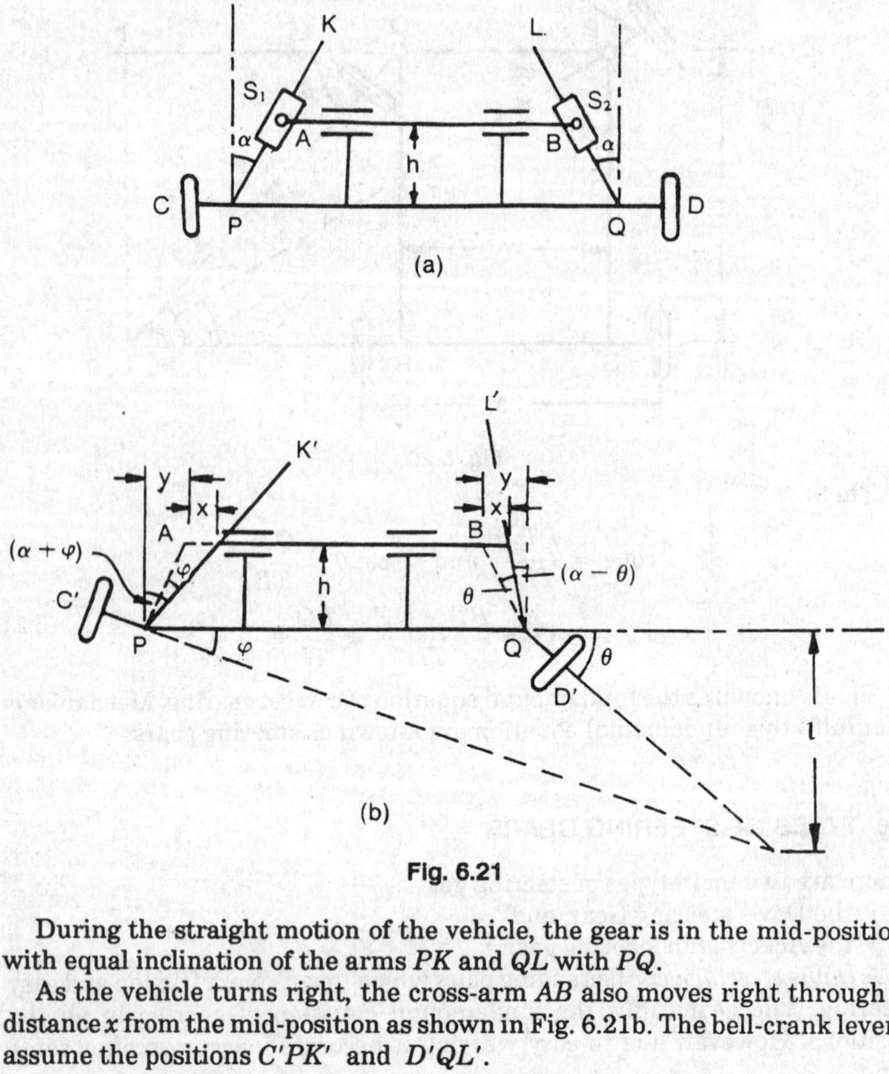

Fig. 6.21

During the straight motion of the vehicle, the gear is in the mid-position with equal inclination of the arms *PK* and *QL* with *PQ*.

As the vehicle turns right, the cross-arm *AB* also moves right through a distance *x* from the mid-position as shown in Fig. 6.21b. The bell-crank levers assume the positions *C′PK′* and *D′QL′*.

Let h = vertical distance between *AB* and *PQ*

$$\tan(\alpha - \theta) = \frac{y-x}{h}$$

$$\frac{\tan\alpha - \tan\theta}{1 + \tan\alpha\tan\theta} = \frac{y-x}{h}$$

$$\frac{\frac{y}{h} - \tan\theta}{1 + \frac{y}{h}\tan\theta} = \frac{y-x}{h} \qquad \left(\tan\alpha = \frac{y}{h}\right)$$

$$\frac{y - h\tan\theta}{xh + y\tan\theta} = \frac{y - x}{h}$$

$$(y - h\tan\theta)h = (h + y\tan\theta)(y - x)$$

$$yh - h^2\tan\theta = hy - hx + y^2\tan\theta - xy\tan\theta$$

$$hx = (y^2 - xy + h^2)\tan\theta$$

$$\tan\theta = \frac{hx}{y^2 - xy + h^2}$$

Also, $\tan(\alpha + \varphi) = \dfrac{y + x}{h}$

and it can be proved that $\tan\varphi = \dfrac{hx}{y^2 + xy + h^2}$

For correct steering action, $\cot\varphi - \cot\theta = \dfrac{w}{l}$

or

$$\frac{y^2 + xy + h^2}{hx} - \frac{y^2 - xy + h^2}{hx} = \frac{w}{l}$$

or

$$\frac{2xy}{hx} = \frac{w}{l}$$

or

$$\frac{y}{h} = \frac{w}{2l}$$

or

$$\tan\alpha = \frac{w}{2l} \tag{6.2}$$

Usual value of w/l is between 0.4 to 0.5 and that of α from 11 to 14 degrees.

Example 6.5 The distance between the steering pivots of a Davis steering gear is 1.3 m. The wheel base is 2.75 m. What will be the inclination of the track arms to the longitudinal axis of the vehicle if it is moving in a straight path?

Solution:

$$w = 1.3 \text{ m} \qquad l = 2.75 \text{ m}$$

$$\tan\alpha = \frac{w}{21} = \frac{1.3}{2 \times 2.75} = 0.236$$

$\therefore$ $$\alpha = 13.3° \text{ or } 13°18'$$

Ackermann Steering Gear

This steering gear consists of a four-link mechanism *PABQ* having four-turning pairs.

As shown in Fig. 6.22, two equal arms *PA* and *QB* are fixed to the stub axles *PC* and *QD* to form two similar bell-crank levers *CPA* and *DQB* pivoted at *P* and *Q* respectively. A cross link *AB* is pin-jointed at the ends to the two bell-crank levers.

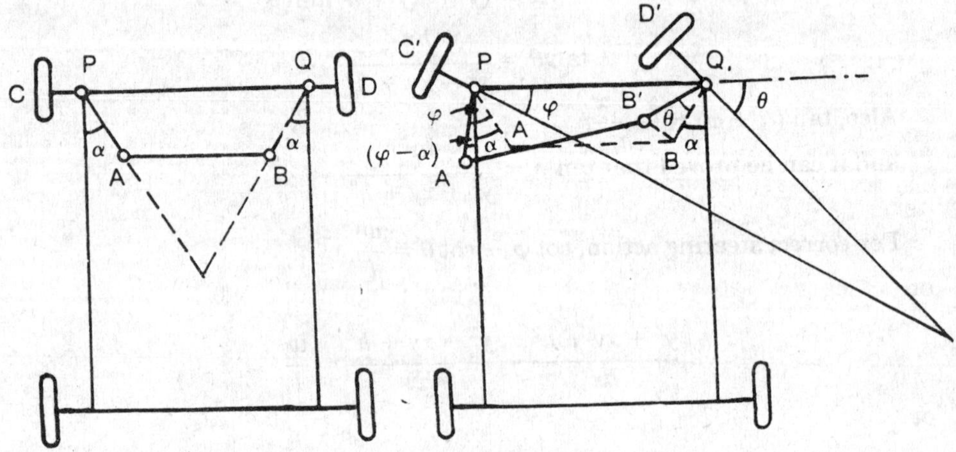

Fig. 6.22

During the straight motion of the vehicle, the gear is in the mid-position with equal inclination of the arms *PA* and *QB* with *PQ*. The cross link *AB* is parallel to *PQ* in this position.

An Ackermann gear does not fulfil the fundamental equation of correct gearing in all the positions but only in three positions. If the values of *PA*, *PQ* and the angle α are known, the mechanism can be drawn to a suitable scale in different positions. The angle φ can be noted for different values of θ. This angle φ found by drawing the gear may be termed as φ_a (φ actual).

Correct or theoretical values of φ corresponding to different values of θ, for the given values of w and l can be calculated from the relation for correct gearing, $\cot\varphi - \cot\theta = w/l$. The angle so obtained may be termed φ_t (φ theoretical).

Comparing φ_a and φ_t, following observations are made:

1. For small values of θ, φ_a is marginally higher than φ_t
2. For larger values of θ, φ_a is lower than φ_t and the difference is substantial.

Thus for larger values of θ or when the vehicle is taking a sharp turn, the wear of the tyres can be more due to slipping. However, to take sharp turns, the vehicle has to be slowed down, which reduces the wear of the tyres. Thus the large difference between φ_a and φ_t does not affect much the life of tyres.

In an Ackermann gear, the instantaneous centre *I* does not lie on the rear axis but on a line parallel to the rear axis at an approximate distance of 0.3 l above it.

Three positions of the correct gearing are

1. when the vehicle moves straight
2. when the vehicle moves at a correct angle to the right, and
3. when the vehicle moves at a correct angle to the left.

In all other positions, pure rolling is not possible due to slipping of the wheel.

Graphically, the two positions of the correct gearing are found by finding $(\cot \varphi - \cot \theta)$ at different positions. The values that give the correct values of w/l ($w/l \approx 0.45$) correspond to correct gearing.

Determination of Angle To determine the angle α, assume that the projections of AB and $A'B'$ on PQ are equal (Fig 6.22). Therefore,

$$\text{Projection of } BB' \text{ on } PQ = \text{Projection of } AA' \text{ on } PQ$$

$$QB\,[\sin(\alpha + \theta) - \sin\alpha] = PA\,[\sin\alpha + \sin(\varphi - \alpha)]$$

or

$$\sin(\alpha + \theta) - \sin\alpha = \sin\alpha + \sin(\varphi - \alpha) \qquad (PA = QB)$$

$$(\sin\alpha\cos\theta + \cos\alpha\sin\theta) - \sin\alpha = \sin\alpha + \sin\varphi\cos\alpha - \cos\varphi\sin\alpha$$

$$\sin\alpha\,(\cos\theta + \cos\varphi - 2) = \cos\alpha(\sin\varphi - \sin\theta)$$

$$\frac{\sin\alpha}{\cos\alpha} = \frac{\sin\varphi - \sin\theta}{\cos\theta + \cos\varphi - 2}$$

$$\tan\alpha = \frac{\sin\varphi - \sin\theta}{\cos\theta + \cos\varphi - 2} \qquad (6.3)$$

where θ and φ are the values of angles for the correct gearing.

6.7 HOOKE'S JOINT

A Hooke's joint (Fig 6.23), commonly known as a *universal joint*, is used to connect two non-parallel and intersecting shafts. It is also used for shafts with angular misalignment. A common application of this joint is in an automobile where it is used to transmit power from the gear box (of the engine) to the rear axle. The driving shaft rotates at a uniform angular speed whereas the driven shaft rotates at a continuously varying angular speed.

The shafts 1 and 2 rotate in the fixed bearings. A complete revolution of the either shaft will cause the other to rotate through a complete revolution in the same time, but with varying angular speed. Each shaft has a fork at its end. The four ends of the two forks are connected by a centre piece, the arms of which rests in the bearings, provided in the fork ends. The centre piece can

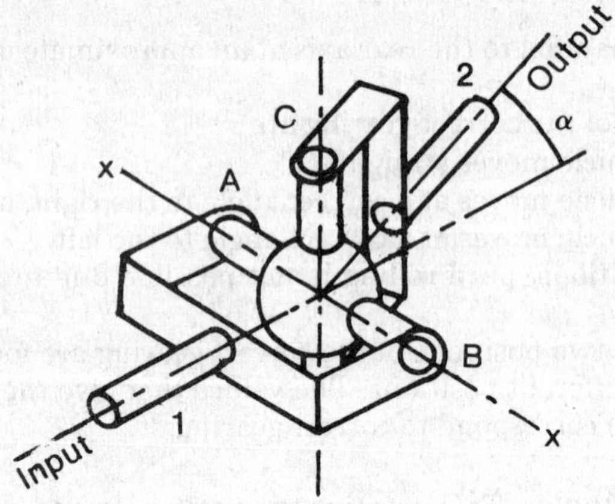

Fig. 6.23

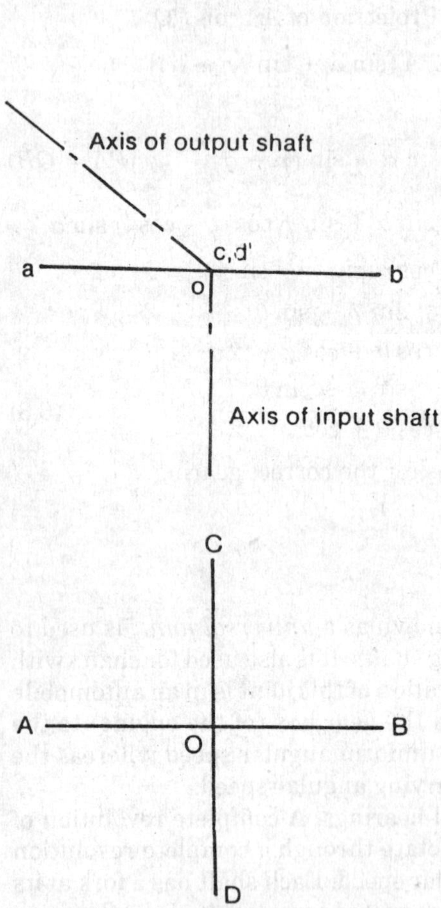

Fig. 6.24

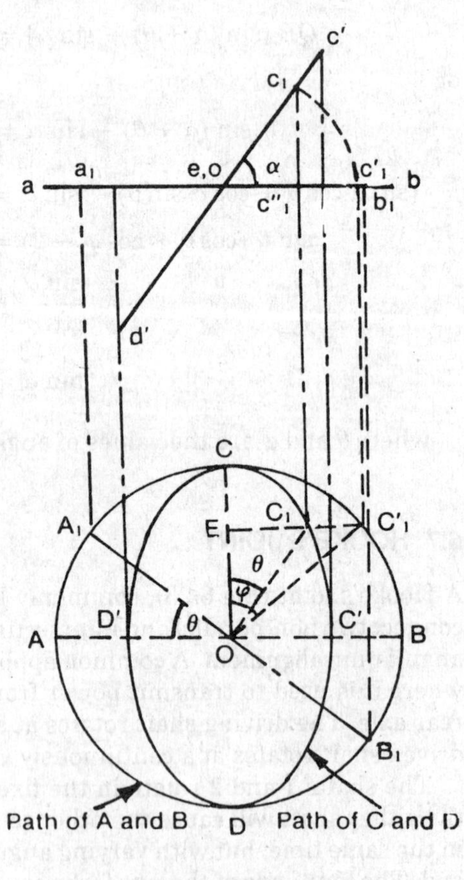

Fig. 6.25

be in the shape of a cross, square or sphere (having four pins or arms). The four arms of the cross are at right angles.

Let two horizontal shafts, the axes of which are at an angle α, be connected by a Hooke's joint. If the joint is looked along the axis of shaft 1, the fork ends of this shaft will be A and B as shown in Fig. 6.24. C and D are the positions assumed by the fork ends of shaft 2. The axis of the shaft 1 is along the perpendicular to the plane of paper at O and that of shaft 2 along OA. When looking from top, c and d, projections of C and D coincide with that of O whereas a and b remain unchanged.

As the shaft 1 is rotated, its fork ends A and B, are rotated in a circle (Fig. 6.25). However, the fork ends C and D of the shaft 2 will move along the path of an ellipse, if looked along the axis of a shaft 1. In the top view, the motion of the fork ends of the shaft 1 is along the line ab whereas that of the shaft 2 on a line $c'd'$ at an angle of α to ab.

Let the shaft 1 rotate through an angle θ so that fork ends assume the positions A_1 and B_1. Now, the angle moved by shaft 2 would also be θ when looking along the axis of shaft 1. Let the fork end C take the position C_1. However, the true angle turned by shaft 2 would be when it is looked along its own axis. In Fig. 6.26, front view of the joint is shown, when looking along the axis of shaft 2. Here C and D move in a circle. The point C_1 lies on a circle at the same height as it is on the ellipse in Fig. 6.25. This gives the true angle φ turned by the shaft 2.

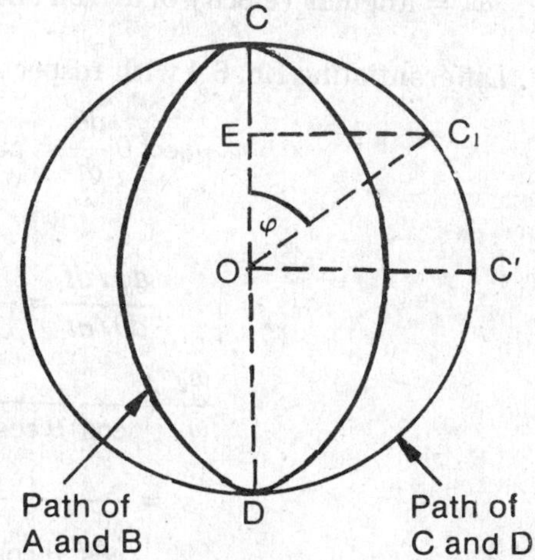

Fig. 6.26

Now,

$$\frac{\tan\varphi}{\tan\theta} = \frac{EC'_1/EO}{EC_1/EO} \qquad \text{(Fig 6.25)}$$

$$= \frac{EC'_1}{EC_1}$$

$$= \frac{ec'_1}{ec''_1} \qquad \text{(Fig 6.25 top view)}$$

$$= \frac{ec_1}{ec''_1}$$

$$= \frac{1}{ec''_1/ec_1}$$

$$= \frac{1}{\cos\alpha}$$

or

$$\tan\theta = \cos\alpha \tan\varphi \tag{6.4}$$

Angular Velocity Ratio

Let ω_1 = angular velocity of driving shaft $\left(= \dfrac{d\theta}{dt}\right)$

ω_2 = angular velocity of driven shaft $\left(= \dfrac{d\varphi}{dt}\right)$

Differentiating Eq. 6.4 with respect to time t,

$$\sec^2\theta \frac{d\theta}{dt} = \cos\alpha \sec^2\varphi \frac{d\varphi}{dt}$$

or

$$\frac{d\varphi/dt}{d\theta/dt} = \frac{\sec^2\theta}{\cos\alpha \sec^2\varphi}$$

$$\frac{\omega_2}{\omega_1} = \frac{1}{\cos^2\theta \cos\alpha\,(1 + \tan^2\varphi)}$$

$$= \frac{1}{\cos^2\theta \cos\alpha \left(1 + \dfrac{\tan^2\theta}{\cos^2\alpha}\right)} \cdots \left(\tan\varphi = \frac{\tan\theta}{\cos\alpha}\right)$$

$$= \frac{1}{\cos^2\theta \cos\alpha \left(1 + \dfrac{\sin^2\theta}{\cos^2\theta \cos^2\alpha}\right)}$$

$$= \frac{\cos^2\theta \cos^2\alpha}{\cos^2\theta \cos\alpha\,(\cos^2\theta \cos^2\alpha + \sin^2\theta)}$$

$$= \frac{\cos\alpha}{\cos^2\theta\,(1 - \sin^2\alpha) + \sin^2\theta}$$

$$= \frac{\cos\alpha}{\cos^2\theta - \cos^2\theta \sin^2\alpha + \sin^2\theta}$$

$$= \frac{\cos\alpha}{1 - \sin^2\alpha \cos^2\theta} \tag{6.5}$$

(i) $\dfrac{\omega_2}{\omega_1}$ is unity when $\dfrac{\cos\alpha}{1 - \sin^2\alpha \cos^2\theta} = 1$

or

$$\cos\alpha = 1 - \sin^2\alpha \cos^2\theta$$

or

$$\cos^2\theta = \frac{1-\cos\alpha}{\sin^2\alpha}$$

$$= \frac{1-\cos\alpha}{1-\cos^2\alpha}$$

$$= \frac{1-\cos\alpha}{(1+\cos\alpha)(1-\cos\alpha)}$$

$$= \frac{1}{1+\cos\alpha}$$

$$= \frac{\sin^2\theta+\cos^2\theta}{1+\cos\alpha}$$

$$= \frac{\cos^2\theta\left(\dfrac{\sin^2\theta}{\cos^2\theta}+1\right)}{1+\cos\alpha}$$

or

$$\frac{\sin^2\theta}{\cos^2\theta}+1 = 1+\cos\alpha$$

or

$$\tan^2\theta = \cos\alpha$$

or

$$\tan\theta = \pm\sqrt{\cos\alpha} \tag{6.6}$$

Thus $\omega_2 = \omega_1$ or the velocities of the driven and the driving shafts are equal when the condition is fulfilled. This is possible once in all the four quadrants for particular values of θ if α is constant.

(ii) $\dfrac{\omega_2}{\omega_1}$ is minimum when the denominator of Eq. (6.5) is maximum, i.e. $(1 - \sin^2\alpha\cos^2\theta)$ is maximum.

This is so when $\cos^2\theta$ is minimum,

or $\theta = 90°$ or $270°$

Then,

$$\frac{\omega_2}{\omega_1} = \cos\alpha \tag{6.7}$$

(iii) $\dfrac{\omega_2}{\omega_1}$ is maximum when the denominator of Eq. (6.5) is minimum, i.e. $(1 - \sin^2\alpha\cos^2\theta)$ is minimum,

or

$$\theta = 0° \text{ or } 180°$$

and $$\frac{\omega_2}{\omega_1} = \frac{\cos\alpha}{1 - \sin^2\alpha} = \frac{\cos\alpha}{\cos^2\alpha} = \frac{1}{\cos\alpha} \tag{6.8}$$

The variation in the speed of the driven shaft corresponding to the rotation of the driving shaft is shown in Fig 6.27. Points '*e*' correspond to the angular displacements of the driving shaft when the angular velocity of the driven shaft is equal to that of the driving shaft. Points 'min' and 'max' correspond to the angular displacements of the driving shaft when the angular speeds of the driven shaft are the minimum and the maximum respectively.

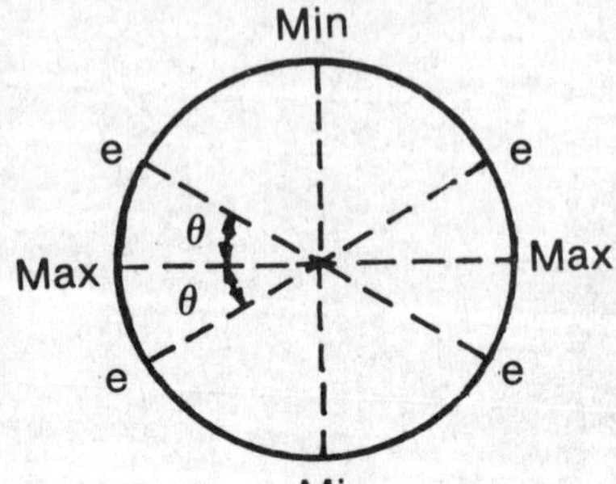

Fig. 6.27

Graphically, the variation of angular velocity of the driven shaft can be represented by an ellipse whereas that of the driving shaft by a circle (Fig 6.28). Such a diagram is known as a *polar velocity diagram.*

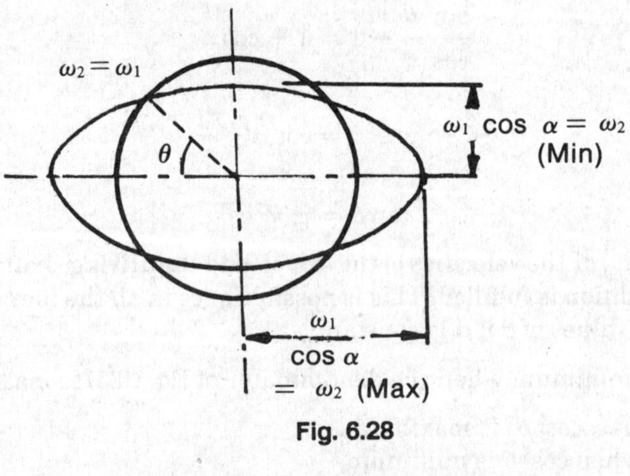

Fig. 6.28

Maximum variation of velocity of the driven shaft of its mean velocity

$$= \frac{\omega_{2\,\max} - \omega_{2\,\min}}{\omega_{\text{mean}}}$$

But ω_{mean} of the driven shaft is equal to the angular velocity ω_1 of the driving shaft as both the shafts complete one revolution in the same period of time.

$$\text{Maximum variation} = \frac{\omega_1/\cos\alpha - \omega_1\cos\alpha}{\omega_1} \tag{6.9}$$

$$= \frac{1 - \cos^2\alpha}{\cos\alpha}$$

$$= \frac{\sin^2 \alpha}{\cos \alpha}$$

$$= \tan \alpha \sin \alpha \tag{6.10}$$

If α is small, i.e. the angle between the axes of the two shafts is small, $\sin \alpha \approx \tan \alpha \approx \alpha$.

$$\text{Maximum variation} \approx \alpha^2$$

Angular Acceleration of Driven Shaft

Differentiating Eq. (6.5) with respect to time (ω_1 = constant)

$$\frac{d\omega_2}{dt} = \omega_1 \frac{d}{dt}\left(\frac{\cos \alpha}{1 - \sin^2 \alpha \cos^2 \theta}\right)$$

or
$$\text{acceleration} = \omega_1 \cdot \frac{d\theta}{dt} \frac{d}{d\theta}\left(\frac{\cos \alpha}{1 - \sin^2 \alpha \cos^2 \theta}\right)$$

$$= \omega_1^2 \cos \alpha \frac{d}{d\theta}(1 - \sin^2 \alpha \cos^2 \theta)^{-1}$$

$$= \omega_1^2 \cos \alpha (-1)(1 - \sin^2 \alpha \cos^2 \theta)^{-2} (-\sin^2 \alpha)(2 \cos \theta)(-\sin \theta)$$

$$= \frac{-\omega_1^2 \cos \alpha \sin^2 \alpha \sin 2\theta}{(1 - \sin^2 \alpha \cos^2 \theta)^2} \tag{6.11}$$

This is maximum or minimum when $\frac{d\,(acc)}{d\theta} = 0$. The resulting expression being very cumbersome, the result can be approximate to

$$\cos 2\theta \approx \frac{2 \sin^2 \alpha}{2 - \sin^2 \alpha} \tag{6.12}$$

This gives the maximum acceleration of the driven shaft corresponding to two values of θ in the second and the fourth quadrants whereas the minimum acceleration (maximum retardation) corresponds to θ values in the first and the third quadrants.

Care is to be taken to keep the angle between the two shafts to the minimum possible and not to attach excessive masses to the driven shaft. Otherwise, very high alternating stresses due to the angular acceleration and retardation will be set up in the parts of the joint, which are undesirable.

Example 6.6 The driving shaft of a Hooke's joint rotates at a uniform speed of 400 rpm. If the maximum variation in speed of the driven shaft is ± 5% of the mean speed, determine the greatest permissible angle between the axes of the shafts. What are the maximum and the minimum speeds of the driven shaft?

Solution:

$$N_1 = 400 \text{ rpm}$$

Maximum variation in speed = 0.1 ...(0.05 + 0.05 = 0.1)

or

$$\frac{1 - \cos^2 \alpha}{\cos \alpha} = 0.1$$

$$1 - \cos^2 \alpha - 0.1 \cos \alpha = 0$$

$$\cos^2 \alpha + 0.1 \cos \alpha = 1$$

$$(\cos \alpha + 0.05)^2 = 1 + (0.05)^2 = 1.0025 = (1.001\,25)^2$$

or

$$\cos \alpha = 1.001\,25 - 0.05 = 0.951\,25$$

$$\alpha = 17.96° \text{ or } 17°58'$$

$$\text{Maximum speed of driven shaft} = \frac{N_1}{\cos \alpha} = \frac{400}{0.951\,25}$$

$$= \underline{420.5 \text{ rpm}}$$

$$\text{Minimum speed of driven shaft} = N_1 \cos \alpha = 400 \times 0.951\,25$$

$$= \underline{380.5 \text{ rpm}}$$

Example 6.7 A Hooke's joint connects two shafts whose axes intersect at 25°. What will be the angle turned by the driving shaft when

(i) the velocity ratio is maximum, minimum and unity
(ii) the acceleration of the driven shaft is maximum, minimum (negative) and zero?

Solution:

(i) (a) ω_2/ω_1 is maximum at $\theta = 0°$ and 180°
(b) ω_2/ω_1 is minimum at $\theta = 90°$ and 270°
(c) ω_2/ω_1 is unity when $\tan \theta = \pm \sqrt{\cos \alpha} = \pm \sqrt{\cos 25°}$

$$= \pm\, 0.952$$

or $\theta = 43°35'$, $136°25'$, $223°35'$ and $316°25'$

(ii) Acceleration of driven shaft is maximum or minimum when

$$\cos 2\theta \approx \frac{2 \sin^2 \alpha}{2 - \sin^2 \alpha} \approx \frac{2 \sin^2 25°}{2 - \sin^2 25°} \approx 0.196$$

or $2\theta \approx 78°42'$, $(360° - 78°42')$, $(360° + 78°42')$, $(720° - 78°42')$
or $\theta \approx 39°21'$, $140°39'$, $219°21'$ and $320°39'$

Now,

$$\text{Acceleration} = \frac{-\omega_1^2 \cos\alpha \sin^2\alpha \sin 2\theta}{(1-\sin^2\alpha\cos^2\theta)^2}$$

Thus acceleration is positive when sin 2θ is negative and is negative when sin 2θ is positive.

Corresponding to four values of 2θ found above, sin 2θ will be +ve, −ve, +ve and −ve respectively.

Maximum acceleration will be at 140°39′ and 320°39′ and minimum acceleration (−ve) will be at 39°21′ and 219°21′.

Acceleration is zero when ω_2/ω_1 is maximum or minimum i.e. at 0°, 90°, 180° and 270°.

or acceleration is zero when 2θ is zero or when 2θ is 0°, 180°, 360°, 540° or when θ is 0°, 90°, 180° and 270°.

Example 6.8 The angle between the axes of two shafts joined by Hooke's joint is 25°. The driving shaft rotates at a uniform speed of 180 rpm. The driven shaft carries a steady load of 7.5 kW. Calculate the mass of the flywheel of the driven shaft if its radius of gyration is 150 mm and the output torque of the driven shaft does not vary by more than 15% of the input shaft.

Solution:

$$\alpha = 25^\circ \qquad N_1 = 180 \text{ rpm}$$

$$P = 7.5\text{ kW} \qquad \omega_1 = \frac{2\pi \times 180}{60} = 6\pi$$

$$k = 0.15\text{ m} \qquad \Delta T = 15\%$$

Maximum torque on the driven shaft will be when the acceleration is maximum, i.e. when

$$\cos 2\theta \approx \frac{2\sin^2\alpha}{2-\sin^2\alpha} \approx \frac{2\sin^2 25^\circ}{2-\sin^2 25^\circ} \approx 0.196$$

or

$$2\theta = 78^\circ 42' \text{ or } 281^\circ 18'$$

$\therefore$

$$\text{Maximum acceleration} = \frac{-\omega_1^2 \cos\alpha \sin^2\alpha \sin 2\theta}{(1-\sin^2\alpha\cos^2\theta)^2}$$

$$= \frac{-(6\pi)^2 \cos 25^\circ \sin^2 25^\circ \sin 281^\circ 18'}{(1-\sin^2 25^\circ \cos^2 140^\circ 39')}$$

$$= 63.14 \text{ rad/s}^2$$

$$P = T\omega_1$$

$$7500 = T \times 6\pi$$

$$\text{input torque, } T = 397.9 \text{ N.m}$$

$$\text{Permissible variation} = \text{torque due to acceleration of driven shaft}$$

$$397.9 \times 0.15 = I\alpha = mk^2 \alpha$$

or

$$397.9 \times 0.15 = m \times (0.15)^2 \times 63.14$$

$$m = \underline{42 \text{ kg}}$$

6.8 DOUBLE HOOKE'S JOINT

In a single Hooke's joint, the speed of the driven shaft is not uniform although the driving shaft rotates at a uniform speed. To get a uniform velocity ratio, a double Hooke's joint has to be used. In a double Hooke's joint, two universal joints and an intermediate shaft are used. If the angular misalignment between each shaft and the intermediate shaft is equal, the driving and the driven shafts remain in exact angular alignment, through the intermediate shaft rotates with the varying speed.

A single Hooke's joint was analysed assuming the axes of the two shafts and the fork of the driving shaft to be horizontal. The results showed that the speed of the driven shaft is the same after an angular displacement of 180°. Therefore, it is immaterial whether the driven shaft makes the angle α with the axis of the driving shaft to its left or its right.

Thus to have a constant velocity ratio:

– the driving and the driven shafts should make equal angles with the intermediate shaft and

– the forks of the intermediate shaft should lie in the same plane

Let γ be the angle turned by the intermediate shaft 3 while the angle turned by the driving shaft 1 and the driven shaft 2 be θ and φ respectively as before (Fig. 6.29).

Then,

$$\tan\theta = \cos\alpha \tan\gamma \qquad \text{(fork of shaft 1 horizontal)}$$

and

$$\tan\varphi = \cos\alpha \tan\gamma \qquad \text{(fork of shaft 2 horizontal)}$$

$$\therefore \qquad \theta = \varphi$$

This type of joint can be used for two intersecting shafts as well as for two parallel shafts.

However, if somehow the forks of the intermediate shafts lie in planes perpendicular to each other, the variation of speed of the driven shaft will be there.

$$\left(\frac{\omega_3}{\omega_1}\right)_{\min} = \cos\alpha \qquad \text{(fork of shaft 1 horizontal)}$$

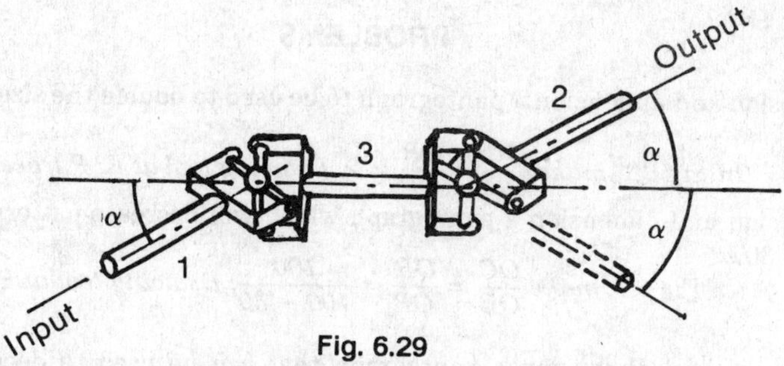

Fig. 6.29

$$\left(\frac{\omega_2}{\omega_3}\right)_{min} = \cos\alpha \qquad \text{(fork of shaft 3 horizontal)}$$

$\therefore$

$$\left(\frac{\omega_2}{\omega_1}\right)_{min} = \cos^2\alpha \tag{6.13}$$

similarly,

$$\left(\frac{\omega_2}{\omega_1}\right)_{max} = \frac{1}{\cos^2\alpha} \tag{6.14}$$

Therefore, the maximum variation (fluctuation) of speed of the driven shaft is from $\cos^2\alpha$ to $1/\cos^2\alpha$.

Example 6.9 The driving shaft of a double Hooke's joint rotates at 400 rpm. The angle of the driving and of the driven shaft with the intermediate shaft is 20°. If somehow the forks of the intermediate shaft lie in planes perpendicular to each other, determine the maximum and the minimum velocities of the driven shaft.

Solution:

$$\omega_{2min} = \omega_1 \cos^2\alpha$$

or

$$N_{2\,min} = N_1\cos^2\alpha = 400 \times \cos^2 20°$$

$$= \underline{353.2 \text{ rpm}}$$

$$N_{2max} = \frac{N_1}{\cos^2\alpha} = \frac{400}{\cos^2 20°}$$

$$= \underline{453 \text{ rpm}}$$

PROBLEMS

6.1 Design and dimension a pantograph to be used to double the size of a pattern.

(In Fig 6.1, make $\frac{OC}{OD} = \frac{OR}{OP} = 2$. *Drawing tool at R; P traces the pattern)*

6.2 Design and dimension a pantograph which will decrease pattern dimensions by 30%.

(In Fig 6.1, make $\frac{OC}{OD} = \frac{OR}{OP} = \frac{100}{100-30}$; *Drawing tool at P; R traces the pattern.)*

6.3 Design and dimension a pantograph that can be used to decrease pattern dimensions by 15%. The fixed pivot should lie between the tracing point and the marking point (tool holder).

(in Fig. 6.1, make $\frac{CD}{OD} = \frac{RP}{OP} = \frac{100}{100-15}$; *P, the fixed pivot; Drawing tool at O; R traces the pattern)*

6.4 In Fig. 6.30, the dimensions of the various links are such that

$$\frac{OA}{OB} = \frac{OE}{OF} = \frac{AC}{BD} = \frac{EC}{FD}$$

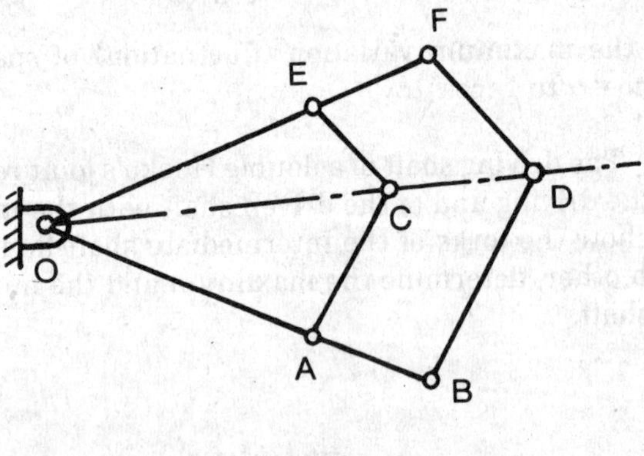

Fig. 6.30

Show that if C traces any path, then D will describe a similar path and vice-versa.

6.5 Figure 6.31 shows a straight-line Watt mechanism. Plot the path of point P and mark and measure the straight line segment of the path of P.

6.6 Figure 6.32 shows a Robert straight line motion in which $ABCD$ is a four-bar linkage. The cranks AB and DC are equal and the connecting rod BC is one-half as long as the line of centres AD. P is a point rigidly attached to the connecting rod and lying on the mid-point of AD when BC is parallel to AD. Show that the point P moves in an approximately straight line for small displacements of the cranks.

(Note: For better results take AB or DC > 0.6 AD)

6.7 In the Robert mechanism (Fig. 6.32) if $AB = BC = CD = AD/2$, locate the point P on the central vertical arm that approximately describes a straight line.

(At a length 1.3 BC below BC)

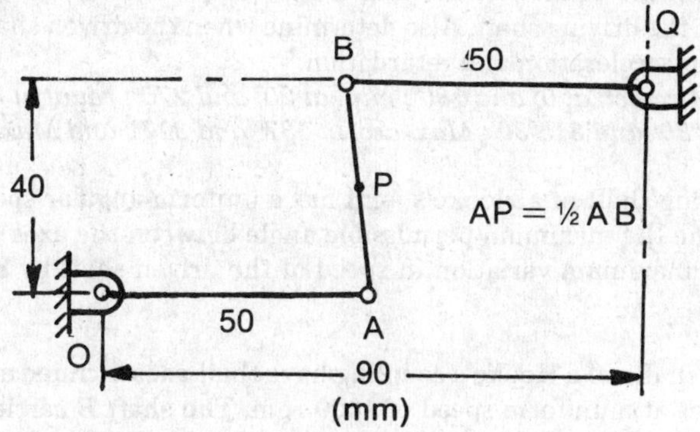

Fig. 6.31

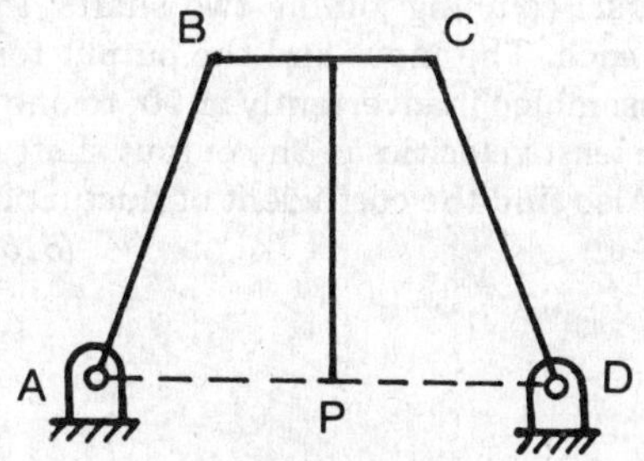

Fig. 6.32

6.8 In a Watt parallel motion (Fig 6.10), the links *OA* and *QB* are perpendicular to the link *AB* in the mean position. The lengths of the moving links are *OA* = 120 mm, *QB* = 200 mm and *AB* = 175 mm.
Locate the position of a point *P* on *AB* to trace approximately a straight line motion. Also trace the locus of *P* for all possible movements.

(AP = 109.3 mm)

6.9 In a Watt mechanism of the type shown in Fig 6.33, the links *OA* and *QB* are perpendicular to the link *AB* in the mean position. If *OA* = 45 mm, *QB* = 90 mm and *AB* = 60 mm, find the point *P* on the link *AB* produced for approximate straight line motion of point *P*.

(AP = 120 mm)

6.10 In a Davis steering gear, the length of the car between axles is 2.4 m and the steering pivots are 1.35 m apart. Determine the inclination of the track arms to the longitudinal axis of the car when the car moves in a straight path.

(15°42′)

6.11 In a Hooke's joint, the angle between the two shafts is 15°. Find the angles turned by the driving

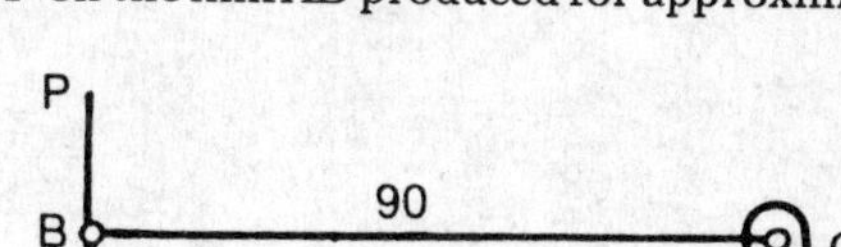

Fig. 6.33

shaft when the velocity of the driven shaft is maximum, minimum and equal to that of the driving shaft. Also determine when the driven shaft will have the maximum acceleration and retardation.

(Max. vel. at 0° and 180°; min. at 90° and 270°; equal at 44°30′, 135°30′, 224°30′ and 315°30′; Max. acc. at 137° and 317°; and Max. ret. at 43° and 223°)

6.12 The driving shaft of a Hooke's joint has a uniform angular speed of 280 rpm. Determine the maximum permissible angle between the axes of the shafts to permit a maximum variation in speed of the driven shaft by 8% of the mean speed.

(22°36′)

6.13 The two shafts of a Hooke's coupling have their axes inclined at 20°. The shaft *A* revolves at a uniform speed of 1000 rpm. The shaft *B* carries a flywheel of mass 30 kg. If the radius of gyration of the flywheel is 100 mm, find the maximum torque in shaft *B*.

(930 N.m)

6.14 In a double universal coupling joining two shafts, the intermediate shaft is inclined at 10° to each. The input and the output forks on the intermediate shaft have been assembled inadvertently at 90° to one another. Determine the maximum and the least velocities of the output shaft if the speed of the input shaft is 500 rpm. Also find the coefficient of fluctuation in speed.

(515.5 rpm; 484.9 rpm; 0.06)

7

CAMS

7.1 INTRODUCTION

A cam is a mechanical member used to impart desired motion to a follower by direct contact. The cam may be rotating or reciprocating whereas the follower may be rotating, reciprocating or oscillating. Complicated output motions which are otherwise difficult to achieve can easily be produced with the help of cams. Cams are widely used in automatic machines, internal combustion engines, machine tools, printing control mechanisms, and so on. They are manufactured usually by die-casting, milling or by punch-presses.

A cam and the follower combination belongs to the category of higher pairs. Necessary elements of a cam mechanism are:

A driver member known as the *cam*
A driven member called the *follower*
A *frame* which supports the cam and guides the follower.

7.2 TYPES OF CAMS

Cams are classified according to:

1. Shape,
2. Follower movement, and
3. Manner of constraint of the follower.

According to Shape

1. Wedge and Flat Cams A wedge cam has a wedge W which, in general, has a translational motion (Figs. 7.1a and b). The follower F can either translate (Fig. 7.1a) or oscillate (Fig. 7.1b). A spring is, usually, used to maintain the contact between the cam and the follower. In Fig. 7.1c, the cam is stationary and the follower constraint or guide G causes the relative motion of the cam and the follower.

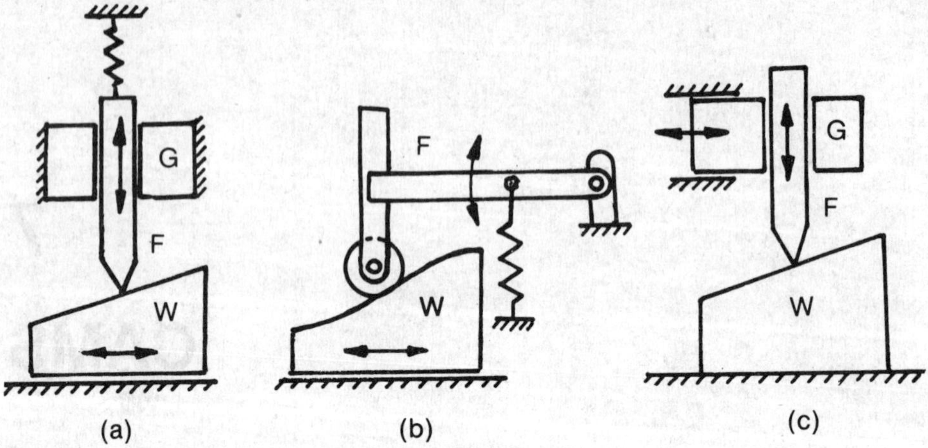

Fig. 7.1

Instead of using a wedge, a flat plate with a groove can also be used. In the groove the follower is held as shown in Fig. 7.2. Thus a positive drive is achieved without the use of a spring.

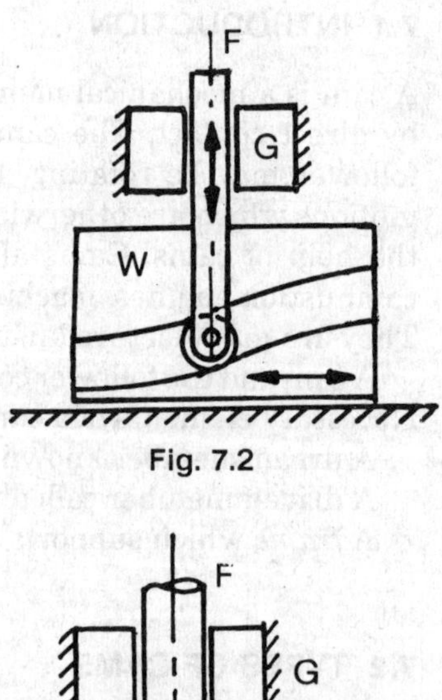

Fig. 7.2

2. *Radial or Disc Cams* A cam in which the follower moves radially from the centre of rotation of the cam is known as a radial or a disc cam (Figs. 7.3 a and b). Radial cams are very popular due to their simplicity and compactness.

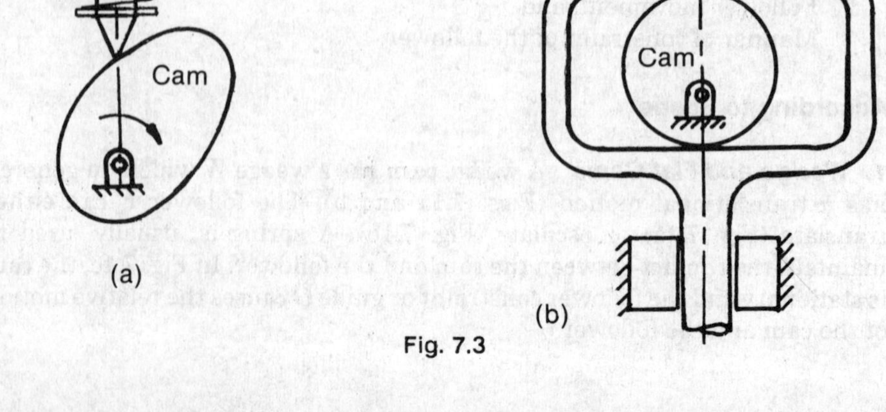

Fig. 7.3

3. Spiral Cams A spiral cam is a face cam in which a groove is cut in the form of a spiral as shown in Fig. 7.4. The spiral groove consists of teeth which mesh with a pin gear follower. The velocity of the follower is proportional to the radial distance of the groove from the axis of the cam.

The use of such a cam is limited as the cam has to reverse the direction to reset the position of the follower. It finds its use in computers.

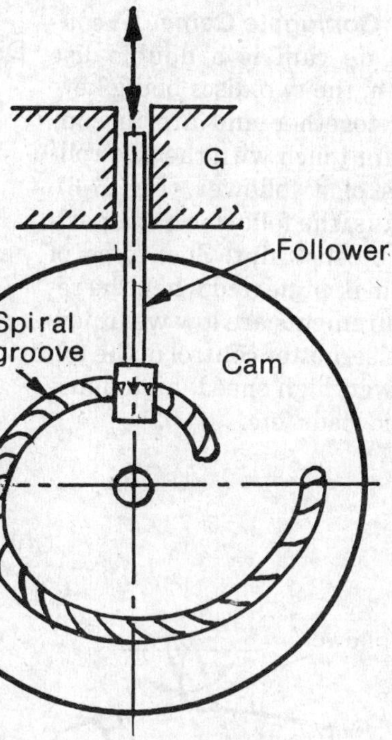

Fig. 7.4

4. Cylindrical Cams In a cylindrical cam, a cylinder which has a circumferential contour cut in the surface, rotates about its axis. The follower motion can be of two types as follows:

In the first type, a groove is cut on the surface of the cam and a roller follower has a constrained (or positive) oscillating motion (Fig. 7.5 a). Another type is an end cam in which end of the cylinder is the working surface (7.5 b). A spring-loaded follower translates along or parallel to the axis of the rotating cylinder.

Cylindrical cams are also known as *barrel* or *drum* cams.

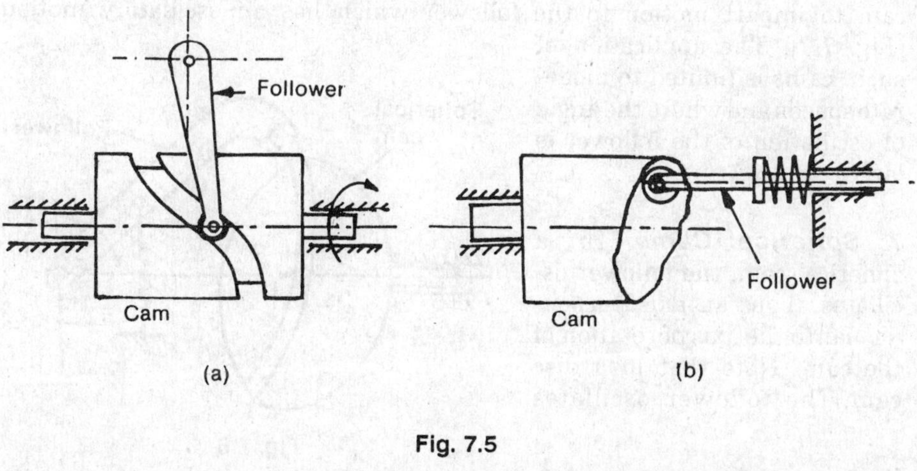

Fig. 7.5

5. Conjugate Cams A conjugate cam is a double-disc cam, the two discs being keyed together and are in constant touch with the two rollers of a follower (Fig. 7.6). Thus, the follower has a positive constraint. Such type of cam is preferred when the requirements are low wear, low noise, better control of the follower, high speed, high dynamic loads, etc.

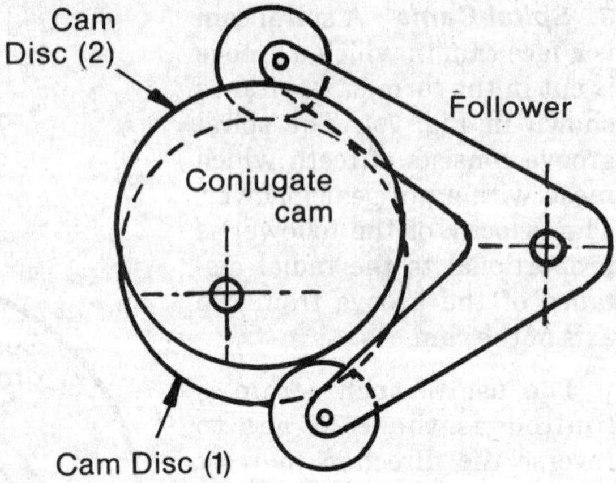

Fig. 7.6

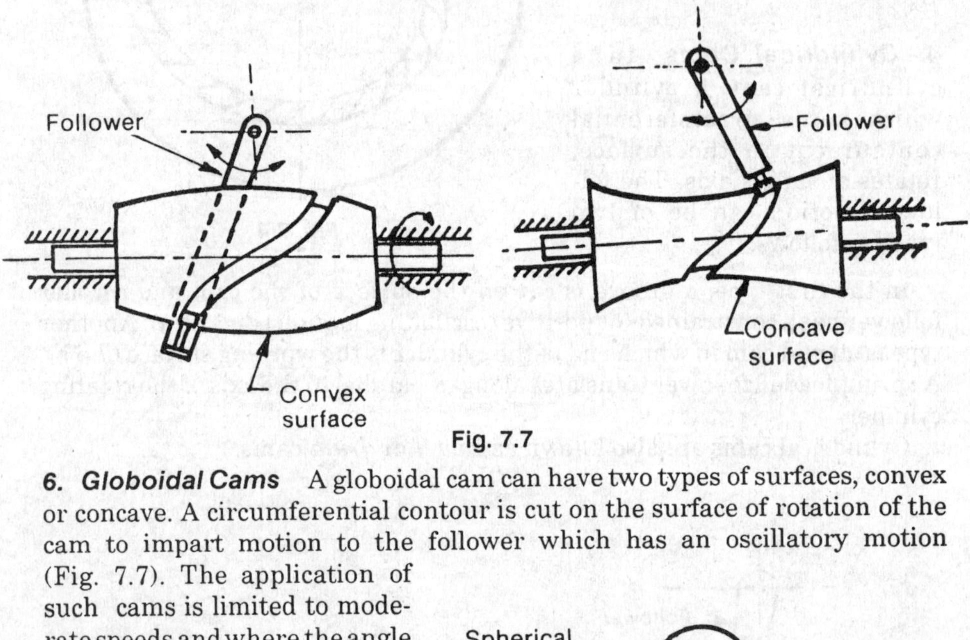

Fig. 7.7

6. Globoidal Cams A globoidal cam can have two types of surfaces, convex or concave. A circumferential contour is cut on the surface of rotation of the cam to impart motion to the follower which has an oscillatory motion (Fig. 7.7). The application of such cams is limited to moderate speeds and where the angle of oscillation of the follower is large.

7. Spherical Cams In a spherical cam, the follower oscillates about an axis perpendicular to the axis of rotation of the cam. Note that in a disc cam, the follower oscillates

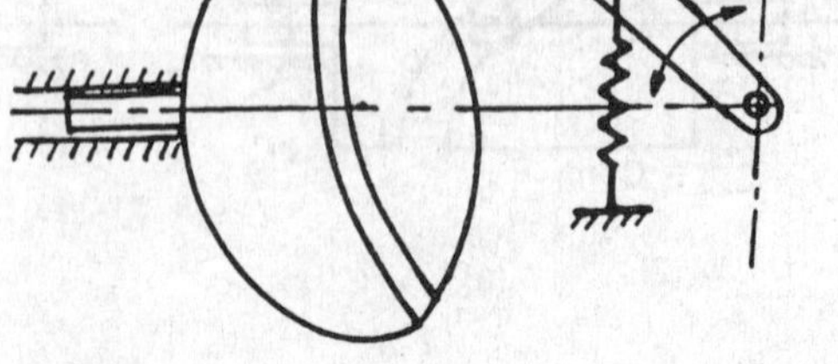

Fig. 7.8

about an axis parallel to the axis of rotation of the cam.

A spherical cam is in the form of a spherical surface which transmits motion to the follower (Fig. 7.8).

According to Follower Movement

The motions of the followers are distinguished from each other by the dwells they have. A *dwell* is the zero displacement or the absence of motion of the follower during the motion of the cam.

Cams are classified according to the motions of the followers in the following ways:

1. ***Rise-Return-Rise (R-R-R)*** In this, there is alternate rise and return of the follower with no periods of dwells (Fig. 7.9 a). Its use is very limited in the industry. The follower has a linear or an angular displacement.

2. ***Dwell-Rise-Return-Dwell (D-R-R- D)*** In such a type of cam, there is rise and return of the follower after a dwell (Fig. 7.9 b). This type is used more frequently than the R-R-R type of cam.

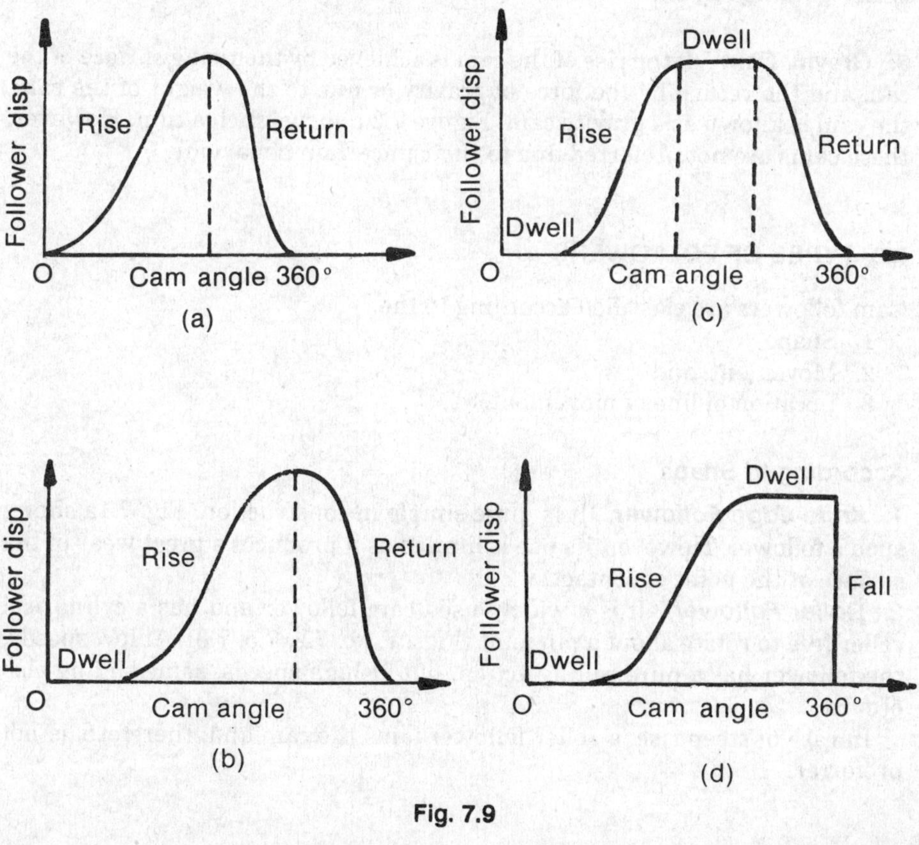

Fig. 7.9

3. Dwell-Rise-Dwell-Return-Dwell (D-R-D-R- D) It is the most widely used type of cam. The dwelling of the cam is followed by rise and dwell and subsequently by return and dwell as shown in Fig. 7.9 c. In case the return of the follower is by a fall (Fig. 7.9 d), the motion may be known as Dwell-Rise-Dwell (D-R-D).

According to Manner of Constraint of the Follower

To reproduce exactly the motion transmitted by the cam to the follower, it is necessary that the two remain in touch at all speeds and at all times. The cams can be classified according to the manner in which this is achieved.

1. Pre-loaded Spring Cam A pre-loaded compression spring is used for the purpose of keeping the contact between the cam and the follower (Figs. 7.1 a and b, 7.3 a, 7.5 b and 7.8).

2. Positive-drive Cam In this type, constant touch between the cam and the follower is maintained by a roller follower operating in the groove of a cam (Figs. 7.2, 7.3 b, 7.4, 7.5 a and 7.7). The follower cannot go out of the groove under the normal working operations. A constrained or positive drive is also obtained by the use of a conjugate cam (Fig. 7.6).

3. Gravity Cam If the rise of the cam is achieved by the rising surface of the cam and the return by the force of gravity or due to the weight of the cam, the cam is known as a gravity cam. Figure 7.2 c shows such a cam. However, these cams are not preferred due to their uncertain behaviour.

7.3 TYPES OF FOLLOWERS

Cam followers are classified according to the:

1. Shape,
2. Movement, and
3. Location of line of movement.

According to Shape

1. Knife-edge Follower It is quite simple in construction. Fig. 7.1a shows such a follower. However, its use is limited as it produces a great wear of the surface at the point of contact.

2. Roller Follower It is a widely used cam follower and has a cylindrical roller free to rotate about a pin joint (Figs. 7.1 b, 7.2, 7.5, 7.8). At low speeds, the follower has a pure rolling action, but at high speeds, some sliding also occurs.

In case o steep rise, a roller follower jams the cam and, therefore, is not preferred.

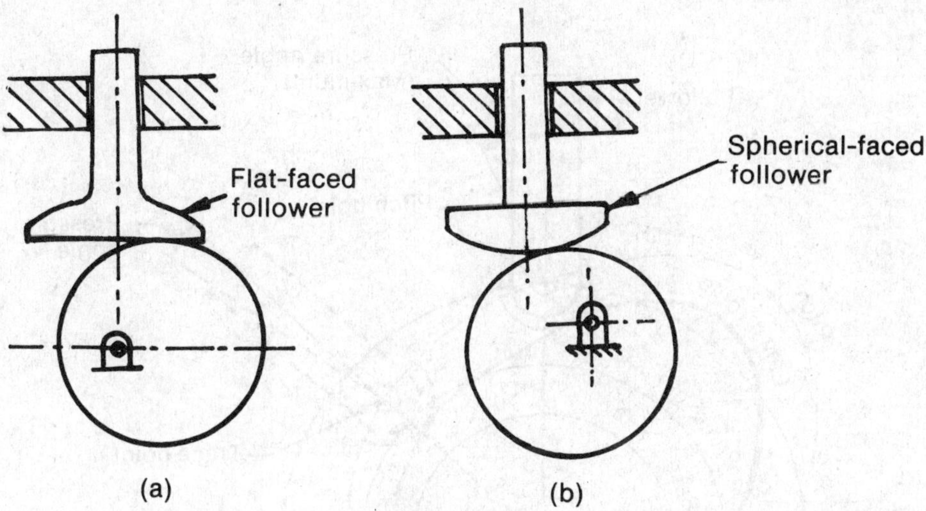

Fig. 7.10

3. Mushroom Follower A mushroom follower (Fig. 7.10) has the advantage that it does not pose the problem of jamming the cam. However, high surface stresses and wear are quite high due to deflection and misalignment if a flat-faced follower is used (Fig. 7.10 a). These disadvantages are reduced if a spherical-faced follower (Fig. 7.10 b) is used instead of a flat-faced follower.

According to Movement

1. Reciprocating Follower In this type, as the cam rotates, the follower reciprocates or translates in the guides (Fig. 7.1a).
2. Oscillating Follower The follower is pivoted at a suitable point on the frame and oscillates as the cam makes the rotary motion (Fig. 7.1b).

According to Location of Line of Movement

1. Radial Follower The follower is known as a radial follower if the line of movement of the follower passes through the centre of rotation of the cam (Figs. 7.3 a and b).

2. Offset Follower If the line of movement of the roller follower is offset from the centre of rotation of the cam, the follower is known as an offset follower (Fig. 7.10 b).

7.4 DEFINITIONS

With reference to Fig. 7.11, some definitions are given below:

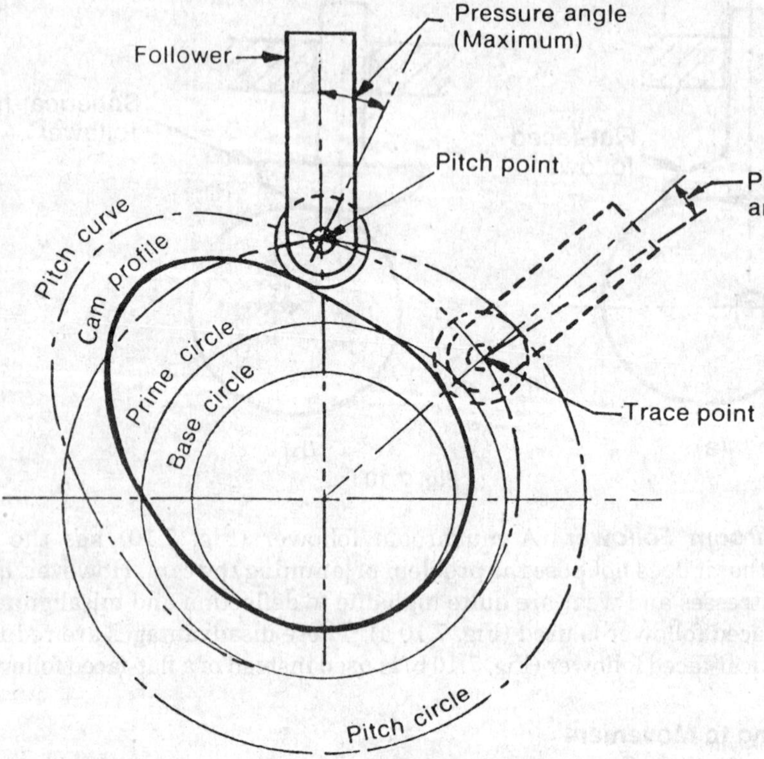

Fig. 7.11

Base Circle It is the smallest circle tangent to the cam profile (contour) drawn from the centre of rotation of a radial cam.

Trace Point It is a reference point on the follower to trace the cam profile such as the knife edge of a knife-edged follower and centre of the roller of a roller follower.

Pitch Curve It is the curve drawn by the trace point assuming that the cam is fixed, and the trace point of the follower rotates around the cam.

Pressure Angle The pressure angle, representing the steepness of the cam profile, is the angle between the normal to the pitch curve at a point and the direction of the follower motion. It varies in magnitude at all instants of the follower motion. A high value of the maximum pressure angle is not desired as it might jam the follower in the bearings.

Pitch Point It is the point on the pitch curve at which the pressure angle is maximum.

Pitch Circle It is the circle passing through the pitch point and concentric with the base circle.

Prime Circle The smallest circle drawn tangent to the pitch curve is known as the prime circle.

7.5 FOLLOWER DISPLACEMENT PROGRAMMING

As a cam rotates about its axis, it imparts a specific motion to the follower which is repeated with each revolution of the cam. Thus it is enough to know the motion of the follower for only one revolution. The motion of the cam can be represented on a graph, the x-axis of which may represent the angular displacements of the cam and the y-axis, the angular or the linear displacements of the follower. The follower displacement is measured from its lowest position and is plotted with the same scale as is to be used in the layout of the cam profile. The following terms are used with reference to the angular motion of the cam:

Angle of Ascent (φ_a) It is the angle through which the cam turns during the time the follower rises.

Angle of Dwell (δ) Angle of dwell is the angle through which the cam turns while the follower remains stationary at the highest or the lowest position.

Angle of Descent (φ_d) It is the angle through which the cam turns during the time the follower returns to the initial position.

Angle of Action Angle of action is the total angle moved by the cam during the time, between the beginning of rise and the end of return of the follower.

To satisfy the given requirements of the follower displacement, a programme can be made keeping in view the following points:

1. In a given specific interval of time, due consideration to the velocity and the acceleration must be given, the effects of which are manifested as inertia loads. The dynamic effects of acceleration, usually, limit the speed of the cams. Moreover, effects of jerks (rate of change of acceleration) in case of high-speed mechanisms produce vibrations of the system, which is undesirable for a follower motion. Though it is very difficult to completely eliminate jerk, efforts are to be made to keep it within tolerable limits.

2. The force exerted by a cam on the follower is always normal to the surface of the cam at the point of contact. The vertical component (**F** cos α) lifts the

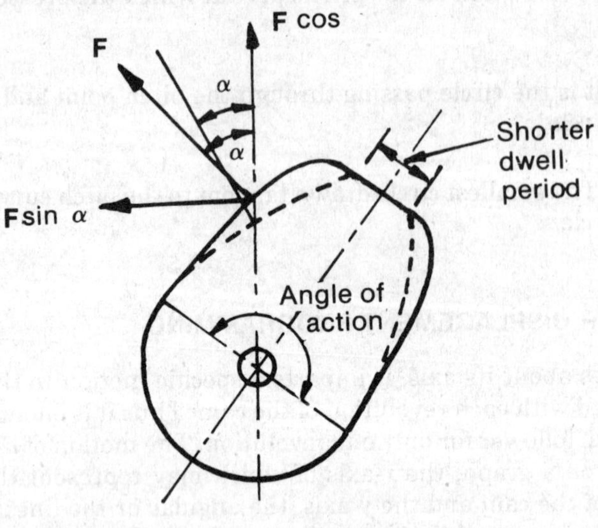

Fig. 7.12

follower whereas the horizontal component (**F** sin α) exerts lateral pressure on the bearing as shown in Fig. 7.12.

In order to reduce the lateral pressure or **F** sin α, α has to be decreased which means making the surface more convex and longer (dotted lines). This results in reduced velocity of the follower and more time for the same rise. This also reduces the dwell period for a fixed angle of action. In internal combustion engines, a shorter dwell period means a smaller period of valve-opening, resulting in less fuel per cycle and lesser power production. Thus the minimum value of α cannot be reduced from a certain value.

3. The size of the base circle controls the pressure angle. As shown in Fig. 7.13, the increase in the base circle diameter increases the length of the arc of the circle upon which the wedge (the raised portion) is to be made. A short wedge for a given rise requires a steep rise or a higher pressure

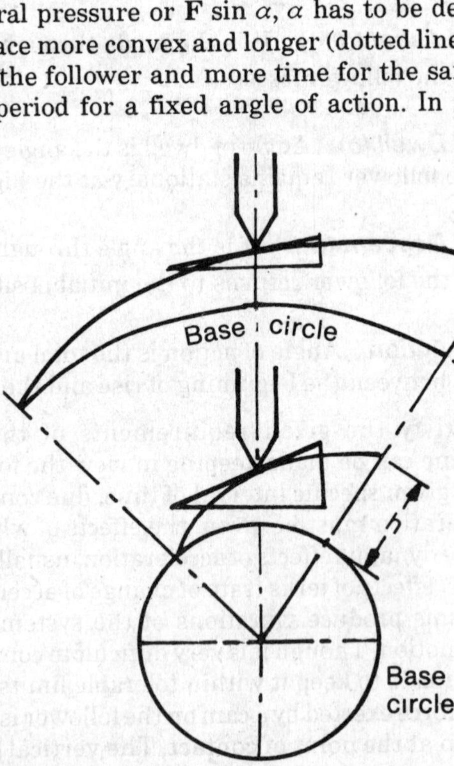

Fig. 7.13

angle, thus increasing the lateral force.

7.6 MOTIONS OF THE FOLLOWER

Though the follower can be made to have any type of desired motion, a knowledge of the existing motion programmes saves time and labour while designing the cams.

Following are some basic displacement programmes.

1. Simple Harmonic Motion (SHM)

This is a popular follower motion and is easy to lay out.

Let s = follower displacement (instantaneous)
h = maximum follower displacement
v = velocity of the follower
f = acceleration of the follower
θ = cam rotation angle (instantaneous)
φ = cam rotation angle for the maximum follower displacement
β = angle on the harmonic circle.

Construction The follower rises through a distance h while the cam turns through an angle φ. Construct the follower displacement curve as follows (Fig. 7.14):

(i) Draw a semicircle with cam rise (or fall) as the diameter. This is, usually, known as the harmonic (semi) circle. Divide this semicircle into n equal arcs (n even).

(ii) Divide the cam displacement interval into n equal divisions.

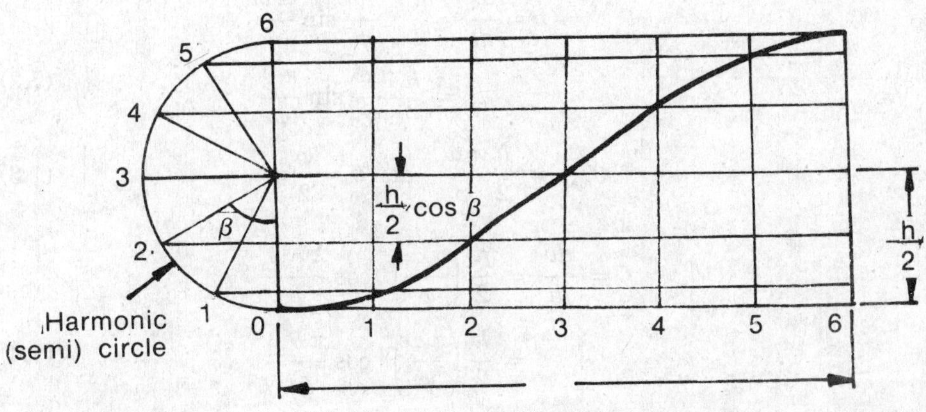

Fig. 7.14

(iii) Project the intercepts of the harmonic semicircle to the corresponding divisions of the cam displacement interval.

(iv) Join the points with a smooth curve to obtain the required harmonic curve.

Displacement At any instant, displacement of the follower is given by,

$$s = \frac{h}{2} - \frac{h}{2}\cos\beta$$

$$= \frac{h}{2}(1 - \cos\beta) \qquad \text{(i)}$$

For the rise (or fall) h of the follower displacement, the cam is rotated through an angle φ whereas a point on the harmonic semicircle traverses an angle π. Thus the cam rotation is proportional to the angle turned by the point on the harmonic semicircle. That is,

$$\beta = \pi\,\frac{\theta}{\varphi}$$

Thus β can be replaced by θ and φ in Eqn (i) above,

$$s = \frac{h}{2}\left(1 - \cos\frac{\pi\theta}{\varphi}\right) \qquad (7.1)$$

The expression is also valid for β more than 90°. In that case $\cos\beta$ or $\cos\pi\theta/\varphi$ becomes negative so that s is again positive and more than $h/2$.

Let ω = Angular velocity of the cam

$\therefore$
$$\theta = \omega t$$

and
$$s = \frac{h}{2}\left(1 - \cos\frac{\pi\omega t}{\varphi}\right)$$

$$v = \frac{ds}{dt} = \frac{h}{2}\,\frac{\pi\omega}{\varphi}\sin\frac{\pi\omega t}{\varphi}$$

$$= \frac{h}{2}\frac{\pi\,\omega}{\varphi}\sin\frac{\pi\,\theta}{\varphi} \qquad (7.2)$$

$$v_{\max} = \frac{h}{2}\frac{\pi\,\omega}{\varphi} \quad \text{at } \theta = \frac{\varphi}{2} \qquad (7.3)$$

$$f = \frac{dv}{dt} = \frac{h}{2}\left(\frac{\pi\,\omega}{\varphi}\right)^2\cos\frac{\pi\omega\, t}{\varphi}$$

$$= \frac{h}{2}\left(\frac{\pi\,\omega}{\varphi}\right)^2\cos\frac{\pi\,\theta}{\varphi} \qquad (7.4)$$

$$f_{\max} = \frac{h}{2}\left(\frac{\pi\,\omega}{\varphi}\right)^2 \quad \text{at } \theta = 0 \qquad (7.5)$$

Let φ_a = Angle of ascent

φ_d = Angle of descent

$\delta_{1,2}$ = Angles of dwells

It can be seen from the plots of Fig. 7.15 that there is an abrupt change of acceleration from zero to maximum at the beginning of the follower motion and also from maximum (negative) to zero at the end of the follower motion when the follower rises. Similar abruption would also be there at the start and end of the return motion. As these abrupt changes result in infinite jerk, vibration and noise etc., the programme should be adopted only for low or moderate cam speeds.

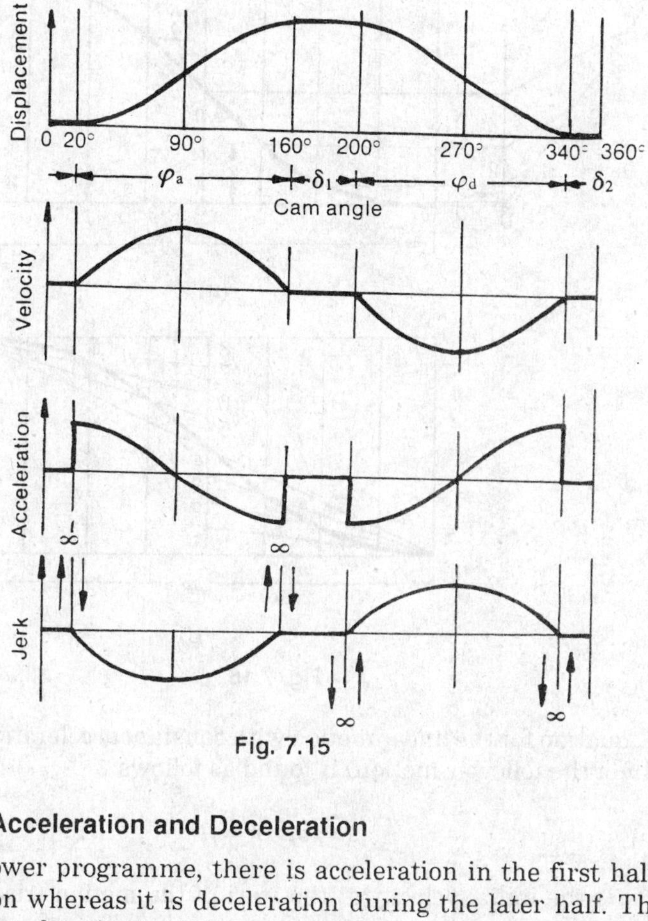

Fig. 7.15

2. Constant Acceleration and Deceleration

In such a follower programme, there is acceleration in the first half of the follower motion whereas it is deceleration during the later half. The magnitude of the acceleration and the deceleration is the same and constant in the two halves.

Construction (Refer to Fig. 7.16 a)

(i) Divide each half of the cam displacement interval into n equal divisions.

(ii) Divide half the follower rise into n^2 equal divisions.

(iii) Project 1^2 displacement interval to the first ordinate of the cam displacement, 2^2 to the second ordinate, 3^2 to the third and so on.

(iv) The second half of the curve is similar to the first half.

Alternatively, divide half of the follower rise at the central ordinate of the cam displacement into n equal divisions (Fig. 17.16 b). Joining the zero point with the first division gives 1^2 on the first ordinate, with the second division 2^2 on the second ordinate and so on.

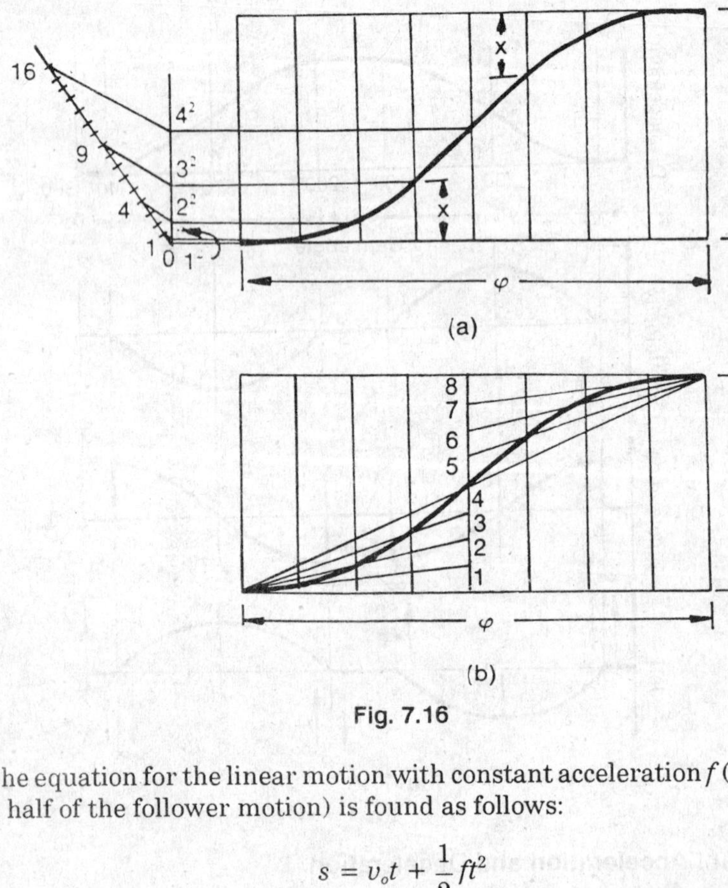

Fig. 7.16

The equation for the linear motion with constant acceleration f (during the first half of the follower motion) is found as follows:

$$s = v_o t + \frac{1}{2} f t^2$$

where v_o is the initial velocity at the start of the motion (rise or fall) and is zero in this case.

$$\therefore \qquad s = \frac{1}{2} f t^2$$

or

$$f = \frac{2s}{t^2} = \text{constant} \tag{7.6}$$

As f is constant during the accelerating period, considering the follower at the midway,

$$s = \frac{h}{2} \quad \text{and} \quad t = \frac{\varphi/2}{\omega}$$

$$\therefore \quad f = \frac{2h/2}{\varphi^2/4\omega^2} = \frac{4\,h\omega^2}{\varphi^2} \tag{7.6a}$$

The velocity is linear during the period and is given by

$$v = \frac{ds}{dt} = \frac{1}{2} \times 2\,ft = ft \tag{7.7}$$

$$= \frac{4h\,\omega^2}{\varphi^2}\frac{\theta}{\omega} \quad (\theta = \omega t)$$

$$= \frac{4h\,\omega}{\varphi^2}\theta \tag{7.7a}$$

The velocity is maximum when θ is maximum or the follower is at the midway, i.e. when $\theta = \varphi/2$

$$v_{\max} = \frac{4h\,\omega}{\varphi^2}\frac{\varphi}{2} = \frac{2h\,\omega}{\varphi} \tag{7.8}$$

During the second half of the follower motion, the follower is decelerated at constant rate so that the velocity reduces to zero at the end.

It can be observed from the plots shown in Fig. 7.17 that there are abrupt changes in the acceleration at the beginning, midway and the end of the follower motion. At midway, an infinite jerk is produced. Thus, this programme of the follower is adopted only up to moderate speeds.

3. Constant Velocity

Constant velocity of the follower implies that the displacement of the follower is proportional to the cam displacement and the slope of the displacement curve is constant (Fig. 7.18).

Displacement of the follower for the angular displacement θ of the cam is given by

$$s = h\frac{\theta}{\varphi} = h\frac{\omega t}{\varphi} \tag{7.9}$$

$$v = \frac{ds}{dt} = \frac{h\omega}{\varphi} = \text{constant} \tag{7.10}$$

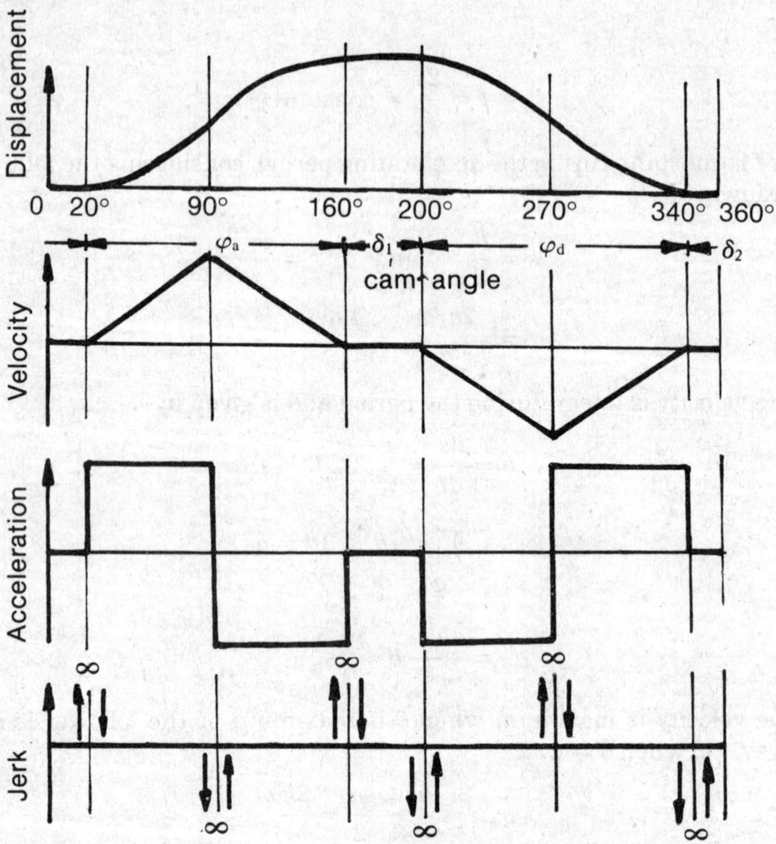

Fig. 7.17

$$f = \frac{dv}{dt} = 0 \tag{7.11}$$

As seen in the plots of Fig. 7.19 a, though acceleration is zero during the rise or the fall of the follower, it is infinite at the beginning and end of the motion as there are abrupt changes in velocity at these points. This results in infinite inertia forces and thus is not suitable from practical point of view.

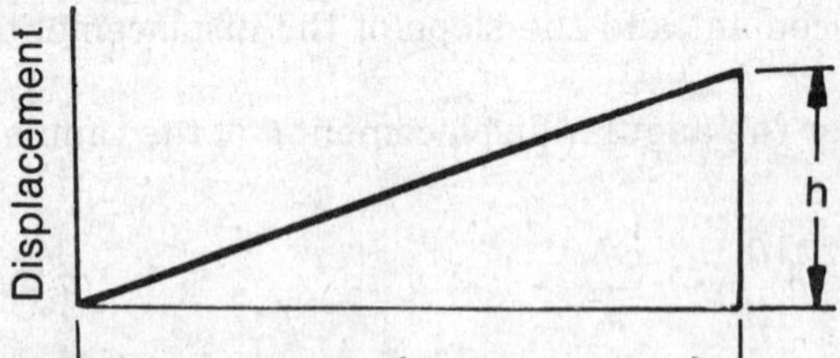

Fig. 7.18

A modified programme for the follower motion can be evolved in which the accelerations are reduced to finite values. This can be done by rounding the sharp corners of the displacement curve so that the velocity changes are

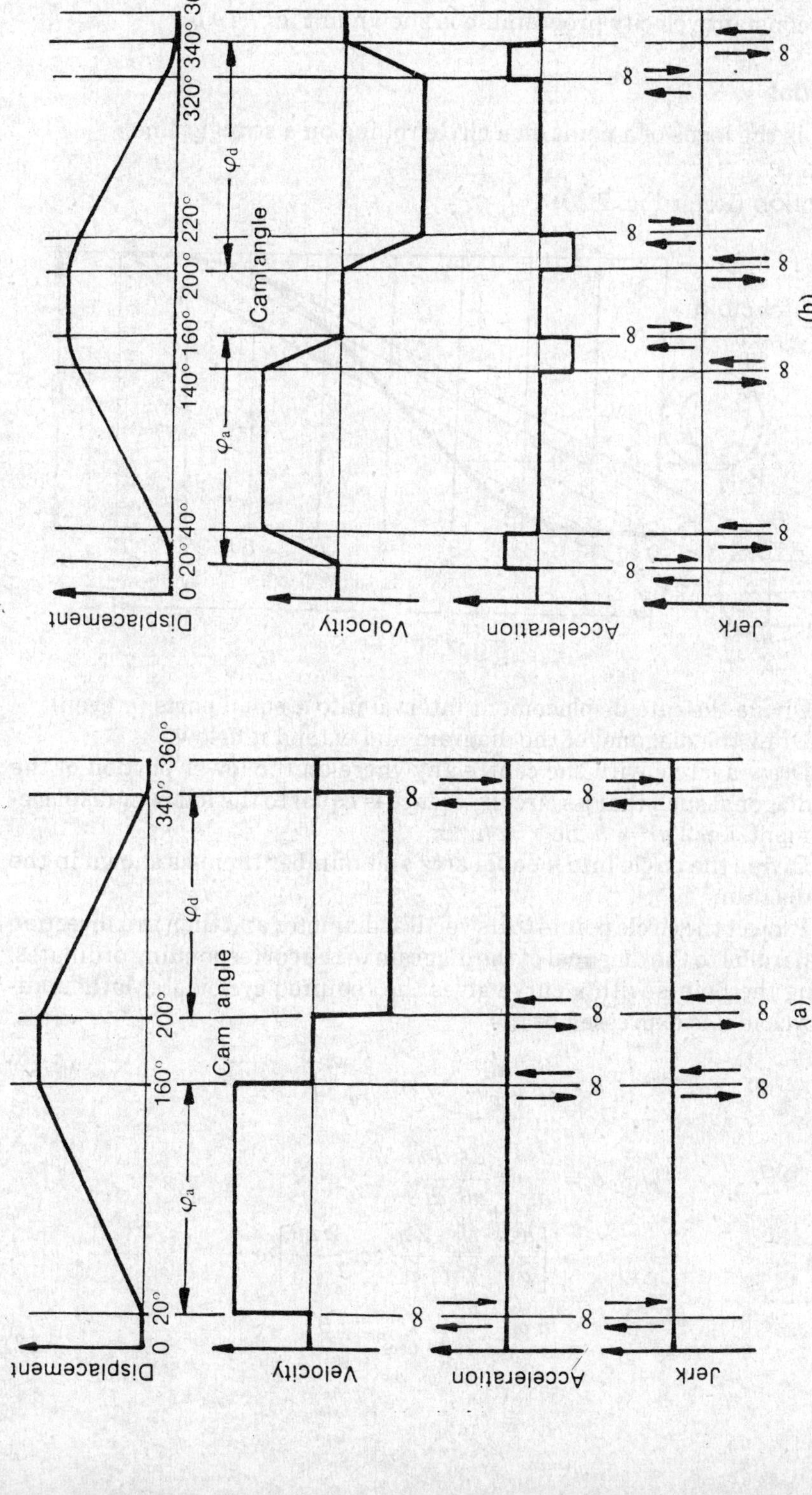
Displacement
Velocity
Acceleration
Jerk
Cam angle
0
20°
40°
140°
160°
200°
220°
320°
340°
360°
φ_a
φ_d
∞
(a)
(b)

Fig. 7.19

gradual at the beginning and end of the follower motion. During these periods, the accelerations may be assumed to be constant and of finite values. A modified constant velocity programme is shown in Fig. 7.19 b.

4. Cycloidal

A cycloid is the locus of a point on a circle rolling on a straight line.

Construction (Refer Fig. 7.20)

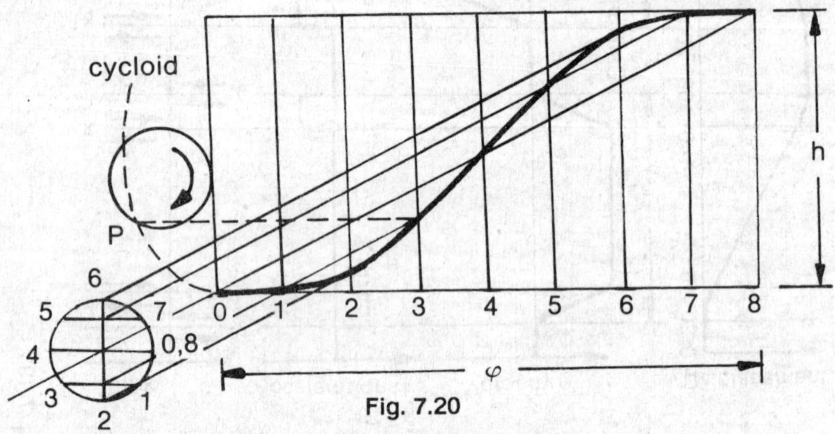

Fig. 7.20

(i) Divide the cam displacement interval into n equal parts (n even).

(ii) Draw the diagonal of the diagram and extend it below.

(iii) Draw a circle with the centre anywhere on the lower portion of the diagonal such that its circumference is equal to the follower displacement, i.e. $2\pi r = h$ or $r = h/2\pi$.

(iv) Divide the circle into n equal arcs and number them as shown in the diagram.

(v) Project the circle points to its vertical diameter and then in a direction parallel to the diagonal of the diagram to the corresponding ordinates.

Joining the points with a curve gives the required cycloidal. Mathematically, a cycloidal is expressed by

$$s = \frac{h}{\pi}\left(\frac{\pi\theta}{\varphi} - \frac{1}{2}\sin\frac{2\pi\theta}{\varphi}\right) \tag{7.12}$$

$$v = \frac{ds}{dt} = \frac{ds}{d\theta}\frac{d\theta}{dt}$$

$$= \left[\frac{h}{\varphi} - \frac{h}{2\pi}\frac{2\pi}{\varphi}\cos\frac{2\pi\theta}{\varphi}\right]\omega$$

$$= \frac{h\omega}{\varphi} - \frac{h\omega}{\varphi}\cos\frac{2\pi\theta}{\varphi} \tag{7.13}$$

$$= \frac{h\omega}{\varphi}\left(1 - \cos\frac{2\pi\theta}{\varphi}\right)$$

$$v_{max} = \frac{2h\omega}{\varphi} \text{ at } \theta = \frac{\varphi}{2} \tag{7.14}$$

$$f = \frac{dv}{dt} = \frac{dv}{d\theta}\frac{d\theta}{dt}$$

$$= \left[\frac{h\omega}{\varphi}\frac{2\pi}{\varphi}\sin\frac{2\pi\theta}{\varphi}\right]\omega$$

$$= \frac{2h\pi\omega^2}{\varphi^2}\sin\frac{2\pi\theta}{\varphi} \tag{7.15}$$

$$f_{max} = \frac{2h\pi\omega^2}{\varphi^2} \text{ at } \theta = \frac{\varphi}{4} \tag{7.16}$$

From the plots of Fig. 7.21, it is observed that there are no abrupt changes in the velocity and the acceleration at any stage of the motion. Thus it is the most ideal programme for high speed follower motion.

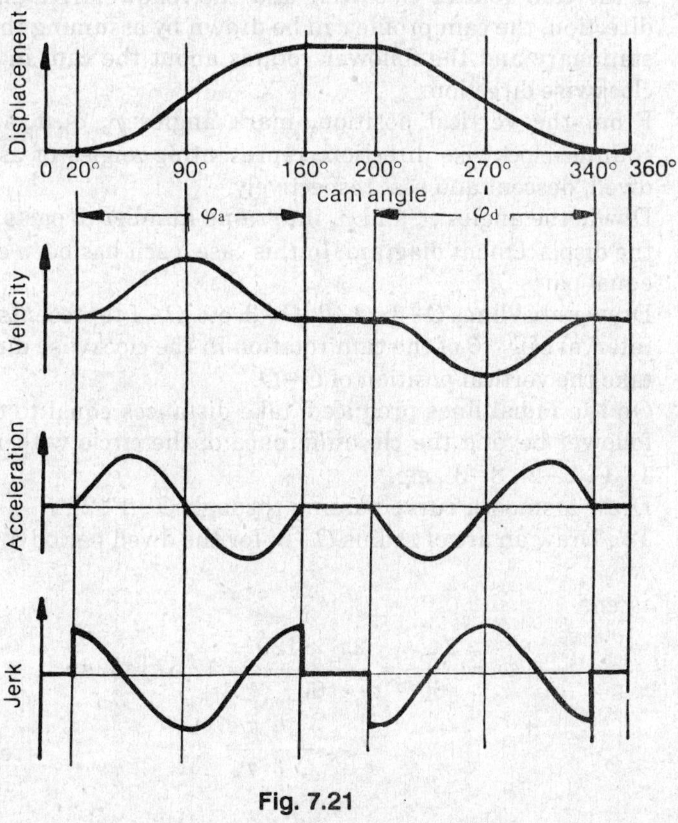

Fig. 7.21

Example 7.1 Draw the profile of a cam operating a knife-edged follower having a lift of 30 mm. The cam raises the follower with SHM for 150° of its rotation followed by a period of dwell for 60°. The follower descends for the next 100° rotation of the cam with uniform velocity, again followed by a dwell period. The cam rotates at a uniform velocity of 120 rpm and has a least radius of 20 mm. What will be the maximum velocity and acceleration of the follower during the lift and the return?

Solution:

$$h = 30 \text{ mm} \qquad \varphi_a = 150°$$
$$N = 120 \text{ rpm} \qquad \delta_1 = 60°$$
$$r_c = 20 \text{ mm} \qquad \varphi_d = 100°$$
$$\delta_2 = (360° - 150° - 100° - 60°) = 50°$$

Draw the displacement diagram of the follower as discussed earlier (Fig. 7.22a) taking a convenient scale. Construct the cam profile as follows (refer Fig. 7.22b):

(i) Draw a circle with radius r_c.

(ii) If the cam rotates clockwise and the follower remains in vertical direction, the cam profile can be drawn by assuming that the cam is stationary and the follower rotates about the cam in the counter-clockwise direction.
From the vertical position, mark angles φ_a, δ_1, φ_d and δ_2 in the counter-clockwise direction, representing angles of ascent, rest or dwell, descent and rest respectively.

(iii) Divide the angles φ_a and φ_d into same number of parts as is done in the displacement diagram. In this case, each has been divided into 6 equal parts.

(iv) Draw radial lines $O-1$, $O-2$, $O-3$, etc. $O-1$ represents that after an interval of $\varphi_d/6$ of the cam rotation in the clockwise direction it will take the vertical position of $O-O'$.

(v) On the radial lines produced, take distances equal to the lift of the follower beyond the circumference of the circle with radius r_c, i.e. $1-1'$, $2-2'$, $3-3'$, etc.

(vi) Draw a smooth curve passing through O', $1'$, $2'$, . . ., $10'$, $11'$ and $12'$. Draw an arc of radius $O-6'$ for the dwell period δ_1.

During ascent

$$\omega = \frac{2\pi N}{60} = \frac{2\pi \times 120}{60} = 12.57 \text{ rad/s}$$

$$v_{\max} = \frac{h}{2}\frac{\pi\omega}{\varphi_a} \qquad \text{[refer Eq. (7.3)]}$$

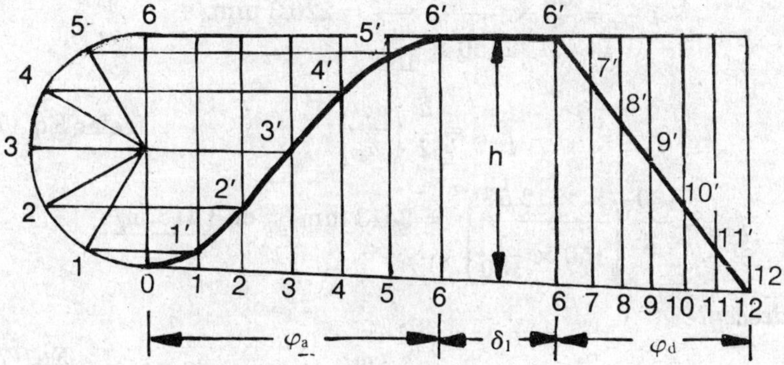
6
5
4
3
2
1
0 1 2 3 4 5 6
1′ 2′ 3′ 4′ 5′ 6′ 6′
h
7′ 8′ 9′ 10′ 11′ 12′
6 7 8 9 10 11 12
φa δ1 φd

(a)

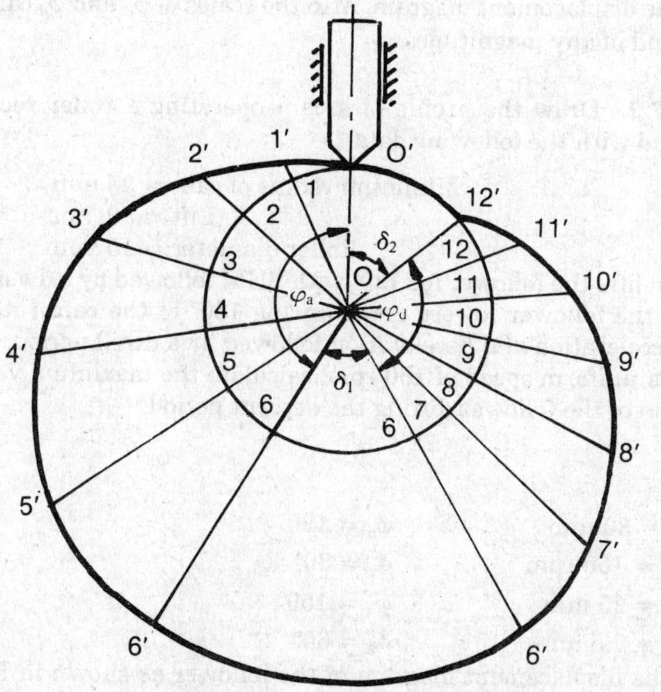
O′
1′ 2′ 3′ 4′ 5′ 6′ 6′ 7′ 8′ 9′ 10′ 11′ 12′
1 2 3 4 5 6 6 7 8 9 10 11 12
O
δ2
φa
φd
δ1

(b)

Fig. 7.22

or

$$v_{max} = \frac{30}{2} \times \frac{\pi \times 12.57}{150 \times \frac{\pi}{180}} = 226.3 \text{ mm/s}$$

$$f_{max} = \frac{h}{2}\left(\frac{\pi\omega}{\varphi_a}\right)^2 \qquad \text{[refer Eq. (7.5)]}$$

or

$$f_{max} = \frac{30}{2}\left(\frac{\pi \times 12.57}{150 \times \frac{\pi}{180}}\right)^2 = 3413 \text{ mm/s}^2 \text{ or } \underline{3.413 \text{ m/s}^2}$$

During descent

$$v_{max} = h\frac{\omega}{\varphi_d} \qquad \text{Eq. (7.10)}$$

or

$$v_{max} = 30 \times \frac{12.57}{100 \times \frac{\pi}{180}} = \underline{216 \text{ mm/s}}$$

$$f_{max} = f = 0 \qquad \text{Eq. (7.11)}$$

Note that to draw the cam profile, it is not necessary that the interval δ_1 is taken in the displacement diagram. Also the scales of φ_a and φ_d can be taken different and of any magnitudes.

Example 7.2 Draw the profile of a cam operating a roller reciprocating follower and with the following data:

Minimum radius of cam = 25 mm
Lift = 30 mm
Roller diameter = 15 mm

The cam lifts the follower for 120° with SHM followed by a dwell period of 30°. Then the follower lowers down during 150° of the cam rotation with uniform acceleration and deceleration followed by a dwell period. If the cam rotates at a uniform speed of 150 rpm, calculate the maximum velocity and acceleration of the follower during the descent period.

Solution:

h = 30 mm φ_a = 120°
N = 150 rpm δ_1 = 30°
r_c = 25 mm φ_d = 150°
r_r = 7.5 mm δ_2 = 60°

Draw the displacement diagram of the follower as shown in Fig. 7.23 a. Consrict the cam profile as described below (Fig. 7.23 b):

(i) Draw a circle with radius $(r_c + r_r)$.

(ii) From the vertical position, mark angles φ_a, δ_1, φ_d and δ_2 in the counter-clockwise direction (assuming that the cam is to rotate in the clockwise direction).

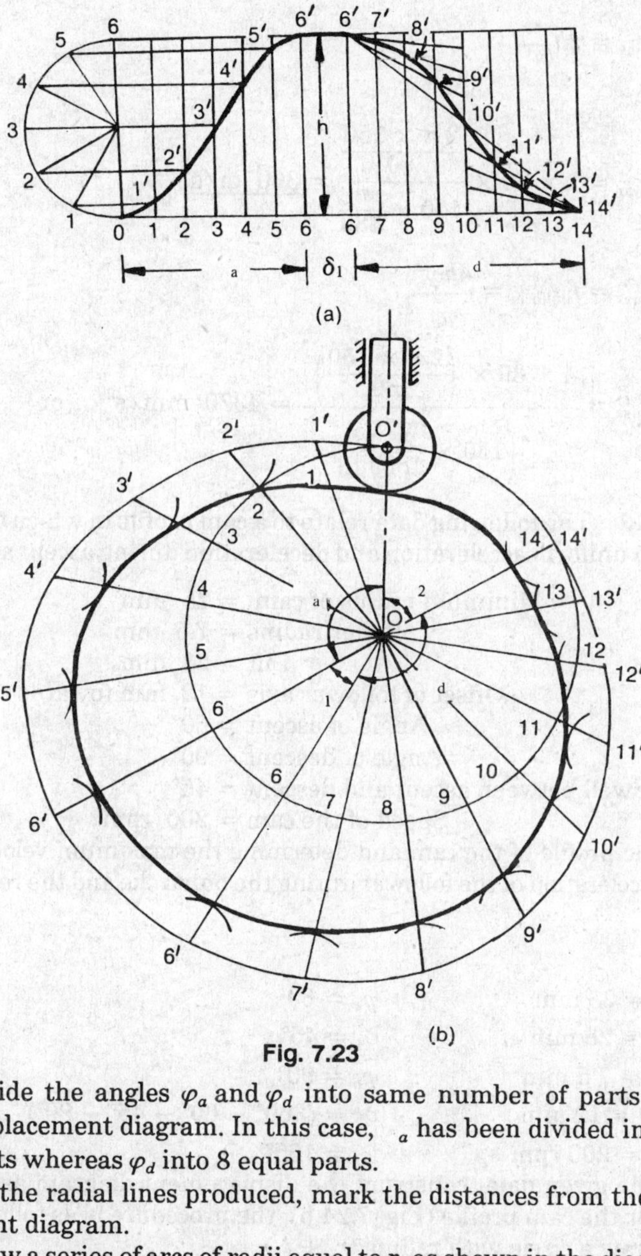

Fig. 7.23

(iii) Divide the angles φ_a and φ_d into same number of parts as in the displacement diagram. In this case, $_a$ has been divided into 6 equal parts whereas φ_d into 8 equal parts.

(iv) On the radial lines produced, mark the distances from the displacement diagram.

(v) Draw a series of arcs of radii equal to r_r as shown in the diagram from the points 1′, 2′, 3′, etc.

(v) Draw a smooth curve tangential to all the arcs which is the required cam profile.

During the descent period, the acceleration and the deceleration is uniform. Therefore, the maximum velocity is at the end of the acceleration period.

$$v_{max} = 2h \frac{\omega}{\varphi_d} \qquad \text{Eq. (7.8)}$$

or

$$v_{max} = 2 \times 30 \times \frac{\frac{2\pi \times 150}{60}}{150 \times \frac{\pi}{180}} = \underline{360 \text{ m/s}}$$

$$f_{max} = f_{uniform} = \frac{4h\omega^2}{\varphi_d^2} \qquad \text{Eq. (7.6)}$$

or

$$f_{max} = \frac{4 \times 30 \times \left(\frac{2\pi \times 150}{60}\right)^2}{\left(150 \times \frac{\pi}{180}\right)^2} = 4320 \text{ mm/s}^2 \quad \text{or} \quad 4.32 \text{ m/s}^2$$

Example 7.3 The following data relate to a cam profile in which the follower moves with uniform acceleration and deceleration during ascent and descent:

Minimum radius of cam = 25 mm
Roller radius = 7.5 mm
Lift = 28 mm
Offset of follower axis = 12 mm towards right
Angle of ascent = 60°
Angle of descent = 90°
Angle of dwell between ascent and descent = 45°
Speed of the cam = 200 rpm

Draw the profile of the cam and determine the maximum velocity and the uniform acceleration of the follower during the outstroke and the return stroke.

Solution:

$h = 28$ mm $\qquad \varphi_a = 60°$
$r_c = 25$ mm $\qquad \delta_1 = 45°$
$r_r = 7.5$ mm $\qquad \varphi_d = 90°$
offset, $x = 12$ mm $\qquad \delta_2 = (360° - 60° - 45° - 90°)$
$N = 200$ rpm $\qquad = 165°$

From the given data, construct the displacement diagram as usual (Fig. 7.24 a). For the cam profile (Fig. 7.24 b), the procedure is as follows:

(i) Draw a circle with radius $(r_c + r_r)$.
(ii) Draw another circle concentric with the previous circle with radius x. If the cam is assumed stationary, the follower will be tangential to this circle in all the positions. Let the initial position be $a - O'$.
(iii) Join $O - O'$. Divide the circle of radius $(r_c + r_r)$ into four parts as usual with angles φ_a, δ_1, φ_d and δ_2 starting from $O - O'$.

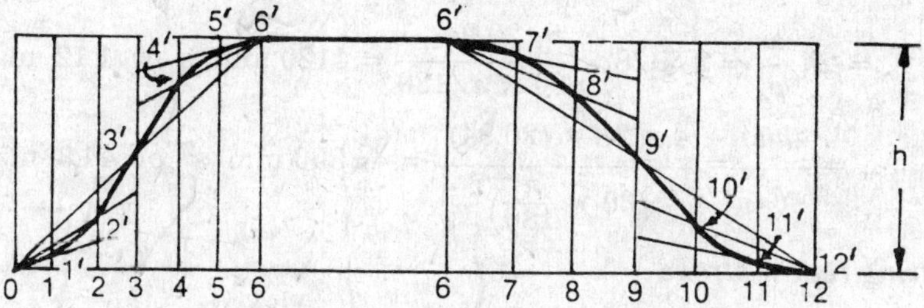

(a)

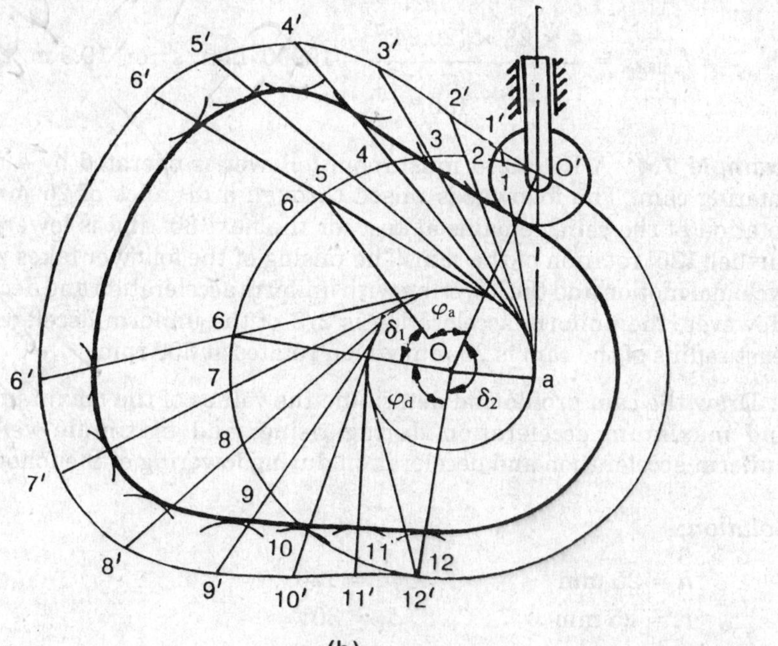

(b)

Fig. 7.24

(iv) Divide the angles φ_a and φ_d into same number of parts as is done in the displacement diagram and obtain the points 1, 2, 3, etc. on the circumference of circle with radius $(r_c + r_r)$.

(v) Draw tangents to the circle with radius x from the points 1, 2, 3, etc.

(vi) On the extension of the tangent lines, mark the distances from the displacement diagram.

(vii) Draw a smooth curve through O', $1'$, $2'$, etc. This is the pitch curve.

(viii) With $1'$, $2'$, $3'$, etc. as centres, draw a series of arcs of radii equal to r_r.

(ix) Draw a smooth curve tangential to all the arcs and obtain the required cam profile.

During outstroke

$$\omega = \frac{2\pi \times 200}{60} = 20.94 \text{ rad/s}$$

$$v_{\max} = 2h\frac{\omega}{\varphi_a} = 2 \times 28 \times \frac{20.94}{60 \times \pi/180} = 1120 \text{ mm/s or } 1.12 \text{ m/s}$$

$$f_{\text{uniform}} = \frac{4h\omega^2}{\varphi_a^2} = \frac{4 \times 28 \times (20.94)^2}{\left(60 \times \frac{\pi}{180}\right)^2} = 44\,800 \text{ mm/s}^2 \text{ or } 44.8 \text{ m/s}^2$$

During return stroke

$$v_{\max} = \frac{2h\,\omega}{\varphi_d} = \frac{2 \times 28 \times 20.94}{90 \times \pi/180} = 747 \text{ mm/s or } 0.747 \text{ m/s}$$

$$f_{\text{uniform}} = \frac{4 \times 28 \times (20.94)^2}{\left(90 \times \frac{\pi}{180}\right)^2} = 19900 \text{ mm/s}^2 \text{ or } 19.9 \text{ m/s}^2$$

Example 7.4 A flat-faced mushroom follower is operated by a uniformly rotating cam. The follower is raised through a distance of 25 mm in 120° rotation of the cam, remains at rest for the next 30° and is lowered during further 120° rotation of the cam. The raising of the follower takes place with cycloidal motion and the lowering with uniform acceleration and deceleration. However, the uniform acceleration is 2/3 of the uniform deceleration. The least radius of the cam is 25 mm which rotates at 300 rpm.

Draw the cam profile and determine the values of the maximum velocity and maximum acceleration during rising, and maximum velocity and uniform acceleration and deceleration during lowering of the follower.

Solution:

$h = 25$ mm	$\varphi_a = 120°$
$r_c = 25$ mm	$\delta_1 = 30°$
$N = 300$ rpm	$\varphi_d = 120°$
	$\delta_2 = 90°$

For uniform acceleration, $v = ft$. Refer Eq. (7.7)

During the return stroke of the follower, initial velocity is zero. Then with a uniform acceleration the maximum velocity is attained, followed by a uniform deceleration that again makes the velocity to be zero.

Thus, if the uniform acceleration is to be 2/3 of the uniform deceleration, the time for acceleration must be 3/2 times the time for deceleration for the same maximum velocity.

Let f' = uniform deceleration (magnitude)

t' = period of deceleration

Then Uniform acceleration, $f = \frac{2}{3}f'$

Period of acceleration,

$$t = \frac{3}{2}t'$$

Displacement, $s = ut + \frac{1}{2}ft^2$

But Initial velocity $u = 0$

Distance moved during acceleration period

$$= \frac{1}{2}\left(\frac{2}{3}f'\right)\left(\frac{3}{2}t'\right)^2$$

$$= \frac{3}{2}\left(\frac{1}{2}f'\,t'^2\right)$$

In words, it means that the distance travelled during acceleration period is 3/2 times the distance travelled during the deceleration.

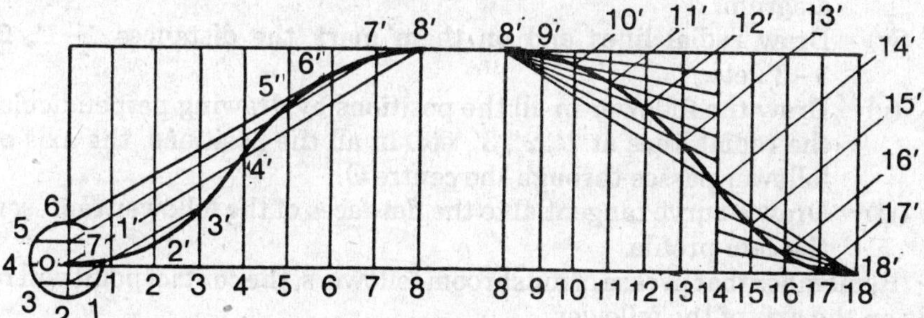

(a)

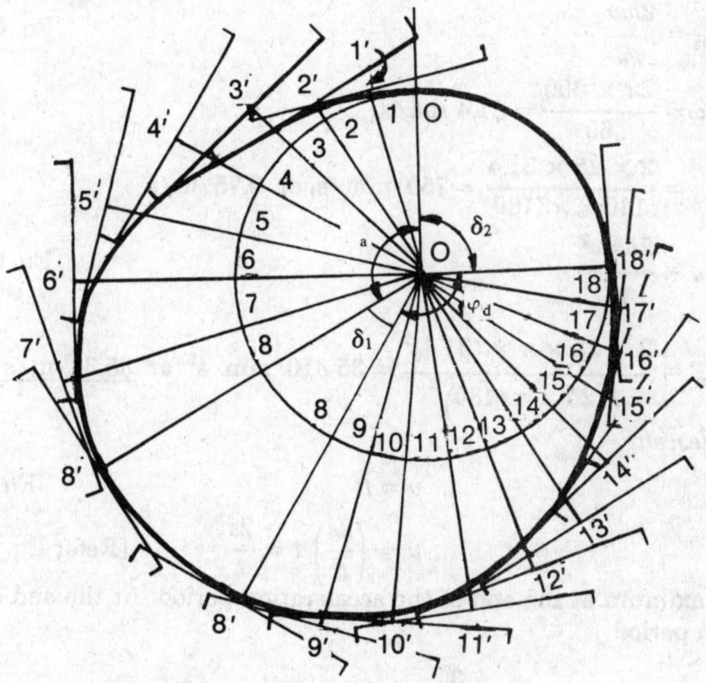

(b)

Fig. 7.25

The displacement diagram has been shown in Fig. 7.25 a. Note that during the return stroke, the time of acceleration and the displacement are 3/2 times of the corresponding values during the deceleration. (Time is measured along the x-axis and displacement along the y-axis.) Thus the time of acceleration is 3/5 of the total time of return and the displacement is 3/5 of the total displacement.

To draw the cam profile, proceed as follows (Fig. 7.25b):

(i) Draw a circle with radius r_c.

(ii) Take angles φ_a, δ_1, φ_d and δ_2 as before (in the counter-clockwise direction if the cam rotation is assumed clockwise).

(iii) Divide φ_a and φ_d into same number of parts as in the displacement diagram.

(iv) Draw radial lines and on them mark the distances $1-1'$, $2-2'$, $3-3'$, etc.

(v) Draw the follower in all the positions by drawing perpendiculars to the radial lines at $1'$, $2'$, $3'$, etc. In all the positions, the axis of the follower passes through the centre O.

(vi) Draw a curve tangential to the flat-faces of the follower representing the cam profile.

Remember that in case of mushroom followers, the contact point will rarely lie on the axis of the follower.

During ascent

$$v_{max} = \frac{2h\omega}{\varphi_a} \qquad \text{Eq. (7.14)}$$

where $\omega = \dfrac{2\pi \times 300}{60} = 31.4$ rad/s

$$v_{max} = \frac{2 \times 25 \times 31.4}{120 \times \pi/180} = 750 \text{ mm/s or } 0.75 \text{ m/s}$$

$$f_{max} = \frac{2h\pi\omega^2}{\varphi_a^2} \qquad \text{Eq. (7.16)}$$

$$f_{max} = \frac{2 \times 25 \times \pi \times (31.4)^2}{(120 \times \pi/180)^2} = 35\,310 \text{ mm/s}^2 \text{ or } \underline{35.31 \text{ m/s}^2}$$

During descent

$$v = ft \qquad \text{Eq. (7.7)}$$

or

$$v = \left(\frac{2s}{t^2}\right) t = \frac{2s}{t} \qquad \text{[Refer Eq. (7.6)]}$$

v will be maximum at the end of the acceleration period. At the end of the acceleration period,

$$s = \frac{3}{5} \times 25 = 15 \text{ mm}$$

and the time taken to travel this distance is found as under,

$$\text{Time for 300 rev.} = 60 \text{ s}$$

$$\text{Time for 1 rev. } (=360°) = \frac{60}{300} = 0.2 \text{ s}$$

$$\text{Time for } \left(\frac{3}{5} \times 120°\right) = 0.2 \times \frac{72}{360} = 0.04 \text{ s}$$

$$v_{max} = \frac{2 \times 15}{0.04} = \underline{750 \text{ mm/s}^2} \text{ or } \underline{0.075 \text{ m/s}^2}$$

$$\text{Uniform acceleration} = \frac{v_{max}}{\text{time}} = \frac{0.75}{0.04} = \underline{18.75 \text{ m/s}^2}$$

$$\text{Uniform deceleration} = 18.75 \times \frac{3}{2} = \underline{28.13 \text{ m/s}^2}$$

Example 7.5 The following data relate to a cam operating an oscillating roller follower:

Minimum diameter of the cam = 44 mm
Diameter of roller = 14 mm
Length of the follower arm = 40 mm
Distance of fulcrum centre from cam centre = 50 mm
Angle of ascent = 75°
Angle of descent = 105°
Angle of dwell for the follower in the highest position = 60°
Angle of oscillation of follower = 28°

Draw the profile of the cam if the ascent and descent both take place with SHM.

Solution:

$r_c = 22$ mm $\quad\quad \theta = 28°$

$r_r = 7$ mm $\quad\quad \varphi_a = 75°$

Follower arm length = 40 mm $\quad\quad \delta_1 = 60°$

$h = \theta \times$ Arm length $\quad\quad \varphi_d = 105°$

$= \left(28° \times \frac{\pi}{180°}\right) \times 40$ $\quad\quad \delta_2 = 120°$

$= 19.5$ mm

The displacement diagram has been shown in Fig. 7.26 a. To draw the cam profile, proceed as follows:

(i) Draw a circle with radius $(r_c + r_r)$ (Fig. 7.26 b).

(ii) Assuming the initial position of the roller centre vertically above the cam centre O, locate the fulcrum centre as its distances from the cam centre and the roller centre (equal to length of follower arm) are known.

(iii) Draw a circle with radius OA and centre at O.

(iv) On the circle through A, starting from OA, take angles φ_a, δ_1 and φ_d as usual.

(v) Divide the angles φ_a and φ_d into same number of parts as is done in the displacement diagram and obtain the points a, b, c, d etc. on the circle through A.

(vi) With centres A, a, b, etc., draw arcs with radii equal to length of the arm.

(vii) Mark distances 1−1′, 2−2′, 3−3′, etc. on these arcs as shown in the diagram. It is on the assumption that for small angular displace-

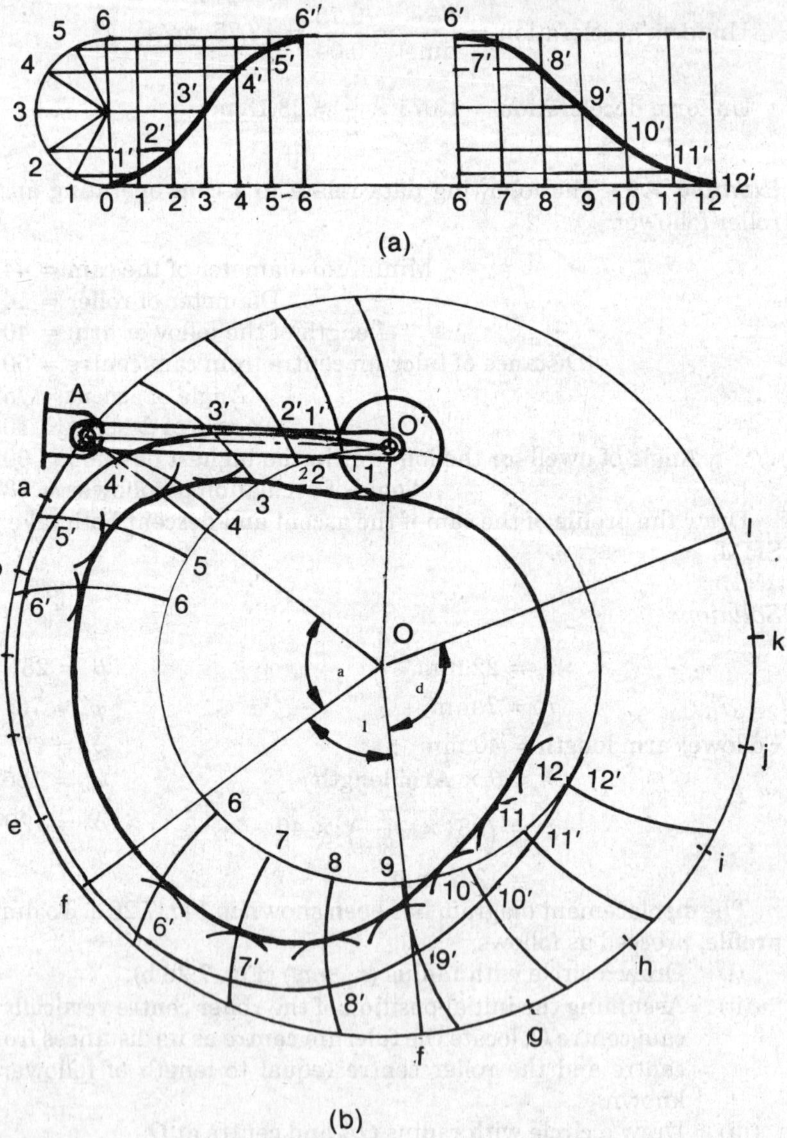

(a)

(b)

Fig. 7.26

ments, the linear displacements on the arcs and on the straight lines are the same.

(viii) With 1′, 2′, 3′, etc. draw a series of arcs of radii equal to r_r.

(ix) Draw a smooth curve tangential to all the arcs and obtain the required cam profile.

7.7 CAMS WITH SPECIFIED CONTOURS

The motion programmes of the follower given in the previous section are helpful in drawing the contour or profile of a cam. In the present section, cams with specified contours will be taken and analysed.

1. Tangent Cam (with Roller Follower)

A tangent cam is symmetrical about the centre line. It has straight flanks (such as AK in Fig. 7.27) with circular nose. The centre of the cam is at O and that of the nose at Q.

Let r_c = least radius of cam

r_n = radius of nose

r_r = radius of roller

r = distance between the cam and the nose centres

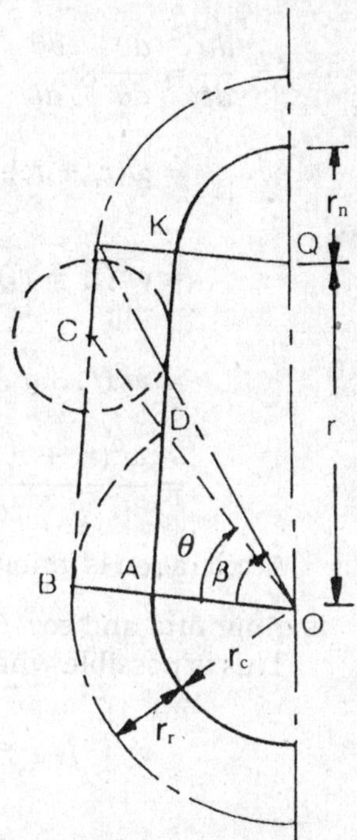

Fig. 7.27

Roller on the Flank When the roller is on the straight flank, the centre of the roller is at C on the pitch profile as shown in Fig. 7.27.

Let θ = angle turned by the cam from the beginning of the follower motion

$$\text{Let } x = OC - OD = OC - OB = \frac{OB}{\cos\theta} - OB = OB\left(\frac{1}{\cos\theta} - 1\right)$$

$$\text{or } x = (r_c + r_r)\left(\frac{1}{\cos\theta} - 1\right) \tag{7.17}$$

$$v = \frac{dx}{dt} = \frac{dx}{d\theta}\cdot\frac{d\theta}{dt} = (r_c + r_r)\left(\frac{\sin\theta}{\cos^2\theta} - 0\right)\omega$$

$$= \omega\,(r_c + r_r)\frac{\sin\theta}{\cos^2\theta} \tag{7.18}$$

$\sin\theta$ increases with the increase in θ whereas $\cos\theta$ decreases. Hence the velocity increases with θ and it is maximum when θ is maximum. This will happen when the point of contact leaves the straight flank.

Let β = angle turned by the cam when the roller leaves the flank.

$$\therefore \quad v_{max} = \omega(r_c + r_r)\frac{\sin\beta}{\cos^2\beta} \quad \text{and} \quad v_{min} = 0 \text{ at } \theta = 0$$

$$f = \frac{dv}{dt} = \frac{dv}{d\theta} \times \frac{d\theta}{dt}$$

$$= \omega(r_c + r_r)\left[\frac{\cos^2\theta\cos\theta - \sin\theta(-2\sin\theta\cos\theta)}{\cos^4\theta}\right]\omega$$

$$= \omega^2(r_c + r_r)\left[\frac{\cos^2\theta + 2\sin^2\theta}{\cos^3\theta}\right]$$

$$= \omega^2(r_c + r_r)\left[\frac{2(\cos^2\theta + \sin^2\theta) - \cos^2\theta}{\cos^3\theta}\right]$$

$$= \frac{\omega^2(r_c + r_r)(2 - \cos^2\theta)}{\cos^3\theta} \qquad (7.19)$$

Acceleration is minimum when $\frac{2 - \cos^2\theta}{\cos^3\theta}$ is minimum, i.e., when $2 - \cos^2\theta$ is minimum and $\cos^3\theta$ is maximum or when $\cos\theta$ is maximum.

This is possible when $\theta = 0°$ or when the roller touches the straight flank.

$$f_{\min} = \omega^2(r_c + r_r) \qquad (7.20)$$

Roller on the Nose Let the centre of the roller be at the point C on the pitch profile over the nose. Let QN be the perpendicular to OC (Fig. 7.28).

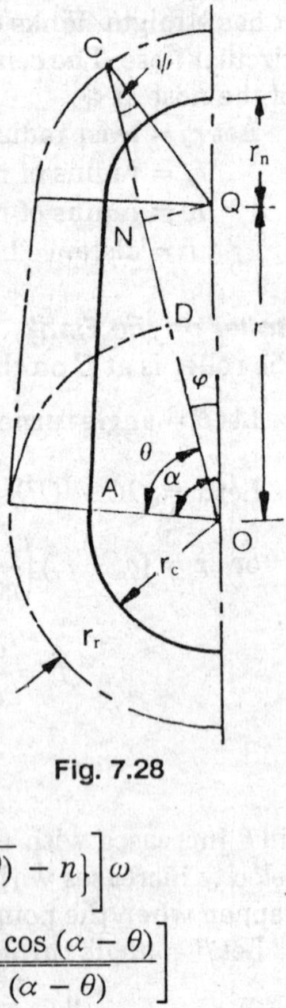

Fig. 7.28

$$x = OC - OD$$

$$= (ON + NC) - OB$$

$$= OQ\cos\varphi + CQ\cos\psi - OB$$

$$= OQ\cos\varphi + CQ\sqrt{1 - \sin^2\psi} - OB$$

$$= OQ\cos\varphi + CQ\sqrt{1 - \frac{(NQ)^2}{(CQ)^2}} - OB$$

$$= r\cos(\alpha - \theta) + \sqrt{l^2 - r^2\sin^2(\alpha - \theta)} - n \qquad (7.21)$$

where $l = CQ = r_n + r_r$

and $n = OB = r_c + r_r$

$$v = \frac{dx}{dt} = \frac{dx\, d\theta}{d\theta\, dt}$$

$$= \left[\frac{d}{d\theta}\left\{r\cos(\alpha - \theta) + \sqrt{(l^2 - r^2\sin^2(\alpha - \theta)} - n\right\}\right]\omega$$

$$= \omega\left[-r\sin(\alpha - \theta)(-1) + \frac{1}{2}\frac{r^2\, 2\sin(\alpha - \theta)\cos(\alpha - \theta)}{\sqrt{l^2 - r^2\sin^2(\alpha - \theta)}}\right]$$

$$= \omega r \left[\sin(\alpha - \theta) + \frac{r \sin 2(\alpha - \theta)}{2\sqrt{l^2 - r^2 \sin^2(\alpha - \theta)}} \right] \quad (7.22)$$

$$f = \frac{dv}{dt} = \frac{dv\, d\theta}{d\theta\, dt} = \omega r \frac{d}{d\theta}\left[\sin(\alpha - \theta) + \frac{r}{2} \frac{\sin 2(\alpha - \theta)}{\sqrt{l^2 - r^2 \sin^2(\alpha - \theta)}} \right] \omega$$

$$= \omega^2 r \frac{d}{d\theta}[\sin(\alpha - \theta)] + \frac{\omega^2 r^2}{2} \frac{d}{d\theta}\left[\sin 2(\alpha - \theta)\left\{l^2 - r^2 \sin^2(\alpha - \theta)\right\}^{-1/2} \right]$$

$$= \omega^2 r[\cos(\alpha - \theta)(-1)] + \frac{\omega^2 r^2}{2}\Big[\sin 2(\alpha - \theta)\left(-\frac{1}{2}\right)$$

$$\left\{l^2 - r^2 \sin^2(\alpha - \theta)\right\}^{-3/2}(-r^2)\left\{2 \sin(\alpha - \theta) \cos(\alpha - \theta)(-1)\right\}$$

$$+ \left\{l^2 - r^2 \sin^2(\alpha - \theta)\right\}^{-1/2} 2 \cos 2(\alpha - \theta)(-1)\Big]$$

$$= \omega^2 r \left[-\cos(\alpha - \theta) - \frac{r^3 \sin^2 2(\alpha - \theta)}{4[l^2 - r^2 \sin^2(\alpha - \theta)]^{3/2}} - \frac{r \cos 2(\alpha - \theta)}{\sqrt{l^2 - r^2 \sin^2(\alpha - \theta)}} \right] \quad (7.23)$$

2. Circular Arc(Convex) Cam (with Flat-faced Follower)

Follower Touching Circular Flank(Fig. 7.29) When the flat-faced follower touches point E on the circular flank, it is lifted through a distance AC.

Let P be the centre of the circular arc of the flank and r_f the radius of the circular flank (= PD = PE).

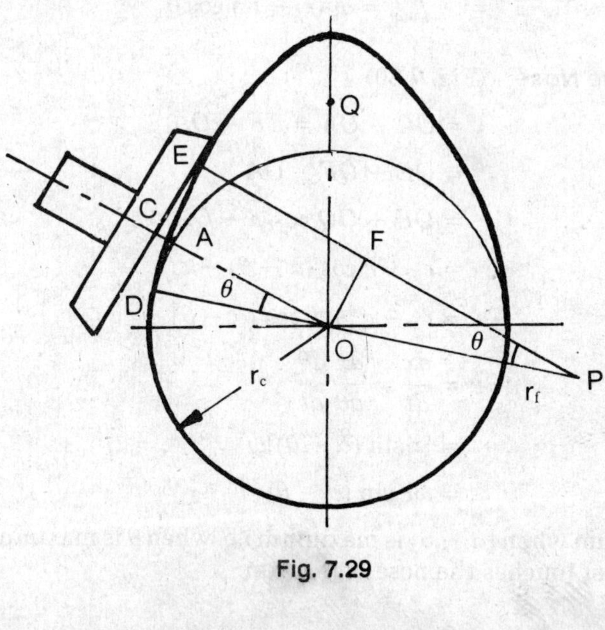

Fig. 7.29

$$x = OC - OA = EF - OA = (PE - PF) - OA$$

$$= PE - OP \cos\theta - OA$$

$$= PE - (PD - OD)\cos\theta - OA$$

or $$x = r_f - (r_f - r_c)\cos\theta - r_c$$

$$= r_f - r_f\cos\theta + r_c\cos\theta - r_c$$

$$= r_f(1 - \cos\theta) - r_c(1 - \cos\theta)$$

$$= (r_f - r_c)(1 - \cos\theta) \tag{7.24}$$

$$v = \frac{dx}{dt} = \frac{dx}{d\theta}\frac{d\theta}{dt}$$

$$= [(r_f - r_c)(\sin\theta)]\,\omega$$

$$= \omega\,(r_f - r_c)\sin\theta \tag{7.25}$$

It is zero when $\theta = 0$, i.e. when the follower starts ascending. It increases with θ and is maximum when θ is maximum or when the follower leaves the flank and $\theta = \beta$.

$$v_{max} = \omega(r_f - r_c)\sin\beta$$

$$f = \frac{dv}{dt} = \frac{dv}{d\theta}\frac{d\theta}{dt} = [(r_f - r_c)\cos\theta]\omega = \omega^2(r_f - r_c)\cos\theta \tag{7.26}$$

It is maximum at the beginning when $\theta = 0$, i.e. when the rise commences.

$$f_{max} = \omega^2(r_f - r_c)$$

$$f_{min} = \omega^2(r_f - r_c)\cos\beta$$

Follower on the Nose (Fig. 7.30)

$$x = OC - OA = EF - OA$$

$$= QE + QF - OA$$

$$= QE + OQ\cos\varphi - OA$$

$$= r_n + r\cos(\alpha - \theta) - r_c$$

$$= r_n - r_c \pm r\cos(\alpha - \theta) \tag{7.27}$$

$$v = \frac{dx}{dt} = \frac{dx}{d\theta}\frac{d\theta}{dt}$$

$$= [r\sin(\alpha - \theta)]\omega$$

$$= \omega r\sin(\alpha - \theta) \tag{7.28}$$

It is maximum when $(\alpha - \theta)$ is maximum i.e. when θ is maximum or when the follower just touches the nose of the cam.

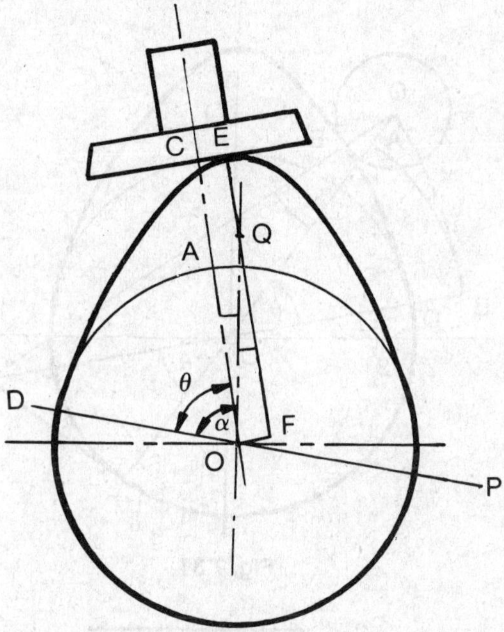

Fig. 7.30

Velocity is minimum when $(\alpha - \theta)$ is minimum or when θ is maximum i.e when the follower is at the apex of the circular nose or $\theta = \alpha$.

$$f = \frac{dv}{dt} = \frac{dv}{d\theta}\frac{d\theta}{dt} = \omega r \cos(\alpha - \theta)(-1)\omega = -\omega^2 r \cos(\alpha - \theta) \qquad (7.29)$$

i.e. it is retardation. It is maximum when $(\alpha - \theta)$ is minimum or when θ is maximum or when the follower is at the apex of the nose.

Similarly, it is minimum at the commencement of the nose travel.

3. Circular Arc (Convex) Cam (with Roller Follower)

Follower on the Flank (Fig. 7.31)

$$x = OC - OA = FC - FO - OB$$

$$= CP \cos\varphi - OP \cos\theta - OB$$

$$= CP \cos\varphi - (DP - DO)\cos\theta - OB$$

or

$$x = (r_f + r_n)\cos\varphi - (r_f - r_c)\cos\theta - (r_c + r_r)$$

where, $\cos\varphi$ is given by,

$$\cos\varphi = \sqrt{1 - \sin^2\varphi}$$

$$= \sqrt{1 - (FP/CP)^2}$$

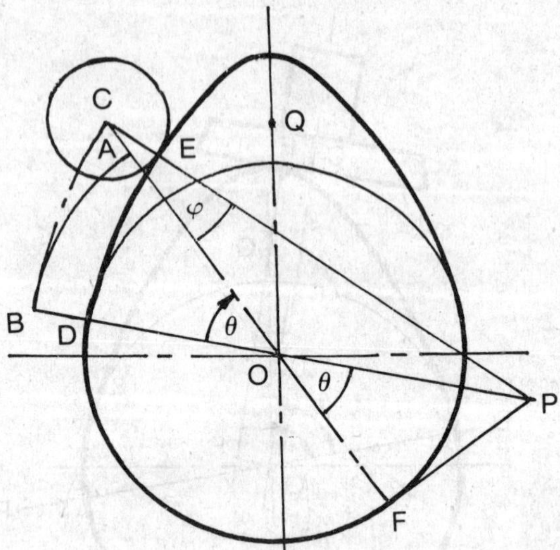

Fig. 7.31

$$= \sqrt{1 - \left(\frac{OP \sin\theta}{CP}\right)^2}$$

$$= \sqrt{1 - \left[\frac{(r_f - r_c)\sin\theta}{(r_f + r_n)}\right]^2}$$

or

$$\cos\varphi = \sqrt{1 - \left(\frac{A\sin\theta}{B}\right)^2} = \frac{1}{B}\sqrt{B^2 - A^2\sin^2\theta}$$

where $A = r_f - r_c$ and $B = r_f + r_n$

$$x = \sqrt{B^2 - A^2\sin^2\theta} - A\cos\theta - (r_c + r_r) \quad (7.30)$$

$$v = \frac{dx}{dt} = \frac{dx\,d\theta}{d\theta\,dt} = \left[\frac{1}{2}(B^2 - A^2\sin^2\theta)^{-1/2}(-A^2\,2\sin\theta\cos\theta) + A\sin\theta\right]\omega$$

$$= \omega A\left[\sin\theta - \frac{A\sin 2\theta}{2\sqrt{B^2 - A^2\sin^2\theta}}\right] \quad (7.31)$$

$$f = \frac{dv}{dt} = \frac{dv}{d\theta}\frac{d\theta}{dt} = \omega^2 A\left[\cos\theta - \frac{A}{2}\left\{\frac{2\cos 2\theta}{\sqrt{B^2 - A^2\sin^2\theta}} + \sin 2\theta\left(-\frac{1}{2}\right)\right.\right.$$

$$\left.\left.(B^2 - A^2\sin^2\theta)^{-3/2}(-A^2)(2\sin\theta\cos\theta)\right\}\right]$$

$$= \omega^2 A\left[\cos\theta - \frac{A\cos 2\theta}{\sqrt{B^2 - A^2\sin^2\theta}} - \frac{A^3\sin 2\theta}{4(B^2 - A^2\sin^2\theta)^{3/2}}\right] \quad (7.32)$$

Follower on the Nose

This case has already been discussed for the tangent cam when the roller follower is on the nose. Same expressions for the displacement, velocity and the acceleration hold good.

Example 7.6 A tangent cam with a base circle diameter of 50 mm operates a roller follower 20 mm in diameter. The line of stroke of the roller follower passes through the axis of the cam. The angle between the tangential faces of the cam is 60°, speed of the cam shaft 200 rpm and the lift of the follower 15 mm. Calculate

(i) the main dimensions of the cam

(ii) the accelerations of the follower at

(a) the beginning of lift

(b) where the roller just touches the nose

(c) the apex of the circular nose.

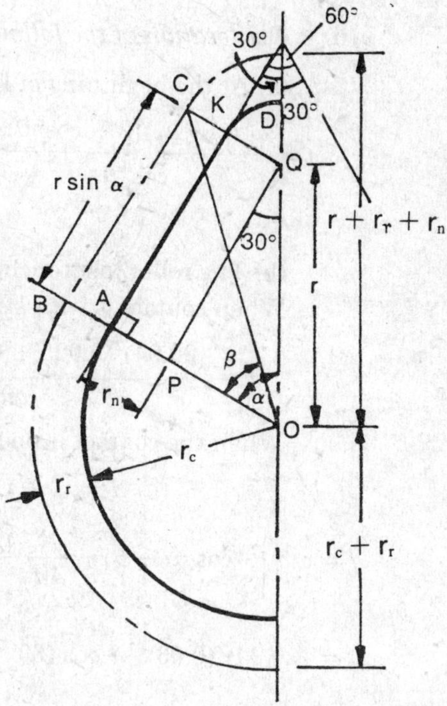

Fig. 7.32

Solution:

$$r_c = 25 \text{ mm} \qquad h = 15 \text{ mm}$$

$$r_r = 10 \text{ mm} \qquad N = 200 \text{ rpm}$$

$$\alpha = (180° - 30° - 90°) = 60°$$

(i) Refer Fig. 7.32,

$$r + r_n + r_r = r_c + r_r + h$$

$$r + r_n = r_c + x = 25 + 15 = 40 \qquad \text{(i)}$$

Also,

$$OP + r_n = r_c$$

or

$$r \cos 60° + r_n = 25$$

or

$$0.5\, r + r_n = 25 \qquad \text{(ii)}$$

Subtracting (ii) from (i),

$$r = 30 \text{ mm}$$

$$r_n = 10 \text{ mm}$$

$$\tan \beta = \frac{r \sin \alpha}{r_c + r_r}$$

$$= \frac{30 \sin 60°}{25 + 10}$$

$$= 0.742$$

$$\therefore \qquad \beta = 36.6° \text{ or } 36°36'$$

(ii) *Acceleration of the follower* $\omega = \dfrac{2\pi \times 200}{60} = 20.94 \text{ rad/s}$

(a) At the beginning of Lift, i.e. roller centre at B, $\theta = 0°$.

$$f = \frac{\omega^2(r_c + r_r)(2 - \cos^2\theta)}{\cos^3\theta} = \frac{(20.94)^2(0.025 + 0.01)(2 - \cos^2 0°)}{\cos^3 0°}$$

$$= \underline{15.35 \text{ m/s}^2}$$

(b) The roller just touches the nose ($\theta = \beta$), i.e. the roller centre at C. When contact is with straight flank, $\theta = 36.6° = \beta$

$$f = \frac{(20.94)^2(0.025 + 0.01)(2 - \cos^2 36.6°)}{\cos^3 36.6°} = \underline{40.2 \text{ m/s}^2}$$

When the contact is on the circular nose,

$$l = r_n + r_r = 10 + 10 = 20 \text{ mm}$$

$$f = \omega^2 r\left[-\cos(\alpha - \theta) - \frac{r^3\sin^2 2(\alpha - \theta)}{4[l^2 - r^2\sin^2(\alpha - \theta)]^{3/2}} - \frac{r\cos 2(\alpha - \theta)}{\sqrt{l^2 - r^2\sin^2(\alpha - \theta)}}\right]$$

$$= (20.94)^2(0.03)\left[-\cos(60° - 36.6°) - \right.$$

$$\left.\frac{(0.03)^3\sin^2 2(60° - 36.6°)}{4[(0.02)^2 - (0.03)^2\sin^2(60° - 36.6°]^{3/2}} - \frac{(0.03)\cos 2(60° - 36.6°)}{\sqrt{(0.02)^2 - (0.03)^2\sin^2(60° - 36.6°)}}\right]$$

$$= 13.15(-0.917 - 0.865 - 1.278)$$

$$= \underline{-40.24 \text{ m/s}^2}$$

(c) When the roller is at the apex of the circular nose, i.e. at D, $\theta = \alpha$, and

$$f = \omega^2 r\left(-1 - \frac{r}{l}\right) = (20.94)^2 \times 0.03\left(-1 - \frac{0.03}{0.02}\right) = \underline{-32.9 \text{ m/s}^2}$$

Example 7.7 The following data relate to a circular cam operating a flat-faced follower:

Least diameter = 40 mm
Lift = 12 mm
Angle of action = 160°
Speed = 500 rpm

If the period of acceleration of the follower is 60% of the retardation during the lift, determine

(i) the main dimensions of the cam
(ii) the acceleration at the main points.

What is the maximum acceleration and deceleration during the lift?

Solution:

$r_c = 20$ mm $\qquad h = 12$ mm

$2\alpha = 160°$ $\qquad N = 500$ rpm

$\alpha = 80°$

During the lifting of the follower, the acceleration takes place when the follower is on the radial flank and the deceleration when the follower is on the nose. When the follower just touches the nose, the follower position will be as shown in Fig. 7.33. OC and PQE are parallel and the angles $\beta = 30°$ and $\varphi = 50°$ so that β is 60% of φ.

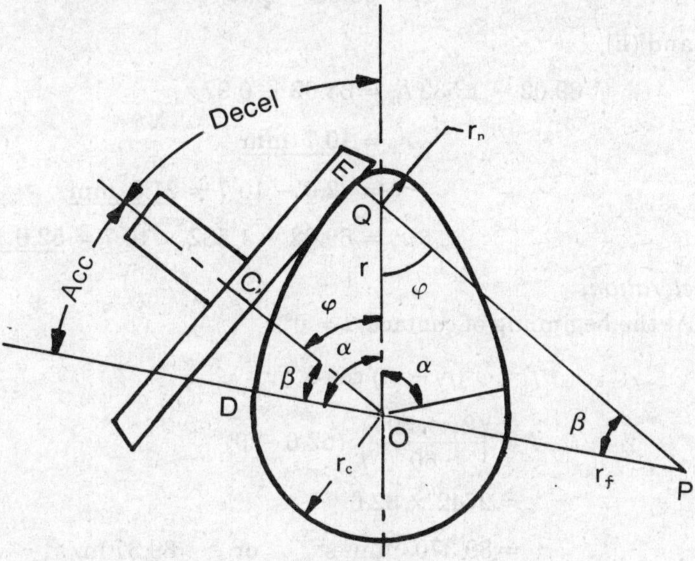

Fig. 7.33

(i) Apply sine rule to $\Delta\, POQ$,

$$\frac{OP}{\sin \varphi} = \frac{PQ}{\sin (180° - \alpha)} = \frac{OQ}{\sin \beta}$$

$$\frac{r_f - r_c}{\sin 50°} = \frac{r_f - r_n}{\sin 100°} = \frac{r}{\sin 30°}$$

But $\qquad r + r_n = r_c + h$

or $\qquad r = r_c + h - r_n$

$$= 20 + 12 - r_n$$

$$= 32 - r_n$$

$$\frac{r_f - 20}{\sin 50°} = \frac{r_f - r_n}{\sin 100°} = \frac{32 - r_n}{\sin 30°}$$

From first and last terms,

$$0.5\, r_f - 10 = 24.51 - 0.766\, r_n$$

or

$$0.5 r_f = 34.51 - 0.766 r_n$$

$$r_f = 69.02 - 1.532 r_n \qquad \text{(i)}$$

From second and last terms,

$$0.5\, r_f - 0.5\, r_n = 31.51 - 0.985\, r_n$$

$$r_f = 63.02 - 0.97\, r_n \qquad \text{(ii)}$$

From (i) and (ii),

$$69.02 - 1.532\, r_n = 63.02 - 0.97\, r_n$$

$$r_n = \underline{10.7 \text{ mm}}$$

$$r = 32.0 - 10.7 = \underline{21.3 \text{ mm}}$$

$$r_f = 69.02 - 1.532 \times 10.7 = \underline{52.6 \text{ mm}}$$

(ii) *Accelerations*

(a) At the beginning of contact, $\theta = 0°$,

$$f = \omega^2 (r_f - r_c) \cos 0°$$

or

$$f = \left(\frac{2\,\pi \times 500}{60}\right)^2 (52.6 - 20)$$

$$= 2742 \times 32.6$$

$$= 89\,370 \text{ mm/s}^2 \quad \text{or} \quad \underline{89.37 \text{ m/s}^2}$$

(b) Contact on circular flank when $\theta = \beta = 30°$,

$$f = 2742 \times 32.6 \cos 30° = 77\,410 \text{ mm/s}^2 \quad \text{or} \quad \underline{77.41 \text{ m/s}^2}$$

(c) Contact on circular nose, when $\theta = \beta = 30°$,

$$f = -\omega^2 r \cos(\alpha - \theta) = -2742 \times 21.3 \cos(80° - 30°)$$

$$= -37\,540 \text{ mm/s}^2 \quad \text{or} \quad \underline{-37.54 \text{ m/s}^2}$$

(d) Contact at the apex of nose, $\theta = \alpha = 80°$,

$$f = -\omega^2 r \cos(80° - 80°) = -2742 \times 21.3$$

$$= -58\,400 \text{ mm/s}^2 \text{ or } \underline{-58.4 \text{ m/s}^2}$$

Maximum acceleration is when the contact is just made with the circular flank; it is 89.37 m/s^2 and the maximum retardation is at the end of the lifting period, i.e. when the contact is at the apex of the nose; it is 58.4 m/s^2.

Example 7.8 In a four-stroke petrol engine, the exhaust valve opens 45° before the t.d.c. and closes 15° after the b.d.c. The valve has a lift of 12 mm. The least radius of the circular arc type cam operating a flat-faced follower is 25 mm. The nose radius is 3 mm. The camshaft rotates at 1500 rpm. Calculate the maximum velocity of the valve and the minimum force exerted by the spring to overcome the inertia of the moving parts that weight 300 g.

Solution:

$r_c = 25$ mm $\qquad N = 1500$ rpm

$h = 12$ mm $\qquad r_n = 3$ mm

$m = 0.3$ kg

Crank rotation during lifting of the exhaust valve =

$$45° + 180° + 15° = 240°$$

In four-stroke engines, the camshaft speed is half that of the crankshaft.

$$\text{Angle of action of the camshaft, } 2\alpha = \frac{240}{2} = 120°$$

$$\therefore \qquad \alpha = \frac{120}{2} = 60°$$

Refer to Fig. 7.33.

$$r + r_n = r_c + h$$

$$\text{or} \qquad r = 25 + 12 - 3 = 34 \text{ mm}$$

$$(PQ)^2 = (OP)^2 + (OQ)^2 - 2(OP)(OQ)\cos\angle POQ$$

$$(r_f - 3)^2 = (r_f - 25)^2 + (34)^2 - 2(r_f - 25)(34)\cos(180° - 60°)$$

$$r_f + 9 - 6r_f = r_f^2 + 625 - 50r_f + 1156 + 34r_f - 85°$$

$$r_f = 92.2 \text{ mm}$$

Applying sine rule to ΔOPQ,

$$\frac{OQ}{\sin\beta} = \frac{PQ}{\sin(180° - \alpha)}$$

$$\text{or} \qquad \frac{r}{\sin\beta} = \frac{r_f - r_n}{\sin 120°}$$

$$\text{or} \qquad \frac{34}{\sin\beta} = \frac{92.2 - 3}{\sin 120°}$$

$$\sin\beta = 0.33 = 19.27°$$

Velocity is maximum when the contact is on the point where the circular flank meets the circular nose.

$$v_{max} = \omega(r_f - r_c) \sin \beta$$

$$= \frac{2\pi \times 1500}{60} (92.2 - 25) \sin 19.27°$$

$$= 157.08 \times 67.2 \times 0.33$$

$$= 3480 \text{ mm/s} \quad \text{or} \quad \underline{3.48 \text{ m/s}}$$

The same result is obtained if relation of Eqn. (7.28) is used, i.e.

$$v_{max} = \omega r \sin (\alpha - \theta)$$

Maximum acceleration is when $\theta = 0$,

$$f_{max} = \omega^2(r_f - r_c) = (157.08)^2(92.2 - 25)$$

$$= 1658\,090 \text{ mm/s}^2 \quad \text{or} \quad 1658.09 \text{ m/s}^2$$

Maximum retardation is when $\alpha - \theta = 0$,

$$f_{max} = \omega^2 r = (157.08)^2 \times 34 = 838\,920 \text{ mm/s}^2 \quad \text{or} \quad \underline{838.92 \text{ m/s}^2}$$

Spring force is needed to maintain contact during the retardation of the follower.

$$\text{Minimum force, } F = m \times f = 0.3 \times 838.92 = \underline{251.7 \text{ N}}$$

PROBLEMS

7.1 Draw the profile of a cam that gives a lift of 40 mm to a rod carrying a 20 mm diameter roller. The axis of the roller passes through the centre of the cam. The least radius of the cam is 50 mm. The rod is to be lifted with simple harmonic motion in a quarter revolution and is to be dropped suddenly at half revolution.
Determine the maximum velocity and maximum acceleration during the lifting. The cam rotates at 60 rpm.

(0.25 m/s; 3.155 m/s²)

7.2 Layout the profile of a cam so that the follower
— is moved outwards through 30 mm during 180° of cam rotation with cycloidal motion
— dwells for 20° of the cam rotation
— returns with uniform velocity during the remaining 160° of the cam rotation.
The base circle diameter of the cam is 28 mm and the roller diameter 8 mm. The axis of the follower is offset by 6 mm to the left. What will be the maximum velocity and acceleration of the follower during the outstroke if the cam rotates at 1500 rpm counter-clockwise.

(3 m/s; 471.2 m/s²)

7.3 Use the following data in drawing the profile of a cam in which a knife-edged follower is raised with uniform acceleration and deceleration and is lowered with simple harmonic motion:

Least radius of cam = 60 mm
Lift of follower = 45 mm
Angle of ascent = 60°
Angle of dwell between ascent and descent = 40°
Angle of descent = 75°

If the cam rotates at 180 rpm, determine the maximum velocity and acceleration during ascent and descent.

(Ascent: 1.6 m/s; 58.3 m/s^2
Descent: 1.02 m/s; 46.05 m/s^2)

7.4 Draw the profile of a cam which is to give oscillatory motion to the follower with uniform angular velocity about its pivot. The base circle diameter is 50 mm, angle of oscillation of the follower 30° and the distance between the cam centre and the pivot of the follower 60 mm. The oscillating lever is 60 mm long with a roller of 8 mm diameter at the end. One oscillation of the follower is completed in one revolution of the cam.

7.5 Set out the profile of a cam to give the following motion to a flat mushroom contact face follower:

— Follower to rise through 24 mm during 150° of cam rotation with SHM.
— Follower to dwell for 30° of the cam rotation.
— Follower to return to the initial position during 90° of the cam rotation with SHM.
— Follower to dwell for the remaining 90° of cam rotation.

Take minimum radius of the cam as 30 mm.

7.6 A cam is required to give motion to a follower fitted with a roller 50 mm in diameter. The lift of the follower is 30 mm and is performed

— with uniform acceleration for 12 mm, the cam turns through 45°.
— with uniform velocity for 12 mm, the cam turns through next 30°.
— with uniform deceleration for the remainder of the lift, the cam turns through next 45°.

The follower falls through immediately with simple harmonic motion while the cam turns through 120°. Then a period of dwell is followed for 120° of the cam angle.

Construct a lift and fall diagram on a cam angle base. Also draw the outline of the cam. The least radius of the cam is 35 mm. The line of motion of the follower passes through the centre of the cam axis.

7.7 The following particulars relate to a symmetrical tangent cam having a roller follower:

Minimum radius of the cam = 40 mm
Lift = 20 mm
Speed = 360 rpm
Roller diameter = 44 mm
Angle of ascent = 60°

Calculate the acceleration of follower at the beginning of lift. Also find its values when the roller just touches the nose and is at the apex of the circular nose. Sketch the variation of displacement, velocity and acceleration during ascent.

(88.12 m/s^2; 164 m/s^2 and $-92.6\ m/s^2$; $-111\ m/s^2$)

7.8 A flat ended valve tappet is operated by a symmetrical cam with circular arcs for flank and nose profiles. The total angle of action is 150°, base circle diameter 125 mm and the lift 25 mm. During the lift, the period of acceleration is half

that of the deceleration. Speed of cam shaft is 1250 rpm. The straight line path of the tappet passes through the cam axis. Find:
(i) Radii of the nose and the flank, and
(ii) Maximum acceleration and deceleration during the lift.

(40 mm, 149 mm; 1476 m/s^2, 813.6 m/s^2)

7.9 In a four-stroke petrol engine, the crank angle is 50° after t.d.c. when the suction valve opens and 53° after b.d.c. when the suction valve closes. The lift is 8 mm, the nose radius 3 mm and the least radius of the cam 18 mm. The shaft rotates at 800 rpm. The cam is of the circular type with a circular nose and flanks while the follower is flat-faced. Determine the maximum velocity and the maximum acceleration and retardation of the valve.
What is the minimum force exerted by the springs to overcome the inertia of moving parts weighing 250 gm.

(1.05 m/s; 187.4 m/s; −168.46 m/s^2; 42.11 N)

8
FRICTION

8.1 INTRODUCTION

When a body slides over another, the motion is resisted by a force called the force of friction. The force arises from the fact that the surfaces, though planed and made smooth, have ridges and depressions that interlock and the relative movement is resisted. Thus the force of friction on a body is parallel to the sliding surfaces and acts in a direction opposite to that of the sliding body.

There are phenomena, where it is necessary to reduce the force of friction whereas in some cases it must be increased. In case of lathe slides, journal bearings, etc. where the power transmitted is reduced due to friction, it has to be decreased by the use of lubricated surfaces. In processes, where the power itself is transmitted through friction, attempts are made to increase it to transmit more power. Examples are friction clutches, belt drives, etc. Even the tightness of a nut and bolt is dependent mainly on the force of friction. Had there been no friction between the nut and the surface on which it is tightened, the nut would loosen off at the moment the spanner is removed after tightening.

8.2 KINDS OF FRICTION

Usually, three kinds of friction, depending upon the conditions of surfaces are considered.

1. Dry Friction

Dry friction is said to occur when there is relative motion between two completely unlubricated surfaces. It is further divided into two types:

(a) *Solid Friction* When the two surfaces have a sliding motion relative to each other, it is called a solid friction (Sec. 8.3).

(b) *Rolling Friction* Friction due to rolling of one surface over another (e.g. ball and roller bearings) is called rolling friction (Sec. 8.10).

2. Skin or Greasy Friction

When the two surfaces in contact have a minute thin layer of lubricant between them, it is known as *skin* or *greasy* friction. Higher spots on the surface break through the lubricant and come in contact with the other surface.

Skin friction is also termed *boundary* friction (Sec. 8.12).

3. Film Friction

When the two surfaces in contact are completely separated by a lubricant, friction will occur due to the resistance of motion between the lubricant and the surfaces in contact with it. This is known as *film* friction or *viscous* friction (Sec. 8.15).

8.3 LAWS OF FRICTION

Experiments have shown that the force of solid friction
- is directly proportional to normal reaction between the two surfaces
- opposes the motion between the surfaces
- depends upon the materials of the two surfaces
- is independent of the area of contact
- is independent of the velocity of sliding.

The last of these laws is not true in the strict sense as it has been found that the friction force decreases slightly with the increase in velocity.

8.4 COEFFICIENT OF FRICTION

Let a body of weight **W** rest on a smooth and dry plane surface. Under the circumstances, the plane surface also exerts a reaction force $\mathbf{R}_n$ on the body which is normal to the plane surface. If the plane surface considered is horizontal, $\mathbf{R}_n$ would be equal and opposite to **W** (Fig. 8.1 a).

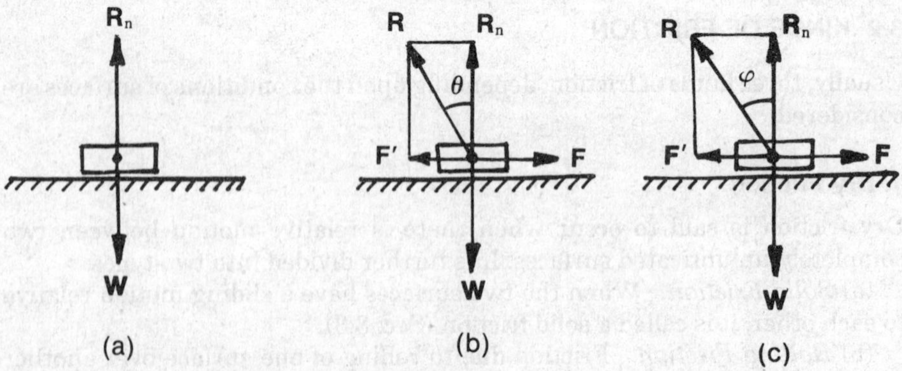

Fig. 8.1

Let a small horizontal force **F** be applied to the body to move it on the surface (Fig. 8.1b). So long the body is unable to move, the equilibrium of the body provides,

$$R_n = W \quad \text{and} \quad F = F'$$

where **F′** is the horizontal force resisting the motion of the body. As the force **F** is increased, the resistive force **F′** also increases accordingly. **F′** and $\mathbf{R}_n$, the friction and the normal reaction forces can also be combined into a single reaction force **R** inclined at an angle θ to the normal. Thus

$$R\cos\theta = W \quad \text{and} \quad R\sin\theta = F$$

At a moment, when the force **F** would just move the body, the value of **F′** or **R** $\sin\theta$ (equal to **F**) is called the static force of friction. Angle θ attains the value φ and the body is in equilibrium under the action of three forces (Fig. 8.1c):

F, in the horizontal direction
W, in the vertical downward direction and
R, at an angle φ with the normal (inclined towards the force of friction).
According to the first law of friction,

$$F' \propto R_n$$

$$= \mu R_n$$

where μ is known as the *coefficient of friction*.
or

$$\mu = \frac{F'}{R_n}$$

Also in Fig. 8.1c, $\tan\varphi = \dfrac{F'}{R_n}$

or

$$\tan\varphi = \frac{\mu R_n}{R_n} = \mu \qquad (8.1)$$

The angle φ is known as the *limiting angle of friction* or simply the *angle of friction.*

Now, if the body moves over the plane surface, it is observed that the friction force will be slightly less than the static friction force. As long as the body moves with a uniform velocity, the force F required for the motion of the body will be equal to the force of friction on the body. However, if the velocity is to increase, additional force will be needed to accelerate the body. Thus while the body is in motion, it can be written that

$$\tan\varphi = \mu$$

where φ is approximately the limiting angle of friction.

Also, no movement is possible until the angle of reaction **R** with the normal becomes equal to the limiting angle of friction or until $\tan\varphi = \mu$.

8.5 INCLINED PLANE

1. Body at Rest

When a body is at rest on an inclined plane making an angle α with the horizontal, the forces acting on the body are (Fig. 8.2):

- **W**, weight of body in downward direction
- $\mathbf{R}_n$, normal reaction
- **F′**, force resisting the motion of body.

From equilibrium conditions,

$W \sin\alpha = F'$ and $W\cos\alpha = R_n$

If the angle of inclination of plane is increased, the body will just slide down the plane of its own when

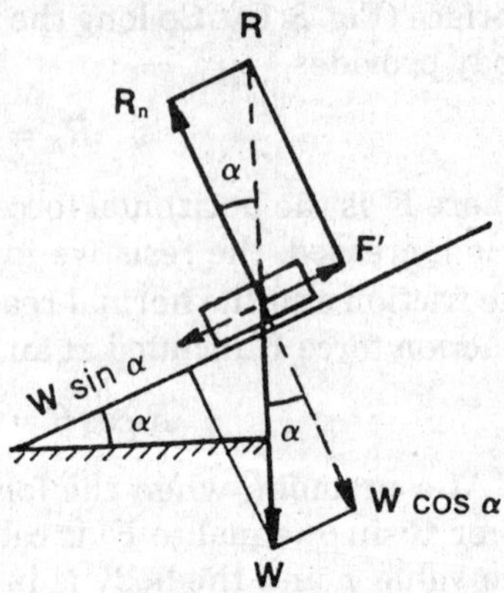

Fig. 8.2

$$W\sin\alpha = F' = \mu R_n = \mu W\cos\alpha$$

or $$\tan\alpha = \mu = \tan\varphi$$

or $$\alpha = \varphi \qquad (8.2)$$

This maximum value of angle of inclination of plane with the horizontal when the body starts sliding of its own is known as the angle of repose or limiting angle of friction.

2. Motion Up the Plane

Consider a body moving up an inclined plane under the action of a force **F** as shown in Fig. 8.3 a.

Let α = angle of inclined plane with the horizontal

θ = angle of force **F** with the direction of **W**.

As the motion is up the plane, the friction force **F′** ($= \mu\,\mathbf{R}_n$) would act downwards along the plane. Combining **F′** and $\mathbf{R}_n$ as before, the body is in equilibrium under the action of forces **F**, **W** and **R** (Fig. 8.3 b).

Applying Lami's theorem, we get

$$\frac{F}{\sin[\pi - (\alpha+\varphi)]} = \frac{W}{\sin[\pi - [\theta - (\alpha+\varphi)]]}$$

$$\frac{F}{\sin(\alpha+\varphi)} = \frac{W}{\sin[\theta - (\alpha+\varphi)]}$$

Alternatively, a triangle of forces can be drawn as shown in Fig. 8.3 c, and applying sine law of forces,

$$\frac{F}{\sin(\alpha+\varphi)} = \frac{W}{\sin[\theta - (\alpha+\varphi)]}$$

Thus, $$F = \frac{W\sin(\alpha+\varphi)}{\sin[\theta-(\alpha+\varphi)]} \qquad (8.3)$$

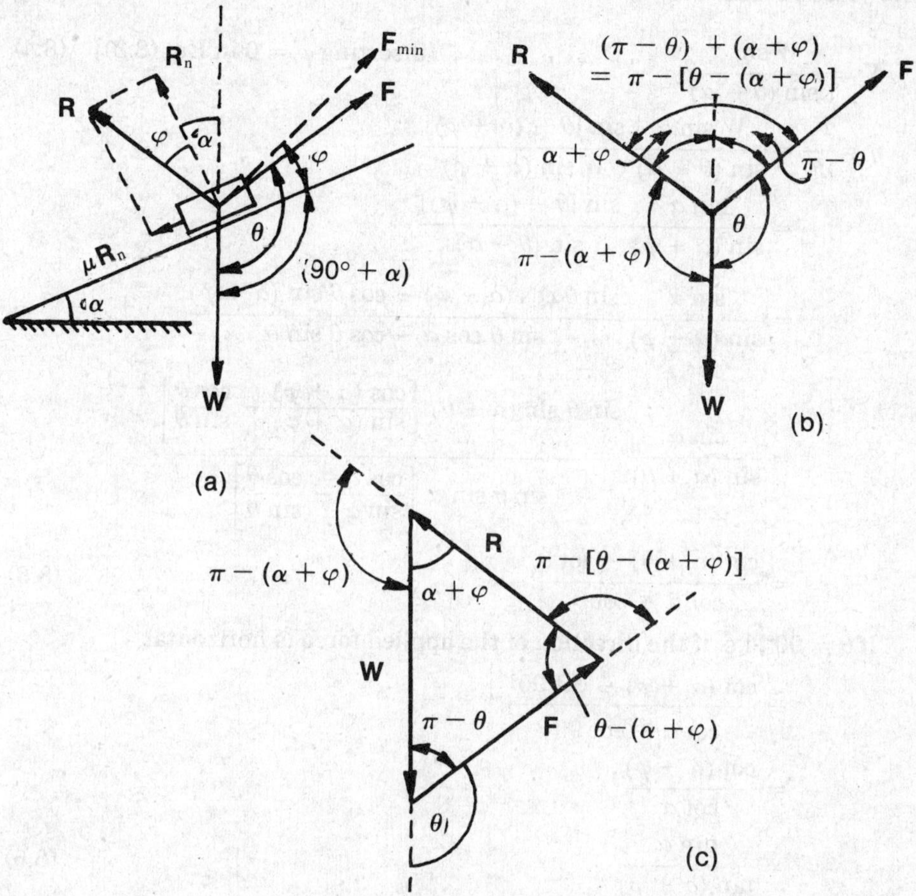

Fig. 8.3

F will be minimum if the denominator on the right-hand side is maximum, i.e.

$$\sin\,[\,\theta - (\alpha + \varphi)] = 1$$

or

$$\theta - \alpha - \varphi = 90^\circ$$

or

$$\theta - (90^\circ + \alpha) = \varphi$$

i.e. the angle between **F** and the inclined plane should be equal to the angle of friction. In that case,

$$F_{\min} = W \sin\,(\alpha + \varphi)$$

Efficiency The efficiency of an inclined plane, when a body slides up the plane, is defined as the ratio of the forces required to move the body without consideration and with consideration of force of friction.

Let F_o = force required to move the body up the plane without friction. In the absence of friction, the force **R** coincides with $\mathbf{R}_n$ and φ is zero, and

$$F_o = \frac{W \sin \alpha}{\sin (\theta - \alpha)} \qquad \text{[inserting } \varphi = 0 \text{ in Eq. (8.3)]} \qquad (8.4)$$

$$\eta = \frac{F_o}{F} = \frac{W \sin \alpha}{\sin (\theta - \alpha)} \frac{\sin [\theta - (\alpha + \varphi)]}{W \sin (\alpha + \varphi)}$$

$$= \frac{\sin \alpha}{\sin (\alpha + \varphi)} \frac{\sin [\theta - (\alpha + \varphi)]}{\sin (\theta - \alpha)}$$

$$= \frac{\sin \alpha}{\sin (\alpha + \varphi)} \frac{\sin \theta \cos (\alpha + \varphi) - \cos \theta \sin (\alpha + \varphi)}{\sin \theta \cos \alpha - \cos \theta \sin \alpha}$$

$$= \frac{\sin \alpha}{\sin (\alpha + \theta)} \frac{\sin \theta \sin (\alpha + \theta) \left[\dfrac{\cos (\alpha + \varphi)}{\sin (\alpha + \varphi)} - \dfrac{\cos \theta}{\sin \theta} \right]}{\sin \theta \sin \alpha \left[\dfrac{\cos \alpha}{\sin \alpha} - \dfrac{\cos \theta}{\sin \theta} \right]}$$

$$= \frac{\cot (\alpha + \theta) - \cot \theta}{\cot \alpha - \cot \theta} \qquad (8.5)$$

If $\theta = 90°$, i.e. if the direction of the applied force is horizontal,

$$= \frac{\cot (\alpha + \varphi) - \cot 90°}{\cot \alpha - \cot 90°}$$

$$= \frac{\cot (\alpha + \varphi)}{\cot \alpha}$$

$$= \frac{\tan \alpha}{\tan (\alpha + \varphi)} \qquad (8.6)$$

3. Motion Down the Plane

When the body moves down the plane, the force of friction $\mathbf{F}'$ $(= \mu \mathbf{R}_n)$ acts in the upwards direction and the reaction $\mathbf{R}$, i.e. the combination of $\mathbf{R}_n$ and $\mathbf{F}'$ is inclined backwards as shown in Fig. 8.4 a. Assume that $\mathbf{F}$ acts downwards.

Applying Lami's theorem as before (Fig. 8.4 b),

$$\frac{F}{\sin [\pi - (\varphi - \alpha)]} = \frac{W}{\sin [\theta + (\varphi - \alpha)]}$$

$$F = \frac{W \sin (\varphi - \alpha)}{\sin [\theta + (\varphi - \alpha)]} \qquad (8.7)$$

The equation suggests that F is positive only for $\varphi > \alpha$ and when $\varphi = \alpha$, the force required to slide the body down is zero, i.e. the body is on the point of moving down under its own weight W.

When $\varphi < \alpha$, i.e. the angle of friction is lesser than the angle of the inclined plane, F will be negative meaning that a force equal to F is to be applied in the opposite direction to resist the motion.

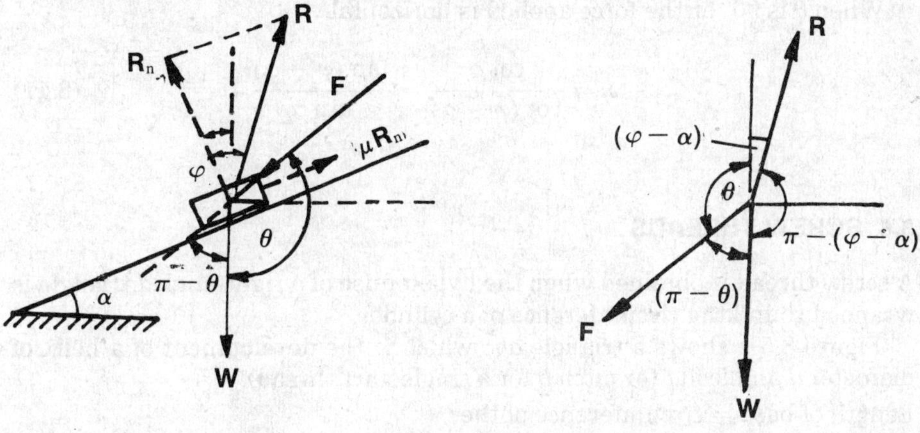

Fig. 8.4

However, for a given value of α, F is minimum when the denominator of Eq. (8.7) is maximum

$$F_{\min} = W \sin(\varphi - \alpha) \tag{8.8}$$

If friction is neglected, i.e. $\varphi = 0$,

$$F_o = \frac{W \sin(-\alpha)}{\sin(\theta - \alpha)} = \frac{-W \sin \alpha}{\sin(\theta - \alpha)} \tag{8.9}$$

The force is negative indicating that in the absence of force of friction, a force in the opposite direction is required to oppose the motion down the plane. This is due to the fact that a component of **W** acts as an effort to move the body in the downward direction.

Efficiency Efficiency of the inclined plane when the body slides down the plane, is defined as the ratio of the forces required to move the body with and without the consideration of force of friction, i.e.,

$$\eta = \frac{F}{F_o} = \frac{W \sin(\varphi - \alpha)}{\sin[\theta + (\varphi - \alpha)]} \frac{\sin(\theta - \alpha)}{W \sin \alpha}$$

$$= \frac{\sin(\varphi - \alpha)}{\sin\theta \cos(\varphi - \alpha) + \cos\theta \sin(\varphi - \alpha)} \frac{\sin\theta \cos\alpha - \cos\theta \sin\alpha}{\sin\alpha}$$

$$= \frac{\sin(\varphi - \alpha)}{\sin\theta \sin(\varphi-\alpha) \left[\dfrac{\cos(\varphi-\alpha)}{\sin(\varphi-\alpha)} + \dfrac{\cos\theta}{\sin\theta}\right]} \frac{\sin\theta \sin\alpha \left[\dfrac{\cos\alpha}{\sin\alpha} - \dfrac{\cos\theta}{\sin\theta}\right]}{\sin\alpha}$$

$$= \frac{\cot \alpha - \cot \theta}{\cot (\varphi - \alpha) + \cot \theta} \tag{8.10}$$

When θ is 90° or the force applied is horizontal,

$$\eta = \frac{\cot \alpha}{\cot (\varphi - \alpha)} = \frac{\tan (\varphi - \alpha)}{\tan \alpha} \tag{8.11}$$

8.6 SCREW THREADS

A screw thread is obtained when the hypotenuse of a right-angled triangle is wrapped round the circumference of a cylinder.

Figure 8.5 a shows a triangle *abc* which is the development of a helix of diameter d and lead l (or pitch p for a singlestart thread).

Length of base = circumference of the cylinder of screw threads

$= \pi d$

Height of triangle $= l$

$$\tan \alpha = \frac{l}{\pi d}$$

where α is the helix angle.

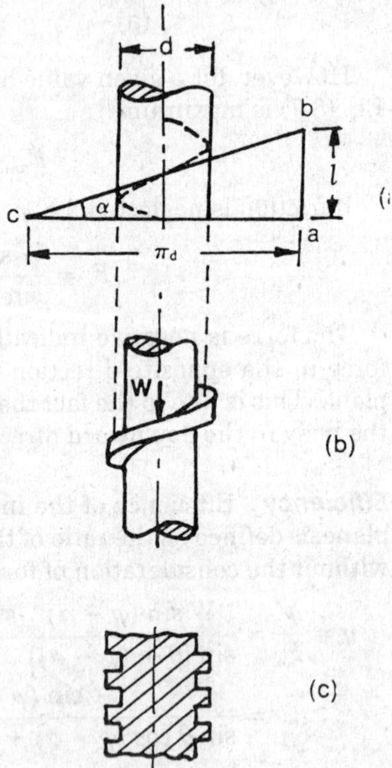

Fig. 8.5

Square Threads

A square-threaded screw used as a jack to raise a load **W** is shown in Fig. 8.5 b. Faces of the square threads in the sectional view (Fig. 8.5 c) are normal to the axis of the spindle. Force **F** acting horizontally is the force at the screw thread required to slide the load **W** up the inclined plane.

From Eq. (8.3) ($\theta = 90°$)

$$F = \frac{W \sin (\alpha + \varphi)}{\sin [90° - (\alpha + \varphi)]}$$

$$= \frac{W \sin (\alpha + \varphi)}{\cos (\alpha + \varphi)}$$

$$= W \tan (\alpha + \varphi) \tag{8.12}$$

$$= W \frac{\tan \alpha + \tan \varphi}{1 - \tan \alpha \tan \varphi}$$

$$= W \frac{\dfrac{l}{\pi\, d} + \mu}{1 - \dfrac{l}{\pi\, d}\mu}$$

$$= W \frac{1 + \mu\pi d}{\pi d - \mu l} \tag{8.13}$$

A bar is, usually fixed to the screw head to use as a lever for the application of force.

Let f = force applied at the end of the bar of length L

Then

$$fL = F\frac{d}{2} = Fr$$

or

$$f = \frac{Fr}{L} = \frac{Wr}{L}\tan(\alpha + \varphi) \tag{8.14}$$

If the weight is to be lowered,

$$F = \frac{W\sin(\varphi - \alpha)}{\sin(90° + (\varphi - \alpha))} \quad \text{[from Eq. (8.7), } \theta = 90°]$$

$$= \frac{W\sin(\varphi - \alpha)}{\cos(\varphi - \alpha)}$$

$$= W\tan(\varphi - \alpha) \tag{8.15}$$

$$f = \frac{Wr}{L}\tan(\varphi - \alpha) \tag{8.15a}$$

$$\text{Screw efficiency} = \frac{\text{work done in lifting the load/rev.}}{\text{work done by the applied force/rev.}}$$

$$= \frac{W \times l}{F \times \pi d}$$

$$= \frac{W}{F} \times \frac{l}{\pi d}$$

$$= \frac{W}{W\tan(\alpha + \varphi)}\tan\alpha$$

$$= \frac{\tan\alpha}{\tan(\alpha + \varphi)} \tag{8.16}$$

This is maximum when $\dfrac{d\eta}{d\alpha} = 0$

or $$\frac{d}{d\alpha}\left[\frac{\tan\alpha}{\tan(\alpha+\varphi)}\right]=0$$

or $$\frac{\sec^2\alpha\tan(\alpha+\varphi)-\sec^2(\alpha+\varphi)\tan\alpha}{\tan^2(\alpha+\varphi)}=0$$

or $$\sec^2\alpha\tan(\alpha+\varphi)-\sec^2(\alpha+\varphi)\tan\alpha=0$$

or $$\frac{\tan(\alpha+\varphi)}{\sec^2(\alpha+\varphi)}=\frac{\tan\alpha}{\sec^2\alpha}$$

or $$\frac{\sin(\alpha+\varphi)}{\cos(\alpha+\varphi)}\cos^2(\alpha+\varphi)=\frac{\sin\alpha}{\cos\alpha}\cos^2\alpha$$

or $$\sin(\alpha+\varphi)\cos(\alpha+\varphi)=\sin\alpha\cos\alpha$$

or $$2\sin(\alpha+\varphi)\cos(\alpha+\varphi)=2\sin\alpha\cos\alpha$$

or $$\sin 2(\alpha+\varphi)=\sin 2\alpha$$

This is possible if either $(\alpha+\varphi)=\alpha$, i.e. $\varphi=0$ (or no friction)

or $$\sin 2(\alpha+\varphi)=\sin(\pi-2\alpha)$$

i.e. $$2(\alpha+\varphi)=\pi-2\alpha$$

$$4\alpha+2\varphi=\pi$$

$$\alpha=\frac{\pi-2\varphi}{4}=45^\circ-\frac{\varphi}{2}$$

Thus the necessary condition for the maximum efficiency is,

$$\alpha=45^\circ-\frac{\varphi}{2} \tag{8.17}$$

Also, $$\eta_{max}=\frac{\tan\left(45^\circ-\frac{\varphi}{2}\right)}{\tan\left(45^\circ-\frac{\varphi}{2}+\varphi\right)}$$

$$=\tan\left(45^\circ-\frac{\varphi}{2}\right)\frac{1}{\tan\left(45^\circ+\frac{\varphi}{2}\right)}$$

$$=\left(\frac{\tan 45^\circ-\tan\frac{\varphi}{2}}{1+\tan 45^\circ\tan\frac{\varphi}{2}}\right)\left(\frac{1-\tan 45^\circ\tan\frac{\varphi}{2}}{\tan 45^\circ+\tan\frac{\varphi}{2}}\right)$$

$$=\frac{\left(1-\tan\frac{\varphi}{2}\right)\left(1-\tan\frac{\varphi}{2}\right)}{\left(1+\tan\frac{\varphi}{2}\right)\left(1+\tan\frac{\varphi}{2}\right)}$$

$$= \frac{\left(1 - \tan\frac{\varphi}{2}\right)^2}{\left(1 + \tan\frac{\varphi}{2}\right)^2}$$

$$= \frac{\left(1 - \frac{\sin\varphi/2}{\cos\varphi/2}\right)^2}{\left(1 + \frac{\sin\varphi/2}{\cos\varphi/2}\right)^2}$$

$$= \frac{\left(\cos\frac{\varphi}{2} - \sin\frac{\varphi}{2}\right)^2}{\left(\cos\frac{\varphi}{2} + \sin\frac{\varphi}{2}\right)^2}$$

$$= \frac{\cos^2\frac{\varphi}{2} + \sin^2\frac{\varphi}{2} - 2\cos\frac{\varphi}{2}\sin\frac{\varphi}{2}}{\cos^2\frac{\varphi}{2} + \sin^2\frac{\varphi}{2} + 2\sin\frac{\varphi}{2}\cos\frac{\varphi}{2}}$$

$$= \frac{1 - 2\sin\frac{\varphi}{2}\cos\frac{\varphi}{2}}{1 + 2\sin\frac{\varphi}{2}\cos\frac{\varphi}{2}}$$

$$= \frac{1 - \sin\varphi}{1 + \sin\varphi} \tag{8.18}$$

$$\text{Mechanical advantage} = \frac{\text{weight lifted}}{\text{force applied}} = \frac{W}{f} = \frac{W}{\frac{Wr}{L}\tan(\alpha + \varphi)}$$

$$= \frac{L}{r}\cot(\alpha + \varphi) \tag{8.19}$$

$$\text{Velocity ratio} = \frac{\text{distance moved by force/rev.}}{\text{distance moved by load/rev.}}$$

or

$$VR = \frac{2\pi L}{l} = \frac{L}{\frac{l}{\pi d}\frac{d}{2}} = \frac{L}{r\tan\alpha} \tag{8.20}$$

Observe that the angle of friction should always be more than the helix angle of the screw. Otherwise, the load will slide down of its own under the weight W. Thus α is not to be more than φ to prevent the nut from turning back. When $\alpha = \varphi$, the nut will be on the point of reversing and

$$\eta = \frac{\tan \varphi}{\tan 2\varphi} \approx \frac{1}{2} \tag{8.21}$$

Thus reversal of the nut is avoided if the efficiency of the thread is less than 50% (approximately).

Note that in case of square threads, as the helix angle α of screw threads is usually very small (3°–8°), the faces of the threads are normal to the axis of spindle and thus the normal reaction $\mathbf{R}_n$ is almost equal to the load $\mathbf{W}$ ($\cos \alpha \approx 1$).

V-Threads

In case of V-threads, the faces are inclined to the axis of spindle even if the helix angle is neglected. Figure 8.6 shows a section through a V-thread in which 2β is the angle between the faces of the thread (α has not been considered thus not shown). If $\mathbf{R}_n$ is the normal reaction, then clearly the axial component of $\mathbf{R}_n$ must be equal to $\mathbf{W}$, i.e.

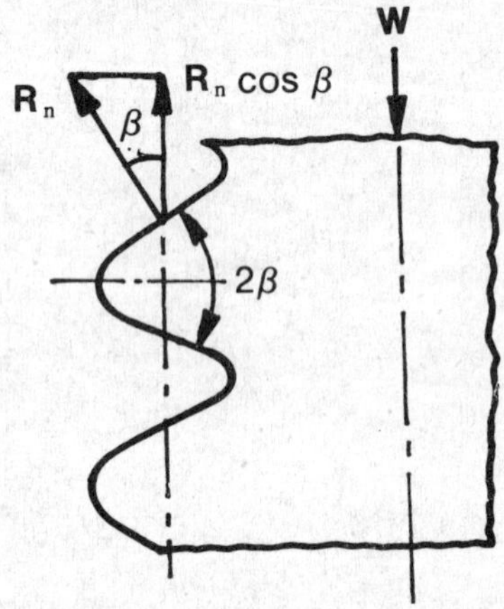

Fig. 8.6

$$W = R_n \cos \beta$$

or
$$R_n = \frac{W}{\cos \beta}$$

Friction force on the surface $= \mu R_n$

$$= \mu \frac{W}{\cos \beta}$$

$$= \frac{\mu}{\cos \beta} W$$

$$= \mu' W \tag{8.22}$$

This shows that the coefficient of friction μ (or $\tan \varphi$) as used in relations for the square threads is to be replaced by μ' or $\mu/\cos \beta$ or $\tan \varphi / \cos \beta$ to adapt them to V-threads.

Example 8.1 The following data relate to a screw jack (Fig. 8.7):

Pitch of the threaded screw = 8 mm
Diameter of the threaded screw = 40 mm
Coefficient of friction between screw and nut = 0.1
Load = 20 kW

Assuming that the load rotates with the screw, determine

(i) the ratio of torques required to raise and lower the load
(ii) the efficiency of the machine.

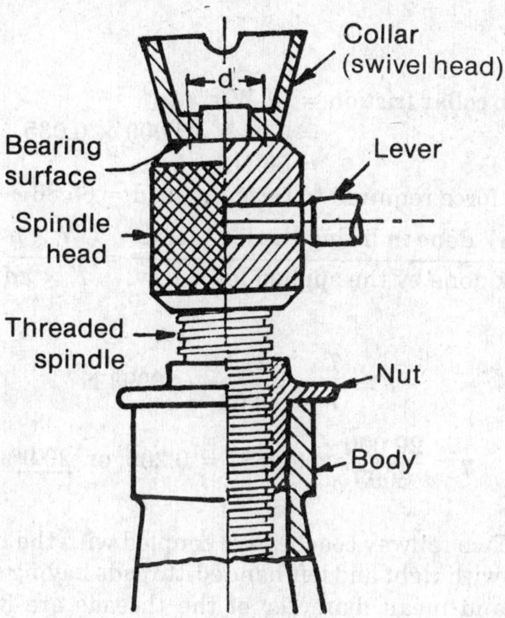

Fig. 8.7

Solution:

$$\tan\alpha = \frac{p}{\pi d} = \frac{8}{\pi \times 40} = 0.0637$$

$$\alpha = 3.64°$$

$\mu = \tan\varphi = 0.1$ or $\mu = 5.71°$

(i) *To raise the load*

$T = Fr = W\tan(\alpha + \varphi)r = 20\,000 \times \tan(3.64° + 5.71°) \times 0.02$

or $T = 65.86$ N.m

To lower the load $T = W\tan(\varphi - \alpha)r$

$= 20\,000 \times \tan(5.71° - 3.64°) \times 0.02$

$= 14.46$ N.m

$$\frac{\text{Torque to raise the load}}{\text{Torque to lower the load}} = \frac{65.86}{14.46} = \underline{4.56}$$

(ii) Efficiency

$$= \frac{\tan\alpha}{\tan(\alpha + \varphi)} = \frac{\tan 3.64°}{\tan(3.64° + 5.71°)} = 0.386 \text{ or } \underline{38.6\%}$$

Example 8.2 If the load in Example 8.1 does not rotate with the screw, but is carried on a swivel head having a bearing diameter of 70 mm, and the coefficient of friction between the swivel head and the spindle is 0.08, what is the total torque required to raise the load and the efficiency?

Solution:

Torque due to collar friction $= (\mu\ W)r$

$$= 0.08 \times 20\,000 \times 0.035$$
$$= 56 \text{ N.m}$$

Total friction force required to raise the load = 65.86 + 56 = 121.86 N.m

$$\eta = \frac{\text{Work done in lifting the load/rev.}}{\text{Work done by the applied force/rev.}} = \frac{W \times p}{F \times \pi d} = \frac{W}{F} \tan \alpha$$

where

$$F = \frac{T}{r} = \frac{121.86}{0.02} = 6093 \text{ N}$$

$$\eta = \frac{20\,000}{6093} \times 0.0637 = 0.209 \text{ or } \underline{20.9\%}$$

Example 8.3 Two railway coaches are coupled with the help of two tie rods of a turn buckle with right and left handed-threads having single-start square threads. Pitch and mean diameter of the threads are 8 mm and 30 mm respectively. What will be the work done in bringing the two coaches closer through a distance of 160 mm against a steady load of 2 kN? Take $\mu = 0.12$.

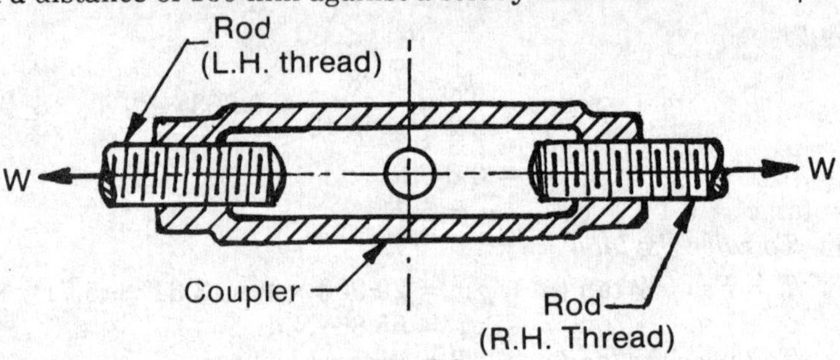

Fig. 8.8

Solution:

Figure 8.8 shows the outline of a turn buckle.

$p = 8$ mm $\quad\quad \mu = 0.12$

$d = 30$ mm $\quad\quad W = 2000$ N

$$\tan \alpha = \frac{p}{\pi d} = \frac{8}{\pi \times 30} = 0.0848 \text{ or } \alpha = 4.85°$$

$$\mu = \tan \varphi = 0.12 \quad \text{or} \quad \varphi = 6.84°$$

Torque on each rod $= Fr = W \tan(\alpha + \varphi) \times r$

$$= 2000 \times \tan(4.85° + 6.84°) \times 0.015$$
$$= 6.21 \text{ N.m}$$

Total torque on the coupling nut = 2 × 6.21 = 12.42 N.m

In one complete revolution of the rod, each coach is moved through a distance equal to pitch.

Number of turns required to move the coaches through a distance of 160 mm = 160/(2 × 8) = 10

$$\text{Work done, } W = T\theta = 12.42 \times 2\pi \times 10 = 780.4 \text{ N.m}$$

$$\eta = \frac{\tan \alpha}{\tan (\alpha + \varphi)} = \frac{\tan 4.85^\circ}{\tan (4.85^\circ + 6.84^\circ)} = 0.41 \text{ or } 41\%$$

8.7 WEDGE

A wedge is used to raise loads like a screw jack. It consists of three sliding pairs as shown in Fig. 8.9 a formed by the frame *A*, wedge *B* and the slider *S*. When a force **F** is applied to the wedge, the slider is raised in the frame guides raising the load.

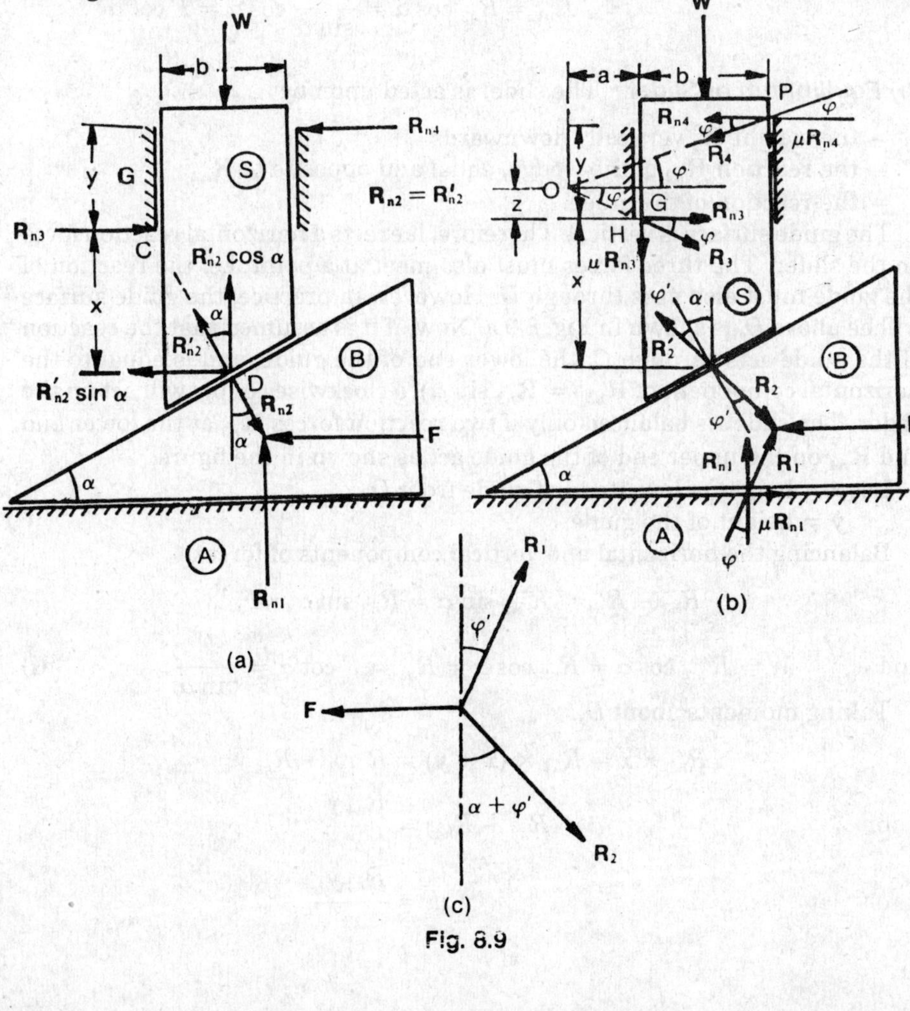

Fig. 8.9

Mechanical efficiency of the wedge is defined as the ratio of the load raised when friction is considered to the load raised when friction is neglected while the force applied is the same.

1. Friction Neglected

(a) Equilibrium of Wedge The wedge is acted upon by (Fig. 8.9 a)

– the horizontal force **F**
– the reaction $\mathbf{R}_{n1}$ of the surface A in the vertical direction
– the reaction $\mathbf{R}_{n2}$ of the slider, normal to slanting surface.

For the equilibrium of the wedge, three forces must meet at a point and be balanced vectorially. Thus $\mathbf{R}_{n2}$ passes through the point of intersection of **F** and $\mathbf{R}_{n1}$.

Also
$$F = R_{n2} \sin \alpha$$
$$R_{n1} = R_{n2} \cos \alpha = \frac{F}{\sin \alpha} \cos \alpha = F \cot \alpha$$

(b) Equilibrium of Slider The slider is acted upon by

– the weight **W**, vertically downwards
– the reaction $\mathbf{R}'_{n2}$ of the wedge, equal and opposite to $\mathbf{R}_{n2}$
– the reaction of the guide.

The guide surface is vertical. Therefore, it exerts a horizontal reaction force on the slider. The three forces must also meet at a point, i.e. the reaction of the guide must also pass through D. However, in practice, the guide surface will be above D as shown in Fig. 8.9 a. Now, if it is assumed that the reaction of the guide acts through C, the lower end of the guide, and is equal to the horizontal component of $\mathbf{R}'_{n2}$ ($= \mathbf{R}'_{n2} \sin \alpha$), a clockwise couple will act on the slider. The slider is balanced only if two reaction forces $\mathbf{R}_{n3}$ at the lower end and $\mathbf{R}_{n4}$ on the upper end of the guide act as shown in the figure.

Let x = height of lower end of guide from D
y = height of the guide.

Balancing the horizontal and vertical components of forces,

$$R_{n3} - R_{n4} = R'_{n2} \sin \alpha = R_{n2} \sin \alpha = F$$

and
$$W = R'_{n2} \cos \alpha = R_{n2} \cos \alpha = R_{n1} = F \cot \alpha = \frac{F}{\tan \alpha} \qquad \text{(i)}$$

Taking moments about D,

$$R_{n3} \times x = R_{n4} \times (x + y) = R_{n4}\, x + R_{n4}\, y$$

or
$$(R_{n3} - R_{n4}) = \frac{R_{n4}\, y}{x}$$

or
$$F = \frac{R_{n4}\, y}{x}$$

or $$R_{n4} = \frac{Fx}{y}$$

and $$R_{n3} = F + F\frac{x}{y} = F\left(1 + \frac{x}{y}\right)$$

2. Friction Considered

Let φ' = Angle of friction between frame A slider S and wedge B

φ = Angle of friction between guide G and slider S

(a) Equilibrium of Wedge The forces acting on the wedge are (Fig. 8.9 b)

– the horizontal force $\mathbf{F}$

– the reaction $\mathbf{R}_1$ inclined at an angle φ' with $\mathbf{R}_{n1}$. The wedge moves towards left and so the friction force acts towards right

– the reaction $\mathbf{R}_2$ inclined at an angle φ' with $\mathbf{R}_{n2}$.

For equilibrium (Fig. 8.9 c),

$$\frac{R_1}{\sin[90° + (\alpha + \varphi')]} = \frac{R_2}{\sin(90° + \varphi')} = \frac{F}{\sin[180° - (\alpha + 2\varphi')]}$$

$$\frac{R_1}{\cos(\alpha + \varphi')} = \frac{R_2}{\cos\varphi'} = \frac{F}{\sin(\alpha + 2\varphi')}$$

$$R_1 = \frac{\cos(\alpha + \varphi')}{\sin(\alpha + 2\varphi')}F$$

and $$R_2 = \frac{\cos\varphi'}{\sin(\alpha + 2\varphi')}F$$

(b) Equilibrium of Slider The slider is acted upon by

– the reaction $\mathbf{R}'_2$ of the wedge, equal and opposite to $\mathbf{R}_2$

– the weight $\mathbf{W}'$, vertically downwards

– the reactions $\mathbf{R}_3$ and $\mathbf{R}_4$ of the guide. The motion of the guide is vertically upwards. Therefore, the friction force acts in the downward direction.

Let $\mathbf{R}_3$ and $\mathbf{R}_4$ intersect at O. The distances a and z are as shown in the diagram.

Taking moments about O,

$$R'_2[\sin(\alpha + \varphi')](x + z) + W'\left(a + \frac{b}{2}\right) - R'_2\cos(\alpha + \varphi')\left(a + \frac{b}{2}\right)$$

$$W' = R'_2\left[\cos(\alpha + \varphi') - \frac{(x + z)\sin(\alpha + \varphi')}{\left(a + \frac{b}{2}\right)}\right]$$

where $R'_2 = R_2 = F\dfrac{\cos\varphi'}{\sin(\alpha + 2\varphi')}$

a and z can be calculated as given below,

$$\tan\varphi = \frac{y-z}{a+b} = \mu$$

$$z = y - a\mu - b\mu$$

In ΔOGH, $\tan\varphi = \frac{z}{a} = \mu$ or $z = a\mu$

$$y - a\mu - b\mu = a\mu$$

or $2a\mu = y - b\mu$

$$a = \frac{y - b\mu}{2\mu} = \frac{y}{2\mu} - \frac{b}{2}$$

$$z = \mu\left(\frac{y}{2\mu} - \frac{b}{2}\right)$$

$$\therefore W' = F\frac{\cos\varphi'}{\sin(\alpha + 2\varphi')}\left[\cos(\alpha+\varphi') - \frac{\left\{x + \mu\left(\frac{y}{2\mu} - \frac{b}{2}\right)\right\}\sin(\alpha+\varphi')}{\frac{y}{2\mu}}\right] \quad \text{(ii)}$$

Dividing (ii) by (i),

$$\eta = \frac{W'}{W} = \frac{\cos\varphi\tan\alpha}{\sin(\alpha+2\varphi')}\left[\cos(\alpha+\varphi') - \mu\left(\frac{2x-\mu b}{y}+1\right)\sin(\alpha+\varphi')\right] \quad (8.23)$$

If x and b are comparatively small,

$$\eta = \frac{\cos\varphi'\tan\alpha}{\sin(\alpha + 2\varphi')}\left[\cos(\alpha + \varphi') - \tan\varphi\sin(\alpha + \varphi')\right]$$

$$= \frac{\cos\varphi'\tan\alpha}{\sin(\alpha + 2\varphi')}\left[\frac{\cos(\alpha + \varphi')\cos\varphi - \sin\varphi\sin(\alpha + \varphi')}{\cos\varphi}\right]$$

$$= \frac{\cos\varphi'\tan\alpha}{\sin(\alpha + 2\varphi')} \times \frac{\cos(\alpha + \varphi + \varphi')}{\cos\varphi} \quad (8.24)$$

If $\varphi = \varphi'$,

$$\eta = \frac{\tan\alpha}{\tan(\alpha + 2\varphi)} \quad (8.25)$$

Example 8.4 For a wedge shown in Fig. 8.9, $\alpha = 30°, \varphi = 6°\ \varphi' = 8°$, $y = 3x = 3b$. Determine the efficiency.

Solution:

$$\mu = \tan\varphi = \tan 6° = 0.105$$

From Eq. (8.23),

$$\eta = \frac{\cos 8°\tan 30°}{\sin(30°+16°)}\left[\cos(30°+8°) - 0.105\left(\frac{2x - 0.105x}{3x}+1\right)\sin(30°+8°)\right]$$

$$= 0.543$$

8.8 PIVOTS AND COLLARS

When a rotating shaft is subjected to an axial load, the thrust (axial force) is taken either by a pivot or a collar. Examples are, the shaft of a steam turbine, propeller shaft of a ship, etc.

Collar Bearing

A collar bearing or simply a collar is provided at any position along the shaft and bears the axial load on a mating surface.

The surface of the collar may be plane (flat) normal to the shaft (Fig. 8.10) or of conical shape (Fig. 8.11).

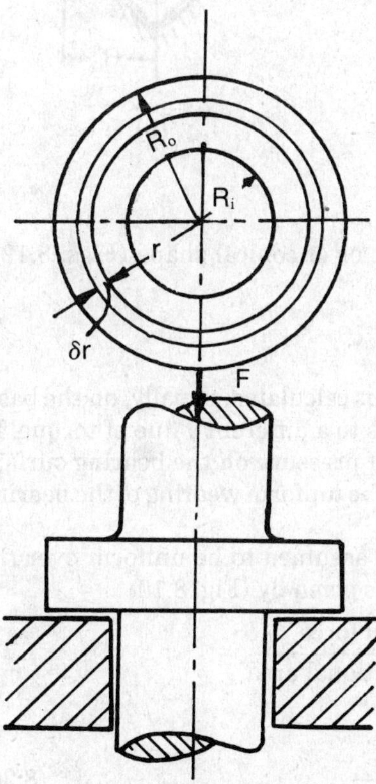

Fig. 8.10

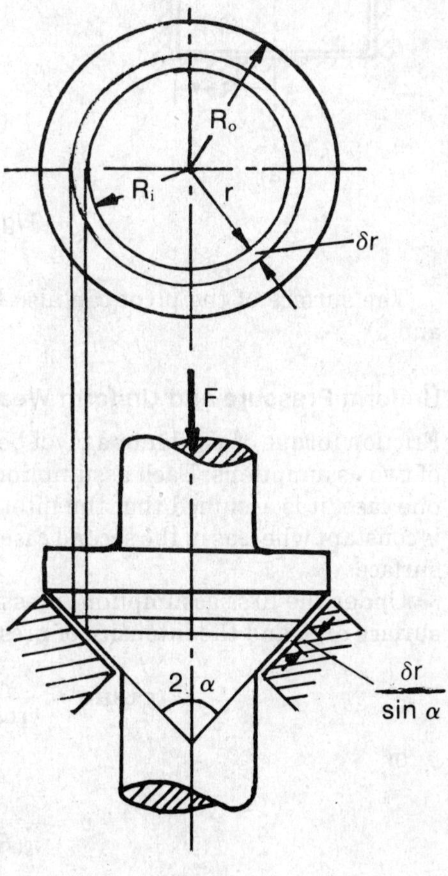

Fig. 8.11

Pivot Bearing

When the axial load is taken by the end of the shaft which is inserted in a recess to bear the thrust, it is called a pivot bearing or simply a pivot. It is also known as footstep bearing.

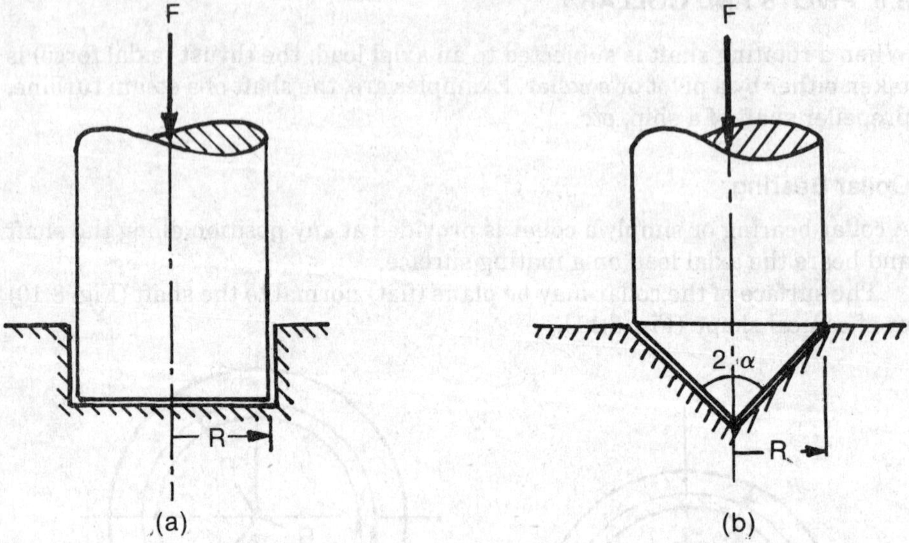

Fig. 8.12

The surface of the pivot can also be flat or of conical shape (Figs. 8.12 a and b).

Uniform Pressure and Uniform Wear

Friction torque of a collar or a pivot bearing is calculated, usually, on the basis of two assumptions. Each assumption leads to a different value of torque. In one case, it is assumed that the intensity of pressure on the bearing surface is constant whereas in the second case, it is the uniform wearing of the bearing surface.

Under the first assumption, pressure is assumed to be uniform over the surface area and the intensity of pressure is given by (Fig. 8.10)

$$\text{Pressure} = \frac{\text{axial force}}{\text{cross-sectional area}}$$

or

$$p = \frac{F}{\pi(R_o^2 - R_i^2)} \tag{8.26}$$

where R_o = outer radius of the collar

R_i = inner radius of the collar

For uniform wear over an area, the intensity of pressure should vary inversely proportional to the elementary areas, i.e. it should decrease with the increase in the elementary area and vice-versa. This can be illustrated by drawing a line with a chalk. In doing so, a little quantity of chalk is worn from the stick. Now, if it is desired that the chalk is worn by the same amount, but the length of the line is doubled, the pressure on the chalk has to be reduced to half that in the previous case. Therefore, for uniform wear, product of the pressure applied and the distance travelled must be constant. For uniform wear of the surface, let

p_1 = normal pressure between two surfaces at radius r_1
p_2 = normal pressure between two surfaces at radius r_2
b = width of the surface at radii r_1 and r_2 (equal width)

$$p_1 \times \text{area at } r_1 = p_2 \times \text{area at } r_2$$

$$p_1 \times 2\pi r_1 \times b = p_2 \times 2\pi r_2 \times b$$

or
$$p_1 r_1 = p_2 r_2$$

or
$$pr = \text{constant} \tag{8.27}$$

Thus in case of uniform weariness of the two surfaces, product of the normal pressure and the corresponding radius must be constant. This means, the pressure is less where the radius is more and vice-versa. Pressure on an elemental area at radius r can be found as given below.

$$\text{Axial force,}\quad F = \int_{R_i}^{R_o} \text{Axial force on the elemental area}$$

$$= \int_{R_i}^{R_o} \text{Pressure on the element} \times \text{Area}$$

$$= \int_{R_i}^{R_o} p \times 2\pi r dr$$

$$= \int_{R_i}^{R_o} \frac{C}{r} \times 2\pi r dr \qquad (p.r = C)$$

$$= \int_{R_i}^{R_o} 2\pi\, c dr$$

$$= (2\pi\, Cr)_{R_i}^{R_o}$$

$$= 2\pi C\,(R_o - R_i)$$

$$= 2\pi\, pr(R_o - R_i)$$

or pressure intensity p at a radius r of the collar,

$$p = \frac{F}{2\pi r\,(R_o - R_i)} \tag{8.28}$$

In a flat pivot, in which $R_i = 0$, the pressure would be infinity at the centre of the bearing ($r = 0$), which cannot be true. Thus, the uniform wear theory has a flaw in it.

Collars and pivots, using the above two theories, have been analysed below.

Collars

(i) Flat Collar

Let p = uniform normal pressure over an area

F = axial thrust

N = speed of the shaft

μ = coefficient of friction between the two surfaces

Consider an element of width δr of the collar at radius r. Friction force on the element (Fig. 8.10),

$$\delta F = \mu \times \text{axial force}$$

$$= \mu \times p \times \text{area of the element}$$

$$= \mu \times p \times 2\pi r \,\delta r$$

Friction torque about the shaft axis,

$$\delta T = \delta F \times r$$

$$= \mu \times p \times 2\pi r \,\delta r \times r$$

$$= 2\mu p \pi r^2 \delta r$$

$$\text{Total friction torque, } T = \int_{R_i}^{R_o} 2\mu p \pi r^2 dr \tag{8.29}$$

(a) With Uniform Pressure Theory Pressure is uniform over the whole area and is given by

$$p = \frac{F}{\pi (R_o^2 - R_i^2)}$$

$$\therefore \quad T = \int_{R_i}^{R_o} 2\mu \pi r^2 \frac{F}{\pi (R_o^2 - R_i^2)} dr$$

$$= \int_{R_i}^{R_o} \frac{2\mu F}{R_o^2 - R_i^2} r^2 \, dr$$

$$= \left[\frac{2\mu F}{R_o^2 - R_i^2} \frac{r^3}{3} \right]_{R_i}^{R_o}$$

$$= \frac{2\mu F (R_o^3 - R_i^3)}{3 (R_o^2 - R_i^2)} \tag{8.30}$$

(b) With Uniform Wear Theory Pressure p at a radius r of the collar is given by

$$p = \frac{F}{2\pi r (R_o - R_i)}$$

$$\therefore \quad T = \int_{R_i}^{R_o} 2\,\mu\,\pi\,r^2 \frac{F}{2\,\pi\,r\,(R_o - R_i)} dr$$

$$= \int_{R_i}^{R_o} \frac{\mu\,Fr}{R_o - R_i} dr$$

$$= \frac{\mu\,F}{2} \frac{(R_o^2 - R_i^2)}{(R_o^2 - R_i)}$$

$$= \frac{\mu\,F}{2} (R_o + R_i) \tag{8.31}$$

$$= \mu\,F \times \text{Mean radius of the collar bearing}$$

(ii) Conical Collar (Frustrum of Cone) Consider an elementary area of width δr at a radius r of the bearing (Fig. 8.11).

Friction force on the element,

$$\delta F = \mu \times p \times \text{Area of the element}$$

$$= \mu \times p \times 2\,\pi\,r \frac{\delta r}{\sin\alpha}$$

Friction torque about the shaft axis,

$$\delta T = \delta F \times r = \frac{2\,\mu\,p\,\pi\,r^2}{\sin\alpha} \delta r$$

Total friction torque,

$$T = \int_{R_i}^{R_o} \frac{2\,\mu\,p\,\pi\,r^2}{\sin\alpha} dr \tag{8.32}$$

(a) With Uniform Pressure Theory

$$T = \int_{R_i}^{R_o} \frac{2\,\mu\,\pi\,r^2}{\sin\alpha} \frac{F}{\pi\,(R_o^2 - R_i^2)} dr$$

$$= \frac{2\,\mu\,F}{\sin\alpha\,(R_o^2 - R_i^2)} \left[\frac{r^3}{3}\right]_{R_i}^{R_o}$$

$$= \frac{2\,\mu\,F}{3\,\sin\alpha} \left[\frac{R_o^3 - R_i^3}{R_o^2 - R_i^2}\right] \tag{8.33}$$

i.e., the torque is increased in the ratio $\frac{1}{\sin\alpha}$ from that for flat collars.

(b) With Uniform Wear Theory

$$T = \int_{R_i}^{R_o} \frac{2\mu\pi r^2}{\sin\alpha} \frac{F}{2\pi r(R_o - R_i)} dr$$

$$= \int_{R_i}^{R_o} \frac{\mu F}{\sin\alpha (R_o - R_i)} r\, dr$$

$$= \frac{\mu F}{\sin\alpha (R_o - R_i)} \left(\frac{r^2}{2}\right)_{R_i}^{R_o}$$

$$= \frac{\mu F}{2\sin\alpha}(R_o + R_i) \tag{8.34}$$

$$= \frac{\mu F}{\sin\alpha} \times \text{mean radius of the bearing}$$

i.e., the torque is increased by $\frac{1}{\sin\alpha}$ times from that for flat collars.

Pivots

Expressions for torque in case of pivots can directly be obtained from the expressions for collars by inserting the values $R_i = 0$ and $R_o = R$.

(i) Flat Pivot

(a) Uniform pressure theory, $T = \frac{2}{3}\mu FR$ (8.35)

(b) Uniform wear theory, $T = \frac{1}{2}\mu FR$ (8.36)

(ii) Conical Pivot

(a) Uniform pressure theory, $T = \frac{2\mu FR}{3\sin\alpha}$ (8.37)

(b) Uniform wear theory, $T = \frac{\mu FR}{2\sin\alpha}$ (8.38)

The above expressions reveal that the value of the friction torque is more when the uniform pressure theory is applied. In practice, however, it has been found that the value of the friction torque lies inbetween that given by the two theories. To be on the safer side, out of the two theories, one is selected on the basis of the use.

A clutch plate transmits torque through the force of friction. Thus, though a clutch will surely be transmitting torque given by the uniform wear theory (lower value), it is not necessary that the clutch can also transmit a torque given by the uniform pressure theory (higher value). Therefore, it is safer to say that the clutch transmits a maximum torque based on the uniform wear theory and design it accordingly. However, the actual torque transmitted will be a little higher.

On the other hand, while calculating the power loss in a bearing, it is to be on the basis of uniform pressure theory, though the actual power loss will be a little less than that calculated.

Example 8.5 In a thrust bearing, the external and the internal diameters of the contacting surfaces are 320 mm and 200 mm respectively. The total axial load is 80 kN and the intensity of pressure is 350 kN/m^2. The shaft rotates at 400 rpm. Taking the coefficient of friction as 0.06, calculate the power lost in overcoming the friction. Also find the number of collars required for the bearing.

Solution:

$$R_o = 0.16 \text{ m} \qquad F = 80 \times 10^3 \text{ N}$$
$$R_i = 0.1 \text{ m} \qquad \mu = 0.06$$
$$N = 400 \text{ rpm} \qquad p = 350 \times 10^3 \text{ N/m}^2$$

Using uniform pressure theory as we are to find the power loss in a bearing.

$$T = \frac{2}{3}\mu F\left(\frac{R_o^3 - R_i^3}{R_o^2 - R_i^2}\right) = \frac{2}{3} \times 0.06 \times 80 \times 10^3 \left[\frac{(0.16)^3 - (0.10)^3}{(0.16)^2 - (0.10)^2}\right]$$
$$= 3200 \times 0.1985$$
$$= 635.12 \text{ N.m}$$

$$P = T\omega = T\frac{2\pi N}{60} = 635.12 \times \frac{2\pi \times 400}{60} = 26\,602 \text{ W or } 26.602 \text{ kW}$$

$$\text{Number of collars} = \frac{\text{total load}}{\text{load per collar}}$$
$$= \frac{F}{p \times \pi (R_o^2 - R_i^2)}$$
$$= \frac{80 \times 10^3}{350 \times 10^3 \times \pi [(0.16)^2 - (0.10)^2]}$$
$$= 4.66 \quad \text{or} \quad \underline{5 \text{ collars}}$$

Example 8.6 A conical pivot with angle of cone as 100°, supports a load of 18 kN. The external radius is 2.5 times the internal radius. The shaft rotates at 150 rpm. If the intensity of pressure is to be 300 kN/m^2 and coefficient of friction as 0.05, what is the power lost in working against the friction.

Solution:

$F = 18$ kN $\qquad R_o = 2.5\,R_i$

$p = 300$ kN/m^2 $\qquad N = 150$ rpm

$\mu = 0.05 \qquad \alpha = 50°$

$$\text{In case of uniform pressure, } p = \frac{F}{\pi\,(R_o^2 - R_i^2)}$$

$$\text{or} \qquad 300 \times 10^3 = \frac{18 \times 10^3}{\pi\,[(2.5R_i)^2 - R_i^2]}$$

$$\text{or} \qquad (2.5R_i)^2 - R_i^2 = \frac{18}{300 \times \pi}$$

$$R_i = 0.0603 \text{ m}$$

$$R_o = 0.0603 \times 2.5 = 0.1508 \text{ m}$$

$$T = \frac{2}{3}\,\frac{\mu F}{\sin\alpha}\left[\frac{R_o^3 - R_i^3}{R_o^2 - R_i^2}\right]$$

$$\text{or} \quad T = \frac{2}{3} \times \frac{0.05 \times 18\,000}{\sin 50°}\left[\frac{(0.1508)^3 - (0.0603)^3}{(0.1508)^2 - (0.0603)^2}\right] = 131.6 \text{ N.m}$$

$$P = T\omega = T\,\frac{2\,\pi\,N}{60} = 131.6 \times \frac{2\,\pi \times 150}{60} = 2067 \text{ W} \quad \text{or} \quad \underline{2.067 \text{ kW}}$$

8.9 FRICTION CLUTCHES

A clutch is a device used to transmit the rotary motion of one shaft to another when desired. The axes of the two shafts are coincident.

In friction clutches, the connection of the engine shaft to the gear box shaft is affected by friction between two or more rotating concentric surfaces. The surfaces can be pressed firmly against one another when engaged and the clutch tends to rotate as a single unit.

1. Disc Clutch (Single-plate Clutch)

A disc clutch consists of a clutch plate attached to a splined hub which is free to slide axially on splines cut on the driven shaft. The clutch plate is made of steel and has a ring of friction lining on each side. The engine shaft supports a rigidly fixed flywheel.

A spring-loaded pressure plate presses the clutch plate firmly against the flywheel when the clutch is engaged. When disengaged, the springs press against a cover attached to the flywheel. Thus both the flywheel and the pressure plate rotate with the input shaft. The movement of the clutch pedal is transferred to the pressure plate through a thrust bearing.

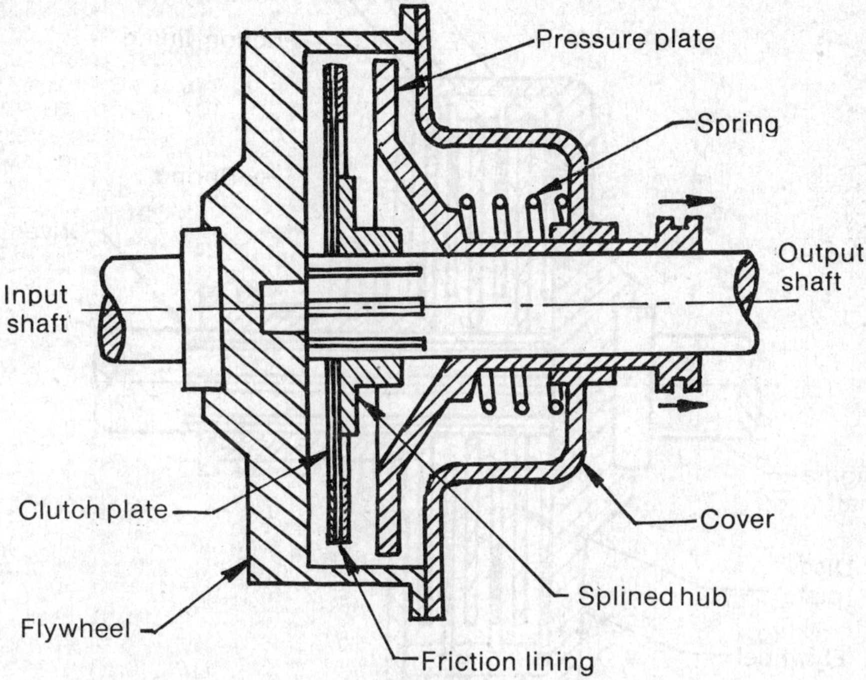

Fig. 8.13

Figure 8.13 shows the pressure plate pulled back by the release levers and the friction linings on the clutch plate are free of contact with the pressure plate or the flywheel. The flywheel rotates without driving the clutch plate and thus the driven shaft.

When the foot is taken off the clutch pedal, the pressure on the thrust bearing is released. As a result, the springs become free to move the pressure plate to bring it in contact with the clutch plate. The clutch plate slides on the splined hub and is tightly gripped between the pressure plate and the flywheel. The friction between the linings on the clutch plate, and the flywheel on one side and the pressure plate on the other, cause the clutch plate and hence the driven shaft to rotate.

In case the resisting torque on the driven shaft exceeds the torque at the clutch, clutch slip will occur.

2. Multi-plate Clutch

In a multi-plate clutch, the number of frictional linings and the metal plates is increased which increases the capacity of the clutch to transmit torque. Figure 8.14 shows a simplified diagram of a multi-plate clutch.

The friction rings are splined on their outer circumference and engage with corresponding splines on the flywheel. They are free to slide axially. The friction material thus rotates with the flywheel and the engine shaft. The number of friction rings depends upon the torque to be transmitted.

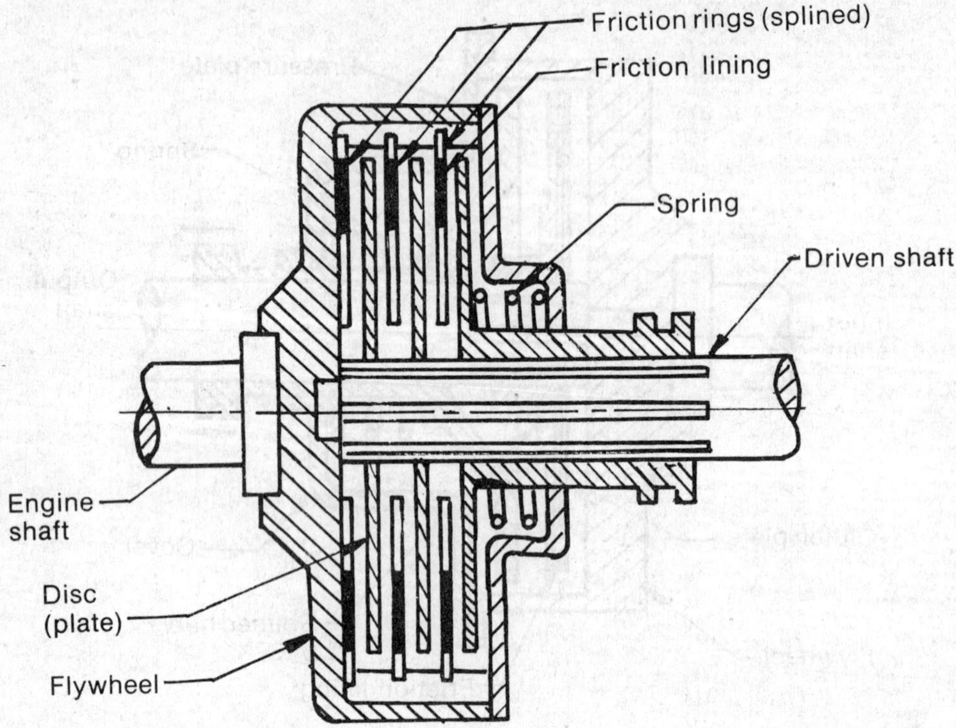

Fig. 8.14

The driven shaft also supports discs on the splines which rotate with the driven shaft and can slide axially. If the actuating force on the pedal is removed, a spring presses the discs into contact with the friction rings and the torque is transmitted between the engine shaft and the driven shaft.

3. Cone Clutch

In a cone clutch (Fig. 8.15), the contact surfaces are in the form of cones. In the engaged position, the friction surfaces of the two cones A and B are in complete contact due to spring pressure that keeps one cone pressed against the other all the time.

When the clutch is engaged, the torque is transmitted from the driving shaft to the driven shaft through the flywheel and the friction cones. For disengaging the clutch, the cone B is pulled back through a lever system against the force of the spring.

The advantage of a cone clutch is that the normal force on the contact surfaces is increased which is equal to the axial force for a disc clutch. If $\mathbf{F}$ is the axial force, $\mathbf{F}_n$ the normal force and the β semi-cone angle of the clutch, then

$$\mathbf{F}_n = \frac{F}{\sin \beta} \tag{8.39}$$

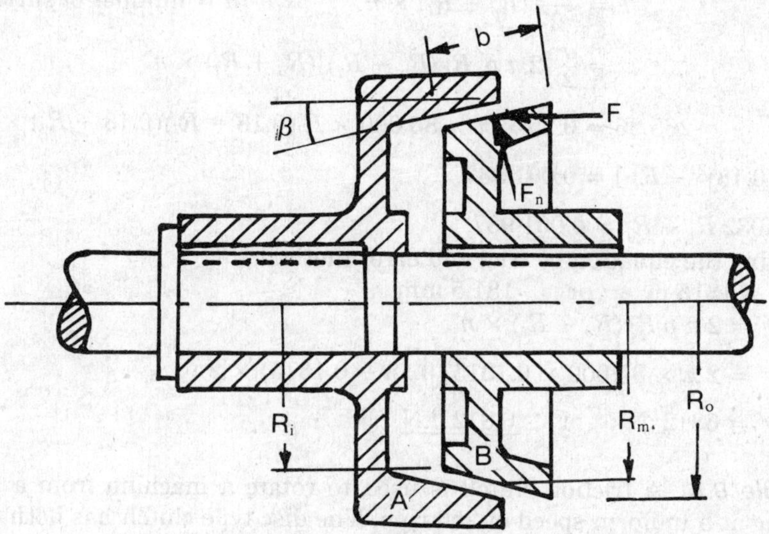

Fig. 8.15

However, cone clutches have become obsolete as small cone angles and exposure to dust and dirt tend to bind the two cones and it becomes difficult to disengage them.

Example 8.7 A single-plate clutch transmits 25 kW at 900 rpm. The maximum pressure intensity between the plates is 85 kN/m^2. The outer diameter of the plate is 360 mm. Both the sides of the plate are effective and the coefficient of friction is 0.25. Determine

(i) the inner diameter of the plate
(ii) the axial force to engage the clutch.

Solution:

P = 25 kW $\qquad\qquad \mu = 0.25$
N = 900 rpm $\qquad\qquad R_o = 0.18$ m

In case of power transmission through a clutch, it is safer to use the expressions obtained by uniform wear theory. In that case maximum pressure is at the inner radius, i.e.,

$$P_i = 85 \text{ kN/m}^2$$

(i) $$P = T\omega$$

or

$$25\,000 = T \times \frac{2\pi \times 900}{60}$$

$$T = 265.26 \text{ N.m}$$

Also $T = \frac{\mu F}{2}(R_o + R_i) \times n$ (n = number of surfaces)

$$= \frac{\mu}{2}[2\pi p_i R_i (R_o - R_i)](R_o + R_i) \times n$$

$$265.26 = 0.25 \times \pi \times 85\,000 \times R_i(0.18 - R_i)(0.18 + R_i) \times 2$$

$R_i\,[(0.18)^2 - R_i^2] = 0.001\,987$

or $0.032\,R_i - R_i^3 = 0.001\,987$

Solving the equation by trial and error method,

$R_i = 0.1315$ m or 131.5 mm

(ii) $F = 2\pi p_i R_i (R_o - R_i) \times n$

$$= 2\pi \times 85\,000 \times 0.1315\,(0.18 - 0.1315) \times 2$$

$$= 6812 \text{ N} \quad \text{or} \quad 6.812 \text{ kN}$$

Example 8.8 A friction clutch is used to rotate a machine from a shaft rotating at a uniform speed of 250 rpm. The disc type clutch has both of its sides effective, the coefficient of friction being 0.3. The outer and the inner diameters of the friction plate are 200 mm and 120 mm respectively. Assuming uniform wear of the clutch, the intensity of pressure is not to be more than 100 kN/m^2. If the moment of inertia of the rotating parts of the machine is 6.5 kg.m^2, determine the time to attain the full speed by the machine and the energy lost in slipping of the clutch.

What will be the intensity of pressure, if the condition of uniform pressure of the clutch is considered? Also determine the ratio of power transmitted with uniform wear to that with uniform pressure.

Solution:

$P_i = 100 \times 10^3$ N/m^2 $\quad$ $R_o = 0.1$ m

$I = 6.5$ kg.m^2 $\quad$ $R_i = 0.06$ m

$\mu = 0.3$ $\quad$ N = 250 rpm

$n = 2$ (both sides effective)

(a) *With uniform wear*

Force/surface, $F = 2\pi p_i R_i (R_o - R_i)$

$$= 2\pi \times 100 \times 10^3 \times 0.06(0.1 - 0.06)$$

$$= 1508 \text{ N}$$

$$T = \frac{\mu F}{2}(R_o + R_i) \times n$$

$$= \frac{0.3 \times 1508}{2}(0.1 + 0.06) \times 2$$

$$= 72.38 \text{ N.m}$$

$$P = T \times \omega = 72.38 \times \frac{2\,\pi \times 250}{60} = 1895 \text{ W}$$

Also $\quad T = I\alpha \qquad$ (α is angular acceleration)

$$72.38 = 6.5 \times \alpha$$

or $\quad \alpha = 11.135 \text{ rad/s}^2$

or $\quad \dfrac{\omega}{t} = 11.135$

or $\quad \dfrac{2\,\pi \times 250}{60 \times t} = 11.135$

or $\quad t = \underline{2.35 \text{ s}}$

Thus the full speed is attained by the machine in 2.35 seconds.

(b) *During the slipping period*

Angle turned by the driving shaft,

$$\theta_1 = \omega t = \frac{2\,\pi N}{60} t = \frac{2\,\pi \times 250}{60} \times 2.35 = 61.52 \text{ rad}$$

Angle turned by the driven shaft, $\theta_2 = \omega_o t + \dfrac{1}{2}\alpha t^2 \qquad (\omega_o = 0)$

$$= 0 + \frac{1}{2} \times 11.135 \times (2.35)^2$$

$$= 30.75 \text{ rad}$$

Energy lost in friction $= T(\theta_1 - \theta_2) = 72.38 \times 30.75$

$$= 2226 \text{ N.m or } 2.226 \text{ kN.m}$$

(c) *With uniform pressure*

$$p = \frac{F}{\pi\,(R_o^2 - R_i^2)} = \frac{1508}{\pi\,(0.1)^2 - (0.06)^2]}$$

$$= 75\,000 \text{ N/m}^2 \text{ or } 75 \text{ kN/m}^2$$

$$T = \frac{2}{3}\mu F \left[\frac{R_o^3 - R_i^3}{R_o^2 - R_i^2}\right] \times n$$

$$= \frac{2}{3} \times 0.3 \times 1508 \left[\frac{(0.1)^3 - (0.06)^3}{(0.1)^2 - (0.06)^2}\right] \times 2 = 73.89 \text{ N.m}$$

$$P = 73.89 \times \frac{2\,\pi \times 250}{60} = 1934 \text{ W}$$

$$\frac{\text{Power with uniform wear}}{\text{Power with uniform pressure}} = \frac{1895}{1934} = \underline{0.98}$$

Example 8.9 A multi-plate disc clutch transmits 55 kW of power at 1 800 rpm. Coefficient of friction for the friction surfaces is 0.1. Axial intensity of pressure is not to exceed 160 kN/m^2. The internal radius is 80 mm and is 0.7 times the external radius. Find the number of plates needed to transmit the required torque.

Solution:

$$p_i = 160 \times 10^3 \text{ N/m}^2 \qquad R_i = 0.08 \text{ m}$$

$$\mu = 0.1 \qquad R_o = \frac{0.08}{0.7} = 0.1143 \text{ m}$$

$$N = 1800 \text{ rpm}$$

$$P = 55 \text{ kW}$$

Assuming uniform wear conditions,

$$F = 2\pi p_i r_i (R_o - R_i)$$

$$= 2\pi \times 160 \times 10^3 \times 0.08(0.1143 - 0.08)$$

$$= 2759 \text{ N}$$

$$T = \frac{1}{2}\mu F(R_o + R_i)$$

$$= \frac{1}{2} \times 0.1 \times 2759 \times (0.1143 + 0.08)$$

$$= 26.78 \text{ N.m/surface}$$

$$\text{Total torque transmitted} = \frac{P}{\omega} = \frac{55\,000}{\dfrac{2\pi \times 1800}{60}} = 291.8 \text{ N.m}$$

$$\text{Number of friction surfaces required} = \frac{291.8}{26.78} = 10.9 \text{ or } 11 \text{ surfaces}$$

In all, there will be 12 plates. 6 plates (rings) revolve with the driving or engine shaft and the other 6 with the driven shaft.

Example 8.10 A torque of 350 N.m is transmitted through a cone clutch having a mean diameter of 300 mm and a semi-cone angle of 15°. The maximum normal pressure at the mean radius is 150 kN/m^2. The coefficient of friction is 0.3. Calculate the width of the contact surface. Also find the axial force to engage the clutch.

Solution:

$$T = 350 \text{ N.m} \qquad R_m = 0.15 \text{ m}$$

$$p = 150 \text{ kN/m}^2 \qquad \mu = 0.3$$

$$\beta = 15°$$

$$T = \mu F_n R_m$$

$$350 = 0.3 \times F_n \times 0.15$$

$$F_n = 7\,778 \text{ N}$$

Also,

$$F_n = (2\pi R_m b)p \qquad \text{[refer Fig. 8.15]}$$

$$7\,778 = (2\pi \times 0.15 \times b) \times 150 \times 10^3$$

$$b = 0.055 \text{ m or } \underline{55 \text{ mm}}$$

Axial force $= F_n \sin\beta = 7\,778 \sin 15° = \underline{2\,012.4 \text{ N}}$

Example 8.11 The semi-cone angle of a cone clutch is 12.5° and the contact surfaces have a mean diameter of 80 mm. Coefficient of friction is 0.32. What is the minimum torque required to produce slipping of the clutch for an axial force of 200 N?

If the clutch is used to connect an electric motor with a stationary flywheel, what is the time needed to attain the full speed and the energy lost during slipping? Motor speed is 900 rpm and the moment of inertia of the flywheel is 0.4 kg.m².

Solution:

$R_m = 0.04$ m $\qquad \mu = 0.32$

$F = 200$ N $\qquad \beta = 12.5°$

$N = 900$ rpm $\qquad I = 0.4$ kg.m²

$$T = I\alpha \qquad (\alpha = \text{angular acceleration})$$

or

$$T = \mu F_n R_m = I\alpha$$

or

$$T = \mu \frac{F}{\sin\beta} R_m = I_\alpha$$

or

$$T = 0.32 \times \frac{200}{\sin 12.5°} \times 0.04 = 0.4\,\alpha$$

$$\alpha = 29.57 \text{ rad/s}^2 \quad \text{and} \quad T = \underline{11.828 \text{ N.m}}$$

or

$$\frac{\omega}{t} = 29.57$$

or

$$\frac{2\pi \times 900}{60t} = 29.57$$

$$t = \underline{3.187 \text{ s}}$$

During slipping period

Angle turned by driving shaft, $\theta_1 = \omega t = \dfrac{2\pi \times 900}{60} \times 3.187$

$$= 300.4 \text{ rad}$$

Angle turned by driven shaft, $\theta_2 = \omega_o t + \frac{1}{2}\alpha t^2$

$$= 0 + \frac{1}{2} \times 29.57 \times (3.187)^2$$

$$= 150.2 \text{ rad.}$$

Energy lost in friction $= T(\theta_1 - \theta_2) = 11.828\,(300.4 - 150.2)$

$$= \underline{1\,777 \text{ N.m}}$$

Alternatively,

$$\text{Energy supplied} = T\omega \times \text{time}$$

$$= 11.828 \times \frac{2\pi \times 900}{60} \times 3.187$$

$$= 3\,553 \text{ N.m}$$

$$\text{Energy of flywheel} = \frac{1}{2}I\omega^2 = \frac{1}{2} \times 0.4 \times \left(\frac{2\pi \times 900}{60}\right)^2 = 1776 \text{ N.m}$$

$$\text{Energy lost} = 3553 - 1776 = 1777 \text{ N.m}$$

Example 8.12 A cone clutch with a semi-cone angle of 15° transmits 10 kW at 600 rpm. The normal pressure intensity between the surfaces in contact is not to exceed 100 kN/m². The width of the friction surfaces is half of the mean diameter. Assume $\mu = 0.25$. Determine

(i) the outer and inner diameters of the plate
(ii) the axial force to engage the clutch.

Solution:

$P = 10$ kW $\qquad p = 100$ kN/m²

$N = 600$ rpm $\qquad \mu = 0.25$

$\beta = 15°$ $\qquad b = \frac{d_m}{2} = R_m$

(i) $b = R_m$

or

$$\frac{R_o - R_i}{\sin\beta} = \frac{R_o + R_i}{2} \qquad \text{[refer Fig. 8.15]}$$

or

$$R_o - R_i = \frac{\sin 15°}{2}(R_o + R_i) = 0.129R_o + 0.129R_i$$

or

$$R_o = 1.296\,R_i$$

Now,

$$T = \frac{P}{\omega} = \frac{10000}{\frac{2\pi \times 600}{60}} = 159 \text{ N.m}$$

Intensity of pressure is maximum at the inner radius (uniform wear theory).

$$T = \frac{\mu F_n}{2}(R_o + R_i) = \frac{\mu F}{2\sin\beta}(R_o + R_i)$$

$$= \frac{\mu}{2\sin\beta}\left[2\pi p_i R_i (R_o - R_i)\right](R_o + R_i)$$

$$= \frac{\mu \pi p_i}{\sin\beta} R_i (R_o^2 - R_i^2)$$

$$159 = \frac{0.25\pi \times 100 \times 10^3}{\sin 15°} R_i\,[(1.296\,R_i)^2 - R_i^2]$$

$$= \frac{0.25\pi \times 100 \times 10^3}{\sin 15°} \times 0.6796\,R_i^3$$

$$R_i = 0.0917 \text{ m} \quad \text{or} \quad \underline{91.7 \text{ mm}}$$

$$R_o = 91.7 \times 1.296 = \underline{118.8 \text{ mm}}$$

$$b = \frac{R_o - R_i}{\sin\beta} = \frac{118.8 - 91.7}{\sin 15°} = 105 \text{ mm} \qquad \text{(if required)}$$

(ii) Axial force, $F = 2\pi p_i R_i (R_o - R_i)$

$$= 2\pi \times 100 \times 10^3 \times 0.0917\,(0.1188 - 0.0917)$$

$$= \underline{1561 \text{ N}}$$

8.10 ROLLING FRICTION

When a ball rolls over a flat surface, the contact is theoretically at a point. Similarly, when a cylinder rolls over the surface, the contact is along a line parallel to its axis. However, the ball or the cylinder possesses weight and due to the pressure of the same, deformation of the flat surface or of the rolling body or of both takes place. The amount of deformation will depend upon the elasticity of the materials in contact and the pressure. With harder surfaces, the deformation is extremely small. The deformation causes the two surfaces to have area of contact rather than point or line contact.

Figure 8.16 shows a ball rolling on a flat surface. The ball is assumed to be made of a very hard material so that its deformation is negligible. The material of the dented surface is stretched at the place of contact as the curved length *AB* is greater than the flat length *AB*.

Now, if the ball is to roll to the right, the material of the flat surface must be stretched to accommodate the curved surface of the ball. The stretching material slides against the surface of the ball which causes frictional resistance. The frictional resistance also occurs when the material left behind again contracts and slides against the surface of the ball. This friction is known as rolling friction.

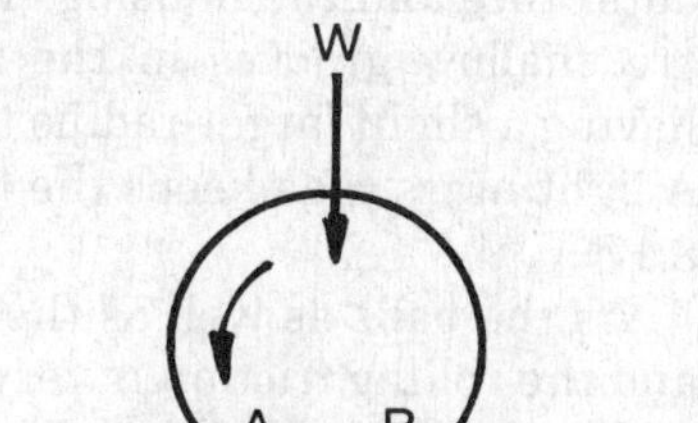

Fig. 8.16

8.11 ANTI-FRICTION BEARINGS

Ordinarily, if a shaft revolves in a bearing, it has a sliding motion. But if balls or rollers are used between the shaft and the journal as shown in Figs. 8.17 and 8.18, then rolling occurs between the journal and the rollers as well as between the rollers and the shaft. Usually, the balls or rollers are made of hardened materials such as chromium steel or chrome-nickel steel.

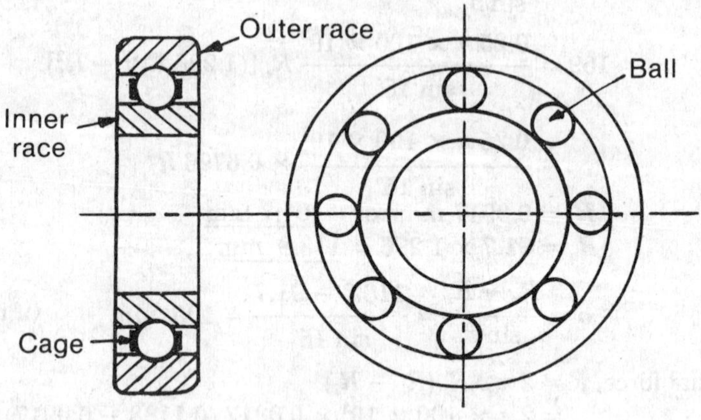

Fig. 8.17

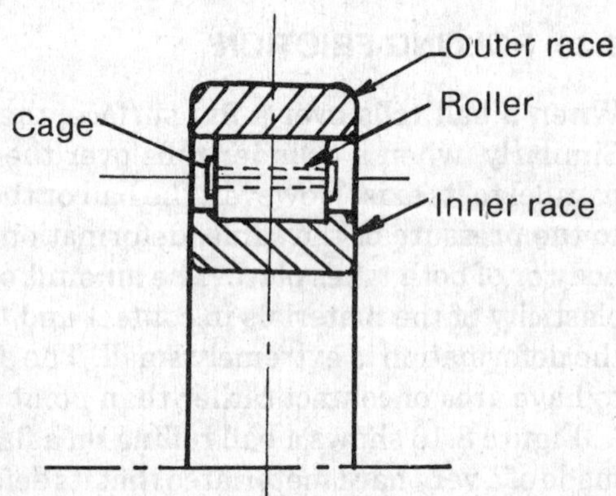

Fig. 8.18

1. Ball Bearings

A ball bearing consists of a number of hardened balls mounted between two hardened races. The inner race is fitted to the shaft and the outer race is a tight fit into the bearing housing. Thus there is no relative movement between the shaft and the inner race, and the outer race and the housing. There are shallow grooves in the races having a slight larger radius than that of the balls to accommodate the balls. A light brass cage keeps the balls at a fixed distance from one another (Fig. 8.17).

As the balls as well as the races are very hard, distortion of each is little and the rolling friction is very low.

The friction of ball bearings is slightly higher when lubricated than when dry. However, a small amount of lubricant is useful to prevent rust formation.

2. Roller Bearings

These are similar to ball bearings where the balls are replaced by hardened cylindrical rollers. These bearings can carry heavier loads.

Figure 8.18 shows a radial roller bearing. This type of bearing will carry only the radial loads or loads perpendicular to the shaft axis. The outer race is plain whereas the inner race has a groove to accommodate the rollers. A cage keeps the rollers at a uniform distance apart.

When the rollers are small and used without a cage, they are known as needles. Such bearings are used to carry heavy loads.

Roller bearings are not used for mis-aligned shafts.

8.12 GREASY FRICTION

If two metallic surfaces are wetted with a small amount of lubricant, a very thin film of the same is formed on each of the surfaces. This thin film is of molecular thickness. It adheres to the surface and is known as adsorbed film. It has been found that when the two surfaces are placed in contact, the coefficient of friction between them is considerably reduced compared with when the surfaces are dry and unlubricated.

The property of a lubricant to form a layer of molecular thickness (adsorbed film) on a metallic surface is known as its oiliness. If two lubricants of equal viscosity are used to lubricate two metallic surfaces under identical conditions, it is found that one reduces the friction more than the other and is said to have greater oiliness.

The friction of two surfaces, when they are wetted with an extreme thin layer of lubricant and metal to metal contact can take place between high spots, is known as *greasy friction* or *boundary friction.*

The laws governing greasy friction are similar to those for solid or dry friction.

8.13 GREASY FRICTION AT A JOURNAL

Greasy or boundary friction occurs in heavily loaded, slow running bearings. In this type of friction, the frictional force is assumed to be proportional to the normal reaction.

When a shaft rests in its bearing (Fig. 8.19 a), the weight of the shaft **W**, acts through its centre of gravity. The reaction of the bearing acts in line with **W** in the vertically upward direction. The shaft rests at the bottom of the bearing at A and metal to metal contact exists between the two.

When a torque is applied to the shaft, it rotates and the seat of pressure creeps or climbs up the bearing in a direction opposite to that of rotation. Metal to metal condition still exists and greasy friction criterion applies as the oil film will be of molecular thickness. The common normal at B between the two surfaces in contact passes through the centre of the shaft.

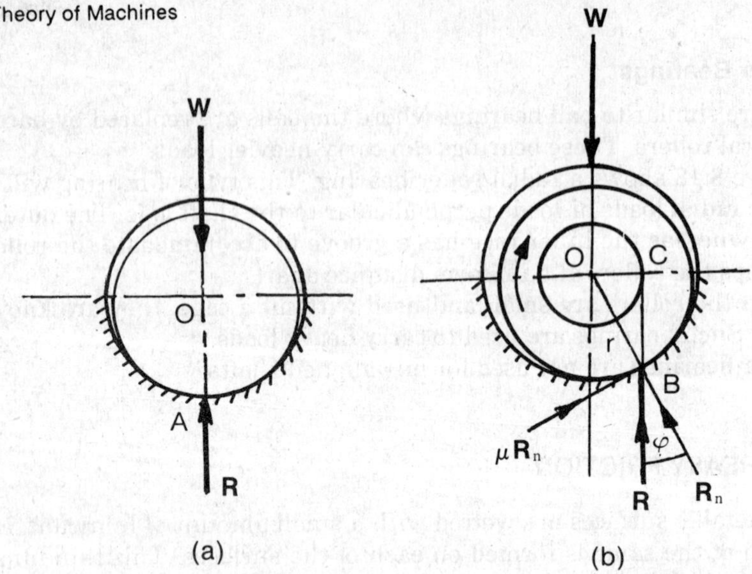

Fig. 8.19

Let $\mathbf{R}_n$ = normal (radial) reaction at B

$\mu\mathbf{R}_n$ = friction force, tangential to the shaft

The normal reaction and the friction force can be combined into a resultant reaction **R** inclined at an angle φ to $\mathbf{R}_n$ (Fig. 8.19 b).

Now the shaft is in equilibrium under the following forces:

(i) Weight **W**, acting vertically downwards, and

(ii) Reaction **R**.

For equilibrium, **R** must act vertically upwards and must equal to **W**. However, the two forces **W** and **R** will be parallel and constitute a couple.

Let OC = Perpendicular to R from O

Friction couple (torque) $= W \times OC = Wr \sin\varphi$ $\qquad (\because \sin\varphi = \frac{OC}{r})$

$\approx Wr \tan\varphi$ $\qquad$ (as φ is small)

$\approx Wr\,\mu$ $\qquad (\mu = \tan\varphi)$

This couple must be equal and opposite to the couple or torque producing motion.

A circle drawn with OC (or $r \sin\varphi \approx r\tan\varphi \approx r\mu$) as radius is known as the *friction circle* of the journal.

Thus the effect of friction is equivalent to displacing the reaction through a distance equal to $r \sin\varphi$ or such that it is tangential to the friction circle.

8.14 FRICTION AXIS OF A LINK

In a pin-jointed mechanism, usually, it is assumed that the resulting thrusts (axial forces) in links act along the longitudinal axes of the links. But as the laws of dry friction are similar to those of greasy friction, the friction at a pin-joint acts in the same way as that of a journal (rotating shaft) rotating in a bearing. In a journal bearing, the resultant force on a journal is tangential to the friction circle. Similarly, in pin-jointed links, the line of thrust on a link is tangential to the friction circles at the pin-joints. The net effect of all this is to shift the axis along which the thrust acts. The new axis is known as the *friction axis* of the link.

1. Slider-crank Mechanism

Figure 8.20 a shows a slider-crank mechanism in which **F** is the thrust on the slider. If the effect of friction is neglected, the force **F** will induce a thrust $\mathbf{F}_c$ in the connecting rod along its axis. However, due to friction, the friction axis will be along a tangent to the friction circles at the joints A and B.

Let r_a = radius of pin at A

r_b = radius of pin at B

μ_a = coefficient of friction at A

μ_b = coefficient of friction at B

Therefore, the radius of friction circle at $A = \mu_a r_a$

The radius of friction circle at $B = \mu_b r_b$

(exactly the two radii will be $r_a \sin \varphi_a$ and $r_b \sin \varphi_b$)

Now, there are four possible ways of drawing a tangent to these circles. To decide about the right one, remember that a friction couple is opposite to the couple or torque producing the motion of a link. Thus while drawing a tangent to a friction circle, see that the friction couple or torque so obtained is opposite to the direction of rotation of the link. Thus the position of right tangent depends upon:

1. the directions of the external forces on the link, and
2. the direction of the motion of the link relative to the links to which it is connected.

Owing to the force **F** on the piston, compressive forces are developed in the connecting rod and the crank rotates in the counter-clockwise direction. At end B, the angle OBA is increasing and thus the rod rotates clockwise relative to the piston. Therefore, the tangent at B must be on the upper side of the friction circle to give a counter-clockwise friction couple. At end A, the angle OAB is decreasing and thus the rod rotates clockwise relative to the crank. Therefore, the tangent at A must be on the lower side of the friction circle to give a counter-clockwise friction couple. The friction axis will be the common tangent to the two friction circles, on the upper side of the circle at B and on the lower side of the circle at A.

Figure 8.20 a also shows one more position of the crank in dotted lines. The direction of F as well as that of the thrust in the connecting rod is the same as before. However, now the angle $OB'A'$ at B' is decreasing. Thus the rod rotates in the counter-clockwise direction relative to the piston and the tangent is on the lower side of the friction circle to give a clockwise friction couple. The tangent at end A' will be as before.

For the positions of the crank shown in Fig. 8.20 b, the direction of F has changed and the rod becomes in tension. The angle $OB'A'$ at B' increases and

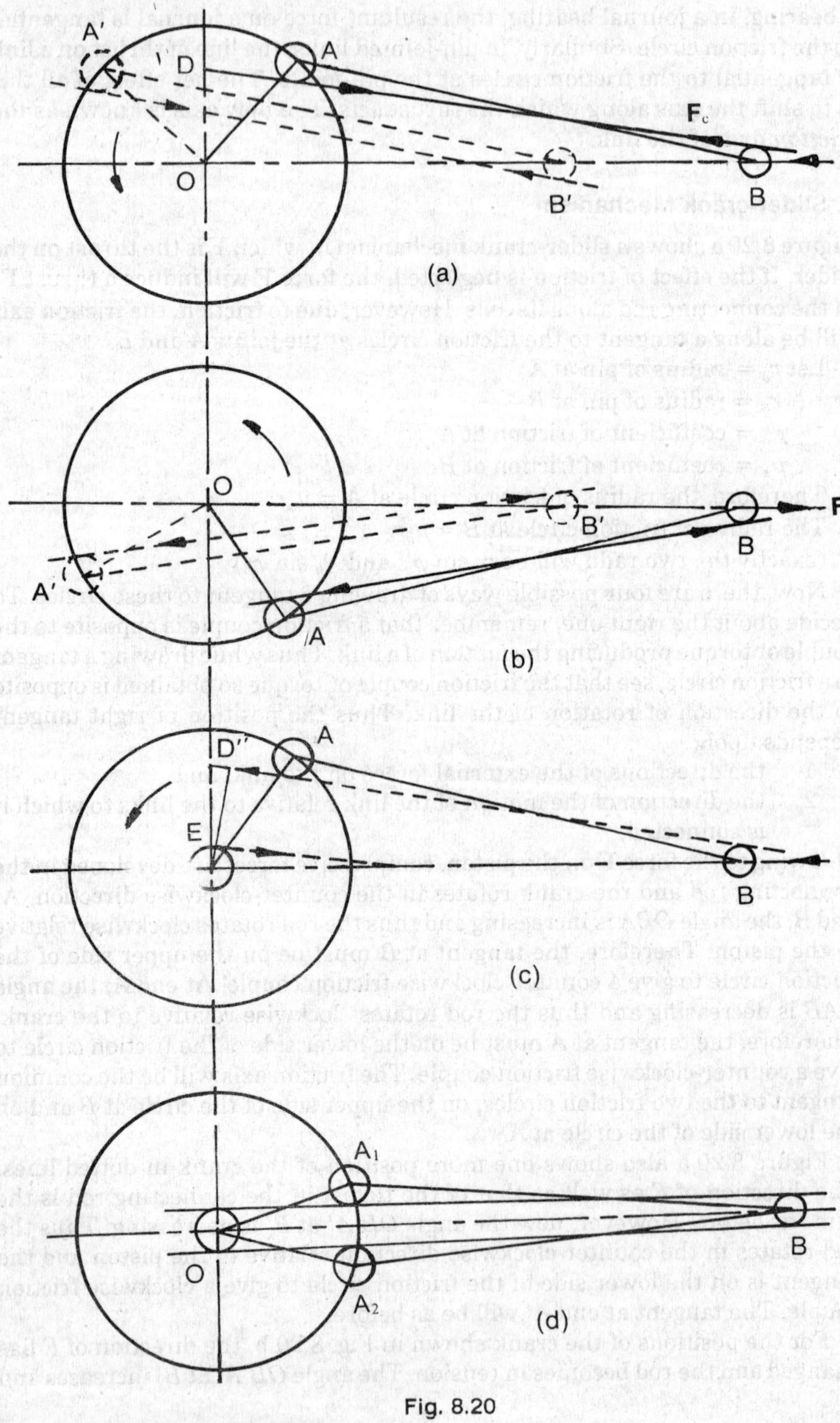

Fig. 8.20

thus the rod rotates in the counter-clockwise direction. The tangent, therefore, is on the upper side of the circle. At A', the angle $OA'B'$ also increases and the rod rotates clockwise relative to the crank. The tangent is on the upper side of the circle to give a counter-clockwise friction couple. In the same way, the tangent can be drawn for the position OA of the crank.

In case a torque **T** turns the crank counter-clockwise and a load **F** is applied to the piston, then the rod will be in tension instead of compression for the positions of Fig. 8.20 a. For the first position of the crank, the tangent at A will be on the upper side and at B on the lower side.

Another method to know the friction axis out of the four possible tangents is to select the one that gives the least intercept from O on a vertical through O such as OD if **F** is the driving force. In case **T** is the driving torque at the crank, the intercept would be maximum.

Reaction at Crankshaft Bearing (Fig. 8.20 c) The crank OA is acted upon by two equal and opposite forces at its two ends O and A forming a couple and a torque. A force equal and opposite to that in the connecting rod acts at the crank end A, the opposite of which acts at end O. As the crankshaft rotates in the counter-clockwise direction, the tangent is on the upper side of the friction circle to give a clockwise friction couple at end O.

The turning moment of the crankshaft = $F_c \times D'E$

where $D'E$ is drawn perpendicular to the friction axis.

Dead Angle and Dead Centre Positions During rotation, when the crank is near the inner dead centre, a position is obtained where the reaction at the crankshaft bearing and the friction axis of the connecting rod become in the same straight line (Fig. 8.20 d). These positions of the crank have been shown as OA_1 and OA_2 and are known as dead centre positions. In these positions, the length $D'E$ is zero and, therefore, the turning moment or the torque transmitted is zero.

The angle A_1OA_2 is known as the dead angle. A corresponding dead angle at the outer dead centre position of the crank can also be found in the same way.

2. Four-bar Mechanism

Consider a four-bar mechanism shown in Fig. 8.21. Let the rotation of the driving link AB be clockwise. In the absence of friction, the driving torque induces compressive axial force in the coupler along its axis BC.

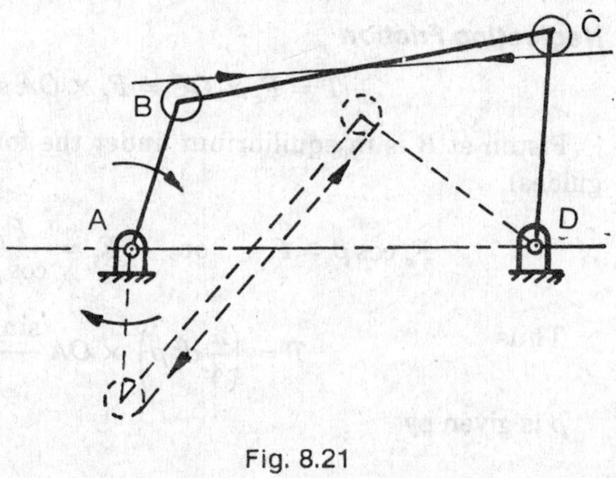

Fig. 8.21

When friction is considered, it is tangential to the friction circles at B and C.

The angle ABC at B is increasing. Thus the coupler rotates in the counter-clockwise direction relative to AB. Therefore, the tangent at B is on the upper side of the circle to give a clockwise friction couple. Similarly at C, the angle BCD is decreasing. Thus BC rotates in the counter-clockwise direction relative to CD. The tangent at C will be on the lower side of the circle so that the resulting friction couple is clockwise. Similarly, the friction axis for the links AB and DC can be drawn.

Example 8.13 **The length of crank of a slider-crank mechanism is 300 mm and of connecting rod 1.25 m. The diameters of the journals at the crosshead, crankpin and the crankshaft are 80 mm, 120 mm and 140 mm respectively. The steam pressure on the piston is 450 kN/m² which has a diameter of 250 mm. Coefficient of friction between the crosshead and the guides is 0.07 and for journals it is 0.05.**

Find the reduction in the turning moment available at the crankshaft due to friction of the crosshead guides and the journals when the crank has rotated 50° from the inner-dead centre.

Solution: Refer to Fig. 8.22 a.

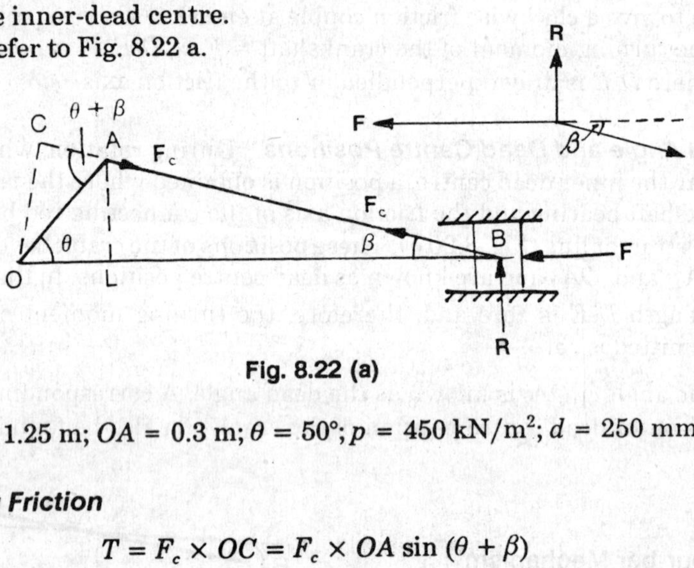

Fig. 8.22 (a)

$$AB = 1.25 \text{ m}; OA = 0.3 \text{ m}; \theta = 50°; p = 450 \text{ kN/m}^2; d = 250 \text{ mm}$$

Neglecting Friction

$$T = F_c \times OC = F_c \times OA \sin(\theta + \beta)$$

Piston at B is in equilibrium under the forces, **F**, $\mathbf{F_c}$ and **R** (reaction of guides).

$$\therefore \quad F_c \cos\beta = F \quad \text{or} \quad F_c = \frac{F}{\cos\beta} = \frac{\frac{\pi}{4} d^2 \times p}{\cos\beta}$$

Thus

$$T = \left(\frac{\pi}{4} d^2 p\right) \times OA \frac{\sin(\theta + \beta)}{\cos\beta}$$

β is given by

$$\beta = \sin^{-}\left(\frac{AL}{AB}\right) = \sin^{-}\left(\frac{OA \sin \theta}{AB}\right) = \sin^{-}\left(\frac{0.3 \sin 50°}{1.25}\right)$$

$$T = \left[\frac{\pi}{4}(0.25)^2 \times 450 \times 10^3\right] \times 0.3 \times \frac{\sin (50° + 10.6°)}{\cos 10.6°}$$

$$= 22089 \times 0.3 \times \frac{\sin 60.6°}{\cos 10.6°}$$

$$= 5873.5 \text{ N.m}$$

Friction Considered Refer to Fig. 8.22 b.

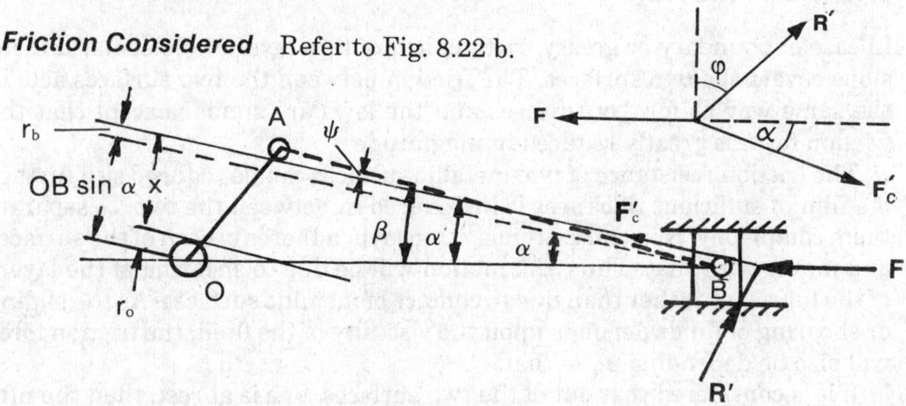

Fig. 8.22 (b)

Radius of friction circle at O (crankshaft), $r_o = \mu r = 0.05 \times \dfrac{0.14}{2}$

$$= 0.0035 \text{ m}$$

Radius of friction circle at A (crankpin), $r_a = 0.05 \times \dfrac{0.12}{2} = 0.003$ m

Radius of friction circle at B (crosshead), $r_b = 0.05 \times \dfrac{0.08}{2} = 0.002$ m

$$OB = OA \cos \theta + AB \cos \beta = 0.3 \cos 50° + 1.25 \cos 10.6°$$

$$= 1.4215 \text{ m}$$

Inclination of the friction axis with OB,

$$\alpha = \beta - \psi = \beta - \sin^{-}\left(\frac{r_a + r_b}{AB}\right) = 10.6° - \sin^{-}\left(\frac{0.003 + 0.002}{1.25}\right)$$

or

$$\alpha = 10.6° - 0.23° = 10.37°$$

The piston at B is in equilibrium under the action of forces $\mathbf{F}$, $\mathbf{F}'_c$ and R'.

$$\varphi = \tan^{-} 0.07 = 4°$$

$$\therefore \quad \frac{F'_c}{\sin (90° + \varphi)} = \frac{F}{\sin (90° + \alpha - \varphi)}$$

or
$$F_c' = \frac{22\,089 \sin (90° + 4°)}{\sin (90° + 10.37° - 4°)} = 22\,172 \text{ N}$$

$$x = OB \sin \alpha + r_b - r_o = 1.4215 \sin 10.37° + 0.002 - 0.0035 = 0.2544 \text{ m}$$

$$T' = F_c' \times x = 22\,172 \times 0.2544 = 5\,640.6 \text{ N.m}$$

Reduction in the torque available $= T - T' = 5\,873.5 - 5\,640.6 = \underline{232.9 \text{ N.m}}$

8.15 FILM FRICTION

In case of boundary or greasy friction, a very thin layer of molecular dimensions covers the two surfaces. The friction between the two surfaces acts in the same way as for dry surfaces and the laws are similar except that the friction force is greatly reduced in magnitude.

The friction resistance of two metallic surfaces can be reduced still further if a film of sufficient thickness is introduced in between the two, to separate them completely. Now, an extremely thin layer adheres to each of the surfaces and moves with that. Thus, the motion will be due to shearing of the layers of the lubricant rather than due to contact of metallic surfaces. As the sliding or shearing of fluids depends upon the viscosity of the fluid, the friction force will also be depending upon that.

If it is considered that out of the two surfaces, one is at rest, then the film of lubricant in contact with the stationary surface will remain at rest and that in contact with the moving surface will move with it. The different layers between the two may be considered to be moving with a speed proportional to its distance from the surface at rest. Thus the force of friction would be the force necessary to slide these adjacent layers over one another.

For a journal rotating in a bearing under the film lubrication conditions, the frictional resistance is found to be

(a) proportional to the area
(b) proportional to the viscosity of the lubricant
(c) proportional to the speed
(d) independent of the pressure
(e) independent of the materials of the journal and the bearing.

The first experiments on film lubrication were carried by Beauchamp Tower in 1883. However, it was Osborne Reynolds who demonstrated that to maintain a film between two surfaces, it is necessary that they are slightly inclined to each other such that a wedge of lubricant is formed between them.

Under greasy friction, in slow running and heavily loaded shafts, the seat of pressure climbs up the bearing in a direction opposite to that of rotation of the shaft and the condition is maintained. However, if the speed is further increased, a film of oil adheres to the surface of the journal which also rotates and becomes of sufficient thickness to lift the journal. The pressure of the oil beneath the journal also rises to support the load.

It is observed that under stable conditions, the journal takes up a new position as shown in Fig. 8.23 a. Two wedges of oil are formed on each side of A, the point of minimum thickness of oil. The point A is also referred as the point of nearest approach. The pressure of the oil in the wedge varies from zero at the point of entrance of the oil to a maximum at a point somewhere near A. On the down side of A, the pressure falls to zero before reaching the point of free discharge.

The pressure built up of the oil, due to decrease in area as it approaches the point A, is enough to bear the load of the bearing and separate the two metallic surfaces. The laws of viscous friction apply in this condition.

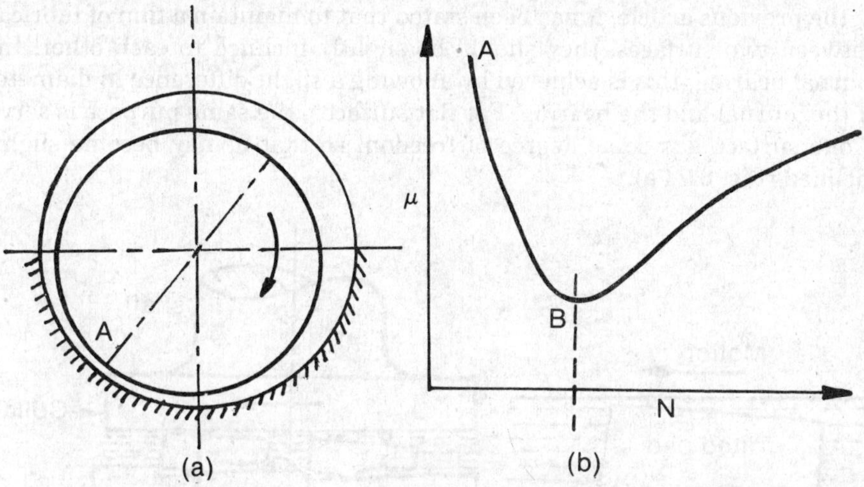

Fig. 8.23

A plot of graph between the coefficient of friction and the speed (Fig. 8.23 b) shows that at low speeds the coefficient of friction falls rapidly as the speed increases. This is because more oil is fed between the surfaces lowering the coefficient of friction. From A to B, the conditions vary from greasy friction at A to film friction at B where a complete oil film is formed. As the speed is further increased, the resistance of the oil film increases, increasing the coefficient of friction.

For clockwise rotation of the shaft, the position of its axis shifts towards right in the range of greasy friction and towards left in that of film friction (Figs. 8.19 b and 8.23 a).

The above discussion may lead to the conclusion that it is possible to have no wear of the two metallic surfaces if film lubrication is adopted. However, in practice, some wear does take place even under the most favourable conditions. The most apparent cause is the failure of the oil film during the starting and stopping of the shaft, when the conditions resemble to that of greasy friction, and metal to metal contact exists. Metal particles of extremely small size are produced through wear which then float in the lubricant. With time, the number of these particles increases which tends to cause further wear. Moreover, the small particles of solid matter tend to destroy the formation of the viscous oil film. Thus in actual practice, the working conditions become that of a combination of film and greasy lubrication. However, by fitting an oil filter to remove the small particles of solid matter may increase the life of a bearing. Otherwise, the abrasive action will increase the wear, increasing the clearances resulting in insufficient pressure rise to lift the bearing surfaces apart. Thus the bearing may fail eventually.

8.16 MITCHELL THRUST BEARING

In the previous article, it has been stated that to maintain a film of lubricant between two surfaces, they should be slightly inclined to each other. In a journal bearing, this is achieved by allowing a slight difference in diameters of the journal and the bearing. For flat surfaces, the same purpose is served if one surface has some degree of freedom so that it may become slightly inclined (Fig. 8.24 a).

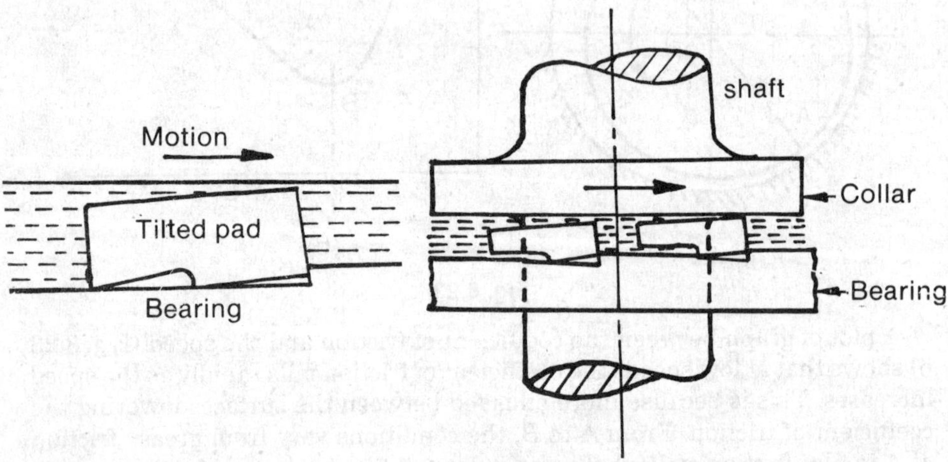

Fig. 8.24

To bear axial forces (thrusts) in shafts, usually, a number of collars, working under dry friction conditions have to be used (Sec. 8.8). In a Mitchell thrust bearing, the condition of film friction, instead of dry friction, is achieved and only one collar is used which considerably reduces the length of the bearing by increasing the allowable pressure. Thus this type of bearing is useful in transmitting very heavy thrusts, e.g. thrust of a ship's propeller to the hull.

A Mitchell thrust bearing consists of a series of metallic pads arranged around a rotating collar fixed to the shaft (Fig. 8.24 b). Each pad is held by the housing of the bearing. Thus the pad is prevented from rotation but is able to tilt about its stepped edge. When the thrust is transmitted to the pads, they can tilt slightly on the edges. The oil carried by the moving collar, is dragged around the pad. Thus an oil film of wedge shape is formed and a considerable pressure is built up to carry the axial load.

The number of pads can be from four to six depending upon the total thrust.

PROBLEMS

8.1 A screw jack is used to raise a load of 5 tonnes (1 tonne = 9.81 kN). The pitch of single start square threads used for the screw is 24 mm. The mean diameter is 72 mm. Determine the force to be applied at the end of a 1.2 m long handle when the load is lifted with constant velocity and rotate with the spindle. Take μ = 0.2. Also find the mechanical efficiency of the jack.

(552.2 N; 33.9%)

8.2 Find the load that can be lifted by applying a force of 220 N at the end of 500 mm long lever of screw jack using single- start square threads. The load does not rotate with the spindle and is carried on a swivel head having a bearing of 100 mm diameter. The pitch of the threads is 10 mm and the root diameter is 50 mm. Coefficient of friction between nut and thread is 0.18 and between spindle and swivel head 0.15.
Find also the efficiency of the jack.

(7.798 kN; 11.28%)

8.3 Determine the mechanical efficiency of a wedge used to raise loads if the angle of wedge is 20° and the coefficient of friction 0.2 between the frame and the wedge and 0.15 between the slider and the guide. The height of the guide is 120 mm and its lower end is 45 mm above the lower point of the axis of the slider which has a width of 50 mm.

(49.64%)

8.4 The shaft of a collar thrust bearing rotates at 200 rpm and carries an end thrust of 10 tonnes. The outer and the inner diameters of the bearing are 480 mm and 280 mm respectively. If the power lost in friction is not to exceed 8 kW, determine the coefficient of friction of the lubricant of the bearing.

(0.02)

8.5 A shaft carrying a load of 12 tonnes and running at 120 rpm has a number of collars integral with it. Shaft diameter is 240 mm and the external diameter of the collars 360 mm. Intensity of uniform pressure is 400 kN/m^2 and the coefficient of friction 0.06. Determine the power absorbed in overcoming the friction and the number of collars required.

(13.49 kW; 6)

8.6 A single plate clutch, having two active surfaces, transmits 10 kW of power and the maximum torque developed is 120 N.m. Axial pressure is not to exceed 100 kN/m^2. Outer diameter of the friction plate is 1.3 times the inner diameter. Determine these diameters and the axial force exerted by the springs. Assume uniform wear and take coefficient of friction as 0.25.

(207 mm, 269 mm)

8.7 A multiple-disc clutch has 6 active friction surfaces. The power transmitted is 20 kW at 400 rpm. Inner and outer radii of the friction surfaces are 90 and 120 mm respectively. Assuming uniform wear with a coefficient of friction 0.3, find the maximum axial intensity of pressure between the discs.

(148.9 kN/m^2)

8.8 Determine the axial force required to engage a cone clutch transmitting 25 kW of power at 750 rpm. Average friction diameter of the cone is 400 mm, semicone angle 10° and coefficient of friction 0.25. Also find the width of the friction cone.

(2.672 kN; 84.4 mm)

8.9 Show that the torque transmitted by a cone clutch when intensity of pressure is uniform is given by

$$\frac{2\,\mu\,W}{3\sin\alpha}\left[\frac{r_o^3 - r_i^3}{r_o^2 - r_i^2}\right]$$

Where W = axial load

μ = coefficient of friction of contact surfaces

α = semi–cone angle

r_o = outer radius of contact surface

r_i = inner radius of contact surface

In a cone clutch with semi-cone angle of 15°, maximum and minimum radii of the contact surface are 120 mm and 80 mm respectively. The speed is 800 rpm and the maximum allowable normal pressure is 150 kN/m^2. Determine the axial load and the power transmitted taking the coefficient of friction as 0.3.

(3.77 kN; 37.097 kW)

8.10 The crank of a steam engine is 250 mm long and the length of the connecting rod is 1 m. The steam pressure at the cross-head is 350 kN/m^2. The diameters of the journals at the crankshaft, crankpin and the crosshead are 180 mm, 140 mm and 100 mm respectively. The piston diameter is 200 mm and the coefficient of friction between the crosshead and guide 0.079 and at the journal 0.06. Determine the reduction in the turning moment available at the crankshaft due to friction of the crosshead guides and the journal at the moment when the crank has rotated through 45° from the inner-dead centre.

(58.85 N.m)

9

BELTS, ROPES AND CHAINS

9.1 INTRODUCTION

Usually, power is transmitted from one shaft to another by means of belts, ropes, chains and gears, the salient features of which are as follows:

1. Belts, ropes and chains are used where the distance between the shafts is large. For small distances, gears are preferred.
2. Belts, ropes and chains are flexible type of connectors, i.e. they are bent easily.
3. The flexibility of belts and ropes is due to the property of their materials whereas chains have a number of small rigid elements having relative motion between the two elements.
4. Belts and ropes transmit power due to friction between them and the pulleys. If the power transmitted exceeds the force of friction, the belt or rope slips over the pulley.
5. Belts and ropes are strained during motion as tensions are developed in them.
6. Owing to slipping and straining action, belts and ropes are not positive type of drives, i.e. their velocity ratios are not constant. On the other hand, chains and gears have constant velocity ratios.

9.2 BELT AND ROPE DRIVES

To transmit power from one shaft to another, pulleys are mounted on the two shafts. The pulleys are then connected by an endless belt or rope passing over the pulleys. The connecting belt or rope is kept in tension so that motion of one pulley is transferred to the other without slip. The speed of the driven shaft can be varied by varying the diametres of the two pulleys.

For an unstreched belt mounted on the pulleys, the outer and the inner faces become in tension and compression respectively (Fig. 9.1). Inbetween there is a neutral section which has no tension or compression. Usually, this is considered at half the thickness of the belt. The effective radius of rotation

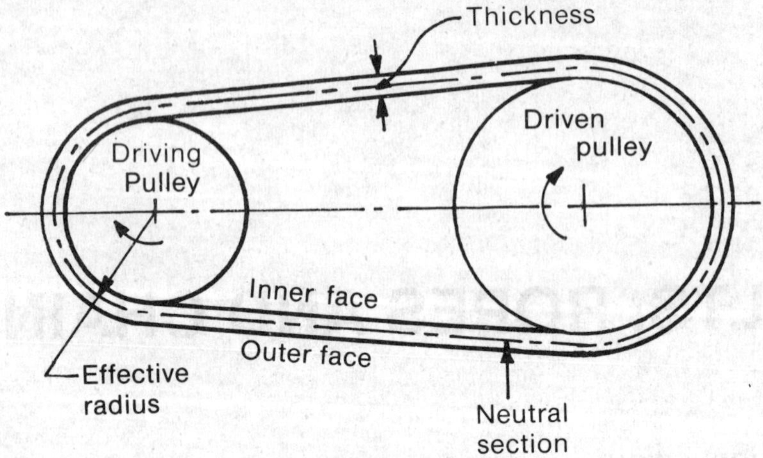

Fig. 9.1

of a pulley is obtained by adding half the belt thickness to the radius of the pulley.

A belt may be of rectangular section, known as a *flat belt* (Fig. 9.2 a) or of trapezoidal section, known as a V-belt (Fig. 9.2 b). In case of a flat belt, the rim of the pulley is slightly *crowned* which helps to keep the belt running centrally on the pulley rim. The groove on the rim of the pulley of a V-belt drive is made deeper to take the advantage of the wedge action. The belt does not touch the bottom of the groove. Owing to wedging action, V-belts need little adjustment and transmit more power, without slip, as compared to flat belts. Also, a multiple V-belt system, using more than one belt in the two

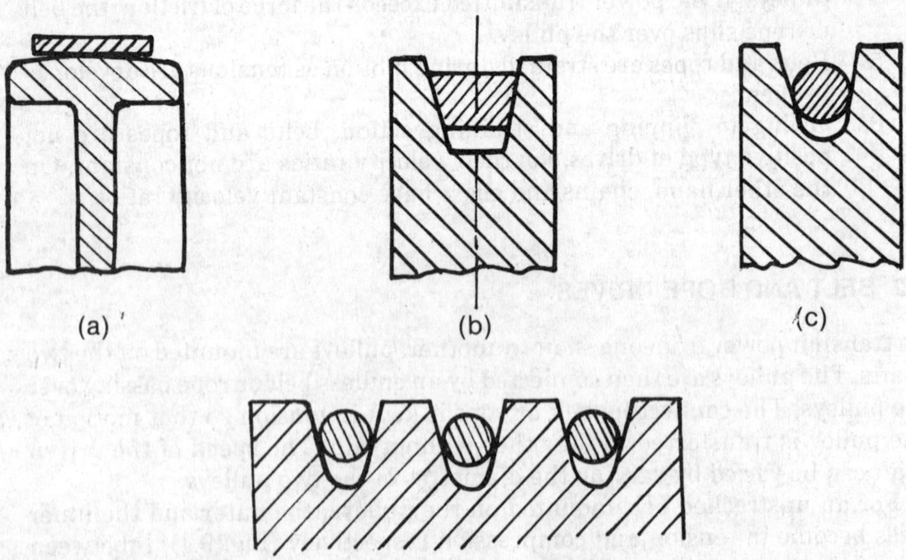

Fig. 9.2

pulleys, can be used to increase the power transmitting capacity. Generally, these are more suitable for shorter centre distances.

For power transmission by ropes, grooved pulleys are used (Fig. 9.2 c). The rope is gripped on its sides as it bends down in the groove reducing the chances of slipping. Pulleys with several grooves can also be employed to increase the capacity of power transmission (Fig. 9.2 d). These may be connected in either of the two ways:

1. Using a continuous rope passing from one pulley to the other and back again to the same pulley in the next groove and so on.
2. Using one rope for each pair of grooves.

The advantage of using continuous rope is that the tension in it is uniformly distributed. However, in case of belt failure, the whole drive is put out of action. Using one rope for each groove poses difficulty in tightening the ropes to the same extent but with the advantage that the system can continue its operation even if a rope fails. The repair can be undertaken when it is convenient.

Rope drives are, usually, preferred for long centre distances between the shafts, ropes being cheaper as compared to belts. These days, however, long distances are avoided and thus the use of ropes has been limited.

9.3 OPEN- AND CROSSED-BELT DRIVES

1. Open-Belt Drive

An open belt drive is used when the driven pulley is desired to be rotated in the same direction as the driving pulley as shown in Fig. 9.1.

Generally, the centre distance for an open-belt drive is 14 to 16 m. If the centre distance is too large, the belt whips, i.e. vibrates in a direction perpendicular to the direction of motion. For very shorter centre distances, the belt slip increases. Both these phenomena limit the use of belt for power transmission.

While transmitting power, one side of the belt is more tightened (known as tight side) than the other (known as slack side). In case of horizontal drives, it is always desired that the tight side is at the lower side of two pulleys. This is because the sag of the belt will be more on the upper side than the lower side. This slightly increases the angles of wrap of the belt on the two pulleys than if the belt had been perfectly straight between the pulleys. In case the tight side is at the upper side, the sag will be greater at the lower side, reducing the angle of wrap and slip could occur earlier. This ultimately affects the power to be transmitted.

2. Crossed-Belt Drive

A crossed-belt drive is adopted when the driven pulley is to be rotated in the opposite direction to that of the driving pulley (Fig. 9.3).

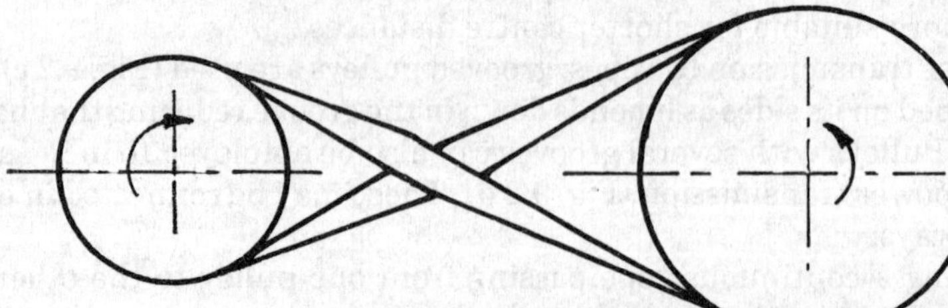

Fig. 9.3

A crossed-belt drive can transmit more power than an open-belt drive as the angle of wrap is more. However, the belt has to bend in two different planes and it wears out more.

9.4 ACTION OF BELT ON PULLEYS

To transmit torque from one shaft to another effectively, it is necessary that the belt does not slip over the pulleys. To understand the action of belt on pulleys, consider a belt resting on the rim of a pulley. The belt is wrapped round the pulley to subtend an angle θ at its centre. The belt is tightened on the pulley by applying an equal amount of pull to the two ends of the belt. This results in an initial tension **T** on each end of the belt (Fig. 9.4 a).

Now, if a small torque or tangential force **F** in the clockwise direction is applied to the driving pulley, it tends to rotate the belt with it. But if the motion of the belt is resisted, the pulley will have a tendency to slip over the belt. Considering the equilibrium of the pulley, it can be said that the slipping of the pulley is prevented by a frictional force **F**′ (=**F**) acting in the counter-clockwise direction. In the same way, it is concluded that the belt is prevented from slipping in the counter-clockwise direction (opposite to that of pulley) due to a frictional force equal to **F** in the clockwise direction (opposite to that on pulley). The frictional force in the belt results in an increased tension on one end of the belt ($\mathbf{T}_1$) and a decrease in tension on the other ($\mathbf{T}_2$) such that $(T_1 - T_2) = F$ (Fig. 9.4 b).

If the tangential force on the pulley is increased, at one stage, it will just cause relative motion (slip) between the belt and the pulley. This will indicate that the frictional force has reached to the limiting value and the force F on the pulley should not be increased further.

Now, consider a belt and pulley arrangement as shown in Fig. 9.4 c. Where a belt with an initial tension **T** passes over driving and driven pulleys. On the driving side, the belt would rotate with the pulley with no relative slip between the two as long as the tangential force on the pulley, F $(= T_1 - T_2)$ is less than the frictional force. On the driven side, the pulley rotate with the belt in the clockwise direction which means that the pulley can have a tendency to slip back in the counter-clockwise direction. This implies that the belt will have

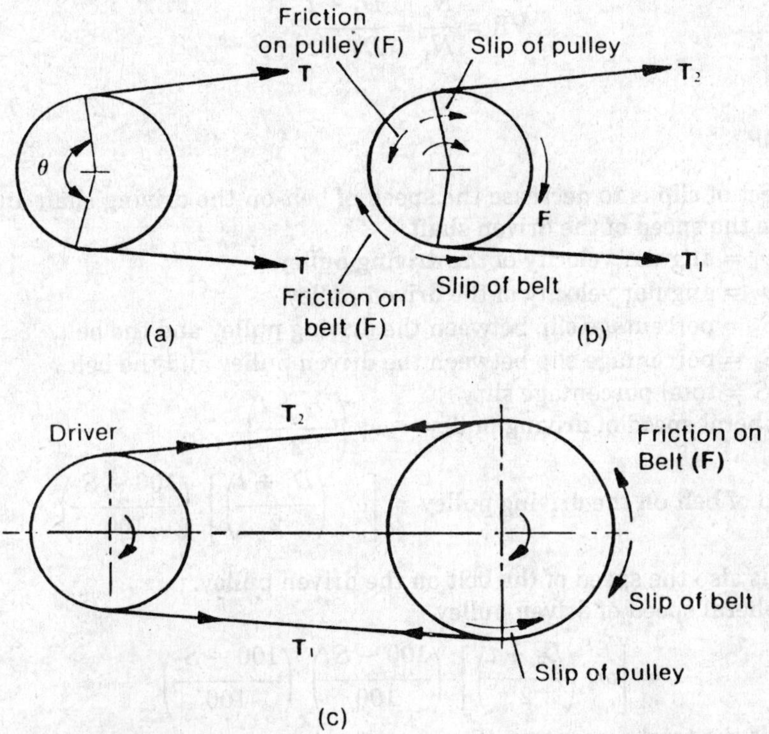

Fig. 9.4

a tendency to slip over the pulley in the clockwise direction meaning a counter-clockwise frictional force equal to $(T_1 - T_2)$.

9.5 VELOCITY RATIO

Velocity ratio is the ratio of speed of the driven pulley to that of the driving pulley.

Let N_1 = rotational speed of the driving pulley
N_2 = rotational speed of the driven pulley
D_1 = diameter of the driving pulley
D_2 = diameter of the driven pulley
t = thickness of the belt

Neglecting any slip between the belt and the pulleys and also considering the belt to be inelastic,

Speed of belt on driving pulley = Speed of belt on driven pulley

$$\left(D_1 + 2\frac{t}{2}\right) N_1 = \left(D_2 + 2\frac{t}{2}\right) N_2$$

or
$$\text{VR} = \frac{N_2}{N_1} = \frac{D_1 + t}{D_2 + t} \tag{9.1}$$

9.6 SLIP

The effect of slip is to decrease the speed of belt on the driving shaft and to decrease the speed of the driven shaft.

Let ω_1 = angular velocity of the driving pulley

ω_2 = angular velocity of the driven pulley

S_1 = percentage slip between the driving pulley and the belt

S_2 = percentage slip between the driven pulley and the belt

S = total percentage slip,

Peripheral speed of driving pulley $= \omega_1 \left(\frac{D_1 + t}{2}\right)$

Speed of belt on the driving pulley $= \left[\omega_1 \left(\frac{D_1 + t}{2}\right)\right] \left(\frac{100 - S_1}{100}\right)$

This is also the speed of the belt on the driven pulley.

Peripheral speed of driven pulley

$$= \left[\omega_1 \left(\frac{D_1 + t}{2}\right)\right] \left(\frac{100 - S_1}{100}\right) \left(\frac{100 - S_2}{100}\right)$$

As S is the total percentage slip,

$$\text{Peripheral speed of driven shaft} = \left[\omega_1 \left(\frac{D_1 + t}{2}\right)\right] \left(\frac{100 - S}{100}\right)$$

$$\left[\omega_1 \left(\frac{D_1 + t}{2}\right)\right] \left(\frac{100 - S_1}{100}\right) \left(\frac{100 - S_2}{100}\right) = \left[\omega_1 \left(\frac{D_1 + t}{2}\right)\right] \left(\frac{100 - S}{100}\right)$$

or
$$\frac{(100 - S_1)(100 - S_2)}{100 \times 100} = \frac{100 - S}{100}$$

or
$$(100 - S_1)(100 - S_2) = 100\,(100 - S)$$

or
$$10\,000 = 100S_2 - 100S_1 + S_1S_2 = 10\,000 - 100S$$

or
$$100\,S = 100\,S_1 + 100\,S_2 - S_1S_2$$

or
$$S = S_1 + S_2 - 0.01\,S_1S_2 \tag{9.2}$$

Effect of slip is to reduce the velocity ratio,

$$\text{VR} = \frac{N_2}{N_1} = \left(\frac{D_1 + t}{D_2 + t}\right) \left(\frac{100 - S}{100}\right) \tag{9.3}$$

Also it is to be remembered that slip will first occur on the pulley with smaller angle of lap i.e. on the smaller pulley.

Example 9.1 A shaft runs at 80 rpm and drives another shaft at 150 rpm through belt drive. The diameter of the driving pulley is 600 mm. Determine the diameter of the driven pulley in the following cases:

(i) Neglecting belt thickness,
(ii) Taking belt thickness as 5 mm,
(iii) Assuming for case (ii) a total slip of 4%, and
(iv) Assuming for case (ii) a slip of 2% on each pulley.

Solution:

$$N_1 = 80 \text{ rpm} \qquad D_1 = 600 \text{ mm}$$
$$N_2 = 150 \text{ rpm}$$

(i) $$\frac{N_2}{N_1} = \frac{D_1}{D_2} \quad \text{or} \quad \frac{150}{80} = \frac{600}{D_2} \quad \text{or} \quad D_2 = \underline{320 \text{ mm}}$$

(ii) $$\frac{N_2}{N_1} = \frac{D_1 + t}{D_2 + t} \quad \text{or} \quad \frac{150}{80} = \frac{600 + 5}{D_2 + 5} \quad \text{or} \quad D_2 = \underline{317.7 \text{ mm}}$$

(iii) $$\frac{N_2}{N_1} = \left(\frac{D_1 + t}{D_2 + t}\right)\left(\frac{100 - S}{100}\right) \quad \text{or} \quad \frac{150}{80} = \left(\frac{600 + 5}{D_2 + 5}\right)\left(\frac{100 - 4}{100}\right)$$

$$D_2 = \underline{304.8 \text{ mm}}$$

(iv) $$\frac{N_2}{N_1} = \left(\frac{D_1 + t}{D_2 + t}\right)\left(\frac{100 - S}{100}\right)$$

where $S = S_1 + S_2 - 0.01\, S_1 S_2$

$= 2 + 2 - 0.01 \times 2 \times 2$

$= 3.96$

$$\frac{150}{80} = \left(\frac{600 + 5}{D_2 + 5}\right)\left(\frac{100 - 3.96}{100}\right)$$

$$D_2 = \underline{304.9 \text{ mm}}$$

9.7 MATERIAL FOR BELTS AND ROPES

Choice of materials for the belts and ropes is influenced by climatic or environmental conditions along with the service requirements. The common materials are as given below:

1. Flat Belts

Usual materials are leather, canvas, cotton and rubber. These belts are used to connect shafts up to 8-10 m apart with speeds as high as 22 m/s.

2. V-Belts

These are made of rubber impregnated fabric with angle of V between 30 to 40 degrees. These are used to connect shafts up to 4 m apart. Speed ratios can be up to 7 to 1 and belt speeds up to 24 m/s.

3. Ropes

The materials for ropes are cotton, hemp, manila or wire. These are used to connect shafts up to 30 m apart, operating speed being less than 3 m/s.

9.8 CROWNING OF PULLEYS

As mentioned in Sec. 9.2, the rim of the pulley of a flat belt drive is slightly crowned to prevent the slipping off the belt from the pulley. The crowning can be in the form of conical surface or a convex surface.

Let somehow a belt has come over the conical portion of the pulley. If it is taken that the belt assumes the position as shown in Fig. 9.5 a, i.e., its centre line remains in a plane, the belt will touch the rim surface at its one edge only. This is impractical. Owing to the pull, the belt always tends to stick to the rim surface. The belt also has a lateral stiffness. Thus a belt has to bend in the way shown in Fig. 9.5 b to be on the conical surface of the pulley.

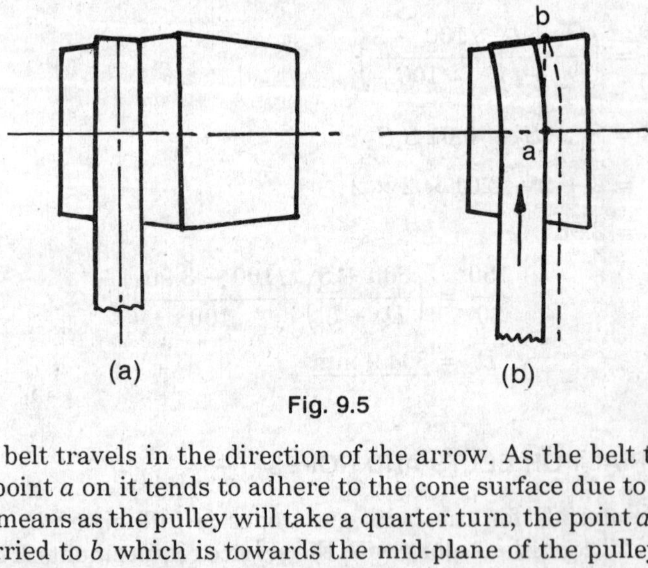

Fig. 9.5

Let the belt travels in the direction of the arrow. As the belt touches the cone, the point a on it tends to adhere to the cone surface due to pull on the belt. This means as the pulley will take a quarter turn, the point a on the belt will be carried to b which is towards the mid-plane of the pulley than that previously occupied by the edge of the belt. But again, the belt cannot be stable on the pulley in the upright position and has to bend to stick to the cone surface, i.e., it will occupy the position shown by dotted lines.

Thus if a pulley is made up of two equal cones or of convex surface, the belt will tend to climb on the slopes and will thus run with its centre line on the mid-plane of the pulley.

The amount of crowning is usually 1/96 of the pulley face width.

9.9 TYPES OF PULLEYS

1. Idler Pulleys

With constant use, the belt is permanently stretched a little in length. This reduces the initial tension in the belt leading to lower power transmission capacity. However, the tension in the belt can be restored to the original value by using an arrangement shown in Fig. 9.6 a.

A bell-crank lever, hinged on the axis of the smaller pulley, supports adjustable weights on its one arm and the axis of a pulley on the other. The pulley is free to rotate on its axis and is known as idler pulley. Owing to weights on one arm of the lever, the pulley exerts pressure on the belt increasing the tension and the angle of contact. Thus life of the belt is increased and power capacity is restored to the previous value.

The pressure force on the belt can be varied by changing the weights on the arm of the lever.

Motion of one shaft can be transmitted to two or more than two shafts by using a number of idler pulleys. This has been illustrated in Fig. 9.6 b.

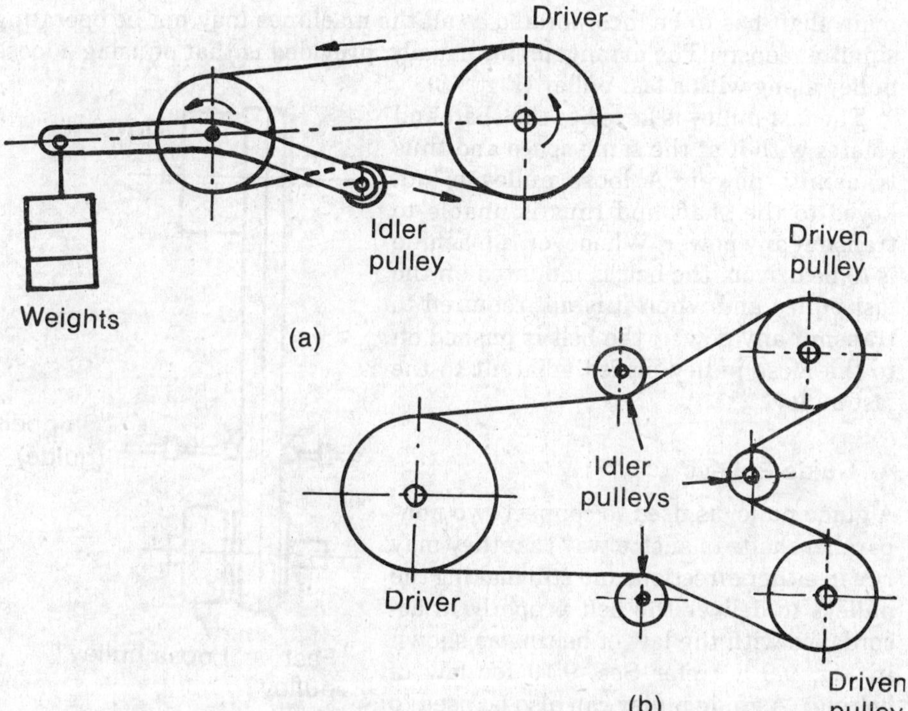

Fig. 9.6

2. Intermediate Pulleys

When it is required to have large velocity ratios, ordinarily, the size of the larger pulley will be quite big. However, by using an intermediate (or countershaft) pulley, the size can be reduced as shown in Fig. 9.7.

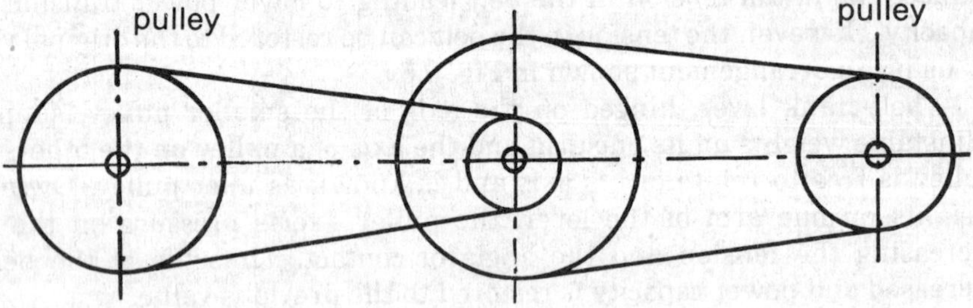

Fig. 9.7

3. Loose and Fast Pulleys

Many times, it is required to drive several machines from a single main shaft. In such cases, some arrangement to link or delink a machine to or from the main shaft has to be incorporated as all the machines may not be operating simultaneously. The arrangement, usually, provided is that of using a loose pulley along with a fast pulley (Fig. 9.8).

The fast pulley is keyed to the shaft and rotates with it at the same speed and thus transmits power. A loose pulley is not keyed to the shaft and thus is unable to transmit any power. Whenever, a machine is to be driven, the belt is mounted on the fast pulley and when it is not required to transmit any power, the belt is pushed on to the loose pulley placed adjacent to the fast pulley.

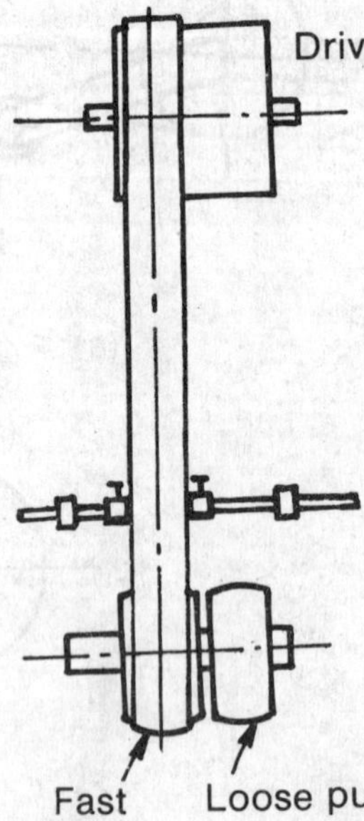

Fig. 9.8

4. Guide Pulleys

A guide pulley is used to connect two non-parallel shafts in such a way that they may run in either direction, and still making the pulleys to deliver the belt properly in accordance with the law of belting as shown in Fig. 9.9 a (refer Sec. 9.10 for law of belting). A guide pulley can also be used to connect even the intersecting shafts as shown in Fig. 9.9 b.

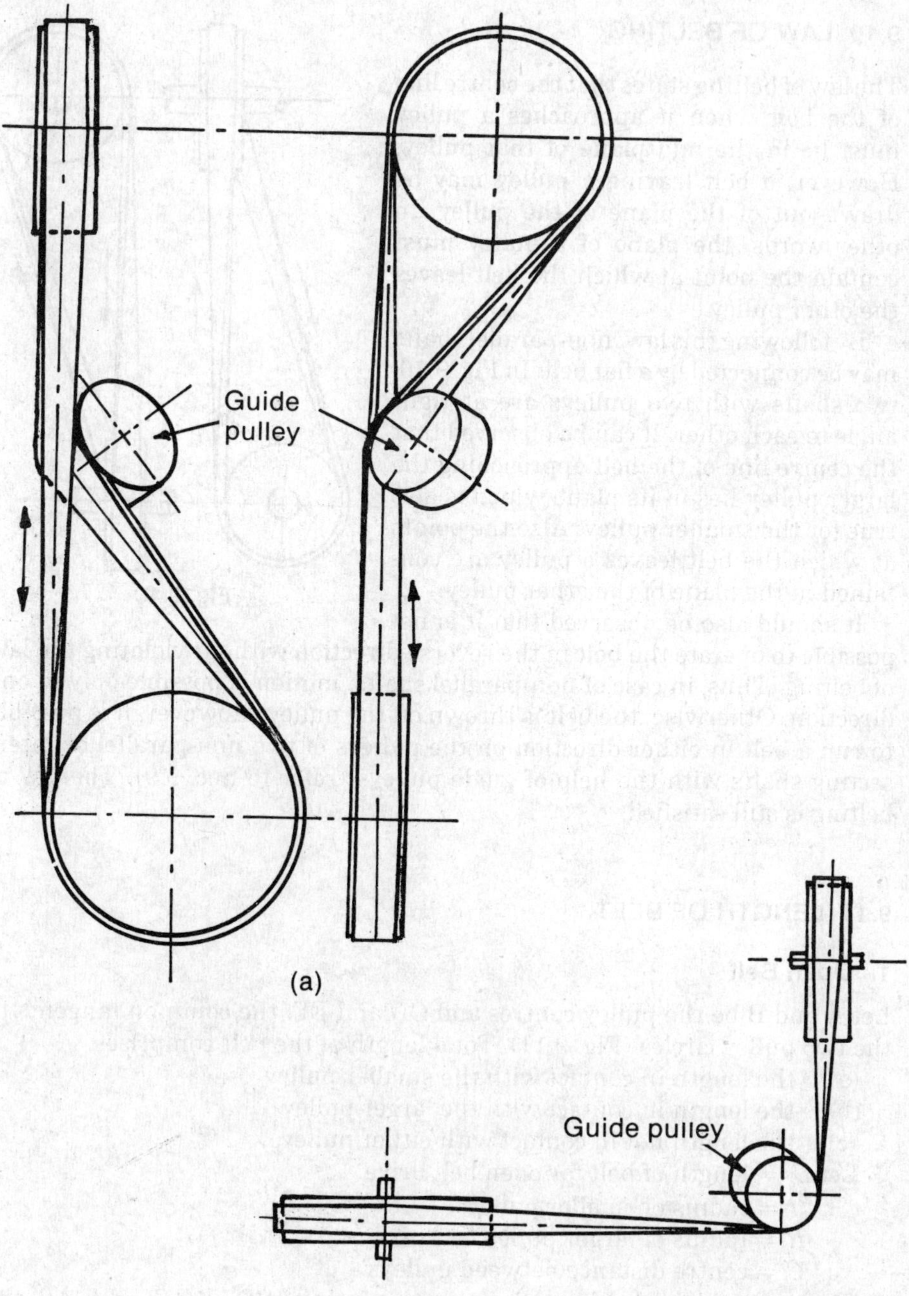

Fig. 9.9

9.10 LAW OF BELTING

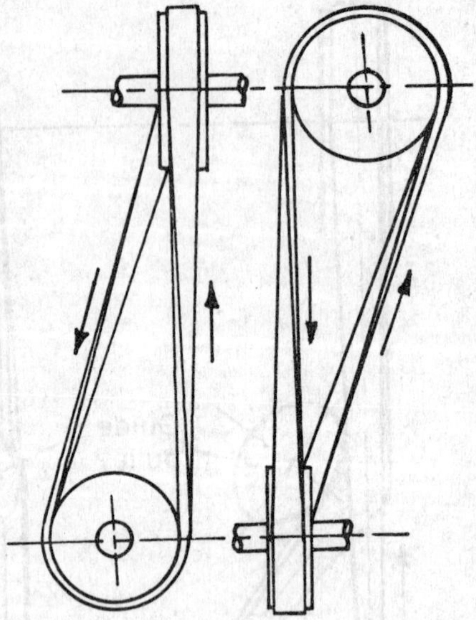

Fig. 9.10

The law of belting states that the centre line of the belt when it approaches a pulley must lie in the mid plane of that pulley. However, a belt leaving a pulley may be drawn out of the plane of the pulley. In other words, the plane of a pulley must contain the point at which the belt leaves the other pulley.

By following this law, non-parallel shafts may be connected by a flat belt. In Fig. 9.10, two shafts with two pulleys are at right angle to each other. It can be observed that the centre line of the belt approaching the larger pulley lies in its plane which is also true for the smaller pulley. Also the points at which the belt leaves a pulley are contained in the plane of the other pulley.

It should also be observed that it is not possible to operate the belt in the reverse direction without violating the law of belting. Thus, in case of non-parallel shafts, motion is possible only in one direction. Otherwise, the belt is thrown off the pulley. However, it is possible to run a belt in either direction on the pulleys of two non-parallel or intersecting shafts with the help of guide pulleys (refer to Sec. 9.9). The law of belting is still satisfied.

9.11 LENGTH OF BELT

1. Open Belt

Let A and B be the pulley centres and CD and EF, the common tangents to the two pulley circles (Fig. 9.11). Total length of the belt comprises:

(a) the length in contact with the smaller pulley

(b) the length in contact with the larger pulley

(c) the length not in contact with either pulley.

Let L_o = length of belt for open belt drive

r = radius of smaller pulley

R = radius of larger pulley

C = centre distance between pulleys

β = angle subtended by each common tangent (CD or EF) with AB, the line of centres of pulleys.

Draw AN parallel to CD so that $\angle BAN = \beta$ and $BN = R - r$.

As CD is tangent to two circles,

AC and BD both are perpendicular to CD or AN.

Now, $AB \perp BK$ and $AN \perp BD$

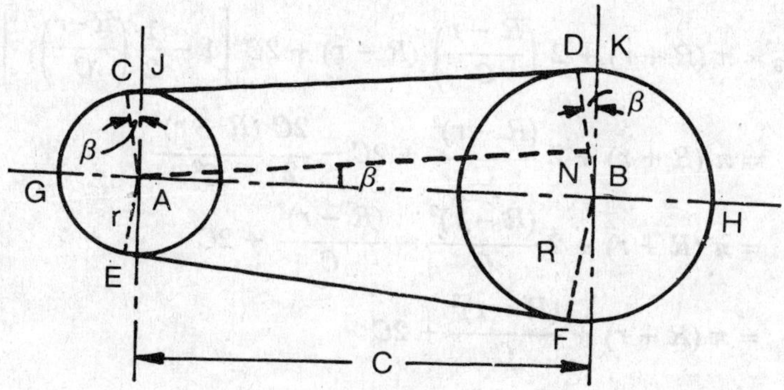

Fig. 9.11

$$\therefore \quad \angle DBK = \angle NAB = \beta$$

Similarly, as $BA \perp AJ$, $NA \perp AC$

$$\angle CAJ = \angle NAB = \beta$$

$$L_o = 2\left[\text{Arc } GC + CD + \text{Arc } DH\right]$$

$$= 2\left[\left(\frac{\pi}{2} - \beta\right) r + AN + \left(\frac{\pi}{2} + \beta\right) R\right]$$

$$= 2\left[\left(\frac{\pi}{2} - \beta\right) r + C\cos\beta + \left(\frac{\pi}{2} + \beta\right) R\right]$$

$$= \pi(R + r) + 2\beta(R - r) + 2C\cos\beta \tag{9.4}$$

This relation gives the exact length of belt required for an open belt drive. In this relation,

$$\beta = \sin^{-}\left(\frac{R - r}{C}\right) \tag{9.5}$$

An approximate relation for the length of belt can also be found in terms of R, r and C eliminating β, if β is small, i.e. if the difference in radii of the two pulleys is small and the centre distance is large.

For small angle of β, $\sin\beta \approx \beta$

$$\therefore \quad \beta = \frac{R - r}{C}$$

and

$$\cos\beta = \sqrt{1 - \sin^2\beta}$$

$$= (1 - \sin^2\beta)^{1/2}$$

$$= \left(1 - \frac{1}{2}\sin^2\beta + \ldots\right) \quad \text{[By binomial theorem]}$$

or

$$\cos\beta = (1 - \frac{1}{2}\beta^2) = 1 - \frac{1}{2}\left(\frac{R - r}{C}\right)^2$$

$$L_o = \pi (R + r) + 2 \left(\frac{R - r}{C}\right) (R - r) + 2C \left[1 - \frac{1}{2}\left(\frac{R-r}{C}\right)^2\right]$$

$$= \pi (R + r) + 2 \frac{(R - r)^2}{C} + 2C - \frac{2C}{2} \frac{(R - r)^2}{C^2}$$

$$= \pi (R + r) + 2 \frac{(R - r)^2}{C} - \frac{(R - r)^2}{C} + 2C$$

$$= \pi (R + r) + \frac{(R - r)^2}{C} + 2C \qquad (9.6)$$

2. Crossed-Belt

As before, let A and B be the pulley centres and CD and EF, the common tangents (crossed) to the two pulley circles (Fig. 9.12).

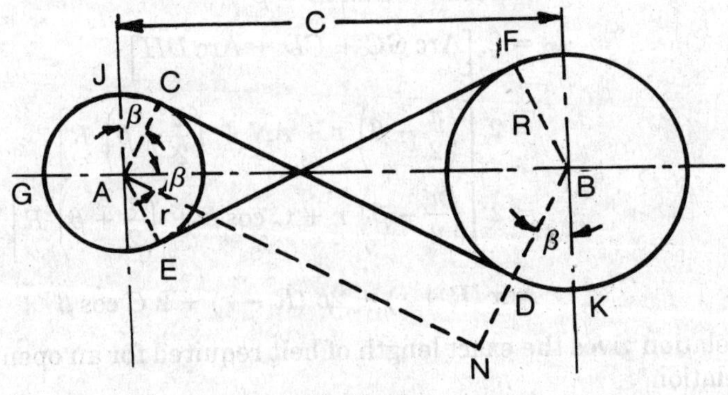

Fig. 9.12

Draw AN parallel to CD meeting BD produced at N so that $\angle BAN = \beta$

We have, $\angle CAJ = \angle DBK = \beta$

Let L_c = length of belt for crossed belt drive

$L_c = 2$ [Arc $GC + CD +$ Arc DH]

$$= 2\left[\left(\frac{\pi}{2} + \beta\right) r + AN + \left(\frac{\pi}{2} + \beta\right) R\right]$$

$$= 2\left[\left(\frac{\pi}{2} + \beta\right) r + C\cos\beta + \left(\frac{\pi}{2} + \beta\right) R\right]$$

$$= (\pi + 2\beta)(R + r) + 2C\cos\beta \qquad (9.7)$$

This is the exact length for a crossed-belt drive where

$$\beta = \sin^{-}\left(\frac{R + r}{C}\right) \qquad (9.8)$$

For small angle of β, $\sin\beta \approx \beta$

$$\therefore \qquad \beta = \frac{R+r}{C}$$

and

$$\cos\beta = \left(1 - \frac{1}{2}\beta^2\right) = 1 - \frac{1}{2}\left(\frac{R+r}{C}\right)^2$$

$$L_c = \left[\pi + 2\left(\frac{R+r}{C}\right)\right](R+r) + 2C\left[1 - \frac{1}{2}\left(\frac{R+r}{C}\right)^2\right]$$

$$= \pi(R+r) + 2\frac{(R+r)^2}{C} + 2C - \frac{(R+r)^2}{C}$$

$$= \pi(R+r) + \frac{(R+r)^2}{C} + 2C \qquad (9.9)$$

This is an approximate relation for the length in terms of R, r and C.

It can be noted that the length of belt depends only on the sum of the pulley radii and the centre distance in case of crossed-belt drive whereas it depends on the sum as well as the difference of the pulley radii apart from the centre distance in case of open-belt drive.

Example 9.2 Two parallel shafts, connected by a crossed-belt, are provided with pulleys 480 mm and 640 mm in diameters. The distance between the centre lines of the shafts is 3 m. Find by how much the length of the belt should be changed if it is desired to alter the direction of rotation of the driven shaft.

Solution:

R = 320 mm
C = 3 m
r = 240 mm

For Cross belt

$$\beta = \sin^{-}\left(\frac{R+r}{C}\right) = \sin^{-}\left(\frac{0.32+0.24}{3}\right) = \sin^{-} 0.1867$$

$$= 10°45' \quad \text{or} \quad 0.1\,878 \text{ rad}$$

$$\cos\beta = 0.9825$$

$$L_c = (\pi + 2\beta)(R+r) + 2C\cos\beta$$

$$= (\pi + 2 \times 0.1878)(0.32 + 0.24) + 2 \times 3 \times 0.9825$$

$$= \underline{7.865 \text{ m}}$$

For Open belt

$$\beta = \sin^{-}\left(\frac{R-r}{C}\right) = \sin^{-}\left(\frac{0.32-0.24}{3}\right) = \sin^{-} 0.0267$$

$$= 1°32'$$

As the angle is very small, the approximate relation can be used.

$$L_o = \pi\,(R+r)+\frac{(R-r)^2}{C} + 2C = \pi\,(0.32+0.24) + \frac{(0.32-0.24)^2}{3} + 2\times 3$$

$$= 7.761 \text{ m}$$

The length of the belt should be reduced by

$$L_c - L_o = 7.865 - 7.761 = 0.104 \text{ m} \quad \text{or} \quad \underline{104 \text{ mm}}$$

9.12 CONE (STEPPED) PULLEYS

Many times, it is required to run the driven shaft at different speeds whereas the driving shaft runs at constant speed which is the speed of the motor. This is facilitated by using a pair of cone or stepped pulleys (Fig. 9.13). A cone pulley has different sets of pulley radii to give varying speeds of the driven shaft. The radii of different steps are so chosen that the same belt can be used at different sets of the cone pulleys.

Let n = speed of the driving shaft (constant)

N_n = speed of the driven shaft when the belt is on nth step

r_n = radius of the nth step of the driving pulley

R_n = radius of the nth step of the driven pulley

The subscript n denotes 1, 2, 3, ... n.

The ratio of speeds of driving to driven shaft is inversely proportional to ratio of their pulley radii, i.e.

$$\frac{N_1}{n} = \frac{r_1}{R_1}$$

Thus to get speed N_1 of the driven shaft from the first pair of steps of the cone pulleys, dimensions of r_1 and R_1 can be chosen convenient to the design.

For second pair of steps,

$$\frac{N_2}{n} = \frac{r_2}{R_2}$$

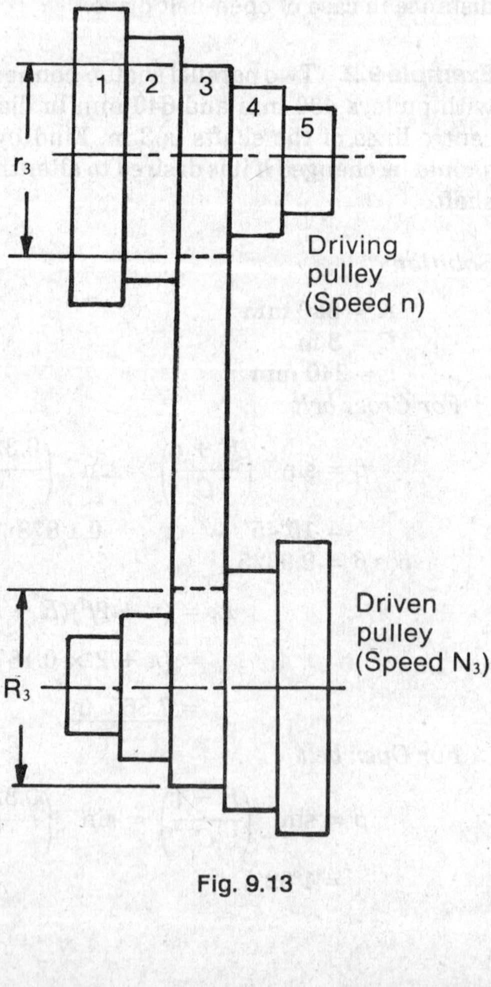

Fig. 9.13

Again some convenient dimensions of r_2 and R_2 can be chosen according to ratio of N_2/n. Similarly, for other pair of steps also, the same procedure can be adopted. For the nth pair,

$$\frac{N_n}{n} = \frac{r_n}{R_n}$$

However, it is always desired that the same belt is used on all the pairs of steps of the cone pulleys. To fulfil this condition, the length of the belt has to be the same for all pairs of steps, i.e.,

$$L_1 = L_2 = \ldots = L_n$$

$$\text{or } (R_1 + r_1) + \frac{(R_1 - r_1)^2}{C} + 2C = (R_2 + r_2) + \frac{(R_2 - r_2)^2}{C} + 2C = \ldots$$

$$= (R_n + r_n) + \frac{(R_n - r_n)^2}{C} + 2C$$

$$\text{or} \quad (R_1 + r_1) + \frac{(R_1 - r_1)^2}{C} = (R_2 + r_2) + \frac{(R_2 - r_2)^2}{C} = \ldots$$

$$= (R_n + r_n) + \frac{(R_n - r_n)^2}{C} \tag{9.10}$$

As R_1 and r_1 have already been chosen for the first pair and ratios $r_2/R_2, \ldots, r_n/R_n$ are known, their values can be easily calculated.

Also it is usual practice to have speeds of the driven shaft in geometrical progression and to make the driving and the driven cones similar.

Let K = ratio of progression of speed

Then $$\frac{N_2}{N_1} = \frac{N_3}{N_2} = \ldots = \frac{N_n}{N_{n-1}} = K$$

$\therefore$ $$N_2 = KN_1$$

$$N_3 = KN_2 = K^2 N_1$$

....

$$N_n = KN_{n-1} = K^{n-1} N_1$$

To obtain these speeds of the driven shaft, the ratio of the radii of different pairs of pulleys can be obtained as under:

$$N_n = K^{n-1} N_1 = K^{n-1} n \frac{r_1}{R_1}$$

or $$\frac{N_n}{n} = K^{n-1} \frac{r_1}{R_1}$$

or $$\frac{r_n}{R_n} = K^{n-1} \frac{r_1}{R_1}$$

Thus, $\frac{r_2}{R_2} = K \frac{r_1}{R_1}$; $\frac{r_3}{R_3} = K^2 \frac{r_1}{R_1}$ and so on.

If the driving and the driven pulleys are to be made similar,

$$\frac{r_n}{R_n} = \frac{r_1}{r_1} \text{ or } K^{n-1}\frac{r_1}{R_1} = \frac{R_1}{r_1} \text{ or } \left(\frac{R_1}{r_1}\right)2 = K^{n-1} \text{ or } \frac{R_1}{r_1} = \sqrt{K^{n-1}} \quad (9.11)$$

This gives the ratio R_1/r_1. After deciding its value, ratios r_2/R_2, r_3/R_3, etc. can be obtained and then from the relation for the length of belt, the values of r_2, R_2 and r_3, R_3 etc. can be obtained.

If cone pulleys are being used for crossed-belt drive, it is very easy to obtain the dimensions of the radii of different pairs of steps. In this case, length of the belt is same if the sums of the radii of different pairs of steps are constant for a given centre distance between the pulleys, i.e.

$$R_1 + r_1 = R_2 + r_2 = \dots = R_n + r_n \quad (9.12)$$

Example 9.3 Design a set of stepped pulleys to drive a machine from a countershaft that runs at 220 rpm. The distance between centres of the two sets of pulleys is 2 m. The diameter of the smallest step on the countershaft is 160 mm. The machine is to run at 80 100 and 130 rpm and should be able to rotate in either direction.

Solution: As the driven shaft is to rotate in either direction, both the cases of a crossed-belt and an open-belt are to be considered.

(i) *For Crossed-belt system*

The smallest step on the countershaft will correspond to the biggest step on the machine shaft (or the minimum speed of the machine shaft).

$n_1 = n_2 = n_3 = 220$ rpm $\qquad r_1 = 80$ mm

$N_1, N_2, N_3 = 80, 100, 130$ rpm respectively

(a) *For First Step:*

$$\frac{R_1}{r_1} = \frac{n_1}{N_1} \text{ or } \frac{R_1}{80} = \frac{220}{80} \text{ or } R_1 = \underline{220 \text{ mm}}$$

(b) *For Second Step:*

$$\frac{R_2}{r_2} = \frac{n_2}{N_2} = \frac{220}{100} \text{ or } R_2 = 2.2\ r_2$$

Also

$$R_2 + r_2 = R_1 + r_1$$

$$2.2\ r_2 + r_2 = 220 + 80$$

$$3.2\ r_2 = 300$$

$$r_2 = \underline{93.75 \text{ mm}}$$

$$R_2 = 93.7 \times 2.2 = \underline{206.3 \text{ mm}}$$

(c) *For Third Step:*

$$\frac{R_3}{r_3} = \frac{220}{130} \quad \text{or} \quad R_3 = 1.69\ r_3$$

Also

$$R_3 + r_3 = R_1 + r_1$$

$$1.69\,r_3 + r_3 = 220 + 80 = 300$$

$$r_3 = \underline{111.5\ \text{mm}}$$

$$R_3 = 111.5 \times 1.69 = \underline{188.5\ \text{mm}}$$

(ii) *For Open-belt System*

(a) *For First Step:* $r_1 = 80$ mm $R_1 = 220$ mm as before

(b) *For Second Step:*

$$\pi\,(R_2 + r_2) + \frac{(R_2 - r_2)^2}{C} = \pi\,(R_1 + r_1) + \frac{(R_1 - r_1)^2}{C}$$

$$\pi\,(2.2\,r_2 + r_2) + \frac{(2.2\,r_2 - r_2)^2}{2} = \pi\,(0.22 + 0.08) + \frac{(0.22 - 0.08)^2}{2}$$

$$10.05\,r_2 + \frac{1.44}{2} r_2^2 = 0.9523$$

$$r_2^2 + 13.958 r_2 = 1.323$$

$$(r_2 + 6.979)^2 = 1.323 + (6.979)^2 = 50.029 = (7.073)^2$$

$$r_2 = 7.073 - 6.979 = 0.094\ \text{m} \quad \text{or} \quad 94\ \text{mm}$$

$$R_2 = 2.2 \times 94 = \underline{206.8\ \text{mm}}$$

(c) *For Third Step*

$$\pi\,(1.69\,r_3 + r_3) + \frac{(1.69\,r_3 - r_3)^2}{2} = 0.9523$$

$$8.451\,r_3 + \frac{0.476}{2} r_3^2 = 0.9523$$

$$r_3^2 + 35.508\,r_3 = 4.001$$

$$(r_3^2 + 17.754)^2 = 4.001 + (17.754)^2 = 319.206 = (17.866)^2$$

$$r_3 = 17.866 - 17.754 = 0.112\ \text{m} \quad \text{or} \quad 112\ \text{mm}$$

$$R_3 = 112 \times 1.69 = \underline{189.3\ \text{mm}}$$

9.13 RATIO OF FRICTION TENSIONS

1. Flat Belt

Let $\mathbf{T}_1$ = tension on tight side

$\mathbf{T}_2$ = tension on slack side

θ = angle of lap of the belt over the pulley

μ = coefficient of friction between the belt and the pulley

Consider a short length of belt subtending an angle $\delta\theta$ at the centre of the pulley (Fig. 9.14).

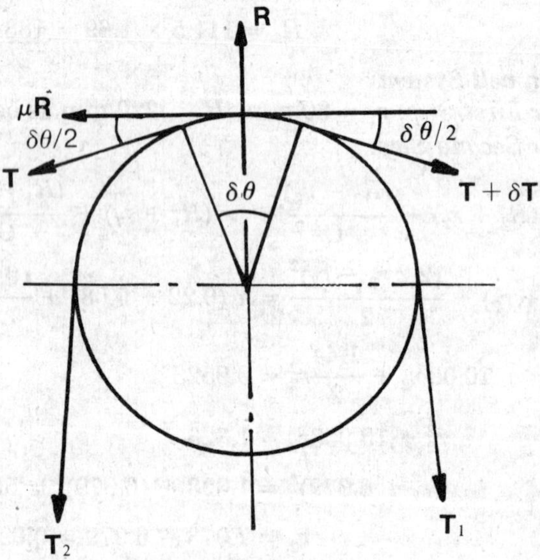

Fig. 9.14

Let $\mathbf{R}$ = normal (radial) reaction between the element length of belt and the pulley

$\mathbf{T}$ = tension on slack side of the element

$\delta\mathbf{T}$ = increase in tension on tight side than that on slack side

$\mathbf{T} + \delta\mathbf{T}$ = tension on tight side of the element

Tensions $\mathbf{T}$ and $(\mathbf{T} + \delta\mathbf{T})$ act in directions perpendicular to the radii drawn at the ends of the element. The friction force $\mu\mathbf{R}$ will act tangentially to the pulley rim resisting the slipping of the elementary belt on the pulley.

Resolving the forces in the tangential direction,

$$\mu R + T \cos\frac{\delta\theta}{2} - (T + \delta T)\cos\frac{\delta\theta}{2} = 0$$

As $\delta\theta$ is small,

$$\cos\frac{\delta\theta}{2} \approx 1$$

$$\therefore \qquad \mu R + T - T - \delta T = 0 \quad \text{or} \quad \delta T = \mu R \qquad \text{(i)}$$

Resolving the forces in the radial direction,

$$R - T \sin \frac{\delta\theta}{2} - (T + \delta T) \sin \frac{\delta\theta}{2} = 0$$

As $\delta\theta$ is small, $\sin \frac{\delta\theta}{2} \approx \frac{\delta\theta}{2}$

$$\therefore \qquad R - T \frac{\delta\theta}{2} - T \frac{\delta\theta}{2} - \frac{\delta T \delta\theta}{2} = 0$$

Neglecting product of two small quantities,

$$R = T\delta\theta \qquad \text{(ii)}$$

From (i) and (ii), $\delta T = \mu T \delta\theta$ or $\frac{\delta T}{T} = \mu \delta\theta$

Integrating between proper limits,

$$\int_{T_2}^{T_1} \frac{dT}{T} = \int_0^{\theta} \mu \, d\theta$$

or

$$\log_e \frac{T_1}{T_2} = \mu\theta$$

or

$$\frac{T_1}{T_2} = e^{\mu\theta} \qquad (9.13)$$

It is to be noted that the above relation is valid only when the belt is on the point of slipping on the pulleys.

2. V-Belt or Rope

In case of V-belt or rope, there are two normal reactions as shown in Fig. 9.15 so that the radial reaction is equal to 2**R** sin α.

Thus total frictional force = $2(\mu R) = 2\mu R$

Resolving the forces tangentially,

$$2\mu R + T \cos \frac{\delta\theta}{2} - (T + \delta T) \cos \frac{\delta\theta}{2} = 0$$

For small angle of $\delta\theta$,

$$\cos \frac{\delta\theta}{2} \approx 1$$

$$\therefore \qquad \delta T = 2\mu R \qquad \text{(iii)}$$

Resolving the forces radially,

$$2R \sin \alpha - T \sin \frac{\delta\theta}{2} - (T + \delta T) \sin \frac{\delta\theta}{2} = 0$$

As $\delta\theta$ is small, $\sin \frac{\delta\theta}{2} \approx \frac{\delta\theta}{2}$

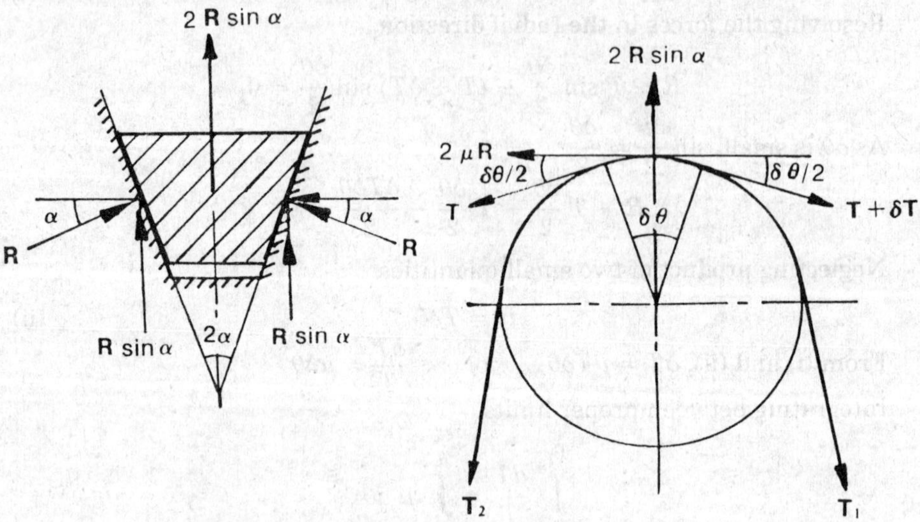

Fig. 9.15

$$2\,R\sin\alpha - T\frac{\delta\theta}{2} - T\frac{\delta^2\theta}{2} = 0$$

or $$R = \frac{T\,\delta\theta}{2\sin\alpha} \qquad \text{(iv)}$$

From (iii) and (iv),

$$\delta T = 2\,\mu\,\frac{T\,\delta\theta}{2\sin\alpha}$$

$$\frac{\delta\,T}{T} = \frac{\mu\delta\theta}{\sin\alpha}$$

Integrating between proper limits,

$$\int_{T_2}^{T_1}\frac{dT}{T} = \int_0^{\theta}\frac{\mu d\theta}{\sin\alpha} \qquad \text{or} \qquad \log_e\frac{T_1}{T_2} = \frac{\mu\theta}{\sin\alpha}$$

or $$\frac{T_1}{T_2} = e^{\mu\theta/\sin\alpha} \qquad (9.14)$$

The expression is similar to that for a flat-belt drive except that μ is replaced by $\mu/\sin\theta$, i.e. the coefficient of friction is increased by $1/\sin\theta$. Thus the ratio T_1/T_2 is far greater in case of V-belts and ropes for the same angle of lap θ and coefficient of friction μ.

Again, it is to be noted that the above expression is derived on the assumption that the belt is on the point of slipping.

9.14 POWER TRANSMITTED

Let T_1 = tension on the tight side
T_2 = tension on the slack side
v = linear velocity of the belt
P = power transmitted
Then,

$$P = \text{net force} \times \text{distance moved/second}$$

$$= (T_1 - T_2) \times v \quad (9.15)$$

This relation gives the power transmitted irrespective of the fact whether the belt is on the point of slipping or not. If it is, the relationship between T_1 and T_2 for a flat belt is given by $T_1/T_2 = e^{\mu\theta}$. If it is not, no particular relation is available to calculate T_1 and T_2.

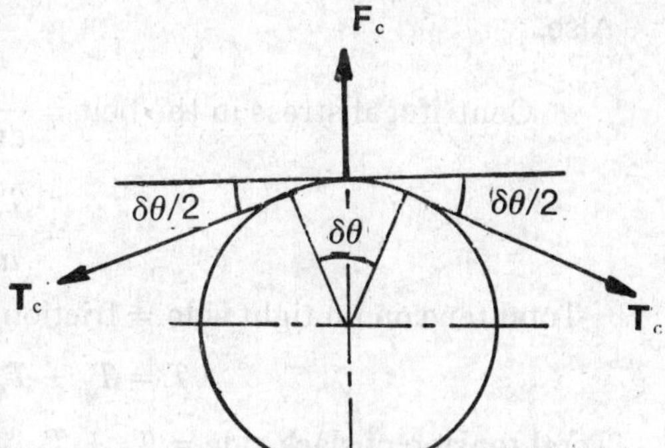

Fig. 9.16

9.15 CENTRIFUGAL EFFECT ON BELTS

While in motion, as a belt passes over a pulley, the centrifugal effect due to its own weight tends to lift the belt from the pulley. Owing to symmetry, the centrifugal force produces equal tensions on the two sides of the belt, i.e. on the tight and the slack sides.

Consider a short element of belt (Fig. 9.16).

Let m = mass per unit length of belt
$\mathbf{T}_c$ = centrifugal tension on tight and slack sides of element
$\mathbf{F}_c$ = centrifugal force on the element
r = radius of the pulley
v = velocity of the belt
$\delta\theta$ = angle of lap of the element over the pulley
$\mathbf{F}_c$ = mass of element × acceleration
= (length of element × mass per unit length) × acceleration

$$= (r\delta\theta \times m) \times \frac{v^2}{r}$$

$$= mv^2\delta\theta \quad \text{(i)}$$

Also,

$$F_c = 2\,T_c \sin\frac{\delta\theta}{2}$$

As $\delta\theta$ is small,

$$\sin\frac{\delta\theta}{2} \approx \frac{\delta\theta}{2}$$

$$F_c = 2\,T_c\,\frac{\delta\theta}{2}$$

$$= T_c\,\delta\theta \tag{ii}$$

From (i) and (ii),

$$T_c\,\delta\theta = mv^2\,\delta\theta$$

or

$$T_c = mv^2 \tag{9.16}$$

Thus centrifugal tension is independent of the tight and slack side tensions and depends only on the velocity of the belt over the pulley.

Also,

$$\text{Centrifugal stress in the belt} = \frac{\text{centrifugal tension}}{\text{area of cross-section of belt}}$$

$$= \frac{T_c}{a}$$

Total tension on tight side = friction tension + centrifugal tension

$$T = T_1 + T_c \tag{9.17}$$

Total tension of slack side = $T_2 + T_c$

It can be shown that the power transmitted is reduced if centrifugal effect is considered for a given value of the total tight side tension T.

(a) *Centrifugal Tension Considered*

Friction tension on tight side = $T - T_c = T_1$

Let T_2 be the friction tension on the slack side.

Then $\dfrac{T_1}{T_2} = e^{\mu\theta} = k$, a constant

or $T_2 = \dfrac{T_1}{k}$

and Power, $P = (T_1 - T_2)v = \left(T_1 - \dfrac{T_1}{k}\right)v = T_1\left(1 - \dfrac{1}{k}\right)v$

(b) *Centrifugal Tension Neglected*

Friction tension on tight side = T

Let T_2' be the friction tension on slack side.

$$\frac{T}{T_2'} = e^{\mu\theta} = k \quad \text{or} \quad T_2' = \frac{T}{k}$$

$$\text{Power, } P = (T - T_2')v = \left(T - \frac{T}{k}\right)v = T\left(1 - \frac{1}{k}\right)v$$

As T_1 is lesser than T, power transmitted is less when centrifugal force is taken into account.

9.16 MAXIMUM POWER TRANSMITTED BY A BELT

If it is desired that a belt transmits maximum possible power, two conditions must be fulfilled simultaneously. They are

1. Larger tension must reach the maximum permissible value for the belt
2. The belt should be on the point of slipping, i.e. maximum frictional force is developed in the belt.

Now,

$$P = (T_1 - T_2)v = T_1\left(1 - \frac{T_2}{T_1}\right)v = T_1\left(1 - \frac{1}{e^{\mu\theta}}\right)v = T_1 kv$$

where $$k = 1 - \frac{1}{e^{\mu\theta}} = \text{constant}$$

or

$$P = (T - T_c)kv$$

$$= kTv - kmv^2\, v$$

$$= kTv - kmv^3$$

The maximum tension T in the belt should not exceed the permissible limit. Hence treating T as constant and differentiating the power with respect to v and equating the same equal to zero.

$$\frac{dP}{dv} = kT - 3\,kmv^2 = 0$$

or $$T = 3mv^2 = 3T_c$$

or $$T_c = \frac{T}{3} \tag{9.18}$$

Therefore, for maximum power to be transmitted centrifugal tension in the belt must be equal to one-third of the maximum allowable belt tension and the belt should be on the point of slipping.

Also,

$$T_1 = T - T_c = T - \frac{T}{3} = \frac{2}{3}T$$

and

$$v_{\text{max}} = \sqrt{\frac{T}{3m}} \tag{9.19}$$

9.17 INITIAL TENSION

When a belt is first fitted to a pair of pulleys, an initial tension T_o is given to the belt when the system is stationary. When transmitting power, the tension on the tight side increases to T_1 and that on slack side decreases to T_2. If it is assumed that the material of the belt is perfectly elastic, i.e. the strain in the belt is proportional to stress in it and the total length of the belt remains unchanged, the tension on the tight side will increase by the same amount as the tension on the slack decreases. If this change in the tension is δT, then

Tension on tight side, $T_1 = T_o + \delta T$

Tension on slack side, $T_2 = T_o - \delta T$

$$\therefore \; T_o = \frac{T_1 + T_2}{2}$$

= mean of the tight and the slack side tensions. (9.20)

Example 9.4 A 100 mm wide and 10 mm thick belt transmits 5 kW between two parallel shafts. The distance between the shaft centres is 1.5 m and the diameter of the smaller pulley is 440 mm. The driving and the driven shafts rotate at 60 rpm and 150 rpm respectively. Find the stress in the belt if the two pulleys are connected by:

(i) an open belt, and (ii) a cross belt. The coefficient of friction is 0.22.

Solution:

$P = 5$ kW $b = 100$ mm

$C = 1.5$ m $t = 10$ mm

$r = 220$ mm $N_1 = 60$ rpm

$\mu = 0.22$ $N_2 = 150$ rpm

$$v = \omega_2\left(r + \frac{t}{2}\right) = \frac{2\pi N_2}{60}\left(r + \frac{t}{2}\right) = \frac{2 \times \pi \times 150}{60}\left(220 + \frac{10}{2}\right)$$

$$= 3535 \text{ mm/s} \quad \text{or} \quad 3.535 \text{ m/s}$$

$$P = (T_1 - T_2)v$$

$$5000 = (T_1 - T_2) = 3.535$$

$$(T_1 - T_2) = 1414.5 \text{ N}$$

(i) *Open-belt Drive*

$$\frac{T_1}{T_2} = e^{\mu\theta}$$

where θ = angle of lap or contact

Remember that in designing the belt drive, the lesser of the angles of contact on the smaller and the larger pulleys has to be considered. Also note that the smaller pulley has lesser angle of contact.

$$\theta = \pi - 2\beta = \pi - 2\sin^{-1}\left(\frac{R-r}{C}\right)$$

$$= \pi - 2\sin^{-1}\left(\frac{220 \times \frac{150}{60} - 220}{1500}\right)$$

$$= \pi - 25.4°$$

$$= \pi - 0.443 \qquad (\beta \text{ is to be in radians})$$

$$= 2.698 \text{ rad.}$$

$$\frac{T_1}{T_2} = e^{0.22 \times 2.698} = 1.81 \quad \text{or} \quad T_1 = 1.81\ T_2$$

But

$$T_1 - T_2 = 1414.5$$

$\therefore$

$$1.81\ T_2 - T_2 = 1414.5$$

$$0.81\ T_2 = 1414.5$$

$$T_2 = 1746.3 \text{ N} \quad \text{and} \quad T_1 = 1746.3 \times 1.81 = 3160.8 \text{ N}$$

$\therefore$ Stress in the belt,

$$f_t = \frac{T_1}{b \times t} = \frac{3160.8}{100 \times 10} = \underline{3.16 \text{ N/mm}^2}$$

(ii) *Cross-belt Drive*

$$\theta = \pi + 2\beta = \pi + 2\sin^{-}\left(\frac{R+r}{C}\right)$$

or

$$\theta = \pi + 2\sin^{-}\left(\frac{220 \times \frac{150}{60} + 220}{1500}\right) = \pi + 61.8°$$

or

$$\theta = \pi + 1.08 = 4.22 \text{ rad}$$

$$\frac{T_1}{T_2} = e^{0.22 \times 4.22} = 2.53$$

$$T_1 - T_2 = 1414.5$$

$$1.53\ T_2 = 1414.5$$

$$T_2 = 924.5 \text{ N}$$

$$T_1 = 924.5 \times 2.53 = 2339 \text{ N}$$

$$f_t = \frac{2339}{100 \times 10} = \underline{2.339 \text{ N/mm}^2}$$

Example 9.5 2.5 kW of power is transmitted by an open-belt drive. The linear velocity of the belt is 2.5 m/s. The angle of lap on the smaller pulley is 165°. The coefficient of friction is 0.3.

Determine the effect on power transmission in the following cases:

(i) Initial tension in the belt is increased by 8%,
(ii) Initial tension in the belt is decreased by 8%,
(iii) Angle of lap is increased by 8% by the use of an idler pulley, for the same speed and the tension on the tight side, and
(iv) Coefficient of friction is increased by 8% by suitable dressing to the friction surface of the belt.

Solution:

$P = 2.5$ kW $\qquad \mu = 0.3$

$\theta = 165°$ $\qquad v = 2.5$ m/s

$$P = (T_1 - T_2)v$$

$$2500 = (T_1 - T_2) \times 2.5$$

$$T_1 - T_2 = 1000 \text{ N}$$

$$\frac{T_1}{T_2} = e^{\mu\theta} = e^{0.33 \times 165\,\pi/180} = 2.37$$

or $$T_1 = 2.37\ T_2$$

$$2.37\ T_2 - T_2 = 100 \quad \text{or} \quad T_2 = 729.9 \text{ N}$$

$$T_1 = 729.9 \times 2.37 = 1729.9 \text{ N}$$

$$\text{Initial tension, } T_0 = \frac{T_1 + T_2}{2} = \frac{1729.9 + 729.9}{2} = 1229.9 \text{ N}$$

(i) When initial tension is increased by 8%:

$$T'_0 = 1229.9 \times 1.08 = 1328.3 \text{ N}$$

or $$\frac{T_1 + T_2}{2} = 1328.3 \quad \text{or} \quad T_1 + T_2 = 2656.6$$

As μ and θ remain unchanged, $e^{\mu\theta}$ or $\frac{T_1}{T_2}$ is same.

$$2.37\ T_2 + T_2 = 2656.6$$

$$T_2 = 788.3 \text{ N}$$

$$T_1 = 1868.3 \text{ N}$$

$$P = (T_1 - T_2)v = (1868.3 - 788.3) \times 2.5 = 2700 \text{ W or } 2.7 \text{ kW}$$

$$\therefore \quad \text{Increase in power} = \frac{2.7 - 2.5}{2.5} = 0.08 \text{ or } 8\%$$

(ii) When initial tension is decreased by 8%:

$$T_0' = 1229.9 \times (1 - 0.08) = 1131.5$$

or $$\frac{T_1 + T_2}{2} = 1131.5 \quad \text{or} \quad T_1 + T_2 = 2263$$

$$3.37\, T_2 = 2263$$

$$T_2 = 671.5 \text{ N}$$

$$T_1 = 1591.5 \text{ N}$$

$$P = (1591.5 - 671.5) \times 2.5 = 2300 \text{ W or } 2.3 \text{ kW}$$

$\therefore$ Decrease in power $= \dfrac{2.5 - 2.3}{2.5} = 0.08$ or 8%

(iii) $\dfrac{T_1}{T_2} = e^{\mu\theta}$

T_1 is the same as before whereas θ increases by 8%.

$$\frac{1729.9}{T_2} = e^{0.3 \times \frac{165 \times 1.08 \times \pi}{180}} = 2.54$$

$$T_2 = 680.5 \text{ N}$$

$$P = (1729.9 - 680.5) \times 2.5 = 2624 \text{ W or } 2.624 \text{ kW}$$

$\therefore$ Increase in power $= \dfrac{2.624 - 2.5}{2.5} = 0.0496$ or 4.96%

(iv) $\dfrac{T_1}{T_2} = e^{\mu\theta} = e^{0.3 \times 1.08 \times \frac{165 \times \pi}{180}} = 2.54$

or $T_1 = 2.54\, T_2$

$$T_1 + T_2 = 1229.9 \times 2 = 2459.8$$

$$T_2 = 694.9 \text{ N}$$

$$T_1 = 694.9 \times 2.54 = 1764.9 \text{ N}$$

$$P = (1764.9 - 694.9) \times 2.5 = 2675 \text{ W or } 2.675 \text{ kW}$$

$\therefore$ Increase in power $= \dfrac{2.675 - 2.5}{2.5} = 0.07$ or 7%

Example 9.6 A belt drive is required to transmit 10 kW from a motor running at 600 rpm. The belt is 12 mm thick and has a mass density of 0.001 gm/mm^3. Safe stress in the belt is not to exceed 2.5 N/mm^2. Diameter of the driving pulley is 250 mm whereas the speed of the driven pulley is 220 rpm. The two shafts are 1.25 m apart. The coefficient of friction is 0.25.

Determine the width of the belt.

Solution:

$$P = 10 \text{ kW} \qquad t = 12 \text{ mm}$$
$$\rho = 0.001 \text{ gm/mm}^3 \qquad R_1 = 125 \text{ mm}$$
$$= 1000 \text{ kg/m}^3 \qquad N_1 = 600 \text{ rpm}$$
$$C = 1.25 \text{ m} \qquad N_2 = 220 \text{ rpm}$$
$$\mu = 0.25 \qquad f_t = 2.5 \text{ N/mm}^2 = 2.5 \times 10^6 \text{ N/m}^2$$

To calculate the width of the belt, we need to know the maximum tension in the belt which is the sum of the tight side tension and the centrifugal tension, i.e., $T = T_1 + T_c$.

Calculation of T_1

$$P = (T_1 - T_2)v$$

where $$v = \omega\left(R_1 + \frac{t}{2}\right) = \frac{2\pi N}{60}\left(R_1 + \frac{t}{2}\right)$$
$$= \frac{2\pi \times 600}{60}\left(125 + \frac{12}{2}\right)$$
$$= 8230 \text{ mm/s or } 8.23 \text{ m/s}$$
$$10000 = (T_1 - T_2) \times 8.23$$

or $$T_1 - T_2 = 1215 \qquad \text{(i)}$$

Also $\frac{T_1}{T_2} = e^{\mu\theta}$ where $\theta = \pi - 2\beta = \pi - 2\sin^{-}\left(\frac{R_2 - R_1}{C}\right)$

or $$\theta = \pi - 2\sin^{-}\left(\frac{125 \times \frac{600}{220} - 125}{1250}\right)$$

or $$\theta = \pi - 19.9° = \pi - 0.347 = 2.79$$

$$\frac{T_1}{T_2} = e^{0.25 \times 2.79} = 2.01 \quad \text{or} \quad T_1 = 2.01\, T_2 \qquad \text{(ii)}$$

From (i) & (ii),

$$2.01\, T_2 - T_2 = 1215$$
$$T_2 = 1203 \text{ N}$$
$$T_1 = 2418 \text{ N}$$

Calculation of T_c

$$T_c = mv^2$$
$$= \text{mass per unit length} \times v^2$$
$$= \text{volume per unit length} \times \text{density} \times v^2$$
$$= (x\text{−sectional area} \times \text{length} \times \text{density}) \times v^2$$
$$= (\text{width} \times \text{thickness} \times \text{length} \times \text{density}) \times v^2$$
$$= b \times 0.012 \times 1 \times 1000 \times (8.23)^2$$
$$= (812.8\, b) \text{ N} \qquad (b \text{ in m})$$

$$T = T_1 + T_c = f_t \times (b \times t)$$

$$2418 + 812.8\,b = 2.5 \times 10^6 \times b \times 0.012$$

$$29\,187\,b = 2418$$

$$b = 0.0828 \text{ m} \quad \text{or} \quad \underline{82.8 \text{ mm}}$$

Example 9.7 The grooves on the pulleys of a multiple rope drive have an angle of 50° and accommodate ropes of 22 mm diameter having a mass of 0.8 kg per meter length for which a safe operating tension 1200 N has been laid down. The two pulleys are of equal size. The drive is designed for maximum power conditions. Speed of both the pulleys is 180 rpm. Assuming coefficient of friction as 0.25, determine the diameters of the pulleys and the number of ropes when the power transmitted is 150 kW.

Solution:

$T = 1200$ N $\qquad P = 150\,000$ W

$m = 0.8$ kg/m length $\qquad \theta = 180°$. . . (two pulleys are of equal size)

$\alpha = \dfrac{50}{2} = 25°$ $\qquad \mu = 0.25$

$N_1 = N_2 = 180$ rpm

Under maximum power conditions,

$$T_1 = \frac{2}{3}T = \frac{2}{3} \times 1200 = 800 \text{ N}$$

$$T_c = T - T_1 = 1200 - 800 = 400 \text{ N}$$

Also $T_c = mv^2$ $\quad$ or $\quad 400 = 0.8\,v^2$ $\quad$ or $\quad v = 22.36$ m/s

or $$v = \frac{\pi D_1 N_1}{60} = 22.36$$

or $$\frac{\pi \times D_1 \times 180}{60} = 22.36$$

$$D_1 = \underline{2.37 \text{ m}}, \quad \text{Also} \quad D_2 = \underline{2.37 \text{ m}}$$

$$\frac{T_1}{T_2} = e^{\mu\theta/\sin\alpha} = e^{0.25 \times \frac{180 \times \pi}{180} \frac{1}{\sin 25°}} = 6.41$$

$$T_2 = \frac{T_1}{6.41} = \frac{800}{6.41} = 124.8 \text{ N}$$

$$P = (T_1 - T_2)vn$$

$$150\,000 = (800 - 124.8) \times 22.36 \times n$$

$$n = 9.94 \text{ say } \underline{10 \text{ ropes}}$$

9.18 CREEP

It is seen that when a belt moves over the driving pulley, tension in the belt decreases from T_1 to T_2. Also it can be assumed that the strain in the belt is proportional to the stress in it.

Let l_1 = stretch in unit length of belt due to T_1

l_2 = stretch in unit length of belt due to T_2

As $$T_1 > T_2,$$

$$l_1 > l_2$$

Thus a length $(1 + l_1)$ of belt will be approaching the driving pulley and a length $(1 + l_2)$ will be leaving it. As $(1 + l_1)$ is greater than $(1 + l_2)$, the belt slips back over the driving pulley. This slip is known as creep of the belt.

On the driven pulley, the belt tension increases from T_2 to T_1. This means a shorter length $(1 + l_2)$ approaches the driving pulley and a greater length $(1 + l_1)$ leaves it. Thus, the belt creeps forward by an amount $(l_1 - l_2)$. This makes the driven pulley to move at a slower speed than the belt.

Thus effect of creep is to slow down the speed of the belt on the driving pulley than that of the rim of the pulley and to reduce the rim velocity of the driven pulley than that of the belt on it. Therefore, net effect of creep is to reduce the speed of the driven pulley than what it would have been without creep and thus reducing the power transmitted.

9.19 CHAINS

A belt or a rope drive does not give a constant velocity ratio due to slip which may occur due to overloads or belt and pulley surfaces becoming slightly greasy. A belt or a rope can be replaced by a chain whenever there is a need to have a constant velocity ratio or a positive drive. A chain is regarded in between the gear drive and the belt drive. Like gears, chains are made of metal and, therefore, occupy lesser space and give constant velocity ratios. Like belts, they are used for longer centre distances.

The disadvantages of a chain drive are that it is heavier as compared to the belt and there is a gradual stretching of it. Thus it increases in length so that from time to time some of its links have to be removed. Lubrication of its parts is also desired.

The wheels over which chains are run, corresponding to the pulleys of a belt drive are known as *sprockets*. The surfaces of sprockets conform to the type of chain used. Usually, a sprocket has projected teeth that fit into the recessess in the chain. Thus the chain passes round the sprockets as a series of chordal links (Fig. 9.17).

Distance between roller centres of two adjacent links is known as *pitch* (p) of the chain. A circle drawn through the roller centres of a wrapped chain

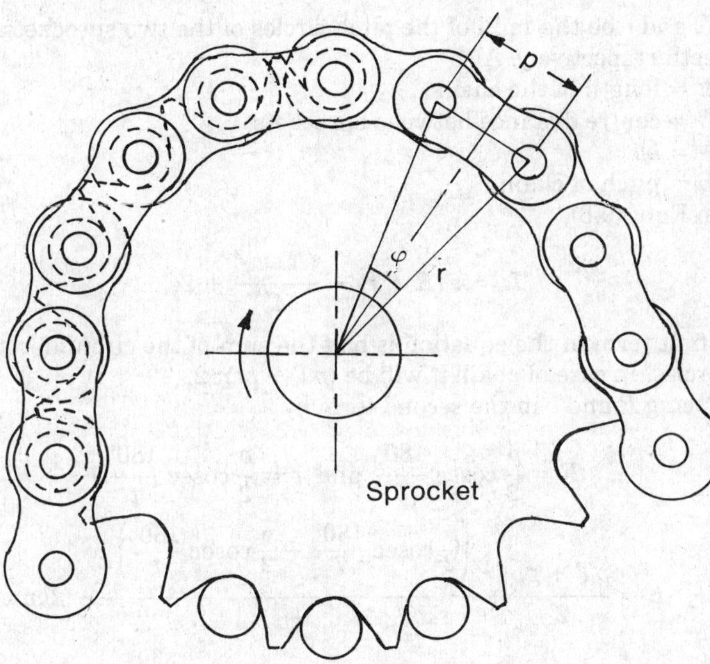

Fig. 9.17

round a sprocket, is called the *pitch circle* and its diameter as pitch circle diameter.

Observe that a chain is wrapped round the sprocket in the form of a pitch polygon and not in the form of a pitch circle.

Let T = number of teeth on a sprocket

φ = angle subtended by chord of a link at the centre of sprocket

r = radius of the pitch circle

$$\text{Then } p = 2\,r \sin\frac{\varphi}{2} = 2\,r\sin\frac{1}{2}\left(\frac{360^\circ}{T}\right) = 2\,r\sin\frac{180^\circ}{T}$$

$$\text{or} \quad r = \frac{p}{2\sin\dfrac{180^\circ}{T}} = \frac{p}{2}\operatorname{cosec}\frac{180^\circ}{T} \tag{9.21}$$

9.20 CHAIN LENGTH

For a given pair of sprockets at a fixed distance apart, the length of the chain may be calculated in the same way as for an open belt. Since the pitch line of a sprocket is a polygon, Eqn. 9.6 will give a length slightly more than the actual length.

Let R and r be the radii of the pitch circles of the two sprockets having T and t teeth respectively. Also

Let L = length of the chain

C = centre distance between sprockets

= kp (assumed)

p = pitch of chain

From Eqn. (9.6),

$$L = \pi (R + r) + \frac{(R-r)^2}{C} + 2C$$

The first term in the equation is half the sum of the circumference of the pitch circles. In case of chain it will be $(pT + pt)/2$.

Replacing R and r in the second term by

$$R = \frac{p}{2} \operatorname{cosec} \frac{180°}{T} \text{ and } r = \frac{p}{2} \operatorname{cosec} \frac{180°}{t}$$

$$L = \frac{pT + pt}{2} + \frac{\left(\frac{p}{2} \operatorname{cosec} \frac{180°}{T} - \frac{p}{2} \operatorname{cosec} \frac{180°}{t}\right)^2}{kp} + 2kp$$

$$= p\left[\frac{T + t}{2} + \frac{\left(\operatorname{cosec} \frac{180°}{T} - \operatorname{cosec} \frac{180°}{t}\right)^2}{4k} + 2k\right] \qquad (9.22)$$

Note that the terms in the square bracket must be an integral number of pitch lengths. In case it is a fraction, it must be rounded off to the next integral number.

9.21 ANGULAR SPEED RATIO

The chain is wrapped round the sprocket in the form of a pitch polygon and not as pitch circle. From Fig. 9.18, it may be observed that the axial line of the chain vibrates between two positions shown by full and dotted lines.

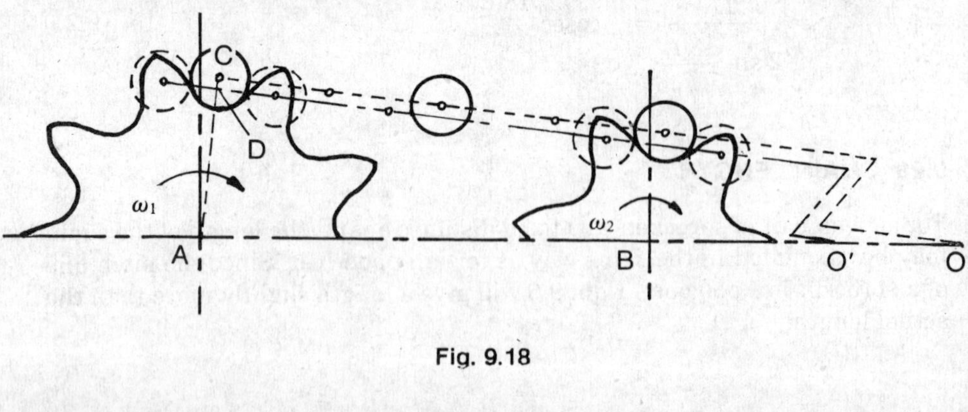

Fig. 9.18

Even if the sprocket rotates at an uniform angular velocity ω, the linear velocity of the chain will be varying from a maximum $\omega\, AC$ to a minimum $\omega\, AD$. Thus the magnitude of the speed variation is the ratio of the distances AC to AD. The variation in the chain speed also causes a variation in the angular speed of the driven sprocket. However, by increasing the number of teeth on the sprocket, the magnitude of the variation in speed may be minimised.

It can be shown that at any instant, if the line of transmission cuts the line of centres at O, the angular velocities of the two sprockets will be in the inverse ratio of the distances of their centres from O, i.e.

$$\frac{\omega_2}{\omega_1} = \frac{OA}{OB}$$

The variation of ω_2 will be between

$$\omega_1 \frac{OA}{OB} \text{ and } \omega_1 \frac{O'A}{O'B}$$

Thus at any instant, the angular velocity of the driven shaft would be changing.

9.22 CLASSIFICATION OF CHAINS

Chains have been classified into hoisting chains, conveyor chains and power transmission chains. Each type has been discussed below:

1. Hoisting Chains

Hoisting chains include oval link and stud link chains. An oval link chain is a common form of hoisting chain (Fig. 9.19 a). It consists of oval links and is also known as coil chain. Such chains are used for lower speeds only.

Figure 9.19 b shows a stud link chain. This does not kink or tangle easily.

2. Conveyor Chains

Conveyor chains may be of detachable or *hook joint* type (Fig. 9.20 a), or of the *closed-end pintle* type (Fig. 9.20 b). The sprocket teeth are so shaped and spaced that the chain should run onto and off the sprocket smoothly and without interference. Such chains are used for low speed agricultural machinery. The material of the links is, usually, malleable cast iron. The motion of the chain is not very smooth.

3. Power Transmission Chains

These chains are made of steel in which the wearing parts are hardened. They are accurately machined and run on carefully designed sprockets.

These are of three types:

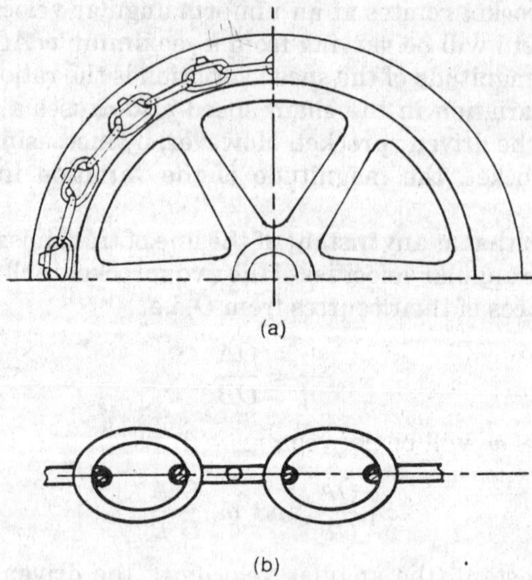

Fig. 9.19

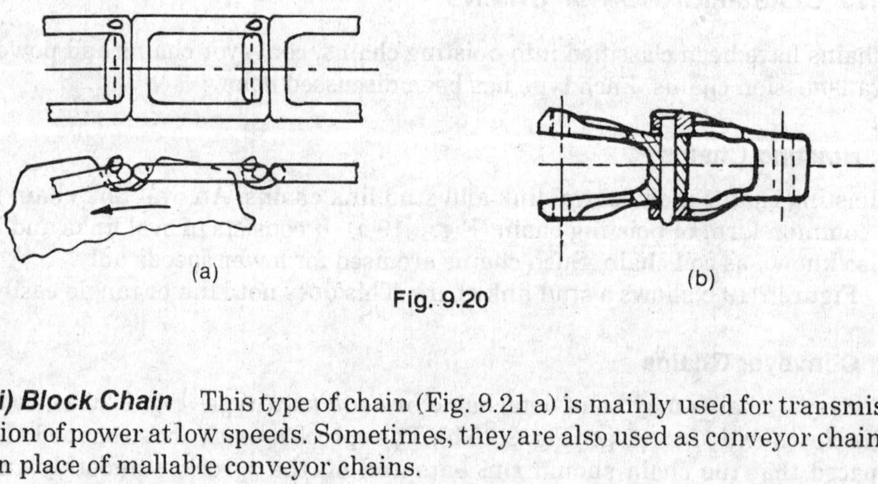

Fig. 9.20

(i) Block Chain This type of chain (Fig. 9.21 a) is mainly used for transmission of power at low speeds. Sometimes, they are also used as conveyor chains in place of mallable conveyor chains.

(ii) Roller Chain A common form of a roller chain is shown in Fig. 9.21 b. A bushing is fixed to the inner link whereas the outer link has a pin fixed to it. There is only sliding motion between the pin and the bushing. The roller is made of a hardened material and is free to turn on the bushing. Figure 9.17 shows this type of chain in place on the sprocket. A good roller chain is quiter and wears less as compared to a block chain.

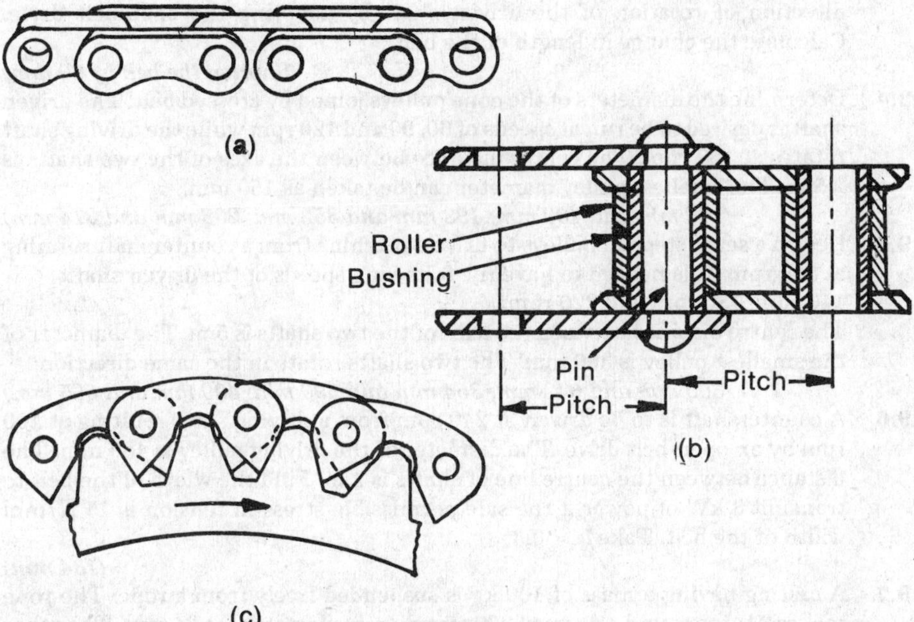

Fig. 9.21

(iii) Silent Chain (Inverted tooth Chain) Though roller chains can run quitely at fairly high speeds, the silent chains or inverted tooth chains are used where maximum quiteness is desired.

Silent chains do not have rollers. The links are so shaped as to engage directly with the sprocket teeth. The included angle is either 60° or 75° (Fig. 9.21 c).

PROBLEMS

9.1 A motor shaft drives a main shaft of a workshop by means of a flat belt, the diameters of the pulleys being 500 mm and 800 mm respectively. Another pulley of diameter 600 mm on the main shaft drives a counter-shaft having a 750 mm diameter pulley. If the speed of the motor is 1600 rpm, find the speed of the countershaft neglecting the thickness of the belt and considering a slip of 4% on each drive.

(737.3 rpm)

9.2 Two pulleys on two shafts are connected by a flat belt. The driving pulley is 250 mm in diameter and runs at 150 rpm. The speed of the driven pulley is to be 90 rpm. The belt is 120 mm wide, 5 mm thick and weighs 1000 kg/m^3. Assuming a slip of 2% between the belt and each pulley, determine the diameter of the driven pulley. Also find the total effective slip.

(416.5 mm; 3.96%)

9.3 The pulleys of two parallel shafts 8 m apart are 600 mm and 800 mm in diameters and are connected by a crossed belt. It is needed to change the direction of rotation of the driven shaft by adopting the open-belt drive. Calculate the change in length of the belt.

(Shorten the belt by 60 mm)

9.4 Determine the diameters of the cone pulleys joined by crossed belt. The driven shaft is desired to be run at speeds of 60, 90 and 120 rpm while the driving shaft rotates at 160 rpm. The centre distance between the axes of the two shafts is 2.5 m. The smallest pulley diameter can be taken as 150 mm.

(150 mm and 400 mm; 198 mm and 352 mm; 236 mm and 314 mm)

9.5 Design a set of stepped pulleys to drive a machine from a countershaft running at 300 rpm. It is needed to have the following speeds of the driven shaft: 140 rpm, 180 rpm and 220 rpm.
The centre distance between the axes of the two shafts is 5 m. The diameter of the smallest pulley is 300 mm. The two shafts rotate in the same direction.

(300 mm and 642 mm; 354 mm and 590 mm; 400 mm and 565 mm)

9.6 A countershaft is to be driven at 240 rpm from a driving shaft rotating at 100 rpm by an open-belt drive. The diameter of the driving pulley is 480 mm. The distance between the centre line of shafts is 2 m. Find the width of the belt to transmit 3 kW of power if the safe permissible stress in tension is 15 N/mm width of the belt. Take $\mu = 0.3$.

(134 mm)

9.7 A casting having a mass of 100 kg is suspended freely from a rope. The rope makes 2 turns round a drum of 300 mm diameter rotating at 24 rpm. The other end of the rope is pulled by a man. Calculate the force required by the man, power to raise the casting and the power supplied by drum run by a prime-mover. Take $\mu = 0.3$.

(222 N; 3.698 kW; 3.613 kW)

9.8 A leather belt transmits 10 kW from a motor running at 600 rpm by an open-belt drive. The diameter of the driven pulley of the motor is 350 mm, centre distance between the pulleys 4 m and speed of the driven pulley 180 rpm. The belt weighs 1100 kg/m^3 and the maximum allowable tension in the belt is 2.5 N/mm^2 · $\mu = 0.25$. Find the width of the belt assuming the thickness to be 10 mm. Neglect the belt thickness to calculate the velocities.

(73.8 mm)

9.9 Determine the maximum power that can be transmitted through a flat belt having the following data:

X-section of the belt = 300 mm × 12 mm
Ratio of friction tensions = 2.2
Maximum permissible tension in belt = 2 N/mm^2
Mass density of the belt material = 0.0011 gm/mm^3

(59.088 kW)

9.10 A V-belt weighing 1.6 kg/m run has an area of cross-section of 750 mm^2. The angle of lap is 165° on the smaller pulley which has a groove angle of 40°. μ = 0.12. The maximum safe stress in the belt is 9.5 N/mm^2. What is the power that can be transmitted by the belt at a speed of 20 m/s.

(82.485 kW)

9.11 A rope drive transmits 40 kW at 120 rpm by using 15 ropes. Angle of lap on the smaller pulley which is 300 mm in diameter is 165°. Coefficient of friction is 0.25. The rope weighs $(50 \times 10^{-6})G^2$ kg per meter length of rope and the working tension is limited to $0.14\ G^2$ N where G is the girth (circumferen... the rope in mm. Determine the initial tension and the diameter of each rope.

(903.8 N; 34.2 mm)

9.12 A man wants to lower an engine weighing 380 kg from a trolley to the ground by using a rope which he passes over a fixed horizontal pipe overhead. The man is capable of controlling the motion with a force of 200 N or less on the free end of the rope. Find the minimum number of times the rope must be passed round the pipe if $\mu = 0.22$.

(2.1 turns)

10

GEARS

10.1 INTRODUCTION

Gears are used to transmit motion from one shaft to another or between a shaft and a slide. This is accomplished by successively engaging teeth.

Gears use no intermediate link or connector and transmit the motion by direct contact. In this method, the surfaces of two bodies make a tangential contact. The two bodies have either a rolling or a sliding motion along the tangent at the point of contact. No motion is possible along the common normal as that will either break the contact or one body will tend to penetrate into the other.

If power transmitted between two shafts is small, motion between them may be obtained by using two plain cylinders or discs 1 and 2 as shown in Fig. 10.1. If there is no slip of one surface relative to the other, a definite motion of 1 can be transmitted to 2 and vice-versa. Such wheels are termed as friction wheels. However as the power transmitted increases, slip occurs between the discs and the motion no longer remains definite.

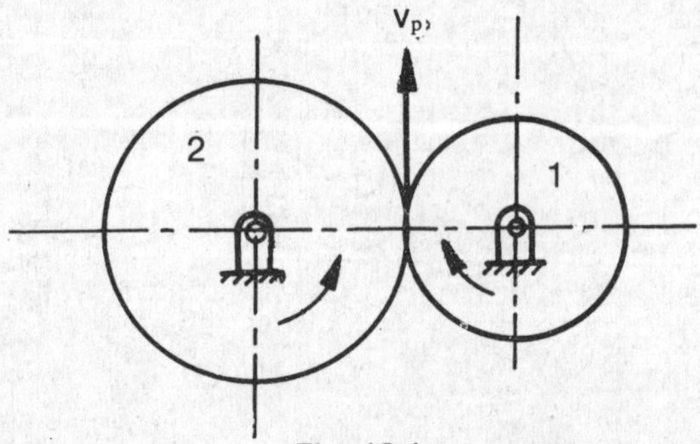

Fig. 10.1

Assuming no sliping of the two surfaces, the following kinematic relationship exists for their linear velocity,

$$v_p = \omega_1 r_1 = \omega_2 r_2$$

$$= 2\pi N_1 r_1 = 2\pi N_2 r_2$$

or

$$\frac{\omega_1}{\omega_2} = \frac{N_1}{N_2} = \frac{r_2}{r_1} \qquad (10.1)$$

where N = angular velocity (rpm)
ω = angular velocity (rad/s)
r = radius of the disc

Subscripts 1 and 2 represent discs 1 and 2 respectively.

The relationship shows that the speeds of the two discs rolling together without slipping are inversely proportional to the radii of the discs.

To transmit a definite motion of one disc to the other or to prevent slip between the surfaces, projections and recesses on the two discs can be made which can mesh with each other. This leads to the formation of teeth on the discs and the motion between the surfaces changes from rolling to sliding. The discs with teeth are known as gears or gear wheels.

It is to be noted that if the disc1 rotates in the clockwise direction, 2 rotates in the counter-clockwise direction and vice-versa.

Although large velocity ratios of the driving and the driven members have been obtained by the use of gears, practically, it is limited to 6 for spur gears and 10 for helical and herringbone gears. To obtain large reductions, two or more pairs of gears are used.

10.2 CLASSIFICATION OF GEARS

Gears can be classified according to the relative positions of their shaft axes as follows:

1. Parallel Shafts

Regardless of the manner of contact, uniform rotary motion between two parallel shafts is equivalent to the rolling of two cylinders, assuming no slipping. Depending upon the teeth of the equivalent cylinders i.e. straight or helical, following are the main types of gears to join parallel shafts:

Spur Gears They have straight teeth parallel to the axes and thus are not subjected to axial thrust due to tooth load (Fig. 10.2 a).

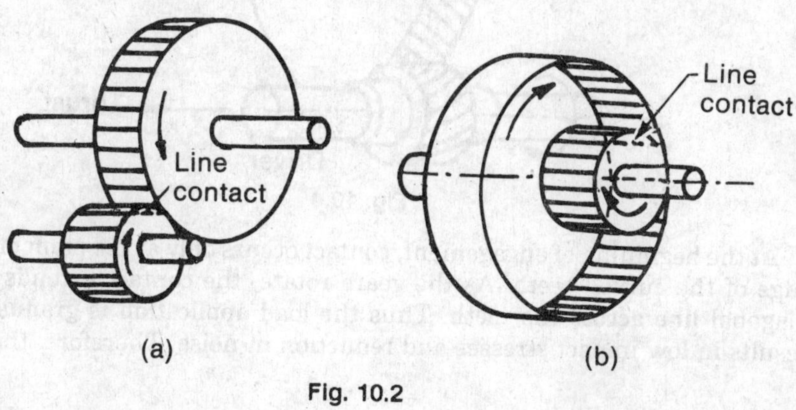

Fig. 10.2

At the time of engagement of the two gears, the contact extends across the entire width on a line parallel to the axes of rotation. This results in sudden application of the load, high impact stresses and excessive noise at high speeds.

Further, if the gears have external teeth on the outer surface of the cylinders, the shafts rotate in the opposite directions (Fig. 10.2 a). In an internal spur gear, the teeth are formed on the inner surface of an annulus ring. An internal gear can mesh with an external pinion (smaller gear) only and the two shafts rotate in the same direction as shown in Fig. 10.2 b.

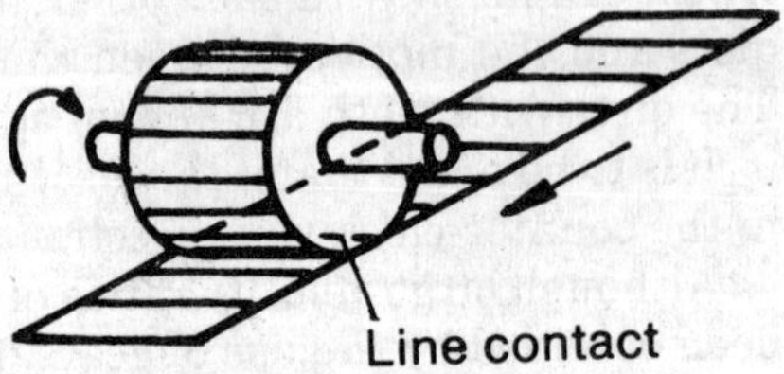

Fig. 10.3

Spur Rack and Pinion Spur rack is a special case of a spur gear where it is made of infinite diameter so that the pitch surface is a plane (Fig. 10.3). The spur rack and pinion combination converts rotary motion into translatory motion, or vice-versa. It is used in a lathe in which the rack transmits motion to the saddle.

Helical Gears or Helical Spur Gears In helical gears, the teeth are curved, each being helical in shape. Two mating gears have the same helix angle, but have teeth of opposite hands (Fig. 10.4).

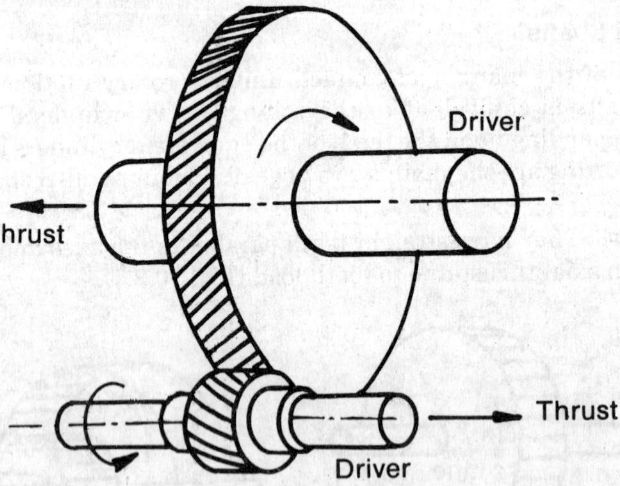

Fig. 10.4

At the beginning of engagement, contact occurs only at the point of leading edge of the curved teeth. As the gears rotate, the contact extends along a diagonal line across the teeth. Thus the load application is gradual which results in low impact stresses and reduction in noise. Therefore, the helical gears can be used at higher velocities than the spur gears and have greater load-carrying capacity.

Helical gears have the disadvantage of having end thrust as there is a force component along the gear axis. The bearings and the assemblies mounting the helical gears must be able to withstand thrust loads.

Double-helical and Herringbone Gears A double-helical gear is equivalent to a pair of helical gears secured together, one having a right-hand helix and the other a left-hand helix. The teeth of the two rows are separated by a groove used for tool runout. Axial thrust which occurs in case of single-helical gears is eliminated in double-helical gears. This is because the axial thrusts of the two rows of teeth cancel each other out. These can be run at high speeds with less noise and vibrations.

If the left and the right inclinations of a double-helical gear meet at a common apex and there is no groove in between, the gear is known as *herringbone* gear (Fig. 10.5).

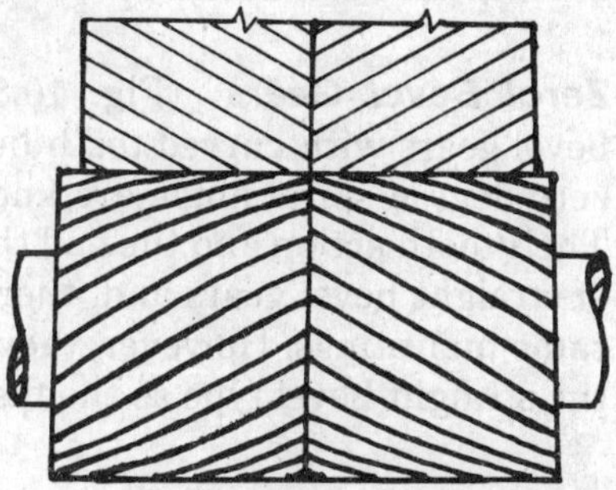

Fig. 10.5

2. Intersecting Shafts

Kinematically, the motion between two intersecting shafts is equivalent to the rolling of two cones, assuming no slipping. The gears, in general, are known as *bevel* gears.

When teeth formed on the cones are straight, the gears are known as straight bevel and when inclined, they are known as *spiral* or *helical bevel.*

Straight Bevel Gears The teeth are straight, radial to the point of intersection of the shaft axes and vary in cross-section throughout their length. Usually, they are used to connect shafts at right angles which run at low speeds (Fig. 10.6). Gears of the same size and connecting two shafts at right angle to each other are known as *mitre* gears.

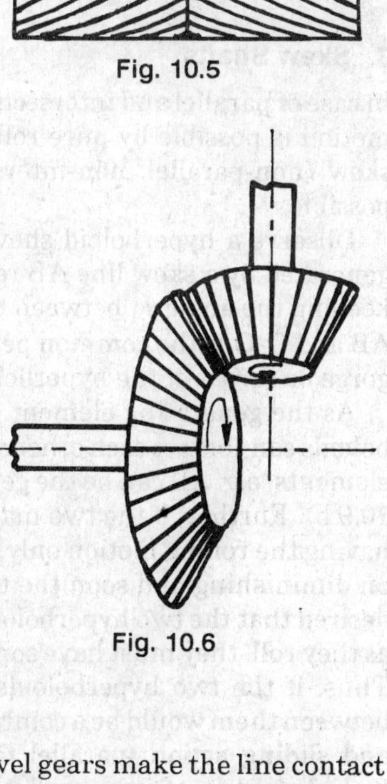

Fig. 10.6

At the beginning of engagement, straight bevel gears make the line contact similar to spur gears. There can also be internal bevel gears analogous to internal spur gears.

Spiral Bevel Gears When the teeth of a bevel gear are inclined at an angle to the face of the bevel, they are known as spiral bevels or helical bevels (Fig. 10.7). They are smoother in action and quieter than straight tooth bevels as there is gradual load application and low impact stresses. Of course, there exists an axial thrust calling for stronger bearings and supporting assemblies.

These are used for the drive to the differential of an automobile.

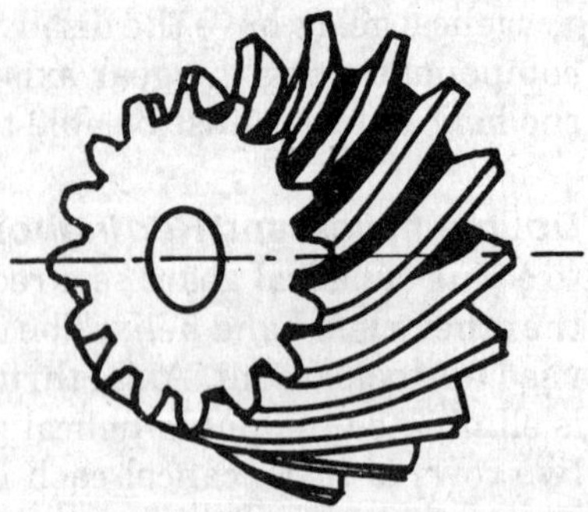

Fig. 10.7

Zerol Bevel Gears (Fig. 10.8) Spiral bevel gears with curved teeth but with a zero degree spiral angle are known as zerol bevel gears. Their tooth action and the end thrust are the same as that of straight bevel gears and, therefore, can be used in the same mountings. However, they are quiter in action than the straight bevel type as the teeth are curved.

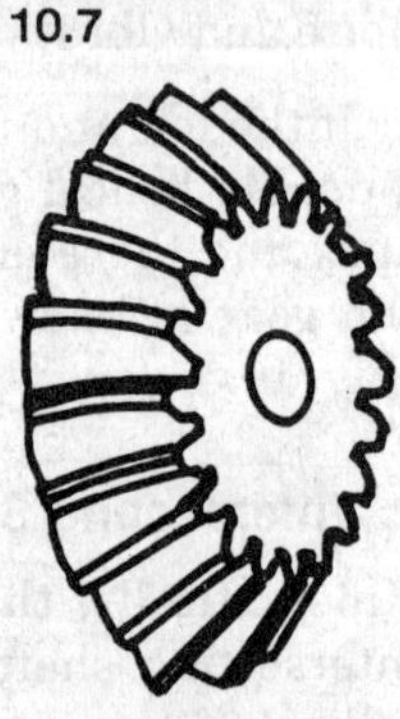

Fig. 10.8

3. Skew Shafts

In case of parallel and intersecting shafts, a uniform rotary motion is possible by pure rolling contact. But in case of skew (non-parallel, non-intersecting) shafts, this is not possible.

Observe a hyperboloid shown in Fig. 10.9 a. It is a surface of revolution generated by a skew line AB revolving around an axis O-O in another plane, keeping the angle ψ_1 between them as constant. Minimum distance between AB and O-O is the common perpendicular CD which is also the radius of the gorge or throat of the hyperboloid.

As the generating element of a hyperboloid is a straight line, two hyperboloids can contact each other on a line common to their respective generating elements, e.g. AB can be the generating element of the two hyperboloids (Fig. 10.9 b). Further, if the two mating hyperboloids are of limited width and are having the rolling motion only, then contact length of their generators will go on diminishing and soon the two could be separated. In other words, if it is desired that the two hyperboloids touch each other on the entire length of AB as they roll, they must have some sliding motion parallel to the line of contact. Thus, if the two hyperboloids rotate on their respective axes, the motion between them would be a combination of rolling (normal to the line of contact) and sliding action (parallel to the line of contact). Teeth are cut on the hyperboloids surfaces parallel to the line of contact to form gears.

Angle between the two shafts will be equal to the sum of the angles of generation of the two hyperboloids.

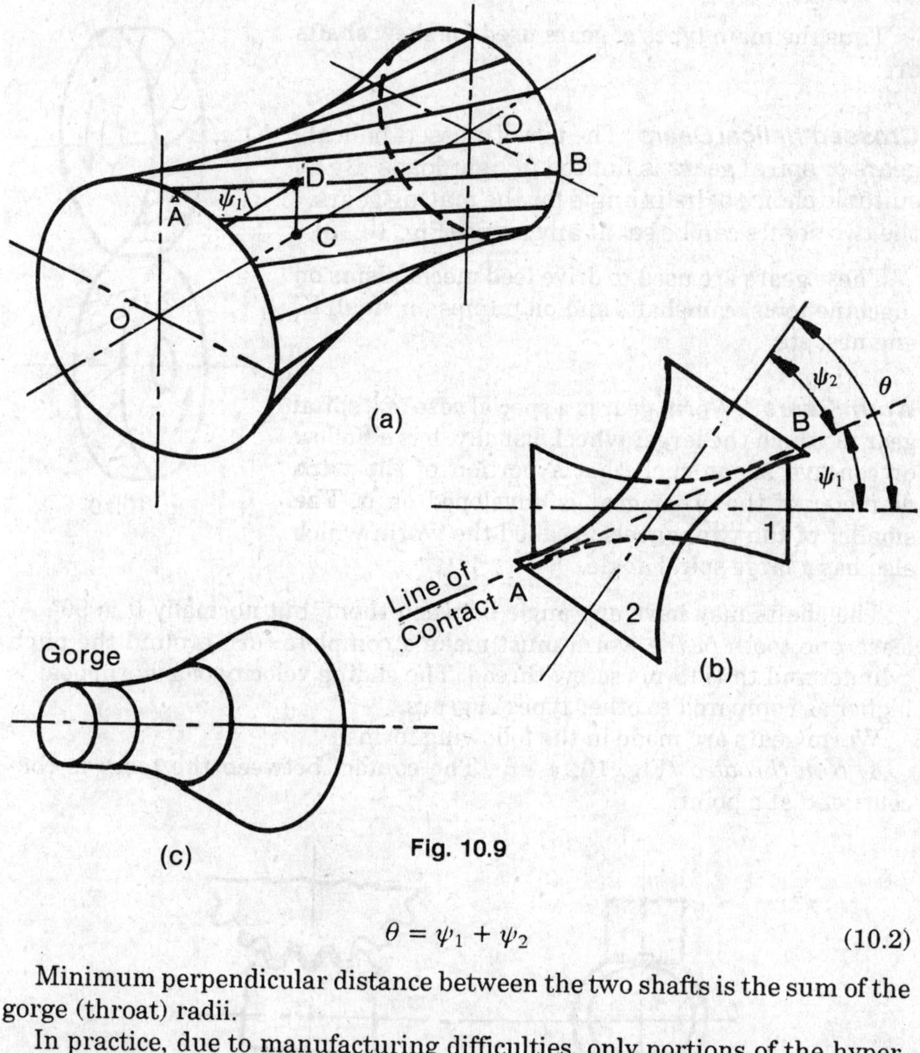

Fig. 10.9

$$\theta = \psi_1 + \psi_2 \tag{10.2}$$

Minimum perpendicular distance between the two shafts is the sum of the gorge (throat) radii.

In practice, due to manufacturing difficulties, only portions of the hyperboloids are used to transmit motion between the skew shafts and that too with approximations as given below:

1. A short segment at the gorge is approximated to a cylinder and the corresponding gear is known as helical or crossed-helical or spiral gear (Fig. 10.9 c). The contact between the two gears is concentrated at a point which limits the capacity.

   For skew shafts with 90° angle between them where high-speed ratios are to be achieved, the helix angle of the pinion (small gear) increases. When the angle exceeds 60°-65° and the number of teeth is less than 3-4, the high-speed pinion is known as *worm* and the mating helical gear as the *gear*.
2. Gears using an end portion of the hyperboloid are known as hypoid gears.

Thus the main types of gears used for skew shafts are:

Crossed-helical Gears The use of crossed-helical gears or spiral gears is limited to light loads. By a suitable choice of helix angle for the mating gears, the two shafts can be set at any angle (Fig. 10.10).

These gears are used to drive feed mechanisms on machine tools, camshafts and oil pumps on small IC engines, etc.

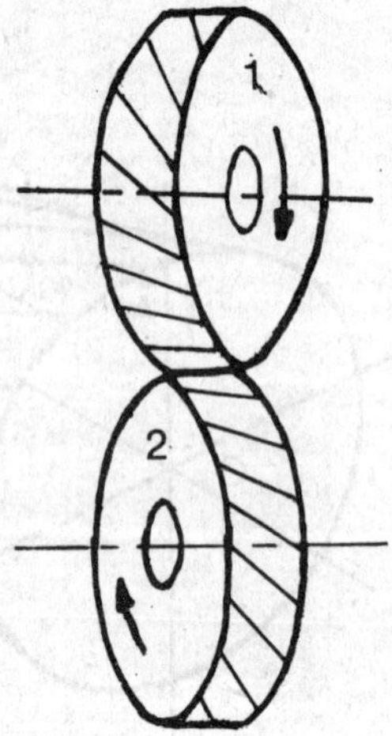

Fig. 10.10

Worm Gears Worm gear is a special case of a spiral gear in which the larger wheel, usually, has a hollow or concave shape such that a portion of the pitch diameter of the other gear is enveloped on it. The smaller of the two wheels is called the worm which also has a large spiral angle.

The shafts may have any angle between them, but normally it is 90°. At least, one tooth of the worm must make a complete turn around the pitch cylinder and thus forms screw thread. The sliding velocity of a worm gear is higher as compared to other types of gears.

Worm gears are made in the following forms:

1. *Non-throated* (Fig. 10.11 a): The contact between the teeth is concentrated at a point.

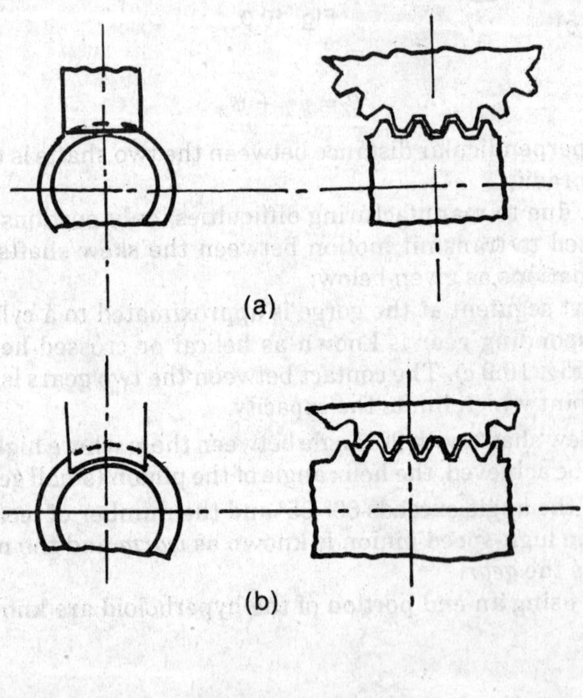

(a)

(b)

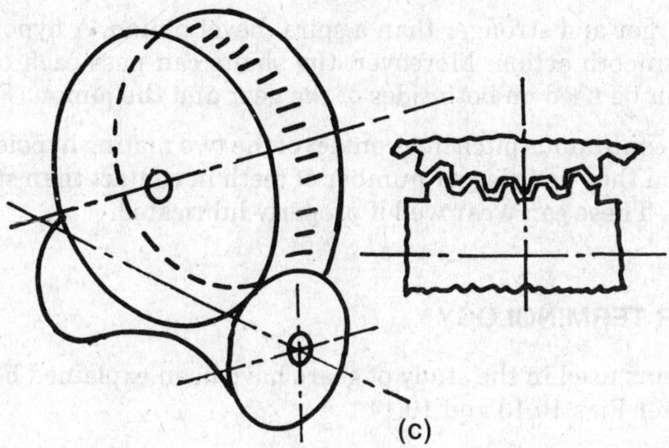

Fig. 10.11

2. *Single-throated* (Fig. 10.11 b): Gear teeth are curved to envelop the worm. There is line contact between the teeth.

3. *Double-throated* (Fig. 10.11 c): There is area contact between the teeth. A worm may be cut with a single or multiple thread cutter.

Hypoid Gears As mentioned earlier, hypoid gears are approximations of hyperboloids though they look like spiral bevel gears (Fig. 10.12 a). A hypoid

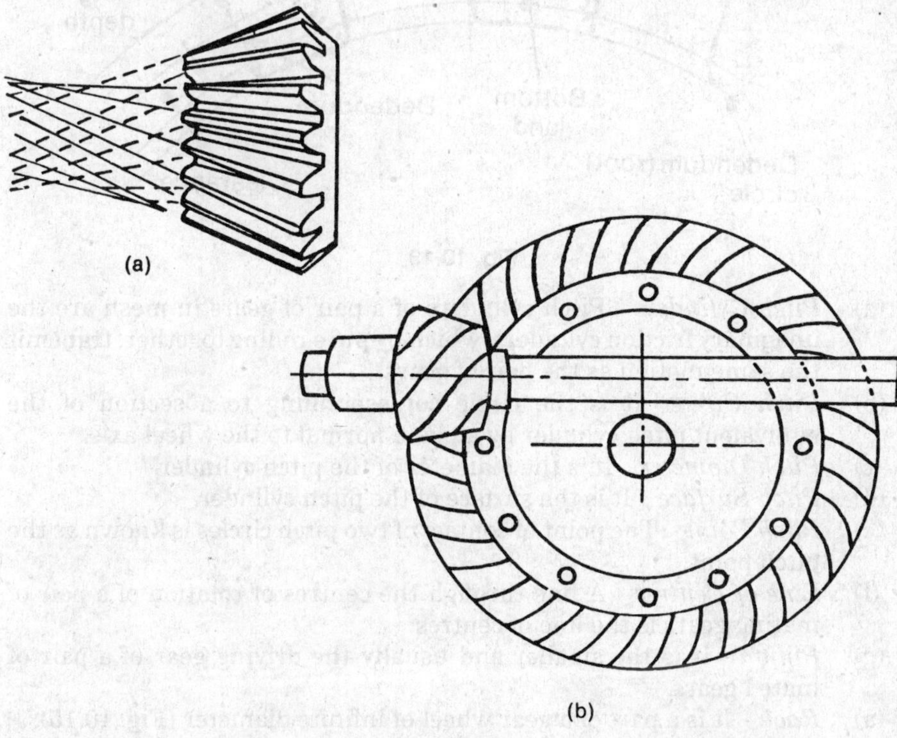

Fig. 10.12

pinion is larger and stronger than a spiral bevel pinion. A hypoid pair has a quite and smooth action. Moreover, the shafts can pass each other so that bearings can be used on both sides of the gear and the pinion (Fig. 10.12 b)

There is continuous pitch line contact of the two mating hypoid gears while in action and they have larger number of teeth in contact than straight tooth bevel gears. These can wear well if properly lubricated.

10.3 GEAR TERMINOLOGY

Various terms used in the study of gears have been explained below:

1. Refer Figs 10.13 and 10.14.

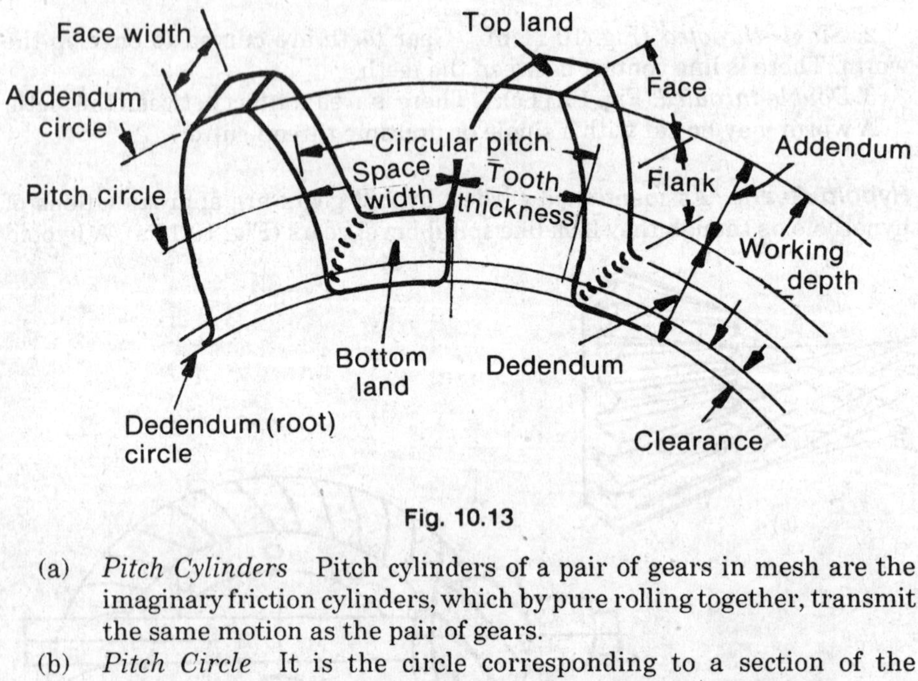

Fig. 10.13

(a) *Pitch Cylinders* Pitch cylinders of a pair of gears in mesh are the imaginary friction cylinders, which by pure rolling together, transmit the same motion as the pair of gears.

(b) *Pitch Circle* It is the circle corresponding to a section of the equivalent pitch cylinder by a plane normal to the wheel axis.

(c) *Pitch Diameter* It is the diameter of the pitch cylinder.

(d) *Pitch Surface* It is the surface of the pitch cylinder.

(e) *Pitch Point* The point of contact of two pitch circles is known as the pitch point.

(f) *Line of Centres* A line through the centres of rotation of a pair of mating gears is the line of centres.

(g) *Pinion* It is the smaller and usually the driving gear of a pair of mated gears.

2. (a) *Rack* It is a part of a gear wheel of infinite diameter (Fig. 10.15).

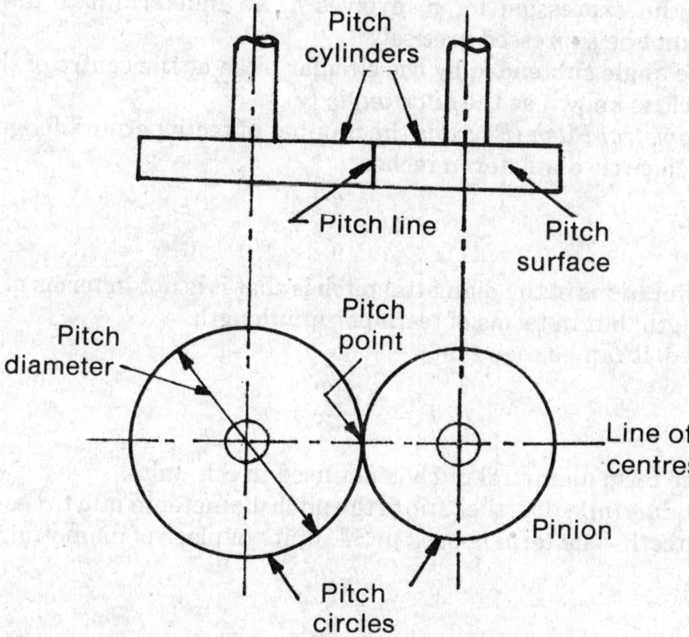

Fig. 10.14

(b) *Pitch Line* It is a part of the pitch circle of a rack and is a straight line (Fig. 10.15).

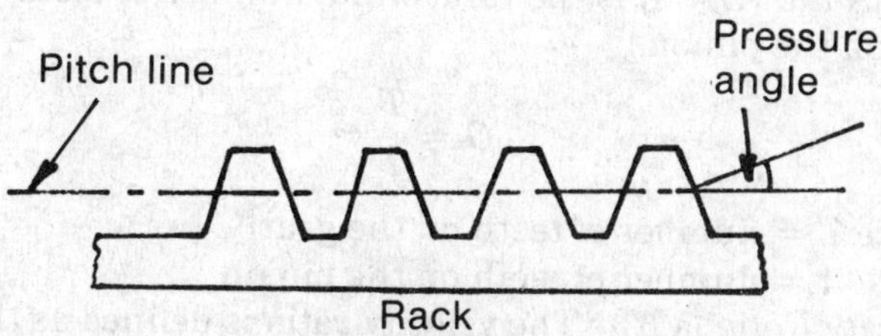

Fig. 10.15

3. *Pitch* Pitch of two mating gears must be same. It is defined as follows:

(a) *Circular Pitch (p)* It is the distance measured along the circumference of the pitch circle from a point on one tooth to the corresponding point on the adjacent tooth (Fig. 10.13).

$$p = \frac{\pi d}{T}$$

where p = circular pitch
d = pitch diameter
T = number of teeth.

As the expression for p involves π, an indeterminate number, p cannot be expressed precisely.
The angle subtended by the circular pitch at the centre of the pitch circle is known as the *pitch angle* (γ).

(b) *Diametral Pitch (P)* It is the number of teeth per unit length of the pitch circle diameter in inches.

$$P = \frac{T}{d}$$

Limitations of the diametral pitch is that it is not in terms of units of length, but in terms of teeth per unit length.
Also, it can be seen that

$$p\,P = \frac{\pi\,d}{T}\,\frac{T}{d} = \pi$$

The term diametral pitch is not used in S.I. units.

(c) *Module (m)* It is the ratio of the pitch diameter in mm to the number of teeth. The term is used in S.I. units in place of diametral pitch.

$$m = \frac{d}{T}$$

Also,

$$p = \frac{\pi\,d}{T} = \pi\,m$$

4. (a) *Gear Ratio (G)* It is the ratio of the number of teeth on the gear to that on the pinion.

$$G = \frac{T}{t}$$

where T = number of teeth on the gear
t = number of teeth on the pinion.

(b) *Velocity Ratio (VR)* The velocity ratio is defined as the ratio of the angular velocity of the follower to the angular velocity of the driving gear.

Let d = pitch diameter
T = number of teeth
ω = angular velocity (rad/s)
N = angular velocity (rpm)
Subscript 1 = driver
2 = follower

$$\text{VR} = \frac{\text{angular velocity of follower}}{\text{angular velocity of driver}}$$

$$= \frac{\omega_2}{\omega_1}$$

$$= \frac{N_2}{N_1} \qquad (\omega = 2\pi N)$$

$$= \frac{d_1}{d_2} \qquad (\because \pi d_1 N_1 = \pi d_2 N_2)$$

$$= \frac{T_1}{T_2} \qquad \left(p = \frac{\pi d_1}{T_1} = \frac{\pi d_2}{T_2}\right) \quad (10.3)$$

5. Refer to Fig. 10.13.

(i) (a) *Addendum Circle* It is a circle passing through the tips of teeth.

(b) *Addendum* It is the radial height of a tooth above the pitch circle. Its standard value is one module.

(c) *Dedendum* or *Root Circle* It is a circle passing through the roots of the teeth.

(d) *Dedendum* It is the radial depth of a tooth below the pitch circle. Its standard value is 1.157 *m*.

(e) *Clearance* Radial difference between the addendum and the dedendum of a tooth. Thus

Addendum circle diameter = $d + 2\,m$

Dedendum circle diameter = $d - 2 \times 1.157\,m$

Clearance = $1.157\,m - m$

$= 0.157\,m$

(ii)(a) *Full Depth of Teeth* It is the total radial depth of the tooth space.

Full depth = Addendum + Dedendum

(b) *Working Depth of Teeth* The maximum depth to which a tooth penetrates into the tooth space of the mating gear is the working depth of teeth.

Working depth = Sum of addendums of the two gears.

(c) *Space Width* It is the width of the tooth space along the pitch circle.

(d) *Tooth Thickness* It is the thickness of the tooth measured along the pitch circle.

(e) *Backlash* It is the difference between the space width and the tooth thickness along the pitch circle.

Backlash = Space width – Tooth thickness

(f) *Face Width* The length of the tooth parallel to the gear axis is the face width.

(iii)(a) *Top Land* It is the surface of the top of the tooth.

(b) *Bottom Land* The surface of the bottom of the tooth between the adjacent fillets.

(c) *Face* Tooth surface between the pitch circle and the top land.

(d) *Flank* Tooth surface between the pitch circle and the bottom land including fillet.

(e) *Fillet* It is the curved portion of the tooth flank at the root circle.

6. Refer to Fig. 10.16.

(i) (a) *Line of Action or Pressure Line* The force, which the driving tooth exerts on the driven tooth, is along a line from the pitch point to the

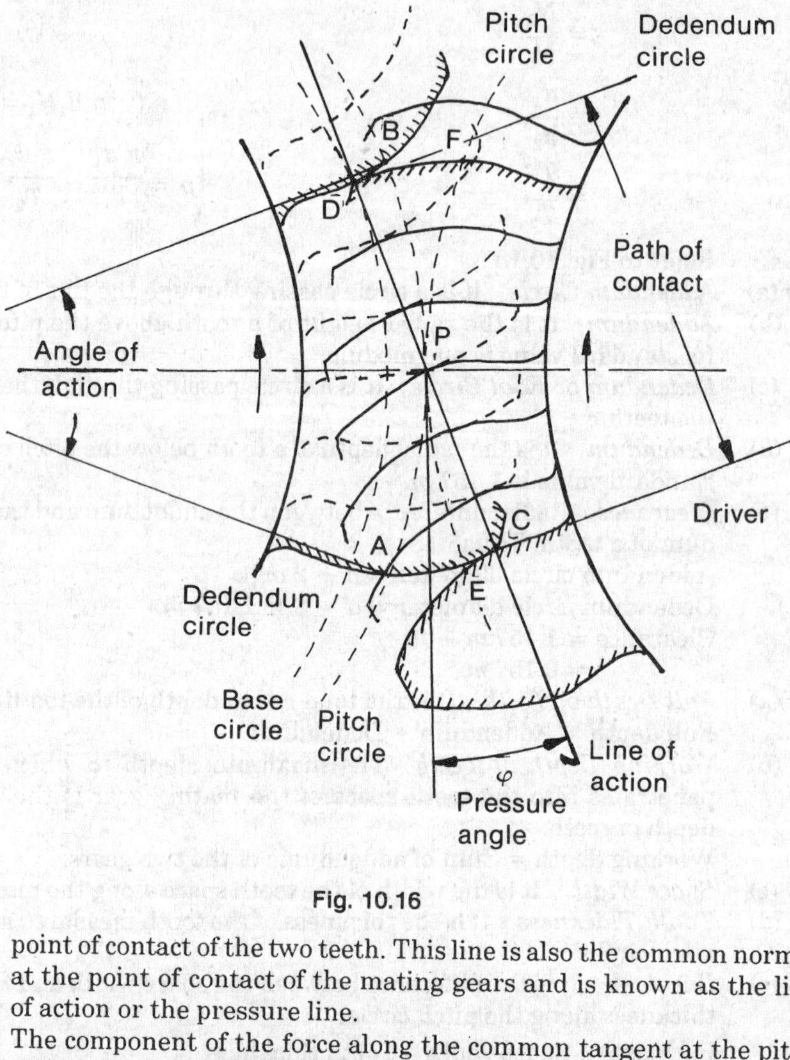

Fig. 10.16

point of contact of the two teeth. This line is also the common normal at the point of contact of the mating gears and is known as the line of action or the pressure line.

The component of the force along the common tangent at the pitch point is responsible for the power transmission.

The component of the force perpendicular to the common tangent at the pitch point produces thrust which tends to make the shafts apart.

(b) *Pressure Angle or Angle of Obliquity* (φ) The angle between the pressure line and the common tangent to the pitch circles is known as the pressure angle or the angle of obliquity.

For more power transmission and lesser pressure on the bearings, pressure angle must be kept small. Standard pressure angles are 20° and 25°. Gears with 14.5° pressure angles have become almost obsolete.

(ii) (a) *Path of Contact or Contact Length* Locus of the point of contact of two mating teeth from the beginning of engagement to the end of engagement is known as the path of contact or the contact length. It is *CD* in the figure. Pitch point *P* is always one point on the path of contact. It can be subdivided as follows:

Path of Approach Portion of the path of contact from the beginning of engagement to the pitch point, i.e. the length *CP*.

Path of Recess Portion of the path of contact from the pitch point to the end of engagement i.e. length *PD*.

(b) *Arc of Contact* Locus of a point on the pitch circle from the beginning to the end of engagement of two mating gears is known as the arc of contact. In fig. 10.16, *APB* or *EPF* is the arc of contact.

It has also been divided into sub-portions.

Arc of Approach It is the portion of the arc of contact from the beginning of engagement to the pitch point, i.e. length *AP* or *EP*.

Arc of Recess Portion of the arc of contact from the pitch point to the end of engagement is the arc of recess i.e. length *PB* or *PF*.

(c) *Angle of Action* (δ) It is the angle turned by a gear from the beginning of engagement to the end of engagement of a pair of teeth i.e. the angle turned by arcs of contact of respective gear wheels. Similarly, angle of approach (α) and angle of recess (β) can be defined.

$$\delta = \alpha + \beta$$

The angles will have different values for the driving and the driven gears.

7. *Contact Ratio* It is the angle of action divided by the pitch angle, i.e.

$$\text{CR} = \frac{\delta}{\gamma} = \frac{\alpha + \beta}{\gamma}$$

Example 10.1 Two spur gears have a velocity ratio of 1/3. The driven gear has 72 teeth of 8 mm module and rotates at 300 rpm. Calculate the number of teeth and the speed of the driver. What will be the pitch line velocities?

Solution:

$$T_2 = 72 \qquad \text{VR} = \frac{1}{3}$$

$$N_2 = 300 \text{ rpm}$$

(i) $$\text{VR} = \frac{N_2}{N_1} = \frac{T_1}{T_2} = \frac{1}{3}$$

$$\frac{300}{N_1} = \frac{1}{3}$$

or $$N_1 = \underline{900 \text{ rpm}}$$

Also $$\frac{T_1}{72} = \frac{1}{3} \quad \text{or} \quad T_1 = 24$$

(ii) Pitch line velocity, $v_p = \omega_1 r_1$ or $\omega_2 r_2$

$$= 2\pi N_1 \times \frac{d_1}{2} \text{ or } 2\pi N_2 \times \frac{d_2}{2}$$

$$= 2\pi N_1 \times \frac{m\,T_1}{2} \text{ or } 2\pi N_2 \times \frac{m\,T_2}{2}$$

$$= 2\pi \times 900 \times \frac{8 \times 24}{2} \text{ or } 2\pi \times 300 \times \frac{8\times 72}{2}$$

$$= 542\,867 \text{ mm/minute}$$

$$= 9047.8 \text{ mm/s or } \underline{9.0478 \text{ m/s}}$$

Example 10.2 The number of teeth of a spur gear is 30 and it rotates at 200 rpm. What will be its circular pitch and the pitch line velocity if it has a module of 2 mm?

$T = 30 \qquad m = 2$ mm

$N = 200$ rpm

$p = \pi m = \pi \times 2 = \underline{6.28 \text{ mm}}$

$$v_p = \omega r = 2\pi N \times \frac{d}{2} = 2\pi N \times \frac{m\,T}{2}$$

$$= \pi \times 200 \times 2 \times 30$$

$$= 37699 \text{ mm/min}$$

$$= \underline{628.3 \text{ mm/s}}$$

10.4 LAW OF GEARING

The law of gearing states the condition which must be fulfilled by the gear tooth profiles to maintain a constant angular velocity ratio between two gears. Figure 10.17 shows two bodies 1 and 2 representing a portion of the two gears in mesh.

A point C on the tooth profile of gear 1 is in contact with a point D on the tooth profile of gear 2. The two curves in contact at point C or D must have a common normal at the point. Let it be $n - n$.

Let ω_1 = instantaneous angular velocity of gear 1 (clockwise)

ω_2 = instantaneous angular velocity of gear 2 (counter-clockwise)

v_c = linear velocity of C

v_d = linear velocity of D

Then $v_c = \omega_1 AC$ in a direction perpendicular to AC or at an angle α to $n - n$

$v_d = \omega_2 BD$ in a direction perpendicular to BD or at an angle β to $n - n$.

Now, if the curved surfaces of the teeth of two gears are to remain in contact, one surface may slide relative to the other along the common tangent $t - t$. The relative motion between the surfaces along the common normal

$n - n$ must be zero to avoid the separation, or the penetration of the two teeth into each other.

Component of v_c along $n - n = v_c \cos \alpha$

Component of v_d along $n - n = v_d \cos \beta$

Relative motion along $n - n = v_c \cos \alpha - v_d \cos \beta$

Draw perpendiculars AE and BF on $n-n$ from points A and B respectively. Then $\angle CAE = \alpha$ and $\angle DBF = \beta$. For proper contact,

$$v_c \cos \alpha - v_d \cos \beta = 0$$

or $$\omega_1 AC \cos \alpha - \omega_2 BD \cos \beta = 0$$

or $$\omega_1 AC \frac{AE}{AC} - \omega_2 BD \frac{BF}{BD} = 0$$

or $$\omega_1 AE - \omega_2 BF = 0$$

or $$\frac{\omega_1}{\omega_2} = \frac{BF}{AE}$$

$$= \frac{BP}{AP} \qquad [\because \Delta s\ AEP \text{ and } BFP \text{ are similar}]$$

Thus, it is seen that the centre line AB is divided at P by the common normal in the inverse ratio of the angular velocities of the two gears. If it is desired that the angular velocities of two gears remain constant, the common normal at the point of contact of the two teeth should always pass through a fixed point P which divides the line of centres in the inverse ratio of angular velocities of two gears.

As seen earlier, P is also the point of contact of two pitch circles which divides the line of centres in the inverse ratio of the angular velocities of the two circles and is the pitch point.

Thus for constant angular velocity ratio of the two gears, the common normal at the point of contact of the two mating teeth must pass through the pitch point.

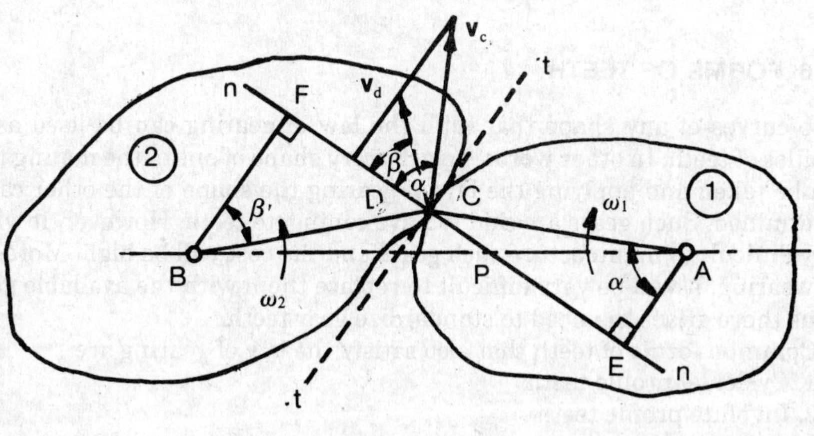

Fig. 10.17

Also as the Δs AEP and BFP are similar,

$$\frac{BP}{AP} = \frac{FP}{EP}$$

or $$\frac{\omega_1}{\omega_2} = \frac{FP}{EP} \quad \text{or} \quad \omega_1\, EP = \omega_2\, FP \tag{10.4}$$

10.5 VELOCITY OF SLIDING

If the curved surfaces of the two teeth of the gears 1 and 2 are to remain in contact, one can have a sliding motion relative to the other along the common tangent $t - t$ at C or D (Fig. 10.17).

Component of v_c along $t - t = v_c \sin\alpha$

Component of v_d along $t - t = v_d \sin\beta$

$$\begin{aligned}
\text{Velocity of sliding} &= v_c \sin\alpha - v_d \sin\beta \\
&= \omega_1\, AC\,\frac{EC}{AC} - \omega_2\, BD\,\frac{FD}{BD} \\
&= \omega_1\, EC - \omega_2\, FD \\
&= \omega_1\,(EP + PC) - \omega_2\,(FP - PD) \\
&= \omega_1\, EP + \omega_1\, PC - \omega_2\, FP + \omega_2\, PC \\
&\qquad (C \text{ and } D \text{ are same points}) \\
&= (\omega_1 + \omega_2)\, PC + \omega_1\, EP - \omega_2\, FP \\
&= (\omega_1 + \omega_2)\, PC \qquad [\omega_1\, EP = \omega_2\, FP, \text{Eq. (10.4)}] \\
&= \text{sum of angular velocities} \times \text{distance between the pitch point and the point of contact}
\end{aligned}$$

10.6 FORMS OF TEETH

Two curves of any shape that fulfil the law of gearing can be used as the profiles of teeth. In other words, an arbitrary shape of one of the mating teeth can be taken and applying the law of gearing the shape of the other can be determined. Such gears are said to have conjugate teeth. However, it will be very difficult to manufacture such gears and the cost will be high. Moreover, on wearing, it will be very difficult to replace them with the available gears. Thus there arises the need to standardize gear teeth.

Common forms of teeth that also satisfy the law of gearing are:

1. Cycloidal profile teeth
2. Involute profile teeth.

10.7 CYCLOIDAL PROFILE TEETH

In this type, the faces of the teeth are epicycloids and the flanks the hypocycloids.

A cycloid is the locus of a point on the circumference of a circle that rolls without slipping on a fixed straight line.

An epicycloid is the locus of a point on the circumference of a circle that rolls without slipping on the circumference of another circle.

A hypocycloid is the locus of a point on the circumference of a circle that rolls without slipping inside the circumference of another circle.

The formation of a cycloidal tooth has been shown in Fig. 10.18. A circle H rolls inside another circle APB (pitch circle). At the start, the point of contact of the two circles is at A. As the circle H rolls inside the pitch circle, the locus of the point A on the circle H traces the path ALP which is a hypocycloid. A small portion of this curve near the pitch circle is used for the flank of the tooth.

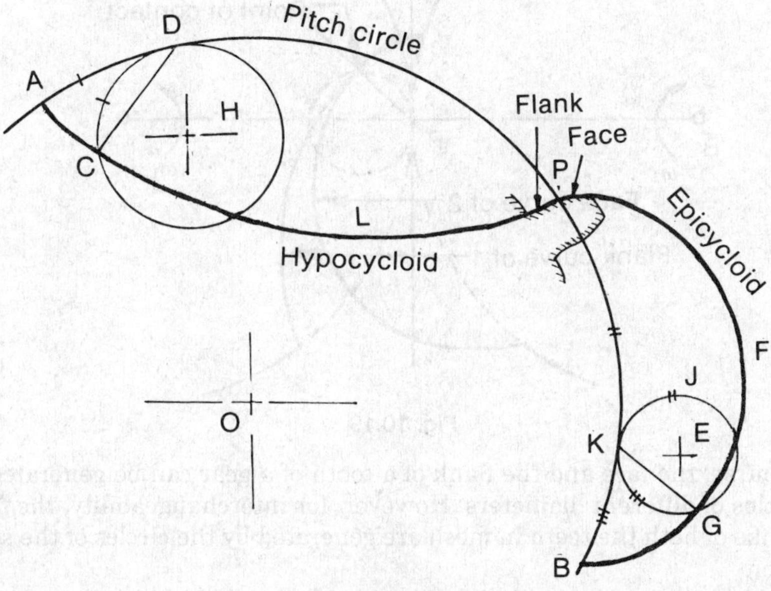

Fig. 10.18

A property of the hypocycloid is that at any instant, the line joining the generating point (A) with the point of contact of the two circles is normal to the hypocycloid, e.g. when the circle H touches the pitch circle at D, the point A is at C and CD is normal to the hypocycloid ALP.

Also, Arc AD = Arc CD (on circle H)

In the same way, if the circle E rolls outside the pitch circle, starting from P, an epicycloid PFB is obtained. Similar to the property of a hypocycloid, the line joining the generating point with the point of contact of the two circles is a normal to the epicycloid, e.g. when the circle E touches the pitch circle at K, the point P is at G and GK is normal to the epicycloid PFB.

Arc PK = Arc KJG (on circle E)

or Arc BK = Arc KG (on circle E)

A small portion of the curve near the pitch circle is used for the face of the tooth.

Meshing of Teeth During meshing of teeth, the face of a tooth on one gear is to mesh with the flank of another tooth on the other gear. Thus for proper meshing, it is necessary that the diameter of the circle generating face of a tooth (on one gear) is the same as the diameter of the circle generating flank of the meshing tooth (on another gear); the pitch circle being the same in the two cases (Fig. 10.19).

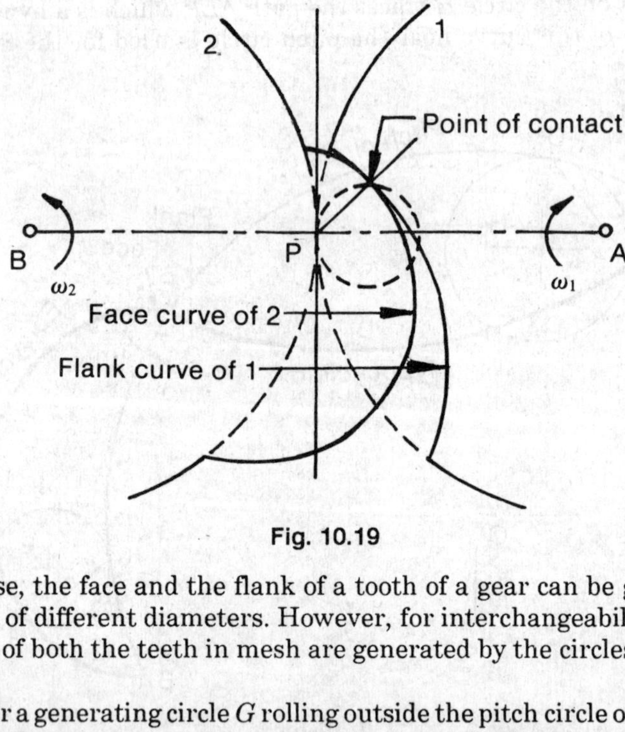

Fig. 10.19

Of course, the face and the flank of a tooth of a gear can be generated by two circles of different diameters. However, for interchangeability, the faces and flanks of both the teeth in mesh are generated by the circles of the same diameter.

Consider a generating circle G rolling outside the pitch circle of gear 2. (Fig. 10.20). It will generate epicycloid, a portion of which is the face of tooth on the gear. Now this face is to mesh with the flank of a tooth on gear 1. This flank will be a portion of the hypocycloid which can be generated by rolling the same generating circle G inside the pitch circle of gear 1.

a_1 is the generating point for the two curves a_1b_1 (epicycloid) and a_1c_1 (hypocycloid). a_1b_1 is generated when the circle G moves in the clockwise direction on the pitch circle of gear 2 and at the start a_1 coincides with b_1. a_1c_1 is generated when the circle G moves clockwise inside the pitch circle of gear 2, and in the beginning a_1 coincides with c_1.

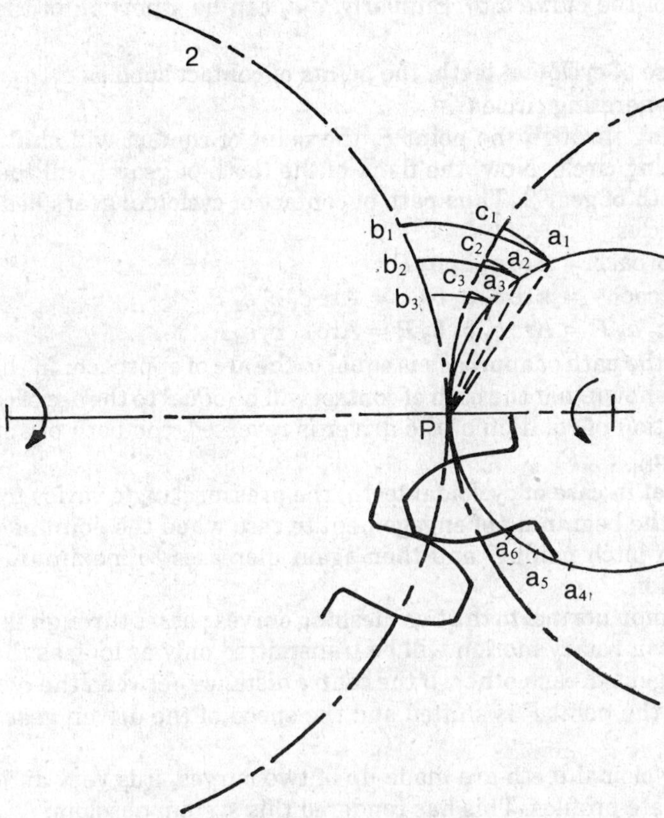

Fig. 10.20

The two pitch circles touch each other at P (pitch point). When the generating circle G touches the pitch circle 2 at P, the generating point of the epicycloid is at a_1 and $a_1 p$ is normal to the face of tooth on the gear 2. Similarly, when G touches the pitch circle 1 at P, the generating point of the hypocycloid is again at a_1 and a_1P is also normal to the flank of tooth on gear 1. Thus if at an instant, a_1P is the common normal to the two profiles of the meshing teeth, the teeth must touch each other tangentially.

According to the law of gearing, the common normal at the point of contact of two mating profiles of the teeth must pass through a fixed point which is also the pitch point. The above discussion shows that the law of gearing is fulfilled in case of cycloidal teeth.

After a little while, let the point of contact of the two mating gears be at a_2. This point is on the generating circle G and if b_2 is considered the start of the epicycloid a_2b_2, and c_2 is considered the start of the hypocycloid a_2c_2, then a_2P will be normal to the two curves a_2b_2 and a_2c_2.

But as the two curves a_1b_1 and a_2b_2 are generated by the same circle rolling outside the same pitch circle, the two curves must be similar. Thus a_2b_2 can be a portion of the curve a_1b_1. Similarly, a_2c_2 can be a portion of the curve a_1c_1.

Thus in case of cycloidal teeth, the points of contact such as a_1, a_2, a_3, . . ., P lie on the generating circle G.

After passing through the point P, the point of contact will shift on the other generating circle. Now, the flank of the tooth of gear 1 will touch the face of the tooth of gear 2. Thus path of contact of cycloidal gears lies on the generating circles.

Path of approach = Arc $a_1\, a_2\, a_3\, P$

Arc of approach = Arc $b_1\, b_2\, b_3\, P$ = Arc $c_1\, c_2\, c_3\, P$

But arc $a_1\, a_2\, a_3\, P$ = Arc $b_1\, b_2\, b_3\, P$ = Arc $c_1\, c_2\, c_3\, P$

Therefore, the path of approach is equal to the arc of approach. In the same way, it can be shown that the path of contact will be equal to the arc of contact.

If the direction of rotation of the driver is reversed, the path of approach will be a_4, a_5, a_6, . . ., P.

Observe that in case of cycloidal teeth, the pressure angle varies from the maximum at the beginning of engagement to zero when the point of contact coincides with pitch point P and then again increases to maximum in the reverse direction.

As the common normal to the two meshing curves passes through the pitch point P, uniform rotary motion will be transmitted only as long as the pitch circles are tangent to each other. If the centre distance between the two pitch circles varies, the point P is shifted and the speed of the driven gear would vary.

Since the cycloidal teeth are made up of two curves, it is very difficult to produce accurate profiles. This has rendered this system obsolete.

10.8 INVOLUTE PROFILE TEETH

An involute is defined as the locus of a point on a straight line which rolls without slipping on the circumference of a circle. Also, it is the path traced out by the end of a piece of taut cord being unwound from the circumference of a circle. The circle on which the straight line rolls or from which the cord is unwound is known as the *base circle.*

Figure 10.21 shows an involute generated by a line rolling over the circumference of a base circle with centre at O. At the start, the tracing point is at A. As the line rolls on the circumference of the circle, the path ABC traced out by the point A is the involute.

Note that as D can be regarded as the instantaneous centre of rotation of B, the motion of B is perpendicular to BD. Since BD is tangent to the base circle, the normal to the involute is a tangent to the base circle.

A short length EF of the involute drawn from A can be utilized to make the profile of an involute tooth. The other side HJ of the tooth has been taken from the involute drawn from G in the reverse direction. The profile of an involute tooth is made up of a single curve, and teeth, usually, are termed as single curve teeth.

Owing to the ease of standardisation and manufacture, and low cost of production, the use of involute teeth has become universal by entirely superseding the cycloidal shape. Only one cutter or tool is necessary to manufacture a complete set of interchangeable gears. The cutter is in the form of a rack as all gears will gear with their corresponding rack. Moreover, the cutters of this form can be made to a higher degree of accuracy as the teeth of an involute rack are straight.

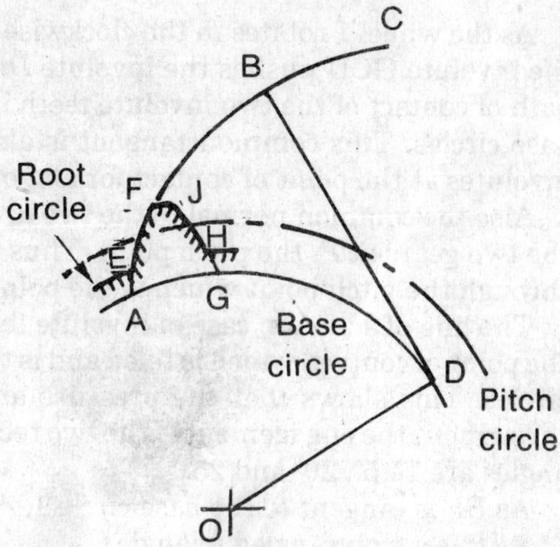

Fig. 10.21

Meshing of Teeth

In Fig. 10.22, two gear wheels 1 and 2 with centres of rotation at *A* and *B* respectively are in contact at *C*. *CE* and *CF* are the tangents to the two base circles 1 and 2 respectively. $t - t$ is the common tangent to the two involutes *DC* and *GCH* of the two meshing teeth. The involute *DC* is traced by rolling line *EF* on the base circle of gear 2 while the involute *GCH* is obtained by rolling line *EF* on the base circle of gear 1.

From the property of the involute, the tangent *CF* to the base circle of gear 2 is normal to the involute *DC* or the tangent $t - t$. Similarly, the tangent *CE* to the base circle of gear 1 is normal to the involute *GC* or the tangent $t - t$. As *CE* and *CF* both are normal to the common tangent $t-t$ at the point *C*; *CE* and *CF* lie on a straight line. *ECF* is thus a straight line.

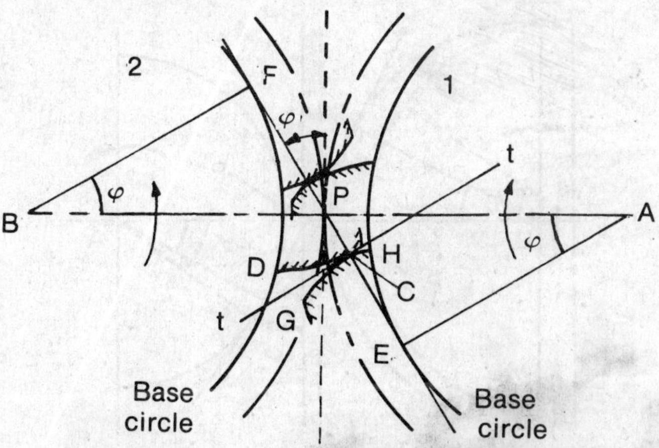

Fig. 10.22

As the wheel 1 rotates in the clockwise direction, the point of contact C on the involute GCH pushes the involute DC along the line CF. Therefore, the path of contact of the two involute teeth is along the common tangent to the base circles. This common tangent is also the common normal to the two involutes at the point of contact for all positions.

Also the common normal to the two involutes divides the line of centres of the two gears at P, the pitch point. Thus the common normal always passes through the pitch point which is the point of contact of two pitch circles.

The line of action in case of involute teeth is along the common normal at the point of contact, which is fixed and is the common tangent to the two base circles. This shows that the pressure angle in this case remains constant throughout the engagement of the two teeth. The usual values of the pressure angles are 14.5°, 20° and 25°.

As EF is tangent to the base circle 1, AE is perpendicular to EF.

AEP is a right-angled triangle.

Also $\angle EAP = \angle\varphi$

$AE = AP\cos\varphi$

Similarly, $BF = BP\cos\varphi$

i.e., [Base circle diameter = Pitch circle diameter × cos φ]

$$\text{Velocity ratio of gears} = \frac{BP}{AP} = \frac{BF}{AE} = \text{constant}$$

Thus for a pair of involute gears, velocity ratio is inversely proportional to the pitch circle diameters as well as base circle diameters.

Any shift in the centres of two gears changes the centre distance. If the involutes are still in contact, the common normal to the two involutes at the point of contact will be the new common tangent to the base circles and its intersection with the line of centres as the new pitch point (Fig. 10.23). It can be judged that the shifting of P does not alter the ratio AP/BP which means the velocity ratio between the two gears remains constant. Of course, in this way there is change in the pressure angle. Altering the centre distance

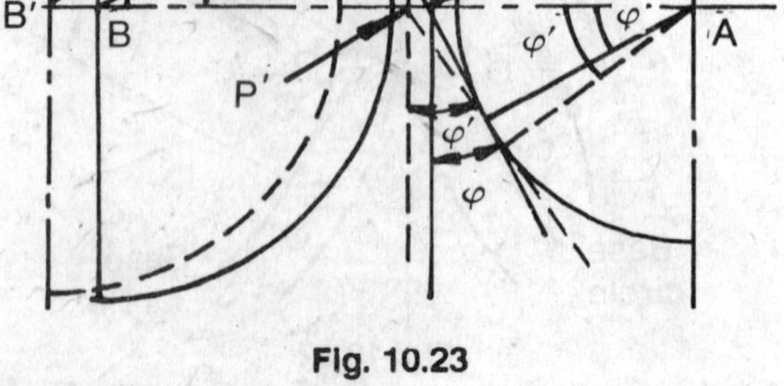

Fig. 10.23

without destroying the correct tooth action is an important property of the involute gears.

Remember in case of involute gears,

1. Points of contact lie on the line of action which is the common tangent to the two base circles.
2. The contact is made when the tip of a tooth of the driven wheel touches the flank of a tooth of the driving wheel and the contact is broken where the tip of the driving wheel touches the flank of the driven wheel.
3. If the direction of angular movement of the wheels is reversed, the points of contact will lie on the other common tangent to the base circles.
4. Initial contact occurs where the addendum circle of the driven wheel intersects the line of action. Final contact occurs at a point where the addendum circle of the driver intersects the line of action.

Example 10.3 The following data relate to two meshing gears:

$$\text{Velocity ratio} = \frac{1}{3}$$

$$\text{Module} = 4 \text{ mm}$$

$$\text{Pressure angle} = 20°$$

$$\text{Centre distance} = 200 \text{ mm}$$

Determine the number of teeth and the base circle radius of the gear wheel.

Solution:

$$\text{VR} = 1/3 \qquad \varphi = 20°$$

$$m = 4 \text{ mm} \qquad C = 200 \text{ mm}$$

(i)
$$\text{VR} = \frac{N_2}{N_1} = \frac{1}{3} = \frac{T_1}{T_2} \quad \text{or} \quad T_2 = 3\,T_1$$

and
$$C = \frac{d_1 + d_2}{2} = \frac{m\,(T_1 + T_2)}{2}$$

or
$$200 = \frac{4\,(T_1 + 3T_1)}{2} = 8\,T_1$$

or
$$T_1 = 25$$

$$T_2 = 25 \times 3 = 75$$

Number of teeth on gear wheel = 75

(ii)
$$d_2 = \text{m}\,T_2 = 4 \times 75 = 300 \text{ mm}$$

$$\text{Base circle radius, } d_{b2} = \frac{d_2}{2}\cos\varphi = \frac{300}{2} \times \cos 20°$$

$$= \underline{141 \text{ mm}}$$

10.9 PATH OF CONTACT

Let two gear wheels with centres A and B be in contact (Fig. 10.24).

The pinion 1 is the driver and is rotating clockwise. The wheel 2 is driven in the counter-clockwise direction. EF is their common tangent to the base circles.

Contact of the two teeth is made where the addendum circle of the wheel meets the line of action EF, i.e. at C and is broken where the addendum circle of the pinion meets the line of action, i.e. at D. CD is then the path of contact.

Let r = pitch circle radius of pinion
R = pitch circle radius of wheel
r_a = addendum circle radius of pinion
R_a = addendum circle radius of wheel.

Path of contact = path of approach + path of recess

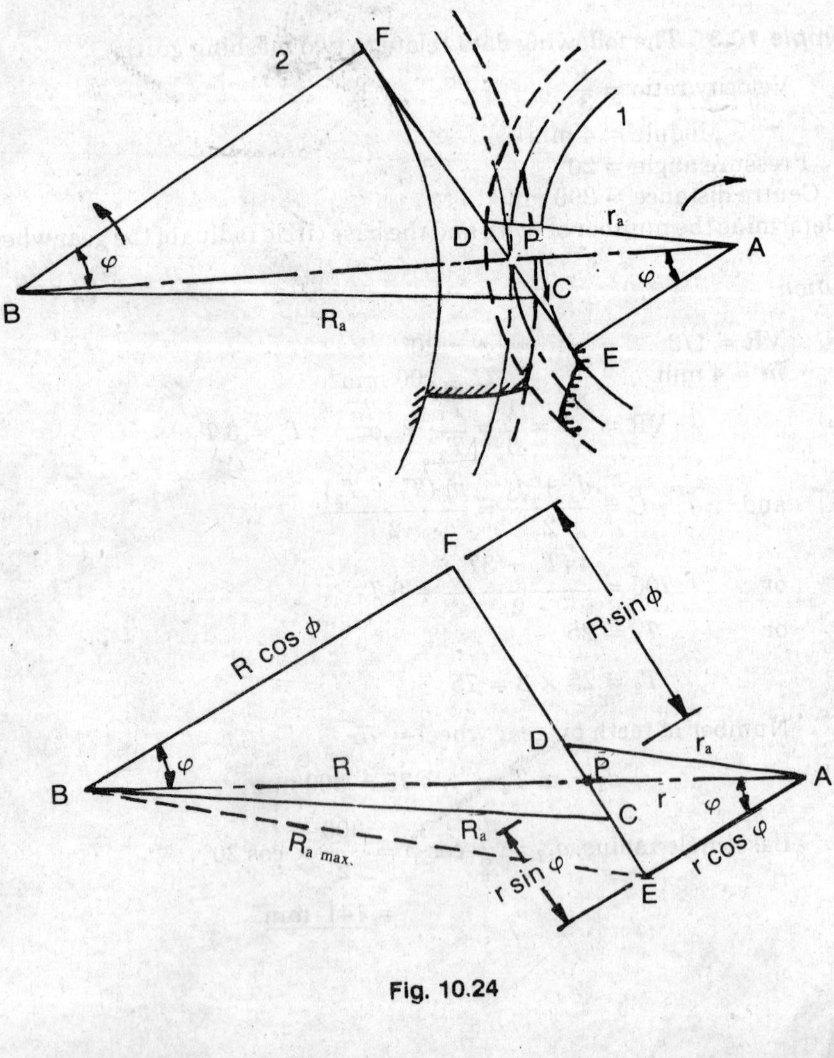

Fig. 10.24

$$CD = CP + PD$$

$$= (CF - PF) + (DE - PE)$$

$$= \left[\sqrt{R_a^2 - R^2\cos^2\varphi} - R\sin\varphi\right] + \left[\sqrt{r_a^2 - r^2\cos^2\varphi} - r\sin\varphi\right]$$

$$= \sqrt{R_a^2 - R^2\cos^2\varphi} + \sqrt{r_a^2 - r^2\cos^2\varphi} - (R + r)\sin\varphi \qquad (10.5)$$

Observe that the path of approach can be found if the dimensions of the driven wheel are known. Similarly, the path of recess is known from the dimensions of the driving wheel (pinion).

10.10 ARC OF CONTACT

Arc of contact is the distance travelled by a point on either pitch circle of the two wheels during the period of contact of a pair of teeth.

In Fig. 10.25, at the beginning of engagement, the driving involute is shown as GH; when the point of contact is at P, it is shown as JK and when at the end of engagement, it is DL. The arc of contact is $P'P''$ and it consists of the arc of approach $P'P$ and the arc of recess PP''.

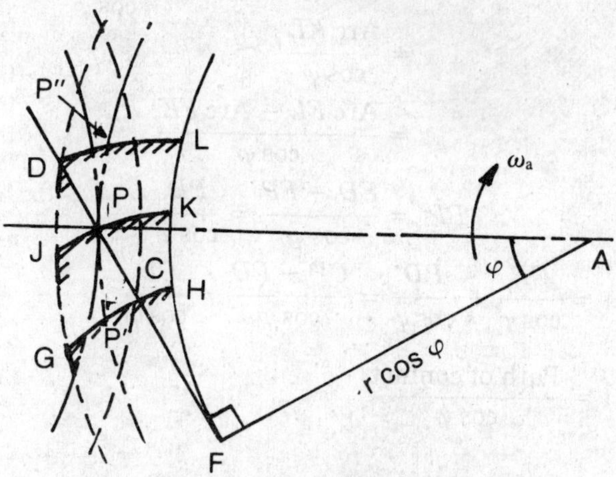

Fig. 10.25

Let the time to traverse the arc of approach is t_a, then

Arc of approach = $P'P$ = Tangential velocity of P' × Time of approach

$$= \omega_a r \times t_a \qquad (t_a = \text{time of approach})$$

$$= \omega_a (r\cos\varphi)\frac{1}{\cos\varphi}t_a$$

$$= (\text{Tang. vel. of } H)\ t_a \frac{1}{\cos\varphi} \qquad (AF = AH)$$

$$= \frac{\text{Arc } HK}{\cos\varphi}$$

$$= \frac{\text{Arc } FK - \text{Arc } FH}{\cos\varphi}$$

$$= \frac{FP - FC}{\cos\varphi}$$

$$= \frac{CP}{\cos\varphi}$$

Arc *FK* is equal to the path *FP* as the point *P* is on the generator *FP* that rolls on the base circle *FHK* to generate the involute *PK*. Similarly, arc *FH* = Path *FC*.

$$\text{Arc of recess} = PP'' = \text{Tang. vel. of P} \times \text{Time of recess}$$

$$= \omega_a r \times t_r \qquad (t_r = \text{time of recess})$$

$$= \omega_a (r\cos\varphi)\frac{1}{\cos\varphi}t_r$$

$$= (\text{Tang. vel. of } K)\ t_r \frac{1}{\cos\varphi}$$

$$= \frac{\text{Arc } KL}{\cos\varphi}$$

$$= \frac{\text{Arc } FL - \text{Arc } FK}{\cos\varphi}$$

or

$$PP'' = \frac{FD - FP}{\cos\varphi} = \frac{PD}{\cos\varphi}$$

$$\text{Arc of contact} = \frac{CP}{\cos\varphi} + \frac{PD}{\cos\varphi} = \frac{CP + PD}{\cos\varphi} = \frac{CD}{\cos\varphi}$$

$$\text{or Arc of contact} = \frac{\text{Path of contact}}{\cos\varphi} \tag{10.6}$$

10.11 NUMBER OF PAIRS OF TEETH IN CONTACT

The arc of contact is the length of the pitch circle traversed by a point on it during the mating of a pair of teeth.

Thus all the teeth lying in between the arc of contact $P'P''$ will be meshing with the teeth on the other wheel.

$$\text{Number of teeth within the arc } P'P'' = \frac{\text{Arc } P'P''}{\text{Circular pitch}}$$

$$n = \frac{CD}{\cos\varphi}\frac{1}{p} \tag{10.7}$$

For continuous transmission of motion, at least one tooth of one wheel must be in contact with another tooth of the second wheel. Therefore, n must be greater than unity.

If n lies between 1 and 2, the number of teeth in contact at any time will not be less than one and never more than two. If n is between 2 and 3, it is never less than two pairs of teeth and not more than three pairs and so on. If n is 1.6, one pair of teeth are always in contact whereas two pairs of teeth are in contact for 60% of the time.

Number of teeth in contact is also given by the contact ratio which is the ratio of the angle of action to the pitch angle.

Example 10.4 Two gears in mesh have a module of 8 mm and a pressure angle of 20°. The larger gear has 57 while the pinion has 23 teeth. If the addenda on pinion and gear wheel are equal to one module, find

(i) the number of pairs of teeth in contact
(ii) the angle of action of the pinion and the gear wheel
(iii) the ratio of the sliding to rolling velocity at
  (a) the beginning of contact
  (b) the pitch point
  (c) the end of contact.

Solution:

$\varphi = 20°$ $\quad\quad T = 57$

$m = 8$ mm $\quad\quad t = 23$

addendum $= 1\,m = 8$ mm

$$R = \frac{mT}{2} = \frac{8 \times 57}{2} = 228 \text{ mm}$$

$$R_a = R + m = 228 + 8 = 236 \text{ mm}$$

$$r = \frac{mt}{2} = \frac{8 \times 23}{2} = 92 \text{ mm}$$

$$r_a = r + m = 92 + 8 = 100 \text{ mm}$$

(i) $$n = \frac{\text{Arc of contact}}{\text{Circular pitch}}$$

$$= \left(\frac{\text{Path of contact}}{\cos\varphi}\right) \times \frac{1}{\pi m}$$

$$= \frac{\text{Path of approach + Path of recess}}{\cos\varphi \times \pi m}$$

$$= \frac{\left[\sqrt{R_a^2 - R^2\cos^2\varphi} - R\sin\varphi\right] + \left[\sqrt{r_a^2 - r^2\cos^2\varphi} - r\sin\varphi\right]}{\cos\varphi \times \pi m}$$

$$= \frac{[\sqrt{(236)^2 - (228)^2 \cos^2 20°} - 228 \sin 20°] + [\sqrt{(100)^2 - (92)^2 \cos^2 20°} - 92 \sin 20°]}{\cos 20° \times \pi \times 8}$$

$$= \frac{20.97 + 18.79}{\cos 20° \times \pi \times 8}$$

$$= 42.31 \times \frac{1}{\pi \times 8}$$

$$= \underline{1.68}$$

(ii) Angle of action = Angle traversed by the arc of contact

$$\therefore \delta_p = \text{Arc of contact} \times \frac{360}{2\,\pi\, r}$$

$$= 42.31 \times \frac{360}{2\,\pi \times 92}$$

$$= 26.3° \quad \text{or} \quad 26°\, 18'$$

$$\delta_g = 42.31 \times \frac{360}{2\,\pi \times 228}$$

$$= 10.63° \quad \text{or} \quad \underline{10°\, 38'}$$

(iii)(a) $$\frac{\text{Sliding velocity}}{\text{Rolling velocity}} = \frac{(\omega_p + \omega_g) \times \text{Path of approach}}{\text{Pitch line velocity } (= \omega_p \times r)}$$

$$= \frac{\left(\omega_p + \frac{23}{57}\,\omega_p\right) \times 20.97}{\omega_p \times 92}$$

$$= \underline{0.32}$$

(b) $$\frac{\text{Sliding velocity}}{\text{Rolling velocity}} = \frac{(\omega_p + \omega_g) \times 0}{\text{Pitch line velocity}} = 0$$

(c) $$\frac{\text{Sliding velocity}}{\text{Rolling velocity}} = \frac{\left(\omega_p + \frac{23}{57}\,\omega_p\right) \times \text{Path of recess}}{\omega_p \times r}$$

$$= \frac{\left(1 + \frac{23}{57}\right) \times 18.79}{92}$$

$$= \underline{0.287}$$

Example 10.5 Two 20° gears have a module pitch of 4 mm. The number of teeth on gear 1 is 40 and on gear 2 is 24. If the gear 2 rotates at 600 rpm, determine the velocity of sliding when the contact is at the tip of the tooth of gear 2. Take addendum equal to one module.

Also find the maximum velocity of sliding.

Solution: Let 1 be the gear wheel and 2 the pinion.

$\varphi = 20°$ $\qquad T = 40$

$N_p = 600$ rpm $\qquad t = 24$

$m = 4$ mm $\qquad$ Addendum = 1 module = 4 mm

$$R = \frac{mT}{2} = \frac{4 \times 40}{2} = 80 \text{ mm}$$

$$R_a = 80 + 4 = 84 \text{ mm}$$

$$r = \frac{mt}{2} = \frac{4 \times 24}{2} = 48 \text{ mm}$$

$$r_a = 48 + 4 = 52 \text{ mm}$$

$$N_g = N_p \times \frac{t}{T} = 600 \times \frac{24}{40} = 360 \text{ rpm}$$

(i) Let pinion (gear 2) be the driver.
The tip of the driving wheel is in contact with a tooth of the driven wheel at the end of engagement. Thus, it is required to find the path of recess which is obtained from the dimensions of the driving wheel.

$$\text{Path of recess} = \sqrt{r_a^2 - (r\cos\varphi)^2} - r\sin\varphi$$

$$= \sqrt{(52)^2 - (48\cos 20°)^2} - 48\sin 20°$$

$$= 9.458 \text{ mm}$$

$$\text{Velocity of sliding} = (\omega_p + \omega_g) \times \text{Path of recess}$$

$$= 2\pi\left(N_p + N_g\right) \times 9.458$$

$$= 2\pi\,(600 + 360) \times 9.458$$

$$= 57\,049 \text{ mm/min}$$

$$= \underline{950.8 \text{ mm/s}}$$

(ii) In case the gear wheel is the driver, the tip of the pinion will be in contact with the flank of a tooth of the gear wheel at the beginning of contact. Thus, it is required to find the distance of the point of contact from the pitch point, i.e. path of approach.
The path of approach is found from the dimensions of the driven wheel which is again pinion.

$$\text{Thus path of approach} = \sqrt{r_a^2 - (r\cos\varphi)^2} - r\sin\varphi$$

$$= 9.458 \text{ mm, as before}$$

and velocity of sliding = 950.8 mm/s
Thus, it is immaterial whether the driver is the gear wheel or the pinion, the velocity of sliding is the same when the contact is at the tip of the pinion.
Maximum velocity of sliding will depend upon the larger path considering any of the wheels to be the driver.
Consider pinion to be the driver.

Path of recess = 9.458 mm

$$\text{Path of approach} = \sqrt{R_a^2 - (R\cos\varphi)^2} - R\sin\varphi$$

$$= \sqrt{(84)^2 - (80\cos 20°)^2} - 80\sin 20°$$

$$= 10.117\ \text{mm}$$

This is also the path of recess if the wheel becomes the driver

$$\text{Maximum velocity of sliding} = (\omega_p + \omega_g) \times \text{Maximum path}$$

$$= 2\pi\,(600 + 360) \times 10.117$$

$$= 61\,024\ \text{mm/min}$$

$$= \underline{1017.1\ \text{mm/s}}$$

10.12 INTERFERENCE IN INVOLUTE GEARS

Power transmission through a pair of teeth is along the line of action or the common normal to the two involutes at the point of contact. The common normal is also a common tangent to the two base circles and passes through the pitch point. At any instant, the portions of the tooth profiles which are in contact must be involute so that the line of action does not deviate. If any of the two surfaces is not involute, the two surfaces would not touch each other tangentially and the transmission of power would not be proper. Mating of two non-conjugate (non-involute) teeth is known as interference because the two teeth do not slide properly and thus rough action and binding occurs. Owing to non-involute profile, the contacting teeth have different velocities which can lock the two gears.

Figure 10.26 shows two gears in mesh. If pinion is the driver, the line of action will be along EF which is the common tangent to base circles of the two gears. Let the addendum radius of the wheel be BC and that of pinion AD. The teeth on the pinion and wheel are engaged at C and are disengaged at D. Now, if the radius of the addendum circle of the pinion is increased, the point D shifts along PF towards F and the point D coincides with F when the radius is equal to AF. Any further increase in the value of this radius, will result in shifting the point of contact inside the base circle of the wheel. Since an involute can exist

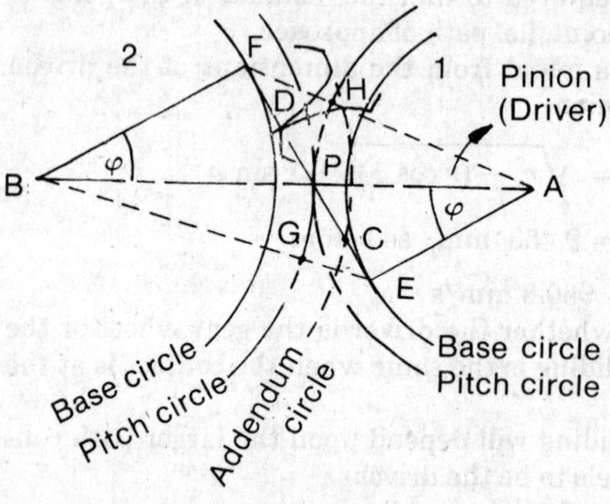

Fig. 10.26

only outside the base circle, therefore, any profile of teeth inside the base circle will be of non-involute type. Usually, a radial profile is adopted for this portion. Thus the profiles in such a case cannot be tangent to each other and the tip of the pinion would try to dig out the flank of the tooth of the wheel. Therefore, interference occurs in the mating of two gears.

Similarly, if the addendum radius of the wheel is made greater than *BE*, the tip of the wheel tooth will be in contact with a portion of the non-involute profile of the pinion tooth for some of the engagement. The conclusion is that to have no interference of the teeth, the addendum circles of the wheel and the pinion must intersect the line of action between *E* and *F*. The points *E* and *F* are called interference points.

Note that to avoid interference, the limiting value of addendum of the wheel is *GE* whereas that of pinion it is *HF* and the latter is clearly greater than the former. Thus if the addenda of the wheel and the pinion are to be equal, the addendum circle of the wheel passes through the limiting point *E* on the line of action before the addendum circle of the pinion passes through the limiting point *F* on the same line. Thus, for equal addenda of the wheel and the pinion, the addendum radius of the wheel decides whether the interference will occur or not.

10.13 MINIMUM NUMBER OF TEETH

As explained in the previous section, the maximum value of the addendum radius of the wheel to avoid interference can be up to *BE*. Referring Fig. 10.24,

$$
\begin{aligned}
(BE)^2 &= (BF)^2 + (FE)^2 \\
&= (BF)^2 + (FP + PE)^2 \\
&= (R\cos\varphi)^2 + (R\sin\varphi + r\sin\varphi)^2 \\
&= R^2\cos^2\varphi + R^2\sin^2\varphi + r^2\sin^2\varphi + 2\,rR\sin^2\varphi \\
&= R^2(\cos^2\varphi + \sin^2\varphi) + \sin^2\varphi(r^2 + 2rR) \\
&= R^2 + (r^2 + 2rR)\sin^2\varphi \\
&= R^2\left[1 + \frac{1}{R^2}(r^2 + 2rR)\sin^2\varphi\right] \\
&= R^2\left[1 + \left(\frac{r^2}{R^2} + \frac{2r}{R}\right)\sin^2\varphi\right] \\
BE &= R\sqrt{1 + \frac{r}{R}\left(\frac{r}{R} + 2\right)\sin^2\varphi}
\end{aligned}
$$

Therefore, maximum value of the addendum of the wheel can be equal to (*BE* – Pitch circle radius)

or

$$a_{w\,\max} = R\sqrt{1+\frac{r}{R}\left(\frac{r}{R}+2\right)\sin^2\varphi} - R$$

$$= R\left[\sqrt{1+\frac{r}{R}\left(\frac{r}{R}+2\right)\sin^2\varphi} - 1\right]$$

Let the adopted value of the addendum in some case be a_w times the module of teeth. Then this adopted value of addendum must be less than the maximum value of addendum to avoid interference.

i.e.,

$$R\left[\sqrt{1+\frac{r}{R}\left(\frac{r}{R}+2\right)\sin^2\varphi} - 1\right] \geq a_w m$$

Let t = number of teeth on the pinion
T = number of teeth on the wheel

$$m = \frac{D}{T} = \frac{d}{t} \quad \text{or} \quad m = \frac{2R}{T} = \frac{2r}{t}$$

or

$$R = \frac{mT}{2} \quad \text{and} \quad r = \frac{mt}{2}$$

Hence,

$$\frac{mT}{2}\left[\sqrt{1+\frac{t}{T}\left(\frac{t}{T}+2\right)\sin^2\varphi} - 1\right] \geq a_w m$$

or

$$\frac{T}{2}\left[\sqrt{1+\frac{1}{G}\left(\frac{1}{G}+2\right)\sin^2\varphi} - 1\right] \geq a_w$$

where $G = \dfrac{T}{t}$ = Gear ratio

or

$$T \geq \frac{2a_w}{\sqrt{1+\frac{1}{G}\left(\frac{1}{G}+2\right)\sin^2\varphi} - 1}$$

In the limit,

$$T = \frac{2a_w}{\sqrt{1+\frac{1}{G}\left(\frac{1}{G}+2\right)\sin^2\varphi} - 1} \tag{10.8}$$

This gives the minimum number of teeth on the wheel for the given values of the gear ratio, the pressure angle and the *addendum coefficient* a_w.

The minimum number of teeth on the pinion are given by,

$$t = \frac{T}{G}$$

For $a_w = 1$ i.e. when the addendum is equal to module,

$$T \geq \frac{2}{\sqrt{1 + \frac{1}{G}\left(\frac{1}{G} + 2\right) \sin^2\varphi} - 1}$$

For equal number of teeth on the pinion and the wheel, $G = 1$ and

$$T_{min} = \frac{2}{\sqrt{1 + 3 \sin^2\varphi} - 1}$$

For a pressure angle of 20°, i.e. $\varphi = 20°$

$$T_{min} = \frac{2}{\sqrt{1 + 3 \sin^2 20°} - 1} = 12.31 \text{ or } 13$$

Thus for two wheels of equal size with 20° pressure angle and addendum equal to one module, the minimum number of teeth on each wheel must be 13 to avoid interference.

10.14 INTERFERENCE BETWEEN RACK AND PINION

Figure 10.27 shows a rack and pinion in which pinion is rotating in the clockwise direction and driving the rack. P is the pitch point and PE is the line of action. Engagement of the rack tooth with the pinion tooth occurs at C. To avoid interference, the maximum addendum of the rack can be increased in such a way that C coincides with E. Thus addendum of the rack must be less than GE.

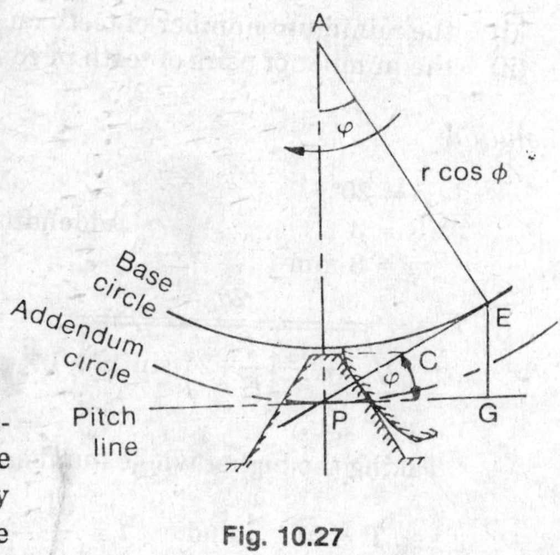

Fig. 10.27

Let the adopted value of the addendum of the rack be $a_r m$ where a_r is the *addendum coefficient* by which the standard value of the addendum has been multiplied.

$$GE = PE \sin\varphi$$

$$= (r \sin\varphi) \sin\varphi$$

$$= r \sin^2\varphi$$

$$= \frac{mt}{2} \sin^2\varphi$$

To avoid interference,

$$GE \geq a_r\, m$$

or

$$\frac{mt}{2}\sin^2\varphi \geq a_r\, m$$

or

$$t \geq \frac{2a_r}{\sin^2\varphi}$$

When $a_r = 1$, i.e. for standard addendum,

$$t_{\min} = \frac{2}{\sin^2\varphi}$$

For 20° pressure angle, $\varphi = 20°$

$$t_{\min} = 17.1 \quad \text{or} \quad 18$$

Thus the number of minimum teeth on the pinion to avoid interference is 18.

Example 10.6 Two 20° involute spur gears mesh externally and give a velocity ratio of 3. Module is 3 mm and the addendum is equal to 1.1 module. If the pinion rotates at 120 rpm, determine

(i) the minimum number of teeth on each wheel to avoid interference
(ii) the number of pairs of teeth in contact.

Solution:

$\varphi = 20°$ $\qquad$ $N_p = 120$ rpm
VR = 3 $\qquad$ Addendum = 1.1 m
$m = 3$ mm $\qquad$ $a_w = 1.1$

(i) $$T = \frac{2a_w}{\sqrt{1+\frac{1}{G}\left(\frac{1}{G}+2\right)\sin^2\varphi} - 1} = \frac{2 \times 1.1}{\sqrt{1+\frac{1}{3}\left(\frac{1}{3}+2\right)\sin^2 20°} - 1}$$

$$= 49.44$$

Taking the higher whole number divisible by the velocity ratio,

i.e., $T = 51$ and $t = \frac{51}{3} = \underline{17}$

(ii) Number of pairs of teeth in contact,

$$n = \frac{\text{Arc of contact}}{\text{Circular pitch}} = \left(\frac{\text{Path of contact}}{\cos\varphi}\right) \times \frac{1}{\pi m}$$

or

$$n = \frac{\sqrt{R_a^2 - R^2\cos^2\varphi} - R\sin\varphi + \sqrt{r_a^2 - r^2\cos^2\varphi} - r\sin\varphi}{\cos\varphi \times \varphi\, m}$$

We have, $R = \frac{mT}{2} = \frac{3 \times 51}{2} = 76.5$ mm

$$R_a = R + 1.1\,m = 76.5 + 1.1 \times 3 = 79.8 \text{ mm}$$

$$r = \frac{mt}{2} = \frac{3 \times 17}{2} = 25.5 \text{ mm}$$

$$r_a = 25.5 + 1.1 \times 3 = 28.8 \text{ mm}$$

$$n = \frac{\sqrt{(79.8)^2 - (76.5 \cos 20°)^2} - 76.5 \sin 20° + \sqrt{(28.8)^2 - (25.5 \cos 20°)^2} - 25.5 \sin 20°}{\cos 20° \times \pi \times 3}$$

$$= \frac{34.646 - 26.165 + 15.977 - 8.720}{\cos 20° \times \pi \times 3}$$

$$= 1.78$$

Thus 1 pair of teeth will always remain in contact whereas for 78% of time 2 pairs of teeth will be in contact.

Example 10.7 Two 20° involute spur gears have a module of 10 mm. The addendum is equal to one module. The larger gear has 40 teeth while the pinion has 20 teeth.

Will the gear interfere with the pinion?

Solution:

$\varphi = 20°$ $\qquad T = 40$

$m = 10$ mm $\qquad t = 20$

Addendum $= 1\,m = 10$ mm

$$R = \frac{mT}{2} = \frac{10 \times 40}{2} = 200 \text{ mm}$$

$$R_a = 200 + 10 = 210 \text{ mm}$$

$$r = \frac{mt}{2} = \frac{10 \times 20}{2} = 100 \text{ mm}$$

$$r_a = 100 + 10 = 110 \text{ mm}$$

Let pinion be the driver (Refer Fig. 10.24),

Path of approach, $PC = \sqrt{R_c^2 - (R \cos \varphi)^2} - R \sin$

$$= \sqrt{(210)^2 - (200 \times \cos 20°)^2} - 200 \sin 20°$$

$$= 25.3 \text{ mm}$$

To avoid interference, maximum length of the path of approach can be PE.

$$PE = r \sin \varphi = 100 \sin 20° = 34.2 \text{ mm}$$

Since the actual path of approach is within the maximum limit, no interference occurs.

Alternative Method Addendum radius of wheel = 210 mm

To avoid interference, maximum addendum radius $R_{a\,max}$ can be equal to BE.

$$\text{i.e., } R_{a\,max} = BE = \sqrt{(BF)^2 + (FP + PE)^2}$$

$$= \sqrt{(R \cos \varphi)^2 + (R \sin \varphi + r \sin \varphi)^2}$$

$$= \sqrt{(200 \cos 20°)^2 + (200 \sin 20° + 100 \sin 20°)^2}$$

$$= 214.1 \text{ mm}$$

As the actual addendum radius of the wheel is lesser than the maximum permissible value of the addendum radius, no interference occurs.

Example 10.8 Two 20° involute spur gears have a module of 10 mm. The addendum is one module. The larger gear has 50 teeth and the pinion 13 teeth. Does the interference occur? If it occurs, to what value should the pressure angle be changed to eliminate interference?

Solution:

$\varphi = 20°$ $\quad\quad T = 50$

$m = 10$ mm $\quad\quad t = 13$

Addendum = $1\,m$ = 10 mm

$$R = \frac{mT}{2} = \frac{10 \times 50}{2} = 250 \text{ mm}$$

$$R_a = 250 + 10 = 260 \text{ mm}$$

$$r = \frac{mt}{2} = \frac{10 \times 13}{2} = 65 \text{ mm}$$

Refer Fig. 10.24,

$$R_{a\,max} = \sqrt{(R \cos \varphi)^2 + (R \sin \varphi + r \sin \varphi)^2}$$

$$= \sqrt{(250 \cos 20°)^2 + (250 \sin 20° + 65 \sin 20°)^2}$$

$$= \sqrt{(250 \cos 20°)^2 + (315 \sin 20°)^2}$$

$$= 258.45 \text{ mm}$$

The actual addendum radius R_a is more than the maximum value $R_{a\,max}$, therefore, interference occurs. The new value of φ can be found by taking $R_{a\,max}$ equal to R_a.

$$\text{i.e.,}\quad 260 = \sqrt{(250\cos\varphi)^2 + (315\sin\varphi)^2}$$

$$\text{or}\quad (260)^2 = (250)^2\cos^2\varphi + (315)^2(1 - \cos^2\varphi)$$

$$= (250)^2\cos^2\varphi + (315)^2 - (315)^2\cos^2\varphi$$

$$\text{or}\quad \cos^2\varphi = \frac{(315)^2 - (260)^2}{(315)^2 - (250)^2} = 0.861$$

$$\cos\varphi = 0.928$$

$$\text{or}\quad \varphi = 21.88^\circ \quad \text{or} \quad 21^\circ 52'$$

Thus if the pressure angle is increased to 21°52′, the interference is avoided.

Example 10.9 Following data relate to two meshing involute gears:

Number of teeth on the gear wheel = 60
Pressure angle = 20°
Gear ratio = 1.5
Speed of the gear wheel = 100 rpm
Module = 8 mm

The addendum on each wheel is such that the path of approach and the path of recess on each side are 40% of the maximum possible length each. Determine the addendum for the pinion and the gear and the length of the arc of contact.

Solution:

$$R = \frac{mT}{2} = \frac{8 \times 60}{2} = 240\text{ mm}$$

$$r = \frac{mt}{2} = \frac{8 \times (60/1.5)}{2} = 160\text{ mm}$$

Refer Fig. 10.24 and let pinion be the driver.
Maximum possible length of path of approach = $r \sin\varphi$
Actual length of path of approach = $0.4 \times r \sin\varphi$
Similarly, actual length of path of recess = $0.4\,R\sin\varphi$
Thus, we have

$$0.4\,r\sin\varphi = \sqrt{R_a^2 - (R\cos\varphi)^2} - R\sin\varphi$$

$$0.4 \times 160\sin 20^\circ = \sqrt{R_a^2 - (240\cos 20^\circ)^2} - 240\sin 20^\circ$$

$$R_a^2 - 50\,862 = 10\,809.8$$

$$R_a^2 = 61\,671.8$$

$$R_a = 248.3 \text{ mm}$$

Addendum of the wheel = 248.3 – 240 = <u>8.3 mm</u>

Also,
$$0.4\,R \sin\varphi = \sqrt{r_a^2 - (r\cos\varphi)^2} - r\sin\varphi$$

$$0.4 \times 240 \sin 20° = \sqrt{r_a^2 - (160 \cos 20°)^2} - 160 \sin 20°$$

$$r_a^2 - 22\,605 = 7666$$

$$r_a^2 = 30\,271$$

$$r_a = 174 \text{ mm}$$

Addendum of the pinion = 174 – 160 = <u>14 mm</u>

$$\text{Arc of contact} = \frac{\text{Path of contact}}{\cos\varphi}$$

$$= 0.4\left(\frac{r\sin\varphi + R\sin\varphi}{\cos\varphi}\right)$$

$$= 0.4 \times (240 + 160)\frac{\sin 20°}{\cos 20°}$$

$$= \underline{58.2 \text{ mm}}$$

10.15 UNDERCUTTING

Figure 10.28 a shows a pinion. A portion of its dedendum falls inside the base circle. The profile of the tooth inside the base circle is radial. If the addendum of the mating gear is more than the limiting value, it interferes with the dedendum of the pinion and the two gears are locked.

Now, if instead of the gear mating with the pinion, a cutting rack having similar teeth is used to cut the teeth in the pinion, the portion of the pinion

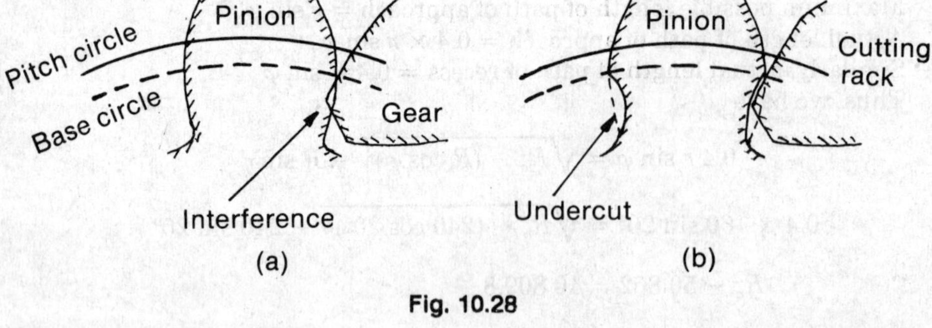

Fig. 10.28

tooth, which would have interfered with the gear, will be removed as shown in Fig. 10.28 b. A gear having its material removed in this manner is said to be undercut and the process, undercutting. In pinion with small number of teeth, this can seriously weaken the tooth. However, when the actual gear meshes with the undercut pinion, no interference occurs.

Undercutting will not take place if the teeth are designed to avoid interference.

10.16 COMPARISON OF CYCLOIDAL AND INVOLUTE TOOTH FORMS

The following table compares the two forms of teeth, the cycloidal and the involute:

	Cycloidal Teeth	*Involute Teeth*
(a)	Pressure angle varies from maximum at the beginning of engagement, reduces to zero at the pitch point and again increases to maximum at the end of engagement resulting in less smooth running of the gears.	Pressure angle is constant throughout the engagement of teeth. This results in smooth running of the gears.
(b)	It involves double curve for the teeth, epicycloid and hypocycloid. This complicates the manufacture.	It involves single curve for the teeth resulting in simplicity of manufacturing and of tools.
(c)	Owing to difficulty to manufacture, these are costlier.	These are simple to manufacture and thus are cheaper.
(d)	Exact centre-distance is required to transmit a constant velocity ratio.	A little variation in the centre distance does not affect the velocity ratio.
(e)	Phenomenon of interference does not occur at all.	Interference can occur if the condition of minimum number of teeth on a gear is not followed.
(f)	The teeth have spreading flanks and thus are stronger.	The teeth have radial flanks and thus are weaker as compared to cycloidal form for the same pitch.
(g)	In this, a convex flank always has contact with a concave face resulting in less wear.	Two convex surfaces are in contact and thus there is more wear.

On careful examination of the above, it can be deduced that the advantages of involute system are more real. Therefore, the use of the involute teeth have become almost universal rendering the cycloidal system obsolete.

10.17 HELICAL AND SPIRAL GEARS

In helical and spiral gears, the teeth are inclined to the axis of a gear. They can be right-handed or left-handed, depending upon the direction in which

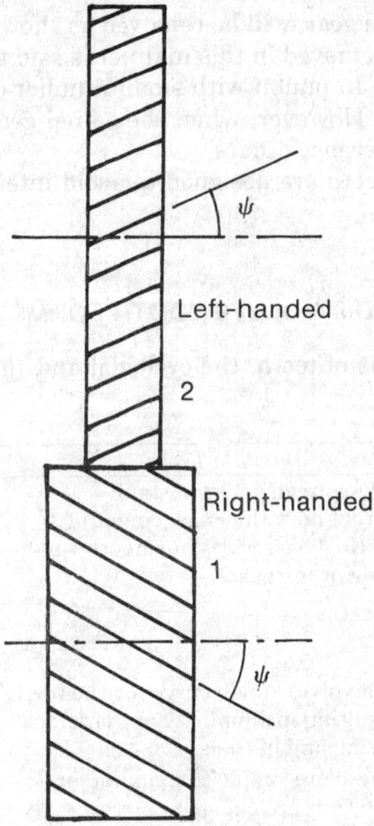

Fig. 10.29

the helix slopes away from the viewer when a gear is viewed parallel to the axis of the gear. In Fig. 10.29, gear 1 is a right-handed helical gear whereas 2 is a left-handed. The two mating gears have parallel axes and equal helix angle ψ. The contact between two teeth on the two gears is first made at one end which extends through the width of the wheel with the rotation of the gears.

Figure 10.30 a shows the same two gears when looking from above. Now, if the helix angle of gear 2 is reduced by a few degrees so that the helix angle of gear 1 is ψ_1 and that of gear 2 it is ψ_2 and it is desired that the teeth of the two gears still mesh with one another tangentially, it is esential to rotate the axis of gear 2 through some angle as shown in Fig. 10.30 b. Let the angle turned by it be θ which is the angle between the axes of the two gears.

From the geometry of the diagram, $\theta = \psi_1 - \psi_2$.

Thus, this is the case of two skew shafts joined with crossed-helical or spiral gears.

When $\psi_1 = \psi_2$, the helix angle is the same as before. Then $\theta = \psi_1 - \psi_2 = 0$, a case of helical gears joining parallel shafts (Fig. 10.30 a).

When $\psi_2 = 0$, i.e. the helix angle of gear 2 is made zero or the gear 2 is a straight spur gear, then $\theta = \psi_1$, i.e. the angle between the axes becomes equal to the helix angle of the gear 1 (Fig. 10.30 c).

In case the helix angle of gear 2 is made negative, i.e. the teeth are made of the same hand as that of gear 1 (Fig. 10.30 d),

$$\theta = \psi_1 - (-\psi_2) = \psi_1 + \psi_2$$

The above discussion leads to the following conclusion:

$$\text{Angle between the shafts, } \theta = \psi_1 + \psi_2 \text{ for gears of same hand} \quad (10.9)$$

$$= \psi_1 - \psi_2 \text{ for gears of opposite hands} \quad (10.10)$$

In case of helical gears for parallel shafts there is a line contact of the teeth through the width. However, gears for skew shafts have a point contact. This can be demonstrated by having two cylinders. When their axes are parallel,

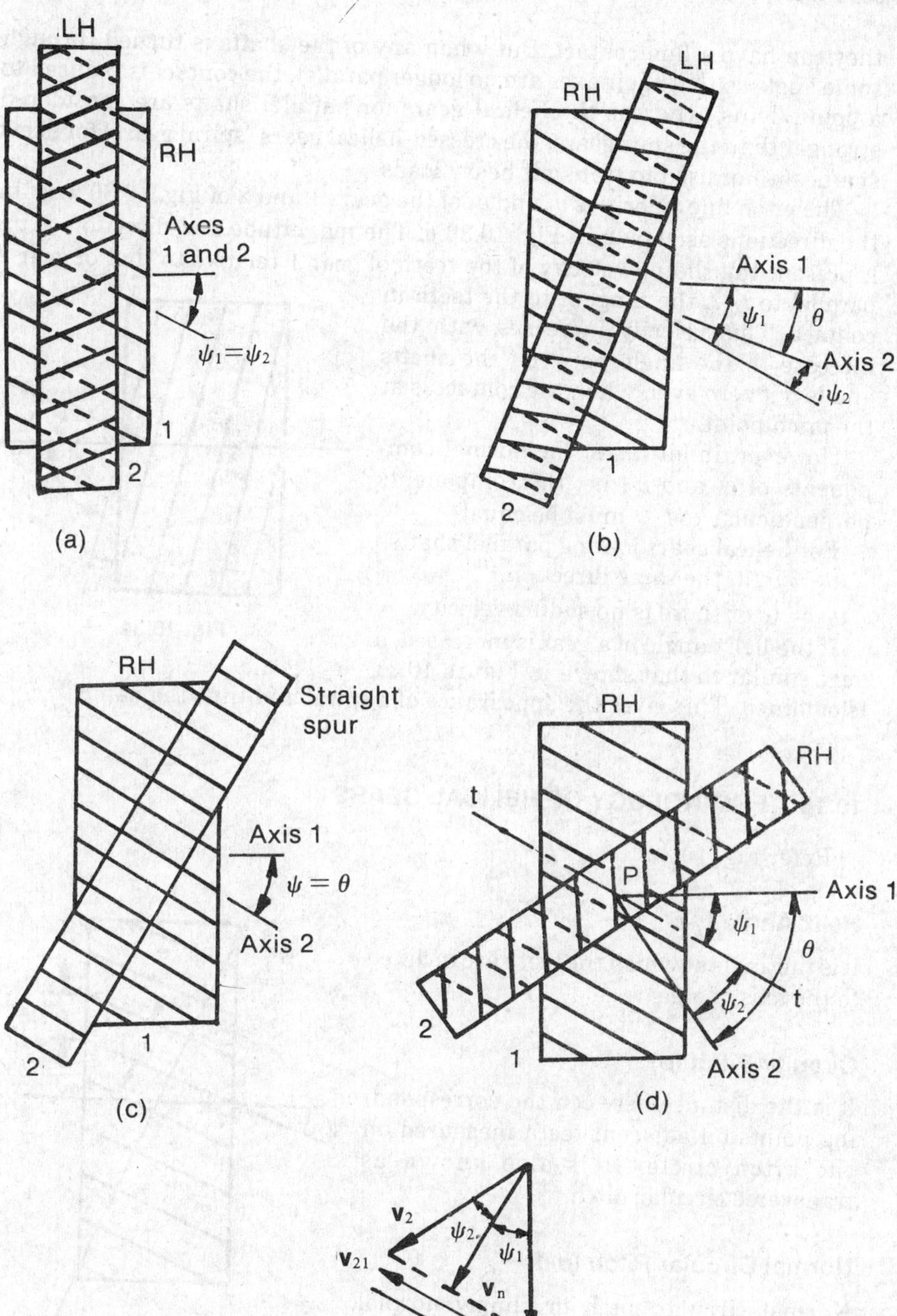
LH
RH
Axes
1 and 2
ψ1=ψ2
1
2
(a)
LH
RH
Axis 1
ψ1
θ
Axis 2
ψ2
1
2
(b)
RH
Straight
spur
Axis 1
ψ = θ
Axis 2
1
2
(c)
RH
RH
t
P
Axis 1
ψ1
θ
t
ψ2
2
1
Axis 2
(d)
v2
ψ2
ψ1
v21
vn
v1
v12
(e)

Fig. 10.30

they can have a line contact. But when any of the shafts is turned through some angle so that their axes are no longer parallel, the contact is reduced to a point. Thus, whereas the helical gears for parallel shafts are considered stronger than the spur gears, the crossed-helical gears (spiral gears) for skew shafts are not used to transmit heavy loads.

The pitch line velocities v_1 and v_2 of the gears 1 and 2 of Fig. 10.30 d act in the directions as shown in Fig. 10.30 e. The magnitude and direction of v_{12} represent the sliding velocity of the teeth of gear 1 relative to that of gear 2 parallel to $t-t$, the tangent to the teeth in contact. This velocity increases with the increase in the angle between the shafts and is not zero even when the contact is at the pitch point.

However, in all cases, the normal components of v_1 and v_2, i.e. the components perpendicular to $t-t$ must be equal.

For helical gears joining parallel shafts,

$v_1 = v_2$ in the same direction.

$v_{12} = 0$ or there is no sliding velocity.

If the helix angle of a gear is increased, a gear similar to that shown in Figure 10.31 is obtained. This gives the appearance of a spiral and thus the name.

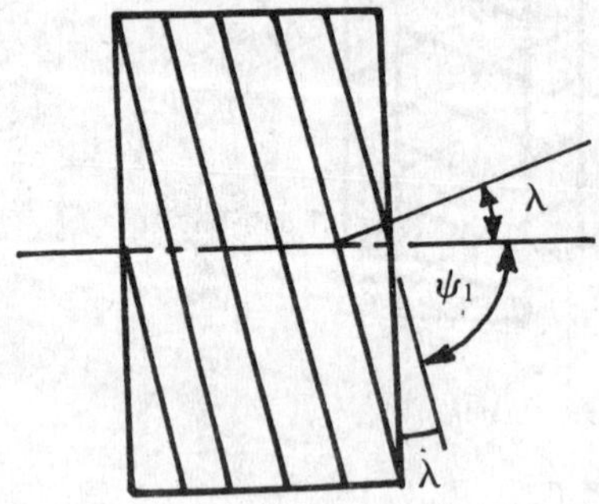

Fig. 10.31

10.18 TERMINOLOGY OF HELICAL GEARS

Refer Fig. 10.32.

Helix Angle (ψ)

It is the angle at which the teeth are inclined to the axis of a gear.

Circular Pitch (p)

It is the distance between the corresponding points on adjacent teeth measured on the pitch circle. It is also known as transverse circular pitch.

Normal Circular Pitch (p_n)

Normal circular pitch or simply normal pitch is the shortest distance measured along the normal to the helix between corresponding points on the adjacent teeth. The normal circular pitch of two mating gears must be same.

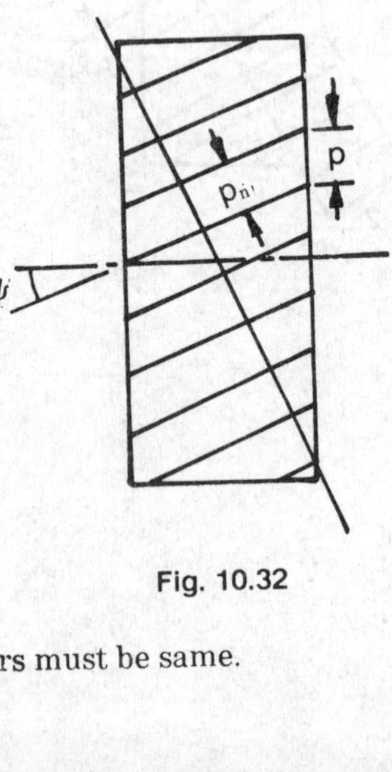

Fig. 10.32

$P_n = p \cos \psi$

Also we have, $p = \pi m$ as for spur gears

$$p_n = \pi m_n$$

and $m_n = m \cos \psi$

10.19 VELOCITY RATIO AND CENTRE DISTANCE OF HELICAL GEARS

Velocity Ratio Refer Fig. 10.30 e,

$$v_n = v_1 \cos \psi_1 = v_2 \cos \psi_2$$

or

$$\frac{v_2}{v_1} = \frac{\cos \psi_1}{\cos \psi_2}$$

$$\text{VR} = \frac{\omega_2}{\omega_1} = \frac{v_2/r_2}{v_1/r_1} = \frac{v_2/d_2}{v_1/d_1} = \frac{d_1}{d_2}\frac{v_2}{v_1}$$

or

$$\text{VR} = \frac{d_1}{d_2}\frac{\cos \psi_1}{\cos \psi_2}$$

$$= \frac{m_1 T_1}{m_2 T_2}\frac{\cos \psi_1}{\cos \psi_2}$$

$$= \frac{m_n/\cos \psi_1}{m_n/\cos \psi_2}\frac{T_1}{T_2}\frac{\cos \psi_1}{\cos \psi_2}$$

$$= \frac{T_1}{T_2} \qquad (10.11)$$

Centre Distance

Let C be the centre distance between two skew shaft axes which is the shortest distance between them.

$$C = r_1 + r_2 = \frac{1}{2}(d_1 + d_2) = \frac{1}{2}(m_1 T_1 + m_2 T_2)$$

or

$$C = \frac{1}{2}\left(\frac{m_n}{\cos \psi_1} T_1 + \frac{m_n}{\cos \psi_2} T_2\right)$$

$$= \frac{m_n}{2}\left(\frac{T_1}{\cos \psi_1} + \frac{T_2}{\cos \psi_2}\right)$$

In case of helical gears for parallel shafts, $\psi_1 = \psi_2 = \psi$

$$C = \frac{m_n}{2 \cos \psi}(T_1 + T_2) \qquad (10.12)$$

10.20 HELICAL GEAR FORCES AND EFFICIENCY

Like all direct contact mechanisms, the force exerted by a helical gear on its mating gear acts normal to the contacting surfaces if friction is neglected. However, a normal force in case of helical gears has three components. Apart from tangential and radial components which are present in spur gears, a third component parallel to the axis of the shaft of the gear also exists. This is known as the axial or the thrust force component. Figure 10.33 shows the normal force and its components acting on a helical gear. The gear shown is the driven gear and the forces are exerted on it by the driving gear.

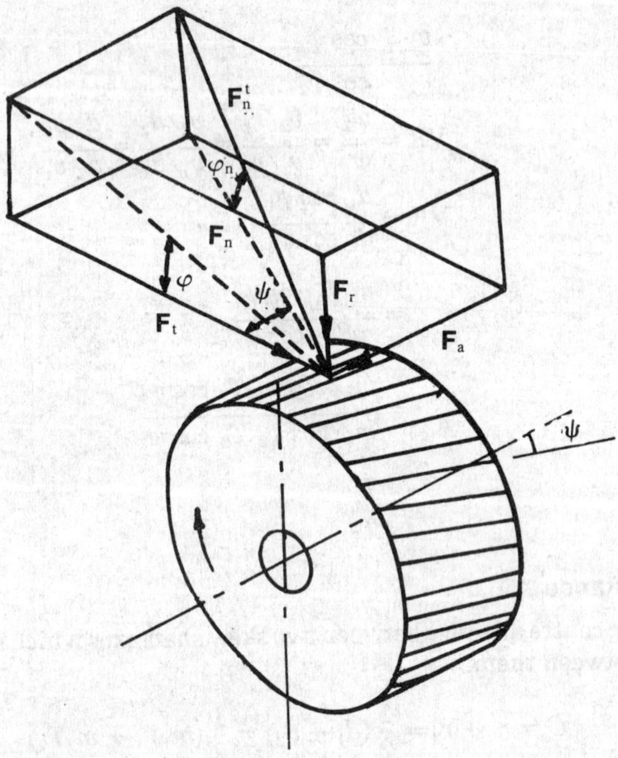

Fig. 10.33

Let F_n^t = total normal force
F_t = tangential force
F_a = axial force
F_r = radial force
F_n = normal force in the plane of F_t and F_a
φ = pressure angle
φ_n = normal pressure angle
ψ = helix angle

Then $F_n = F_n^t \cos \varphi_n$ and $F_r = F_n^t \sin \varphi_n$

$F_t = F_n \cos \psi$ and $F_a = F_n \sin \psi$

Efficiency of Spiral and Helical Gears

In spiral or crossed helical gears, the sliding action between the surfaces acts chiefly along the tangent to the pitch helix. The friction force is equal to μF_n and acts in a direction opposite to the direction of sliding of the gear surface.

Two mating spiral gears 1 and 2 are shown separately in Figs. 10.34 a and b along with the forces acting on them. Gear 1 is driving the gear 2.

Let F_{t1} = tangential force acting on gear wheel 1

F_{t2} = tangential force acting on gear wheel 2

$F_{n1} = F_{n2} = F_n$ = normal reaction force between two surfaces in contact.

The direction of the sliding velocity of gear 2 relative to that of gear 1, $\mathbf{v}_{21}$ has been shown in Fig. 10.34 c. The friction force $\mu\ \mathbf{F}_{n2}$ acts in the opposite direction. $\mathbf{F}_{n2}$ and $\mu\ \mathbf{F}_{n2}$ combine into one reaction force $\mathbf{F}_2$ inclined at an angle φ with the normal reaction, where φ is the angle of friction.

$$F_{t2} = F_2 \cos(\psi_2 + \varphi) \quad \text{and} \quad F_{t1} = F_1 \cos(\psi_1 - \varphi)$$

But as $\quad F_n = F_{n1} = F_{n2} \quad$ and $\quad \mu F_{n1} = \mu F_{n2}$

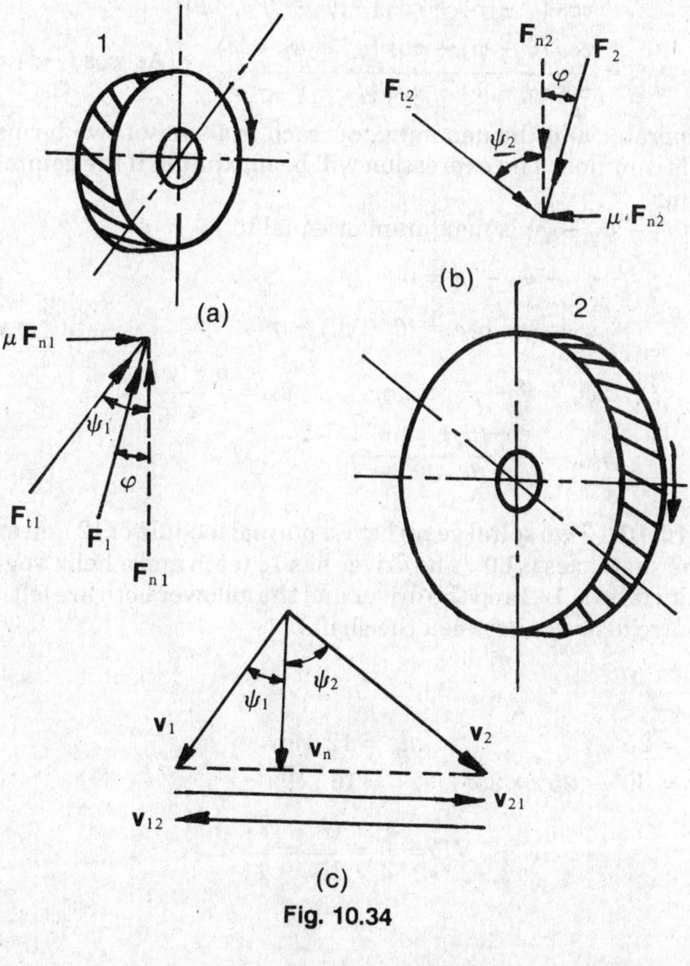

Fig. 10.34

$$F_1 = F_2 = F$$

or $\quad F_{t1} = F\cos(\psi_1 - \varphi) \quad$ and $\quad F_{t2} = F\cos(\psi_2 + \varphi)$

$$\text{Input} = F_{t1} \times v_1$$

$$\text{Output} = F_{t2} \times v_2$$

$$\text{Efficiency, } \eta = \frac{F_{t2} \times v_2}{F_{t1} \times v_1}$$

$$= \frac{\cos(\psi_2 + \varphi)\cos\psi_1}{\cos(\psi_1 - \varphi)\cos\psi_2} \qquad \left(\frac{v_2}{v_1} = \frac{\cos\psi_1}{\cos\psi_2}\right) \qquad (10.13)$$

$$= \frac{2\cos\psi_1\cos(\psi_2 + \varphi)}{2\cos\psi_2\cos(\psi_1 - \varphi)}$$

$$= \frac{\cos(\psi_1 + \psi_2 + \varphi) + \cos(\psi_1 - \psi_2 - \varphi)}{\cos(\psi_2 + \psi_1 - \varphi) + \cos(\psi_2 - \psi_1 + \varphi)}$$

$$= \frac{\cos(\theta + \varphi) + \cos(\psi_1 - \psi_2 - \varphi)}{\cos(\theta - \varphi) + \cos[-(\psi_1 - \psi_2 - \varphi)]}$$

$$= \frac{\cos(\theta + \varphi) + \cos(\psi_1 - \psi_2 - \varphi)}{\cos(\theta - \varphi) + \cos(\psi_1 - \psi_2 - \varphi)} \qquad [\text{As } \cos(-\alpha) = \cos\alpha]$$

The numerator and the denominator, each is a sum of two terms, out of which one is common. The expression will be maximum if the common term is maximum.

i.e., $\cos(\psi_1 - \psi_2 - \varphi)$ is maximum or equal to 1.

or $\quad \psi_1 - \psi_2 - \varphi = 0$

or $\quad \psi_1 = \psi_2 + \varphi = (\theta - \psi_1) + \varphi \qquad (\theta = \psi_1 + \psi_2)$

or $\quad 2\psi_1 = \theta + \varphi \quad$ or $\quad \psi_1 = \dfrac{\theta + \varphi}{2}$

and $\quad \eta_{max} = \dfrac{\cos(\theta + \varphi) + 1}{\cos(\theta - \varphi) + 1}$

Example 10.10 Two spiral gears have a normal module of 12 mm and angle between the shaft axes is 60°. The driver has 16 teeth and a helix angle of 25°. If the velocity ratio is 1/2 and the driver and the follower both are left-handed, find the centre distance between the shafts.

Solution:

$\psi_1 = 25°$ $\qquad m_n = 12$ mm

$\psi_2 = 60° - 25° = 35°$ $\qquad T_1 = 16$

$$\text{VR} = \frac{1}{2} = \frac{N_2}{N_1} = \frac{T_1}{T_2}$$

$$T_2 = \frac{T_1}{VR} = \frac{16}{1/2} = 32$$

$$C = \frac{m_n}{2}\left(\frac{T_1}{\cos\psi_1} + \frac{T_2}{\cos\psi_2}\right) = \frac{12}{2}\left(\frac{16}{\cos 25°} + \frac{32}{\cos 35°}\right)$$

or $\quad C = \underline{340.3 \text{ mm}}$

Example 10.11 The centre distance between two meshing spiral gears is 260 mm and the angle between the shafts 65°. The normal circular pitch is 14 mm and the gear ratio 2.5. The driven gear has a helix angle of 35°. Find

(i) the number of teeth on each wheel
(ii) the exact centre distance
(iii) the efficiency assuming the friction angle to be 5.5°.

Solution:

$\psi_2 = 35°$ $\qquad G = 2.5$

$\psi_1 = 65° - 35° = 30°$ $\qquad C = 260$ mm

$p_n = 14$ mm $\qquad \varphi = 5.5°$

Let the gear with smaller number of teeth be the driver.

$$G = \frac{T_2}{T_1} = 2.5 \quad \text{or} \quad T_2 = 2.5\,T_1$$

$$C = \frac{p_n}{2\pi}\left(\frac{T_1}{\cos\psi_1} + \frac{T_2}{\cos\psi_2}\right) \quad \left(m_n = \frac{p_n}{\pi}\right)$$

$$260 = \frac{14}{2\pi}\left(\frac{T_1}{\cos 30°} + \frac{2.5T_1}{\cos 35°}\right) = 9.373\,T_1$$

or $\quad T_1 = 27.74$

Take $\quad T_1 = \underline{28}$

Then $\quad T_2 = 2.5 \times 28 = \underline{70}$

$$C_{exact} = \frac{14}{2\pi}\left(\frac{28}{\cos 30°} + \frac{70}{\cos 35°}\right) = \underline{262.4 \text{ mm}}$$

$$\eta = \frac{\cos(\psi_2 + \varphi)}{\cos(\psi_1 - \varphi)}\frac{\cos\psi_1}{\cos\psi_2} = \frac{\cos(35° + 5.5°)}{\cos(30° - 5.5°)}\frac{\cos 30°}{\cos 35°}$$

$$= \underline{0.883}$$

Example 10.12 Two left-handed helical gears connect two shafts 60° apart. The normal module is 6 mm. The larger gear has 70 teeth and the velocity ratio is 1/2. The centre distance is 370 mm. Find the helix angles of the two gears.

Solution:

$$\theta = 60^\circ \qquad m_n = 6 \text{ mm}$$

$$T_2 = 70 \qquad C = 370 \text{ mm}$$

$$\psi_2 = 60^\circ - \psi_1$$

$$\text{VR} = \frac{N_2}{N_1} = \frac{T_1}{T_2}$$

or $$T_1 = \text{VR} \times T_2 = \frac{1}{2} \times 70 = 35$$

$$C = \frac{m_n}{2}\left[\frac{T_1}{\cos \psi_1} + \frac{T_2}{\cos (60^\circ - \psi_1)}\right]$$

$$370 = \frac{6}{2}\left[\frac{35}{\cos \psi_1} + \frac{70}{\cos (60^\circ - \psi_1)}\right]$$

By trial and error, $\psi_1 = 26^\circ$, $\psi_2 = 34^\circ$

Example 10.13 The following data relate to two spiral gears in mesh:

Shaft angle	= 90°
Centre distance	= 160 mm (approx.)
Normal circular pitch	= 8 mm
Gear ratio	= 3
Friction angle	= 5°

For maximum efficiency of the drive, determine

(i) the spiral angles of the teeth
(ii) the number of teeth
(iii) the centre distance (exact)
(iv) the pitch diameters
(v) the efficiency

Solution:

(i) $$\psi_1 = \frac{\theta + \varphi}{2} \quad \text{for maximum efficiency}$$

$$= \frac{90^\circ + 5^\circ}{2}$$

$$= 47.5^\circ$$

$$\psi_2 = 90^\circ - 47.5^\circ = 42.5^\circ$$

(ii) $$C = \frac{p_n}{2\pi}\left(\frac{T_1}{\cos \psi_1} + \frac{T_2}{\cos \psi_2}\right)$$

$$160 = \frac{8}{2\pi}\left(\frac{T_1}{\cos 47.5^\circ} + \frac{3T_1}{\cos 42.5^\circ}\right) = 7.065\, T_1$$

or $T_1 = 22.65$

let T_1 to be 23

$$T_2 = 3\, T_1 = 69$$

(iii) $$C_{\text{exact}} = \frac{8}{2\pi}\left(\frac{23}{\cos 47.5^\circ} + \frac{69}{\cos 42.5^\circ}\right) = \underline{162.5 \text{ mm}}$$

(iv) $$d_1 = \frac{P_1 T_1}{\pi} = \frac{p_n}{\cos \psi_1}\frac{T_1}{\pi} = \frac{8}{\cos 47.5^\circ} \times \frac{23}{\pi} = \underline{86.7 \text{ mm}}$$

$$d_2 = \frac{8}{\cos 42.5^\circ} \times \frac{69}{\pi} = 238.3 \text{ mm}$$

(v) $$\eta_{\max} = \frac{\cos(\theta + \varphi) + 1}{\cos(\theta - \varphi) + 1} = \frac{\cos(90^\circ + 5^\circ) + 1}{\cos(90^\circ - 5^\circ) + 1} = \underline{0.84}$$

10.21 WORM AND WORM GEAR

To accomplish large speed reduction in skew shafts, spiral gears with a small driver and a larger follower are required. Also the load transmitted through these gears is limited. To transmit a little higher load than with the usual spiral gears, use of worm and worm gears (throated type) can be made (Fig. 10.11b and c). Usually, worm and worm gears are used to connect two skew shafts at right angle to each other. The axial length of the worm is increased so that at least one or two threads (teeth) complete the circle on it (Fig. 10.31).

A worm can be a single, double or triple start if one, two or three threads are traversed on the worm for one tooth advancement of the gear wheel. Figs. 10.35 a and b show a single start and a double start worm respectively.

10.22 TERMINOLOGY OF WORM GEARS

Refer Fig. 10.35,

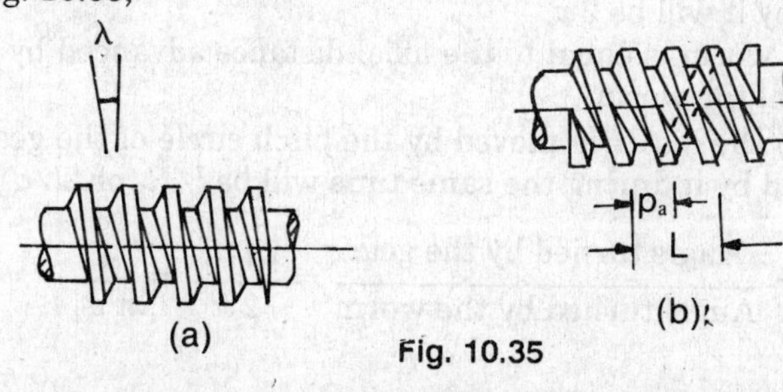

Fig. 10.35

(i) Axial Pitch (p_a)

It is the distance between corresponding points on adjacent teeth measured along the direction of the axis.

(ii) Lead (L)

The distance by which a helix advances along the axis of the gear for one turn around is known as lead.

In a single helix, the axial pitch is equal to lead, in a double helix, this is one-half the lead, in a triple helix, one third of lead and so on.

(iii) Lead Angle (λ)

It is the angle at which the teeth are inclined to the normal to the axis of rotation. Obviously, the lead angle is the complement of the helix angle.

i.e., $\psi + \lambda = 90°$

In case of worms, the lead angle is very small and the helix angle approaches 90°.

As the shaft axes of worm and worm gear are at 90°,

$$\psi_1 + \psi_2 = 90° \qquad \text{(1 denotes worm)}$$

$$(90° - \lambda_1) + \psi_2 = 90°$$

or $$\lambda_1 = \psi_2$$

i.e., lead angle of worm = helix angle of the gear wheel

Also, p_n of worm = p_n of wheel

$$p_{a1} \cos \lambda_1 = p_2 \cos \psi_2$$

But $$\lambda_1 = \psi_2$$

$\therefore$ $$p_{a1} = p_2$$

i.e. axial pitch of worm = circular pitch of wheel

10.23 VELOCITY RATIO AND CENTRE DISTANCE OF WORM GEARS

Velocity Ratio

As a worm may be multistart, the velocity is not calculated from the number of teeth.

Assume that a worm rotates through one revolution about its axis. Then the angle turned by it will be 2π.

The lead of the worm is equal to the axial distance advanced by a thread in one revolution of the worm.

The lead is also the distance moved by the pitch circle of the gear wheel. Thus, angle turned by it during the same time will be l/R_2 or $2l/d_2$.

$$\text{VR} = \frac{\text{Angle turned by the gear}}{\text{Angle turned by the worm}} = \frac{2\,l/d_2}{2\,\pi} = \frac{l}{\pi\,d_2} \qquad (10.15)$$

Centre Distance

$$C = \frac{m_n}{2}\left(\frac{T_1}{\cos\psi_1} + \frac{T_2}{\cos\psi_2}\right)$$

$$= \frac{m_2 \cos\psi_2}{2}\frac{1}{\cos\psi_2}\left(\frac{\cos\psi_2}{\cos\psi_1}\cdot T_1 + T_2\right)$$

$$= \frac{m_2}{2}\left[\frac{\cos\lambda_1}{\cos(90° - \lambda_1)} T_1 + T_2\right] \qquad [\psi_2 = \lambda_1,\ \psi_1 = 90° - \lambda_1]$$

$$= \frac{m_2}{2}\left[\frac{\cos\lambda_1}{\sin\lambda_1} T_1 + T_2\right]$$

$$= \frac{m_2}{2}\,[T_1 \cot\lambda_1 + T_2] \tag{10.16}$$

10.24 EFFICIENCY OF WORM GEARS

$$\eta = \frac{\cos(\psi_2 + \varphi)\cos\psi_1}{\cos(\psi_1 - \varphi)\cos\psi_2} \qquad \text{[Refer Eq. (10.13)]}$$

$$= \frac{\cos(\lambda_1 + \varphi)}{\cos(90° - \lambda - \varphi)}\frac{\cos(90° - \lambda_1)}{\cos\lambda_1}$$

$$= \frac{\cos(\lambda_1 + \varphi)}{\cos[90° - (\lambda_1 + \varphi)]}\frac{\sin\lambda_1}{\cos\lambda_1}$$

$$= \frac{\cos(\lambda_1 + \varphi)}{\sin(\lambda_1 + \varphi)}\tan\lambda_1$$

$$= \frac{\tan\lambda_1}{\tan(\lambda_1 + \varphi)} \tag{10.17}$$

$$\eta_{\max} = \frac{\cos(\theta + \varphi) + 1}{\cos(\theta - \varphi) + 1} = \frac{\cos(90° + \varphi) + 1}{\cos(90° - \varphi) + 1} \qquad (\theta = 90°)$$

$$= \frac{1 - \sin\varphi}{1 + \sin\varphi} \tag{10.17a}$$

If the gear wheel is the driver, it can be deduced that

$$\eta = \frac{\tan(\lambda_1 - \varphi)}{\tan\lambda_1}$$

and

$$\eta_{\max} = \frac{1 - \sin\varphi}{1 + \sin\varphi}$$

Example 10.14 **A two start worm rotating at 800 rpm drives a 26-tooth worm gear. The worm has a pitch diameter of 54 mm and a pitch of 18 mm. If coefficient of friction (μ) is 0.06, find**

(i) the helix angle of worm
(ii) the speed of gear
(iii) the centre distance
(iv) the lead angle for maximum efficiency
(v) the efficiency
(vi) the maximum efficiency

Solution:

$N_1 = 800$ rpm $\qquad T_2 = 26$

$\mu = \tan\varphi = 0.06 \qquad d_1 = 54$ mm

$\varphi = 3.43° \; (= \tan^{-1} 0.06) \qquad p_1 = 18$ mm

(i) Unwrap one thread of the worm,

$$\tan\lambda_1 = \frac{\text{Lead}}{\text{Pitch circumference}} = \frac{2p}{\pi d_1} \quad \text{(for two-start worm)}$$

$$\text{or } \tan\lambda_1 = \frac{2 \times 18}{\pi \times 54} = 0.212$$

$$\lambda_1 = 11.98° \quad \text{or} \quad 11°59'$$

Helix angle $\psi_1 = 90° - \lambda_1 = \underline{78°01'}$

(ii) Pitch of wheel = Axial pitch of worm = 18 mm

$$\therefore \quad p_2 = \frac{\pi d_2}{T_2}$$

$$\text{or} \quad 18 = \frac{\pi d_2}{26}$$

$$\text{or} \quad d_2 = 149 \text{ mm}$$

$$\text{VR} = \frac{\omega_2}{\omega_1} = \frac{N_2}{N_1} = \frac{\text{Lead}}{\pi d_2} = \frac{2 \times 18}{\pi \times 149}$$

$$\text{or} \quad N_2 = \frac{2 \times 18}{\pi \times 149} \times 800 = \underline{61.5 \text{ rpm}}$$

$$\text{Alternatively, } \frac{N_2}{N_1} = \frac{T_1}{T_2}$$

$$N_2 = \frac{2}{26} \times 800 = \underline{61.5 \text{ rpm}}$$

(iii) $$C = \frac{m_2}{2}(T_1 \cot\lambda_1 + T_2) = \frac{p_2}{2\pi}(T_1 \cot\lambda_1 + T_2)$$

$$= \frac{18}{2\pi}(2 \cot 11.98° + 26) = \underline{101.4 \text{ mm}}$$

(iv) For maximum efficiency, $\psi_1 = \dfrac{\theta + \varphi}{2}$

or $$90° - \lambda_1 = \frac{90° + \varphi}{2}$$

or $$\lambda_1 = 45° - \frac{3.43°}{2}$$

$$= 43.29° \quad \text{or} \quad \underline{43°18'}$$

(v) $$\eta = \frac{\tan \lambda_1}{\tan (\lambda_1 + \varphi)}$$

$$= \frac{\tan 11.98°}{\tan (11.98° + 3.43°)} \qquad (\tan^{-1} 0.06 = 3.43°)$$

$$= \underline{0.77}$$

(vi) $$\eta_{\max} = \frac{1 - \sin \varphi}{1 + \sin \varphi} = \frac{1 - \sin 3.43°}{1 + \sin 3.43°} = \underline{0.887}$$

10.25 BEVEL GEARS

To have a gear drive between two intersecting shafts, bevel gears are used. Kinematically, bevel gears are equivalent to rolling cones. Some of the common terms used in bevel gears are illustrated in Fig. 10.36 a.

Let γ_g, γ_p = pitch angles of gear and pinion respectively

r_g, r_p = pitch radii of gear and pinion respectively.

The pitch cones for two mating external bevel gears are shown in Fig. 10.36 b. We have,

$$\sin \gamma_g = \frac{r_g}{OP} = \frac{r_g}{r_p/\sin \gamma_p} = \frac{r_g}{r_p} \sin (\theta - \gamma_g)$$

or $$\sin \gamma_g = \frac{r_g}{r_p} (\sin \theta \cos \gamma_g - \cos \theta \cos \gamma_g)$$

Dividing both sides by $\cos \gamma_g$,

$$\tan \gamma_g = \frac{r_g}{r_p} (\sin \theta - \cos \theta \tan \gamma_g)$$

or $$\frac{r_p}{r_g} \tan \gamma_g = \sin \theta - \cos \theta \tan \gamma_g$$

$$\tan \gamma_g = \frac{\sin \theta}{\dfrac{r_p}{r_g} + \cos \theta}$$

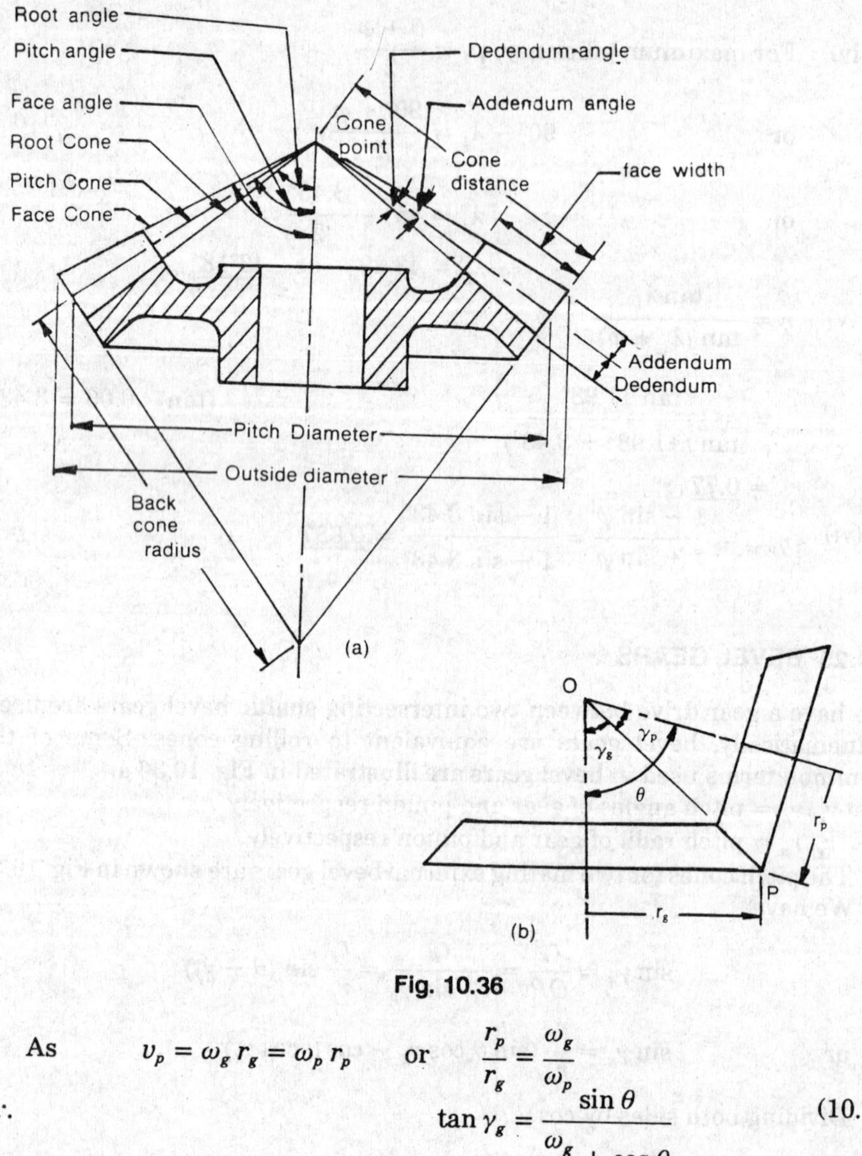

Fig. 10.36

As $\quad v_p = \omega_g r_g = \omega_p r_p \quad$ or $\quad \dfrac{r_p}{r_g} = \dfrac{\omega_g}{\omega_p}$

$\therefore$

$$\tan \gamma_g = \frac{\sin \theta}{\dfrac{\omega_g}{\omega_p} + \cos \theta} \tag{10.18}$$

Similarly,

$$\tan \gamma_p = \frac{\sin \theta}{\dfrac{\omega_p}{\omega_p} + \cos \theta} \tag{10.19}$$

Example 10.15 A pair of bevel gears is mounted on two intersecting shafts whose shaft angle is 72°. The velocity ratio of the gears is 2. Find the pitch angles.

Solution: As the velocity ratio is more than 1, the gear is the driver,

$$\tan \gamma_g = \frac{\sin \theta}{\frac{\omega_g}{\omega_p} + \cos \theta} = \frac{\sin 72°}{\frac{1}{2} + \cos 72°} = 1.176$$

$$\gamma_g = 49.61° \quad \text{or} \quad 49°37'$$

$$\tan \gamma_p = \frac{\sin \theta}{\frac{\omega_p}{\omega_g} + \cos \theta} = \frac{\sin 72°}{2 + \cos 72°} = 0.412$$

$$\gamma_p = 22.39° \quad \text{or} \quad \underline{22°23'}$$

As a check, $\gamma_g + \gamma_p = 49°37' + 22°23' = 72°$

PROBLEMS

10.1 The velocity ratio of two spur gears in mesh is 0.4 and the centre distance 75mm. For a module of 1.2 mm, find the number of teeth of the gears. What will be the pitch line velocity if the pinion speed is 800 rpm? Also find the speed of the gear wheel.

(36,90; 1810 mm/s; 320 rpm)

10.2 A spur gear has 30 teeth and a module of 1.4 mm. It rotates at 360 rpm. Determine its circular pitch and pitch line velocity.

(4.4 mm; 791.7 mm/s)

10.3 Two meshing spur gears with 20° pressure angle have a module of 4mm. The centre distance is 220mm and the number of teeth on the pinion 40. To what value should the centre distance be increased so that the pressure angle is increased to 22°?

(221.89 mm)

10.4 Two gears in mesh have a module of 10mm and a pressure angle of 25°. The pinion has 20 teeth and the gear has 52. The addendum on both the gears is equal to one module. Determine

(i) the number of pairs of teeth in contact

(ii) the angles of action of the pinion and the wheel

(iii) the ratio of the sliding velocity to the rolling velocity at the pitch point and at the beginning and end of engagement.

(1.475; δ_p = 26°36', δg = 10°13'; zero, 0.304, 0.278)

10.5 The number of teeth on the gear and the pinion of two spur gears in mesh are 30 and 18 respectively. Both the gears have a module of 6 mm and a pressure angle of 20°. If the pinion rotates at 400 rpm, what will be the sliding velocity at the moment the tip of the tooth of pinion has contact with the gear flank? Take addendum equal to one module. Also find the maximum velocity of sliding.

(908 mm/s; 981.5 mm/s)

10.6 Two 20° involute spur gears have a module of 6 mm. The larger wheel has 36 teeth and the pinion 16 teeth. If the addendum be equal to one module, will the interference occur? What will be the effect if the number of teeth on the pinion are reduced to 14?

(No; interference occurs)

10.7 Following data refer to two meshing gears having 20° involute teeth:

Number of teeth of gear wheel = 52
Number of teeth of pinion = 20
Speed of pinion = 360 rpm
Module = 8 mm

If addendum of each gear is such that the path of approach and path of recess are half of their maximum possible values, determine the addendum for the gear and the pinion and the length of arc of contact.

(5.07 mm; 18.04 mm; 52.4 mm)

10.8 Determine the minimum number of teeth and the arc of contact (in terms of module) to avoid interference in the following cases:

(a) Gear ratio is unity
(b) Gear ratio is 3
(c) Pinion gears with a rack.

Addendum of the teeth is 0.88 module and the power component is 0.94 times the normal thrust.

(11, 3.96 m; 14, 4.50 m; 16, 4.24 m)

10.9 Two 20° involute spur gears having a velocity ratio of 2.5 mesh externally. Module is 4mm and the addendum is equal to 1.23 module. Pinion rotates at 150 rpm. Find

(i) the minimum number of teeth on each wheel to avoid interference.
(ii) the number of pairs of teeth in contact.

(45, 18; 1.632)

10.10 If the angle of obliquity of a pair of gear wheels is 20°, and the arc of approach or recess not less than the pitch, what will be the least number of teeth on the pinion?

(9)

10.11 The centre distance between two meshing spiral gears is 150mm and the angle between the shafts 60°. The gear ratio is 2 and the normal circular pitch 10mm. The driven gear has a helix angle of 25°, determine

(i) the number of teeth on each wheel
(ii) the exact centre distance
(iii) the efficiency if the friction angle is 4°.

(27, 54; 152.3 mm; 91.5)

10.12 Two right-handed helical gears connect two shafts 70° apart. The larger gear has 50 teeth and the smaller 20. If the centre distance is 180mm, determine the helix angle of the gears. The normal module is 4mm.

(27°54′; 42°06′)

10.13 Angle between two shafts is 90°. They are joined by two spiral gears having a normal circular pitch of 6 mm and gear ratio of 2. If the approximate distance between the shafts be 200 mm and the friction angle 6°, determine for the maximum efficiency of the drive

(a) the number of teeth
(b) the centre distance (exact)
(c) the pitch diameters
(d) the efficiency.

(48, 42; 199.85 mm; 142.7 mm, 257 mm; 0.81)

10.14 A three-start worm has a pitch diameter of 80 mm and a pitch of 20 mm. It rotates at 600 rpm and drives a 40-tooth worm gear. If coefficient of friction is 0.05, find
(i) the helix angle of the worm
(ii) the speed of the gear
(iii) the centre distance
(iv) the efficiency and maximum efficiency.

(76°56′; 45 rpm; 167.3 mm; 0.817; 0.905)

11

GEAR TRAINS

11.1 INTRODUCTION

A gear train is a combination of gears used to transmit motion from one shaft to another. It becomes necessary when it is required to obtain large speed reduction within a small space. The following are the main types of gear trains:

1. Simple gear train
2. Compound gear train
3. Reverted gear train
4. Planetary gear train

11.2 SIMPLE GEAR TRAIN

A series of gears, capable of receiving and transmitting motion from one gear to another is called a simple gear train. In it, all the gear axes remain fixed relative to the frame and each gear is on a separate shaft (Fig. 11.1).

In a simple gear train,

1. A pair of mated external gears always move in opposite directions.

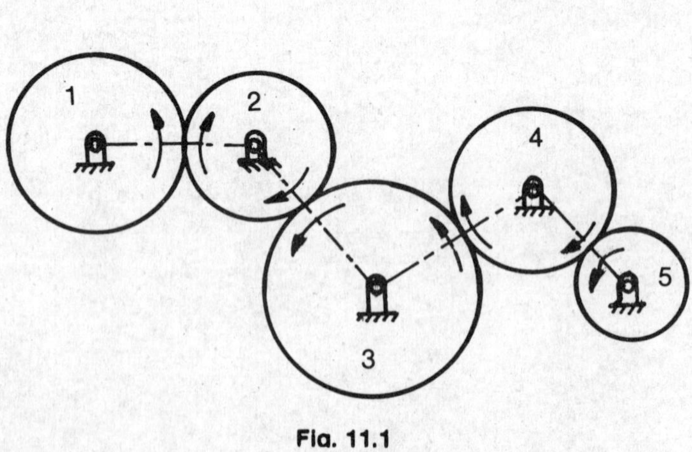

Fig. 11.1

2. All odd numbered gears move in one direction and all even numbered gears in the opposite direction. For example, gears 1, 3, 5, etc. move in the counter-clockwise direction.

3. Speed ratio, the ratio of the speed of the driving to that of the driven shaft, is negative when the input and the output gears rotate in the opposite directions and it is positive when the two rotate in the same direction. Reverse of the speed ratio is known as the train value of the gear train.

4. All the gears can be in a straight line or arranged in a zig-zag manner. A simple gear train can also have bevel gears.

Let T = number of teeth on a gear

N = speed of a gear in rpm.

Refer to Fig. 11.1,

$$\frac{N_2}{N_1} = \frac{T_1}{T_2} \qquad \left[\text{Also } \frac{\omega_2}{\omega_1} = \frac{2\pi N_2}{2\pi N_1} = \frac{N_2}{N_1}\right]$$

and

$$\frac{N_3}{N_2} = \frac{T_2}{T_3}, \quad \frac{N_4}{N_3} = \frac{T_3}{T_4} \quad \text{and} \quad \frac{N_5}{N_4} = \frac{T_4}{T_5}$$

Multiplying,

$$\frac{N_2}{N_1} \times \frac{N_3}{N_2} \times \frac{N_4}{N_3} \times \frac{N_5}{N_4} = \frac{T_1}{T_2} \times \frac{T_2}{T_3} \times \frac{T_3}{T_4} \times \frac{T_4}{T_5}$$

or

$$\text{Train value } \frac{N_5}{N_1} = \frac{T_1}{T_5} = \frac{\text{number of teeth on driving gear}}{\text{number of teeth on driven gear}}$$

$$\text{Speed ratio} = \frac{1}{\text{train value}}$$

or

$$\frac{N_1}{N_5} = \frac{T_5}{T_1} \tag{11.1}$$

Thus it is seen that the intermediate gears have no effect on the speed ratio and, therefore, they are known as idlers.

11.3 COMPOUND GEAR TRAIN

When a series of gears are connected in such a way that two or more gears rotate about an axis with the same angular velocity, it is known as compound gear train. In this type, some of the intermediate shafts, i.e. other than the input and the output shafts, carry more than one gear as shown in Fig. 11.2.

If gear 1 is the driver, then

$$\frac{N_2}{N_1} = \frac{T_1}{T_2}, \quad \frac{N_4}{N_3} = \frac{T_3}{T_4} \quad \text{and} \quad \frac{N_6}{N_5} = \frac{T_5}{T_6}$$

or

$$\frac{N_2}{N_1} \times \frac{N_4}{N_3} \times \frac{N_6}{N_5} = \frac{T_1}{T_2} \times \frac{T_3}{T_4} \times \frac{T_5}{T_6}$$

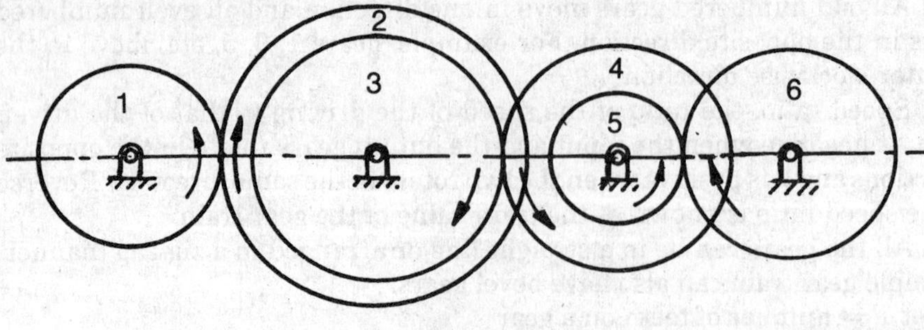

Fig. 11.2

or
$$\frac{N_2}{N_1} \times \frac{N_4}{N_2} \times \frac{N_6}{N_4} = \frac{T_1}{T_2} \times \frac{T_3}{T_4} \times \frac{T_5}{T_6}$$

$$\frac{N_6}{N_1} = \frac{T_1 T_3 T_5}{T_2 T_4 T_6} \qquad (11.2)$$

$$\text{or Train value} = \frac{\text{product of number of teeth on driving gears}}{\text{product of number of teeth on driven gears}}$$

11.4 REVERTED GEAR TRAIN

If the axes of the first and the last wheels of a compound gear coincide, it is called a reverted gear train. Such an arrangement is used in clocks and in simple lathes where 'back gear' is used to give a slow speed to the chuck.

Referring to Fig. 11.3,

$$\frac{N_4}{N_1} = \frac{\text{product of number of teeth on driving gears}}{\text{product of number of teeth on driven gears}}$$

$$= \frac{T_1 T_3}{T_2 T_4} \qquad (11.3)$$

Also, If r is the pitch circle radius of a gear,

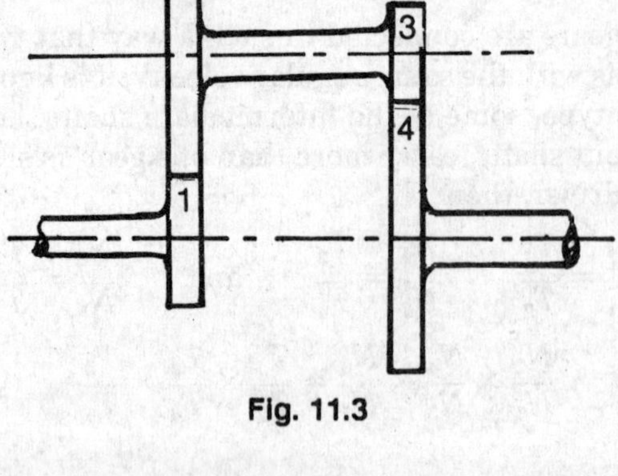

Fig. 11.3

$$r_1 + r_2 = r_3 + r_4 \tag{11.4}$$

11.5 PLANETARY OR EPICYCLIC GEAR TRAIN

When there exists a relative motion of axes in a gear train, it is called a planetary or an epicyclic gear train (or simply epicyclic gear or train). Thus in an epicyclic train, the axis of at least one of the gears also moves relative to the frame.

Consider two gear wheels S and P, the axes of which are connected by an arm a (Fig. 11.4). If the arm a is fixed, the wheels S and P constitute a simple train. However, if the wheel S is fixed so that the arm can rotate about the axis of S, the wheel P would also move around S. Therefore, it is an epicyclic train.

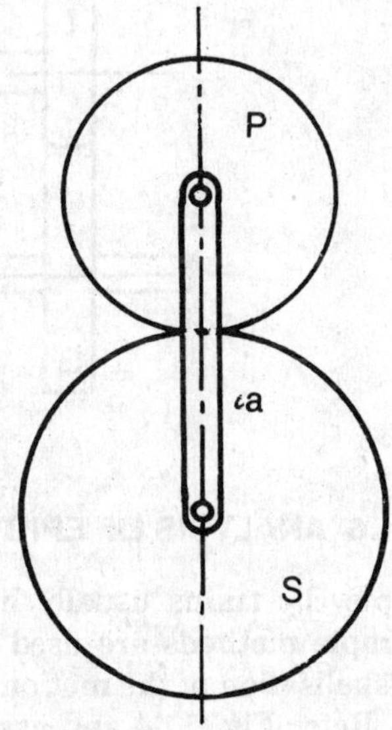

Fig. 11.4

Usually, the wheel P is known as the epicyclic wheel. The term epicyclic emerges from the fact that the wheel P rolls outside another wheel S and traces an epicyclic path. It is also possible that the fixed wheel is annular and another wheel rolls inside it. In that case, the path traced will be a hypocycloid. However, it has become customary to call all gears, in which one of the axes rotates about a fixed axis, as epicyclic gears.

Large speed reductions are possible with epicyclic gears and if the fixed wheel is annular, a more compact unit could be obtained. Important applications of epicyclic gears are in transmission, computing devices, and so on.

In an epicyclic gear, one wheel is, usually, fixed as in the above case. However, it is not necessary at all and the wheel S can have rotations in any direction about its axis i.e. clockwise or counter-clockwise. Figure 11.5 shows an epicyclic gear with no fixed wheel. The epicyclic gear consists of wheels B, C, D, E and F and the arm a. Wheels G and H are merely the drivers; G drives the arm a through the wheel I whereas H drives the wheel C.

In general, gear trains have two degrees of freedom. This means to obtain a controlled motion of the output, the train must have two inputs. In Fig. 11.5, two inputs, to the wheel C and the arm a result in a definite motion of the wheel F or of the output shaft. However, number of inputs can be reduced to one, if one wheel of the train is fixed. That amounts to reducing the speed of that gear wheel to zero.

In dealing with epicyclic gears, distinction must be made between wheels which are part of the epicyclic train and those which are not. In gear train of Fig. 11.5, G and H cannot be the parts of the epicyclic train. Also the speed of the arm a will be the same as that of wheel I.

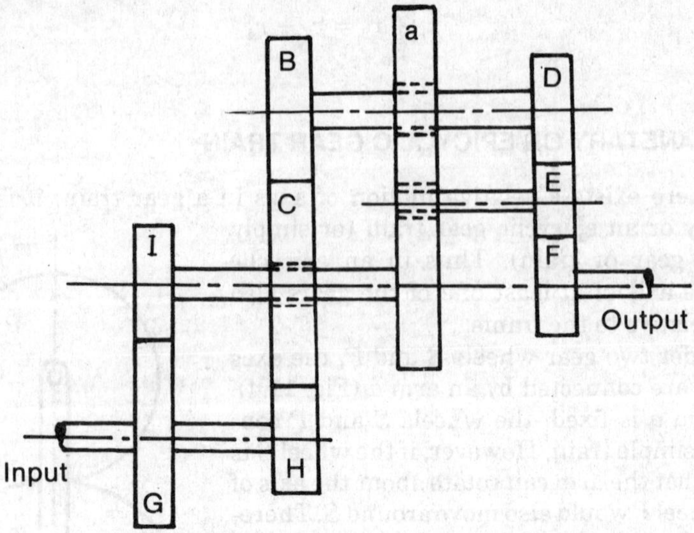

Fig. 11.5

11.6 ANALYSIS OF EPICYCLIC GEAR TRAIN

Epicyclic trains usually have complex motions. Therefore, comparatively simple methods are used to analyse them which do not require accurate visualisation of the motions.

Refer Fig. 11.4 and assume that the arm a is fixed. Turn S through x revolutions in the clockwise direction. Assuming clockwise motion of a wheel as positive and counter-clockwise as negative.

Revolutions made by $a = 0$ (Arm a fixed)
Revolutions made by $S = x$
Revolutions made by $P = -(T_S/T_P)\,x$

Now, if the mechanism is locked together and turned through a number of revolutions, the relative motions between a, S and P will not alter. Let the locked system is turned through y revolutions in the clockwise direction. Then

Revolutions made by $a = y$
Revolutions made by $S = y + x$
Revolutions made by $P = y - \left(\dfrac{T_S}{T_P}\right) x$

This implies that if the arm a turns through y revolutions and S through $(y + x)$ revolutions in the same direction, then P will rotate through $[y - (T_S/T_P)x]$ revolutions in space or relative to the fixed axis of S. Thus if revolutions made by any of the two elements are known, x and y can be solved and the revolutions made by the third can be determined. Thus, the procedure can be summarised as follows:

1. Lock the arm and assume the other wheels free to rotate.

2. Turn any convenient gear through one revolution in the clockwise direction and record the number of revolutions made by each of the other wheels.
3. Multiply all the above recordings by x and write the same in the second row. This is equivalent to the statement that the chosen wheel is given x revolutions in the clockwise direction keeping the arm fixed.
4. Add y to all the quantities in the second row and make the recordings in the third row. This amounts to the fact that by locking the whole system, it is turned through y revolutions in the clockwise direction. Thus the arm makes y revolutions, the chosen wheel $(y + x)$ revolutions and so on.
5. Apply the given conditions and find the values of x and y. Having known x and y, the revolutions made by any of the wheels can be known.

The above procedure is also illustrated by the following table for Fig. 11.4.

Line	*Action*	*Revs. of a*	*Revs. of S*	*Revs. of P*
1.	a fixed, S +1 rev.	0	1	$-\frac{T_S}{T_P}$
2.	a fixed, $S + x$ rev.	0	x	$-\frac{T_S}{T_P}x$
3.	Add y	y	$y + x$	$y - \frac{T_S}{T_P}x$

Note that the number of revolutions of the wheel P given in the third row of the table is the number of revolutions in space or relative to the fixed axis of S and not about its own axis.

Consider a system shown in Fig. 11.6 in which a is an arm which can rotate about a fixed axis O, P is a wheel which has its axis at the other end of a. For the moment, assume the wheel P to be fixed to the arm a. Now turn the arm through 1/4 revolution, i.e. from position 1 to the position 2. It will be observed that the wheel P has also turned through 1/4 revolution. But the rotation of P is not about its own axis, as it is fixed to the arm. Similarly, it can be seen that P also rotates through one revolution as the arm turns through one revolution. However, the rotation of P is in space or about the axis of rotation of the arm and not about its own axis. Thus, if the arm makes y revolutions about O the wheel P also rotates through y revolutions. This suggests that the number of revolutions of P about its own axis can be obtained by subtracting the number of revolutions of the arm from the total number of revolutions of P, or

Rev. of P about its own axis = total revs. about axis of arm − revs. of the arm

$$= \left(y - \frac{T_S}{T_P}x\right) - y$$

$$= -\frac{T_S}{T_P}x \qquad (11.5)$$

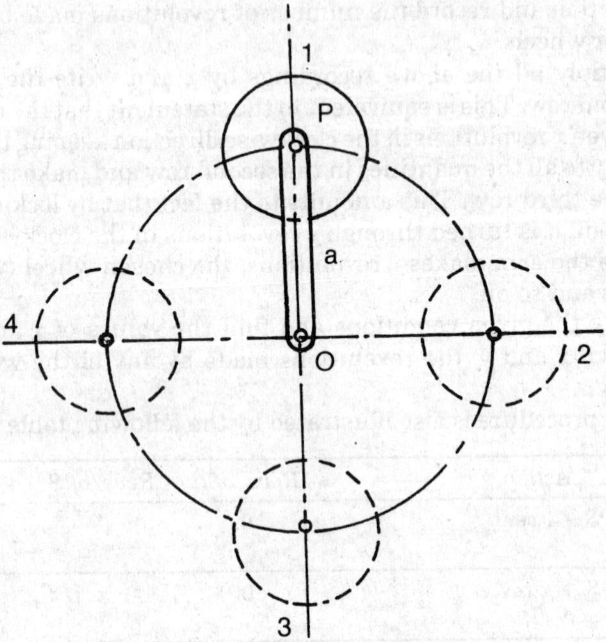

Fig. 11.6

Relative Velocity Method

Angular vel. of S = angular vel. of S rel. to a + angular vel. of a

or $\omega_S = \omega_{Sa} + \omega_a$

or $N_S = N_{Sa} + N_a$

Similarly, $N_P = -N_{Pa} + N_a$ (N_S and N_P are to be in opposite directions)

$\therefore$ $N_{Sa} = N_S - N_a$

and $N_{Pa} = N_a - N_P$

or $\dfrac{N_{Sa}}{N_{Pa}} = \dfrac{N_S - N_a}{N_a - N_P}$

or $\dfrac{T_P}{T_S} = -\dfrac{N_S - N_a}{N_P - N_a}$

Usually, T_P/T_S and out of N_S, N_a and N_P, two are known and the third is calculated.

Example 11.1 The annulus A in the gear shown in Fig. 11.7 rotates at 300 rpm about the axis of the fixed wheel S which has 80 teeth. The three-armed spider (only one arm a is shown in Fig. 11.7a) is driven at 180 rpm. Determine the number of teeth required on the wheel P.

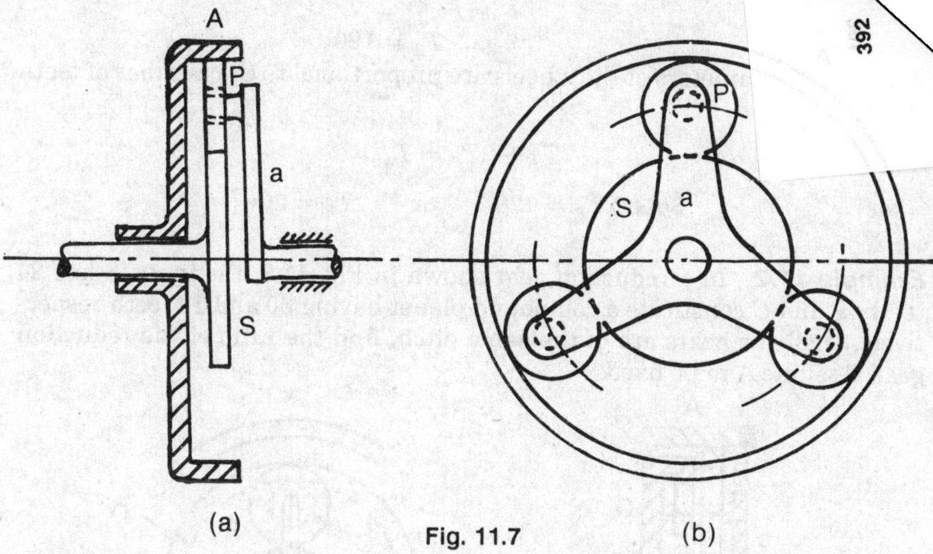

Fig. 11.7

Solution:

$N_S = 0$ $\qquad$ $N_A = 300$ rpm

$N_a = 180$ rpm $\qquad$ $T_S = 80$

Action	*a*	*S*	*P*	*A*
'**a**' fixed, $S + 1$ rev.	0	1	$-\dfrac{80}{T_P}$	$-\dfrac{80}{T_P} \times \dfrac{T_P}{T_A} = -\dfrac{80}{T_A}$
'a' fixed, $S + x$ rev.	0	x	$-\dfrac{80x}{T_P}$	$-\dfrac{80x}{T_A}$
All given y revs.	y	$y + x$	$y - \dfrac{80x}{T_P}$	$y - \dfrac{80x}{T_A}$

From given conditions,

$$N_a = y = 180 \qquad \text{(i)}$$

$$N_S = y + x = 0 \qquad \text{(ii)}$$

$$N_A = y - \frac{80x}{T_A} = 300 \qquad \text{(iii)}$$

From (i) and (ii), $x = -y = -180$

Solving (iii),

$$180 - \frac{80(-180)}{T_A} = 300$$

or

$$\frac{14\,400}{T_A} = 120$$

or $$T_A = 120$$

The pitch diameters of the wheels are proportional to the number of teeth on them.

$$T_S + 2T_P = T_A$$

or $$80 + 2T_P = 120 \quad \text{or} \quad T_P = 20$$

Example 11.2 In a reduction gear shown in Fig. 11.8, the input S has 24 teeth. P and C constitute a compound planet having 30 and 18 teeth respectively. If all the gears are of the same pitch, find the ratio of the reduction gear. Assume A to be fixed.

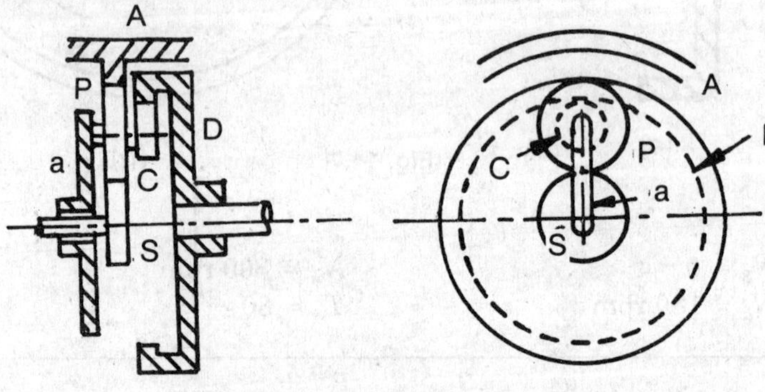

Fig. 11.8

Solution:

$T_S = 24$
$T_P = 30$
$T_C = 18$
$N_A = 0$

Pitch diameters of wheels are proportional to the number of teeth on them.

$$T_A = 2\left[\frac{T_S}{2} + T_P\right] = 2\left[\frac{24}{2} + 30\right] = 84$$

$$T_D = 2\left[\frac{T_S}{2} + \frac{T_P}{2} + \frac{T_C}{2}\right] = 2\left[\frac{24}{2} + \frac{30}{2} + \frac{18}{2}\right] = 72$$

Action	*a*	*S*	*P/C*	*A*	*D*
a fixed, S + 1 rev.	0	1	$-\frac{24}{30}$	$-\frac{24}{30} \times \frac{30}{84}$	$-\frac{24}{30} \times \frac{18}{72}$
a fixed, S + x rev.	0	x	$-\frac{4x}{5}$	$\frac{-2x}{7}$	$-\frac{x}{5}$
Add y	y	$y + x$	$y - \frac{4x}{5}$	$y - \frac{2x}{7}$	$y - \frac{x}{5}$

From given conditions,

$$N_A = y - \frac{2x}{7} = 0 \quad \text{or} \quad y = \frac{2x}{7}$$

$$\therefore \quad \frac{N_S}{N_D} = \frac{y+x}{y-\frac{x}{5}} = \frac{\frac{2x}{7}+x}{\frac{2x}{7}-\frac{x}{5}} = \frac{9}{7} \times \frac{35}{3} = 15$$

Example 11.3 Figure 11.9 shows a port indicator for a twin-screw ship. It is found that the pointer P remains stationary if the propellers run at the same speed and drive the gears *C* and *D* in the same direction through equal gears *A* and *B*. If the number of teeth on *G* and *F* are 24 and 50 respectively, find the ratio of the number of teeth on *C* to that on *D*.

What will be the speed of the pointer if *B* runs at 5% faster than *A* and if the speed of *C* is 100 rpm?

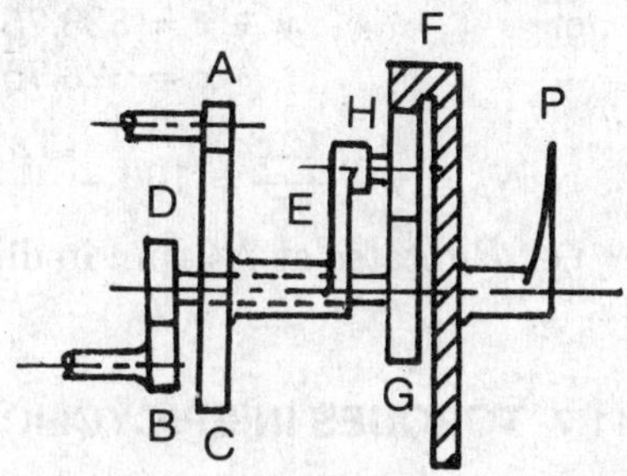

Fig. 11.9

Solution:

$T_G = 24 \qquad T_F = 50 \qquad N_F = N_P = 0$

$$\text{Now, } T_F = 2\left[\frac{T_G}{2} + T_H\right]$$

$$\text{or} \quad 50 = 2\left[\frac{24}{2} + T_H\right]$$

$$\text{or} \quad T_H = \frac{50-24}{2} = 13$$

Action	*C/E*	*D/G*	*H*	*F/P*
C fixed, *D* + 1 rev.	0	1	$-\frac{24}{13}$	$-\frac{24}{13} \times \frac{13}{50}$
C fixed, *D* + *x* rev.	0	x	$-\frac{24x}{13}$	$-\frac{12x}{25}$
Add *y*	y	$y + x$	$y - \frac{24x}{3}$	$y - \frac{12x}{25}$

Given,

$$N_F = y - \frac{12x}{25} = 0 \quad \text{or} \quad y = \frac{12x}{25}$$

$$\therefore \quad \frac{T_C}{T_D} = \frac{N_D}{N_C} = \frac{y+x}{y} = \frac{\frac{12x}{25}+x}{\frac{12x}{25}} = \frac{37}{12}$$

When B and C rotate at different speeds,

$$N_C = y = 100$$

$$N_A = 100 \times \frac{T_C}{T_A}$$

$$N_B = \left(100 \times \frac{T_C}{T_A}\right) \times 1.05$$

$$N_D = N_B \times \frac{T_B}{T_D} = \left(100 \times \frac{T_C}{T_A} \times 1.05\right) \times \frac{T_B}{T_D}$$

$$= 100 \times 1.05 \times \frac{T_C}{T_D} \qquad (T_A = T_B \text{ as } N_A = N_B)$$

$$= 100 \times 1.05 \times \frac{37}{12}$$

or $$y + x = 323.75$$

$$x = 323.75 - 100 = 223.75$$

$$\therefore\ N_P = y - \frac{12x}{25} = 100 - \frac{12 \times 223.75}{25} = \underline{-7.4 \text{ rpm}}$$

i.e., P rotates at 7.4 rpm in direction opposite to that of C.

11.7 TORQUES IN EPICYCLIC TRAINS

Torques are transmitted from one element to another when a geared system transmits power. Assume that all the wheels of a gear train rotate at uniform speeds, i.e. accelerations are not involved. Also each wheel is in equilibrium under the action of torques acting on it.

Let N_S, N_a, N_P and N_A be the speeds and $\mathbf{T}_S$, $\mathbf{T}_a$, $\mathbf{T}_P$ and $\mathbf{T}_A$ the external torques transmitted by S, a, P and A respectively (Fig. 11.7).

We have,

$$\mathbf{T} = 0 \qquad (T \text{ is the torque})$$

or $$\mathbf{T}_S + \mathbf{T}_a + \mathbf{T}_P + \mathbf{T}_A = 0 \qquad (11.7)$$

Now S and a are connected to machinery outside the system and thus transmit external torques. The planet P can rotate on its own pin fixed to a but is not connected to anything outside. Therefore, it does not transmit external torque. The annulus A is either locked by an external torque or transmits power or torque either to or from the system through external teeth.

Therefore, Eq. (11.7) becomes,

$$\mathbf{T}_S + \mathbf{T}_a + 0 + \mathbf{T}_A = 0$$

or $$\mathbf{T}_S + \mathbf{T}_a + \mathbf{T}_A = 0 \qquad (11.8)$$

If A is fixed, $\mathbf{T}_A$ is usually known as the braking or the fixing torque. Out of $\mathbf{T}_S$ and $\mathbf{T}_a$ one will be the driving torque and the other, the output or the resisting torque.

Assuming no losses in power transmission,

$$\Sigma\, \mathbf{T}\omega = 0$$

or

$$\Sigma\, \mathbf{T}N = 0$$

or

$$\mathbf{T}_S N_S + \mathbf{T}_a N_a + \mathbf{T}_A N_A = 0 \qquad (11.9)$$

If A is fixed, $N_A = 0$

and

$$\mathbf{T}_S N_S + \mathbf{T}_a N_a = 0 \qquad (11.10)$$

Proper directions of speeds and torques must be taken into account.

Example 11.4 The number of teeth in the gear shown in Fig. 11.10 are as follows:

$$T_S = 18,\ T_P = 24,\ T_C = 12,\ T_A = 72$$

P and C form a compound gear carried by the arm a and the annular gear A is held stationary. Determine the speed of the output at a. Also find the holding torque required on A if 5 kW is delivered to S at 800 rpm with an efficiency of 94%.

In case the annulus A rotates at 100 rpm in the same direction as S, what will be the new speed of a?

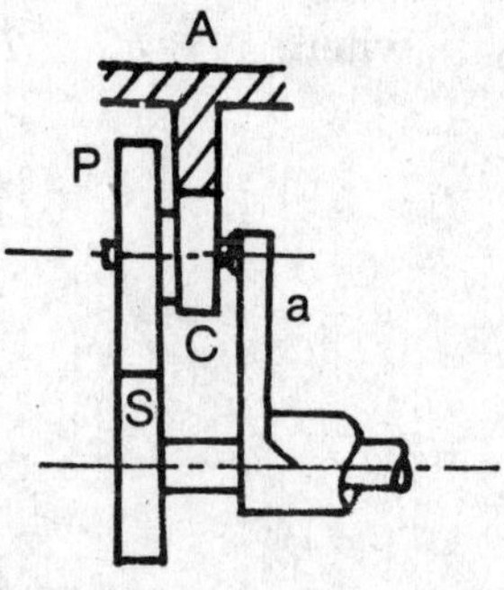

Fig. 11.10

Solution:

$T_S = 18$ $\quad N_S = 800$ rpm

$T_P = 24$ $\quad \eta = 0.94$

$T_C = 12$ $\quad N_A = 0$ in first case

$\qquad\qquad = 100$ rpm in second case

$T_A = 72$ $\quad P = 5$ kW

Action	*a*	*S*	*P/C*	*A*
'*a*' fixed, *S* + 1 rev.	0	1	$-\frac{18}{24}$	$-\frac{18}{24} \times \frac{12}{72}$
'*a*' fixed, *S* + *x* rev.	0	x	$-\frac{3x}{4}$	$-\frac{x}{8}$
Add *y*	y	$y + x$	$y - \frac{3x}{4}$	$y - \frac{x}{8}$

Given conditions (*1st case*)

$$N_A = y - \frac{x}{8} = 0 \quad \text{or} \quad y = \frac{x}{8}$$

$\therefore \qquad N_S = y + x = \frac{x}{8} + x = 800$

or $\qquad \frac{9}{8}x = 800$

or $\qquad x = 711$

$\therefore \qquad y = 88.9$

$\therefore$ Speed of $\qquad a = y = \underline{88.9 \text{ rpm}}$

Now

$$\Sigma \mathbf{T}\ N = 0$$

or $$\mathbf{T}_S N_S + \mathbf{T}_a N_a + \mathbf{T}_A N_A = 0$$

where $$T_S = \frac{\text{Power input}}{\frac{2\pi N}{60}} = \frac{5000}{\frac{2\pi \times 800}{60}} = 59.7 \text{ N.m}$$

$$T_a = \text{Theoretical torque at output}$$

$$= \frac{\text{Actual torque at output}}{\text{Efficiency of power transmission}}$$

$$= \frac{T'_a}{\eta}$$

$\therefore$ $$59.7 \times 800 + \frac{T'_a}{0.94} \times 88.9 + T_A \times 0 = 0$$

or

$$T'_a = -504.9 \text{ N.m}$$

Also,

$$T_S + T'_a + T_A = 0 \qquad \text{[Eq. (11.8)]}$$

or $\qquad 59.7 - 504.9 + T_A = 0$

or $\qquad T_A = \underline{-445.2 \text{ N.m}}$

2nd case:

$$N_A = y - \frac{x}{8} = 100 \quad \text{or} \quad y = 100 + \frac{x}{8}$$

$$N_S = y + x = 100 + \frac{x}{8} + x = 800$$

or $\qquad \frac{9x}{8} = 700$

or $\qquad x = 622.2$

$$y = 100 + \frac{622.2}{8} = 177.8$$

$\therefore \qquad$ New speed of arm $a = y = \underline{177.8 \text{ rpm}}$

Example 11.5 In the epicyclic gear train shown in Fig. 11.11, a gear C which has teeth cut internally and externally is free to rotate on an arm driven by

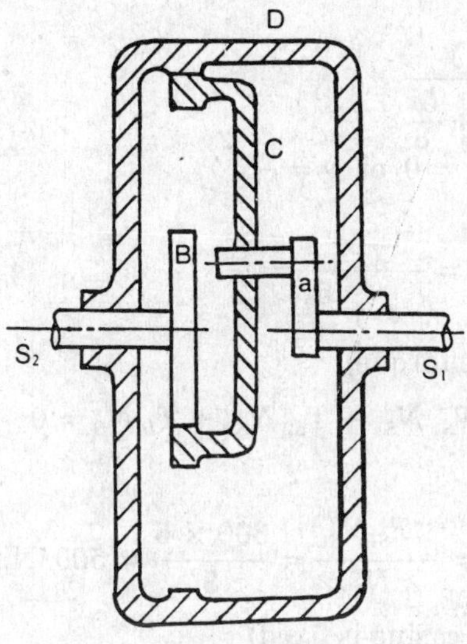

Fig. 11.11

shaft S_1. It meshes externally with the casing D and internally with the pinion B. The gears have the following number of teeth:

$$T_B = 24,\ T_C = 32 \text{ and } 40,\ T_D = 48$$

Find the velocity ratio between

(i) S_1 and S_2 when D is fixed

(ii) S_1 and D when S_2 is fixed.

What will be the torque required to fix the casing D if a torque of 300 N.m is applied to the shaft S_1?

Solution: Complete the following table.

Action	a/S_1	B/S_2	C	D
'a' fixed, B + 1 rev.	0	1	$\frac{24}{32}$	$\frac{24}{32} \times \frac{40}{48}$
'a' fixed, B + x rev.	0	x	$\frac{3x}{4}$	$\frac{5x}{8}$
Add y	y	$y + x$	$y + \frac{3x}{4}$	$y + \frac{5x}{8}$

(i) From the given conditions,

$$N_D = y + \frac{5x}{8} = 0 \quad \text{or} \quad y = -\frac{5x}{8}$$

$$\frac{N_{S1}}{N_{S2}} = \frac{N_a}{N_B} = \frac{y}{y+x} = \frac{-5x/8}{-\frac{5x}{8} + x} = -\frac{5}{3}$$

(ii) $$\frac{N_{S1}}{N_D} = \frac{N_a}{N_D} = \frac{y}{y + \frac{5x}{8}}$$

But $N_B = y + x = 0$ or $y = -x$

$$\therefore \quad \frac{N_{S1}}{N_D} = \frac{-x}{-x + \frac{5x}{8}} = \frac{8}{3}$$

If T denotes the torque,

$$\mathbf{T}_{S1} N_{S1} + \mathbf{T}_{S2} N_{S2} + \mathbf{T}_D N_D = 0$$

or

$$T_{S2} = -\frac{T_{S1}N_{S1}}{N_{S2}} = \frac{300 \times 5}{3} = 500 \text{ N.m}$$

($N_D = 0$, as the casing is fixed)

Also,

$$\mathbf{T}_{S1} + \mathbf{T}_{S2} + \mathbf{T}_D = 0$$

$$300 + 500 + T_D = 0$$

or $$T_D = \underline{-800 \text{ N.m}}$$

11.8 SUN AND PLANET GEAR

When an annular wheel A is added to the epicyclic gear train of Fig. 11.4, the combination is usually, referred to as sun and planet gear (Fig. 11.12). The annular wheel gears with the wheel P which can rotate freely on the arm a. The wheels S and P are generally called the sun and the planet wheels respectively due to analogy of motion of a planet around the sun.

Fig. 11.12

In general, S, A and a are free to rotate independently of each other. It is also possible that either S of A are fixed. If A is fixed, S will be the driving member and if S is fixed, A will be the driving member. In each case the driven member is the arm a.

Let N_S = speed of the sun wheel S

N_A = speed of the annular wheel A

N_a = speed of the arm a

T_S = number of teeth on S

T_A = number of teeth on A

Then the table can be prepared as usual.

Action	Arm a	S	P	A
'a' fixed, S + 1 rev.	0	1	$-\frac{T_S}{T_P}$	$\left(-\frac{T_S}{T_P} \times \frac{T_P}{T_A}\right) = -\frac{T_S}{T_A}$
'a' fixed, S + x rev.	0	x	$-\frac{T_S}{T_P}x$	$-\frac{T_S}{T_A}x$
All given y rev. (add y)	y	$y + x$	$y - \frac{T_S}{T_P}x$	$y - \frac{T_S}{T_A}x$

$$\therefore \quad x + y = N_S \tag{i}$$

and
$$y - \frac{T_S}{T_A}x = N_A \tag{ii}$$

Subtracting (ii) from (i),

$$x\left(1 + \frac{T_S}{T_A}\right) = N_S - N_A$$

or
$$x\left(\frac{T_A + T_S}{T_A}\right) = N_S - N_A$$

or
$$x = \left(\frac{N_S - N_A}{T_A + T_S}\right)T_A$$

and
$$y = N_S - x$$

$$= N_S - \frac{N_S T_A - N_A T_A}{T_A + T_S}$$

$$= \frac{N_S T_A + N_S T_S - N_S T_A + N_A T_A}{T_A + T_S}$$

$$N_a = y = \frac{N_S T_S + N_A T_A}{T_S + T_A} \tag{11.11}$$

If the sun wheel S is fixed, $N_S = 0$

Speed of the arm,

$$N_a = \frac{N_A T_A}{T_S + T_A} \quad \text{or} \quad \frac{N_a}{N_A} = \frac{1}{T_S/T_A + 1}$$

If the annular wheel A is fixed, $N_A = 0$.

Speed of the arm,

$$N_a = \frac{N_S T_S}{T_S + T_A} \quad \text{or} \quad \frac{N_a}{N_S} = \frac{T_S/T_A}{1 + T_S/T_A}$$

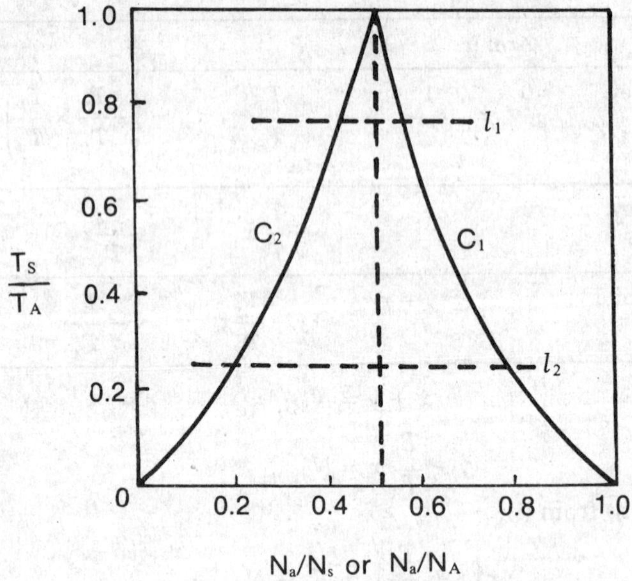

Fig. 11.13

The number of teeth on the sun wheel can vary from 0 to T_A, i.e. from zero to the number of teeth on the annular wheel. Therefore, the ratio of the number of teeth on the sun wheel to that on the annular wheel, i.e. T_S/T_A can vary from 0 to 1. If a graph T_S/T_A vs. N_a/N_A (S is fixed) is plotted, the curve C_1 is obtained (Fig. 11.13) which shows that the sun and the planet gear always acts as a reduction gear. The speed of the arm decreases from N_A to $1/2\,N_A$ as the number of teeth of the sun wheel increases from zero to T_A. Similarly, if A is fixed and T_S/T_A vs. N_a/N_S is plotted, the curve C_2 is obtained. This shows that it again acts as a reduction gear in which the speed of the arm varies from zero to $1/2N_S$.

In both cases, the direction of the arm is the same as that of the driving member.

Horizontal dotted lines l_1 and l_2 show the practical limits of the ratio of T_S/T_A and of the corresponding speeds of the arm, where middle portion shows the range.

Example 11.6 Figure 11.14 shows a compound gear in which an input torque of 150 N.m is given to shaft B at 1000 rpm. The sun and planet gears are all of the same diameter and pitch. What will be the speed and the torque at the output shaft C assuming an efficiency of 97%?

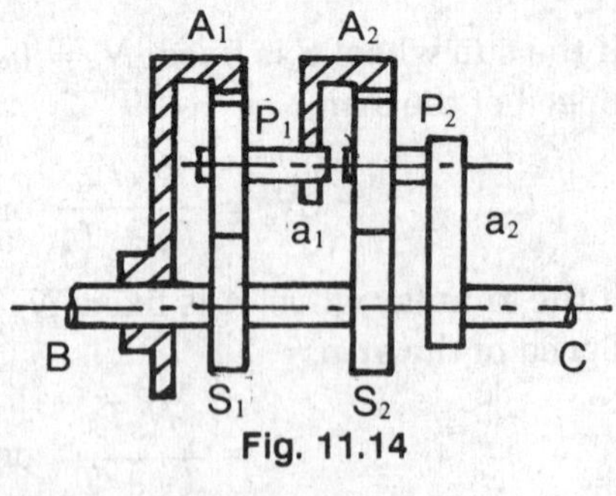

Fig. 11.14

Also find the torque required to hold stationary the annulus A_1.

Solution:

$$N_{A2} = N_{a1}, \quad N_{S1} = N_{S2} = 1000 \text{ rpm}$$

Speed of arm a_1,

$$N_{a1} = \frac{N_{A1}T_{A1} + N_{S1}T_{S1}}{T_{A1} + T_{S1}} = \frac{N_{S1}T_{S1}}{T_{A1} + T_{A1}} \qquad (N_{A1} = 0)$$

Similarly,

$$N_{a2} = \frac{N_{A2}T_{A2} + N_{S2}T_{S2}}{T_{A2} + T_{S2}}$$

All the sun and the planet gears are of the same diameter and pitch,

$$\therefore \quad T_{A1} = 2\left[\frac{T_{S1}}{2} + T_{P1}\right] = 2\left[\frac{T_{S1}}{2} + T_{S1}\right] = 3T_{S1}$$

Thus,

$$T_{A1} = T_{A2} = 3T_{S1} = 3T_{S2} = 3T_{P1} = 3T_{P2}$$

$$\text{Then } N_{a1} = \frac{N_{S1}T_{S1}}{T_{A1} + T_{S1}} = \frac{N_{S1}T_{S1}}{3T_{S1} + T_{S1}} = \frac{N_{S1}}{4}$$

$$\text{and } \quad N_{a2} = \frac{3N_{A2} + N_{S2}}{4}$$

$$= \frac{3N_{a1} + N_{S1}}{4} \qquad (N_{A2} = N_{a1})$$

$$= \frac{3N_{S1}/4 + N_{S1}}{4}$$

or

$$N_{a2} = \frac{7}{16} \times 1000 = 437.5 \text{ rpm}$$

$$\text{Speed of shaft } C = \underline{437.5 \text{ rpm}}$$

If T denotes the torque,

$$T_S N_S + \frac{T_{a2}N_{a2}}{\eta} = 0$$

$$150 \times 1000 + \frac{T_{a2} \times 437.5}{0.97} = 0$$

$$T_{a2} = \underline{-332.6 \text{ N.m}}$$

Also

$$T_{S1} + T_{a2} + T_{A1} = 0$$

$$150 - 332.6 + T_{A1} = 0$$

or

$$\text{Holding torque, } T_{A1} = \underline{182.6 \text{ N.m}}$$

in the same direction as the input torque.

11.9 BEVEL EPICYCLIC GEAR

The methods used for the solution of epicyclic trains of the spur gears are also valid for epicyclic trains consisting of bevel wheels. However, for wheels whose axes are inclined to the main axis, the terms clockwise and counter-clockwise are not applicable. So, in the tabular method of solution, plus or minus sign is omitted for these wheels and also the addition of y is not convenient to make as the rotation is about a different axis.

Example 11.7 Figure 11.15 shows a Humpage gear used in a lathe headstock. The number of teeth on the wheels B, C, D, E and F are 18, 54, 22, 44 and 72 respectively. If the shaft G rotates at 200 rpm, what will be the speed of shaft H when (i) F is held stationary and (ii) it is rotated at 300 rpm opposite to G?

Solution:

$$T_B = 18 \qquad T_E = 44$$
$$T_C = 54 \qquad T_F = 72$$
$$T_D = 22 \qquad N_G = N_B = 200 \text{ rpm}$$

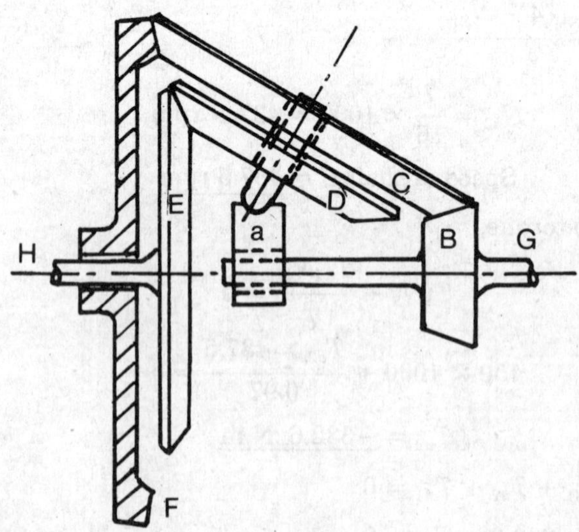

Fig. 11.15

Note that the axis of the wheels C and D is inclined to the main axis (of wheels B, E and F). Therefore, it is usual not to affix any sign to the direction of rotation of these wheels as mentioned earlier. However, they influence the direction of rotation of the other wheels.

Action	a	B/G	C/D	E/H	F
'a' fixed, B + 1 rev.	0	1	$\frac{18}{54}$	$-\frac{18}{54}\times\frac{22}{44}$	$-\frac{18}{54}\times\frac{54}{72}$
'a' fixed, B + x rev.	0	x	$\frac{x}{3}$	$-\frac{x}{6}$	$-\frac{x}{4}$
Add y	y	$y + x$	$y+\frac{x}{3}$	$y-\frac{x}{6}$	$y-\frac{x}{4}$

From given conditions,

$$N_G = y + x = 200 \quad \text{or} \quad y = 200 - x$$

and $$N_F = y - \frac{x}{4} = 0$$

or $$200 - x - \frac{x}{4} = 0$$

or $$x = 160$$

$$y = 200 - 160 = 40$$

$\therefore$ $$N_H = y - \frac{x}{6} = 40 - \frac{160}{6} = \underline{13.3 \text{ rpm}}$$

In the second case,

$$N_G = y + x = 200 \quad \text{or} \quad y = 200 - x$$

$$N_F = y - \frac{x}{4} = -30$$

or $$200 - x - \frac{x}{4} = -30$$

or $$x = 184$$

$$y = 200 - 184 = 16$$

$\therefore$ $$N_A = y - \frac{x}{6} = 16 - \frac{184}{6} = \underline{-13.3 \text{ rpm}}$$

Thus in the second case, shaft H rotates in the opposite direction to that of G.

Example 11.8 Figure 11.16 shows a train of bevel gears. The wheel E is fixed whereas the wheels B and G are keyed to the driving and the driven shafts respectively. The wheels C, D and F are keyed to the inclined shaft which is supported on the arm a. The arm is free to rotate about the common axis of the driving and driven shafts. The number of teeth on the wheels B,

C, D, E, F and G are 15, 45, 45, 135, 40 and 100 respectively. Find the ratio of the driving and the driven shaft speeds.

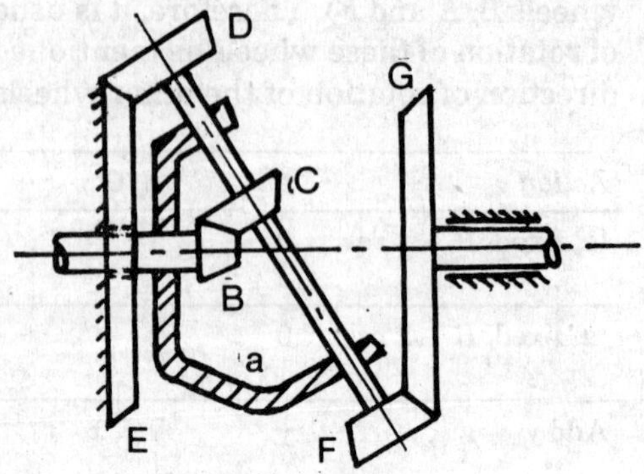

Fig. 11.16

Solution:

$$T_B = 15 \qquad T_E = 135$$
$$T_C = 45 \qquad T_F = 40$$
$$T_D = 45 \qquad T_G = 100$$

From the figure, it is clear that the train consists of two epicyclic trains. The train BCDE causes the arm a to turn and the arm causes G to turn through train $EDFG$.

Considering the first train $BCDE$,

Action	*a*	*B*	*C/D*	*E*
'a' fixed, B + 1 rev.	0	1	$\frac{15}{45}$	$-\frac{15}{45} \times \frac{45}{135}$
'a' fixed, B + x rev.	0	x	$\frac{x}{3}$	$-\frac{x}{9}$
Add y	y	$y + x$	$y + \frac{x}{3}$	$y - \frac{x}{9}$

From the given condition,

$$N_E = y - \frac{x}{9} = 0 \quad \text{or} \quad y = \frac{x}{9}$$

$$\therefore \qquad \frac{N_a}{N_B} = \frac{y}{y+x} = \frac{\frac{x}{9}}{\frac{x}{9} + x} = \frac{1}{10}$$

or $$N_a = \frac{N_B}{10}$$

Considering the second train $EDFG$,

Action	*a*	*E*	*D/F*	*G*
'a' fixed, E + 1 rev.	0	1	$\frac{135}{45}$	$\frac{135}{45} \times \frac{40}{100}$
'a' fixed, E + x rev.	0	x	$3x$	$\frac{6x}{5}$
Add y	y	$y + x$	$y + 3x$	$y + \frac{6x}{5}$

We have,

$$N_E = y + x = 0 \qquad \text{or} \qquad y = -x$$

Also

$$N_a = y = \frac{N_B}{10}$$

$\therefore$

$$x = -\frac{N_B}{10}$$

Hence,

$$\frac{N_G}{N_B} = \frac{y + \frac{6x}{5}}{N_B} = \frac{\frac{N_B}{10} - \frac{6N_B}{5 \times 10}}{N_B} = -\frac{1}{50}$$

Thus, B must turn 50 times to turn G once in the opposite direction.

11.10 COMPOUND EPICYCLIC GEAR

When an epicyclic gear consists of a number of epicyclic gears (sun and planet gears) in series such that the pin of the arm of first epicyclic gear drives an element of another epicyclic gear, it is known as a compound epicyclic gear.

Figure 11.17 shows a compound epicyclic gear. It consists of three simple epicyclic gears namely $A_1a_1P_1S_1$, $A_2a_2P_2S_2$ and $A_3a_3P_3S_3$. The planet P_1 of the first epicyclic gear rotates freely on the pin carried by the arm a_1. As the arm a_1 is integral with the annulus A_2 of the second epicyclic gear and the sun wheel of the third, the pin of the arm a_1 also drives A_2 and S_3, i.e. the annulus and the sun wheel of the second and the third epicyclic gears respectively, at the same speed and direction as its own about the axis of the arm. Also the

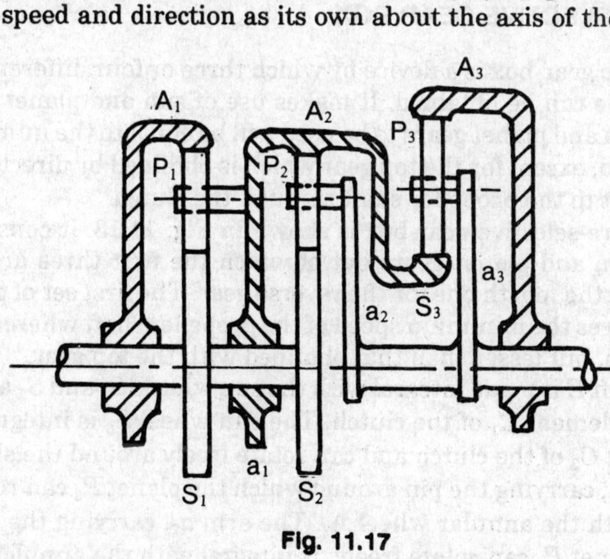

Fig. 11.17

sun wheel of the first is integral with that of second and the arm of the second with that of the third.

To analyse a compound epicyclic gear, each epicyclic gear or sun and planet gear is treated separately. Thus for the first epicyclic gear,

Speed of the arm,

$$N_{a1} = \frac{N_{S1}T_{S1} + N_{A1}T_{A1}}{T_{S1} + T_{A1}} \tag{11.12}$$

Thus if the speeds N_{S1} and N_{A1} are known (if A_1 is fixed, $N_{A1} = 0$), N_{a1} can be calculated.

Similarly, for the second sun and planet gear,

$$N_{a2} = \frac{N_{S2}T_{S2} + N_{A2}T_{A2}}{T_{S2} + T_{A2}}$$

As A_2 is integral with a_1 and S_2 with S_1,

$$\therefore \quad N_{A2} = N_{a1} \quad \text{and} \quad N_{S2} = N_{S1}$$

Thus N_{a2} can be known.

In the same way,

$$N_{a3} = \frac{N_{S3}T_{S3} + N_{A3}T_{A3}}{T_{S3} + T_{A3}}$$

a_3 is integral with a_2 and S_3 with a_1.

$$\therefore \quad N_{a3} = N_{a2} \quad \text{and} \quad N_{S3} = N_{a1}$$

So, N_{A3} can be calculated.

11.11 PRE-SELECTIVE GEAR BOX

A pre-selective gear-box is a device by which three or four different speeds of the automobile can be obtained. It makes use of sun and planet gears. The number of sun and planet gears to be used will be equal to the number of gear ratios required, except for the top gear which is obtained by direct drive from the crankshaft to the propeller shaft through the clutch.

A typical pre-selective gear-box is shown in Fig. 11.18. It consists of four sets of the sun and planet gears out of which the first three are for speed reduction and the fourth one for the reverse gear. The first set of the sun and planet gear gives the minimum speed of the propeller shaft whereas the third the maximum, but lesser than that obtained with the top gear.

Engine shaft E is made integral with the sun wheels S_1 and S_2 and is keyed to the inner element C_1 of the clutch. The sun wheel S_3 is integral with the outer element C_2 of the clutch and can rotate freely around the shaft E.

The arm a_3, carrying the pin around which the planet P_3 can rotate freely, is integral with the annular wheel A_2. The arm a_2, carrying the pin around which the planet P_2 can rotate freely, is integral with the annular wheels A_1

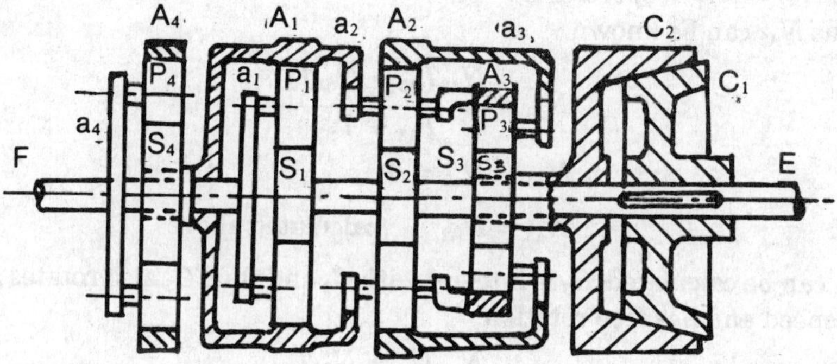

Fig. 11.18

and A_3. The arm a_1, carrying the pin for the planet P_1, is integral with the propeller shaft F and the arm a_4. The arm a_4 carries a pin for the planet P_4.

The sun wheel S_4 is integral with the annular wheel A_1 and can rotate freely around the shaft F.

All the sun wheels gear with their annular wheels through their respective planet gears.

Brake bands (not shown in the figure) are provided around the annular wheels A_1, A_2, A_4 and the element C_2 of the clutch by which any of these can be locked in turn. During the locking of any of these, the two elements C_1 and C_2 of the clutch are disengaged automatically. The clutch is engaged only for the direct drive (top gear).

First Gear

To engage the first gear, A_1 is locked and the clutch is disengaged. Consider the first set of the sun and planet gear,

Speed of arm,

$$N_{a1} = \frac{N_{A1}T_{A1} + N_{S1}T_{S1}}{T_{A1} + T_{S1}}$$

where N_{S1} = speed of the sun wheel S_1

= speed of the propeller shaft E

N_{A1} = speed of the annular wheel A_1

= 0

Therefore, N_{a1} can be known which is also the speed of the shaft F.

The speed of the other gears which rotate freely can also be known as follows:

For the second set of the sun and planet gear,

$$N_{a2} = \frac{N_{A2}T_{A2} + N_{S2}T_{S2}}{T_{A2} + T_{S2}}$$

where $N_{a2} = N_{A1} = 0$
and $N_{S2} = N_{S1}$ = speed of E
Thus N_{A2} can be known.

$$N_{a3} = \frac{N_{A3}T_{A3} + N_{S3}T_{S3}}{T_{A3} + T_{S3}}$$

$$N_{A3} = N_{a2} = 0$$

and $N_{a3} = N_{A2}$ calculated above

N_{S3} can be calculated. S_3 is integral with C_2 and thus C_2 also rotates at the same speed and has free rotation.

$$N_{a4} = \frac{N_{A4}T_{A4} + N_{S4}T_{S4}}{T_{A4} + T_{S4}}$$

where $N_{a4} = N_{a1}$
and $N_{S4} = N_{A1} = 0$
Thus N_{A4} can be calculated. The annular wheel A_4 has free rotation.

Second Gear

To engage the second gear, A_2 is locked and the clutch is disengaged.

$$N_{a2} = \frac{N_{A2}T_{A2} + N_{S2}T_{S2}}{T_{A2} + T_{S2}}$$

where $N_{A2} = 0$
and N_{S2} = speed of E
Thus N_{a2} can be known.

$$N_{a1} = \frac{N_{A1}T_{A1} + N_{S1}T_{S1}}{T_{A1} + T_{S1}}$$

where $N_{A1} = N_{a2}$
and $N_{S1} = N_E$

N_{a1}, which is also the speed of the propeller shaft, can be known.

Free rotations of A_4 and C_2 can also be known in a similar way as in case of first gear.

Third Gear

Lock C_2 and disengage the clutch. The operation also locks the sun wheel S_3.

$$N_{a3} = \frac{N_{A3}T_{A3} + N_{S3}T_{S3}}{T_{A3} + T_{S3}} = \frac{N_{A3}T_{A3}}{T_{A3} + T_{S3}} \qquad (N_{S3} = 0)$$

and $$N_{a2} = \frac{N_{A2}T_{A2} + N_{S2}T_{S2}}{T_{A2} + T_{S2}}$$

In the above two equations,

$$N_{a3} = N_{A2},\ N_{a2} = N_{A3} \quad \text{and} \quad N_{S2} = N_E$$

Thus N_{a2} can be calculated. From N_{a2}, the value of N_{a1} can be calculated in the way it is done for the second gear which further provides the speed of the propeller shaft.

Fourth (Top) Gear

All the brake bands are freed. C_1 and C_2 are engaged. Thus S_1, S_2 and S_3 all rotate at the engine speed and it can be shown that the whole system rotates as a complete unit at the engine speed. Thus speed of the propeller shaft is equal to that of the engine shaft.

Reverse Gear

The clutch is disengaged and A_4 is locked.

We have,

$$N_{a4} = \frac{N_{A4}T_{A4} + N_{S4}T_{S4}}{T_{A4} + T_{S4}}$$

$$= \frac{N_{S4}T_{S4}}{T_{A4} + T_{S4}} \qquad (T_{A4} = 0)$$

and

$$N_{a1} = \frac{N_{A1}T_{A1} + N_{S1}T_{S1}}{T_{A1} + T_{S1}}$$

In the two equations,

$$N_{a4} = N_{a1}$$

$$N_{S4} = N_{A1}$$

and

$$N_{S1} = N_E$$

Thus N_{a1} or N_{a4} can be known which is the speed of the propeller shaft.

Assuming roughly that $T_{A1} = T_{A4}$ and $T_{S1} = T_{S4}$,

$$N_{S4}T_{S4} = N_{A1}T_{A1} + N_{S1}T_{S1} = N_{S4}T_{A1} + N_{S1}T_{S1}$$

$$N_{S4}\,(T_{S4} - T_{A1}) = N_{S1}T_{S1}$$

$$N_{S4} = \frac{N_{S1} - T_{S1}}{T_{S4} - T_{A1}}$$

As T_{A1} is greater than T_{S4}, N_{S4} will be negative. Thus N_{a4} is negative, which shows that the propeller shaft rotates in the reverse direction.

Example 11.9 In the pre-selective gear-box shown in Fig. 11.18, number of teeth are:

$T_{A1} = T_{A2} = 72$	$T_{S1} = T_{S2} = 22$
$T_{A3} = 63$	$T_{S3} = 19$
$T_{A4} = 81$	$T_{S4} = 37$

If the input shaft E rotates at a uniform speed of 800 rpm, determine the speeds of the output shaft F when different gears are engaged.

Solution: *When the first gear is engaged,*

$$N_{a1} = \frac{N_{S1}T_{S1}}{T_{A1} + T_{S1}} = \frac{800 \times 22}{72 + 22} = 187.2$$

or $$N_F = \underline{187.2 \text{ rpm}}$$

When the second gear is engaged,

$$N_{a2} = \frac{N_{S2}T_{S2}}{T_{A2} + T_{S2}} = \frac{800 \times 22}{72 + 22} = 187.2$$

$$N_{a1} = \frac{N_{A1}T_{A1} + N_{S1}T_{S1}}{T_{A1} + T_{S1}} = \frac{N_{a2}T_{A1} + N_{S1}T_{S1}}{T_{A1} + T_{S1}}$$

$$= \frac{187.2 \times 72 + 800 \times 22}{72 + 22}$$

or $$N_F = N_{a1} = \underline{330.6 \text{ rpm}}$$

When the third gear is engaged,

$$N_{a3} = \frac{N_{A3}T_{S3}}{T_{A3} + T_{S3}} = \frac{63N_{A3}}{63 + 19} = \frac{63N_{A3}}{82}$$

or $$N_{A3} = \frac{82N_{a3}}{63}$$

$$N_{a2} = \frac{N_{A2}T_{A2} + N_{S2}T_{S2}}{T_{A2} + T_{S2}}$$

$$= \frac{N_{a3}T_{A2} + N_{S2}T_{S2}}{T_{A2} + T_{S2}} \qquad (N_{A2} = N_{a3})$$

$$= \frac{\frac{63N_{A3}}{82}\cdot T_{A2} + N_{S2}T_{S2}}{T_{A2} + T_{S2}}$$

$$= \frac{\frac{63N_{a2}}{82}\cdot T_{A2} + N_{S2}T_{S2}}{T_{A2} + T_{S2}} \qquad (N_{a2} = N_{A3})$$

$$= \frac{\frac{63N_{a2}}{82} \times 72 + 800 \times 22}{72 + 22}$$

or $$94\,N_{a2} = 55.32\,N_{a2} + 17\,600$$

or $$38.68\,N_{a2} = 17\,600$$

or $$N_{a2} = 455 = N_{A1}$$

$$N_{a1} = \frac{N_{A1}T_{A1} + N_{S1}T_{S1}}{T_{A1} + T_{S1}} = \frac{455 \times 72 + 800 \times 22}{94}$$

or $$N_F = N_{a1} = \underline{535.7 \text{ rpm}}$$

When the fourth gear is engaged,

$$N_F = N_E = \underline{800 \text{ rpm}}$$

In the reverse gear,

$$N_{a4} = \frac{N_{S4}T_{S4}}{T_{A4} + T_{S4}} \qquad (N_{A4} = 0)$$

or $$N_{a4} = \frac{37N_{S4}}{81 + 37} = \frac{37N_{S4}}{118}$$

$$N_{a1} = \frac{N_{A1} \times 72 + 800 \times 22}{94}$$

But

$$N_{a4} = N_{a1} \quad \text{and} \quad N_{S4} = N_{A1}$$

$$\frac{37N_{S4}}{118} = \frac{72N_{S4} + 17600}{94}$$

or $$0.452\, N_{S4} = -187.23$$

or $$N_{S4} = -413.9$$

$$N_F = N_{a4} = -\frac{37 \times 413.9}{118} = -\underline{129.8 \text{ rpm}}$$

11.12 DIFFERENTIAL GEAR

When a vehicle takes a turn, the outer wheels must travel farther than the inner wheels. In automobiles, the front wheels can rotate freely on their axes and thus can adapt themselves to the conditions. However, both rear wheels are driven by the engine through gearing. Therefore, some sort of automatic device is necessary so that the two rear wheels are driven at slightly different speeds. This is accomplished by fitting a differential gear on the rear axle.

The fact that an epicyclic gear has two degrees of freedom has been utilised in the differential gear of an automobile. It permits the two wheels to rotate at the same speed when driving in straight while allowing the wheels to rotate at different speeds when taking a turn. Thus a differential gear is a device which adds or subtracts angular displacements.

Figure 11.19 shows the arrangement of gears in the differential of an automobile. The shaft S is driven by the engine through the gear-box and has a bevel pinion A keyed to it. Bevel pinion A meshes with a bevel wheel B which turns loosely on the hub of gear C. Shafts S_1 and S_2 form the rear axle to which are fixed the rear wheels of the automobile. Gears C and D are keyed to the shafts S_1 and S_2 respectively. C and D gear with equal bevel pinions E and F which are free to rotate on their respective axes. The wheel B carries two brackets that support the bearings for gears E and F.

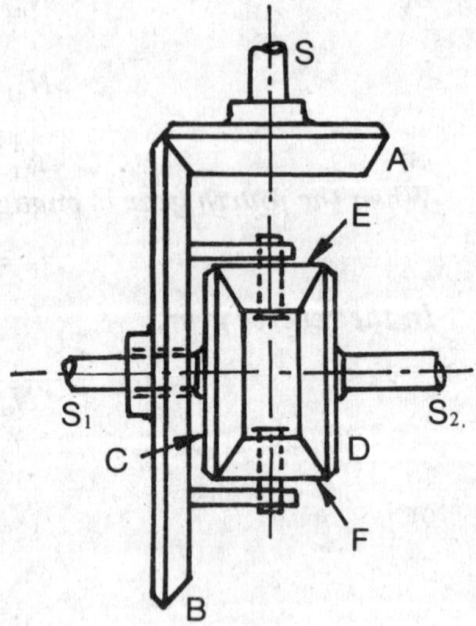

Fig. 11.19

When the automobile moves in a straight path, the bevel pinion A drives the wheel B. The whole differential acts as one unit and rotates with the bevel wheel B so that the wheels C and D rotate with the same speed and in the same direction as B. There is no relative motion between gears C and D, and E and F. Gears E and F also do not rotate about their own axes. They act just like keys to transmit motion from B to C and D. Thus C and D, which are keyed to the shafts that carry the wheels rotate at the same speed as B.

When a turn is taken, E and F rotate about their own axes and the system works as an epicyclic gear giving two outputs at C and D with one input at B. Prepare a table as below:

Action	*B*	*C/S*$_1$	*D/S*$_2$	*E/F*
B fixed, *C* + 1 rev.	0	1	−1	$\frac{T_C}{T_E}$
B fixed, *C* + *x* rev.	0	x	$-x$	$\frac{T_C}{T_E}x$
Add *y*	y	$y + x$	$y - x$	$\frac{T_C}{T_E}x + y$

With reference to the above table, it is seen that the speed of B is the arithmetical mean of the speeds of C and D, because $y = \{(y + x) + (y - x)\}/2$. This shows that while taking a turn, if the speed of C decreases than that of B, there will be a corresponding increase in the speed of D.

PROBLEMS

11.1 An epicyclic gear consists of a pinion, a wheel of 40 teeth and an annulus with 84 internal teeth concentric with the wheel. The pinion gears with the wheel and the annulus. The arm that carries the axis of the pinion rotates at 100 rpm. If the annulus is fixed, find the speed of the wheel; if wheel is fixed, find the speed of the annulus.

(310 rpm; 147.6 rpm)

11.2 In an epicyclic gear shown in Fig. 11.7, the pitch circle diameter of the annulus *A* is to be approximately 324 mm and the module 6 mm. When the annulus is stationary, the three-armed spider makes one revolution for every five revolutions of the wheel *S*. Find the number of teeth for all the wheels and exact pitch circle diameter of the annulus. If a torque of 30 N.m is applied to the shaft carrying *S*, determine the fixing torque of the annulus.

($T_S = 14$, $T_P = 21$, $T_A = 56$; 120 N.m)

11.3 Determine a suitable train of wheels to satisfy the requirements of a clock, the minute hand of which is fixed to a spindle and the hour hand to a sleeve rotating freely on the same spindle. The pitch is the same for all the wheels and each wheel has at least 11 teeth. The total number of teeth should be as small as possible.

(12, 48 and 15, 45)

11.4 The pinion *S* (Fig. 11.8) has 15 teeth, and is rigidly fixed to a motor shaft. The wheel *P* has 20 teeth and gears with *S* and also with a fixed annulus wheel *A*. The pinion *C* has 15 teeth and is fixed to the wheel *P*. *C* gears with the annular wheel *D*, which is keyed to a machine shaft. *P* and *C* can rotate together on a pin carried by an arm which rotates about the shaft on which *S* is fixed. Find the speed of the machine shaft if the motor rotates at 1000 rpm.

(37.5 rpm in the same direction as S)

11.5 In an epicyclic gear (Fig. 11.7), the wheel *A* has also external teeth and is driven by a pinion *B*. Number of teeth on *S* = number of teeth on *B* = one fourth of number of teeth on *A*. The wheel *A* has same number of teeth internally and externally.

Determine the speed of the driven shaft fixed to the arm *a*, when *B* makes 600 rpm in the same direction as *S* whose speed is 400 rpm. Also find the speed and direction of rotation of *B* if the driven shaft rotates at the same speed as above but in the opposite direction.

(40 rpm in a direction opposite to S; 200 rpm in direction of S)

11.6 Figure 11.20 shows an epicyclic train known as Ferguson's paradox. The gears have number of teeth as indicated. Gear 1 is fixed to the frame and is stationary. The arm *a* and the gears 2 and 3 are free to rotate on the shafts. The pitch circle diameters of all are the same so that the planet gear *P* meshes with them all. Find the number of revolutions of gears 2 and 3 for one revolution of arm *a*.

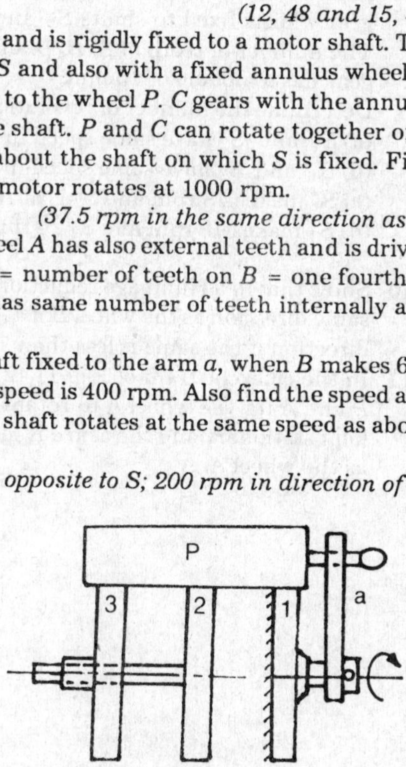

Fig. 11.20

(2 makes 1/81 revs. in the same direction and 3 makes 1/79 revs in the opposite direction of arm 'a')

Note: If the three gears 1, 2 and 3 have the same diameter, their pitches must be slightly different and, theoretically, the drive will not be perfect.

11.7 In the epicyclic gear of Fig. 11.12, $T_S = 40, T_P = 20$ and $T_A = 80$. If the sun gear rotates at 150 rpm counter-clockwise and the annular ring clockwise at 400 rpm, find the arm speed.

(216.7 rpm clockwise)

11.8 In an epicyclic gear (Fig. 11.21), the wheel A fixed to S_1 has 30 teeth and rotates at 500 rpm. B gears with A and is fixed rigidly to C, both being free to rotate on S_2. The wheels B, C and D have 50, 70 and 90 teeth respectively. If D rotates at 80 rpm in a direction opposite to that of A, find the speed of the shaft S_2.

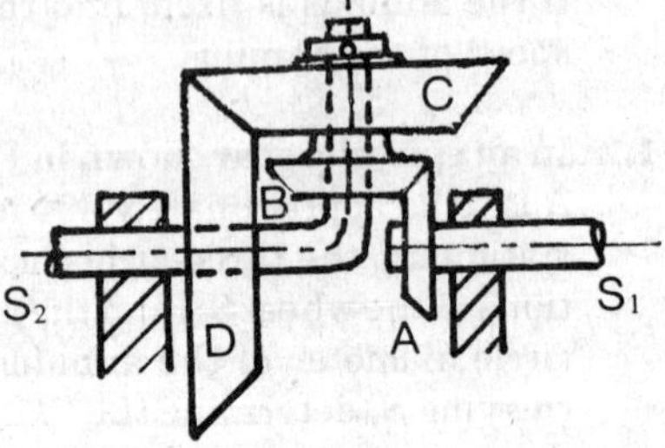

Fig. 11.21

(104.5 rpm in same direction)

11.9 In an indexing mechanism of a milling machine (Fig. 11.19), the drive is from gear wheels fixed to shafts S_1 and S_2 to the bevel A through the gear train. The number of teeth of A, B, D and F are 30, 60, 28 and 24 respectively. Each gear has a module of 10 mm.

Determine the number of revolutions of S (or A) for one revolution of S_1 when:

(a) S_1 and S_2 have same speed in the same direction

(b) S_1 and S_2 have same speed in the opposite directions

(c) S_1 makes 48 rpm and S_2 is at rest

(d) S_1 makes 48 rpm and S_2 24 rpm in the same direction.

(2; zero; 48 rpm; 72 rpm)

11.10 Show that in a Humpage reduction gear (Fig. 11.15), the wheel E rotates in the same direction as the wheel B if T_C/T_D is more than T_F/T_E and in the opposite direction if the same is less than T_F/T_E.

11.11 In the epicyclic train of Fig. 11.12, determine the numbers of teeth on wheels P and A for the wheel A to rotate 130 times if the wheel S having 60 teeth, is kept stationary and the arm a is made to rotate 100 times in the same direction as the wheel A.

(70; 200)

12

STATIC FORCE ANALYSIS

12.1 INTRODUCTION

In the design of machine mechanisms, it is imperative to know the magnitudes as well as the directions of forces transmitted from the input to the output. The analysis helps in selecting proper sizes of the machine components to withstand the stresses developed in them. If proper sizes are not selected, the components may fail during the machine operation. On the other hand, if the members are designed to have more strength than required, the machine may not be able to compete with others due to more cost, weight, size, etc.

If components of a machine accelerate, inertia forces are produced due to their masses. However, if the magnitudes of these forces are small compared to the externally applied loads, they can be neglected while analysing the mechanism. Such an analysis is known as *static-force analysis*. For example, in lifting cranes, the bucket load and the static weight loads may be quite high relative to any dynamic loads due to accelerating masses, and thus static-force analysis is justified.

When the inertia effect due to the mass of the components is also considered, it is called *dynamic-force analysis*. This will be discussed in detail in Chapter 13.

12.2 STATIC EQUILIBRIUM

A body is in static equilibrium if it remains in its state of rest or motion. If the body is at rest, it tends to remain at rest and if it is in motion, it tends to keep the motion. In static equilibrium

– the vector sum of all the forces acting on the body is zero and
– the vector sum of all the moments about any arbitrary point is zero.

Mathematically,

$$\Sigma \mathbf{F} = 0 \tag{12.1}$$

$$\Sigma \mathbf{T} = 0 \tag{12.2}$$

In a planer system, forces can be described by two-dimensional vectors and, therefore,

$$\Sigma \mathbf{F}_x = 0 \tag{12.3}$$

$$\Sigma \mathbf{F}_y = 0 \tag{12.4}$$

$$\Sigma \mathbf{T}_z = 0 \tag{12.5}$$

12.3 EQUILIBRIUM OF TWO-FORCE AND THREE-FORCE MEMBERS

A member under the action of two forces will be in equilibrium if

– the forces are of the same magnitude,

– the forces act along the same line, and

– the forces are in opposite directions.

Figure 12.1 shows such a member.

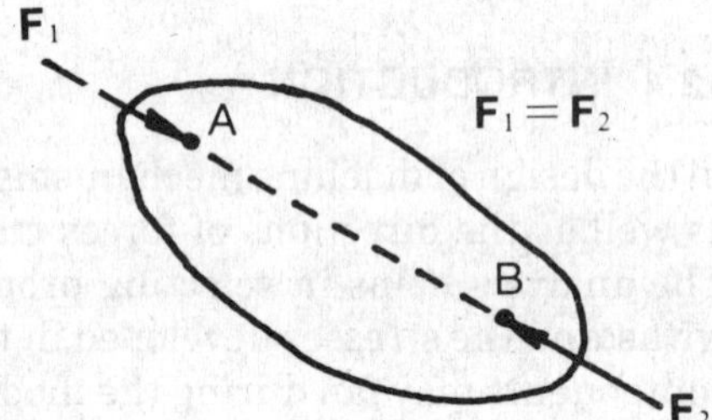

Fig. 12.1

A member under the action of three forces will be in equilibrium if:

– the resultant of the forces is zero, and

– the lines of action of the forces intersect at a point (known as point of concurrency).

Figure 12.2 a shows a member acted upon by three forces $\mathbf{F}_1$, $\mathbf{F}_2$ and $\mathbf{F}_3$ and is in equilibrium as the lines of action of forces intersect at one point O and the resultant is zero. This is verified by adding the forces vectorially (Fig. 12.2 b). As the head of the last vector $\mathbf{F}_3$ meets the tail of the first vector $\mathbf{F}_1$, the resultant is zero. It is not necessary to add the three vectors in order to obtain the resultant as is shown in Fig. 12.2 c in which $\mathbf{F}_2$ is added to $\mathbf{F}_3$ and then $\mathbf{F}_1$ is taken.

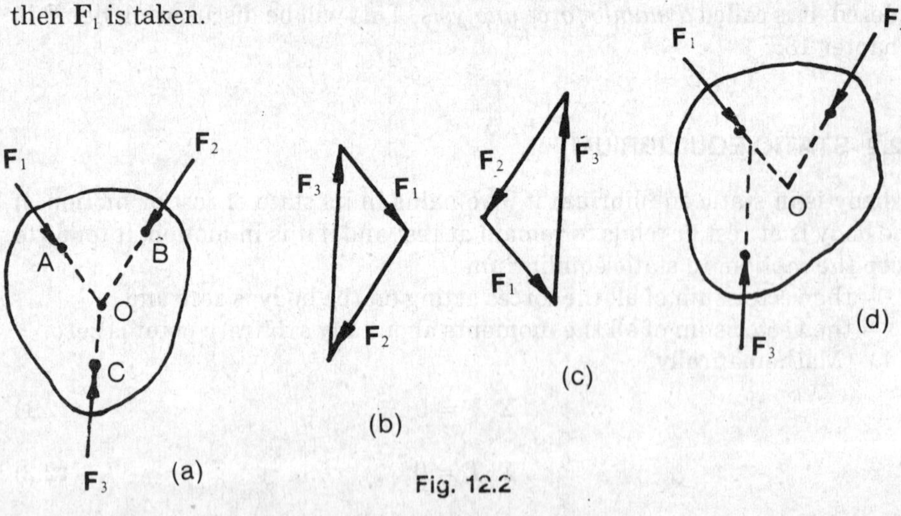

Fig. 12.2

Figure 12.2 d shows a case where the magnitudes and directions of the forces are the same as before, but the lines of action of the forces do not intersect at one point. Thus the member is not in equilibrium.

Consider a member in equilibrium in which force $\mathbf{F}_1$ is completely known, $\mathbf{F}_2$ known in direction only and $\mathbf{F}_3$ is completely unknown. The points of applications of $\mathbf{F}_1$, $\mathbf{F}_2$ and $\mathbf{F}_3$ are *A*, *B* and *C* respectively. To solve such a problem, first find the point of concurrency *O* from the two forces with known directions, i.e. from $\mathbf{F}_1$ and $\mathbf{F}_2$. Joining *O* with *C* gives the line of action of the third force $\mathbf{F}_3$. To know the magnitudes of the forces $\mathbf{F}_2$ and $\mathbf{F}_3$, take a vector of proper magnitude and direction to represent the force $\mathbf{F}_1$. From its two ends draw lines parallel to lines of action of the forces $\mathbf{F}_2$ and $\mathbf{F}_3$ forming a force triangle (Figs. 12.2 b or c). Mark arrowheads on $\mathbf{F}_2$ and $\mathbf{F}_3$ so that $\mathbf{F}_1$, $\mathbf{F}_2$ and $\mathbf{F}_3$ are in the same order.

If the lines of action of two forces are parallel, then the point of concurrency lies at infinity and, therefore, the third force is also parallel to the first two.

12.4 MEMBER WITH TWO FORCES AND A TORQUE

A member under the action of two forces and an applied torque will be in equilibrium if:

– the forces are equal in magnitude, parallel in direction and opposite in sense and

– the forces form a couple which is equal and opposite to the applied torque.

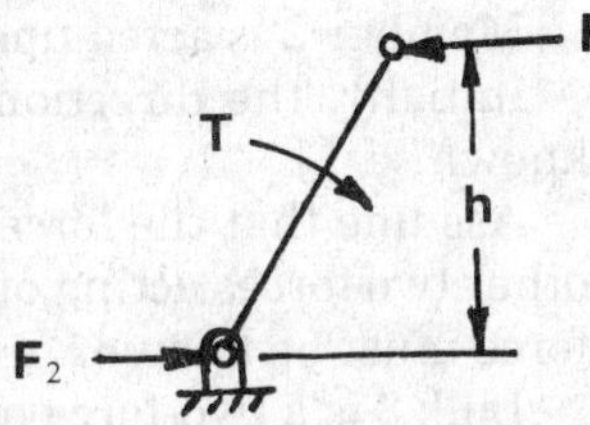

Fig. 12.3

Figure 12.3 shows a member acted upon by two equal forces $\mathbf{F}_1$ and $\mathbf{F}_2$ and an applied torque $\mathbf{T}$. For equilibrium,

$$T = F_1 \times h = F_2 \times h \tag{12.6}$$

where T, F_1 and F_2 are the magnitudes of $\mathbf{T}$, $\mathbf{F}_1$ and $\mathbf{F}_2$ respectively. $\mathbf{T}$ is clockwise whereas the couple formed by $\mathbf{F}_1$ and $\mathbf{F}_2$ is counter-clockwise.

12.5 FORCE CONVENTION

The force exerted by member *i* on member *j* is represented by $\mathbf{F}_{ij}$.

12.6 FREE BODY DIAGRAMS

A free body diagram is a sketch or diagram of a part isolated from the mechanism in order to determine the nature of forces acting on it.

Figure 12.4 a shows a four-link mechanism. The free-body diagrams of its members 2, 3 and 4 are shown in Figs. 12.4 b, c and d respectively. Each member is in equilibrium individually.

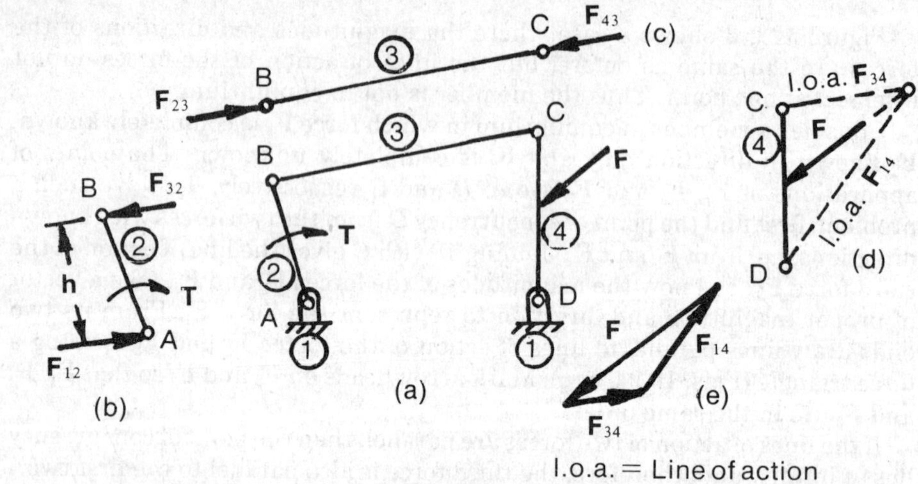

Fig. 12.4

Member 4 is acted upon by three forces $\mathbf{F}$, $\mathbf{F}_{34}$ and $\mathbf{F}_{14}$.

Member 3 is acted upon by two forces $\mathbf{F}_{23}$ and $\mathbf{F}_{43}$.

Member 2 is acted upon by two forces $\mathbf{F}_{32}$ and $\mathbf{F}_{12}$ and a torque $\mathbf{T}$.

Initially, the direction and the sense of some of the forces may not be known.

Assume that the force $\mathbf{F}$ on member 4 is known completely. To know the other two forces acting on this member completely, the direction of one more force must be known.

Link 3 is a two-force member and for its equilibrium $\mathbf{F}_{23}$ and $\mathbf{F}_{43}$ must act along BC. Thus $\mathbf{F}_{34}$, being equal and opposite to $\mathbf{F}_{43}$, also acts along BC. For member 4 to be in equilibrium, $\mathbf{F}_{14}$ passes through the intersection of $\mathbf{F}$ and $\mathbf{F}_{34}$. By drawing a force triangle ($\mathbf{F}$ is completely known), magnitudes of $\mathbf{F}_{14}$ and $\mathbf{F}_{34}$ can be known (Fig. 12.4 e).

Now $F_{34} = F_{43} = F_{23} = F_{32}$

Member 2 will be in equilibrium if $\mathbf{F}_{12}$ is equal, parallel and opposite to $\mathbf{F}_{32}$ and

$$T = F_{12} \times h = F_{32} \times h$$

12.7 SUPERPOSITION

In linear systems, if a number of loads act on a system of forces, the net effect is equal to the superposition of the effects of the individual loads taken one at a time. A linear system is one in which the output force is directly proportional to the input force, i.e. in mechanisms where coulomb or dry friction is neglected.

Example 12.1 Figure 12.5 a shows a quaternary link $ABCD$ under the action of forces $\mathbf{F}_1$, $\mathbf{F}_2$, $\mathbf{F}_3$ and $\mathbf{F}_4$ acting at A, B, C and D respectively. The link is in

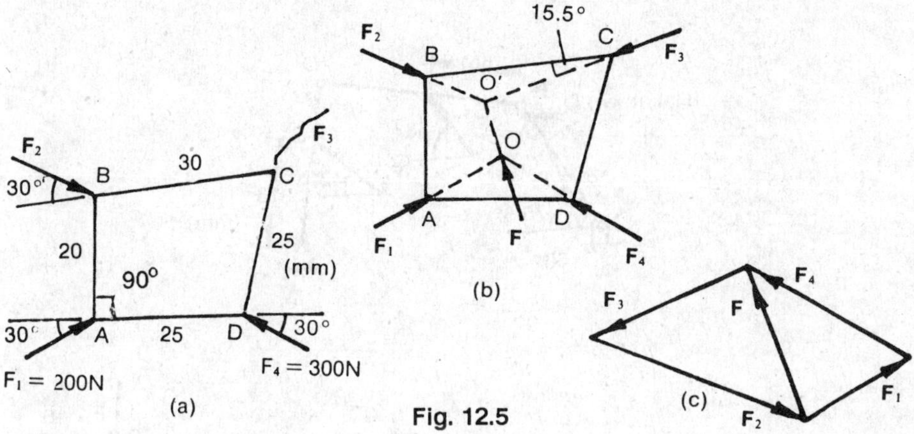

Fig. 12.5

static equilibrium. Determine the magnitudes of the forces $\mathbf{F}_2$ and $\mathbf{F}_3$ and the direction of $\mathbf{F}_3$.

Solution: The forces $\mathbf{F}_1$ and $\mathbf{F}_4$ can be combined into a single force $\mathbf{F}$ by obtaining their resultant (Figs. 12.5b and c). The force $\mathbf{F}$ acts through O, the point where lines of action of $\mathbf{F}_1$ and $\mathbf{F}_4$ meet.

Now, the four-force member $ABCD$ is reduced to a three-force member under the action of forces $\mathbf{F}$ (completely known), $\mathbf{F}_2$ (only the direction known) and $\mathbf{F}_3$ (completely unknown).

Let $\mathbf{F}$ and $\mathbf{F}_2$ meet at O'. Then CO' is the line of action of force $\mathbf{F}_3$. By completing the force triangle, obtain the magnitude of $\mathbf{F}_2$ and $\mathbf{F}_3$.

Magnitude of $\mathbf{F}_2$ = $\underline{380 \text{ N}}$

Magnitude of $\mathbf{F}_3$ = $\underline{284 \text{ N}}$

Line of action of force $\mathbf{F}_3$ makes an angle of 15.5° with CB.

Example 12.2 A four-link mechanism is subjected to following external forces (Fig. 12.6a):

Link	*Length*	*Force*	*Magnitude*	*Point of application of force (r)*
AB (2)	500 mm	$\mathbf{F}_2$	80 ∠73.5°N	325 mm from *A*
BC (3)	660 mm	$\mathbf{F}_3$	144 ∠58°N	297 mm from *B*
CD (4)	560 mm	$\mathbf{F}_4$	60 ∠42°N	373 mm from *D*
AD (1)	1000 mm	–	(Fixed link)	

Determine the shaft torque $\mathbf{T}_2$ on the input link AB for static equilibrium of the mechanism. Also find the forces on the bearings A, B, C and D.

Solution: The solution of the stated problem (Fig. 12.6a) can be obtained by superposition of the solutions of subproblems a, b and c.

Subproblem a (Fig. 12.6a) Neglecting forces $\mathbf{F}_3$ and $\mathbf{F}_4$.

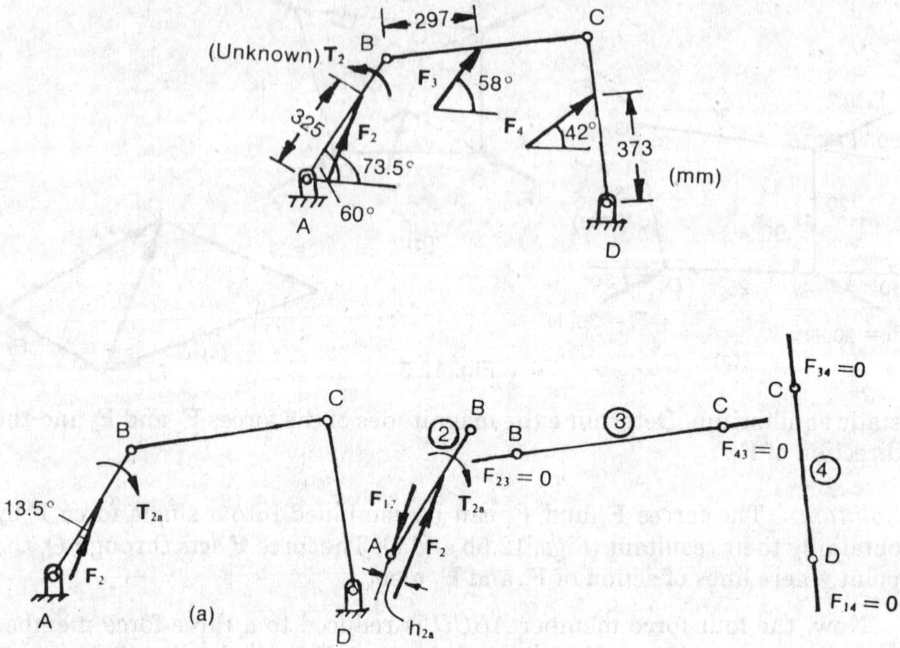

Fig. 12.6 a

Links 3 and 4 are both two-force members. Therefore, $\mathbf{F}_{43}$ can be along *BC* and $\mathbf{F}_{34}$ along *DC*. But $\mathbf{F}_{43}$ must be equal and opposite to $\mathbf{F}_{34}$.

Thus, both $\mathbf{F}_{43}$ and $\mathbf{F}_{34}$ must be zero.

Also $F_{43} = F_{23} = F_{32} = 0$

Hence link 2 is inequilibrium under the action of two forces $\mathbf{F}_2$ and $\mathbf{F}_{12}$ ($F_{12} = F_2$) and torque $\mathbf{T}_{2a}$.

$$T_{2a} = F_2 \times h_{2a} = 80 \times 0.325 \sin 13.5° = 6 \text{ N.m clockwise}$$

Subproblem b (Fig. 12.6b) Neglecting forces $\mathbf{F}_2$ and $\mathbf{F}_4$.

Link 4 is a two-force member.

$\therefore$ $\mathbf{F}_{34} = \mathbf{F}_{14}$, magnitudes unknown, directions parallel to *DC*

Link 3 is a three-force member in which $\mathbf{F}_3$ is completely known, only the direction of $\mathbf{F}_{43}$ is known (parallel to *DC*) and $\mathbf{F}_{23}$ is completely unknown. If the lines of action of $\mathbf{F}_3$ and $\mathbf{F}_{43}$ meet at *O*, the line of action of $\mathbf{F}_{23}$ will be along *OB*. Draw the force triangle ($\mathbf{F}_3$ is completely known) by taking $\mathbf{F}_3$ to a suitable scale and two lines parallel to lines of action of $\mathbf{F}_{23}$ and $\mathbf{F}_{43}$. Mark arrowheads on $\mathbf{F}_{23}$ and $\mathbf{F}_{43}$ to know the directions.

$$F_{43} = 50 \text{ N}$$

Also,
$$F_{43} = F_{34} = F_{14} = 50 \text{ N}$$

$$F_{23} = 108 \text{ N}$$

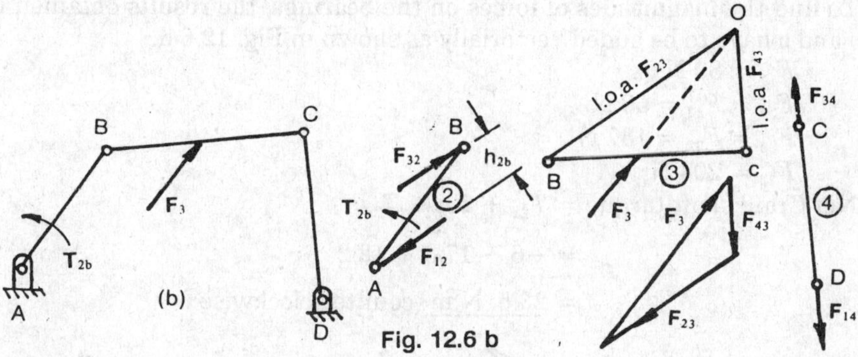

Fig. 12.6 b

and $$F_{23} = F_{32} = 108\text{ N}$$

Link 2 is subjected to two forces and a torque $\mathbf{T}_{2b}$.

For equilibrium, $\mathbf{F}_{12}$ is equal, parallel and opposite to $\mathbf{F}_{32}$.

$$T_{2b} = F_{32} \times h_{2b} = 108 \times 0.16 = 17.3\text{ N.m counter-clockwise.}$$

Subproblem c (Fig. 12.6c) Neglecting forces $\mathbf{F}_2$ and $\mathbf{F}_3$.

Link 4 is a three-force member in which only one force $\mathbf{F}_4$ is known. However, line of action of $\mathbf{F}_{34}$ can be obtained from the equilibrium of link 3 which is a two-force member. $\mathbf{F}_{34}$ will be equal and opposite to $\mathbf{F}_{43}$ which is along *BC*. If $\mathbf{F}_4$ and $\mathbf{F}_{34}$ intersect at *O*, then line of action of $\mathbf{F}_{14}$ will be along *OD*. Draw the force triangle ($\mathbf{F}_4$ is completely known) and obtain the magnitudes of forces $\mathbf{F}_{34}$ and $\mathbf{F}_{14}$.

$$F_{14} = 34.8\text{ N}$$

$$F_{34} = F_{43} = F_{23} = F_{32} = 33\text{ N}$$

Link 2 is subjected to two forces and torque $\mathbf{T}_{2c}$.

For equilibrium,

$$F_{12} = F_{32}$$

$$T_{2c} = F_{32} \times h_{2c} = 33 \times 0.37 = 12.2\text{ N.m counter-clockwise.}$$

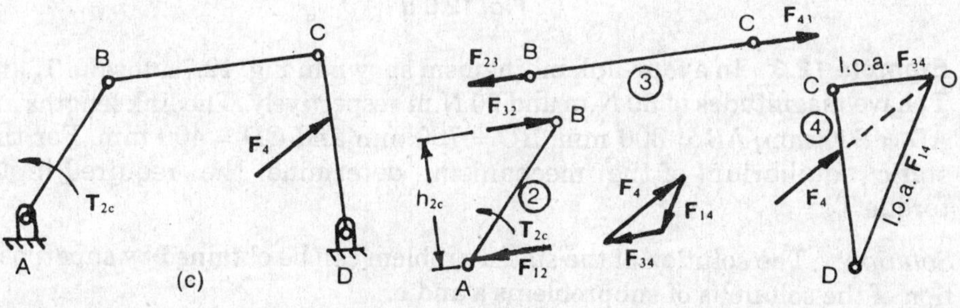

Fig. 12.6 c

To find the magnitudes of forces on the bearings, the results obtained in a, b and c have to be added vectorially as shown in Fig. 12.6 d.

$F_{14} = 80$ N
$F_{34} = F_{43} = 60$ N
$F_{23} = F_{32} = 137$ N
$F_{12} = 204$ N

Net Crankshaft torque $= T_{2a} + T_{2b} + T_{2c}$

$= -6 + 17.3 + 12.2$

$=$ 23.5 N.m counter-clockwise

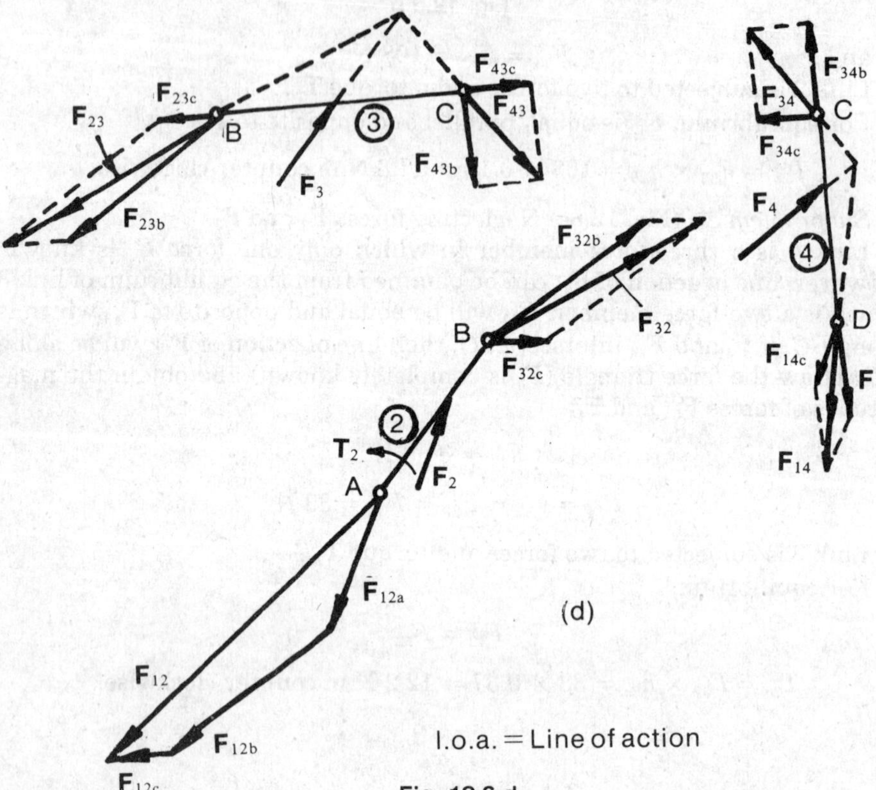

Fig. 12.6 d

Example 12.3 In a four-link mechanism shown in Fig. 12.7a, torque $\mathbf{T}_3$ and $\mathbf{T}_4$ have magnitudes of 30 N.m and 20 N.m respectively. The link lengths are AD = 800 mm, AB = 300 mm, BC = 700 mm and CD = 400 mm. For the static equilibrium of the mechanism, determine the required input torque $\mathbf{T}_2$.

Solution: The solution of the stated problem can be obtained by superposition of the solutions of subproblems a and b.

Subproblem a (Fig. 12.7a) Neglecting torque $\mathbf{T}_3$.

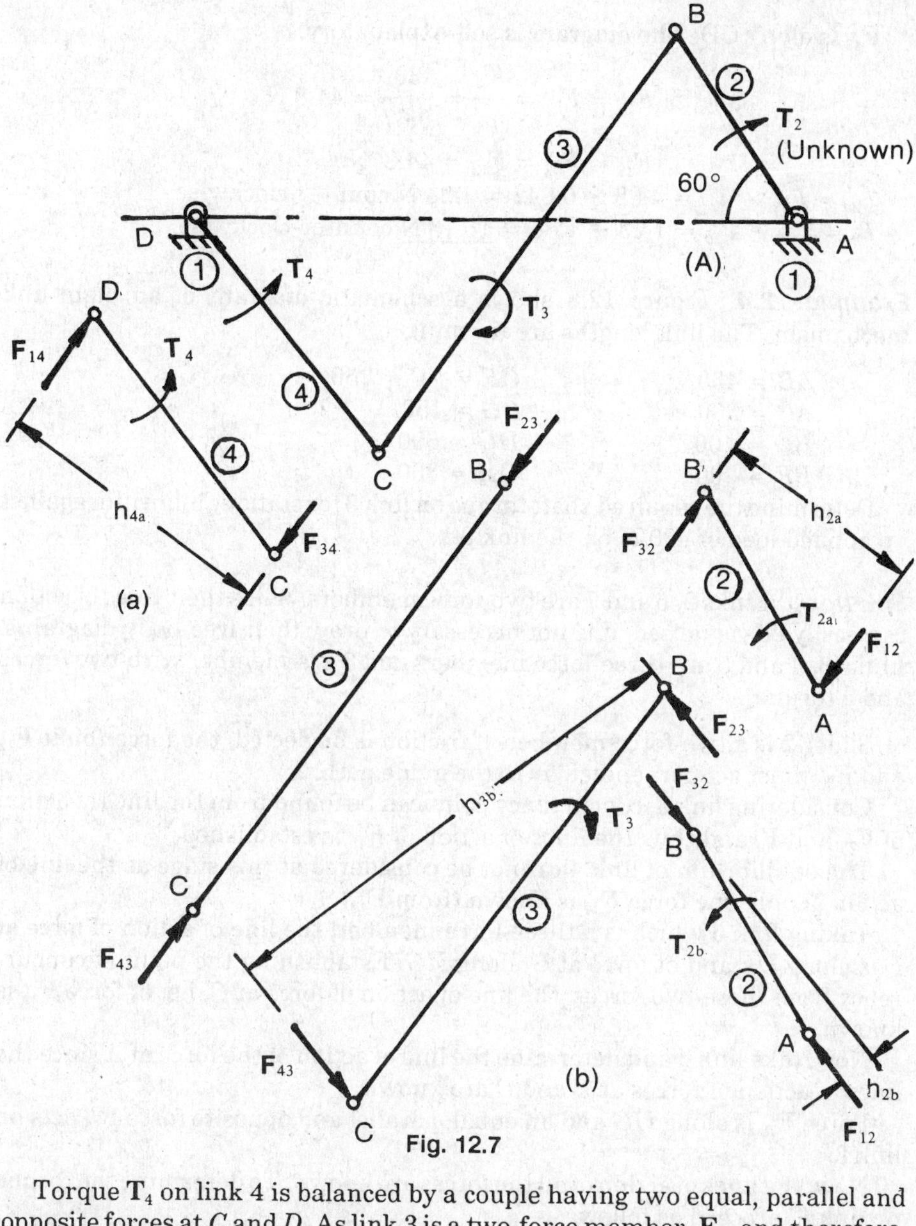

Fig. 12.7

Torque $\mathbf{T}_4$ on link 4 is balanced by a couple having two equal, parallel and opposite forces at C and D. As link 3 is a two-force member, $\mathbf{F}_{43}$ and therefore, $\mathbf{F}_{34}$ and $\mathbf{F}_{14}$ will be parallel to BC.

$$F_{34} = F_{14} = \frac{T_4}{h_{4a}} = \frac{20}{0.383} = 52.2 \text{ N}$$

and $$F_{34} = F_{43} = F_{23} = F_{32} = F_{12} = 52.2 \text{ N}$$

$$T_{2a} = F_{32} \times h_{2a} = 52.2 \times 0.274 = 14.3 \text{ N.m counter-clockwise.}$$

Subproblem b (Fig. 12.7b) Neglecting torque $\mathbf{T}_4$.

$\mathbf{F}_{43}$ is along CD. The diagram is self-explanatory.

$$F_{43} = F_{23} = \frac{T_3}{h_{3b}} = \frac{30}{0.67} = 44.8 \text{ N}$$

$$F_{23} = F_{32} = F_{12} = 44.8 \text{ N}$$

$$T_{2b} = F_{32} \times h_{2b} = 44.8 \times 0.042 = 1.88 \text{ N counter-clockwise}$$

$$T_2 = T_{2a} + T_{2b} = 14.3 + 1.88 = \underline{16.18 \text{ N}} \text{ counter-clockwise.}$$

Example 12.4 Figure 12.8 shows a schematic diagram of an eight-link mechanism. The link lengths are (in mm),

$AB = 450$	$OF = FC = 250$
$AC = 300$	$CG = 150$
$BD = 400$	$HG = 600$
$BE = 200$	$QH = 300$

Determine the required shaft torque on link 8 for static equilibrium against an applied load of 400 N on the link 3.

Solution: Links 2, 6 and 7 are two-force members. Since their lines of action can easily be visualised, it is not necessary to draw their free-body diagrams. Links 3, 4 and 5 are three-force members and 8 is a member with two forces and a torque.

Slider 2 is a two-force member. If friction is neglected, the forces on it $\mathbf{F}_{12}$ and $\mathbf{F}_{32}$ must act perpendicular to the guide path.

Considering link 3, concurrency point can be found from the lines of action of $\mathbf{F}_{23}$ and $\mathbf{F}$, and thus the lines of action of $\mathbf{F}_{43}$ is established.

The equilibrium of link 4 cannot be considered at this stage as the line of action of only one force $\mathbf{F}_{34}$ is known (from $\mathbf{F}_{43}$).

Taking link 5 which is a three-force member, the line of action of force at F is along OF and of force at G along HG. Establishing the point of concurrency from these two forces, the line of action of force at C, i.e. of force $\mathbf{F}_{45}$ is known.

Now, take link 4 and determine the line of action of the force at A since the lines of action of forces at B and C are known.

Force $\mathbf{F}_{78}$ is along HG and an equal, parallel and opposite force $\mathbf{F}_{18}$ acts on link 8.

Now, the lines of action of all the forces are known. To determine the torque on link 8, proceed as follows:

Construct a force diagram for the forces on link 3 ($\mathbf{F}$ is completely known) and find $\mathbf{F}_{43}$ (thus $\mathbf{F}_{34}$ is known).

Draw a force diagram for the forces on member 4 ($\mathbf{F}_{34}$ is completely known) and find $\mathbf{F}_{54}$ (thus $\mathbf{F}_{45}$ is known).

Draw a force diagram for the forces on member 5 ($\mathbf{F}_{45}$ is completely known) and find $\mathbf{F}_{75}$ (thus $\mathbf{F}_{57}$ is known).

Now $F_{57} = F_{87} = F_{78} = F_{18}$

$T_8 = F_{78} \times h = 75 \times 240 = \underline{18\,000 \text{ N.mm}}$ or $\underline{18 \text{ N.m}}$ clockwise

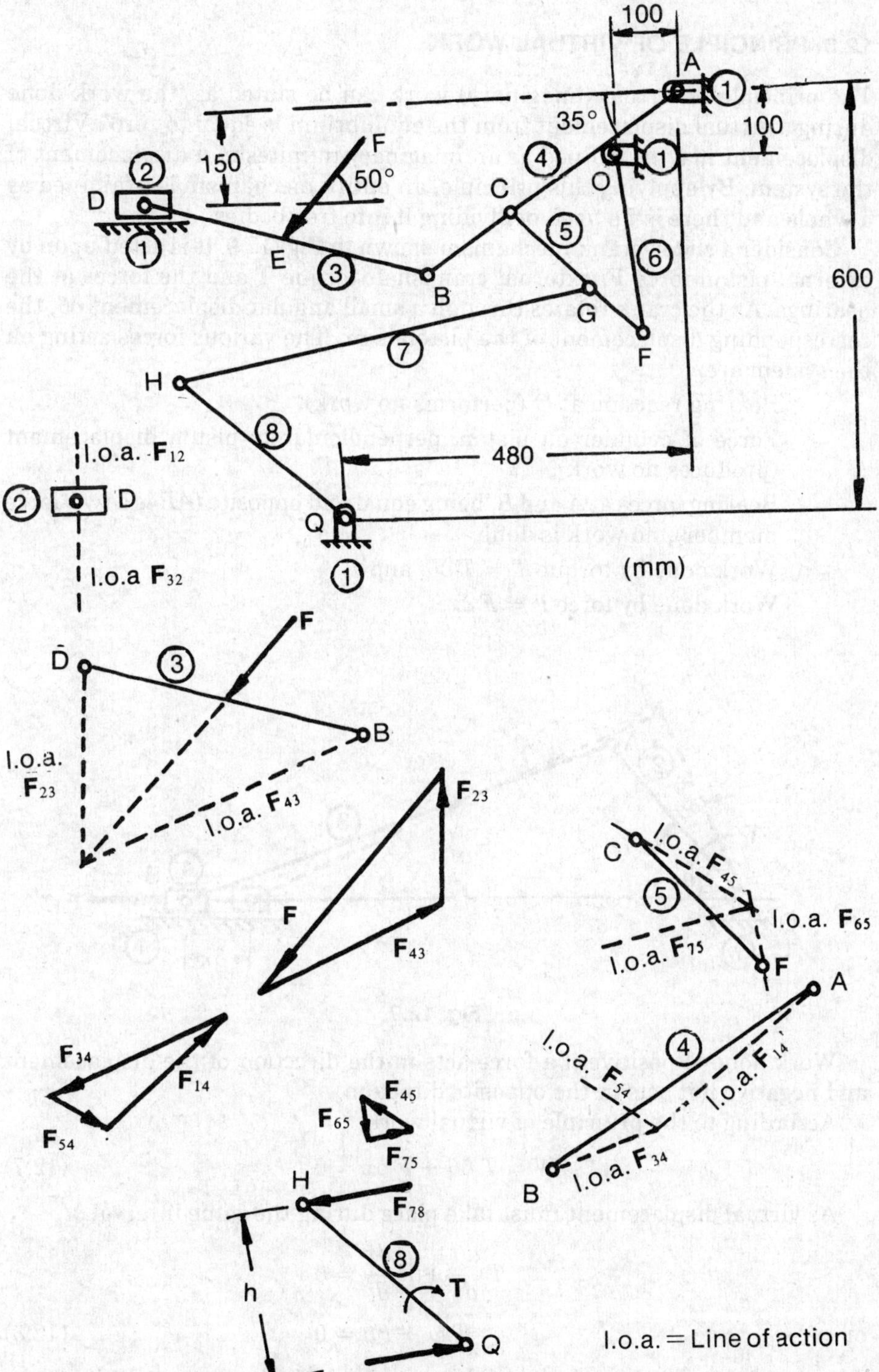
100
A
1
100
35°
600
150
F
4
O
1
2
50°
D
C
1
E
5
6
3
B
G
F
7
H
8
480
l.o.a. F12
2
D
Q
1
l.o.a F32
1
(mm)
F
D
3
B
l.o.a.
F23
l.o.a. F43
F23
F
F43
C
l.o.a. F45
5
l.o.a. F65
l.o.a. F75
F
A
F34
F14
l.o.a. F54
4
l.o.a. F14
F65
F45
F54
F75
B
l.o.a. F34
H
F78
8
h
T
Q
F18
l.o.a. = Line of action
Fig. 12.8

12.8 PRINCIPLE OF VIRTUAL WORK

The principle of virtual (imaginary) work can be stated as 'the work done during a virtual displacement from the equilibrium is equal to zero'. Virtual displacement may be defined as an imaginary infinitesimal displacement of the system. By applying this principle, an entire mechanism is examined as a whole and there is no need of dividing it into free bodies.

Consider a slider-crank mechanism shown in Fig. 12.9. It is acted upon by external piston force **F**, external crankshaft torque **T** and the forces at the bearings. As the crank rotates through a small angular displacement $\delta\theta$, the corresponding displacement of the piston is δx. The various forces acting on the system are:

- Bearing reaction at O (performs no work),
- Force of cylinder on piston, perpendicular to piston displacement (produces no work),
- Bearing forces at A and B, being equal and opposite (AB is a two-force member), no work is done,
- Work done by torque $T = T\,\delta\theta$, and
- Work done by force $F = F\,\delta x$.

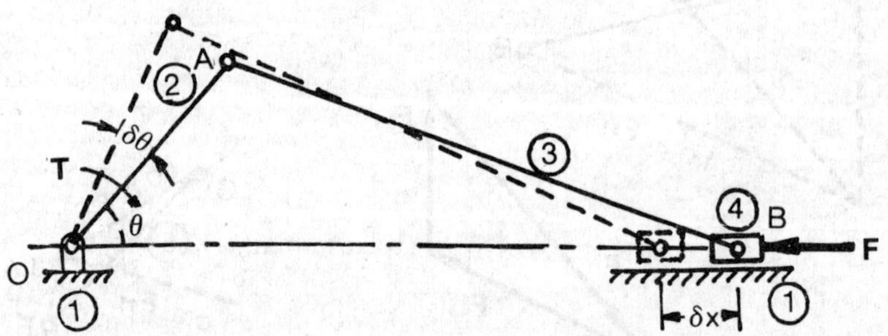

Fig. 12.9

Work done is positive if a force acts in the direction of the displacement and negative if it acts in the opposite direction.

According to the principle of virtual work,

$$W = T\,\delta\theta + F\,\delta x = 0 \tag{12.7}$$

As virtual displacement must take place during the same interval δt,

$$\therefore \quad T\frac{d\theta}{dt} + F\frac{dx}{dt} = 0$$

or

$$T\,\omega + Fv = 0 \tag{12.8}$$

where ω is the angular velocity of the crank and v is the linear velocity of the piston.

$$T = -\frac{F}{\omega} v$$

The negative sign indicates that for equilibrium, **T** must be applied in the opposite direction to the angular displacement.

Example 12.5 Solve Example 12.2 by using the principle of virtual work.

Solution: Assume that the link AB has an instantaneous angular velocity of ω rad/s counter-clockwise. Then $v_b = 0.5\ \omega$ m/s.

From the configuration diagram (Fig. 12.10 a) draw the velocity diagram (Fig. 12.10 b). Locate the points *E*, *F* and *G* on the velocity diagram and locate the velocity vectors for the same. Take their components parallel and perpendicular to the direction of forces.

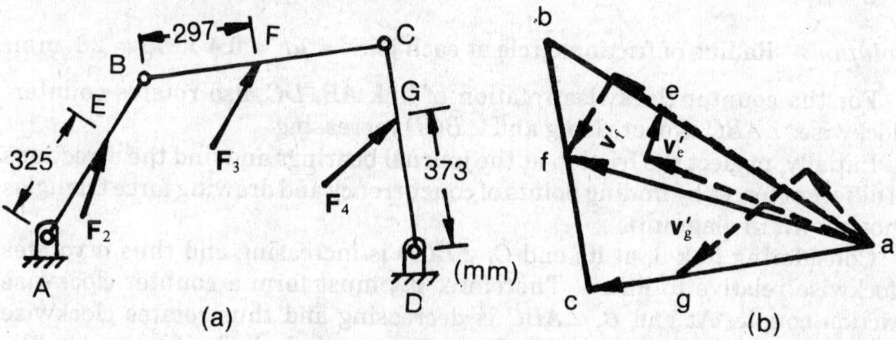

Fig. 12.10

$v_e' = 0.0745$ rad/s (parallel to $\mathbf{F}_2$)

$v_f' = 0.124$ rad/s (parallel to $\mathbf{F}_3$)

$v_g' = 0.205$ rad/s (parallel to $\mathbf{F}_4$)

Assuming *T* to be counter-clockwise and applying the principle of virtual work,

$$T \times \omega + F_2 \times 0.0745\ \omega - F_3 \times 0.124\ \omega - F_4 \times 0.205\ \omega = 0$$

or $$T + 80 \times 0.0745 - 144 \times 0.124 - 60 \times 0.205 = 0$$

or $$T = -6 + 17.3 + 12.3 = 23.5 \text{ N.m counter-clockwise}$$

12.9 FRICTION IN MECHANISMS

When two members of a mechanism move relative to each other, friction occurs at the joints. The presence of friction increases the energy requirements of a machine.

Friction at the bearing is taken into account by drawing friction circles and at the sliding pairs by considering the angle of friction. For details refer Secs. 8.13 and 8.14.

Example 12.6 In a four-link mechanism $ABCD$,

$AB = 350$ mm $\quad DE = 150$ mm

$BC = 500$ mm $\quad \angle DAB = 60°$ (AD is the fixed link).

$CD = 400$ mm

$AD = 700$ mm

A force of 35 N acts at E on link DC as shown in Fig. 12.11 a. Determine the force on the link AB required at the midpoint in the direction shown in the diagram for the static equilibrium of the mechanism. The coefficient of friction is 0.4 for each revolving pair. Assume impending motion of AB to be counter-clockwise. Radius of each journal is 50 mm.

Also find the torque on AB for its impending clockwise motion. (A very high value of coefficient of friction has been assumed to obtain a clear diagram.)

Solution: Radius of friction circle at each joint = $\mu r = 0.4 \times 50 = 20$ mm.

For the counter-clockwise rotation of link AB, DC also rotates counter-clockwise, $\angle ABC$ is decreasing and $\angle BCD$ increasing.

Initially, neglect the friction at the journal bearings and find the directions of different forces by finding points of concurrency and drawing force triangles (not shown in diagram).

Considering link 3, at its end C, $\angle BCD$ is increasing and thus it rotates clockwise relative to link 4. Therefore, $\mathbf{F}_{43}$ must form a counter-clockwise friction couple. At end B, $\angle ABC$ is decreasing and thus rotates clockwise relative to link 2. Therefore, $\mathbf{F}_{23}$ forms a counter-clockwise friction couple. The friction axis for the coupler BC is the common tangent to the two friction circles.

Now consider the link 4. The line of action of force $\mathbf{F}_{34}$ will be opposite to that of $\mathbf{F}_{43}$. Intersection of this line with the line of action of $\mathbf{F}_4$ gives the point of concurrency O for the forces acting on link 4. As the link 4 rotates counter-clockwise, the tangent to the friction circle at D drawn from point O is such that a clockwise friction couple is obtained.

By drawing a force triangle for the forces acting on link 4 ($\mathbf{F}_4$ is completely known), $\mathbf{F}_{34}$ is obtained.

$$F_{34} = F_{43} = F_{23} = F_{32}$$

The point of concurrency for the forces acting on link 2 is at O' which is the intersection of $\mathbf{F}_{32}$ and $\mathbf{F}_2$. As the link 1 rotates counter-clockwise, draw a tangent to the friction circle at A from O' such that a clockwise friction couple is obtained.

Draw a force diagram for the forces acting on link 2($\mathbf{F}_{32}$ is completely known) and obtain the value of $\mathbf{F}_2$.

$$F_2 = 20.3 \text{ N}$$

When the motion of AB is clockwise, DC also moves clockwise. For the equilibrium of link 4, the friction couples at D and C are to be counter-clock-

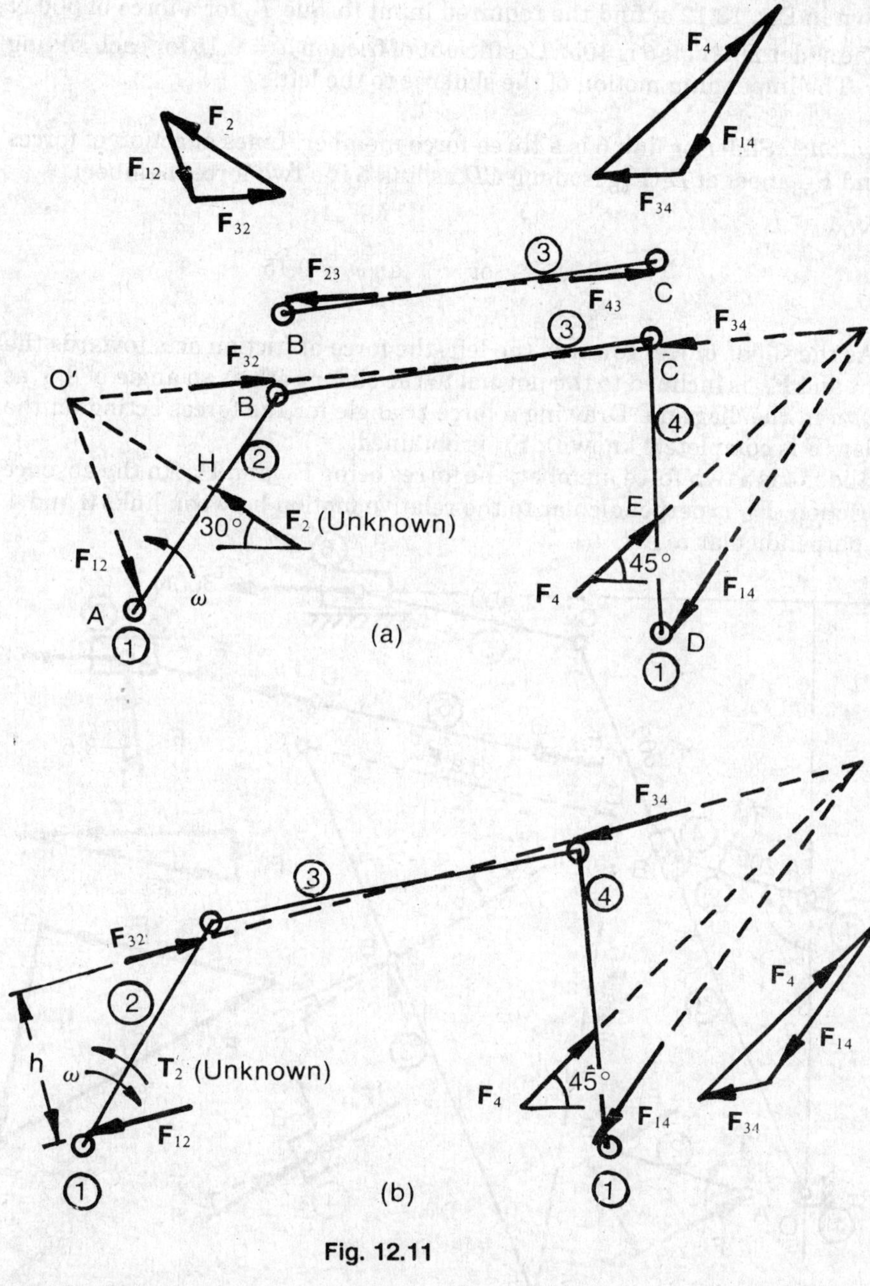

Fig. 12.11

wise. For the equilibrium of link 2, friction couples at *A* and *B* are also to be counter-clockwise. Obtain $\mathbf{F}_{32}$ in the manner discussed above and shown in Fig. 12.11b. $\mathbf{F}_{12}$ will be equal, parallel and opposite to $\mathbf{F}_{32}$.

$$T_2 = F_{32} \times h = 8.6 \times 208 = 1789 \text{ N.mm} \quad \text{or} \quad \underline{1.789 \text{ N.m}}$$

Example 12.7 For the static equilibrium of the quick-return mechanism shown in Fig. 12.12 a, find the required input torque T_2 for a force of 300 N on the slider D. Angle θ is 105°. Coefficient of friction $\mu = 0.15$ for each sliding pair. The impending motion of the slider is to the left.

Solution: Slider or link 6 is a three-force member. Lines of action of forces **F** and $\mathbf{F}_{56}$ meet at D ($\mathbf{F}_{56}$ is along CD as link 5 is a two-force member).

Now,

$$\mu = 0.15 \qquad \text{or} \qquad \tan\varphi = 0.15$$

$$\therefore \qquad \varphi = 8.5°$$

As the slider moves towards the left, the force of friction acts towards the right and $\mathbf{F}_{16}$ is inclined to the normal to the slider path at an angle of 8.5° as shown in the diagram. Drawing a force triangle for the forces acting on the slider (**F** is completely known), $\mathbf{F}_{56}$ is obtained.

Slider 4 is a two-force member, the forces being $\mathbf{F}_{34}$ and $\mathbf{F}_{24}$. In the absence of friction, $\mathbf{F}_{34}$ is perpendicular to the relative motion between links 3 and 4, i.e. perpendicular to AB.

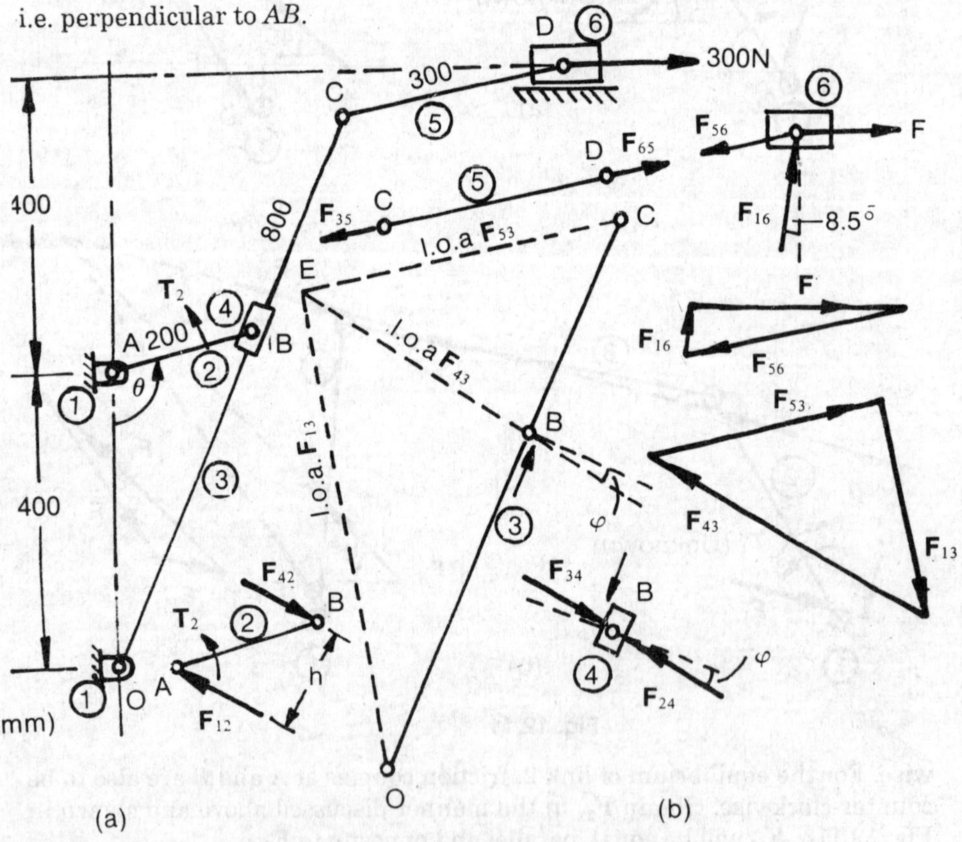

Fig. 12.12

Considering the equilibrium of link 3, $\mathbf{F}_{53}$ acts along CD and $\mathbf{F}_{43}$ at an angle of 8.5° to the perpendicular to AB at B. Note that the force of friction is to be upwards on link 3 as it moves downwards relative to the slider 4. Joining of $\mathbf{F}_{43}$ and $\mathbf{F}_{53}$ gives the point of concurrency E for the forces acting on link 3. Join EO which is the line of action of force $\mathbf{F}_{13}$.

Draw the force triangle for the forces acting on link 3 ($\mathbf{F}_{53}$ is completely known) and obtain the value of $\mathbf{F}_{43}$.

$$F_{43} = F_{34} = F_{24} = F_{42} = F_{12}$$

$$T = F_{42} \times h = 437 \times 147$$

$$= 64\,239 \text{ N.mm} \quad \text{or} \quad \underline{64.239 \text{ N.m}} \text{ counter-clockwise}$$

Example 12.8 Solve Example 8.12 using graphical method. Take coefficient of friction for the journals as 0.4 instead of 0.05. (A fictitious high value of coefficient of friction has been assumed to have friction circles of reasonable diameter to facilitate drawing on a smaller scale.)

Solution: Figure 12.13 a shows the solution of the problem neglecting friction. From the force triangle for the forces on the slider,

$$F_{34} = 22\,500 \text{ N}$$

Now, $F_{34} = F_{43} = F_{23} = F_{32}$

$$T = F_{32} \times h = 22\,500 \times 0.261 = 5872.5 \text{ N.m clockwise}$$

When friction is considered (Fig. 12.13b),

$$\text{Radius of friction circle at } O = 0.4 \times \frac{140}{2} = 28 \text{ mm}$$

$$\text{Radius of friction circle at } A = 0.4 \times \frac{120}{2} = 24 \text{ mm}$$

$$\text{Radius of friction circle at } B = 0.4 \times \frac{80}{2} = 16 \text{ mm}$$

As the crank moves counter-clockwise, $\angle OAB$ decreases. AB rotates clockwise relative to OA. Thus tangent at A is to be such that a counter-clockwise friction couple is obtained.

At B, $\angle OBA$ is increasing. Therefore, BA rotates clockwise relative to the piston. Thus the tangent to the friction circle is to be such that it gives a counter-clockwise friction couple.

For the sliding pair, $\varphi = \tan^{-} 0.07 = 4°$

The point of intersection of $\mathbf{F}_{34}$ and $\mathbf{F}$ gives the point of concurrency for the forces on the slider. Force $\mathbf{F}_{14}$, i.e. the reaction of the guide, is inclined to the perpendicular to the slider path, and passes through the point of concurrency.

By drawing a force triangle for the forces acting on the slider, $\mathbf{F}_{34}$ is obtained.

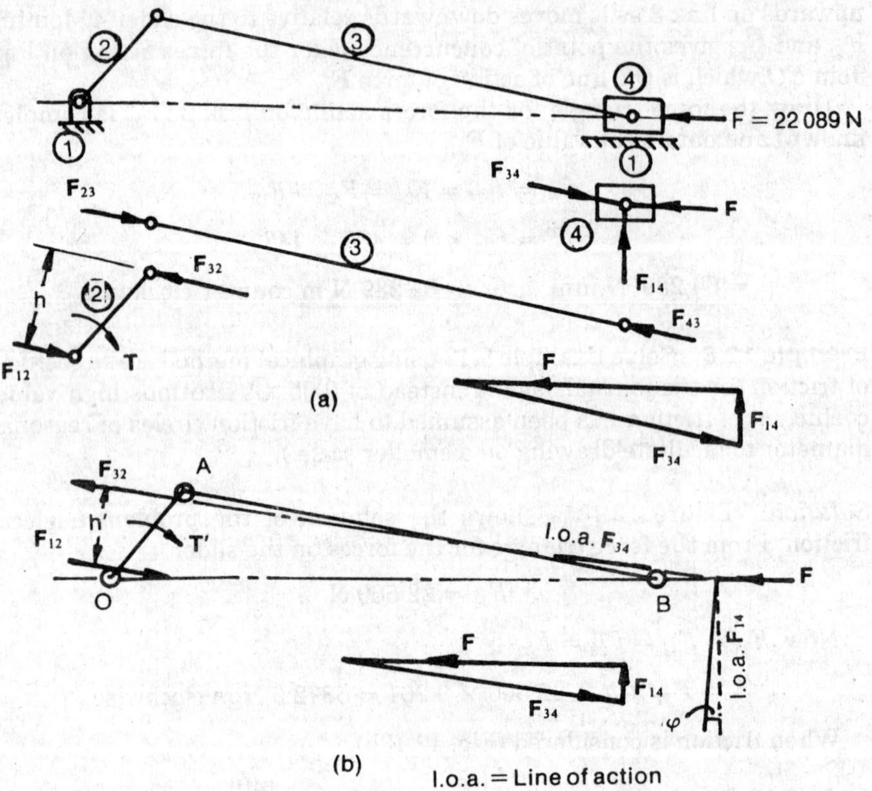

Fig. 12.13

$$F_{34} = F_{43} = F_{23} = F_{32} = 22\ 200 \text{ N}$$

The force at A is equal, parallel and opposite to $\mathbf{F}_{32}$ and tangent to the friction circle such that a clockwise friction couple is obtained.

$$T' = F_{32} \times h' = 22\ 200 \times 0.202 = 4484 \text{ N.m} \qquad \text{clockwise}$$

PROBLEMS

12.1 The dimensions of a four-link mechanism are: AB = 400 mm, BC = 600 mm, CD = 500 mm, AD = 900 mm, and $\angle DAB = 60°$. AD is the fixed link. E is a point on link BC such that BE = 400 mm and CE = 300 mm (BEC clockwise).
A force of 150 $\angle 45°$ N acts on DC at a distance of 250 mm from D. Another force of magnitude 100 $\angle 180°$ N acts at point E. Find the required input torque on link AB for static equilibrium of the mechanism.

(0.35 N.m clockwise)

12.2 Determine the required input torque on the crank of a slider-crank mechanism for the static equilibrium when the applied piston load is 1500 N. The lengths of the crank and the connecting rod are 40 mm and 100 mm respectively and the crank has turned through 45° from the inner-dead centre.

(55 N)

12.3 For the mechanism shown in Fig. 12.14, find the required input torque for the static equilibrium. The lengths *OA* and *AB* are 250 mm and 650 mm respectively. F = 500 N.

(67.5 N.m clockwise)

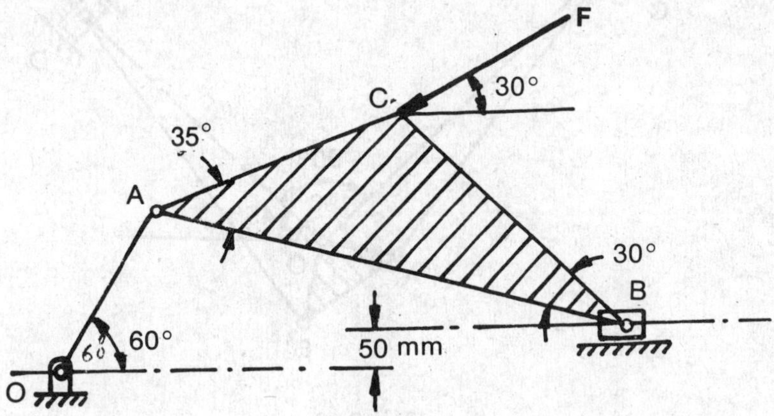

Fig. 12.14

12.4 For the static equilibrium of the mechanism of Fig. 12.15, find the required input torque. The dimensions are:
AB = 150 mm, BC = AD = 500 mm, DC = 300 mm, CE = 100 mm and EF = 450 mm.

(46.7 N.m clockwise)

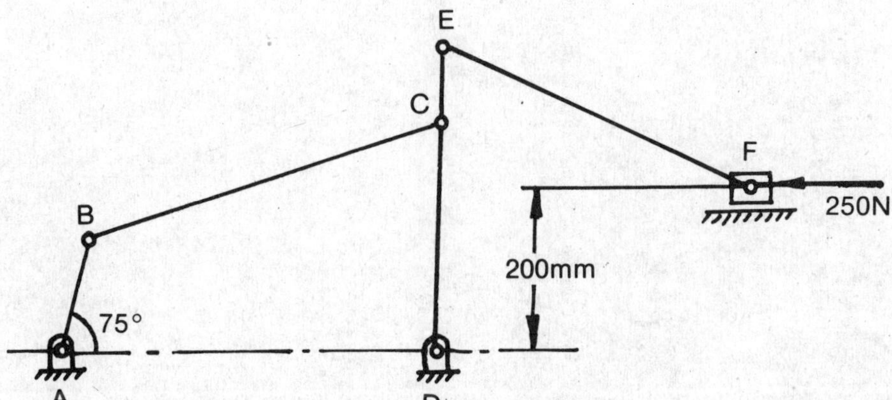

Fig. 12.15

12.5 A two-cylinder engine shown in Fig. 12.16 is in static equilibrium. The dimensions are $OA = OB = 50$ mm, $AC = BD = 250$ mm, $\angle AOB = 90°$. Determine the torque on the crank OAB.

(101.5 N.m clockwise)

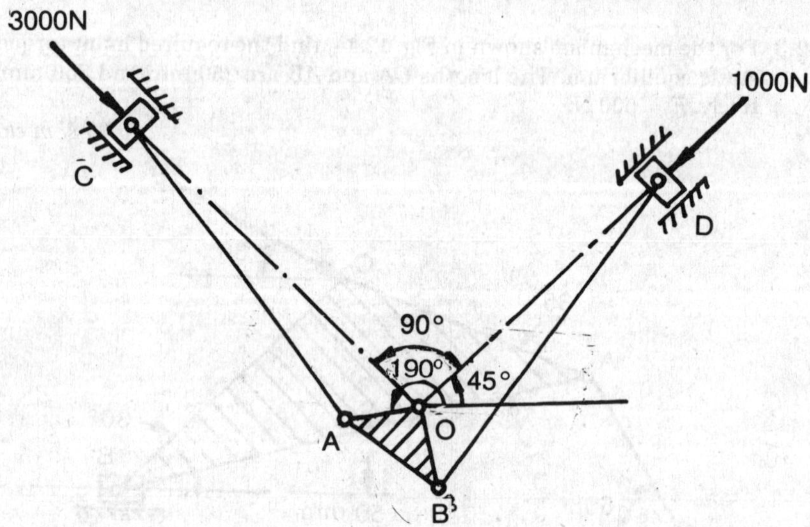

Fig. 12.16

13

DYNAMIC FORCE ANALYSIS

13.1 INTRODUCTION

Dynamic forces are associated with accelerating masses. As all machines have some accelerating parts, dynamic forces are always present when the machines operate. In situations where dynamic forces are dominant or comparable with magnitudes of external forces and operating speeds are high, dynamic analysis has to be carried out. For example, in case of rotors which rotate at speeds more than 80 000 rpm, even the slightest eccentricity of the centre of mass from the axis of rotation produces very high dynamic forces. This may lead to vibrations, wear, noise or even machine failure.

13.2 D'ALEMBERT'S PRINCIPLE

D'Alembert's principle states that the inertia forces and couples, and the external forces and torques on a body together give statical equilibrium.

Inertia is a property of matter by virtue of which a body resists any change in velocity.

$$\text{Inertia force } \mathbf{F}_i = -m\mathbf{f}_g \tag{13.1}$$

where m = mass of body

$\mathbf{f}_g$ = acceleration of centre of mass of the body.

Negative sign indicates that the force acts in the opposite direction to that of the acceleration. The force acts through centre of mass of the body.

Similarly, an inertia couple resists any change in the angular velocity.

Inertia couple,

$$\mathbf{C}_i = -I_g\alpha \tag{13.2}$$

where I_g = moment of inertia about an axis passing through centre of mass G and perpendicular to plane of rotation of the body

α = angular acceleration of the body.

Let $\mathbf{F}_1$, $\mathbf{F}_2$, $\mathbf{F}_3$, etc. = external forces on the body

$\mathbf{T}_{g1}$, $\mathbf{T}_{g2}$, $\mathbf{T}_{g3}$, etc. = external torques on the body about centre of mass G.

According to D' Alembert's principle, the vector sum of forces and torques (or couples) has to be zero, i.e.

$$\mathbf{F}_1 + \mathbf{F}_2 + \mathbf{F}_3 + \ldots + \mathbf{F}_i = 0 \tag{13.3}$$

and

$$\mathbf{T}_{g1} + \mathbf{T}_{g2} + \mathbf{T}_{g3} + \ldots \mathbf{C}_i = 0 \tag{13.4}$$

These equations are similar to equations of a body in static equilibrium, i.e. $\Sigma\mathbf{F} = 0$ and $\Sigma\mathbf{T} = 0$.

This suggests that first the magnitudes and the directions of inertia forces and couples can be determined, after which they can be treated just like static loads on the mechanism. Thus a dynamic analysis problem is reduced to one requiring static analysis.

13.3 EQUIVALENT OFFSET INERTIA FORCE

The moment of a force about a point is known as torque.

The moment of a couple about any arbitrary point is the same and is equal to magnitude of one of the two equal, parallel and opposite forces and the perpendicular distance between them. The resultant force of a couple is zero.

In plane motion, the inertia force acts on a body through its centre of mass along with the couple.

In graphical solutions, it is possible to replace inertia force and inertia couple by an equivalent offset inertia force which can account for both. This is done by displacing the line of action of the inertia force from the centre of mass. The perpendicular displacement h of the force from the centre of mass is such that the torque so produced is equal to the inertia couple acting on the body.

i.e.

$$T_i = C_i$$

or

$$F_i \times h = C_i$$

or

$$h = \frac{C_i}{F_i} = \frac{-I_g\alpha}{-mf_g} = \frac{mk^2\alpha}{mf_g} = \frac{k^2\alpha}{f_g} \tag{13.5}$$

h is taken in such a way that the force produces a moment, about the centre of mass, which is opposite in sense to angular acceleration α.

13.4 DYNAMIC ANALYSIS OF FOUR-LINK MECHANISM

Adopt the following procedure:

1. Draw the velocity and acceleration diagrams of the mechanism from the configuration diagram by usual methods.
2. Determine the linear accelerations of centres of masses of various links, and also their angular accelerations.

3. Calculate the inertia forces and inertia couples from the relations $\mathbf{F}_i = -m\mathbf{f}_g$ and $\mathbf{C}_i = -I_g\alpha$.
4. Replace $\mathbf{F}_i$ with equivalent offset inertia force to take into account $\mathbf{F}_i$ as well as $\mathbf{C}_i$.
5. Assume equivalent offset inertia forces on the link as static forces and analyse the mechanism by any of the methods outlined in Chapter 12.

Example 13.1 The dimensions of a four-link mechanism are

AB = 500 mm, BC = 660 mm, CD = 560 mm and AD = 1000 mm.

The link AB has an angular velocity of 10.5 rad/s counter-clockwise and an angular retardation of 26 rad/s^2 at the instant when it makes an angle of 60° with AD, the fixed link.

The mass of the links BC and CD is 4.2 kg/m length. The link AB has a mass of 3.54 kg, the centre of which lies at 200 mm from A and a moment of inertia of 88 500 kg.mm^2.

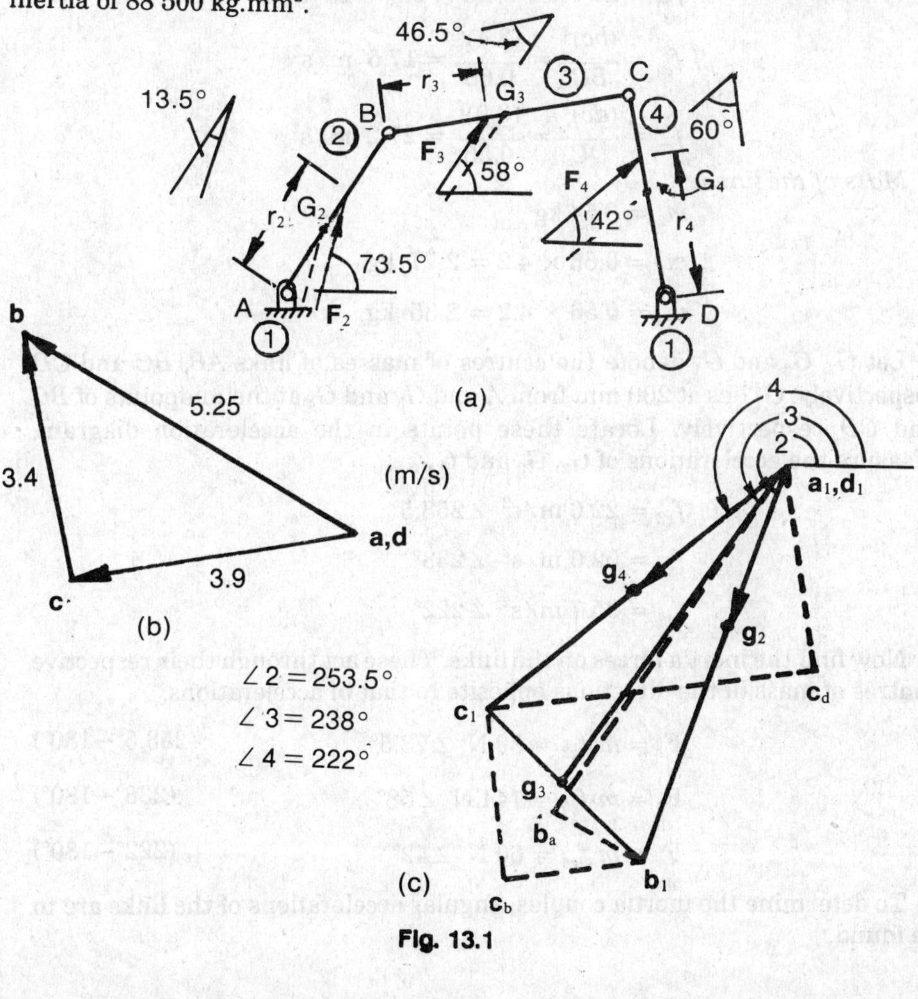

Fig. 13.1

Neglecting gravity and friction effects, determine the instantaneous value of the drive torque required to be applied on AB to overcome the inertia forces.

Solution: Draw the configuration diagram *ABCD* of the mechanism to a suitable scale (Fig. 13.1 a). The velocity and acceleration diagrams of the same have been shown in Figs. 13.1 b and c (also refer to Examples 2.1 and 3.1).

From velocity diagram,

$$v_b \text{ or } ab = \omega_{ba} \times AB = 10.5 \times 0.5 = 5.25 \text{ m/s}$$

$$v_{cb} \text{ or } bc = 3.4 \text{ m/s} \quad \text{and} \quad v_c \text{ or } dc = 3.9 \text{ m/s}$$

From acceleration diagram,

$$f_{ba}^c = \frac{(ab)^2}{AB} = \frac{(5.25)^2}{0.5} = 55.1 \text{ m/s}^2$$

$$f_{ba}^t = \alpha \times AB = 26 \times 0.5 = 13 \text{ m/s}^2$$

$$f_{cb}^c = \frac{(bc)^2}{BC} = \frac{(3.4)^2}{0.66} = 17.5 \text{ m/s}^2$$

$$f_{cd}^c = \frac{(dc)^2}{DC} = \frac{(3.9)^2}{0.56} = 27.2 \text{ m/s}^2$$

Mass of the links

$$m_2 = 3.54 \text{ kg}$$

$$m_3 = 0.66 \times 4.2 = 2.77 \text{ kg}$$

$$m_4 = 0.56 \times 4.2 = 2.35 \text{ kg}$$

Let G_2, G_3 and G_4 denote the centres of masses of links *AB*, *BC* and *CD* respectively. G_2 lies at 200 mm from *A* and G_3 and G_4 at the midpoints of *BC* and *CD* respectively. Locate these points in the acceleration diagram. Measure the accelerations of G_2, G_3 and G_4.

$$f_{g2} = 22.6 \text{ m/s}^2 \ \angle 253.5°$$

$$f_{g3} = 52.0 \text{ m/s}^2 \ \angle 238°$$

$$f_{g4} = 25.7 \text{ m/s}^2 \ \angle 222°$$

Now find the inertia forces on the links. These act through their respective centres of mass in the directions opposite to that of accelerations.

$$\mathbf{F}_2 = m_2 f_{g2} = 80 \text{ N} \ \angle 73.5° \qquad (253.5° - 180°)$$

$$\mathbf{F}_3 = m_3 f_{g3} = 144 \text{ N} \ \angle 58° \qquad (238° - 180°)$$

$$\mathbf{F}_4 = m_4 f_{g4} = 60 \text{ N} \ \angle 42° \qquad (222° - 180°)$$

To determine the inertia couples, angular accelerations of the links are to be found.

$\alpha_2 = 26 \text{ rad/s}^2$ clockwise

$$\alpha_3 = \frac{f^t_{cb}}{CB} = \frac{22.5}{0.66} = 34.1 \text{ rad/s}^2 \text{ counter-clockwise}$$

$$\alpha_4 = \frac{f^t_{cd}}{CD} = \frac{44.3}{0.56} = 79.1 \text{ rad/s}^2 \text{ counter-clockwise}$$

Then $C_i = I_g\alpha$

However, the inertia couples can be taken into account by replacing the inertia forces with equivalent offset inertia forces.

Now,

$$k_2^2 = \frac{I_g}{m_2} = \frac{88\,500}{3.54} = 25\,000 \text{ mm}^2$$

Links 3 and 4 have uniform cross-sections,

$$k_3^2 = \frac{l^2}{12} = \frac{(660)^2}{12} = 36\,300 \text{ mm}^2$$

$$k_4^2 = \frac{l^2}{12} = \frac{(560)^2}{12} = 26\,133 \text{ mm}^2$$

and

$$h_2 = \frac{k^2\alpha}{f_{g2}} = \frac{25\,000 \times 26}{22\,600} = 28.8 \text{ mm}$$

$$h_3 = \frac{36\,300 \times 34.1}{52\,000} = 23.8 \text{ mm}$$

$$h_4 = \frac{26\,133 \times 79.1}{25\,700} = 80.4 \text{ mm}$$

Also,

$$r_2 = 200 + \frac{28.8}{\sin 13.5°} = 325 \text{ mm}$$

$$r_3 = 330 - \frac{23.8}{\sin 46.5°} = 297 \text{ mm}$$

$$r_4 = 280 + \frac{80.3}{\sin 60°} = 373 \text{ mm}$$

An inertia couple acts in a direction opposite to that of the angular acceleration. Thus offsets h_2, h_3 and h_4 are to be such that the required inertia couples are set up. For example, angular acceleration of link 2 is clockwise (being retardation), therefore, inertia couple must be counter-clockwise. Links 2 and 3 have counter-clockwise accelerations and thus the inertia couples are to be clockwise.

Now, assume equivalent offset inertia forces on the links as static forces and solve. This has been done in Examples 12.2 and 12.5.

The required input torque = 23.5 N.m (counter-clockwise)

13.5 DYNAMIC ANALYSIS OF SLIDER-CRANK MECHANISM

The steps outlined for dynamic analysis of a four-link mechanism also hold good for slider-crank mechanism and the analysis can be carried out in exactly the same manner.

However, an analytical approach is also being described in detail in the following sections.

13.6 VELOCITY AND ACCELERATION OF PISTON

Figure 13.2 shows a slider-crank mechanism in which the crank *OA* rotates in the clockwise direction. l and r are the lengths of the connecting rod and the crank respectively.

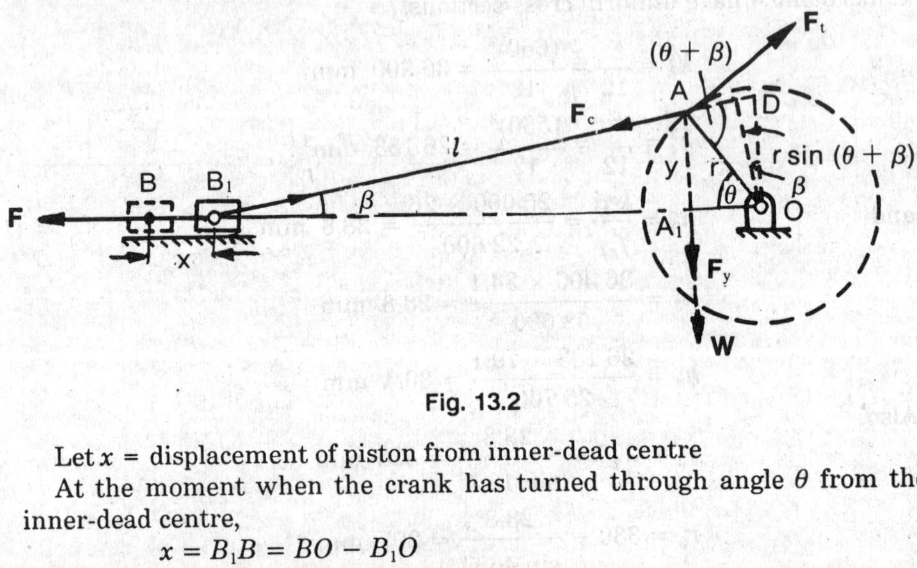

Fig. 13.2

Let x = displacement of piston from inner-dead centre

At the moment when the crank has turned through angle θ from the inner-dead centre,

$$
\begin{aligned}
x &= B_1B = BO - B_1O \\
&= BO - (B_1A_1 + A_1O) \\
&= (l + r) - (l \cos\beta + r\cos\theta) \\
&= (nr + r) - (nr\cos\beta + r\cos\theta) \qquad \text{(taking } l/r = n) \\
&= r[(n + 1) - (n\cos\beta + \cos\theta)] \qquad (13.6)
\end{aligned}
$$

$$
\begin{aligned}
\text{where } \cos\beta &= \sqrt{1 - \sin^2\beta} \\
&= \sqrt{1 - \frac{y^2}{l^2}} \\
&= \sqrt{1 - \frac{(r\sin\theta)^2}{l^2}} \\
&= \sqrt{1 - \frac{\sin^2\theta}{n^2}}
\end{aligned}
$$

$$= \frac{1}{n}\sqrt{n^2 - \sin^2\theta}$$

$$\therefore \quad x = r\,[(n + 1) - (\sqrt{n^2 - \sin^2\theta} + \cos\theta)\,]$$

$$= r\,[(1 - \cos\theta) + (n - \sqrt{n^2 - \sin^2\theta}\,] \qquad (13.7)$$

If the connecting rod is very large as compared to crank, n^2 will be large and the maximum value of $\sin^2\theta$ can be unity. Then $\sqrt{n^2 - \sin^2\theta}$ will be approaching $\sqrt{n^2}$ or n, and

$$x = r\,(1 - \cos\theta) \qquad (13.8)$$

This is the expression for a simple harmonic motion. Thus the piston executes a simple harmonic motion when the connecting rod is large.

Velocity of Piston

$$v = \frac{dx}{dt} = \frac{dx\,d\theta}{d\theta\,dt}$$

$$= \frac{d}{d\theta}\left[r\left\{(1 - \cos\theta) + n - (n^2 - \sin^2\theta)^{1/2}\right\}\right]\frac{d\theta}{dt}$$

$$= r\left[(0 + \sin\theta) + 0 - \frac{1}{2}(n^2 - \sin^2\theta)^{-1/2}(-2\sin\theta\cos\theta)\right]\omega$$

$$= r\omega\left[\sin\theta + \frac{\sin 2\theta}{2\sqrt{n^2 - \sin^2\theta}}\right] \qquad (13.9)$$

If n^2 is large compared to $\sin^2\theta$,

$$v = r\omega\left[\sin\theta + \frac{\sin 2\theta}{2n}\right] \qquad (13.10)$$

If $\dfrac{\sin 2\theta}{2n}$ can be neglected (when n is quite large),

$$v = r\omega\sin\theta \qquad (13.11)$$

Acceleration of Piston

$$f = \frac{dv}{dt} = \frac{dv\,d\theta}{d\theta\,dt}$$

$$= \frac{d}{d\theta}\left[r\omega\left(\sin\theta + \frac{\sin 2\theta}{2n}\right)\right]\omega$$

$$= r\omega\left(\cos\theta + \frac{2\cos 2\theta}{2n}\right)\omega$$

$$= r\omega^2\left(\cos\theta + \frac{\cos 2\theta}{n}\right) \qquad (13.12)$$

If n is very very large,

$$f = r\omega^2 \cos\theta \text{ as in case of SHM} \tag{13.13}$$

When $\theta = 0°$ i.e. at IDC, $f = r\omega^2\left(1 + \frac{1}{n}\right)$

When $\theta = 180°$, i.e. at ODC, $f = r\omega^2\left(-1 + \frac{1}{n}\right)$

At $\theta = 180°$, when the direction of motion is reversed,

$$f = r\omega^2\left(1 - \frac{1}{n}\right) \tag{13.14}$$

Note that this expression of acceleration has been obtained by differentiating the approximate expression for the velocity. It is, usually, very cumbersome to differentiate the exact expression for velocity. However, this gives satisfactory results.

13.7 ANGULAR VELOCITY AND ANGULAR ACCELERATION OF CONNECTING ROD

$$\sin\beta = \frac{\sin\theta}{n}$$

Differentiating with respect to time,

$$\cos\beta\,\frac{d\beta}{dt} = \frac{1}{n}\cos\theta\,\frac{d\theta}{dt}$$

$$\frac{d\beta}{dt} = \frac{\cos\theta}{n\cos\beta}\,\omega$$

or

$$\omega_c = \omega\,\frac{\cos\theta}{n\,\frac{1}{n}\sqrt{n^2 - \sin^2\theta}} \tag{13.15}$$

where ω_c is the angular velocity of the connecting rod

$$= \omega\,\frac{\cos\theta}{\sqrt{n^2 - \sin^2\theta}}$$

Let α_c = angular acceleration of the connecting rod

$$= \frac{d\omega_c}{dt} = \frac{d\omega_c d\theta}{d\theta\ dt}$$

$$= \omega\,\frac{d}{d\theta}[\cos\theta(n^2 - \sin^2\theta)^{-1/2}]\omega$$

$$= \omega^2[-\cos\theta\,\frac{1}{2}(n^2-\sin^2\theta)^{-3/2}(-2\sin\theta\cos\theta)+(n^2-\sin^2\theta)^{-1/2}(-\sin\theta)]$$

$$= \omega^2 \sin\theta \left[\frac{\cos^2\theta - (n^2 - \sin^2\theta)}{(n^2 - \sin^2\theta)^{3/2}}\right]$$

$$= -\omega^2 \sin\theta \left[\frac{n^2 - 1}{(n^2 - \sin^2\theta)^{3/2}}\right] \tag{13.16}$$

The negative sign indicates that the sense of angular acceleration of the rod is such that it tends to reduce the angle β.

13.8 PISTON EFFORT (EFFECTIVE DRIVING FORCE)

Piston effort is termed as the net or effective force applied on the piston. In reciprocating engines, the reciprocating masses accelerate during the first half of the stroke and the inertia force due to the same tends to resist the motion. Thus the net force on the piston is decreased. During the later half of the stroke, the reciprocating masses decelerate and the inertia force opposes this deceleration or the inertia force acts in the direction of the applied gas pressure and thus increases the effective force on the piston.

In a vertical engine, the weight of the reciprocating masses assists the piston during the outstroke (down stroke), thus increasing the piston effort by an amount equal to the weight of the piston. During the instroke (upstroke), piston effort is decreased by the same amount.

Let A_1 = area of the cover end

A_2 = area of the piston end

p_1 = pressure on the cover end

p_2 = pressure on the piston end

m = mass of the reciprocating parts

Force on the piston, $F_p = p_1A_1 - p_2A_2$ (13.17)

Inertia force, $F_b = mf = mr\omega^2\left(\cos\theta + \dfrac{\cos 2\theta}{n}\right)$ (13.18)

Net (effective) force on the piston, $F = F_p - F_b$ (13.19)

13.9 CRANK EFFORT

Crank effort is the net effort (force) applied at the crankpin perpendicular to the crank which gives the required turning moment on the crankshaft.

Let F_t = crank effort

F_c = force on the connecting rod

As

$$F_t \times r = F_c\, r \sin(\theta + \beta) \qquad \text{(refer to Fig. 13.2)}$$

$$\therefore \quad F_t = F_c \sin(\theta + \beta)$$

$$= \frac{F}{\cos\beta}\sin(\theta+\beta) \quad (\because F_c \cos\beta = F) \tag{13.20}$$

13.10 TURNING MOMENT ON CRANKSHAFT

$$T = F_t \times r$$

$$= \frac{F}{\cos\beta}\sin(\theta+\beta) \times r$$

$$= \frac{Fr}{\cos\beta}(\sin\theta\cos\beta + \cos\theta\sin\beta)$$

$$= Fr\left(\sin\theta + \cos\theta\sin\beta\frac{1}{\cos\beta}\right)$$

$$= Fr\left(\sin\theta + \cos\theta\frac{\sin\theta}{n}\frac{1}{\frac{1}{n}\sqrt{n^2-\sin^2\theta}}\right)$$

$$= Fr\left(\sin\theta + \frac{2\sin\theta\cos\theta}{2\sqrt{n^2-\sin^2\theta}}\right)$$

$$= Fr\left(\sin\theta + \frac{\sin 2\theta}{2\sqrt{n^2-\sin^2\theta}}\right) \tag{13.21}$$

Also as $r\sin(\theta+\beta) = OD\cos\beta$

$$T = F_t \times r$$

$$= \frac{F}{\cos\beta} r\sin(\theta+\beta) \qquad \text{[from (13.20)]}$$

$$= \frac{F}{\cos\beta}(OD\cos\beta)$$

$$= F \times OD \tag{13.22}$$

13.11 INERTIA OF CONNECTING ROD

The inertia force due to the reciprocating mass at B is equivalent to an inertia torque at the crankshaft given by Equation 13.21. Apart from this, it is also desired to consider the inertia of the connecting rod. As the motion of the connecting rod is not linear, its inertia cannot be found as such. The same can be taken into account by considering its mass in two parts at two points as follows.

The mass of the connecting rod can be replaced by two point masses at two points if it is ensured that the two masses together have the same dynamical properties as before. The two members of Fig. 13.3 will be dynamically similar if

(i) the sum of the two masses is equal to the total mass
(ii) the combined centre of mass coincides with that of the rod
(iii) the moment of inertia of two point masses about perpendicular axis through their combined centre of mass is equal to that of the rod.

Let m be the total mass of the rod and one of the masses be located at the small end B. Let the second mass be placed at D and

$$m_b = \text{mass at B}$$

$$m_d = \text{mass at D}$$

Take $$BG = b \text{ and } DG = d$$

Then

$$m_b + m_d = m \quad \text{(i)}$$

and

$$m_b\, b = m_d\, d \quad \text{(ii)}$$

From (i) and (ii)

$$m_b + \left(m_b \frac{b}{d}\right) = m$$

or $$m_b \left(1 + \frac{b}{d}\right) = m$$

or $$m_b \left(\frac{b+d}{d}\right) = m$$

or $$m_b = m \frac{d}{b+d}$$

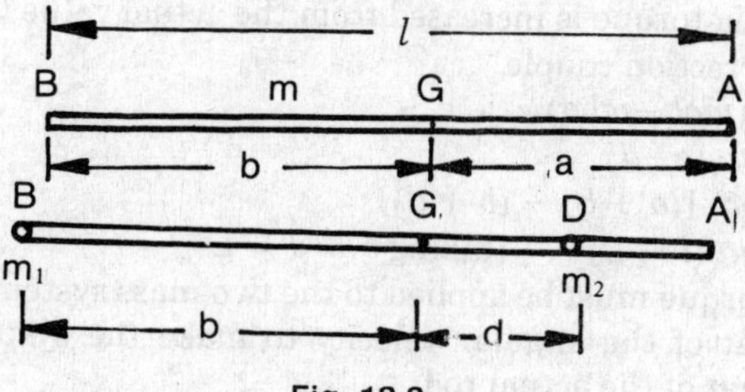

Fig. 13.3

Similarly,

$$m_d = m \frac{b}{b+d}$$

Also, $I = m_b b^2 + m_d d^2$

$$= m\frac{d}{b+d}b^2 + m\frac{b}{b+d}d^2$$

$$= m\,bd\left(\frac{b+d}{b+d}\right)$$

$$= mbd \tag{13.23}$$

Let k = radius of gyration of the connecting rod about an axis through centre of mass G perpendicular to the plane of motion.

Then $mk^2 = mbd$

or

$$k^2 = bd \tag{13.24}$$

Comparing this result with that of an equivalent length of a simple pendulum, it can be observed that $b + d$ $(=L)$ is the equivalent length if the rod is suspended from point B, and D is the centre of oscillation or percussion.

However, in the analysis of the connecting rod, it is much more convenient if the two point masses are considered to be located at the centres of the two end bearings, i.e. at A and B.

Let m_a = mass at A, distance $AG = a$

Then $m_a + m_b = m$

$$m_a = m\frac{b}{a+b} = m\frac{b}{l} \qquad (l = \text{length of rod})$$

$$m_b = m\frac{a}{a+b} = m\frac{a}{l}$$

$$I' = mab$$

Assuming $a > d$, $I' > I$

This means that by considering the two masses at A and B instead of at D and B, the inertia torque is increased from the actual value $(T = I\alpha_c)$

Let ΔT = correction couple

$= \alpha_c\,(mab - mbd)$

$= mb\alpha_c(a-d)$

$= mb\alpha_c\,[(a + b) - (b + d)]$

$= mb\alpha_c(l - L)$ (taking $b + d = L$) (13.25)

This much torque must be applied to the two mass system in the opposite direction to that of the angular velocity to make the system dynamically equivalent to that of the actual rod.

The correction couple will be produced by two equal, parallel and opposite forces F_y acting at the gudgen pin and crankpin ends perpendicular to the line of stroke (Fig. 13.2). Force at B is taken by the reaction of guides.

Torque at crankshaft due to force at A or correction torque, T_c

$$= F_y \times r \cos\theta$$

$$= \frac{\Delta T}{l \cos\beta} \times r \cos\theta \qquad (\because T = F_y l \cos\beta)$$

$$= \frac{mb\alpha_c(l - L)}{l/r} \frac{\cos\theta}{\cos\beta}$$

$$= mb\alpha_c(l - L) \frac{\cos\theta}{n \frac{1}{n}\sqrt{n^2 - \sin^2\theta}}$$

$$= mb\alpha_c(l - L) \frac{\cos\theta}{\sqrt{n^2 - \sin^2\theta}} \tag{13.26}$$

Also due to the weight of the mass at A, a torque is exerted on the crankshaft which is given by

$$T_a = (m_a g) r \cos\theta \tag{13.27}$$

In case of vertical engines, a torque is also exerted on the crankshaft due to weight of mass at B and the expression will be similar to Eq. (13.21), i.e.,

$$T_b = (m_b g)\, r \left(\sin\theta + \frac{\sin 2\theta}{2\sqrt{n^2 - \sin^2\theta}} \right) \tag{13.28}$$

The net torque on the crankshaft will be the algebraic sum of the torques T, T_c, T_a and T_b.

13.12 INERTIA FORCE IN RECIPROCATING ENGINES (GRAPHICAL METHOD)

The inertia forces in reciprocating engines can be obtained graphically as follows (Fig. 13.4):

1. Draw the acceleration diagram by Klein's construction (Refer Sec. 3.8). Remember that the acceleration diagram is turned through 180° from the actual diagram and, therefore, the directions of accelerations are towards O.
2. Replace the mass of the connecting rod by dynamically equivalent system of two masses. If one mass is placed at B, the other will be at D given by $d = k^2/b$, where k is the radius of gyration and b and d are the distances of centre of mass from B and D respectively.
   Point D can also be obtained graphically. Draw $GE \perp AB$ at G and take $GE = k$. Make angle $BED = 90°$, and obtain point D on AB.
3. Obtain the accelerations of points G and D from the acceleration diagram by locating the points g_1 and d_1 on Ab_1 which represents the total acceleration of the connecting rod.
   As Ad_1/AD will be equal to Ab_1/AB and Ag_1/AG equal to Ab_1/AB, Dd_1 and Gg_1 can be drawn parallel to OB. Thus d_1O and g_1O represent accelerations of points D and G respectively.

4. The acceleration of the mass at B is along BO. Therefore, the inertia force due to this mass acts along OB.

The acceleration of the mass at D is parallel to d_1O and, therefore, the inertia force due to this mass acts in the opposite direction through D. Draw a line parallel to Od_1 through D to represent the direction of the inertia force.

Let the lines of action of the two forces due to masses at B and D meet at L. Then the resultant of the forces which is the total inertia force of the connecting rod and is parallel to Og_1 must also pass through point L. Therefore, draw a line parallel to Og_1 through L to represent the direction of the inertia force of connecting rod.

Now, the connecting rod is under the action of following forces:

– Inertia force of reciprocating parts $\mathbf{F}_b$ along OB,
– The reaction of the guide $\mathbf{R}_b$,
– Inertia force of the connecting rod $\mathbf{F}_i$
– The weight of the connecting rod $\mathbf{W}$ (= mg)
– Tangential force $\mathbf{F}_t$ at the crank pin
– Radial force $\mathbf{F}_r$ at the crank pin along OA.

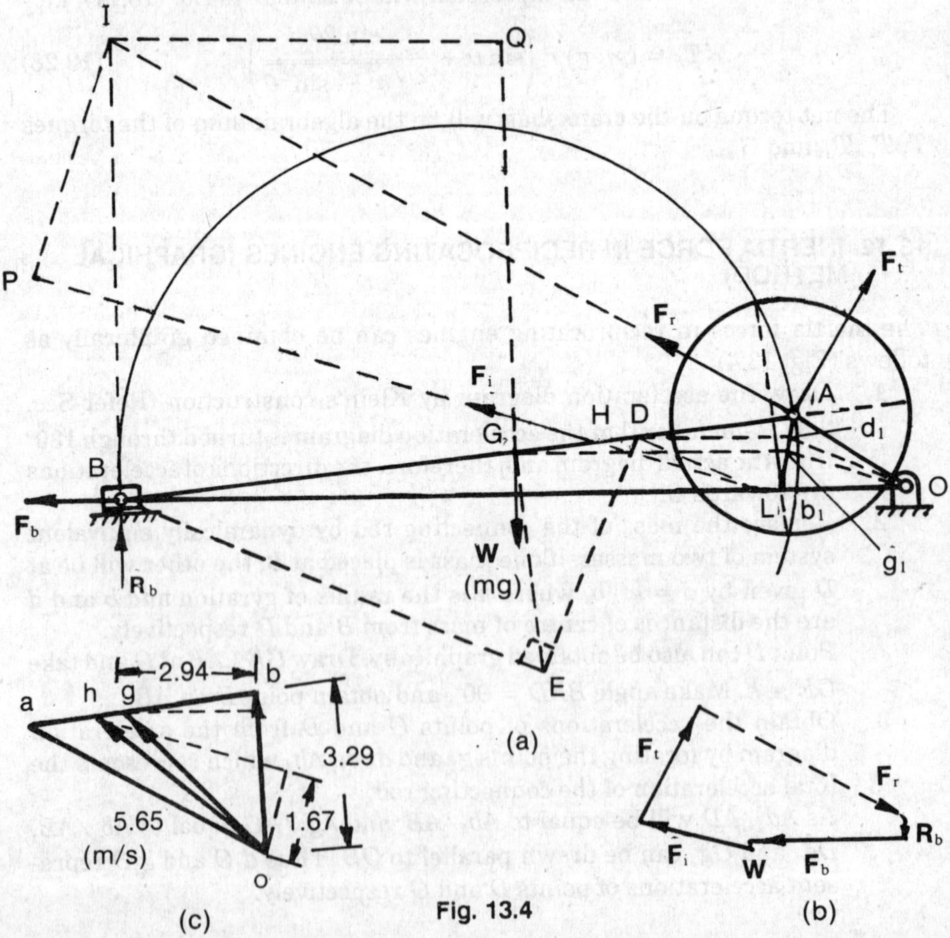

Fig. 13.4

Produce the lines of action of $\mathbf{F}_i$ and $\mathbf{R}_b$ to meet at I, the instantaneous centre of the connecting rod. Draw IP and IQ perpendicular to the lines of action of $\mathbf{F}_i$ and weight $\mathbf{W}$ respectively.

For the equilibrium of the connecting rod, take moments about I,

$$F_t \times IA = F_b \times IB + F_i \times IP + mg \times IQ \tag{13.29}$$

Obtain the value of F_t from it and draw the force polygon to find the values of $\mathbf{F}_r$ and $\mathbf{R}_b$ (Fig. 13.4b).

Torque on the crankshaft to overcome the inertia of the moving parts = $F_t \times OA$.

Example 13.2 The following data relate to the connecting rod of a reciprocating engine:

Mass = 50 kg
Distance between bearing centres = 900 mm
Diameter of big end bearing = 100 mm
Diameter of small end bearing = 80 mm
Time of oscillation when the connecting rod is suspended from
big end = 1.7 s
small end = 1.85 s

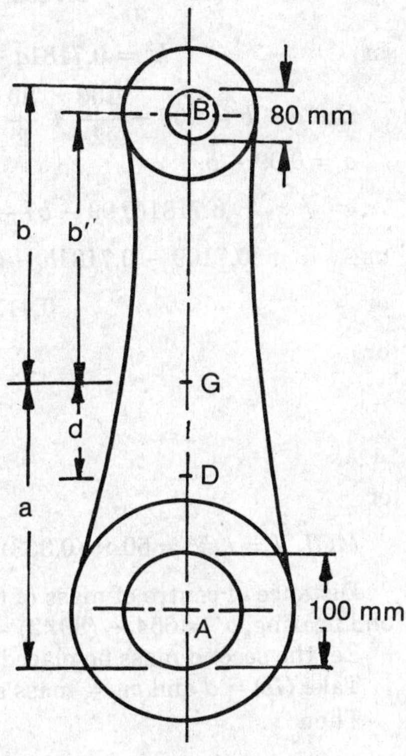

Fig. 13.5

Determine:
(i) the radius of gyration k of the rod about an axis through centre of mass perpendicular to the plane of oscillation,
(ii) the moment of inertia of the rod about the same axis, and
(iii) the dynamically equivalent system of the connecting rod comprising two masses, one at the small end bearing centre.

Solution: Refer Fig. 13.5.

Let L_a = length of equivalent simple pendulum when suspended from the top of big end bearing
L_b = length of equivalent simple pendulum when suspended from the top of small end bearing
a = distance of centre of mass G from top of big-end bearing
b = distance of centre of mass G from top of small-end bearing.

$$t_a = 2\pi\sqrt{\frac{L_a}{g}} \quad \text{and} \quad t_b = 2\pi\sqrt{\frac{L_b}{g}}$$

or $$1.7 = 2\pi\sqrt{\frac{L_a}{9.81}} \quad \text{and} \quad 1.85 = 2\pi\sqrt{\frac{L_b}{9.81}}$$

or $$L_a = 0.7181 \text{ m} \quad \text{and} \quad L_b = 0.8505 \text{ m}$$

or $$a + \frac{k^2}{a} = 0.7181 \quad \text{and} \quad b + \frac{k^2}{b} = 0.8505$$

or $$k^2 = 0.7181a - a^2 = 0.8505\,b - b^2$$

But $a + b = 900 + \frac{100}{2} + \frac{80}{2} = 990 \text{ mm} = 0.99 \text{ m}$

$a = 0.99 - b$

$\therefore$ $$0.7181(0.99 - b) - (0.99 - b)^2 = 0.8505\,b - b^2$$

or $$0.7109 - 0.7181b - (0.9801 + b^2 - 1.98\,b) = 8505\,b - b^2$$

or $$0.4115\,b = 0.2692$$

or $$b = 0.654 \text{ m}$$

$$a = 0.99 - 0.654 = 0.336 \text{ m}$$

$$k^2 = 0.8505 \times 0.654 - (0.654)^2 = 0.1285$$

or $$k = \underline{0.358 \text{ m}}$$

MOI, $I = mk^2 = 50 \times (0.358)^2 = \underline{6.4 \text{ kg.m}^2}$

Distance of centre of mass of the connecting rod from the centre of small end bearing, $b' = 654 - (80/2) = 614$ mm

Let the second mass be placed at D.

Take $GD = d$ and m_d = mass at D

Then

$$d = \frac{k^2}{b'} = \frac{0.1285}{0.614} = 0.209 \text{ m}$$

$$m_d = \frac{m \times b'}{b' + d} = \frac{50 \times 0.614}{0.614 + 0.209} = 37.3 \text{ kg}$$

$$m_b' = 50 - 37.3 = \underline{12.7 \text{ kg}}$$ (m_b' = mass at the small end bearing centre)

Example 13.3 The following data relate to a horizontal reciprocating engine:

Mass of reciprocating parts = 120 kg
Crank length = 90 mm
Engine speed = 600 rpm

Connecting rod: Mass = 90 kg
Length between centres = 450 mm

Distance of centre of mass from big end centre = 180 mm
Radius of gyration about an axis through centre of mass = 150 mm
Find the magnitude and the direction of the inertia torque on the crankshaft when the crank has turned 30° from the inner-dead centre.

Solution: Draw the configuration diagram *OAB* of the engine mechanism to a convenient scale (Fig. 13.4) and its velocity and acceleration diagrams by Klein's construction.

$$\omega = \frac{2\pi N}{60} = \frac{2\pi \times 600}{60} = 62.8 \text{ rad/s}$$

$$v_a = \omega r = 62.8 \times 0.09 = 5.65 \text{ m/s}$$

$$f_a = \omega^2 r = (62.8)^2 \times 0.09 = 355 \text{ m/s}^2$$

Locate the points *B* and *G* on the acceleration diagram and measure b_1O and g_1O.

$$f_b = b_1O = 343.2 \text{ m/s}^2 \quad \text{and} \quad f_g = g_1O = 345 \text{ m/s}^2$$

$$\therefore \quad F_b = m_b \times f_b = 120 \times 343.2 = 41\,186 \text{ N}$$

$$F_i = m \times f_g = 90 \times 345 = 31\,050 \text{ N}$$

Complete the diagram of Fig. 13.4 as discussed in Sec. 13.12. Taking moments about *I*,

$$F_t \times IA = F_b \times IB + F_i \times IP + mg \times IQ$$

$$F_t \times 515 = 41\,186 \times 300 + 31\,050 \times 152 + 90 \times 9.81 \times 268$$

$$F_t = 23\,992 + 9164 + 459.5$$

$$\therefore \quad T = F_t \times r = 2159.3 + 824.8 + 41.3 = 3025.4 \text{ N.m}$$

Instead of taking moments about the I-centre, principle of virtual work (Sec. 12.8) can also be applied to obtain the torque.

On the velocity diagram, locate the points *b*, *h* and *g* corresponding to *B*, *H* and *G* respectively and take the components of velocities in the directions of forces F_b, F_i and *mg* (Fig. 13.4c). In Klein's construction, the velocity diagram in turned through 90°. Then

$$T \times \omega = F_b \times v_b + F_i \times v_h + mg \times v_g$$

$$T \times 62.8 = 41186 \times 3.29 + 31050 \times 1.67 + 90 \times 9.81 \times 2.94$$

$$T = 2157.6 + 825.7 + 41.3$$

$$= \underline{3024.6 \text{ N.m}}$$

Analytical Method Divide the mass of the rod into two parts.

$$\text{Mass at crank pin, } m_a = 90 \times \left(\frac{450-180}{450}\right) = 54 \text{ kg}$$

Mass at gudgeon pin, $m_b = 90 - 54 = 36$ kg

Total mass of reciprocating parts, $m = 120 + 36 = 156$ kg

$$F = mr\,\omega^2 \left(\cos\theta + \frac{\cos 2\theta}{n}\right) \qquad \text{[Eq. (13.18)]}$$

$$= 156 \times 0.09 \times (62.8)^2 \left(\cos 30° + \frac{\cos 60°}{5}\right)$$

$$= 53\,490 \text{ N}$$

$$T = Fr\left(\sin\theta + \frac{\sin 2\theta}{2\sqrt{n^2 - \sin^2\theta}}\right) \qquad \text{[Eq. (13.21)]}$$

$$= 53\,490 \times 0.09 \left(\sin 30° + \frac{\sin 60°}{2\sqrt{(5)^2 - \sin^2 30°}}\right)$$

$$= 2826 \text{ N.m} \quad \text{counter-clockwise}$$

$$\Delta T = m\alpha_c b(l - L) \qquad \text{[Eq. (13.25)]}$$

where $b = 450 - 180 = 270$ mm

$l = 450$ mm

and $L = b + \dfrac{k^2}{b} = 270 + \dfrac{(150)^2}{270} = 353.3$ mm

$$\alpha_c = -\omega^2 \sin\theta \left[\frac{n^2 - 1}{(n^2 - \sin^2\theta)^{3/2}}\right] \qquad \text{[Eq. (13.16)]}$$

$$= -(62.8)^2 \sin 30° \left[\frac{5^2 - 1}{(25 - \sin^2 30°)^{3/2}}\right]$$

$$= -384.7 \text{ rad/s}^2$$

$\therefore$ $\Delta T = 90 \times (-384.7) \times 0.27 \times (0.45 - 0.3533)$

$= -903.97$ N.m

$\therefore$ $$\text{Correction torque} = \Delta T \frac{\cos\theta}{\sqrt{n^2 - \sin^2\theta}}$$

$$= -903.97 \times \frac{\cos 30°}{\sqrt{25 - \sin^2 30°}}$$

$$= -157.4 \text{ N.m}$$

Correction torque is to be deducted from the inertia torque on the crankshaft.

Torque due to weight of mass at A, $T_a = (m_a\,g)\, r\cos\theta$

$= 54 \times 9.81 \times 0.09 \times \cos 30°$

$= 41.3$ N.m counter-clockwise

$\therefore$ Total inertia torque on the crankshaft $= 2826 - (-157.4) + 41.3$

$= \underline{3024.7 \text{ N.m}}$ counter-clockwise

Example 13.4 The connecting rod of a vertical reciprocating engine is 2 m long between centres and it weighs 250 kg. The mass centre is 800 mm from the big end bearing. When suspended as a pendulum from the gudgen pin axis, it makes 8 complete oscillations in 22 seconds. Calculate the radius of gyration of the rod about an axis through its mass centre. The crank is 400 mm long and rotates at 200 rpm.

When the crank has turned through 40° from the top dead centre and the piston is moving downwards, analytically find the inertia torque exerted on the crankshaft.

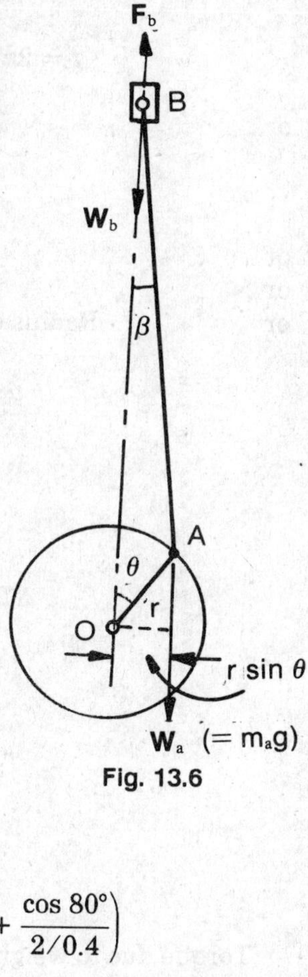

Fig. 13.6

Solution: Refer Fig. 13.6.

Divide the mass of the rod into two parts,

Mass at the crank pin,

$$m_a = 250 \times \frac{2.0 - 0.8}{2.0} = 150 \text{ kg}$$

Mass at the gudgeon pin,

$$m_b = 250 - 150 = 100 \text{ kg}$$

$$F = mr\,\omega^2 \left(\cos\theta + \frac{\cos 2\theta}{n}\right)$$

$$= 100 \times 0.4 \times \left(\frac{2\pi \times 200}{60}\right)^2 \left(\cos 40° + \frac{\cos 80°}{2/0.4}\right)$$

$$= 100 \times 0.4 \times 438.6 \times 0.8$$

$$= 14\,049 \text{ N}$$

$$T = Fr\left(\sin\theta + \frac{\sin 2\theta}{2\sqrt{n^2 - \sin^2\theta}}\right)$$

$$= 14\,049 \times 0.4 \left(\sin 40° + \frac{\sin 80°}{2\sqrt{25 - \sin^2 40°}}\right)$$

$$= 14\,049 \times 0.4 \times 0.7421$$

$$= 4170.3 \text{ N.m counter-clockwise}$$

We have,

$$b + \frac{k^2}{b} = L$$

where $b = 2 - 0.8 = 1.2$ m and L can be found from

$$t = 2\pi\sqrt{\frac{L}{g}} \quad \text{or} \quad \frac{22}{8} = 2\pi\sqrt{\frac{L}{9.81}}$$

or $$L = 1.88 \text{ m}$$

$\therefore$ $$1.2 + \frac{k^2}{1.2} = 1.88$$

or $$k^2 = 0.816$$

or $$k = 0.903$$

or $$\text{Radius of gyration} = 903 \text{ mm}$$

$$\alpha_c = -\omega^2 \sin\theta \left[\frac{n^2 - 1}{(n^2 - \sin^2\theta)^{3/2}}\right]$$

$$= -438.6 \sin 40° \left[\frac{25 - 1}{(25 - \sin^2 40°)^{3/2}}\right]$$

$$= -55.5 \text{ rad/s}^2$$

$$\Delta T = m\alpha_c b(l - L)$$

$$= 250 \times (-55.5) \times 1.2 \times (2.0 - 1.88)$$

$$= -1998 \text{ N.m}$$

$$T_c = \Delta T \frac{\cos\theta}{\sqrt{n^2 - \sin^2\theta}}$$

$$= -1998 \times \frac{\cos 40°}{\sqrt{25 - \sin^2 40°}}$$

$$= -308.7 \text{ N.m}$$

Torque due to weight of mass at A, $T_a = m_a gr \sin\theta$

$$= 150 \times 9.81 \times 0.4 \sin 40°$$

$$= 378.3 \text{ N.m clockwise}$$

Torque due to weight of mass at B,

$$T_b = m_b gr \left(\sin\theta + \frac{\sin 2\theta}{2\sqrt{25 - \sin^2 40°}}\right)$$

$$= 100 \times 9.81 \times 0.4 \times 0.7421$$

$$= 291.2 \text{ N.m clockwise}$$

Total inertia torque on crankshaft $= 4170.3 - (-308.7) - 378.3 - 291.2$

$= \underline{3809.5 \text{ N.m}}$

13.13 TURNING-MOMENT DIAGRAMS

During one revolution of the crankshaft of a steam engine or IC engine, the torque on it varies and is given by

$$T = F_t \times r$$

$$= Fr\left(\sin\theta + \frac{\sin 2\theta}{2\sqrt{n^2 - \sin^2\theta}}\right) \qquad \text{[refer to Eq. (13.21)]}$$

where F is the net piston effort.

A plot of T vs. θ is known as turning-moment diagram. The inertia effect of the connecting rod is, usually, ignored while drawing these diagrams, but can be taken into account if desired.

As $T = F_t \times r$, a plot of F_t vs. θ (known as *crank effort diagram*) is identical to a turning-moment diagram.

Turning-moment diagrams for different types of engines are being given below:

1. Single-cylinder Double-acting Steam engine

Figure 13.7 shows a turning-moment diagram for a single-cylinder double acting steam engine. Crank angle θ is represented along the x-axis and the turning-moment along the y-axis. It can be observed that during the outstroke (*ogp*) the turning moment is maximum when the crank angle is a little less than 90° and zero when the crank angle is zero and 180°. A somewhat similar turning-moment diagram is obtained during the instroke (*pkq*).

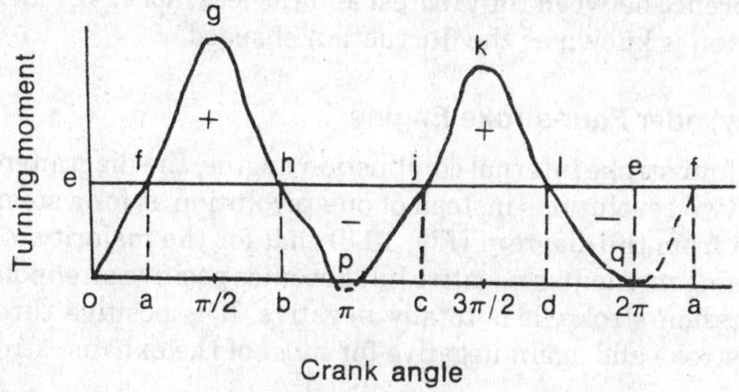

Fig. 13.7

Note that the area of the turning-moment diagram is proportional to the work done per revolution as the work is the product of turning-moment and the angle turned.

The mean torque against which the engine works is given by

$$oe = \frac{\text{Area } ogpkq}{2\pi}$$

where *oe* is the mean torque and is the mean height of the turning-moment diagram.

When the crank turns from angle *oa* to *ob* (Fig. 13.7), the work done by the engine is represented by area *afghb*. But the work done against the resisting torque is represented by the area *afhb*. Thus the engine has done more work than what has been taken from it. The excess work is represented by the area *fgh*. This excess work increases the speed of the engine and is stored in the flywheel.

During the crank travel from *ob* to *oc*, the work needed for the external resistance is proportional to *bhjc* whereas the work produced by the engine is represented by the area under *hpj*. Thus during this period, more work has been taken from the engine than that is produced. The loss is made up by the flywheel which gives up some of its energy and the speed decreases during this period.

Similarly, during the period of crank travel from *oc* to *od*, excess work is again developed and is stored in the flywheel and the speed of the engine increases. During the crank travel from *od* to *oa*, the loss of work is made up by the flywheel and the speed again decreases.

The areas *fgh*, *hpj*, *jkl* and *lqf* represent fluctuations of energy of the flywheel. When the crank is at *b*, the flywheel has absorbed energy while the crank has moved from *a* to *b* and thereby, the speed of the engine is maximum. At *c*, the flywheel has given out energy while the crank has moved from *b* to *c* and thus the engine has a minimum speed. Similarly, the engine speed is again maximum at *d* and minimum at *a*. Thus there are two maximum and two minimum speeds for the turning-moment diagram.

The greatest speed is the greater of the two maximum speeds and the least speed is the lesser of the two minimum speeds.

The difference between the greatest and the least speeds of the engine over one revolution is known as the fluctuation of speed.

2. Single-cylinder Four-stroke Engine

In case of a four stroke internal combustion engine, the diagram repeats itself after every two revolutions instead of one revolution as for a steam engine. It can be seen from the diagram (Fig. 13.8) that for the majority of the suction stroke, turning moment is negative but becomes positive after point *p*. During the compression stroke, it is totally negative. It is positive throughout the expansion stroke and again negative for most of the exhaust stroke.

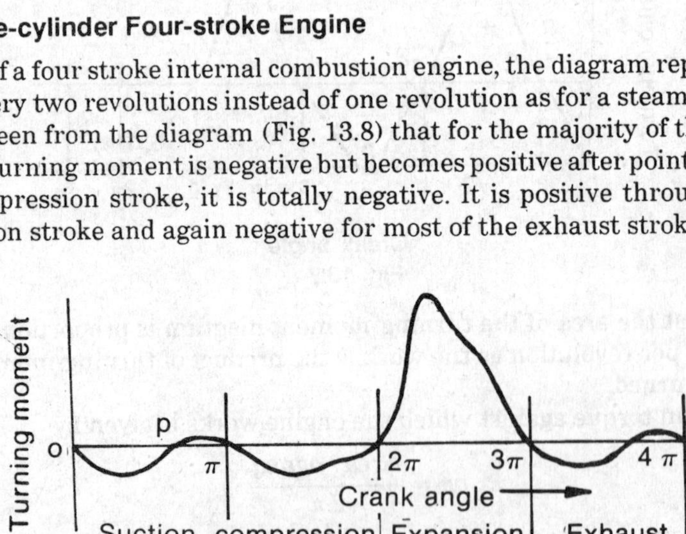

Fig. 13.8

3. Multi-cylinder Engines

As observed in the foregoing paragraphs, the turning-moment diagram for a single-cylinder engine varies considerably and a greater variation of the same is observed in case of four-stroke, single-cylinder engine. For engines with more than one cylinder, the total crankshaft torque at any instant is given by the sum of the torques developed by each cylinder at the instant. For example, if an engine has two cylinders with cranks at 90°, the resultant turning moment diagram has a less variation than that for a single cylinder. In a three-cylinder engine having its cranks at 120°, the variation is still less.

Figure 13.9 shows the turning moment diagram for a multicylinder engine. The mean torque line *ab* intersects the turning moment curve at *c, d, e, f, g* and *h*. The area under the wavy curve is equal to the area *oabk*. As discussed earlier, the speed of the engine will be maximum when the crank positions correspond to *d, f* and *h*, and minimum corresponding to *c, e* and *g*.

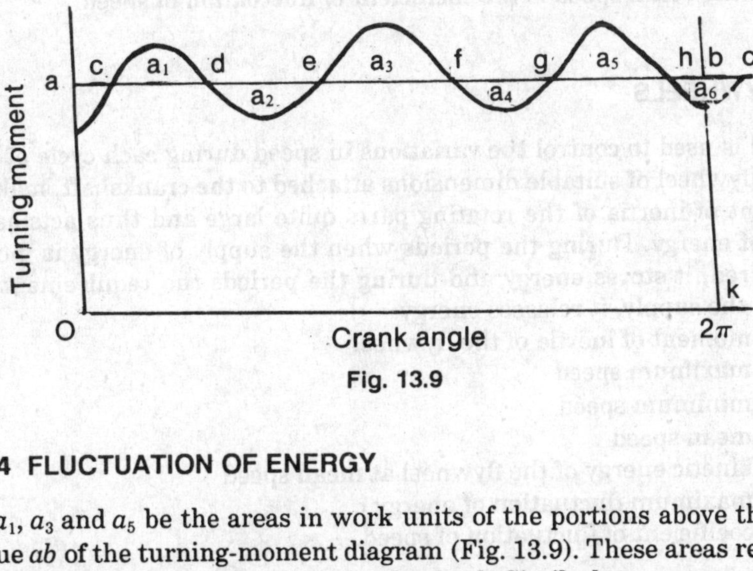

Fig. 13.9

13.14 FLUCTUATION OF ENERGY

Let a_1, a_3 and a_5 be the areas in work units of the portions above the mean torque *ab* of the turning-moment diagram (Fig. 13.9). These areas represent quantities of energies added to the flywheel. Similarly, areas a_2, a_4 and a_6 below *ab* represent quantities of energies taken from the flywheel.

The energies of the flywheel corresponding to positions of the crank are as follows:

Crank position	Flywheel energy
c	E
d	$E + a_1$
e	$E + a_1 - a_2$
f	$E + a_1 - a_2 + a_3$
g	$E + a_1 - a_2 + a_3 - a_4$
h	$E + a_1 - a_2 + a_3 - a_4 + a_5$
c	$E + a_1 - a_2 + a_3 - a_4 + a_5 - a_6$

From the two values of the energies of the flywheel corresponding to the position c, it is concluded that

$$a_1 - a_2 + a_3 - a_4 + a_5 - a_6 = 0$$

The greatest of these energies is the maximum kinetic energy of the flywheel and for the corresponding crank position, the speed is maximum.

The least of these energies is the least kinetic energy of the flywheel and for the corresponding crank position, the speed is minimum.

The difference between the maximum and minimum kinetic energies of the flywheel is known as the maximum fluctuation of energy whereas the ratio of this maximum fluctuation of energy to the work done per cycle is defined as the coefficient of fluctuation of energy.

The difference between the greatest speed and the least speed is known as the maximum fluctuation of speed and the ratio of the maximum fluctuation of speed to the mean speed is the coefficient of fluctuation of speed.

13.15 FLYWHEELS

A flywheel is used to control the variations in speed during each cycle of an engine. A flywheel of suitable dimensions attached to the crankshaft, makes the moment of inertia of the rotating parts quite large and thus acts as a reservoir of energy. During the periods when the supply of energy is more than required, it stores energy and during the periods the requirement is more than the supply, it releases energy.

Let I = moment of inertia of the flywheel
ω_1 = maximum speed
ω_2 = minimum speed
ω = mean speed
E = kinetic energy of the flywheel at mean speed
e = maximum fluctuation of energy
K = coefficient of fluctuation of speed.

$$\begin{aligned}
e &= \frac{1}{2} I\omega_1^2 - \frac{1}{2} I\omega_2^2 \\
&= \frac{1}{2} I\,(\omega_1^2 - \omega_2^2) \\
&= I\left(\frac{\omega_1 + \omega_2}{2}\right)(\omega_1 - \omega_2) \\
&= I\,\omega\,(\omega_1 - \omega_2) \\
&= I\,\omega^2\left(\frac{\omega_1 - \omega_2}{\omega}\right) \\
&= I\,\omega^2 K
\end{aligned}$$

or $$K = \frac{e}{I\omega^2} = \frac{e}{2 \times \frac{1}{2} I\omega^2} = \frac{e}{2E} \tag{13.30}$$

Example 13.5 A punching machine carries out 6 holes per minute. Each hole of 40 mm diameter in 35 mm thick plate requires 8 N.m of energy/mm^2 of the sheared area. The punch has a stroke of 95 mm. Find the power of the motor required if the mean speed of the flywheel is 20 m/s.

If total fluctuation of speed is not to exceed 3% of the mean speed, determine the mass of the flywheel.

Solution:

$d = 40$ mm $\quad K = 0.03$

$t = 35$ mm $\quad$ Stroke $= 95$ mm

$v = 20$ m/s

$$\begin{aligned}
\text{Power, } P &= \text{Energy required for punching work/s} \\
&= \frac{1}{60} (\text{Energy required/minute}) \\
&= \frac{1}{60} (\text{Energy required/hole} \times \text{No. of holes/minute}) \\
&= \frac{1}{60} [(\pi dt \times 8) \times 6] \\
&= \frac{1}{60} [(\pi \times 40 \times 35 \times 8) \times 6] \\
&= \frac{1}{60} (35\,186 \times 6) \\
&= 3519 \text{ N.m/s} \quad \text{or} \quad 3519 \text{ W} \quad \text{or} \quad 3.519 \text{ kW}
\end{aligned}$$

The punch travels a distance of 190 mm (upstroke + downstroke) in 10 seconds. (6 holes are punched in 1 minute).

∴ Actual time required to punch a hole in 35 mm thick plate

$$= \frac{10}{190} \times 35 = 1.842 \text{ s}$$

Energy required/hole or energy supplied by the motor in 10 seconds

$$= \pi\, dt \times 8 = 35\,186 \text{ N.m}$$

$$\text{Energy supplied by the motor in 1.842 s} = \frac{35\,186}{10} \times 1.842 = 6481 \text{ N.m}$$

$$\text{Energy supplied by the flywheel} = 35\,186 - 6481 = 28\,705 \text{ N.m}$$

$$K = \frac{e}{2E}$$

or $$0.03 = \frac{28\,705}{2E}$$

or $$E = 478\,417$$

or $$\frac{1}{2} mv^2 = 478\,417$$

or $$\frac{1}{2} m\,(20)^2 = 478\,417$$

or $$m = \underline{2392 \text{ kg}}$$

Example 13.6 **The turning moment diagram for a multicylinder engine has been drawn to a vertical scale of 1 mm = 650 N.m and a horizontal scale of 1 mm = 4.5°. The areas above and below the mean torque line are −28, +380, −260, +310, −300, +242, −380, +265 and −229 mm².**

The fluctuation of speed is limited to ±1.8% of the mean speed which is 400 rpm. Density of the rim material is 7000 kg/m³ and width of the rim is 4.5 times its thickness. The centrifugal stress (hoop stress) in the rim material is limited to 6 N/mm². Neglecting the effect of the boss and arms, determine the diameter and cross-section of the flywheel rim.

Solution:

$\rho = 7000 \text{ kg/m}^3$ $\quad$ $f = 6 \times 10^6 \text{ N/m}^2$

$N = 400$ rpm $\quad$ $K = 0.018$

$b = 4.5t$

$f = \rho v^2$

$$6 \times 10^6 = 7000 \times v^2$$

$$v = 29.28 \text{ m/s}$$

or $$\frac{\pi\, dn}{60} = \frac{\pi \times d \times 400}{60} = 29.28$$

or $$d = 1.398 \text{ m}$$

Let flywheel KE at $a = E$ (refer Fig. 13.10)

at $b = E - 28$

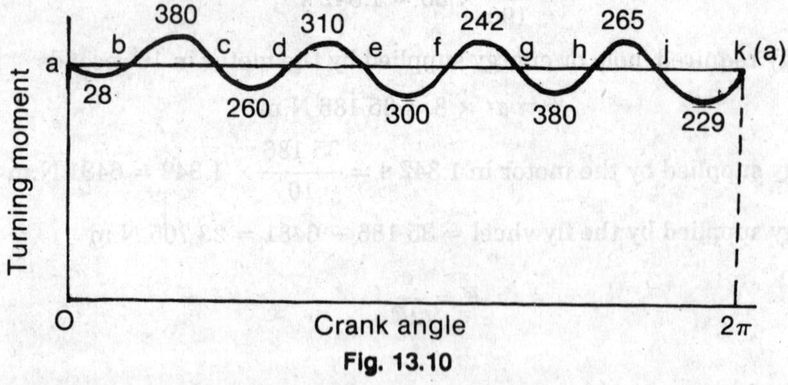

Fig. 13.10

$$\begin{aligned}
\text{at } c &= E - 28 + 380 = E + 352\\
\text{at } d &= E + 352 - 260 = E + 92\\
\text{at } e &= E + 92 + 310 = E + 402\\
\text{at } f &= E + 402 - 300 = E + 102\\
\text{at } g &= E + 102 + 242 = E + 344\\
\text{at } h &= E + 344 - 380 = E - 36\\
\text{at } j &= E - 36 + 265 = E + 229\\
\text{at } k &= E + 229 - 229 = E
\end{aligned}$$

$$\text{Maximum energy} = E + 402 \qquad (\text{at } e)$$

$$\text{Minimum energy} = E - 36 \qquad (\text{at } h)$$

Maximum fluctuation of energy,

$$\begin{aligned}
e_{max} &= (E + 402) - (E - 36) \times \text{Hor. scale} \times \text{Vert. scale}\\
&= 438 \times \left(4.5 \times \frac{\pi}{180}\right) \times 650\\
&= 22\,360 \text{ N.m}
\end{aligned}$$

$$K = \frac{e}{mk^2\omega^2}$$

$$0.018 = \frac{22\,360}{m\left(\frac{1.398}{2}\right)^2\left(\frac{2\pi \times 400}{60}\right)^2}$$

$$m = 1449 \text{ kg}$$

or $\text{Density} \times \text{Volume} = 1449$

or $\rho \times (\pi d) \times t \times 4.5t = 1449$

or $7000 \times \pi \times 1.398 \times t \times 4.5t = 1449$

$$t = 0.1023 \text{ m or } 102.3 \text{ mm}$$

$$b = 4.5 \times 102.3 = 460.5 \text{ mm}$$

Example 13.7 The torque delivered by a two-stroke engine is represented by

$$T = (1000 + 300 \sin 2\theta - 500 \cos 2\theta) \text{ N.m}$$

where θ is the angle turned by the crank from the inner-dead centre. The engine speed is 250 rpm. The mass of the flywheel is 400 kg and radius of gyration 400 mm. Determine

(i) the power developed
(ii) the total percentage fluctuation of speed
(iii) the angular acceleration of flywheel when the crank has rotated through an angle of 60° from the inner-dead centre
(iv) the maximum angular acceleration and retardation of the flywheel.

Solution: The expression for torque being a function of 2θ, the cycle is repeated every 180° of the crank rotation.

(i)
$$T_{mean} = \frac{1}{\pi}\int_0^{\pi} T\,d\theta$$
$$= \frac{1}{\pi}\int_0^{\pi} (1000 + 300 \sin 2\theta - 500 \cos 2\theta)\,d\theta$$
$$= \frac{1}{\pi}\left[1000\,\theta - \frac{300}{2}\cos 2\theta - \frac{500}{2}\sin 2\theta\right]_0^{\pi}$$
$$= \frac{1}{\pi}[(1000\,\pi - 150 \cos 2\pi - 0) - (0 - 150 - 0)]$$
$$= 1000 \text{ N.m}$$
$$P = T\omega = 1000 \times \frac{2\pi \times 250}{60} = 26\,180 \text{ W or } \underline{26.18 \text{ kW}}$$

(ii) At any instant, $\Delta T = T - T_{mean}$
$$= (1000 + 300 \sin 2\theta - 500 \cos 2\theta) - 1000$$
$$= 300 \sin 2\theta - 500 \cos 2\theta$$

ΔT is zero, when $300 \sin 2\theta - 500 \cos 2\theta = 0$

or $300 \sin 2\theta = 500 \cos 2\theta$

or $\tan 2\theta = \frac{5}{3}$

or $2\theta = 59°$ or $239°$

$\theta = 29.5°$ or $119.5°$

$$e_{max} = \int_{29.5°}^{119.5°} \Delta T\,d\theta$$
$$= \int_{29.5°}^{119.5°} (300 \sin 2\theta - 500 \cos 2\theta)\,d\theta$$
$$= [-150 \cos 2\theta - 250 \sin 2\theta]_{29.5°}^{119.5°}$$
$$= 583.1 \text{ N.m}$$
$$K = \frac{e}{mk^2\omega^2}$$
$$= \frac{583.1}{400 \times (0.4)^2 \times \left(\frac{2\pi \times 250}{60}\right)^2}$$
$$= 0.013\,29 \quad \text{or} \quad 1.329\%$$

(iii) Acceleration is produced by excess or deficit torque at any moment.
$$\Delta T = 300 \sin 2\theta - 500 \cos 2\theta$$

When $\theta = 60°$,

$$\Delta T = 509.8 \text{ N.m}$$

or $$I\alpha = mk^2\alpha = 509.8$$

or $$400 \times (0.4)^2 \times \alpha = 509.8$$

or $$\alpha = 7.966 \text{ rad/s}^2$$

(iv) For ΔT_{max} and ΔT_{min},

$$\frac{d}{d\theta}(\Delta T) = \frac{d}{d\theta}(300 \sin 2\theta - 500 \cos 2\theta)d\theta = 0$$

or

$$2 \times 300 \cos 2\theta + 2 \times 500 \sin 2\theta = 0$$

or $$600 \cos 2\theta = -1000 \sin 2\theta$$

or $$\tan 2\theta = -0.6$$

or $$2\theta = 149.04^\circ \text{ and } 329.04^\circ$$

When $2\theta = 149.04^\circ$, $T = 1583.1$ N.m, $\Delta T = 583.1$ N.m

When $2\theta = 329.04^\circ$, $T = 416.9$ N.m, $\Delta T = -583.1$ N.m

As values of ΔT at maximum and minimum torque T are same, maximum acceleration is equal to maximum retardation.

or $$\Delta T = mk^2\alpha = 583.1$$

or $$400 \times (0.4)^2 \times \alpha = 583.1$$

Maximum acceleration or retardation, $\alpha = 9.11$ rad/s^2

Example 13.8 A machine is coupled to a two-stroke engine which produces a torque of $(800 + 180 \sin 3\theta)$ N.m, where θ is the crank angle. The mean engine speed is 400 rpm. The flywheel and the other rotating parts attached to the engine have a mass of 350 kg at a radius of gyration of 200 mm. Calculate

(i) the power of the engine

(ii) the total fluctuation of speed of the flywheel when

(a) the resisting torque is constant

(b) the resisting torque is $(800 + 80 \sin \theta)$ N.m.

Solution:

$m = 350$ kg $\qquad N = 400$ rpm

k = 220 mm $\qquad \omega = \dfrac{2\pi \times 400}{60} = 41.89$ rad/s

The expression for torque being a function of 3θ, the cycle is repeated after every 120° of the crank rotation.

(i)
$$T_{mean} = \frac{1}{2\pi/3}\int_0^{2\pi/3} T\,d\theta$$
$$= \frac{3}{2\pi}\int_0^{2\pi/3} (800 + 180\sin 3\theta)\,d\theta$$
$$= \frac{3}{2\pi}\left[800\,\theta - \frac{180}{3}\cos 3\theta\right]_0^{2\pi/3}$$
$$= 800 \text{ N.m}$$
$$P = T\omega = 800 \times 41.89 = 33\,512 \text{ W}$$

(ii) (a) At any instant, $\Delta = T - T_{mean}$
$$= 800 + 180\sin 3\theta - 800$$
$$= 180\sin 3\theta$$

ΔT is zero when $180\sin 3\theta = 0$

or when $\sin 3\theta = 0$

or $3\theta = 0° \text{ or } 180°$

or $\theta = 0° \text{ or } 60°$

$$e_{max} = \int_{0°}^{60°} \Delta T\,dt$$
$$= \int_{0°}^{60°} (180\sin 3\theta)\,d\theta$$
$$= \left[-\frac{180\cos 3\theta}{3}\right]_{0°}^{60°}$$
$$= 120 \text{ N.m}$$
$$K = \frac{e}{mk^2\omega^2} = \frac{120}{350 \times (0.22)^2 \times (41.89)^2} = 0.00404 \text{ or } \underline{0.404\%}$$

(b)
$$\Delta T = T \text{ of engine} - T \text{ of machine}$$
$$= (800 + 180\sin 3\theta) - (800 + 80\sin\theta)$$
$$= 180\sin 3\theta - 80\sin\theta$$

ΔT is zero when $180\sin 3\theta - 80\sin\theta = 0$

or $180\sin 3\theta = 80\sin\theta$

or $180\,(3\sin\theta - 4\sin^3\theta) = 80\sin\theta$

or $3 - 4\sin^2\theta = \frac{80}{180} = 0.389$

or $\sin^2\theta = 0.639$

or $$\sin\theta = \pm\, 0.799$$

or $$\theta = \pm\, 53° \text{ and } \pm\, 127°$$

$$e_{max} = \int_{53°}^{127°} \Delta T\, dt = \int_{53°}^{127°} (180 \sin 3\theta - 80 \sin\theta)\, d\theta$$

$$= \left[-\frac{180 \cos 3\theta}{3} + 80 \cos\theta\right]_{53°}^{127°}$$

$$= -60 \cos 381° + 80 \cos 127° + 60 \cos 159° - 80 \cos 53°$$

$$= -208.7 \text{ N.m}$$

$$K = \frac{e}{mk^2\omega^2} = \frac{208.7}{350 \times (0.22)^2 \times (41.89)^2} = 0.007\,03 \text{ or } \underline{0.703\%}$$

Example 13.9 In a machine, the intermittent operations demand the torque to be applied as follows:

– During the first half revolution, the torque increases uniformly from 800 N.m to 3000 N.m
– During next one revolution, the torque remains constant
– During next one revolution, the torque decreases uniformly from 3000 N.m to 800 N.m
– During last 1½ revolution, the torque remains constant.

Thus a cycle is completed in 4 revolutions. The motor to which the machine is coupled exerts a constant torque at a mean speed of 250 rpm. A flywheel of mass 1800 kg and radius of gyration of 500 mm is fitted to the shaft. Determine

(i) the power of the motor
(ii) the total fluctuation of speed of the machine shaft.

Solution:

m = 1800 kg N = 250 rpm
k = 500 mm

(i) Refer to Fig. 13.11.
Torque for one complete cycle, T = area $OABCDEF$

or T = Area $OAEF$ + Area ABL + Area $LBCM$ + Area MCD

$$= 8\pi \times 800 + \frac{\pi \times 2200}{2} + 2\pi \times 2200 + \frac{2\pi \times 2200}{2}$$

$$= 14\,100\,\pi \text{ N.m}$$

$$T_{mean} = \frac{14\,100\,\pi}{8\,\pi} = 1762.5 \text{ N.m}$$

$$P = T_m\,\omega = 1762.5 \times \frac{2\pi \times 250}{60} = 46\,142 \text{ W or } \underline{46.142 \text{ kW}}$$

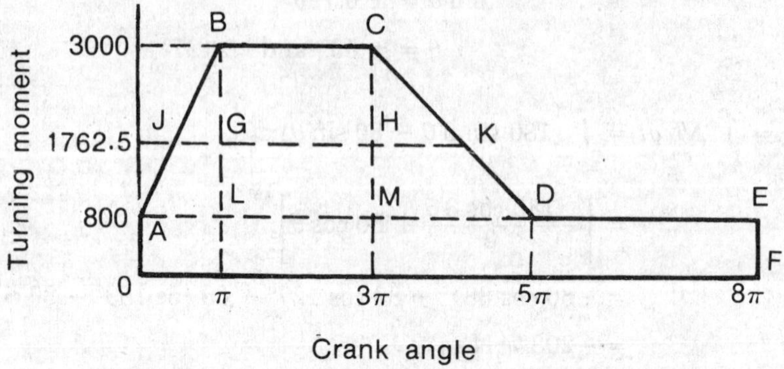

Fig. 13.11

(ii) $JG = AL \times \dfrac{BG}{BL} = \pi \times \dfrac{3000 - 1762.5}{3000 - 800} = 1.767$

$$HK = MD \times \frac{CH}{CM} = 2\,\pi \times \frac{3000 - 1762.5}{3000 - 800} = 3.534$$

Fluctuation of energy is equal to the area above the mean torque line.

$$e = \text{Area } JBCK = \text{area } JBG + \text{area } GBCH + \text{area } HCK$$

$$= (3000 - 1762.5)\left[\frac{1.767}{2} + 2\pi + \frac{3.534}{2}\right]$$

$$= 11\ 055 \text{ N.m}$$

$$K = \frac{e}{mk^2\omega^2}$$

$$= \frac{11\ 055}{1800 \times (0.5)^2 \times \left(\dfrac{2\pi \times 250}{60}\right)^2}$$

$$= 0.0358 \text{ or } 3.58\%$$

PROBLEMS

13.1 In a four-link mechanism $ABCD$, the link AB revolves with an angular velocity of 10 rad/s and angular acceleration of 25 rad/s^2 at the instant when it makes an angle of 45° with AD, the fixed link. The lengths of the links are:
$AB = CD = 800$ mm, $BC = 1000$ mm, and $AD = 1500$ mm.
The mass of the links is 4 kg/m length. Determine the torque required to overcome the inertia forces, neglecting the gravitational effects. Assume all links to be of uniform cross-sections.

(112.2 N.m)

13.2 The following data relate to a four-link mechanism:

Link	*Length*	*Mass*	*MOI about an axis through centre of mass*
AB	60 mm	0.2 kg	80 kg.mm^2
BC	200 mm	0.4 kg	1600 kg.mm^2
CD	100 mm	0.6 kg	400 kg.mm^2
AD	140 mm		

AD is the fixed link. The centres of mass for the links *BC* and *CD* lie at their midpoints whereas the centre of mass for link *AB* lies at *A*. Find the drive torque on the link *AB* at the instant when it rotates at an angular velocity of 47.5 rad/s counter-clockwise and ∠*DAB* 135°. Neglect gravity effects.

(1.675 N.m clockwise)

13.3 The effective steam pressure on the piston of a vertical steam engine is 20 N when the crank is 40° from the inner-dead centre on the down stroke. The crank length is 300 mm and the connecting rod length 1200 mm. The diameter of the cylinder is 800 mm. What will be the torque on the crank-shaft if the engine speed is 300 rpm and the mass of the reciprocating parts 250 kg?

(9916 N.m)

13.4 The length of the connecting rod of a gas engine is 500 mm and its centre of gravity lies at 165 mm from the crank pin centre. The rod has a mass of 80 kg and a radius of gyration of 182 mm about an axis through the centre of mass. The stroke of piston is 225 mm and the crank speed is 300 rpm. Determine the inertia force on the crankshaft when the crank has turned (a) 30° and (b) 135° from the inner-dead centre.

(302.3 N.m; 226.7 N.m)

13.5 The connecting rod of an IC engine is 450 mm long and has a mass of 2 kg. The centre of mass of the rod is 300 mm from the small end and its radius of gyration about an axis through this centre is 175 mm. The mass of the piston and the gudgen pin is 2.5 kg and the stroke is 300 mm. The cylinder diameter is 115 mm. Determine the magnitude and the direction of the torque applied on the crank shaft when the crank is at 40° and the piston is moving away from the inner-dead centre under an effective gas pressure of 2 N/mm^2. The engine speed is 1000 rpm.

(1014 N.m)

13.6 The connecting rod of a vertical high-speed engine is 600 mm long between centres and has a mass of 3 kg. Its centre of mass lies at 200 mm from the big-end bearing. When suspended as a pendulum from the gudgen pin axis, it makes 45 complete oscillations in 30 seconds. The piston stroke is 250 mm. The mass of the reciprocating parts is 1.2 kg. Determine the inertia torque on the crankshaft when the crank makes an angle of 140° with top-dead centre. The engine speed is 1500 rpm.

(361.7 N.m)

13.7 A constant torque 2.5 kW motor drives a rivetting machine. The mass of the moving parts including the flywheel is 125 kg at 700 mm radius. One rivetting operation absorbs 10 000 J of energy and takes one second. Speed of the flywheel is 240 rpm before rivetting. Determine:

(i) the number of rivets closed per hour, and

(ii) the reduction in speed after the riveting operation.

(900; 52.2 rpm)

13.8 The turning-moment diagram for a petrol engine is drawn to a vertical scale of 1 mm to 6 N.m and a horizontal scale of 1 mm to 1°. The turning moment repeats itself after every half revolution of engine. The areas above and below the mean torque line are:
305, 710, 50, 350, 980 and 275 mm^2.
The rotating parts amount to a mass of 40 kg at a radius of gyration of 140 mm. Calculate the coefficient of fluctuation of speed if the speed of the engine is 1500 rpm.

(0.55%)

13.9 The cranks of a three-cylinder single-acting engine are set equally at 120°. The engine speed is 540 rpm. The turning-moment diagram for each cylinder is a triangle for the power stroke with a maximum torque of 100 N.m at 60° after dead-centre of the corresponding crank. On the return stroke, the torque is sensibly zero. Determine
(a) the power developed
(b) the coefficient of fluctuation of speed if the flywheel has a mass of 7.5 kg with a radius of gyration of 65 mm
(c) the coefficient of fluctuation of energy
(d) the maximum angular acceleration of the flywheel.

(4.24 kW; 12.9%; 2.78%; 7.89 rad/s^2)

13.10 A certain machine requires a torque of $(1500 + 200 \sin \theta)$ N.m to drive it, where θ is the angle of rotation of the shaft. The machine is directly coupled to an engine which produces a torque of $(1500 + 250 \sin 2\theta)$ N.m. The flywheel and the other rotating parts attached to the engine has a mass of 300 kg at a radius of gyration of 200 mm. If the mean speed is 200 rpm, find
(a) the fluctuation of energy
(b) the total percentage fluctuation of speed
(c) the maximum and the minimum angular acceleration of the flywheel and the corresponding shaft positions.

(518.4 N.m; 9.85%; 9.07 rad/s^2, 34°21′; 34.6 rad/s^2, 127°16′)

14

BALANCING

14.1 INTRODUCTION

Often, an unbalance of forces is produced in rotary or reciprocating machinery due to the inertia forces associated with the moving masses. Balancing is the process of designing or modifying machinery so that the unbalance is reduced to an acceptable level and if possible is eliminated entirely.

A particle or mass moving in a circular path experiences a centripetal acceleration and a force is required to produce it. An equal and opposite force acting radially outwards acts on the axis of rotation and is known as centrifugal force (Fig. 14.1a). This is a disturbing force on the axis of rotation, the magnitude of which remains constant but the direction changes with the rotation of the mass.

In a revolving rotor, the centrifugal force remains balanced as long as the centre of the mass of rotor lies on the axis of the shaft. When the centre of mass does not lie on the axis or there is an eccentricity, an unbalanced force is produced (Fig. 14.1b). This type of unbalance is very common. For example, in steam turbine rotors, engine crankshafts, rotory compressors, centrifugal pumps, etc.

Most of the serious problems encountered in high-speed machinery are the direct result of unbalanced forces. These forces exerted on the frame by the

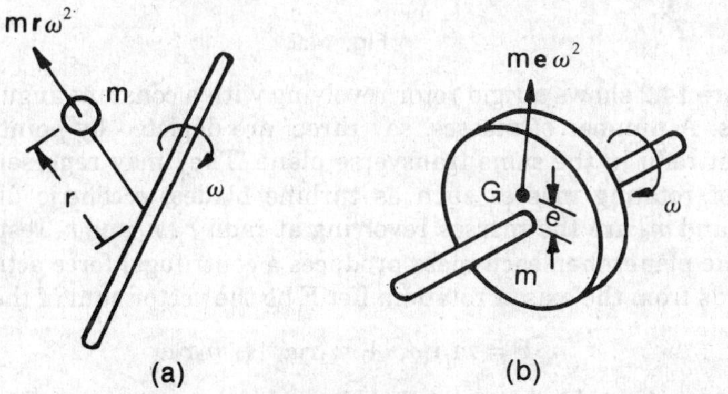

Fig. 14.1

moving machine members are time varying, impart vibratory motion to the frame and produce noise. Also there are human discomfort and detrimental effects on the machine performance and the structural integrity of the machine foundation.

The most common approach to balancing is by redistributing the mass which may be accomplished by addition or removal of mass from various machine members.

There are two basic types of unbalance, rotating unbalance and reciprocating unbalance, which may occur separately or in combination.

14.2 STATIC BALANCING

A system of rotating masses is said to be in static balance if the combined mass centre of the system lies on the axis of rotation.

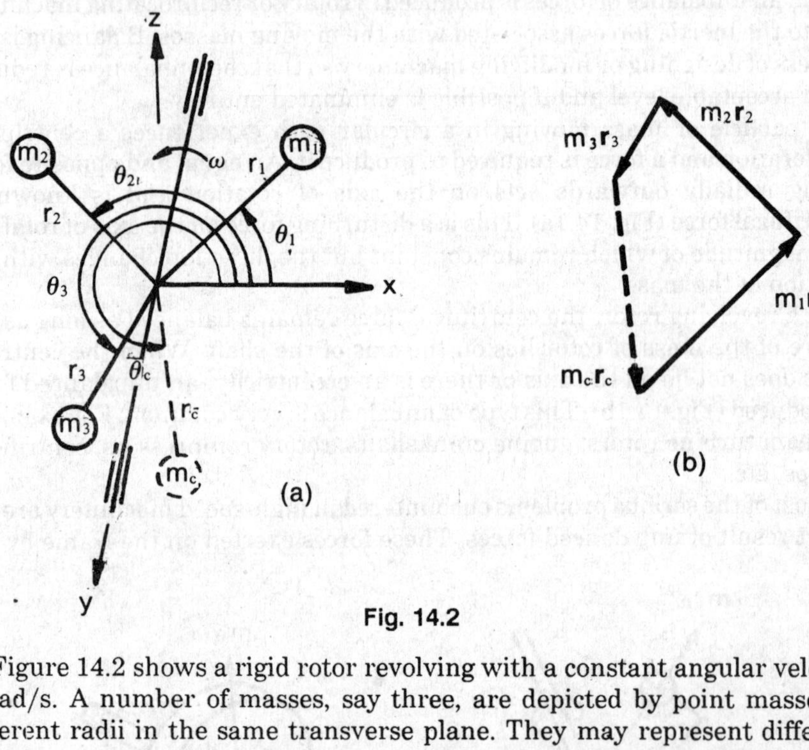

Fig. 14.2

Figure 14.2 shows a rigid rotor revolving with a constant angular velocity ω rad/s. A number of masses, say three, are depicted by point masses at different radii in the same transverse plane. They may represent different kinds of rotating masses such as turbine blades, eccentric discs, etc. If m_1, m_2 and m_3 are the masses revolving at radii r_1, r_2 and r_3 respectively in the same plane, then each mass produces a centrifugal force acting radially outwards from the axis of rotation. Let $\mathbf{F}$ be the vector sum of these forces,

$$\mathbf{F} = m_1\mathbf{r}_1\omega^2 + m_2\mathbf{r}_2\omega^2 + m_3\mathbf{r}_3\omega^2$$

The rotor is said to be statically balanced if the vector sum $\mathbf{F}$ is zero.

If **F** is not zero, i.e. the rotor is unbalanced, then introduce a counterweight (balance weight) of mass m_c at radius $\mathbf{r}_c$ to balance the rotor so that

$$m_1\mathbf{r}_1\omega^2 + m_2\mathbf{r}_2\omega^2 + m_3\mathbf{r}_3\omega^2 + m_c\mathbf{r}_c\omega^2 = 0 \tag{14.1}$$

or

$$m_1\mathbf{r}_1 + m_2\mathbf{r}_2 + m_3\mathbf{r}_3 + m_c\mathbf{r}_c = 0 \tag{14.1a}$$

The magnitude of either m_c or $\mathbf{r}_c$ may be selected and of the other can be calculated.

In general, if $\Sigma m\mathbf{r}$ is the vector sum of $m_1\mathbf{r}_1$, $m_2\mathbf{r}_2$, $m_3\mathbf{r}_3$, $m_4\mathbf{r}_4$, etc., then

$$\Sigma\, m\mathbf{r} + m_c\mathbf{r}_c = 0 \tag{14.2}$$

The equation can be solved either mathematically or graphically. To solve it mathematically, divide each force into its x and z components,

i.e., $\Sigma\, mr\cos\theta + m_c r_c\cos\theta_c = 0$

and $\Sigma\, mr\sin\theta + m_c r_c\sin\theta_c = 0$

or

$$m_c r_c\cos\theta_c = -\Sigma mr\cos\theta \tag{i}$$

and

$$m_c r_c\sin\theta_c = -\Sigma mr\sin\theta \tag{ii}$$

Squaring and adding (i) and (ii),

$$m_c r_c = \sqrt{(\Sigma mr\cos\theta)^2 + (\Sigma mr\sin\theta)^2} \tag{14.3}$$

Dividing (ii) by (i),

$$\tan\theta_c = \frac{-\Sigma mr\sin\theta}{-\Sigma mr\cos\theta} \tag{14.4}$$

The signs of numerator and denominator of this function identify the quadrant of the angle.

In graphical solution, vectors, $m_1\mathbf{r}_1$, $m_2\mathbf{r}_2$, $m_3\mathbf{r}_3$, etc. are added. If they close in a loop, the system is balanced. Otherwise, the closing vector will be giving $m_c\mathbf{r}_c$. Its direction identifies the angular position of the countermass relative to the other masses.

Example 14.1 Figure 14.2a shows a rotor having the following properties.

$m_1 = 4$ kg	$r_1 = 75$ mm	$\theta_1 = 45°$
$m_2 = 3$ kg	$r_2 = 85$ mm	$\theta_2 = 135°$
$m_3 = 2.5$ kg	$r_3 = 50$ mm	$\theta_3 = 240°$

Determine the amount of the countermass at a radial distance of 75 mm required for the static balance.

Solution:

$$m_c r_c = [(\Sigma mr \cos\theta)^2 + (\Sigma mr \sin\theta)^2]^{1/2}$$

$$m_1 r_1 = 300,\ m_2 r_2 = 255,\ m_3 r_3 = 125$$

$$\therefore \quad m_c r_c = \begin{bmatrix} 300 \cos 45° + 255 \cos 135° + 125 \cos 240°)^2 \\ +(300 \sin 45° + 255 \sin 135° + 125 \cos 240°)^2 \end{bmatrix}^{1/2}$$

$$\text{or} \quad m_c \times 75 = [(-30.68)^2 + (284.2)^2]^{1/2}$$

$$= 285.8 \text{ kg.mm}$$

$$\text{or} \quad m_c = \underline{3.81 \text{ kg}}$$

$$\tan\theta_c = \frac{-284.2}{-(-30.68)} = \frac{-284.2}{+30.68} = -9.26$$

$$\therefore \quad \theta_c = \underline{276°12'}$$

θ_c lies in the fourth quadrant ($\because$ numerator is negative and denominator is positive).

The graphical solution has been carried out in Fig. 14.2b.

14.3 DYNAMIC BALANCING

When several masses rotate in different planes, the centrifugal forces, in addition to being out of balance, also form couples. A system of rotating masses is in dynamic balance when there does not exist any resultant centrifugal force as well as resultant couple.

In the work that follows, the products of mr and mrl (instead of $mr\,\omega^2$ and $mrl\,\omega^2$), usually, have been referred as force and couple respectively as it is more convenient to draw force and couple polygone with these quantities.

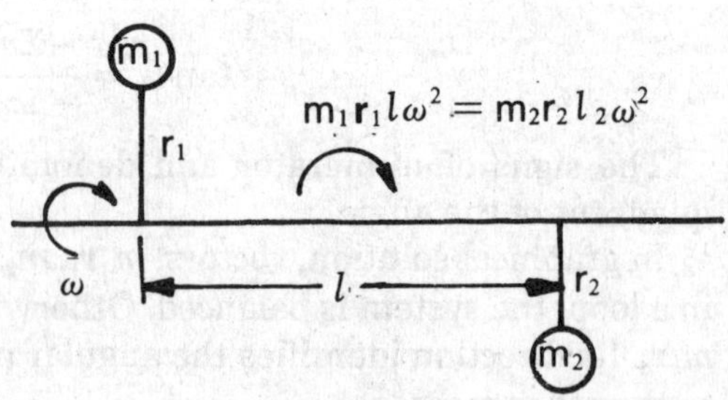

Fig. 14.3

If m_1 and m_2 are two masses (Fig. 14.3) revolving diametrically opposite to each other in different planes such that $m_1 r_1 = m_2 r_2$, the centrifugal forces are balanced, but an unbalanced couple of magnitude $m_1 r_1 l$ $(= m_2 r_2 l)$ is introduced. The couple acts in a plane that contains the axis of rotation and the two masses. Thus the couple is of constant magnitude but variable direction.

14.4 TRANSFERENCE OF A FORCE FROM ONE PLANE TO ANOTHER

Let m be the mass at radius r rotating in a plane at a distance l from another plane (Fig. 14.4). The equilibrium of the system does not change if two equal and opposite forces $F_1 = F_2$ (= mr) are added in the latter plane. The net effect would be a single force $\mathbf{F}_1$ (=mr) in the second plane having the direction of the original force along with a couple mrl formed by the forces $m\mathbf{r}$ and $\mathbf{F}_2$ in a plane containing these forces and the shaft. As the moment of a couple is the same about any point in its plane (equal to the product of one of the forces and the arm), the couple may be assumed to rotate the shaft about point O.

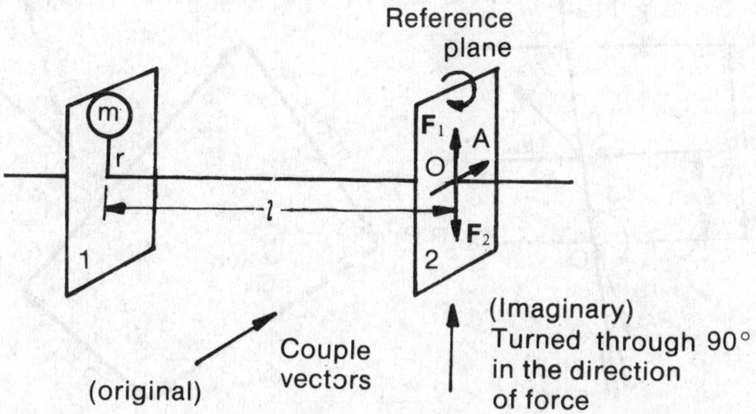

Fig. 14.4

Axis of rotation of the couple is thus a line OA drawn perpendicular to the shaft through O. A line drawn parallel to the axis and to a suitable scale can represent the couple vectorially, the sense of rotation of which is given by the right hand corkscrew rule i.e. for a clockwise couple, the direction is to be away from the viewer. However, in balancing problems, it becomes convenient if the couple vectors are drawn by turning them through 90°, i.e. by drawing them parallel to the force vectors.

A plane passing through a point such as O and perpendicular to the axis of the shaft is called a reference plane. Other masses acting in different planes can be transferred to the reference plane in a similar manner as discussed above.

14.5 BALANCING OF SEVERAL MASSES IN DIFFERENT PLANES

Let there be a rotor revolving with a uniform angular velocity ω (Fig. 14.5). m_1, m_2 and m_3 are the masses attached to the rotor at radii $\mathbf{r}_1$, $\mathbf{r}_2$ and $\mathbf{r}_3$ respectively. The masses m_1, m_2 and m_3 rotate in planes 1, 2 and 3 respectively.

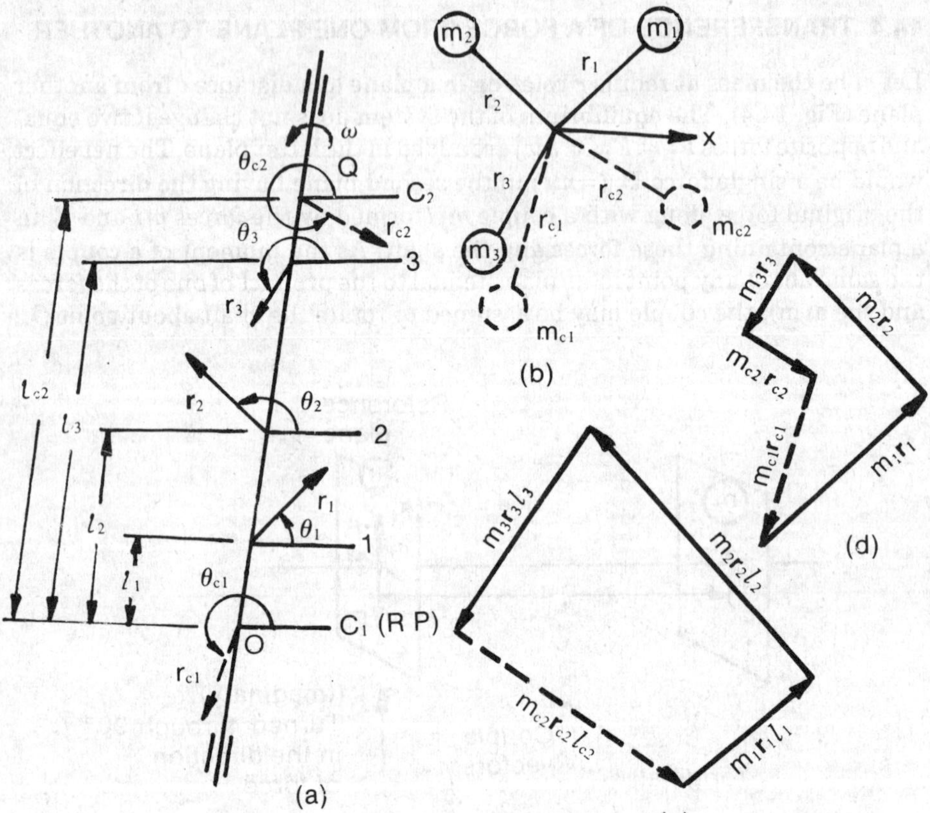

Fig. 14.5

Choose a reference plane at O so that the distance of the planes 1, 2 and 3 from O are l_1, l_2 and l_3 respectively.

Transference of each unbalanced force to the reference plane introduces the like number of forces and couples.

The unbalanced forces in the reference plane are $m_1\mathbf{r}_1$, $m_2\mathbf{r}_2$ and $m_3\mathbf{r}_3$ acting radially outwards.

The unbalanced couples on the rotor are $m_1\mathbf{r}_1 l_1$, $m_2\mathbf{r}_2 l_2$ and $m_3\mathbf{r}_3 l_3$ which may be represented by vectors parallel to the respective force vectors, i.e. parallel to the respective radii of m_1, m_2 and m_3.

For complete balancing of the rotor, the resultant force and the resultant couple both should be zero.

i.e

$$m_1\mathbf{r}_1\,\omega^2 + m_2\mathbf{r}_2\,\omega^2 + m_3\mathbf{r}_3\,\omega^2 = 0 \tag{14.5}$$

and

$$m_1\mathbf{r}_1 l_1\,\omega^2 + m_2\mathbf{r}_2 l_2\,\omega^2 + m_3\mathbf{r}_3 l_3\,\omega^2 = 0 \tag{14.6}$$

If the Eqs. (14.5) and (14.6) are not satisfied, then there are unbalanced forces and couples. A mass placed in the reference plane may satisfy the force equation but the couple equation is satisfied only by two equal forces in different planes. Thus, in general, two planes are needed to balance a system of rotating masses.

Therefore, in order to satisfy Eqs (14.5) and (14.6), introduce two counter-masses m_{c1} and m_{c2} at radii r_{c1} and r_{c2} respectively. Then Eq. (14.5) may be written as

$$m_1\mathbf{r}_1\,\omega^2 + m_2\mathbf{r}_2\omega^2 + m_3\mathbf{r}_3\,\omega^2 + m_{c1}\mathbf{r}_{c1}\omega^2 + m_{c2}\mathbf{r}_{c2}\,\omega^2 = 0 \qquad (14.7)$$

or

$$m_1\mathbf{r}_1 + m_2\mathbf{r}_2 + m_3\mathbf{r}_3 + m_{c1}\mathbf{r}_{c1} + m_{c2}\mathbf{r}_{c2} = 0 \qquad (14.7a)$$

In general,

$$\Sigma m\mathbf{r} + m_{c1}\mathbf{r}_{c1} + m_{c2}\mathbf{r}_{c2} = 0 \qquad (14.8)$$

Let the two countermasses be placed in transverse planes at axial locations O and Q i.e. the countermass m_{c1} be placed in the reference plane and the distance of the plane of m_{c2} be l_{c2} from the reference plane.

Equation (14.6) modifies to (taking moments about O)

$$m_1\mathbf{r}_1l_1\,\omega^2 + m_2\mathbf{r}_2l_2\,\omega^2 + m_3\mathbf{r}_3l_3\,\omega^2 + m_{c2}\mathbf{r}_{c2}l_{c2}\,\omega^2 = 0 \qquad (14.9)$$

or

$$m_1\mathbf{r}_1l_1 + m_2\mathbf{r}_2l_2 + m_3\mathbf{r}_3l_3 + m_{c2}\mathbf{r}_{c2}l_{c2} = 0 \qquad (14.9a)$$

In general,

$$\Sigma m\mathbf{r}l + m_{c2}\mathbf{r}_{c2}l_{c2} = 0 \qquad (14.10)$$

Thus Eqs. (14.8) and (14.10) are the necessary conditions for dynamic balancing of the rotor. Again the equations can be solved mathematically or graphically.

Dividing Eq. (14.10) into component form

$$\Sigma mrl\cos\theta + m_{c2}r_{c2}l_{c2}\cos\theta_{c2} = 0$$

and

$$\Sigma mrl\sin\theta + m_{c2}r_{c2}l_{c2}\sin\theta_{c2} = 0$$

or

$$m_{c2}r_{c2}l_{c2}\cos\theta_{c2} = -\Sigma mrl\cos\theta \qquad \text{(i)}$$

and

$$m_{c2}r_{c2}l_{c2}\sin\theta_{c2} = -\Sigma mrl\sin\theta \qquad \text{(ii)}$$

Squaring and adding (i) and (ii)

$$m_{c2}r_{c2}l_{c2} = \sqrt{(\Sigma mrl\cos\theta)^2 + (\Sigma mrl\sin\theta)^2} \qquad (14.11)$$

Dividing (ii) by (i),

$$\tan \theta_{c2} = \frac{-\Sigma mrl \sin \theta}{-\Sigma mrl \cos \theta} \tag{14.12}$$

After obtaining the values of m_{c2} and θ_{c2} from the above equations, solve Eq. (14.8) by taking its components,

$$m_{c1}r_{c1} = \sqrt{(\Sigma mr \cos \theta + m_{c2}r_{c2} \cos \theta_{c2})^2 + (\Sigma mr \cos \theta + m_{c2}r_{c2} \sin \theta_{c2})^2} \tag{14.13}$$

and

$$\tan \theta_{c1} = \frac{-(\Sigma mr \sin \theta + m_{c2}r_{c2} \sin \theta_{c2})}{-(\Sigma mr \cos \theta + m_{c2}r_{c2} \cos \theta_{c2})} \tag{14.14}$$

To solve Eqs (14.8) and (14.10) graphically, Eq. (14.10) is solved first by adding the known vectors and considering each vector parallel to the radial line of the mass. Then the closing vector will be $m_{c2}\mathbf{r}_{c2}l_{c2}$, the direction of which specifies the angular position of the countermass m_{c2} in the plane at point Q. Then solve Eq. (14.8) by adding the known vectors (along with the vector $m_{c2}\mathbf{r}_{c2}$). The closing vector is $m_{c1}\mathbf{r}_{c1}$, identifying the magnitude and the direction of the countermass m_{c1}.

Example 14.2 Figures 14.5 a and b shows a rotor having the following properties:

$m_1 = 4$ kg	$r_1 = 75$ mm	$\theta_1 = 45°$	$r_{c1} = 75$ mm
$m_2 = 3$ kg	$r_2 = 85$ mm	$\theta_2 = 135°$	$r_{c2} = 40$ mm
$m_3 = 2.5$ kg	$r_3 = 50$ mm	$\theta_3 = 240°$	

The shaft length is 800 mm between bearings and the axial distances as shown in the figure are

$$l_1 = 150 \text{ mm}, \; l_2 = 350 \text{ mm} \quad \text{and} \quad l_3 = 525 \text{ mm}$$

Determine the countermasses for complete balance if these are to be located at 75 mm from each bearing.

Solution: $l_{c2} = (800 - 75 \times 2) = 650$ mm

From Eq. (14.11)

$$m_{c2}r_{c2}l_{c2} = [\,(\Sigma mrl \cos \theta)^2 + (\Sigma mrl \sin \theta)^2]^{1/2}$$

$$m_1r_1l_1 = 4 \times 75 \times 150 = 45\,000 \qquad m_1r_1 = 4 \times 75 = 300$$
$$m_2r_2l_2 = 3 \times 85 \times 350 = 89\,250 \qquad m_2r_2 = 3 \times 85 = 255$$
$$m_3r_3l_3 = 2.5 \times 50 \times 525 = 65\,625 \qquad m_3r_3 = 2.5 \times 50 = 125$$

$$\therefore\; m_{c2}r_{c2}l_{c2} = \begin{bmatrix} (45\,000 \cos 45° + 89\,250 \cos 135° + 65\,625 \cos 240°)^2 \\ + (45\,000 \sin 45° + 89\,250 \sin 135° + 65\,625 \sin 240°)^2 \end{bmatrix}^{1/2}$$
$$= [(-64\,102)^2 + (38\,096)^2]^{1/2}$$

or $$m_{c2} \times 40 \times 650 = 74\,568$$

$$m_{c2} = \underline{2.868 \text{ kg}}$$

$$\tan \theta_{c2} = \frac{-38\,096}{-(-64\,102)} = -0.594$$

$$\theta_{c2} = 329.3° \quad \text{or} \quad \underline{329°18'}$$

From Eq. (14.13)

$$m_{c1}r_{c1} = \left[\begin{array}{l}(300 \cos 45° + 255 \cos 135° + 125 \cos 240° + 2.868 \\ \quad \times 40 \cos 329.3°)^2 + (300 \sin 45° + 255 \sin 135° \\ \quad + 125 \sin 240° + 2.868 \times 40 \sin 329.3°)^2\end{array}\right]^{1/2}$$

$$m_{c1} \times 75 = [(67.96)^2 + (225.62)^2]^{1/2} = 235.63$$

$$m_{c1} = \underline{3.14 \text{ kg}}$$

$$\tan \theta_{c1} = \frac{-225.62}{-67.96} = 3.32; \ \theta_{c1} = 253.2° \text{ or } \underline{253°12'}$$

The graphical solution has also been shown in Figs. 14.5 c and d. From Fig. 14.5c,

$$m_{c2}r_{c2}l_{c2} = 74\,000, \qquad \therefore m_{c2} = \frac{74\,000}{40 \times 650} = 2.846 \text{ kg at } \underline{329°}$$

From Fig. 14.5d,

$$m_{c1}r_{c1} = 235, \qquad \therefore m_{c1} = \frac{235}{75} = 3.13 \text{ kg at } \underline{253°}$$

Example 14.3 Four masses A, B, C and D are completely balanced. Masses C and D make angles of 90° and 195° respectively with B in the same sense. The rotating masses have following properties:

$m_b = 25$ kg	$r_a = 150$ mm
$m_c = 40$ kg	$r_b = 200$ mm
$m_d = 35$ kg	$r_c = 100$ mm
	$r_d = 180$ mm

Planes B and C are 250 mm apart. Determine
(i) the mass A and its angular position
(ii) the positions of planes A and D.

Solution: Refer Fig. 14.6a.

$$m_br_b = 25 \times 200 = 5000$$
$$m_cr_c = 40 \times 100 = 4000$$
$$m_dr_d = 35 \times 180 = 6300$$

For complete balance,

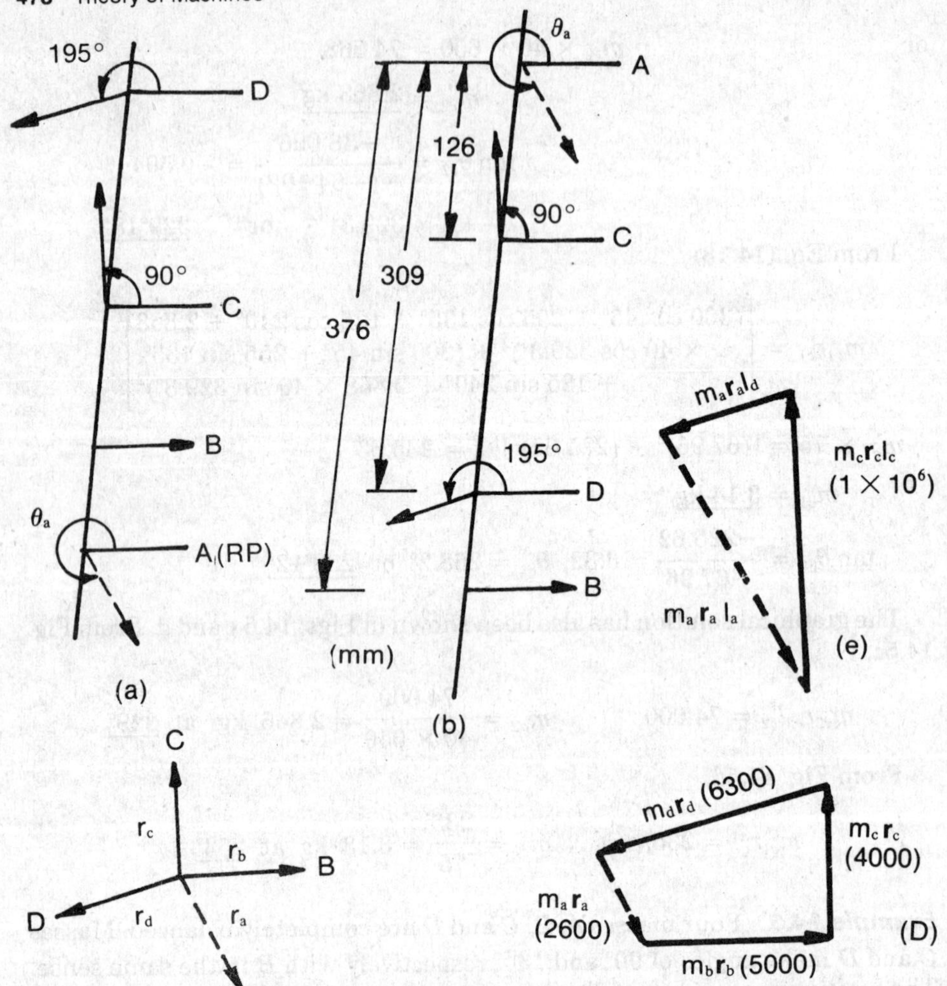

Fig. 14.6

$$\Sigma mr \cos\theta = 0 \quad \text{and} \quad \Sigma mr \sin\theta = 0$$

i.e., $m_a \times 150 \times \cos\theta_a + 5000 \cos 0° + 4000 \cos 90° + 6300 \cos 195° = 0$

or $\quad m_a \times 150 \times \cos\theta_a + 5000 + 0 - 6085 = 0$

or $\quad 150\, m_a \cos\theta_a = 1085 \quad$ (i)

and $\quad m_a \times 150 \times \sin\theta_a + 5000 \sin 0° + 4000 \sin 90° + 6300 \sin 195° = 0$

or $\quad m_a \times 150 \times \sin\theta_a + 0 + 4000 - 1631 = 0$

or

$$150\, m_a \sin\theta_a = -2369 \quad \text{(ii)}$$

Squaring and adding (i) and (ii),

$$22\,500\, m_a^2 = (1085)^2 + (-2369)^2$$

or $$m_a^2 = 30\,175$$

or $$m_a = \underline{17.37 \text{ kg}}$$

Dividing (ii) by (i),

$$\tan\theta_a = \frac{-236.9}{108.5} = -2.184$$

or $$\theta_a = 294.6° \quad \text{or} \quad \underline{294°36'}$$

For complete balance, the couple equations are,

$$\Sigma mrl \cos\theta = 0 \qquad \text{and} \qquad \Sigma mrl \sin\theta = 0$$

Taking A as the reference plane,

$$5000\, l_b \cos 0° + 4000\, l_c \cos 90° + 6300\, l_d \cos 195° = 0$$

or $$5000\, l_b = 6085\, l_d$$

or $$l_b = 1.217\, l_d$$

and $$5000\, l_b \sin 0° + 4000\, l_c \sin 90° + 6300\, l_d \sin 195° = 0$$

or $$4000\, l_c = 1631\, l_d$$

or $$l_c = 0.4078\, l_d$$

or $$l_b + 250 = 0.4078\, l_d$$

or $$1.217\, l_d + 250 = 0.4078\, l_d$$

or $$0.8092\, l_d = -250$$

or $$l_d = \underline{-309 \text{ mm}}$$

$$l_b = 1.217\, l_d = 1.217 \times (-309) = \underline{-376 \text{ mm}}$$

$$l_c = l_b + 250 = -376 + 250 = \underline{-126 \text{ mm}}$$

The correct positions of the planes have been shown in Figs. 14.3 b and c.

To solve the problem graphically, $m_a\mathbf{r}_a$ is obtained from the vector sum of $m_b\mathbf{r}_b$, $m_c\mathbf{r}_c$ and $m_d\mathbf{r}_d$ (Fig. 14.3c). On measuring,

$$m_a r_a = 2600, \quad \therefore\ m_a = \frac{2600}{150} = \underline{17.3 \text{ kg}} \text{ and } \theta_a = \underline{294.5°}$$

Now, $m_c r_c l_c = 4000 \times 250 = 1 \times 10^6$, taking B as the reference plane. Take the vector $m_c\mathbf{r}_c l_c$ and from its two ends, draw lines parallel to $m_a\mathbf{r}_a$ and $m_d\mathbf{r}_d$, thus forming a triangle. Measuring the two sides,

$$m_a\mathbf{r}_a l_a = 985\,000, \qquad l_a = \frac{985000}{17.3 \times 150} = \underline{379 \text{ mm}}$$

$$m_d \mathbf{r}_d l_d = 437000, \qquad l_d = \frac{437000}{6300} = \underline{69 \text{ mm}}$$

l_a and l_d establish the relative positions of the planes.

14.6 BALANCING OF RECIPROCATING MASS

Acceleration of the reciprocating mass of a slider-crank mechanism is given by (Refer Equation 13.12)

$$f = r\,\omega^2 \left(\cos\theta + \frac{\cos 2\theta}{n}\right)$$

Therefore, the force required to accelerate mass m is

$$F = mr\,\omega^2 \left(\cos\theta + \frac{\cos 2\theta}{n}\right)$$

$$= mr\,\omega^2 \cos\theta + mr\,\omega^2 \frac{\cos 2\theta}{n} \qquad (14.15)$$

$mr\omega^2 \cos\theta$ is called the *primary accelerating force* and $mr\omega^2 \dfrac{\cos 2\theta}{n}$ is called the *secondary accelerating force.*

Maximum value of the primary force $= mr\omega^2$

Maximum value of the secondary force $= \dfrac{mr\omega^2}{n}$

As n is, usually, much greater than unity, the secondary force is small compared with the primary force and can be safely neglected for slow speeed engines.

In the slider-crank mechanism OAB of Fig. 14.7, the mass at B of the reciprocating parts is accelerating. Thus a primary force is required to accelerate the reciprocating mass. This force acts in a direction from B to A. The reaction of this force tends to move the frame in a direction from O to B. To prevent this, an attempt has to be made to balance this reaction. The usual approach is by addition of a rotating countermass at radius r directly opposite the crank which however, provides only a partial balance. This countermass

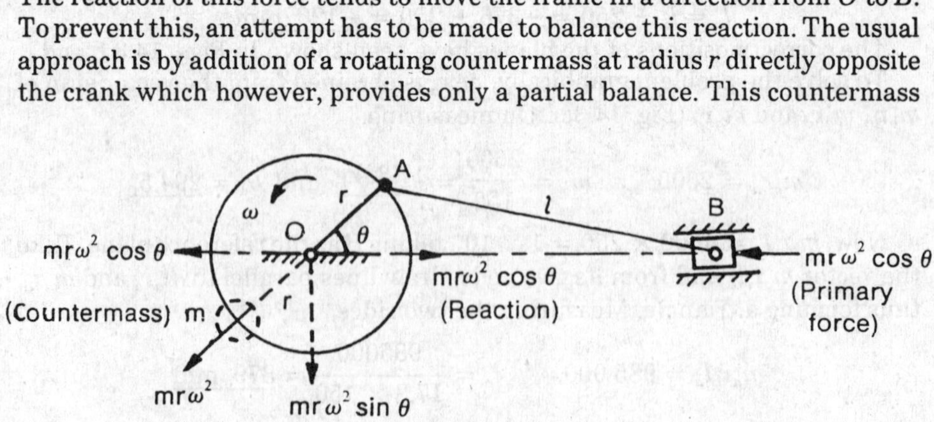

Fig. 14.7

is in addition to the mass used to balance the rotating unbalance due to the mass at the crank pin.

Figure 14.7 also shows the reciprocating mechanism with a countermass m at the radial distance r. The horizontal component of the centrifugal force due to the balancing mass is $mr\omega^2 \cos \theta$ in the line of stroke. This neutralises the unbalanced reciprocating force. But the rotating mass also has a component $mr\omega^2 \sin \theta$ perpendicular to the line of stroke which remains unbalanced. The unbalanced force is zero at the ends of the stroke when $\theta = 0°$ or $180°$ and maximum at the middle when $\theta = 90°$. The magnitude of the unbalanced force remains the same i.e. equal to $mr\omega^2$. Thus instead of slide to and fro on its mounting, the mechanism tends to jump up and down.

To minimise the effect of the unbalanced force, a compromise is, usually, made, i.e. 2/3 of the reciprocating mass is balanced (or a value between one-half and three-quarters). If c is the fraction of the reciprocating mass thus balanced, then

Primary force balanced by the mass $= cmr\omega^2 \cos \theta$

Primary force unbalanced by the mass $= (1-c)mr\omega^2 \cos \theta$

Vertical component of centrifugal force which remains unbalanced

$$= cmr\omega^2 \sin \theta$$

In fact, in reciprocating engines, unbalanced forces in the direction of line of stroke are more dangerous than the forces perpendicular to the line of stroke.

Resultant unbalanced force at any instant

$$= \sqrt{[(1 - c)mr\omega^2 \cos \theta]^2 + [cmr\omega^2 \sin \theta]^2} \tag{14.16}$$

The resultant unbalanced force is minimum when $c = 1/2$.

The method just discussed above to balance the disturbing effect of a reciprocating mass is just equivalent to as if a revolving mass at the crankpin is completely balanced by providing a countermass at the same radius diametrically opposite the crank. Thus if m_p is the mass at the crankpin and c is the fraction of the reciprocating mass m to be balanced, the mass at the crankpin may be considered as $cm + m_p$ which is to be completely balanced.

Example 14.4 The following data relate to a single-cylinder reciprocating engine:

Mass of reciprocating parts	= 40 kg
Mass of revolving parts	= 30 kg at 180 mm radius
Speed	= 150 rpm
Stroke	= 350 mm

If 60% of the reciprocating parts and all the revolving parts are to be balanced, determine

(i) the balance mass required at a radius of 320 mm

(ii) the unbalanced force when the crank has turned 45° from the top-dead centre.

Solution:

$$\omega = \frac{2\pi \times 150}{60} = 15.7 \text{ rad/s}$$

$$r = \frac{350}{2} = 175 \text{ mm}$$

(i) Mass to be balanced at the crankpin $= cm + m_p$

$$= 0.6 \times 40 + 30$$

$$= 54 \text{ kg}$$

$$m_c r_c = mr$$

$$m_c \times 320 = 54 \times 180$$

$$m_c = \underline{30.375 \text{ kg}}$$

(ii) Unbalanced force (at $\theta = 45°$)

$$= \sqrt{[(1 - c)mr\,\omega^2 \cos\theta]^2 + (cmr\,\omega^2 \sin\theta)^2}$$

$$= \sqrt{[(1 - 0.6) \times 40 \times 0.175 \times (15.7)^2 \cos 45°]^2 + [0.6 \times 40 \times 0.175 \times (15.7)^2 \sin 45°]^2}$$

$$= \underline{880.7 \text{ N}}$$

14.7 BALANCING OF LOCOMOTIVES

Locomotives are of two types, coupled and uncoupled. If two or more pairs of wheels are coupled together to increase the adhesive force between the wheels and the track, it is called a coupled locomotive. Otherwise, it is an uncoupled locomotive.

Locomotives usually have two cylinders. If the cylinders are mounted between the wheels, it is called an inside cylinder locomotive and if the cylinders are outside the wheels, it is an outside cylinder locomotive. The cranks of the two cylinders are set at 90° to each other so that the engine can be started easily after stopping in any position. Balance masses are placed on the wheels in both types.

In coupled locomotives, wheels are coupled by connecting their crankpins with coupling rods. As the coupling rod revolves with the crankpin, its proportionate mass can be considered as a revolving mass which can be completely balanced.

Thus whereas in uncoupled locomotives, there are four planes for consideration, two of the cylinders and two of the driving wheels, in coupled locomotives there are six planes, two of cylinders, two of coupling rods and two of the wheels. The planes which contain the coupling rod masses lie outside the planes that contain the balance (counter) masses. Also, in case of coupled locomotives, the mass required to balance the reciprocating parts is distributed among all the wheels which are coupled. This results in a reduced hammer-blow (Refer Sec. 14.8).

Uncoupled locomotives have become obsolete nowadays.

14.8 EFFECTS OF PARTIAL BALANCING IN LOCOMOTIVES

1. Hammer-blow

Hammer-blow is the maximum vertical unbalanced force caused by the mass provided to balance the reciprocating masses. Its value is $mr\omega^2$. Thus, it varies as square of the speed. At high speeds, force of the hammer-blow could exceed the static load on the wheels and the wheels can be lifted off the rail when the direction of the hammer-blow will be vertically upwards.

2. Variation of Tractive Force

A variation in the tractive force (effort) of an engine is caused by the unbalanced portion of the primary force which acts along the line of stroke of a locomotive engine.

If c is the fraction of the reciprocating mass that is balanced, then

Unbalanced primary force for cylinder 1 $= (1 - c)mr\omega^2 \cos\theta$

Unbalanced primary force for cylinder 2 $= (1 - c)mr\omega^2 \cos(90° + \theta)$

$$= -(1 - c)mr\omega^2 \sin\theta$$

Total unbalanced primary force or the variation in the tractive force

$$= (1 - c)mr\omega^2 (\cos\theta - \sin\theta)$$

This is maximum when $(\cos\theta - \sin\theta)$ is maximum,

or when $$\frac{d}{d\theta}(\cos\theta - \sin\theta) = 0$$

or $$-\sin\theta - \cos\theta = 0$$

or $$\sin\theta = -\cos\theta$$

or $$\tan\theta = -1$$

or $$\theta = 135° \text{ or } 315°$$

When $$\theta = 135°,$$

Maximum variation in tractive force

$$= (1 - c)mr\omega^2 (\cos 135° - \sin 135°)$$

$$= (1 - c)mr\omega^2 \left(\frac{1}{\sqrt{2}} - \frac{1}{\sqrt{2}}\right)$$

$$= -\sqrt{2}\,(1 - c)mr\omega^2$$

When $\theta = 315°$

Maximum variation in tractive force

$$= (1 - c)mr\omega^2 (\cos 315° - \sin 315°)$$

$= (1 - c)mr\omega^2 \left(\frac{1}{\sqrt{2}} + \frac{1}{\sqrt{2}}\right)$

$= \sqrt{2}\,(1 - c)mr\omega^2$

Thus maximum variation

$= \pm\sqrt{2}\,(1 - c)mr\omega^2 \qquad (14.17)$

3. Swaying Couple

Unbalanced primary forces along the lines of stroke are separated by a distance l apart and thus constitute a couple (Fig. 14.8). This tends to make the leading wheels sway from side to side.

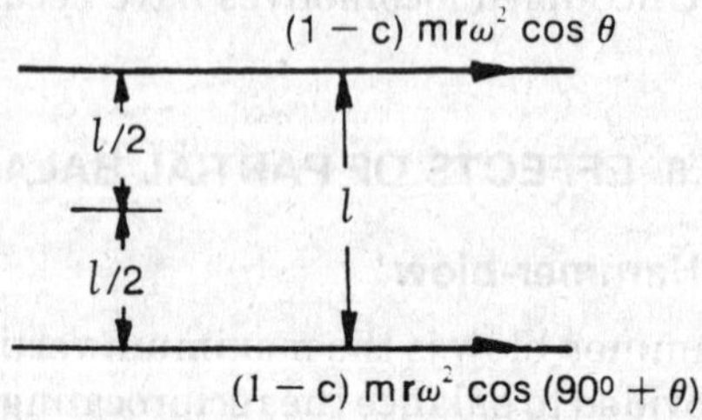

Fig. 14.8

$$\text{Swaying couple} = \text{moments of forces about the engine centre line}$$
$$= [(1-c)mr\omega^2 \cos\theta]\frac{l}{2} - [(1-c)mr\omega^2 \cos(90° + \theta)]\frac{l}{2}$$
$$= (1-c)mr\omega^2 (\cos\theta + \sin\theta)\frac{l}{2}$$

This is maximum when $(\cos\theta + \sin\theta)$ is maximum.

i.e., when $\frac{d}{dt}(\cos\theta + \sin\theta) = 0$

or $\quad -\sin\theta + \cos\theta = 0$

or $\quad \sin\theta = \cos\theta$

or $\quad \tan\theta = 1$

or $\quad \theta = 45° \text{ or } 225°$

When $\theta = 45°$, maximum swaying couple $= \frac{1}{\sqrt{2}}(1-c)mr\omega^2 l$

When $\theta = 225°$, maximum swaying couple $= -\frac{1}{\sqrt{2}}(1-c)mr\omega^2 l$

Thus, maximum swaying couple $= \pm\frac{1}{\sqrt{2}}(1-c)mr\omega^2 l \qquad (14.18)$

Example 14.5 **A four-cylinder steam engine is in complete primary balance. The arrangement of the reciprocating masses in different planes is as shown in Fig. 14.9a. The stroke of each piston is $2r$ mm. Determine the reciprocating mass of the L.P. cylinder and the relative crank positions.**

Solution: Take 2 as the reference plane and $\theta_3 = 0°$

$m_1r_1l_1 = 380\,r \times (-1.3) = -494r$
$m_3r_3l_3 = 590\,r \times 2.8 = 1652r$
$m_4r_4l_4 = 480\,r \times 4.1 = 1968r$

$$-494\, r \cos \theta_1 + 1652\, r \cos 0° + 1968\, r \cos \theta_4 = 0$$

or

$$494 \cos \theta_1 = 1652 + 1968 \cos \theta_4 \qquad \text{(i)}$$

and
$$-494\, r \sin \theta_1 + 1652\, r \sin 0° + 1968\, r \sin \theta_4 = 0$$

or
$$494 \sin \theta_1 = 1968 \sin \theta_4 \qquad \text{(ii)}$$

Squaring and adding (i) and (ii),

$$(494)^2 = (1652 + 1968 \cos \theta_4)^2 + (1968 \sin \theta_4)^2$$

$$= (1652)^2 + (1968)^2 \cos^2 \theta_4 + 2 \times 1652 \times 1968 \cos \theta_4 + (1968)^2 \sin^2 \theta_4$$

$$= (1652)^2 + (1968)^2 + 2 \times 1652 \times 1968 \cos \theta_4$$

$$\cos \theta_4 = -0.978$$

or
$$\theta_4 = 167.9° \text{ or } 192.1°$$

Choosing one value, say $\theta_4 = \underline{167.9°}$,

Dividing (ii) by (i), $\tan \theta_1 = \dfrac{1968 \sin 167.9°}{1652 + 1968 \cos 167.9°}$

$$= \frac{+412.53}{-272.28}$$

$$= -1.515$$

$$\theta_1 = \underline{123.4°}$$

Writing the force equation, (r is common),

$$380 \cos 123.4° + m_2 \cos \theta_2 + 590 \cos 0° + 480 \cos 167.9° = 0$$

or

$$m_2 \cos \theta_2 = 88.5 \qquad \text{(iii)}$$

and
$$380 \sin 123.4° + m_2 \sin \theta_2 + 590 \sin 0° + 480 \sin 167.9° = 0$$

or

$$m_2 \sin \theta_2 = -417.9 \qquad \text{(iv)}$$

Squaring and adding (iii) and (iv), $m_2 = \underline{427.1 \text{ kg}}$

Dividing (iii) by (iv), $\tan \theta_2 = \dfrac{-417.9}{+88.5} = -4.72$

or
$$\theta_2 = \underline{282°}$$

Had we chosen $\theta_4 = 192.1°$, a different set of values of m_2, θ_1 and θ_2 would have come.

To solve the problem graphically, draw the couple polygon (triangle) as shown in Fig. 14.9 b from the three known values. This provides the relative directions of the masses m_1, m_3 and m_4. Now, complete the force polygon (Fig. 14.9 c) and obtain the magnitude and direction of m_2. The results obtained are $\theta_4 = 168°$, $\theta_1 = 124°$, $\theta_2 = 282°$.

$$m_2 r_2 = 428\, r$$

$$m_2 = \frac{428\, r}{r} = \underline{428 \text{ kg}}$$

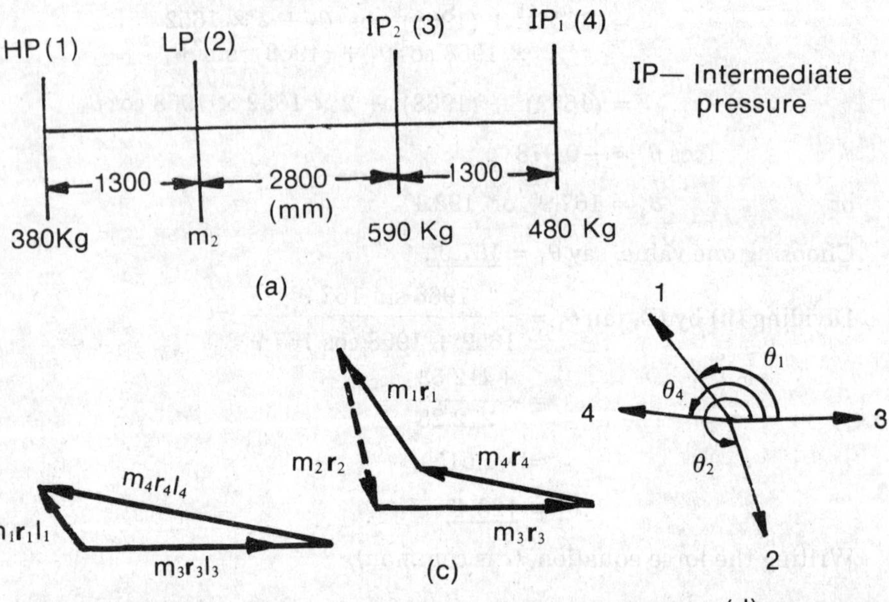

Fig. 14.9

Note that the couple triangle can be drawn in more than one way, thus providing a different set of solution. Also $m_1 \mathbf{r}_1 l_1$ is negative and, therefore, its direction is reversed in the diagram.

Example 14.6 The following data refer to a two cylinder uncoupled locomotive:

Rotating mass per cylinder	= 280 kg
Reciprocating mass per cylinder	= 300 kg
Distance between wheels	= 1400 mm
Distance between cylinder centres	= 600 mm
Diameter of treads of driving wheels	= 1800 mm
Crank radius	= 300 mm

Radius of centre of balance mass	= 620 mm
Locomotive speed	= 50 km/hr
Angle between cylinder cranks	= 90°
Dead load on each wheel	= 3.5 tonne

Determine

(i) the balancing mass required in the planes of driving wheels if whole of the revolving and two-third of the reciprocating mass are to be balanced.

(ii) the swaying couple

(iii) the variation in the tractive force

(iv) the maximum and minimum pressure on the rails

(v) the maximum speed of locomotive without lifting the wheels from the rails.

Solution: Total mass to be balanced $= m_p + cm$

$$= 280 + \frac{2}{3} \times 300$$

$$= 480 \text{ kg}$$

(i) Taking 1 as the reference plane and angle $\theta_2 = 0°$(Fig. 14.10). Writing the couple equations,

$$m_2 r_2 l_2 \cos\theta_2 + m_3 r_3 l_3 \cos\theta_3 + m_4 r_4 l_4 \cos\theta_4 = 0$$

or $$480 \times 300 \times 400 \cos 0° + 480 \times 300 \times 1000 \cos 90° + m_4 \times 620 \times 1400 \cos\theta_4 = 0$$

or $$m_4 \cos\theta_4 = -66.36 \qquad \text{(i)}$$

and $$m_2 r_2 l_2 \sin\theta_2 + m_3 r_3 l_3 \sin\theta_3 + m_4 r_4 l_4 \sin\theta_4 = 0$$

or $$480 \times 300 \times 400 \sin 0° + 480 \times 300 \times 1000 \sin 90° + m_4 \times 620 \times 1400 \sin\theta_4 = 0$$

or

$$m_4 \sin\theta_4 = -165.9 \qquad \text{(ii)}$$

Squaring and adding (i) and (ii), $m_4 = \underline{178.7 \text{ kg}}$

Dividing (ii) by (i), $\tan\theta_4 = \dfrac{-165.9}{-66.36} = 2.5$

$$\theta_4 = \underline{248.2°}$$

Taking 4 as the reference plane and writing the couple equations,

$$m_2 r_2 l_2 \cos\theta_2 + m_3 r_3 l_3 \cos\theta_3 + m_1 r_1 l_1 \cos\theta_1 = 0$$

$$480 \times 300 \times 1000 \cos 0° + 480 \times 300 \times 400 \cos 90° + m_1 \times 620 \times 1400 \sin\theta_1 = 0$$

or

$$m_1 \cos \theta_1 = -165.9 \qquad \text{(iii)}$$

Similarly,

$$m_1 \sin \theta_1 = -66.36 \qquad \text{(iv)}$$

From (iii) and (iv), $m_1 = \underline{178.7 \text{ kg}} = m_4$

$$\tan \theta_1 = \frac{-66.36}{-165.9} = 0.4 \text{ or } \theta_1 = \underline{201.8°}$$

The treatment shows that the magnitude of m_1 could have directly been written equal to m_4.

(ii) $\omega = \dfrac{50 \times 1000 \times 1000}{60 \times 60} \times \dfrac{1}{\frac{1800}{2}} = 15.43 \text{ rad/s}$

$$\text{Swaying couple} = \pm \frac{1}{\sqrt{2}} (1-c) m r \omega^2 l$$

$$= \pm \frac{1}{\sqrt{2}} \left(1 - \frac{2}{3}\right) \times 300 \times 0.3 \times (15.43)^2 \times 0.6$$

$$= \underline{3030.3 \text{ N.m}}$$

(iii) Variation in tractive force $= \pm\sqrt{2}\,(1-c) m r \omega^2$

$$= \pm\sqrt{2}\left(1 - \frac{2}{3}\right) \times 300 \times 0.3 \times (15.43)^2$$

$$= \underline{10\ 100 \text{ N}}$$

(iv) Balance mass for reciprocating parts only

$$= 178.7 \times \frac{\frac{2}{3} \times 300}{480} = 74.46 \text{ kg}$$

$$\text{Hammer-blow} = m r \omega^2$$
$$= 74.46 \times 0.62 \times (15.43)^2$$
$$= 10\ 991 \text{ N}$$

Dead load = 3.5 × 1000 × 9.81 = 34 335 N

Maximum pressure on rails = 34 335 + 10 991 = <u>45 326 N</u>

Minimum pressure on rails = 34 335 – 10 991= <u>23 344 N</u>

(v) Maximum speed of the locomotive without lifting the wheels from the rails will be when the dead load becomes equal to the hammer-blow.

i.e., $74.46 \times 0.62 \times \omega^2 = 34335$

or $\omega = 27.27 \text{ rad/s}$

$$\text{Velocity of wheels} = \omega r = \left(27.27 \times \frac{1.80}{2}\right) \text{ m/s}$$

$$= \left(27.27 \times \frac{1.8}{2} \times \frac{60 \times 60}{1000}\right) \text{ km/hr}$$

$$= \underline{88.36 \text{ km/hr}}$$

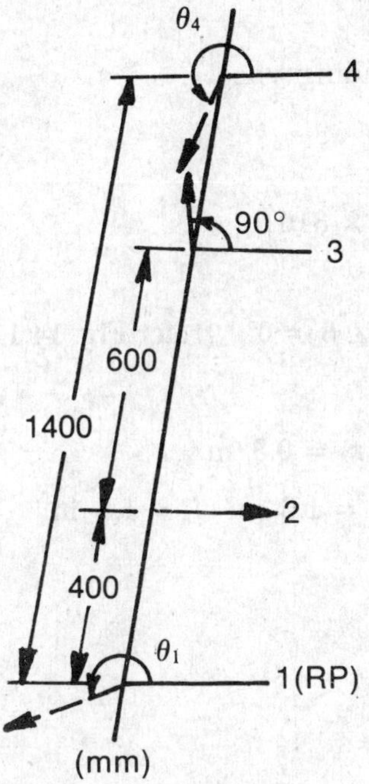

Fig. 14.10

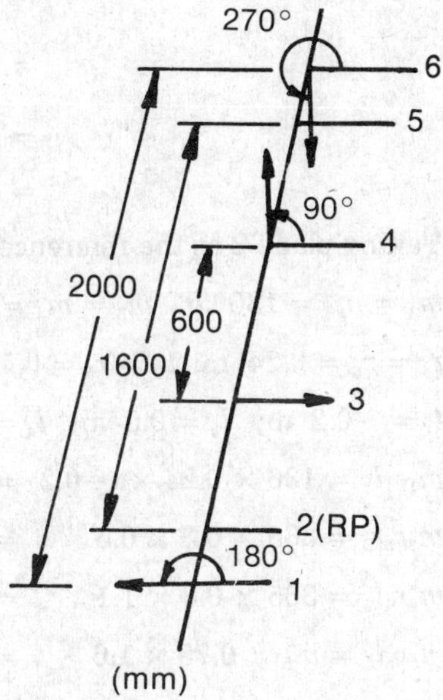

Fig. 14.11

Example 14.7 The following data refer to a four-coupled wheel locomotive with two inside cylinders:

Pitch of cylinders	= 600 mm
Reciprocating mass/cylinder	= 315 kg
Revolving mass/cylinder	= 260 kg
Distance between driving wheels	= 1.6 m
Distance between coupling rods	= 2 m
Diameter of driving wheels	= 1.9 m
Revolving parts for each coupling rod crank	= 130 kg
Engine crank radius	= 300 mm
Coupling rod crank radius	= 240 mm
Distance of centre of balance mass in planes of driving wheels from axle centre	= 750 mm
Angle between engine cranks	= 90°
Angle between coupling rod crank with adjacent engine crank	= 180°

The balanced mass required for the reciprocating parts is equally divided between each pair of coupled wheels. Determine.

(i) the magnitude and position of the balance mass required to balance two-third of reciprocating and whole of the revolving parts

(ii) the hammer-blow and the maximum variation of tractive force when the locomotive speed is 80 km/hr.

Solution:

Leading wheels Balance mass on each leading wheel

$$= m_p + \frac{1}{2} cm$$
$$= 260 + \frac{1}{2}\left(\frac{2}{3} \times 315\right)$$
$$= 365 \text{ kg}$$

Taking plane 2 as the reference plane and $\angle \theta_3 = 0°$ [refer Fig. 14.11]

$m_1 = m_6 = 136$ kg; $m_3 = m_4 = 365$ kg

$r_1 = r_6 = 0.24$ m; $r_2 = r_5 = 0.75$ m; $r_3 = r_4 = 0.3$ m

$l_1 = -0.2$ m; $l_3 = 0.5$ m; $l_4 = 1.1$ m; $l_5 = 1.6$ m; $l_6 = 1.8$ m

$m_1r_1l_1 = 130 \times 0.24 \times (-0.2) = -6.24$

$m_3r_3l_3 = 365 \times 0.3 \times 0.5 = 54.75$

$m_4r_4l_4 = 365 \times 0.3 \times 1.1 = 120.45$

$m_5r_5l_5 = m_5 \times 0.75 \times 1.6 = 1.2\, m_5$

$m_6r_6l_6 = 130 \times 0.24 \times 1.8 = 56.16$

$$1.2m_5 = \begin{bmatrix} (-6.24 \cos 180° + 54.75 \cos 0° + 120.45 \cos 90° \\ + 56.16 \cos 270°)^2 + (-6.24 \sin 180° + 54.75 \sin 0° \\ + 120.45 \sin 90° + 56.16 \sin 270°)^2 \end{bmatrix}^{1/2}$$

$$= [(60.99)^2 + (64.29)^2]^{1/2}$$
$$= 88.62$$
$$m_5 = \underline{73.85 \text{ kg}}$$
$$\tan \theta_5 = \frac{-64.29}{-60.99} = 1.054 \text{ or } \theta_5 = \underline{226.5°}$$

From symmetry of the system, $m_2 = m_5 = \underline{73.85 \text{ kg}}$

and $\tan \theta_2 = \frac{-60.99}{-64.29} = 0.949$ or $\theta_2 = \underline{223.5°}$

Trailing Wheels The arrangement remains the same except that only half of the required reciprocating masses have to be balanced at the cranks.

i.e., $m_3 = m_4 = \frac{1}{2}\left(\frac{2}{3} \times 315\right) = 105$ kg

Then, $m_3r_3l_3 = 105 \times 0.3 \times 0.5 = 15.75$

and $$m_4 r_4 l_4 = 105 \times 0.3 \times 1.1 = 34.65$$

$$1.2\, m_5 = \left[\begin{array}{r}(-6.24 \cos 180° + 15.75 \cos 0° + 34.65 \cos 90° \\ + 56.16 \cos 270°)^2 + (-6.24 \sin 180° + 15.75 \sin 0° \\ + 34.65 \sin 90° + 56.16 \sin 270°)^2\end{array}\right]^{1/2}$$

$$= [(21.99)^2 + (-21.51)^2]^{1/2}$$

$$= 30.76$$

$$m_5 = 25.63 \text{ kg}$$

$$\tan \theta_5 = \frac{-21.51}{-21.99} = \frac{+21.51}{-21.99} = -0.978$$

or $\theta_5 = \underline{135.4°}$

By symmetry, $m_2 = m_5 = \underline{25.63 \text{ kg}}$

and $\tan \theta_2 = \dfrac{-21.99}{+21.51} = -1.022$ or $\theta_2 = \underline{314.4°}$

(ii) Hammer-blow = $mr\omega^2$

where m is the balance mass for reciprocating parts only and neglecting m_1 and m_6 in the above calculations.

Thus $m_1 r_1 l_1 = m_6 r_6 l_6 = 0$

$$1.2\, m_5 = \left[\begin{array}{c}(15.75 \cos 0° + 34.65 \cos 90°)^2 \\ +(15.75 \sin 0° + 34.65 \sin 90°)^2\end{array}\right]^{1/2}$$

$$= [(15.75)^2 + (34.65)^2]^{1/2}$$

$$= 38.06$$

$$m_5 = 31.72 \text{ kg}$$

$$\omega = \frac{80 \times 1000}{60 \times 60} \times \frac{1}{\frac{1.9}{2}} = 23.39 \text{ rad/s}$$

Hammer-blow = $31.72 \times 0.75 \times (23.39)^2 = 13\,015$ N

Maximum variation of tractive force

$$= \pm\sqrt{2}(1-c)mr\omega^2$$

$$= \pm\sqrt{2}\left(1 - \frac{2}{3}\right) \times 315 \times 0.3 \times (23.39)^2$$

$$= \pm \underline{24\,372 \text{ N}}$$

14.9 SECONDARY BALANCING

It was stated earlier that the secondary acceleration force is defined as

$$\text{Secondary force} = mr\omega^2 \frac{\cos 2\theta}{n} \tag{14.19}$$

Its frequency is twice that of the primary force and the magnitude $1/n$ times the magnitude of the primary force.

The expression can also be written as $mr(2\omega)^2 \dfrac{\cos 2\theta}{4n}$

Now, consider two cranks of an engine. One actual one and the other imaginary, with the following specifications:

	Actual	Imaginary
Angular velocity	ω	2ω
Length of crank	r	$\dfrac{r}{4n}$
Mass at the crank pin	m	m

Thus when the actual crank has turned through an angle $\theta = \omega t$, the imaginary crank would have turned an angle of $2\theta = 2\,\omega t$

$$\text{Centrifugal force induced in the imaginary crank} = \frac{mr(2\omega)^2}{4n}$$

$$\text{Component of this force along line of stroke} = \frac{mr(2\omega)^2}{4n}\cos 2\theta$$

Thus the effect of the secondary force is equivalent to an imaginary crank of length $r/4n$ rotating at double the angular velocity i.e. twice of the engine speed.

The imaginary crank coincides with the actual crank at inner top-dead centre. At other times, it makes an angle with the line of stroke equal to twice that of the engine crank.

The secondary couple about a reference plane is given by the multiplication of the secondary force with the distance l of the plane from the reference plane.

Complete Balancing of Reciprocating Parts

From the foregoing discusion, it is concluded that for complete balancing of the reciprocating parts, the following conditions must be fulfilled:

1. Primary forces must balance i.e. primary force polygon is enclosed.
2. Primary couples must balance, i.e. primary couple polygon is enclosed.
3. Secondary forces must balance i.e. secondary force polygon is enclosed.
4. Secondary couples must balance i.e. secondary couple polygon is enclosed.

Usually, it is not possible to satisfy all the above conditions fully for a multicylinder engine. Mostly some unbalanced force or couple would exist in the reciprocating engines.

14.10 BALANCING OF INLINE ENGINES

If a reciprocating mass is transferred to the crankpin, the axial component parallel to the cylinder axis of the resulting centrifugal force represents the primary unbalanced force.

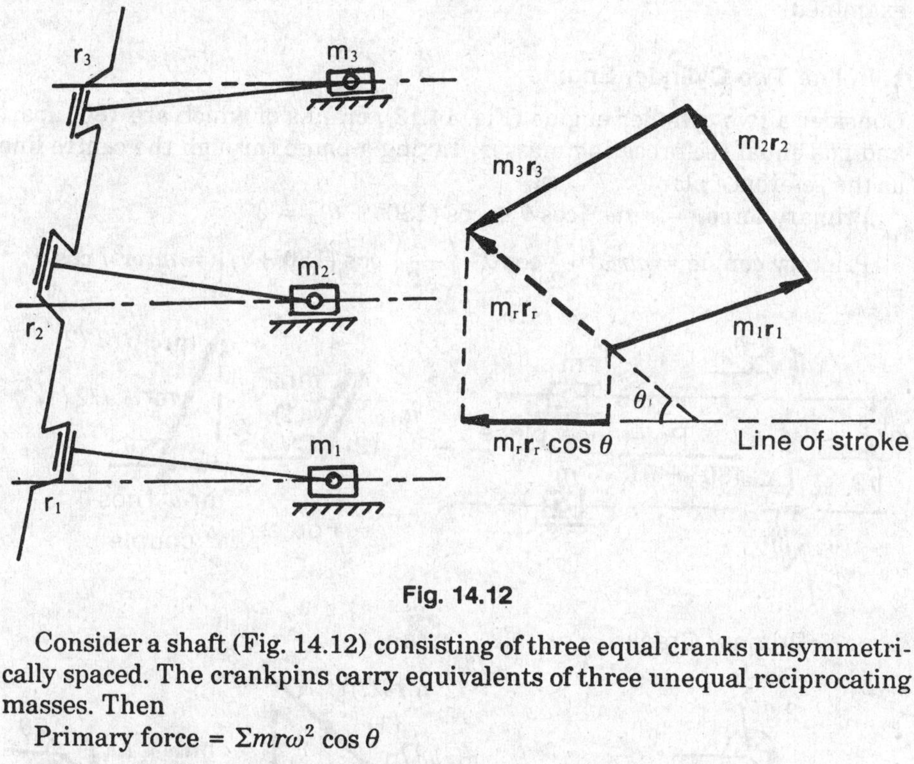

Fig. 14.12

Consider a shaft (Fig. 14.12) consisting of three equal cranks unsymmetrically spaced. The crankpins carry equivalents of three unequal reciprocating masses. Then

Primary force = $\Sigma mr\omega^2 \cos\theta$

Primary couple = $\Sigma mr\omega^2 \, l \cos\theta$

$$\text{Secondary force} = \Sigma mr \frac{(2\omega)^2}{4n} \cos 2\theta = \Sigma mr \frac{\omega^2}{n} \cos 2\theta$$

$$\text{Secondary couple} = \Sigma mr \frac{(2\omega)^2}{4n} l \cos 2\theta = \Sigma mr \frac{\omega^2}{n} l \cos 2\theta$$

In order to solve the above equations graphically, first draw the $\Sigma mr \cos\theta$ polygon (ω^2 is common to all forces). Then the axial component of the resultant force ($\mathbf{F}_r \cos\theta$) multiplied by ω^2 provides the primary unbalanced force on the system at that moment. This unbalanced force is zero when $\theta = 90°$ and a maximum when $\theta = 0°$.

In case the force polygon encloses, the resultant as well as the axial component will always be zero and thus the system will be in primary balance. Then $\Sigma F_{ph} = 0$ and $\Sigma F_{pv} = 0$.

To find the secondary unbalance force, first find the positions of the imaginary secondary cranks. Then transfer the reciprocating masses at the crankpins. Draw the mr polygon. Measure the resulting axial component and multiply the same by $(2\omega)^2/4n$ or ω^2/n to get the secondary force.

In the same way primary and secondary couple ($mr\,l$) polygons can be drawn for primary and secondary couples.

In the following paragraphs, some multi-crank arrangements have been examined.

1. In-line Two Cylinder Engine

Consider a two cylinder engine (Fig. 14.13), cranks of which are 180° apart and has equal reciprocating masses. Taking a plane through the centre line as the reference plane,

$$\text{Primary force} = mr\omega^2\,[\cos\theta + \cos(180° + \theta)] = 0$$

$$\text{Primary couple} = mr\omega^2\left[\frac{l}{2}\cos\theta + \left(-\frac{l}{2}\right)\cos(180° + \theta)\right] = mr\omega^2\, l\cos\theta$$

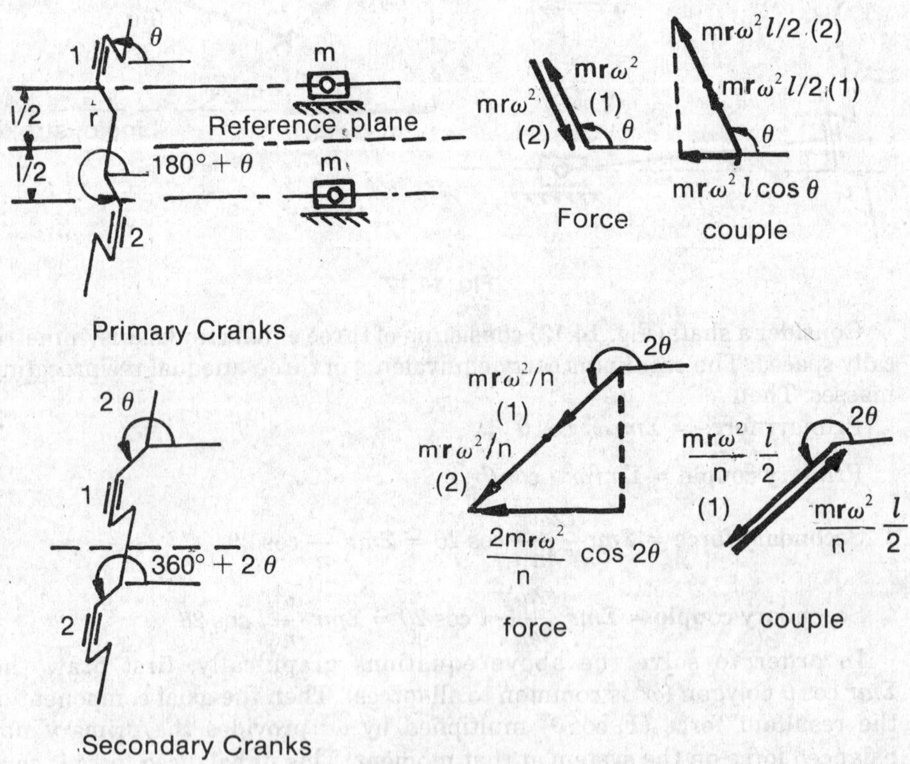

Fig. 14.13

maximum values are $mr\,\omega^2\,l$ at $\theta = 0°$ and $180°$

$$\text{Secondary force} = \frac{mr\,\omega^2}{n}\left[\cos 2\theta + \cos(360° + 2\theta)\right] = 2\,\frac{mr\,\omega^2}{n}\cos 2\theta$$

Maximum values are $\dfrac{2mr\,\omega^2}{n}$ when $2\theta = 0°, 180°, 360°$ and $540°$

or $\theta = 0°, 90°, 180°$ and $270°$

$$\text{Secondary couple} = \frac{mr\,\omega^2}{n}\left[\frac{l}{2}\cos 2\theta + \left(-\frac{l}{2}\right)\cos(360° + 2\theta)\right] = 0$$

Graphical solution has also been shown in Fig. 14.13 which is self-explanatory.

2. In-line Four Cylinder Four-stroke Engine

Such an engine has two outer as well as inner cranks (throws) in line. The inner throws are at 180° to the outer throws. Thus the angular positions for the cranks are: θ for the first, $(180° + \theta)$ for the second, $(180° + \theta)$ for the third and θ for the fourth (Fig. 14.14).

for convenience, choose a plane passing through the middle bearing about which the arrangement is symmetrical as the reference plane.

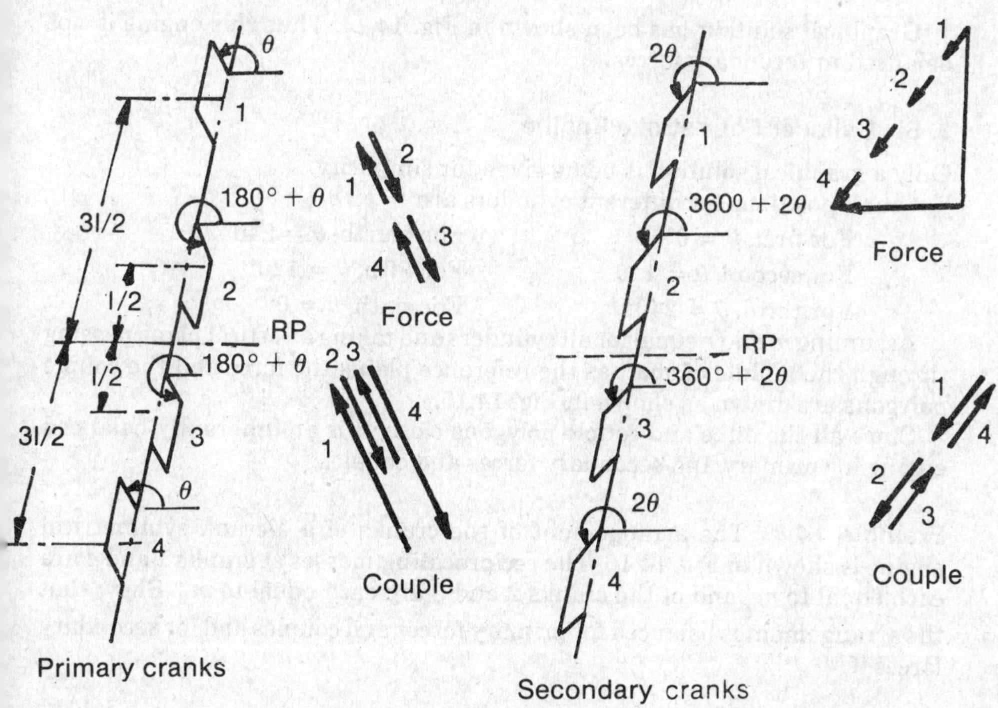

Fig. 14.14

Primary force

$$= mr\omega^2 [\cos\theta + \cos(180° + \theta) + \cos(180° + \theta) + \cos\theta] = 0$$

Primary couple

$$= mr\omega^2 \left[\frac{3l}{2}\cos\theta + \frac{l}{2}\cos(180° + \theta) + \left(-\frac{l}{2}\right)\cos(180° + \theta) + \left(-\frac{3l}{2}\right)\cos\theta\right]$$

$= 0$

Secondary force

$$= \frac{mr\omega^2}{n}[\cos 2\theta + \cos(360° + 2\theta) + \cos(360° + 2\theta) + \cos 2\theta)]$$

$$= \frac{4mr\omega^2}{n}\cos 2\theta$$

Maximum value $= \dfrac{4mr\omega^2}{n}$ at $2\theta = 0°, 180°, 360°$ and $540°$ or $\theta = 0°, 90°, 180°$ and $270°$

Secondary couple

$$= \frac{mr\omega^2}{n}\left[\left(\frac{3l}{2}\right)\cos 2\theta + \frac{l}{2}\cos(360° + 2\theta) + \left(-\frac{l}{2}\right)\cos(360° + 2\theta) + \left(-\frac{3l}{2}\right)\cos 2\theta\right]$$

$= 0$

Graphical solution has been shown in Fig. 14.14. Thus this engine is not balanced in secondary forces.

3. Six Cylinder Four-stroke Engine

Only a graphical solution is being given for simplicity.

Crank positions for different cylinders are

For first, $\theta = 0°$	For fourth, $\theta = 240°$
For second, $\theta = 120°$	For fifth, $\theta = 120°$
For third, $\theta = 240°$	For sixth, $\theta = 0°$

Assuming m and r equal for all cylinders and taking a vertical plane passing through the middle of shaft as the reference plane, the force and the couple polygons are drawn as shown in Fig. 14.15.

Since all the force and couple polygons close, it is an inherently balanced engine for primary and secondary forces and couples.

Example 14.8 The arrangement of the cranks of a 4-crank symmetrical engine is shown in Fig. 14.16. The reciprocating masses at cranks 1 and 4 are each equal to m_1 and of the cranks 2 and 3 are each equal to m_2. Show that the arrangement is balanced for primary forces and couples and for secondary forces if

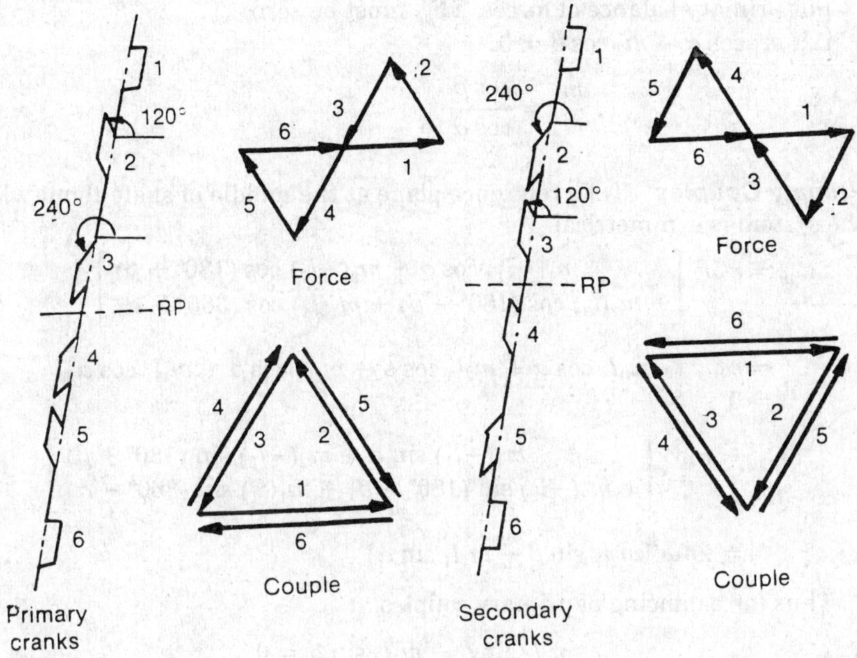

Fig. 14.15

$$\frac{m_1}{m_2} = \frac{\cos\beta}{\cos\alpha}, \frac{l_1}{l_2} = \frac{\tan\beta}{\tan\alpha}, \cos\alpha\cos\beta = \frac{1}{2}$$

Determine also the magnitude of the out of balance secondary couple if the system rotates at ω rad/s.

Solution: Considering horizontal and vertical components of primary and secondary forces and couples,

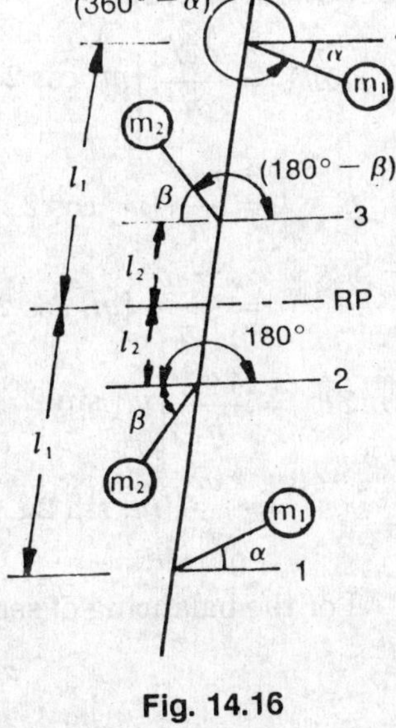

Fig. 14.16

Primary Forces

$$\Sigma F_{ph} = r\omega^2 \left[\begin{array}{r} m_1 \cos\alpha + m_2 \cos(180° + \beta) \\ + m_2 \cos(180° - \beta) \\ + m_1 \cos(360° - \alpha) \end{array} \right]$$

$$= 2r\,\omega^2\,[m_1 \cos\alpha - m_2 \cos\beta]$$

$$\Sigma F_{pv} = r\,\omega^2 \left[\begin{array}{r} m_1 \sin\alpha + m_2 \sin(180° + \beta) + m_2 \sin(180° - \beta) \\ + m_1 \sin(360° - \alpha) \end{array}\right]$$

$$= 0$$

For primary balance of forces, $\Sigma \mathbf{F}_{ph}$ must be zero,

i.e., $m_1 \cos\alpha - m_2 \cos\beta = 0$

or
$$\frac{m_1}{m_2} = \frac{\cos\beta}{\cos\alpha}$$

Primary Couples Take reference plane at the middle of shaft about which the system is symmetrical.

$$\Sigma C_{ph} = r\,\omega^2 \left[\begin{array}{r} m_1(-l_1)\cos\alpha + m_2(-l_2)\cos(180° + \beta) \\ + m_2(l_2)\cos(180° - \beta) + m_1(l_1)\cos(360° - \alpha) \end{array}\right]$$

$$= r\,\omega^2 \left[-m_1 l_1 \cos\alpha + m_2 l_2 \cos\beta - m_2 l_2 \cos\beta + m_1 l_1 \cos\alpha\right]$$

$$= 0$$

$$C_{pv} = r\,\omega^2 \left[\begin{array}{r} m_1(-l_1)\sin\alpha + m_2(-l_2)\sin(180° + \beta) \\ + m_2(-l_2)\sin(180° - \beta) + m_1(l_1)\sin(360° - \alpha) \end{array}\right]$$

$$= 2r\,\omega^2 [m_2 l_2 \sin\beta - m_1 l_1 \sin\alpha]$$

Thus for balancing of primary couples,

$$m_2 l_2 \sin\beta - m_1 l_1 \sin\alpha = 0$$

or
$$\frac{l_1}{l_2} = \frac{m_2}{m_1}\frac{\sin\beta}{\sin\alpha} = \frac{\cos\alpha}{\cos\beta}\frac{\sin\beta}{\sin\alpha} = \frac{\tan\beta}{\tan\alpha}$$

Secondary Forces

$$\Sigma F_{sh} = \frac{r\,\omega^2}{n}\left[\begin{array}{r} m_1 \cos 2\alpha + m_2 \cos 2(180° + \beta) + m_2 \cos 2(180° - \beta) \\ + m_1 \cos 2(360° - \alpha) \end{array}\right]$$

$$= \frac{r\omega^2}{n}[m_1 \cos 2\alpha + m_2 \cos 2\beta + m_2 \cos 2\beta + m_1 \cos 2\alpha]$$

$$= \frac{2r\,\omega^2}{n}[m_1 \cos 2\alpha + m_2 \cos 2\beta]$$

$$\Sigma F_{sv} = \frac{r\,\omega^2}{n}\left[\begin{array}{r} m_1 \sin 2\alpha + m_2 \sin 2(180° + \beta) + m_2 \sin 2(180° - \beta) \\ + m_1 \sin 2(360° - \alpha) \end{array}\right]$$

$$= \frac{r\,\omega^2}{n}[m_1 \sin 2\alpha + m_2 \sin 2\beta - m_2 \sin 2\beta - m_1 \sin 2\alpha]$$

$$= 0$$

For the balancing of secondary forces,

$$m_1 \cos 2\alpha + m_2 \cos 2\beta = 0$$

or $\frac{m_1}{m_2}\cos\alpha + \cos 2\beta = 0$

or $\frac{\cos\beta}{\cos\alpha}(2\cos^2\alpha - 1) + (2\cos^2\beta - 1) = 0$

or $2\cos\beta\cos^2\alpha = \cos\beta + 2\cos^2\beta\cos\alpha - \cos\alpha = 0$

or $2\cos\beta\cos\alpha(\cos\alpha + \cos\beta) - (\cos\alpha + \cos\beta) = 0$

or $(2\cos\alpha + \cos\beta)(2\cos\beta\cos\alpha - 1) = 0$

As $\cos\alpha + \cos\beta \neq 0$,

$\therefore \quad 2\cos\beta\cos\alpha - 1 = 0$

or $\cos\alpha\cos\beta = \frac{1}{2}$

Secondary Couples

$$\Sigma C_{sh} = \frac{r\omega^2}{n}\begin{bmatrix} m_1(-l_1)\cos 2\alpha + m_2(-l_2)\cos 2(180° + \beta) \\ + m_2(l_2)\cos 2(180° - \beta) + m_1(l_1)\cos 2(360° - \alpha) \end{bmatrix}$$

$$= \frac{r\omega^2}{n}[-m_1 l_1 \cos 2 - m_2 l_2 \cos 2 + m_2 l_2 \cos 2 + m_1 l_1 \cos 2]$$

$$= 0$$

$$\Sigma C_{sv} = \frac{r\omega^2}{n}\begin{bmatrix} m_1(-l_1)\sin 2\alpha + m_2(-l_2)\sin 2(180° + \beta) \\ + m^2(l_2)\sin 2(180° - \beta) + m_1(l_1)\sin 2(360° - \alpha) \end{bmatrix}$$

$$= \frac{2r\omega^2}{n}[-m_1 l_1 \sin 2\alpha - m_2 l_2 \sin 2\beta]$$

Out of balance secondary couple $= -\frac{2r\omega^2}{n}[m_1 l_1 \sin 2\alpha + m_2 l_2 \sin 2\beta]$

Example 14.9 The cranks of a four-cylinder marine oil engine are arranged at angular intervals of 90°. The engine speed is 70 rpm and the reciprocating mass per cylinder is 800 kg. The inner cranks are 1m apart and are symmetrically arranged between the outer cranks which are 2.6 m apart. Each crank is 400 mm long.

Determine the firing order of the cylinders for the best balance of reciprocating masses and also the magnitude of the unbalanced primary couple for that arrangement.

Solution:

$m = 800$ kg $\qquad N = 70$ rpm

$r = 0.4$ m $\qquad \omega = \frac{2\pi \times 70}{60} = 7.33$ rad/s

$$mr\omega^2 = 800 \times 0.4 \times (7.33)^2 = 17\,195$$

There are four cranks. They can be used in six different arrangements as shown in Fig. 14.17 a. It can be observed that in all the cases, primary forces are always balanced. Primary couples in each case will be as under:

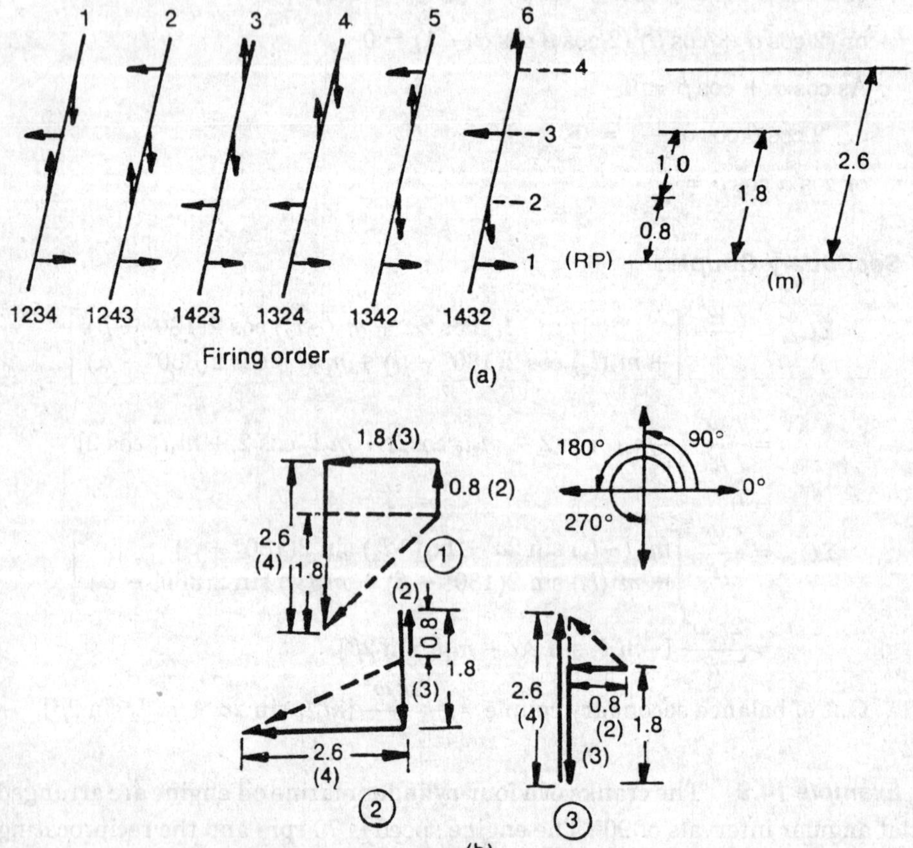

Fig. 14.17

Taking 1 as the reference plane,

$$C_{p1} = mr\omega^2 \sqrt{(-l_3)^2 + (l_2 - l_4)^2}$$

$$= 17\,195 \sqrt{(-1.8)^2 + (0.8 - 2.6)^2}$$

$$= 43\,761 \text{ N.m}$$

$C_{p6} = C_{p1} = 43\,761$ N.m, only l_2 and l_4 are interchanged.

$$C_{p2} = mr\omega^2 \sqrt{(-l_4)^2 + (l_2 - l_3)^2} = 17\,195 \sqrt{(-2.6)^2 + (0.8 - 1.8)^2}$$

$$= 47\,905 \text{ N.m}$$

$C_{p5} = C_{p2} = 47\,905$ N.m, l_2 and l_3 are interchanged.

$$C_{p3} = mr\omega^3 \sqrt{(-l_2)^2 + (l_4 - l_3)^2} = 17\,195 \sqrt{(-0.8)^2 + (2.6 - 1.8)^2}$$

$$= 19\,448 \text{ N.m}$$

$C_{p4} = C_{p3} = 19\,448$ N.m, l_4 and l_3 are interchanged.

Thus the best arrangement is of 3rd and 4th as shown in Fig. 14.17. The firing orders are 1423 and 1324 respectively.

Unbalanced couple = 19 448 N.m

Graphical solution has also been shown in Fig. 14.17.

Example 14.10 The intermediate cranks of a four-cylinder symmetrical engine, which is in complete primary balance, are at 90° to each other and each has a reciprocating mass of 400 kg. The centre distance between intermediate cranks is 600 mm and between extreme cranks 1800 mm. Lengths of the connecting rods and the cranks are 900 mm and 200 mm respectively. Calculate the masses fixed to the extreme cranks with their relative angular positions. Also find the magnitude of the secondary forces and couples about the centre line of the system if the engine speed is 150 rpm.

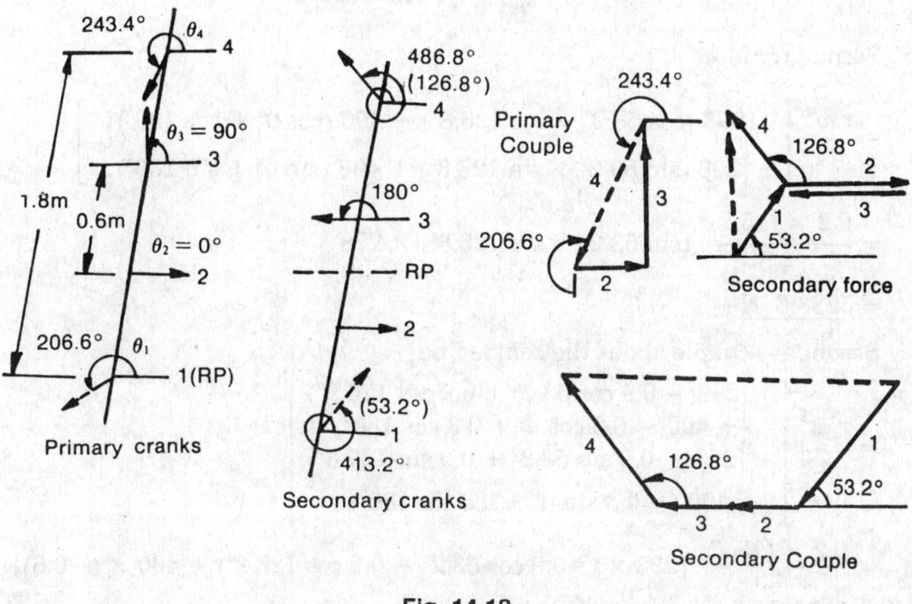

Fig. 14.18

Solution: Refer Fig. 14.18.

$l = 0.9$ m $\qquad m_2 = m_3 = 400$ kg

$r = 0.2$ m $\qquad n = \dfrac{0.9}{0.2} = 4.5$

The engine is in complete primary balance. Taking 1 as the reference plane,

$$m_2 r_2 l_2 = 400 \times 0.2 \times 0.6 = 48$$

$$m_3 r_3 l_3 = 400 \times 0.2 \times 1.2 = 96$$

$$m_4 r_4 l_4 = m_4 \times 0.2 \times 1.8 = 0.36\, m_4$$

$$0.36\, m_4 = \sqrt{(48 \cos 0° + 96 \cos 90°)^2 + (48 \sin 0° + 96 \sin 90°)^2}$$

$$= \sqrt{(48)^2 + (96)^2}$$

$$= 107.33$$

$$m_4 = \underline{298 \text{ kg}}$$

$$\tan \theta_4 = \frac{-96}{-48} = 2;\ \theta_4 = \underline{243.4°}$$

By symmetry, $m_1 = \underline{298 \text{ kg}}$

$$\text{and } \tan \theta_1 = \frac{-48}{-96} = 0.5;\ \theta_1 = 206.6°$$

The positions of the cranks for secondary forces and couples will be such that the angles are doubled (Fig. 14.18).

$$\omega = \frac{2\pi \times 150}{60} = 15.7 \text{ rad/s}$$

Secondary force

$$= \frac{r\omega^2}{n}\left[\begin{array}{l}\{298(\cos 53.2° + \cos 126.8°) + 400(\cos 0° + \cos 180°)\}^2 \\ + \{298(\sin 53.2° + \sin 126.8°) + 400(\sin 0° + \sin 180°)\}^2\end{array}\right]^{1/2}$$

$$= \frac{0.2 \times (15.7)^2}{4.5}(\sin 53.2° + \sin 126.8°) \times 298$$

$$= \underline{5233.6 \text{ N}}$$

Secondary couple about the centre line

$$= \frac{r\omega^2}{n}\left[\begin{array}{l}\{298(-0.9 \cos 53.2° + 0.9 \cos 126.8°) \\ \quad + 400(-0.3 \cos 0° + 0.3 \cos 180°)^2\} \\ + \{298(-0.9 \sin 53.2° + 0.9 \sin 126.8°) \\ \quad + 400(-0.3 \sin 0° + 0.3 \sin 180°)\}^2\end{array}\right]^{1/2}$$

$$= \frac{0.2 \times (15.7)^2}{4.5}[298 \times (-0.9 \cos 53.2° + 0.9 \cos 126.8°) + 400 \times (-0.6)]$$

$$= \underline{6155 \text{ N.m}}$$

14.11 BALANCING OF V-ENGINES

In V-engines, a common crank OA is operated by two connecting rods OB_1 and OB_2. Figure 14.19 shows a symmetrical two cylinder V-engine, the centre lines of which are inclined at angle α to the x-axis.

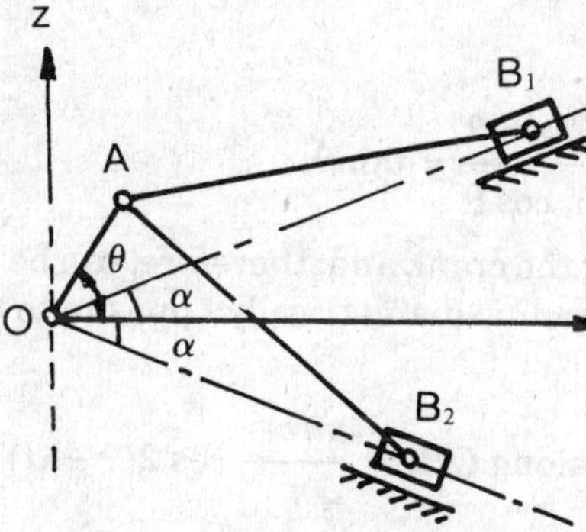

Fig. 14.19

Let θ be the angle moved by the crank from the x-axis.

Primary force of 1 along line of stroke $OB_1 = mr\omega^2 \cos(\theta - \alpha)$

Primary force of 1 along x-axis $= mr\omega^2 \cos(\theta - \alpha)\cos\alpha$

Primary force of 2 along line of stroke $OB_2 = mr\omega^2 \cos(\theta + \alpha)$

Primary force of 2 along x-axis $= mr\omega^2 \cos(\theta + \alpha)\cos\alpha$

Total primary force along x-axis

$$= mr\omega^2 \cos\alpha[\cos(\theta - \alpha) + \cos(\theta + \alpha)]$$

$$= mr\omega^2 \cos\alpha\,[(\cos\theta\cos\alpha + \sin\theta\sin\alpha) + (\cos\theta\cos\alpha - \sin\theta\sin\alpha)]$$

$$= mr\omega^2 \cos\alpha\; 2\cos\theta\cos\alpha$$

$$= 2mr\omega^2 \cos^2\alpha\cos\theta$$

Similarly, total primary force along z-axis

$$mr\omega^2 [\cos(\theta - \alpha)\sin\alpha - \cos(\theta + \alpha)\sin\alpha]$$

$$= mr\omega^2 \sin\alpha \begin{bmatrix}(\cos\theta\cos\alpha + \sin\theta\sin\alpha) \\ -(\cos\theta\cos\alpha - \sin\theta\sin\alpha)\end{bmatrix}$$

$$= mr\omega^2 \sin\alpha\; 2\sin\theta\sin\alpha$$

$$= 2mr\omega^2 \sin^2\alpha\sin\theta$$

Resultant primary force

$$= \sqrt{(2mr\omega^2\cos^2\alpha\cos\theta)^2 + (2mr\omega^2\sin^2\alpha\sin\theta)^2}$$

$$= 2mr\omega^2\sqrt{(\cos^2\alpha\cos\theta)^2 + (\sin^2\alpha\sin\theta)^2}$$

It will be at an angle β with the x-axis, given by

$$\tan\beta = \frac{\sin^2\alpha \sin\theta}{\cos^2\alpha \cos\theta}$$

If $2\alpha = 90°$, Resultant force

$$= 2mr\omega^2 \sqrt{(\cos^2 45° \cos\theta)^2 + (\sin^2 45° \sin\theta)^2}$$

$$= 2mr\omega^2$$

$$\tan\beta = \frac{\sin^2 45° \sin\theta}{\cos^2 45° \cos\theta} = \tan\theta$$

i.e., $\beta = \theta$ or it acts along the crank and, therefore, can be completely balanced by a mass at a suitable radius diametrically opposite to the crank such that $m_r r_r = mr$.

Secondary forces

Secondary force of 1 along $OB_1 = \dfrac{mr\omega^2}{n} \cos 2(\theta - \alpha)$

Secondary force of 2 along x-axis $= \dfrac{mr\omega^2}{n} \cos 2(\theta - \alpha) \cos\alpha$

Secondary force of 2 along $OB_2 = \dfrac{mr\omega^2}{n} \cos 2(\theta + \alpha) \cos\alpha$

Secondary force of 2 along x-axis $= \dfrac{mr\omega^2}{n} \cos 2(\theta + \alpha) \cos\alpha$

Total secondary force along x-axis

$$= \frac{mr\omega^2}{n} \cos\alpha\, [\cos 2(\theta - \alpha) + \cos 2(\theta + \alpha)]$$

$$= \frac{mr\omega^2}{n} \cos\alpha\, [(\cos 2\theta \cos 2\alpha + \sin 2\theta \sin 2\alpha) + (\cos 2\theta \cos 2\alpha - \sin 2\theta \sin 2\alpha)]$$

$$= \frac{2mr\omega^2}{n} \cos\alpha \cos 2\theta \cos 2\alpha$$

Similarly, secondary force along z-axis $= \dfrac{2mr\omega^2}{n} \sin\alpha \sin 2\theta \sin 2\alpha$

Resultant secondary force

$$= \frac{2mr\omega^2}{n} \sqrt{(\cos\alpha \cos 2\theta \cos 2\alpha)^2 + (\sin\alpha \sin 2\theta \sin 2\alpha)^2}$$

$$\tan\beta' = \frac{\sin\alpha \sin 2\theta \sin 2\alpha}{\cos\alpha \cos 2\theta \cos 2\alpha}$$

If $2\alpha = 90°$ or $\alpha = 45°$,

$$\text{Secondary force} = \frac{2mr\omega^2}{n} \sqrt{\left(\frac{\sin 2\theta}{\sqrt{2}}\right)^2}$$

$$= \sqrt{2}\, \frac{mr\omega^2}{n} \sin 2\theta$$

$$\tan \beta' = \infty, \quad \beta' = 90^\circ$$

This means that the force acts along *z*-axis and is a harmonic force and special methods are needed to balance it.

PROBLEMS

14.1 The rotor shown in Fig. 14.2a has the following properties:

$m_1 = 3$ kg $\quad r_1 = 30$ mm $\quad \theta_1 = 30^\circ$

$m_2 = 4$ kg $\quad r_2 = 20$ mm $\quad \theta_2 = 120^\circ$

$m_3 = 2$ kg $\quad r_3 = 25$ mm $\quad \theta_3 = 270^\circ$

Find the amount of the countermass at a radial distance of 35 mm for the static balance.

(2.13 kg; 239°26')

14.2 The rotor shown in Fig. 14.5a has the following properties:

$m_1 = 3$ kg $\quad r_1 = 30$ mm $\quad \theta_1 = 30^\circ \quad l_1 = 100$ mm

$m_2 = 4$ kg $\quad r_2 = 20$ mm $\quad \theta_2 = 120^\circ \quad l_2 = 300$ mm

$m_3 = 2$ kg $\quad r_3 = 25$ mm $\quad \theta_3 = 270^\circ \quad l_3 = 600$ mm

$r_{c1} = 35$ mm and $r_{c2} = 20$ mm

l_1, l_2 and l_3 are the distances from bearing 1. The axial distance between the bearings is 500 mm. Determine the countermass to be placed in the planes of m_1 and a mid-plane of m_2 and m_3 for the complete balance.

(m_{c1} = 1.96 kg, 54°19'; m_{c2} = 2.195 kg, 229°01')

14.3 A rotor has the following properties:

Mass	Magnitude	Radius	Angle	Axial distance from 1st mass
1	9 kg	100 mm	0°	
2	7 kg	120 mm	60°	160 mm
3	8 kg	140 mm	135°	320 mm
4	6 kg	120 mm	270°	560 mm

If the shaft is balanced by two countermasses located at 100 mm radii and revolving in planes midway of planes 1 and 2, and midway of 3 and 4, determine the magnitudes of the masses and their respective angular positions.

(6.65 kg, 25°; 5.2 kg, 275°)

14.4 Four masses *A*, *B*, *C* and *D* are completely balanced. Masses *C* and *D* make angles of 90° and 210° respectively with *B* in the same sense. The planes containing *B* and *C* are 300 mm apart. Masses *A*, *B*, *C* and *D* can be assumed to be concentrated at radii of 360, 480, 240 and 300 mm respectively. The masses *B*, *C* and *D* are 15 kg, 25 kg and 20 kg respectively. Determine

(i) the mass *A* and its angular position

(ii) the positions of planes *A* and *D*.

(40 kg, 236°; A is 978 mm towards right and D is 374 mm towards left of plane B)

14.5 A single cylinder reciprocating engine has a reciprocating mass of 60 kg. The crank rotates at 60 rpm and the stroke is 320 mm. Mass of the revolving parts at 160 mm radius is 40 kg. If two-thirds of the reciprocating parts and the whole

of the reciprocating parts and the whole of the revolving parts are to be balanced, determine
(i) the balance mass required at a radius of 350 mm
(ii) the unbalanced force when the crank has turned 50° from the top-dead centre.

(36.57 kg; 3644 N)

14.6 The cranks of a three-cylinder locomotive are set at 120°. The reciprocating masses are 450 kg for the inside cylinder and 390 kg for each outside cylinder. The pitch of the cylinders is 1.2 m and the stroke of each piston 500 mm. The planes of rotation of the balance masses are 960 mm from the inside cylinder. If 40% of the reciprocating masses are to be balanced, determine the magnitude and the position of the balancing masses required at a radial distance of 500 mm, and the hammer-blow per wheel when the axle rotates at 350 rpm.

(86.25 kg each at 24° and 215°; 57.93 kN)

14.7 The firing order of a six-cylinder vertical four-stroke in-line engine is 1, 4, 2, 6, 3, 5. The piston stroke is 80 mm and the length of each connecting rod is 180 mm. The pitch distances between the cylinder centre lines are 80 mm, 80 mm, 120 mm, 80 mm and 80 mm respectively. The reciprocating mass per cylinder is 1.2 kg and the engine speed is 2400 rpm. Determine the out of balance primary and secondary forces and couples on the engine taking a plane midway between the cylinder 3 and 4 as the reference plane.

(Completely balanced engine; no out of balance primary and secondary forces and couples)

14.8 A four-cylinder engine is arranged as shown in Fig. 14.16. The reciprocating masses in planes 1 and 4 are each 142 kg and in planes 2 and 3 are each 200 kg. If the crank radii are 400 mm each, the speed 200 rpm and the length of the connecting rod 1.6 m, determine the magnitude of primary and secondary forces and couples. Given that $\alpha = 25°$, $\beta = 50°$, $l_1 = 1.28$ m and $l_2 = 0.5$ m.

(Primary forces and couples are zero; secondary force = 4959 N; Secondary couple = 20.56 kN.m)

15

BRAKES AND DYNAMOMETERS

15.1 INTRODUCTION

A brake is an appliance used to apply frictional resistance to a moving body to stop or retard it by absorbing its kinetic energy. In general, in all types of motion, there is always some amount of resistance which retards the motion and is sufficient to bring the body to rest. However, the time taken and the distance covered in this process is usually too large. By providing brakes, the external resistance is considerably increased and the period of retardation shortened.

A dynamometer is a brake incorporating a device to measure the frictional resistance applied. This is used to determine the power developed by the machine, while maintaining its speed at the rated value.

The functional difference between a clutch and a brake is that a clutch connects two moving members of a machine, whereas a brake connects a moving member to a stationary member.

15.2 TYPES OF BRAKES

The following are the main types of mechanical brakes

(i) Block or shoe brake
(ii) Band brake
(iii) Band and block brake
(iv) Internal expanding shoe brake

15.3 BLOCK OR SHOE BRAKE

A block or shoe brake consists of a block or shoe which is pressed against a rotating drum. The force on the drum is increased by using a lever (Fig. 15.1a). If only one block is used for the purpose, a side thrust on the bearing of the shaft supporting the drum will act. This can be prevented by using two blocks on the two sides of the drum (Fig. 15.1b). This also doubles the braking torque.

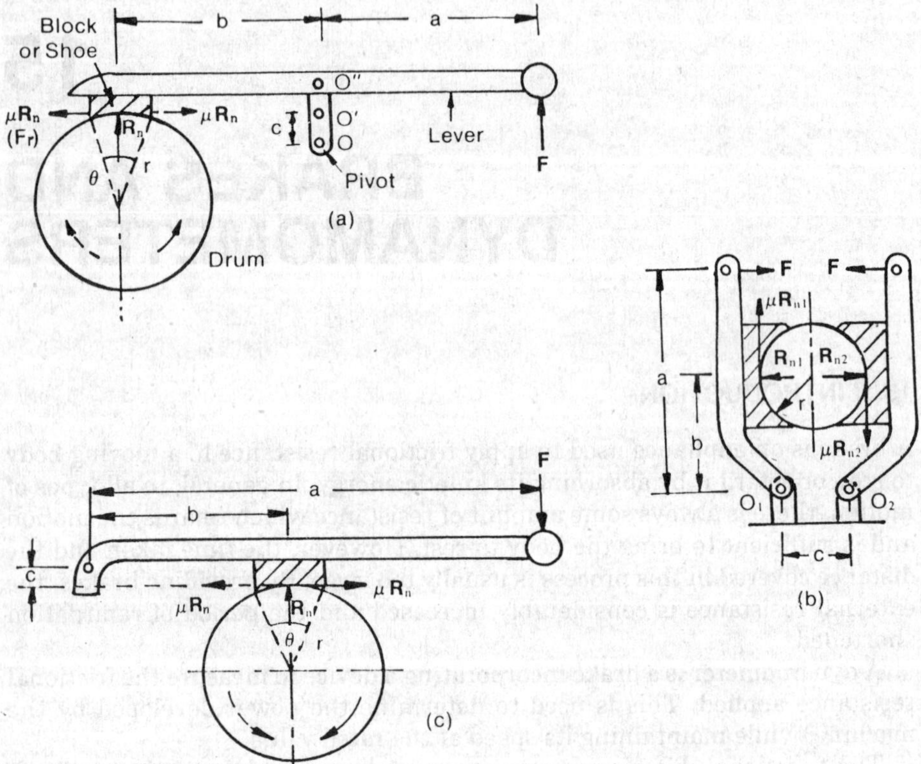

Fig. 15.1

A material softer than that of the drum or the rim of the wheel is used to make the blocks so that these can be replaced easily on wearing. Wood and rubber are used for light and slow vehicles and cast steel for heavy and fast ones.

Let r = radius of the drum

μ = coefficient of friction

F_r = radial force applied on the drum (not shown in the figure)

R_n = normal reaction on the block($= F_r$)

F = force applied at the lever end

F_f = frictional force = μR_n

Assuming that the normal reaction R_n and the frictional force F_f act at the mid-point of the block,

Braking torque on the drum = frictional force × radius

or

$$T_B = \mu R_n \times r \tag{15.1}$$

To obtain R_n, consider the equilibrium of the block as follows.

The direction of the frictional force on the drum is to be opposite to that of its rotation while on the block it is in the same direction. Taking moments about the pivot O (Fig. 15.1a),

$$F \times a - R_n \times b + \mu R_n \times c = 0$$

$$R_n = \frac{Fa}{b - \mu c} \tag{15.2}$$

Also

$$F = R_n \left(\frac{b - \mu c}{a}\right) \tag{15.3}$$

When $b = \mu c$, $F = 0$ which implies that the force needed to apply the brake is virtually zero, or that once contact is made between the block and the drum, the brake is applied itself. Such a brake is known as a *self-locking* brake.

As the moment of the force F_f about O is in the same direction as that of the applied force F, F_f aids in applying the brake. Such a brake is known as a *self-energised brake.*

If the rotation of the drum is reversed, i.e. it is made clockwise $F = R_n$ $[(b + \mu c)/a]$ which shows that required force F will be far greater than what it would be when the drum rotates counter-clockwise.

If the pivot lies on the line of action of F_f, i.e. at O', $c = 0$ and $F = R_n \dfrac{a}{b}$, which is valid for clockwise as well as for counter-clockwise rotation.

If c is made negative. i.e. if the pivot is at O'',

$$F = R_n \left(\frac{b + \mu c}{a}\right) \quad \text{for counter-clockwise rotation}$$

and
$$F = R_n \left(\frac{b - \mu c}{a}\right) \quad \text{for clockwise rotation}$$

In case the pivot is provided on the same side of the applied force and the block as shown in Fig. 15.1c, the equilibrium condition can be considered accordingly.

In the above treatment, it is assumed that the normal reaction and the frictional force act at the mid-point of the block. However, this is true only for small angles of contact. When the angle of contact is more than 40°, the normal pressure is less at the ends than at the centre. In that case, μ has to be replaced by an equivalent coefficient of friction μ given by

$$\mu' = \mu \left(\frac{4 \sin\left(\frac{\theta}{2}\right)}{\theta + \sin \theta}\right)$$

Example 15.1 The diameter of the brake drum of a single block brake shown in Fig. 15.1c is 1m. It sustains 240 N.m of torque at 400 rpm. The coefficient of friction is 0.32. Determine the required force to be applied when the rotation of the drum is : (a) clockwise (b) counter-clockwise, and the angle of contact: (i) 35°, and (ii) 100°. Given that a = 800 mm, b = 150 mm and c = 25 mm. Also find the new values of c for self-locking of the brake.

Solution:

T_B = 240 N.m, r = 0.5 m

(i) Angle of contact 35°:

$$\mu = 0.32$$

$$T_B = \mu R_n r$$

or $$240 = 0.32 \times R_n \times 0.5$$

$$R_n = 1500 \text{ N}$$

(a) Rotation clockwise

$$F \cdot a - R_n b - \mu R_n c = 0$$

$$F \times 0.8 - 1500 \times 0.15 - 0.32 \times 1500 \times 0.025 = 0$$

$$F \times 0.8 - 225 - 12 = 0$$

$$F = \underline{296.25 \text{ N}}$$

(b) Rotation counter-clockwise

$$F \times 0.8 - 1500 \times 0.15 + 0.32 \times 1500 \times 0.025 = 0$$

$$F = \underline{266.25 \text{ N}}$$

(c) For self-locking, F is to be zero. This is possible for counter-clockwise rotation of the drum, i.e. when

$$0 - 1500 \times 0.15 + 0.32 \times 1500 \times c = 0$$

or $$c = \frac{0.15}{0.32} = 0.469 \text{ m or } \underline{469 \text{ mm}}$$

(ii) Angle of contact 100°:

$$\mu' = \mu \left(\frac{4 \sin(\theta/2)}{\theta + \sin \theta}\right) = 0.32 \left(\frac{4 \sin 50°}{100 \times \frac{\pi}{180} + \sin 100°}\right) = 0.36$$

$$T_B = \mu' R_n r$$

$$240 = 0.36 \times R_n \times 0.5$$

$$R_n = 1333 \text{ N}$$

(a) Rotation clockwise

$$F \times 0.8 - 1333 \times 0.15 - 0.36 \times 1333 \times 0.025 = 0$$

$$0.8 F - 200 - 12 = 0$$

$$F = \underline{265 \text{ N}}$$

(b) Rotation counter-clockwise

$$F \times 0.8 - 1333 \times 0.15 + 0.36 \times 1333 \times 0.025 = 0$$

$$0.8\,F - 200 + 12 = 0$$

$$F = \underline{235\text{ N}}$$

(c) For self-locking

$$0 - 1333 \times 0.15 + 0.36 \times 1333 \times c = 0.$$

or

$$c = \frac{0.15}{0.36} = 0.417\text{ m or } \underline{417\text{ mm}}$$

(Note: 400 rpm is the superfluous data in the problem)

Example 15.2 A bicycle and rider, travelling at 12 km/hr on a level road, have a mass of 105 kg. A brake is applied to the rear wheel which is 800 mm in diameter. The pressure applied on the brake is 80 N and the coefficient of friction 0.06. Find the distance covered by the bicycle and number of turns of its wheel before coming to rest.

Solution:

$m = 105$ kg $d = 0.8$ m

$v = \frac{12000}{3600} = 3.333$ m/s $F_r = 80\text{ N} = R_n$

$\mu = 0.06$

Let s = Distance covered by the bicycle before it comes to rest.
Work done against friction = KE of the bicycle and the rider

$$\mu\, R_n\, s = \frac{1}{2} m\, v^2$$

$$0.06 \times 80 \times s = \frac{1}{2} \times 105 \times (3.333)^2$$

$$s = \underline{121.5\text{ m}}$$

$$\pi\, d\, n = s$$

or

$$\pi \times 0.8 \times n = 121.5$$

$$n = \underline{48.4\text{ revolutions}}$$

Example 15.3 A spring-operated pivoted shoe brake shown in Fig. 15.1b is used for a wheel diameter of 500 mm. The angle of contact is 90° and the coefficient of friction 0.3. The force F applied by the spring on each arm is 5 kN. Determine the brake torque on the wheel. Take a = 1 m, b = 0.4 m and c = 0.1 m.

Solution:

$\mu = 0.3 \qquad F = 5000 \text{ N}$
$r = 0.25$

$$\mu' = \mu \left(\frac{4 \sin (\theta/2)}{\theta + \sin \theta}\right) = 0.3 \left(\frac{4 \sin 45°}{(\pi/2) + \sin 90°}\right) = 0.33$$

For the left-hand side block, taking moments about O_1,

$$F \times 1 - R_{n1} \times 0.4 + \mu' R_{n1} \times (0.25 - 0.05) = 0$$

$$5000 \times 1 - R_{n1} \times 0.4 + 0.33 \times R_{n1} \times 0.2 = 0$$

$$R_{n1} = 14\,970 \text{ N}$$

For the right-hand side block, taking moments about O_2,

$$5000 \times 1 - R_{n2} \times 0.4 - 0.33\, R_{n2} \times 0.2 = 0$$

$$R_{n2} = 10\,730 \text{ N}$$

Maximum braking torque, $T_B = \mu' (R_{n1} + R_{n2})\, r$

$$= 0.33\,(14\,970 + 10\,730) \times 0.25$$

$$= 2120.2 \text{ N.m}$$

15.4 BAND BRAKE

It consists of a rope, belt or flexible steel band (lined with friction material) which is pressed against the external surface of a cylindrical drum when the brake is applied. The force is applied at the free end of a lever (Fig. 15.2a).

Brake torque on the drum = $(T_1 - T_2)\, r$

where r is the effective radius of the drum.

The ratio of the tight and the slack side tensions is given by $T_1/T_2 = e^{\mu\theta}$ (Refer Sec. 9.13) on the assumption that the band is on the point of slipping on the drum.

The effectiveness of the force F depends upon

- the direction of rotation of the drum
- the ratio of lengths a and b
- the direction of the applied force F

To apply the brake to the rotating drum, the band has to be tightened on the drum. This is possible if

(i) F is applied in the downward direction when $a > b$.
(ii) F is applied in the upward direction when $a < b$.

If the force applied is not as above, the band is further loosened on the drum which means no braking effect is possible.

(i) $a > b$, F Downwards

(a) Rotation counter-clockwise For counter-clockwise rotation of the drum, the tight and the slack sides of the band will be as shown in Fig. 15.2a.

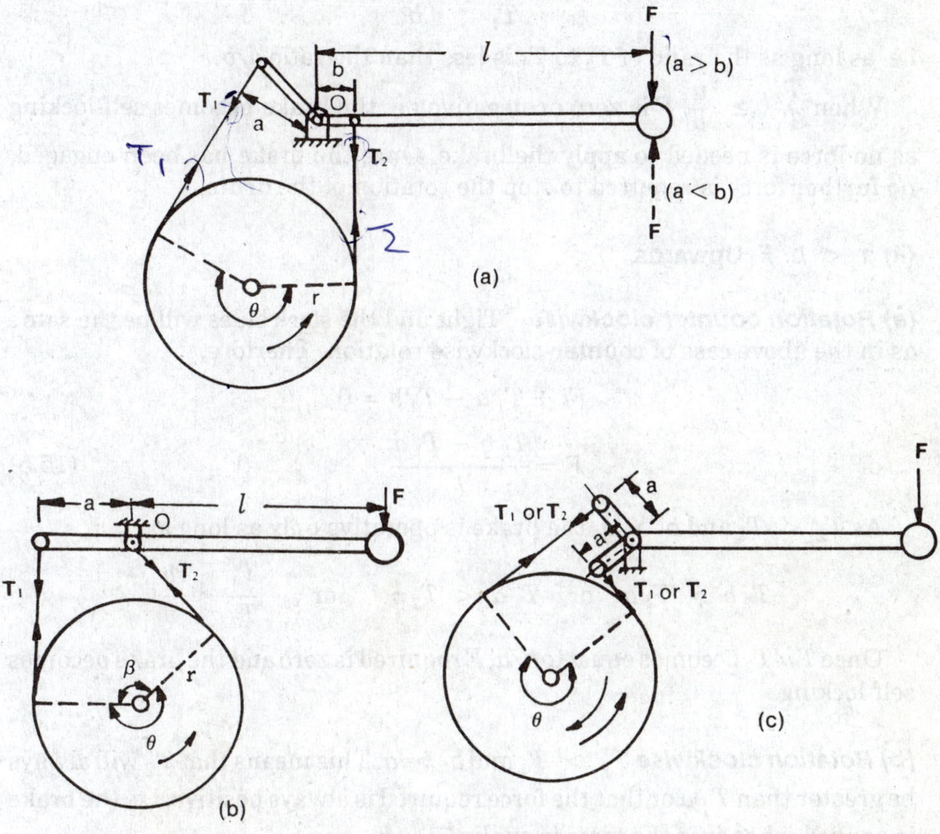

Fig. 15.2

Considering the forces acting on the lever and taking moments about the pivot,

$$F l - T_1 a + T_2 b = 0$$

or
$$F = \frac{T_1 a - T_2 b}{l} \tag{15.4}$$

As $T_1 > T_2$ and $a > b$ under all conditions, the effectiveness of the brake will depend upon the force F.

(b) Rotation clockwise In this case, the tight and the slack sides are reversed, i.e. T_2 becomes greater than T_1.

Then $T_1 < T_2$ and $a > b$.

The brake will be effective as long as $T_1\,a$ is greater than $T_2\,b$

or $$T_2\,b < T_1\,a$$

or $$\frac{T_2}{T_1} < \frac{a}{b}$$

i.e. as long as the ratio of T_2 to T_1 is less than the ratio a/b.

When $\frac{T_2}{T_1} \geq \frac{a}{b}$, F is zero or negative i.e. the brake becomes self-locking as no force is needed to apply the brake. Once the brake has been engaged, no further force is required to stop the rotation of the drum.

(ii) $a < b$ F Upwards

(a) Rotation counter-clockwise Tight and the slack sides will be the same as in the above case of counter-clockwise rotation. Therfore,

$$Fl + T_1\,a - T_2\,b = 0$$

or $$F = \frac{T_2\,b - T_1\,a}{l} \tag{15.5}$$

As $T_2 < T_1$ and $b > a$, the brake is operative only as long as

$$T_2\,b > T_1\,a \quad \text{or} \quad T_1\,a < T_2\,b \quad \text{or} \quad \frac{T_1}{T_2} < \frac{b}{a}.$$

Once T_1/T_2 becomes equal to b/a, F required is zero and the brake becomes self-locking.

(b) Rotation clockwise $T_2 > T_1$ and $b > a$. This means that T_2 will always be greater than $T_1\,a$ or that the force required is always positive, i.e. the brake is applied whenever the force is applied.

(iii) When $a = b$, the band connot be tightened and thus the brake cannot be applied.

(iv) The band brake just discussed is known as a differential band brake. However, if either a or b is made zero, a simple band brake is obtained. If $b = 0$ (Fig. 15.2b) and F downwards,

$$F\,l - T_1\,a = 0$$

or $$F = T_1\,\frac{a}{l} \tag{15.6}$$

Similarly, the force can be found for the other cases.

Note that such a brake can neither have self-energizing properties nor it can be self-locked.

(v) The brake is said to be more effective when maximum braking force is applied with the least effort F.

For case (i), when $a > b$ and F is downwards, the force (effort) F required is less when the rotation is clockwise (assuming that $[(T_2/T_1) < a/b]$).

For case (ii), when $a < b$ and F is upwards, F required is less when the rotation is counter-clockwise (assuming that $[(T_1/T_2) < (a/b)]$).

Thus, for the given arrangement of the differential brake, it is more effective when

(a) $a > b$, F downwards, rotation clockwise

(b) $a < b$, F upwards, rotation counter-clockwise

(vi) The advantage of self-locking is taken in hoists and conveyers where motion is permissible in only one direction. If the motion gets reversed somehow, the self-locking is engaged which can be released only by reversing the applied force.

(vii) It is seen in (v) that a differential band brake is more effective only in one direction of rotation of the drum. However, a two-way band brake can also be designed which is equally effective for both the directions of rotation of the drum (Fig. 15.2c). In such a brake, the two lever arms are made equal.

For both directions of rotation of the drum,

$$F\,l - T_1\,a - T_2\,a = 0$$

or

$$F = (T_1 + T_2)\,\frac{a}{l} \tag{15.7}$$

Example 15.4 A differential band brake (Fig. 15.2a) has a drum of diameter 800 mm. The two ends of the band are fixed to the pins on the opposite sides of the fulcrum of the lever at distances of 40 mm and 200 mm from the fulcrum. The angle of contact is 270° and the coefficient of friction 0.2. Determine the brake torque when a force of 600 N is applied to the lever at a distance of 800 mm from the fulcrum.

Solution:

$F = 600$ N $\qquad l = 800$ mm

$r = 400$ mm $\qquad$ assume $a = 40$ mm

$\theta = 270°$ $\qquad$ and $b = 200$ mm

$\mu = 0.2$

As $a < b$, F must act upwards to apply the brake. Also the brake is more effective when the drum rotates counter-clockwise.

$$\frac{T_1}{T_2} = e^{\mu\theta} = e^{0.2 \times 270 \times \frac{\pi}{180}} = 2.57$$

Taking moments about the fulcrum O,

$$F\,l + T_1\,a - T_2\,b = 0$$

$$600 \times 800 + 2.57\,T_2 \times 40 - T_2 \times 200 = 0$$

or $$600 \times 800 = T_2 (200 - 2.57 \times 40)$$

$$T_2 = 4938 \text{ N}$$

$$T_1 = 4938 \times 2.57 = 12\,691 \text{ N}$$

Braking torque, $T_B = (T_1 - T_2)\, r$

$$= (12\,691 - 4938) \times 0.4.$$

$$= \underline{3101 \text{ N.m}}$$

The braking torque when the drum rotates clockwise can be calculated as follows:

$$\frac{T_2}{T_1} = 2.57$$

Taking moments about O,

$$600 \times 800 + T_1 \times 40 - 2.57\, T_1 \times 200 = 0$$

or $$T_1 = 1012.7 \text{ N}$$

and $$T_2 = 1012.7 \times 2.57 = 2602.5 \text{ N}$$

$$T_B = (2602.5 - 1012.7) \times 0.4$$

$$= \underline{636 \text{ N.m}}$$

Example 15.5 A simple band brake (Fig. 15.2b) is applied to a shaft carrying a flywheel of mass 250 kg and of radius of gyration 300 mm. The shaft speed is 200 rpm. The drum diameter is 200 mm and the coefficient of friction 0.25. The dimensions a and l are 100 mm and 280 mm respectively and the angle $\beta = 135°$. Determine

(i) the brake torque when a force of 120 N is applied at the lever end
(ii) the number of turns of the flywheel before it comes to rest
(iii) the time taken by the flywheel to come to rest.

Solution:

$m = 250$ kg	$\mu = 0.25$
$k = 300$ mm	$r = 100$ mm
$N = 200$ rpm	$a = 100$ mm
$\beta = 135°$	$l = 280$ mm

(i) $$\theta = 360° - 135° = 225°$$

or $$\theta = 225 \times \frac{\pi}{180} = 3.93 \text{ rad}$$

$$\frac{T_1}{T_2} = e^{\mu\theta} = e^{0.25 \times 3.93} = 2.67$$

Taking moments about O,

$$F \times l - T_1 \times a = 0$$

$$120 \times 280 - T_1 \times 100 = 0$$

$$T_1 = 336 \text{ N}$$

$$T_2 = \frac{336}{2.67} = 125.8 \text{ N}$$

$$T_B = (336 - 125.8) \times 0.1 = \underline{21 \text{ N.m}}$$

(ii) KE of the flywheel $= \frac{1}{2} I \omega^2 = \frac{1}{2} mk^2 \left(\frac{2 \pi N}{60}\right)^2$

$$= \frac{1}{2} \times 250 \times (0.3)^2 \times \left(\frac{2 \pi \times 200}{60}\right)^2$$

$$= 4935 \text{ N.m}$$

Let the KE be used to overcome the work done by the braking torque in n revolutions. Then

$$T_B \times \text{Angular displacement} = KE \text{ of flywheel}$$

$$21 \times 2 \pi n = 4935$$

$$n = \underline{37.4 \text{ revolutions}}$$

(iii) time taken $= \frac{n}{N} = \frac{37.4}{200}$ min. $= \frac{37.4}{200/60} = \underline{11.2 \text{ s}}$

Example 15.6 A crane is required to support a load of 1.2 tonnes on the rope round its barrel of 400 mm diameter (Fig. 15.3). The brake drum diameter is 600 mm, the angle of contact of the band brake 275° and the coefficient of friction 0.22. Determine the force required at the end of the lever to support the load. Take a = 150 mm and l = 750 mm.

Solution:

$W = (1.2 \times 1000 \times 9.81)$ N
$R = 300$ mm
$r = 200$ mm
$\mu = 0.22$
$\theta = 275°$

$$\frac{T_1}{T_2} = e^{\mu\theta} = e^{0.22 \times 275 \times \frac{\pi}{180}}$$

$$= 2.87$$

For equilibrium,

$$(T_1 - T_2) R = W \times r$$

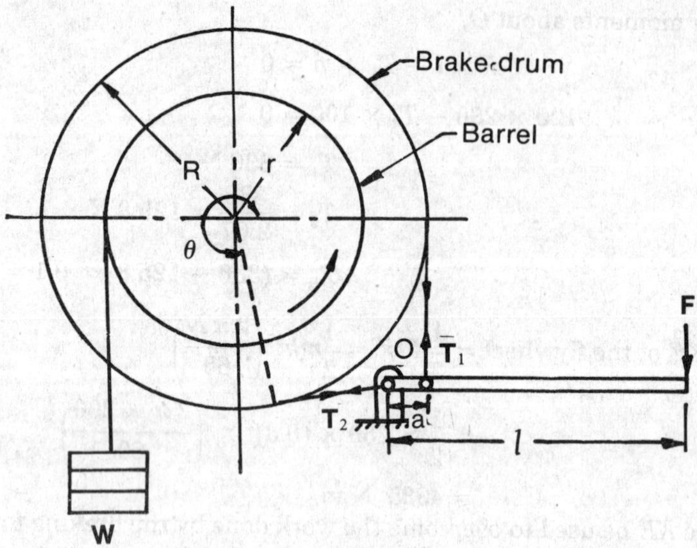

Fig. 15.3

or $$(2.87\,T_2 - T_2) \times 300 = (1.2 \times 1000 \times 9.81) \times 200$$

$$T_2 = 4197 \text{ N.m}$$

$$T_1 = 4197 \times 2.87 = \underline{12\,045 \text{ N.m}}$$

Taking moments about O,

$$F \times l - T_1 \times a = 0$$

or $$F \times 750 - 12\,045 \times 150 = 0$$

or $$F = \underline{2409 \text{ N}}$$

15.5 BAND AND BLOCK BRAKE

A band and block brake consists of a number of wooden blocks secured inside a flexible steel band. When the brake is applied, the blocks are pressed against the drum. Two sides of the band become tight and slack as usual. Wooden blocks have a higher coefficient of friction, thus increasing the effectiveness of the brake. Also, such blocks can be easily replaced on being worn out (Fig. 15.4a).

Each block subtends a small angle 2θ at the centre of the drum. The frictional force on the blocks acts in the direction of rotation of the drum. For n blocks on the brake,

Let T_0 = tension on the slack side

T_1 = tension on the tight side after one block

T_2 = tension on the tight side after two block

....

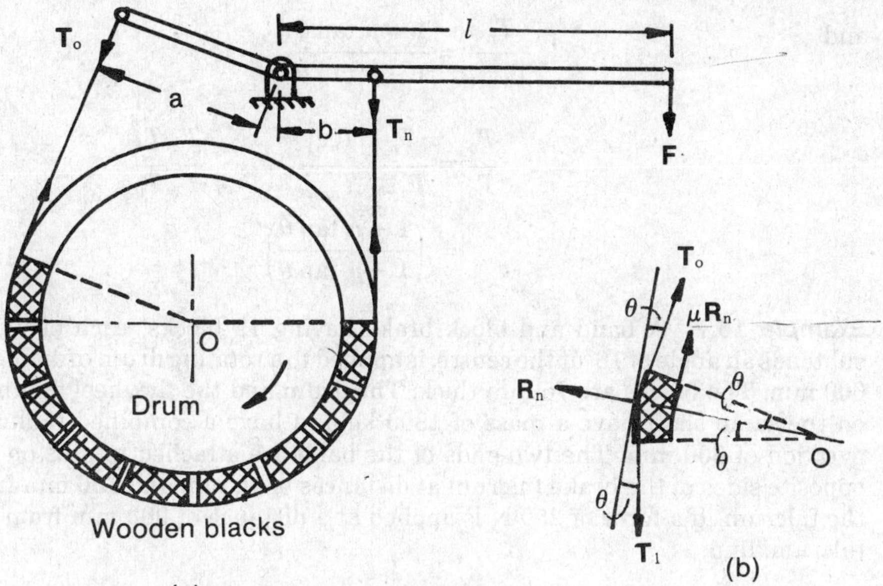

Fig. 15.4

T_n = tension on the tight side after n blocks

μ = coefficient of friction

R_n = normal reaction on the block

The forces on one block of the brake are shown in Fig. 15.4b.

For equilibrium,

$$(T_1 - T_0)\cos\theta = \mu R_n$$

$$(T_1 + T_0)\sin\theta = R_n$$

or
$$\frac{T_1 - T_0}{T_1 + T_0}\frac{1}{\tan\theta} = \mu$$

or
$$\frac{T_1 - T_0}{T_1 + T_0} = \frac{\mu\tan\theta}{1}$$

or
$$\frac{(T_1 - T_0) + (T_1 + T_0)}{(T_1 - T_0) - (T_1 + T_0)} = \frac{\mu\tan\theta + 1}{\mu\tan\theta - 1}$$

$$\frac{2T_1}{2T_0} = -\frac{1 + \mu\tan\theta}{1 - \mu\tan\theta}$$

$$\frac{T_1}{T_0} = \frac{1 + \mu\tan\theta}{1 - \mu\tan\theta}$$

Similarly,
$$\frac{T_2}{T_1} = \frac{1 + \mu\tan\theta}{1 - \mu\tan\theta}$$ and so on.

....

....

and
$$\frac{T_n}{T_{n-1}} = \frac{1 + \mu \tan\theta}{1 - \mu \tan\theta}$$

$$\frac{T_n}{T_0} = \frac{T_n}{T_{n-1}} \frac{T_{n-1}}{T_{n-2}} \cdots \frac{T_2}{T_1} \frac{T_1}{T_0}$$

$$= \left(\frac{1 + \mu \tan\theta}{1 - \mu \tan\theta}\right)^n \tag{15.8}$$

Example 15.7 A band and block brake having 12 blocks, each of which subtends an angle of 16° at the centre, is applied to a rotating drum of diameter 600 mm. The blocks are 75 mm thick. The drum and the flywheel mounted on the same shaft have a mass of 1800 kg and have a combined radius of gyration of 600 mm. The two ends of the band are attached to pins on the opposite sides of the brake fulcrum at distances of 40 mm and 150 mm from the fulcrum. If a force of 250 N is applied at a distance of 900 mm from the fulcrum, find:

(i) the maximum braking torque
(ii) the angular retardation of the drum
(iii) the time taken by the system to be stationary from the rated speed of 300 rpm.

Take coefficient of friction between the blocks and the drum as 0.3.

Solution: Refer to Fig. 15.4.

(i)
$$\frac{T_{12}}{T_0} = \left(\frac{1 + \mu \tan\theta}{1 - \mu \tan\theta}\right)^n = \left(\frac{1 + 0.3 \tan 8°}{1 - 0.3 \tan 8°}\right)^{12} = 2.752.$$

Assume $a = 150$ mm and $b = 40$ mm

As $a > b$, F must be downwards and rotation clockwise for maximum braking torque. Taking moments about the fulcrum,

$$F \times l - T_0 a + T_{12} b = 0$$

$$250 \times 900 - T_0 \times 150 + 2.752\, T_0 \times 40 = 0$$

$$T_0 (150 - 2.752 \times 40) = 250 \times 900$$

$$T_0 = 5636 \text{ N}$$

$$T_{12} = 5636 \times 2.752 = 15\,511 \text{ N}$$

Maximum braking torque, $T_B = (T_{12} - T_0) \times \dfrac{d}{2}$

$$= (15\,511 - 5636) \times \left(\frac{0.6 + 0.075 \times 2}{2}\right)$$

$$= \underline{3703 \text{ N.m}}$$

(ii) $$T_B = I\,\alpha = mk^2\,\alpha$$

$$3703 = 1800 \times (0.6)^2 \times \alpha$$

$$\alpha = \underline{5.71\ \text{rad/s}^2}$$

(iii) Initial angular speed, $\omega_0 = \dfrac{2\,\pi \times 300}{60} = 31.4$ rad/s

Final angular speed, $\omega = 0$.

$\therefore$ $$\omega = \omega_0 - \alpha\,t \quad (\alpha \text{ negative due to retardation})$$

or $$0 = 31.4 - 5.71\,t$$

$$t = \underline{5.5\ \text{s}}$$

15.6 INTERNAL EXPANDING SHOE BRAKE

Earlier, automobiles used band brakes which were exposed to dirt and water. Their heat dissipation capacity was also poor. These days, band brakes have been replaced by internally expanding shoe brakes having at least one self-energising shoe per wheel. This results in tremendous friction, giving great braking power without excessive use of pedal pressure.

Figure 15.5 shows an internal shoe automobile brake. It consists of two semi-circular shoes which are lined with a friction material such as ferodo. The shoes press against the inner flange of the drum when the brakes are applied. Under normal running of the vehicle, the drum rotates freely as the outer diameter of the shoes is a little less than the internal diameter of the drum.

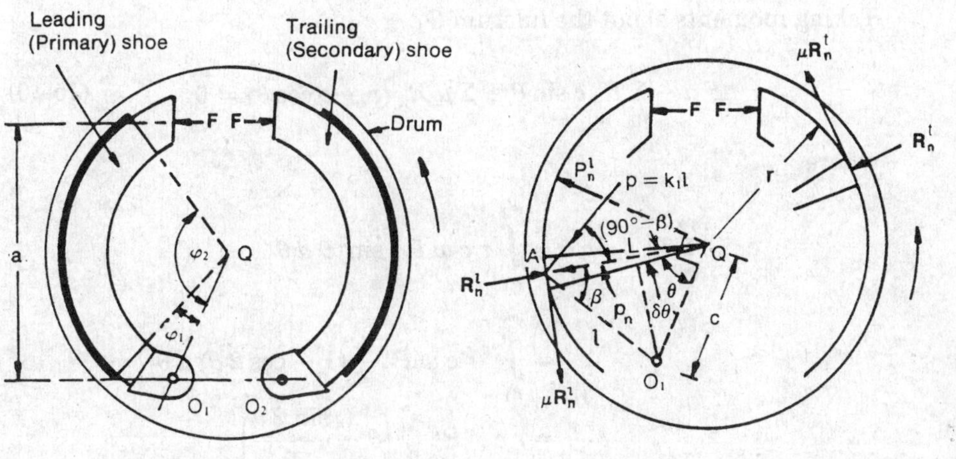

Fig. 15.5

The actuating force F is usually applied by two equal-diameter pistons in a common hydraulic cylinder and is applied equally in magnitude to each shoe. For the shown direction of the drum rotation, the left shoe is known as the leading or forward shoe and the right as the trailing or rear shoe.

Assuming that each shoe is rigid as compared to the friction surface, the pressure p at any point A on the contact surface will be proportional to its distance l from the pivots.

Considering the leading shoe,

$p \alpha l = k_1 l$ where k_1 is a constant

The direction of p is perpendicular to OA.

The normal pressure, $p_n = k_1 l \cos (90° - \beta) = k_1\, l \sin \beta$

$= k_1 c \sin \theta \qquad (OL = l \sin \beta = c \sin \theta)$

$= k_2 \sin \theta$ where k_2 is another constant

p_n is maximum when $\theta = 90°$

Let P_n^l = Maximum intensity of normal pressure on the leading shoe.

$$p_{n\ max} = P_n^l = k_2 \sin 90° = k_2$$

or
$$p_n = P_n^l \sin \theta \tag{15.9}$$

Let w = width of brake lining

μ = coefficient of friction

Consider a small element of brake lining on the leading shoe that makes an angle $\delta\theta$ at the centre.

Normal reaction on the differential surface,

$$R_n^l = \text{Area} \times \text{Pressure}$$

$$= (r\, \delta\theta\, w)\, p_n$$

$$= r\, \delta\theta\, w\, P_n^l \sin \theta$$

Taking moments about the fulcrum O_1,

$$F a - \sum_{\varphi_1}^{\varphi_2} R_n^l\, c \sin \theta + \sum_{\varphi_1}^{\varphi_2} \mu\, R_n^l\, (r - c \cos \theta) = 0 \tag{15.10}$$

where

$$\sum_{\varphi_1}^{\varphi_2} R_n^l\, c \sin \theta = \int_{\varphi_1}^{\varphi_2} r\, c\, w\, P_n^l \sin^2 \theta\, d\theta$$

$$= \int_{\varphi_1}^{\varphi_2} r\, c\, w\, P_n^l\, \frac{1}{2} (1 - \cos 2\theta)\, d\theta$$

$$= r\, c\, w\, P_n^l\, \frac{1}{2} \left(\theta - \frac{\sin 2\theta}{2}\right)_{\varphi_1}^{\varphi_2}$$

$$= \frac{r\, c\, w\, P_n^l}{4} (2\varphi_2 - 2\varphi_1 - \sin 2\varphi_2 + \sin 2\varphi_1)$$

and

$$\sum_{\varphi_1}^{\varphi_2} \mu R_n^l (r - c\cos\theta) = \int_{\varphi_1}^{\varphi_2} \mu r^2 w P_n^l \sin\theta \, d\theta - \int_{\varphi_l}^{\varphi_2} \mu r c w P_n^l \sin\theta \cos\theta \, d\theta$$

$$= \mu r^2 w P_n^l (-\cos\theta)_{\varphi_1}^{\varphi_2} - \int_{\varphi_1}^{\varphi_2} \mu r c w P_n^l \frac{1}{2} \sin 2\theta \, d\theta$$

$$= \mu r^2 w P_n^l (\cos\varphi_1 - \cos\varphi_2) - \mu r c w P_n^l \frac{1}{2} \left(\frac{-\cos 2\theta}{2} \right)_{\varphi_1}^{\varphi_2}$$

$$= \frac{\mu r w P_n^l}{4} [4r (\cos\varphi_1 - \cos\varphi_2) - c (\cos 2\varphi_1 - \cos 2\varphi_2]$$

Taking moments about the fulcrum O_2 for the trailing shoe,

$$F a - \sum_{\varphi_1}^{\varphi_2} R_n^t c \sin\theta - \sum_{\varphi_1}^{\varphi_2} \mu R_n^t (r - c\cos\theta) = 0$$

where $\sum_{\varphi_1}^{\varphi_2} R_n^t c \sin\theta = \dfrac{r c w P_n^t}{4} [2\varphi_2 - 2\varphi_1 - \sin 2\varphi_2 + \sin 2\varphi_1]$

and

$$\sum_{\varphi_1}^{\varphi_1} \mu R_n^t (r - c\cos\theta) = \frac{\mu r w P_n^t}{4} [4r (\cos\varphi_1 - \cos\varphi_2) - c (\cos 2\varphi_1 - \cos 2\varphi_2)]$$

Thus P_n^l and P_n^t, the maximum pressure intensities on the leading and the trailing shoes, can be determined.

Braking torque, $T_B = \sum_{\varphi_1}^{\varphi_2} \mu R_n^l r + \sum_{\varphi_1}^{\varphi_2} \mu R_n^t r$

$$= \int_{\varphi_1}^{\varphi_2} \mu r^2 w P_n^l \sin\theta \, d\theta + \int_{\varphi_1}^{\varphi_2} \mu r^2 w P_n^t \sin\theta \, d\theta$$

$$= r^2 \mu w (P_n^l + P_n^t) (-\cos\varphi)_{\varphi_1}^{\varphi_2}$$

$$= r^2 \mu w (P_n^l + P_n^t) (\cos\varphi_1 - \cos\varphi_2) \tag{15.11}$$

Note that for the same applied force F on each shoe, P_n^l is not equal to P_n^t and $P_n^l > P_n^t$. Usually, more than 50% of the total braking torque is supplied by the leading shoe.

Also note that the leading shoe is self-energizing whereas the trailing shoe is not. This is because the friction forces acting on the leading shoe help the applied force F, and that on the trailing shoe oppose it. On reversing the direction of drum rotation, the right shoe will become self-energizing whereas the left will not be so any longer.

In Eq. 15.10, if the third term exceeds the second term on the LHS, F will be negative and the brake becomes self-locking. A brake should be self-energizing but not self-locking. The amount of self-energizing is measured by the ratio of the friction moment and the normal reaction moment, i.e. the ratio of the third term to the second term. When this ratio is equal to or more than unity, the brake is self-locking. When the ratio is less than unity (more than zero), the brake is self-energizing.

Example 15.8 The following data refer to an internal expanding shoe brake shown in Fig. 15.5:

Force F on each shoe = 180 N
Coefficient of friction, μ = 0.3
Internal radius of the brake drum, r = 150 mm
Width of the brake lining, w = 40 mm
Distance: a = 200 mm c = 120 mm
Angles: $\varphi_1 = 30°$ $\varphi_2 = 135°$

Determine the bracking torque applied when the drum rotates (i) counter-clockwise, and (ii) clockwise

Solution:

(i)
$$F a - \int_{\varphi_1}^{\varphi_2} R_n^l\, c \sin\theta + \int_{\varphi_1}^{\varphi_2} \mu R_n^t (r - c\cos\theta) = 0$$

$$180\times 0.2 - \frac{0.15\times 0.12\times 0.04\times P_n^l}{4}\left(2\times 135\times\frac{\pi}{180} - 2\times 30\times\frac{\pi}{180} - \sin 270° + \sin 60°\right)$$

$$+\frac{0.3\times 0.15\times 0.04\times P_n^l}{4}\left[4\times 0.15(\cos 30° - \cos 135°) - 0.12(\cos 60° - \cos 270°)\right]$$

$$36 - 0.000\,996\, P_n^l + 0.000\,398\, P_n^l = 0$$

$$P_n^l = 60\,201 \text{ Nm}^2$$

For the trailing shoe,

$$36 - 0.000\,996\, P_n^t - 0.000\,398\, P_n^t = 0$$

$$P_n^t = 25\,825 \text{ Nm}^2$$

Braking torque, $T_B = r^2 \mu\, w\, (P_n^l + P_n^t)(\cos\varphi_1 - \cos\varphi_2)$

$$= (0.15)^2 \times 0.3 \times 0.04\,(60\,201 + 25\,825)(\cos 30° - \cos 135°)$$

$$= \underline{36.54 \text{ N.m}}$$

(ii) When the rotation is reversed, P_n^l and P_n^t are interchanged and thus the braking torque is the same.

15.7 EFFECT OF BRAKING

Consider a vehicle moving up an inclined plane (Fig. 15.6).

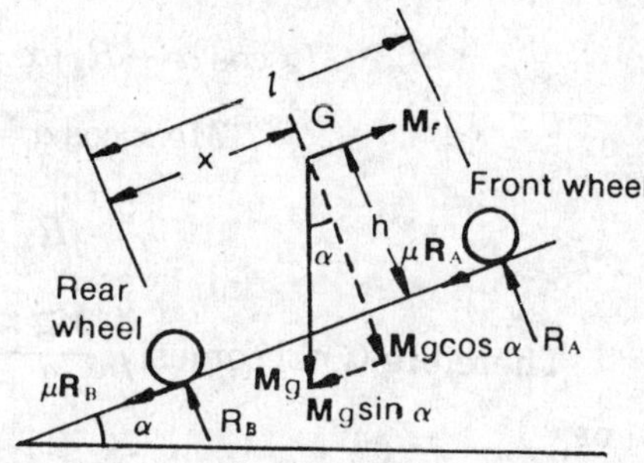

Fig. 15.6

Brakes applied to rear wheels only

For equilibrium,

$$R_A + R_B = Mg \cos \alpha \quad \text{(i)}$$

$$\mu R_B + Mg \sin \alpha = M f \quad \text{(ii)}$$

where f is the retardation of the vehicle.

Taking moments about G, the centre of mass of the vehicle,

$$R_B x + \mu R_B \times h - R_A (l - x) = 0 \quad \text{(iii)}$$

From (i), $$R_A = Mg \cos \alpha - R_B$$

∴ (iii) becomes,

$$R_B x + \mu R_B \times h - (Mg \cos \alpha - R_B)(l - x) = 0$$

or $$R_B (x + \mu h + l - x) = Mg \cos \alpha \, (l - x)$$

or $$R_B = \frac{Mg \cos \alpha \, (l - x)}{l + \mu h}$$

(iv) becomes, $$\mu \frac{Mg \cos \alpha \, (l - x)}{l + \mu h} + Mg \sin \alpha = M f$$

or $$f = g \cos \alpha \left[\frac{\mu (l - x)}{l + \mu h} + \tan \alpha \right] \quad (15.12)$$

On a level road, $\alpha = 0$, and so

$$f = g \frac{\mu (l - x)}{l + \mu h} \quad (15.13)$$

When the vehicle moves down a plane,

$$f = g \cos \alpha \left[\frac{\mu (l - x)}{l + \mu h} - \tan \alpha \right] \quad (15.14)$$

Brakes applied to front wheels only

$$R_A + R_B = Mg \cos \alpha \quad \text{(iv)}$$

$$\mu R_A + Mg \sin \alpha = M f \quad \text{(v)}$$

Taking moments about G,

$$R_B x + \mu R_A \times h - R_A (l - x) \quad \text{(vi)}$$

From (iv) and (vi),

$$(Mg \cos \alpha - R_A) x + \mu R_A \times h - R_A (l - x) = 0$$

or

$$Mg\, x \cos \alpha = R_a (x - \mu h + l - x)$$

$$R_A = \frac{Mg\, x \cos \alpha}{l - \mu h}$$

Therefore (v) becomes, $\mu \dfrac{Mg\, x \cos \alpha}{l - \mu h} + Mg \sin \alpha = M f$

or

$$f = g \cos \alpha \left[\frac{\mu\, x}{l - \mu h} + \tan \alpha \right] \quad (15.15)$$

On a level road, $\alpha = 0$, and therefore

$$f = g \frac{\mu x}{l - \mu h} \quad (15.16)$$

On a down plane,

$$f = g \cos \alpha \left(\frac{\mu\, x}{l - \mu h} - \tan \alpha \right) \quad (15.17)$$

Brakes applied to all four wheels

$$R_A + R_B = Mg \cos \alpha \quad \text{(vii)}$$

$$\mu R_A + \mu R_B + Mg \sin \alpha = M f \quad \text{(viii)}$$

or

$$\mu (R_A + R_B) + Mg \sin \alpha = M f$$

or

$$\mu Mg \cos \alpha + Mg \sin \alpha = M f$$

or

$$f = g \cos \alpha (\mu + \tan \alpha) \quad (15.18)$$

On a level road, $\alpha = 0$. Therefore

$$f = g \mu \quad (15.19)$$

On a down plane,

$$f = g \cos \alpha (\mu - \tan \alpha) \quad (15.20)$$

15.8 TYPES OF DYNAMOMETERS

There are mainly two types of dynamometers:

(i) *Absorption Dynamometers* In this type, the work done is converted into heat by friction while being measured. They can be used for the measurement of moderate powers only. Examples are: Prony brake dynamometer and Rope brake dynamometer.

(ii) *Transmission Dynamometers* In this type, the work is not absorbed in the process, but is utilised after the measurement. Examples are the belt transmission dynamometer and the Torsion dynamometer.

15.9 PRONY BRAKE DYNAMOMETER

A prony brake dynamometer consists of two wooden blocks clamped together on a revolving pulley carrying a lever (Fig. 15.7). The friction between the blocks and the pulley tends to rotate the blocks in the direction of rotation of the shaft. However, the weight due to suspended mass at the end of the lever prevents this tendency. The grip of the blocks over the pulley is adjusted using the bolts of the clamp until the engine runs at the required speed. The mass added to the scale pan is such that the arm remains horizontal in the equilibrium position. the power of the engine is thus absorbed by the friction.

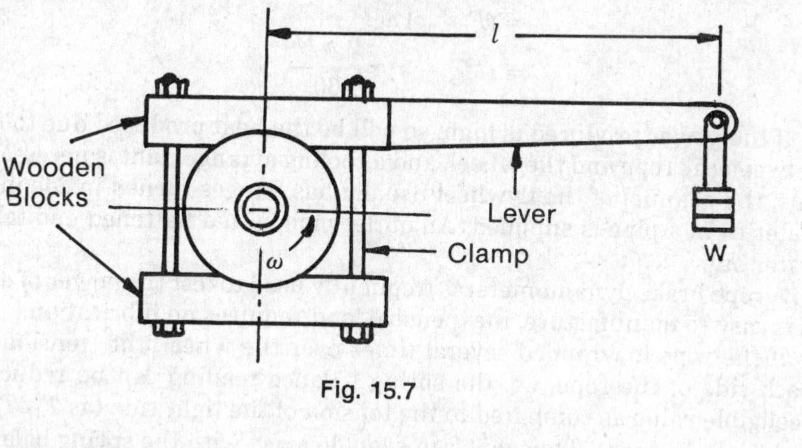

Fig. 15.7

Frictional torque $= W\,l = Mg\,l$

Power of the machine under test $= T\,\omega = Mg\,l\,\dfrac{2\,\pi\,N}{60}$

$= MNk$

where k is a constant for a particular brake.

Note that the expression for power is independent of the size of the pulley and the coefficient of friction.

15.10 ROPE BRAKE DYNAMOMETER

In a rope brake dynamometer (Fig. 15.8), a rope is wrapped over the rim of a pulley keyed to the shaft of the engine. The diameter of the rope depends upon the power of the machine. The spacing of the ropes on the pulley is done by

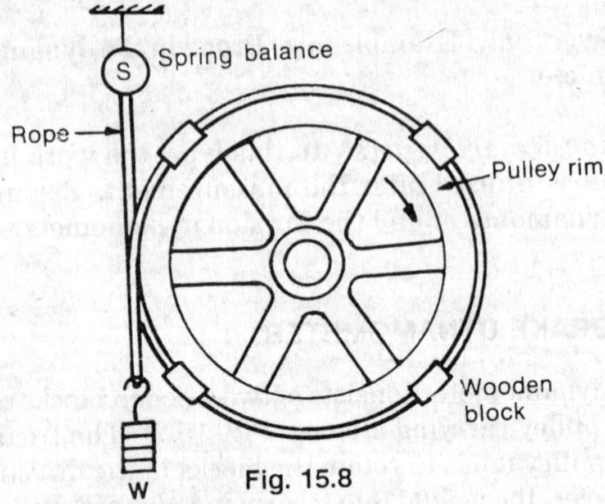

Fig. 15.8

3 or 4 U-shaped wooden blocks which also prevent the rope from slipping off the pulley. The upper end of the rope is attached to a spring balance whereas the lower end supports the weight of suspended mass.

$$\text{Power of the machine} = T\,\omega$$
$$= (F_t \times r)\,\omega$$
$$= (Mg - s)\,r\,\frac{2\pi N}{60}$$

If the power produced is high, so will be the heat produced due to friction between the rope and the wheel, and a cooling arrangement is necessary. For this, the channel of the flywheel usually has flanges turned inside in which water from a pipe is supplied. An outlet pipe with a flattened end takes the water out.

A rope brake dynamometer is frequently used to test the power of engines. It is easy to manufacture, inexpensive, and requires no lubrication.

If the rope is wrapped several times over the wheel, the tension on the slack side of the rope, i.e. the spring balance reading can be reduced to a negligible value as compared to the tension of the tight side (as $T_1/T_2 = e^{\mu\theta}$ and θ is increased). Thus, one can even do away with the spring balance.

15.11 BELT TRANSMISSION DYNAMOMETER

The belt transmission dynamometer occupies a prominent position among transmission dynamometers. When a belt transmits power from one pulley to another, there exists a difference in tensions between the tight and slack sides. A dynamometer measures directly the difference in tensions $(T_1 - T_2)$ while the belt is running.

Figure 15.9 shows a Tatham dynamometer. A continuous belt runs over the driving and the driven pulleys through two intermediate pulleys. The intermediate pulleys are fixed to a lever having a fulcrum at the midpoint of the two pulley centres. The weight of a suspended mass at one end of the lever

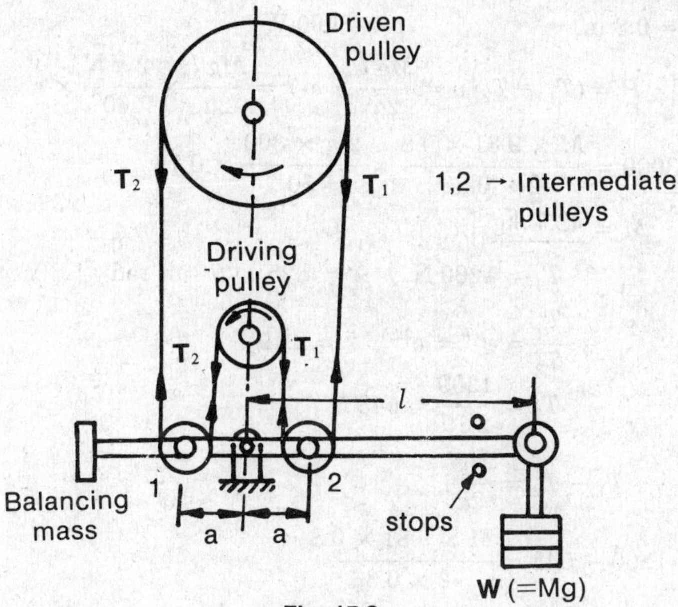

Fig. 15.9

balances the difference in tensions of the tight and the slack sides. As the lever is not pivoted at its midpoint, a mass at the other end is used for its initial equilibrium. Two stops, one on each side of the lever arm, are used to limit the motion of the lever.

Taking moments about the fulcrum,

$$Mg\,l - 2\,T_1\,a + 2\,T_2\,a = 0$$

$$Mg\,l - 2a\,(T_1 - T_2) = 0$$

$$T_1 - T_2 = \frac{Mg\,l}{2a}$$

Power, $P = (T_1 - T_2)\,v$

where v = belt speed in metres per second.

Example 15.9 In a belt transmission dynamometer, the driving pulley rotates at 300 rpm. The distance between the centre of the driving pulley and the dead mass is 800 mm. The diameter of each of the driving as well as the intermediate pulleys is equal to 360 mm. Find the value of the dead mass required to maintain the lever in a horizontal position when the power transmitted is 3 kW. Also find its value when the belt just begins to slip on the driving pulley, μ being 0.25 and the maximum tension in the belt 1200 N.

Solution:

N = 300 rpm $\qquad a$ = 0.36 m

$l = 0.8$ m $\qquad\qquad p = 3000$ W

(i) $$P = (T_1 - T_2)\,v = \frac{Mg\,l}{2a} \times \omega\, r = \frac{Mg\,l}{2a} \times \frac{2\pi N}{60} \times r$$

$$3000 = \frac{M \times 9.81 \times 0.8}{2 \times 0.36} \times \frac{2\pi \times 300}{60} \times 0.18$$

$$M = \underline{48.7 \text{ kg}}$$

(ii) $T_1 = 1200$ N $\quad \mu = 0.25 \quad \theta = \pi$ rad

$$\frac{T_1}{T_2} = e^{\mu\theta} = e^{0.25 \times \pi} = 2.19$$

$$T_2 = \frac{1200}{2.19} = 548 \text{ N}$$

$$T_1 - T_2 = \frac{Mgl}{2a}$$

$$1200 - 548 = \frac{M \times 9.81 \times 0.8}{2 \times 0.36}$$

$$M = \underline{59.8 \text{ kg}}$$

15.12 BEVIS-GIBSON TORSION DYNAMOMETER

A Bevis-Gibson torsion dynamometer consists of two discs A and B, a lamp and a movable torque finder arranged as shown in Fig. 15.10a. The two discs are similar and are fixed to the shaft at a fixed distance from each other. Thus the two discs revolve with the shaft. The lamp is masked and fixed on the bearing of the shaft. The torque finder has an eyepiece capable of moving circumferentially. Each disc has a small radial slot near its periphery. Similar slots are also made at the same radius on the mask of the lamp and on the torque finder.

When the shaft rotates freely and does not transmit any torque, all the four slots are in a line and a ray of light from the lamp can be seen through the eyepiece after every revolution. When torque is transmitted, the shaft twists and the slot in the disc B shifts its position. The ray of light can no longer pass through the four slots. However, if the eyepiece is moved circumferentially by an amount equal to the displacement, the flash will again be visible once in each revolution of the shaft. The eyepiece is moved by a micrometer spindle. The angle of twist may be measured up to one hundredth of a degree.

In case the torque is varied during each revolution of the shaft as in reciprocating engines and it is required to measure the angle of twist at different angular positions, then each disc can be perforated with several slots arranged in the form of a spiral at varying radii (Fig. 15.10b). The lamp and the torque finder have to be moved radially to and from the shaft so that they come opposite each pair of slots in the discs.

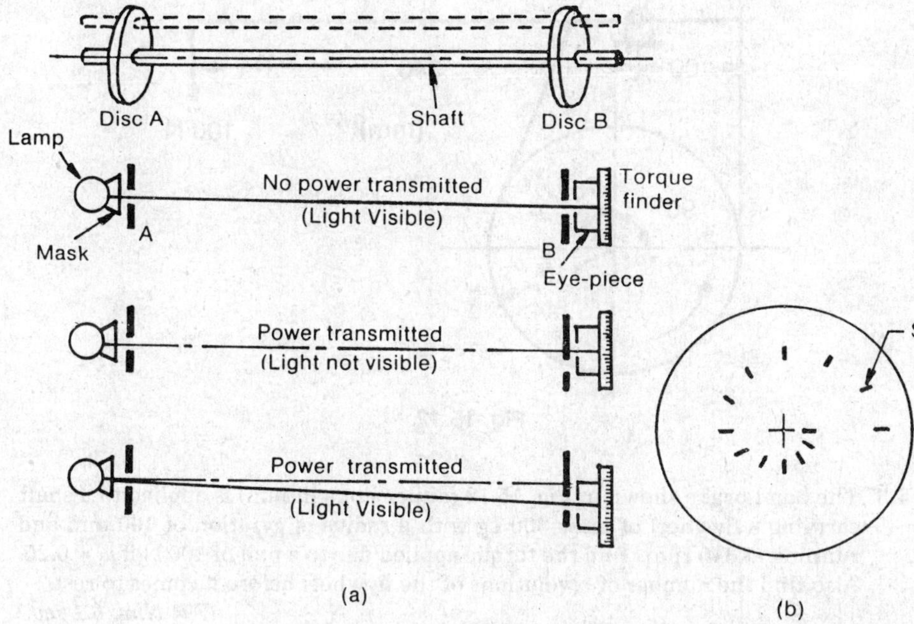

Fig. 15.10

PROBLEMS

15.1 In a brake shoe applied to a drum Fig. 15.11, the radius of the drum is 80 mm and the coefficient of friction at the brake lining 0.3. For the counter-clockwise rotation of the drum, determine the braking torque due to a force of 400 N applied at the end of the lever.

(219 N.m)

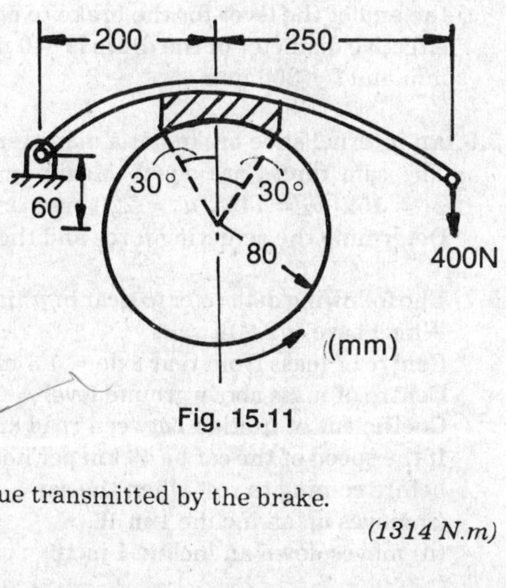

Fig. 15.11

15.2 For the shoe brake shown in Fig.15.1b, the diameter of the brake drum is 400 mm and the angle of contact 96°. The applied force is 3 kN on each arm and the coefficient of friction between the drum and the lining is 0.35. Taking the dimensions a = 500 mm, b = 200 mm and c = 60 mm, determine the maximum torque transmitted by the brake.

(1314 N.m)

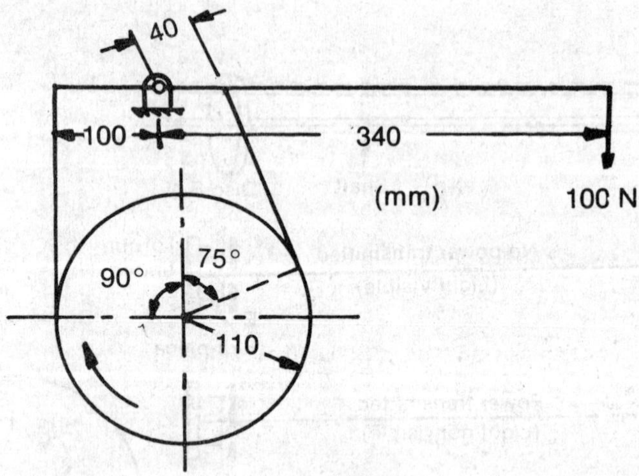

Fig. 15.12

15.3 The band brake shown in Fig. 15.12 (dimensions in mm) is applied to a shaft carrying a flywheel of mass 300 kg with a radius of gyration of 400 mm and running at 340 rpm. find the torque applied due to a pull of 100 N if μ = 0.25. Also find the number of revolutions of the flywheel before it comes to rest.

(794 N.m; 6.1 rev.)

15.4 A crane is used to support a load of 1.1 tonne on the rope round its barrel of 360 mm diameter (Fig. 15.3). The brake drum diameter is 560 mm, the angle of contact 300° and the coefficient of friction between the band and the drum 0.22. What will be the force F required at the end of the lever? Take a = 150 mm and l = 800 mm.

(2663 N)

15.5 In a band and block brake (Fig. 15.4) having 10 blocks, each block subtends an angle of 15° at the centre of wheel. Determine the maximum force required at the end of the lever for the brake to absorb 250 kW of power at 280 rpm. The effective diameter of the drum is 840 mm. Take μ = 0.35, a = 200 mm, b = 40 mm and l = 300 mm.

(4440 N)

15.6 An internal shoe brake has a diameter of 320 mm and width of 30 mm. The cam forces are equal. Maximum pressure is not to exceed 80 kN/m^2. φ_1 = 15°, φ_2 = 145°, a = 220 mm, c = 125 mm and μ = 0.32 (Fig. 15.5) Determine the actuating force and the brake torque.

(175.7 N; 48.04 N.m)

15.7 The following data refer to a car in which brakes are applied to the front wheels:

Wheel base = 2.8 m

Centre of mass from rear axle = 1.3 m

Centre of mass above ground level = 0.96 m

Coefficient of friction between road and tyres = 0.4

If the speed of the car be 40 km per hour, find the distance travelled by the car before coming to rest when the car

(i) moves up an incline 1 in 16

(ii) moves down an incline 1 in 16

(iii) moves on a level track

(22.5 m; 40.75 m; 29.03 m)

15.8 The following data refer to a laboratory experiment with rope brake:
Diameter of the flywheel = 1m
Diameter of the rope = 10 mm
Dead weight on the brake = 50 kg
Speed of the engine = 180 rpm
Spring balance reading = 120 N
Find the power of the engine.

(3527 W)

15.9 In a belt transmission dynamometer (Fig. 15.9), the diameters of the driving and driven pulleys are 0.36 m and 0.8 m respectively. The power transmitted from the driving to the driven shaft is 20 kW. The speed of the driving shaft is 500 rpm. If l = 1.2 m and a = 400 mm, determine the weight on the lever.

(144.2 kg)

16

GOVERNORS

16.1 INTRODUCTION

The function of a governor is to maintain the speed of an engine within specified limits whenever there is a variation of load. In general, the speed of an engine varies in two ways – during each revolution (cyclic variation) and over a number of revolutions. In the former case, it is due to variation in the output torque of the engine during a cycle and can be regulated by mounting a suitable flywheel on the shaft. In the latter case, it is due to variation of load upon the engine and requires a governor to maintain the speed. The operation of a flywheel is continuous whereas that of a governor is more or less intermittent. A flywheel may not be used if there is no undesirable cyclic fluctuation of the energy output, but a governor is essential for all types of engines as it adjusts the supply according to the demand.

If the load on the shaft increases, the speed of the engine decreases unless the supply of fuel is increased by opening the throttle valve. On the other hand, if the load on the shaft decreases, the speed of the engine increases unless the fuel supply is decreased by closing the valve sufficiently to slow the engine to its original speed. The throttle valve is operated by the governor through a mechanism for the purpose.

16.2 TYPES OF GOVERNORS

Governors can broadly be classified into two types.

(i) Centrifugal Governor

This is the more common type. Its action depends on the change of speed. It has a pair of masses, known as governor balls, which rotate at a distance from the axis of rotation. The action of the governor depends upon the centrifugal effects produced by the masses. With the increase in the speed, the balls tend to rotate at a greater radius from the axis and through a suitable linkage the throttle valve is made to close to the required extent. When the speed decreases, the balls rotate at a smaller radius and the valve is opened according to the requirement.

(ii) Inertia Governor

In this type, the positions of the balls are affected by the forces set up by an angular acceleration or deceleration of the given spindle in addition to centrifugal forces on the balls. Using suitable linkages and springs, the change in position of the balls is made to open or close the throttle valve.

Thus, whereas the valve is operated by the actual change of engine speed in the case of centrifugal governors, it is by the rate of change of speed in case of inertia governors. Therefore, the response of inertia governors is faster than that of centrifugal types.

16.3 WATT GOVERNOR (SIMPLE CONICAL GOVERNOR)

Figure 16.1 shows three forms of a simple centrifugal or a Watt governor. In this, a pair of balls (masses) are attached to a spindle with the help of links. In Fig. 16.1a, the upper links are pinned at point O. In Fig. 16.1b, the upper links are connected by a horizontal link and the governor is known as the *open-arm* type Watt governor. On extending the upper arms, they still meet at O. in Fig. 16.1c, the upper links cross the spindle and are connected by a horizontal link and is known as a *crossed-arm* Watt governor. In this type also, the two links intersect at O. The lower links in every case are fixed to a sleeve free to move on the vertical spindle.

As the spindle rotates, the balls take up a position depending upon the speed of the spindle. If it lowers, they move near to the axis due to reduction

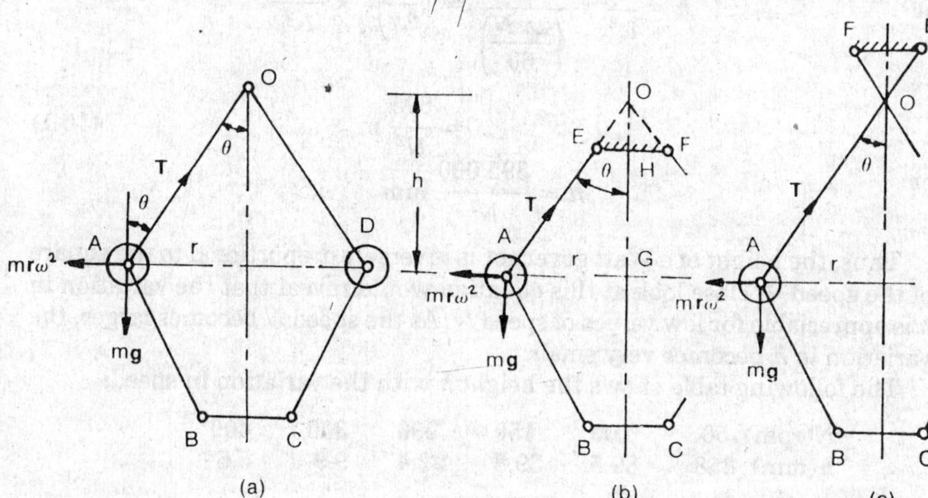

Fig. 16.1

in the centrifugal force on the balls and the ability of the sleeve to slide on the spindle. The movement of the sleeve is further taken to the throttle of the engine by means of a suitable linkage to decrease or increase the fuel supply.

The vertical distance from the plane (horizontal) of rotation of the balls to the point of intersection of the upper arms along the axis of the spindle is called the height h of the governor. The height of the governor decreases with increase in speed, and increases with decrease in speed.

Let m = mass of each ball
w = weight of each ball $(= mg)$
ω = angular velocity of the balls, arms and the sleeve
T = tension in the arm
r = radial distance of ball-center from spindle-axis.

Assuming the links to be massless and neglecting the friction of the sleeve, the mass m at A is in static equilibrium under the action of

– Weight $\mathbf{w}(= mg)$
– Centrifugal force $m\mathbf{r}\,\omega^2$
– Tension $\mathbf{T}$ in the upper link

If the sleeve is massless and also friction is neglected, the lower links will be tension free.

The equilibrium of the mass provides

$$T\cos\theta = mg \text{ and } T\sin\theta = mr\omega^2$$

$$\therefore \quad \tan\theta = \frac{mr\omega^2}{mg} = \frac{r\omega^2}{g}$$

or $$\frac{r}{h} = \frac{r\omega^2}{g}$$

or $$h = \frac{g}{\omega^2} = \frac{g}{\left(\frac{2\pi N}{60}\right)^2} = \left(\frac{60}{2\pi}\right)^2 \times \frac{9.81}{N^2}$$

$$= \frac{895}{N^2}\text{ m} \qquad (16.1)$$

or $$h = \frac{895\,000}{N^2}\text{ mm}$$

Thus, the height of a Watt governor is inversely proportional to the square of the speed. A close look at this equation would reveal that the variation in h is appreciable for low values of speed N. As the speed N becomes larger, the variation in h becomes very small.

The following table shows the height h with the variation in speed:

N(rpm)	50	100	150	200	300	400
h(mm)	358	89.5	39.8	22.4	9.9	5.6

This shows that this type of governor is unsuitable for high speeds. However, this drawback has been overcome by loading the governor with a dead weight or by means of a spring. Such governors have been discussed in the sections that follow.

Example 16.1 In an open-arm type governor (Fig. 16.1b), AE = 400 mm, EF = 50 mm and angle θ = 35°. Determine the percentage change in speed when θ decreases to 30°.

Solution:

$$h = GO = GH + HO = AE \cos\theta + EH \cot\theta$$

$$h = 400 \cos 35° + 25 \cot 35°$$

$$363.4 \text{ mm}$$

$$h' = 400 \cos 30° + 25 \cot 30°$$

$$389.7 \text{ mm}$$

$$\text{Now, } h = \frac{g}{\omega^2} \text{ and } h' = \frac{g}{\omega'^2}$$

$$\therefore \quad \frac{\omega'}{\omega} = \sqrt{\frac{h}{h'}} = \sqrt{\frac{363.4}{389.7}} = 0.966$$

Decrease in speed = (1 – 0.966) × 100 = <u>3.44%</u>
Alternatively,

$$N = \sqrt{\frac{895000}{h}} = \sqrt{\frac{895000}{363.4}} = 49.63 \text{ rpm}$$

$$N' = \sqrt{\frac{895000}{389.7}} = 47.92 \text{ rpm}$$

$$\text{Decrease} = \frac{N - N'}{N} \times 100 = \frac{49.63 - 47.92}{49.63} = \underline{3.44\%}$$

16.4 PORTER GOVERNOR

If the sleeve of a Watt governor is loaded with a heavy mass, it becomes a Porter governor (Fig. 16.2a)

Let M = mass of the sleeve

m = mass of each ball

f = force of friction at the sleeve

The force of friction always acts in a direction opposite to that of the motion. Thus when the sleeve moves up, the force of friction acts in the downward direction and the downward force acting on the sleeve is $(Mg + f)$. Similarly, when the sleeve moves down, the force on the sleeve will be $(Mg - f)$. In

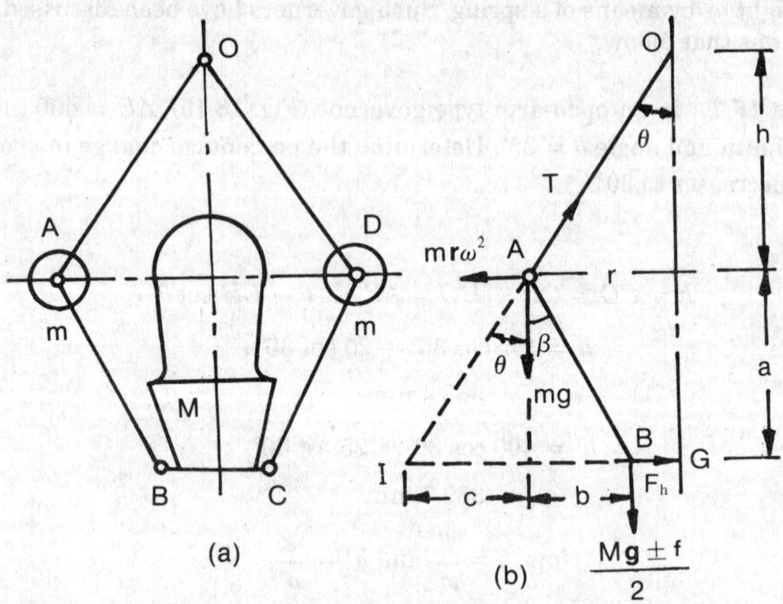

Fig. 16.2

general, the net force acting on the sleeve is $(Mg \pm f)$ depending upon whether the sleeve moves upwards or downwards.

Forces acting on the sleeve and on each ball have been shown in Fig. 16.2b.

Let h = height of the governor

r = distance of the centre of each ball from axis of rotation

As BAO is kinematically equivalent to a slider-crank mechanism with B as slider (vertical motion), the instantaneous centre of rotation of the link AB is at I (Refer Sec. 2.11) for the given configuration of the governor.

Considering the equilibrium of the left-hand half of the governor and taking moments about I.

$$mr\omega^2 a = mg\, c + \frac{Mg \pm f}{2}(c + b)$$

or
$$mr\omega^2 = mg\frac{c}{a} + \frac{Mg \pm f}{2}\left(\frac{c}{a} + \frac{b}{a}\right)$$

$$= mg \tan\theta + \frac{Mg \pm f}{2}(\tan\theta + \tan\beta)$$

$$= \tan\theta\left[mg + \frac{Mg \pm f}{2}(1 + k)\right] \quad (16.2)$$

where
$$k = \frac{\tan\beta}{\tan\theta}$$

The height of the governor will be given by

$$mr\omega^2 = \frac{r}{h}\left[mg + \frac{Mg \pm f}{2}(1 + k)\right]$$

or
$$h = \frac{g}{\omega^2} + \frac{(Mg \pm f)(1 + k)}{2m\,\omega^2} \qquad (16.3)$$

This equation would provide two values of ω for the same height of the governor. The phenomenon can be explained as below.

First assume that the sleeve has just moved down. This means that the force acting on the sleeve is $(Mg - f)$ downwards. Now, if the speed of the engine increases, the balls would tend to move away from the axis, but now as the friction has to act in the downward direction, the resistance to the motion would be $(Mg + f)$. Thus until the speed rises to such a value as to overcome this resistance, the sleeve will not move. In the same way, when the sleeve has moved up and the speed decreases, the resistance to the sleeve movement would be only $(Mg - f)$. Thus, until the speed reduces to such a value as to give a force equal to $(Mg - f)$, the sleeve will not move.

Thus for a given value of h, the governor is insensitive between two values of ω given by Eq. (16.3).

Example 16.2 In a Porter governor, each of the four arms is 400 mm long. The upper arms are pivoted on the axis of the sleeve whereas the lower arms are attached to the sleeve at a distance of 45 mm from the axis. Each ball has a mass of 8 kg and the load on the sleeve is 60 kg. What will be the equilibrium speeds for the two extreme radii of 250 mm and 300 mm of rotation of the governor balls?

Solution:

$m = 8$ kg $\qquad BG = 45$ mm
$M = 60$ kg $\qquad OA = 400$ mm

We have,

$$mr\,\omega^2 = \tan\theta\left[mg + \frac{Mg}{2}(1 + k)\right] \qquad (f = 0)$$

(i) When $r = 250$ mm

$$\tan\theta = \frac{r}{h} = \frac{r}{\sqrt{(OA)^2 - r^2}} = \frac{250}{\sqrt{(400)^2 - (250)^2}} = 0.8$$

$$k = \frac{\tan\beta}{\tan\theta} = \frac{b/a}{\tan\theta}$$

As $b = 250 - 45 = 205$ mm,

$$a = \sqrt{(AB)^2 - (b)^2} = \sqrt{(400)^2 - (205)^2} = 343.4 \text{ mm}$$

$$k = \frac{205/343.4}{0.8} = 0.746$$

$$\therefore \quad 3 \times 0.25 \times \omega^2 = 0.8 \left[8 \times 9.81 + \frac{60 \times 9.81}{2} (1 + 0.746) \right]$$

$$2\,\omega^2 = 0.8\,(78.48 + 513.85)$$

$$\omega^2 = 237$$

$$\omega = \frac{2\,\pi N}{60} = 15.39$$

$$N = \underline{147 \text{ rpm}}$$

(ii) When r = 300 mm,

$$\tan\theta = \frac{300}{\sqrt{(400)^2 - (300)^2}} = 1.134$$

$$b = 300 - 45 = 255 \text{ mm}$$

$$a = \sqrt{(400)^2 - (255)^2} = 308.2 \text{ mm}$$

$$k = \frac{\tan\beta}{\tan\theta} = \frac{b/a}{\tan\theta} = \frac{255/308.2}{1.134} = 0.73$$

$$\therefore \quad 8 \times 0.3 \times \omega^2 = 1.134 \left[8 \times 9.81 + \frac{60 \times 9.81}{2} (1 + 0.73) \right]$$

$$2.4\,\omega^2 = 1.134\,(78.48 + 509.14)$$

$$\omega^2 = 277.6$$

$$\omega = \frac{2\,\pi N}{60} = 16.66$$

$$N = \underline{159.1 \text{ rpm}}$$

Also, range of speed = 159.1 – 147 = 12.1 rpm

Example 16.3 Each arm of a Porter governor is 250 mm long. The upper and lower arms are pivoted to links of 40 mm and 50 mm respectively from the axis of rotation. Each ball has a mass of 5 kg and the sleeve mass is 50 kg. The force of friction on the sleeve of the mechanism is 40 N. Determine the range of speed of the governor for extreme radii of rotation of 125 mm and 150 mm.

Solution: Refer Fig. 16.3.

m = 5 kg	$AB = AE$ = 250 mm
M = 50 kg	BG = 50 mm
f = 40 N	EH = 40 mm

(i) When r = 125 mm,

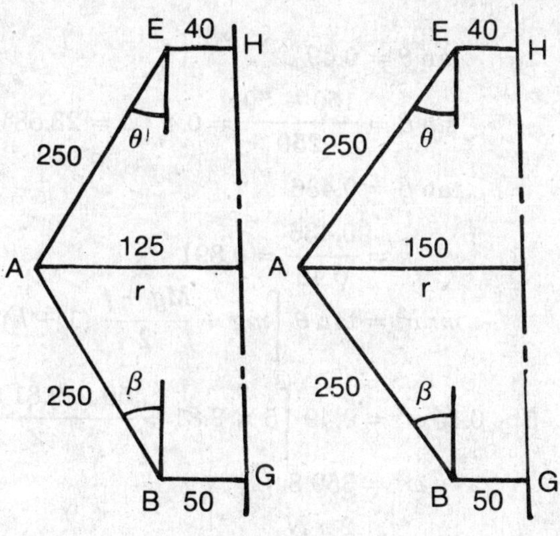

Fig. 16.3

$$\sin\theta = \frac{125 - 40}{250} = 0.34 \quad \theta = 19.88°$$

$$\tan\theta = \tan 19.88° = 0.362$$

$$\sin\beta = \frac{125 - 50}{50} = 0.3 \quad \beta = 17.46°$$

$$\tan\beta = \tan 17.46° = 0.315$$

$$k = \frac{\tan\beta}{\tan\theta} = 0.87$$

As the radii decrease, the sleeve moves down and the force of friction f acts upwards.

$$mr\omega^2 = \tan\theta\left[mg + \frac{Mg - f}{2}(1 + k)\right]$$

$$5 \times 0.125\,\omega^2 = 0.362\left[5 \times 9.81 + \frac{50 \times 9.81 - 40}{2}(1 + 0.87)\right]$$

$$\omega^2 = 272.4$$

$$\omega = \frac{2\pi N}{60} = 16.5$$

$$N_{min} = 157.6 \text{ rpm}$$

(ii) When r = 150 mm,

$$\sin\theta = \frac{150 - 40}{250} = 0.44 \quad \theta = 26.1°$$

$$\tan\theta = 0.49$$

$$\sin\beta = \frac{150 - 50}{250} = 0.4 \quad \beta = 23.58°$$

$$\tan\beta = 0.436$$

$$k = \frac{0.436}{0.49} = 0.891$$

$$mr\omega^2 = \tan\theta\left[mg + \frac{Mg + f}{2}(1 + k)\right] \text{ (sleeve moves up)}$$

$$5 \times 0.15\,\omega^2 = 0.49\left[5 \times 9.81 + \frac{50 \times 9.81 + 40}{2}(1 + 0.891)\right]$$

$$\omega^2 = 359.8$$

$$\omega = \frac{2\pi N}{60} = 18.97$$

$$N_{max} = 181.1 \text{ rpm}$$

Range of speed = 157.6 rpm to 181.1 rpm = 23.5 rpm

16.5 PROELL GOVERNOR

A Porter governor is known as a Proell governor if the two balls (masses) are fixed on the upward extensions of the lower links which are in the form of bent links *BAE* and *CDF* (Fig. 16.4a).

Considering the equilibrium of the link *BAE* which is under the action of

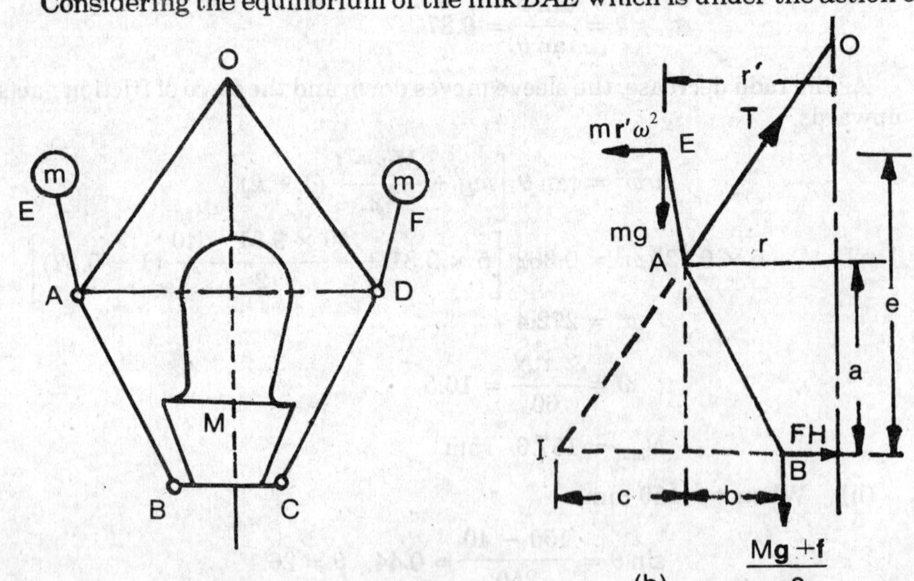

Fig. 16.4

- the weight of the ball, mg
- the centrifugal force, $mr'\omega^2$
- the tension in the link AO
- the horizontal reaction of the sleeve.
- the weight of sleeve and friction, $\frac{1}{2}(Mg \pm f)$.

As before, I is the instantaneous centre of the link BAE.

Taking moments about I,

$$mr'\omega^2 e = mg\,(c + r - r') + \frac{Mg \pm f}{2}(c + b)$$

where b, c, d and r are the dimensions as indicated in the diagram.

$$mr'\omega^2 = \frac{1}{e}\left[mg\,(c + r - r') + \frac{Mg \pm f}{2}(c + b)\right] \tag{16.4}$$

In the position when AE is vertical, i.e. neglecting its obliquity,

$$mr\omega^2 = \frac{1}{e}\left[mg\,c + \frac{Mg \pm f}{2}(c + b)\right]$$

$$= \frac{a}{e}\left[mg\frac{c}{a} + \frac{Mg \pm f}{2}\left(\frac{c}{a} + \frac{b}{a}\right)\right]$$

$$= \frac{a}{e}\left[mg\tan\theta + \frac{Mg \pm f}{2}(\tan\theta + \tan\beta)\right]$$

$$= \frac{a}{e}\tan\theta\left[mg + \frac{Mg \pm f}{2}(1 + k)\right] \tag{16.5}$$

Example 16.4 The mass of each ball of a Proell governor is 7.5 kg and the load on the sleeve is 80 kg. Each of the arms is 300 mm long. The upper arms are pivoted on the axis of rotation whereas the lower arms are pivoted to links of 40 mm from the axis of rotation. The extensions of the lower arms to which the balls are attached are 100 mm long and are parallel to the governor axis at the minimum radius. Determine the equilibrium speeds corresponding to extreme radii of 180 mm and 240 mm.

Solution: When AE is vertical, $r' = r = 180$ mm (Fig. 16.5a).

$$mr\,\omega^2 = \frac{a}{e}\tan\theta\left[mg + \frac{Mg}{2}(1 + k)\right] \text{ (friction neglected)}$$

We have,

$$a = \sqrt{(300)^2 - (180 - 40)^2} = 265.3 \text{ mm}$$

$$e = 265.3 + 100 = 365.3 \text{ mm}$$

$$\sin\theta = \frac{180}{300} = 0.6;\ \theta = 36.87°$$

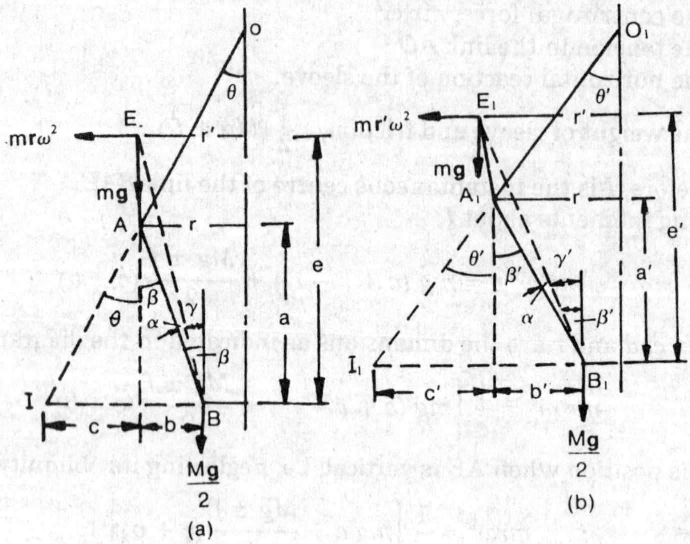

Fig. 16.5

$$\tan\theta = 0.75$$

$$\sin\beta = \frac{180 - 40}{300} = 0.467 \qquad \beta = 27.82°$$

$$\tan\beta = 0.528$$

$$k = \frac{\tan\beta}{\tan\theta} = \frac{0.528}{0.75} = 0.704$$

$$7.5 \times 0.18 \times \omega^2 = \frac{0.2653}{0.3653} \times 0.75 \left[7.5 \times 9.81 + \frac{80 \times 9.81}{2}(1 + 0.704)\right]$$

$$\omega^2 = 299.5$$

$$\omega = \frac{2\pi N}{60} = 17.305$$

$$\underline{N = 165.3 \text{ rpm}}$$

$$BE = \sqrt{e^2 + b^2} = \sqrt{(365.3)^2 + (140)^2} = 391.2 \text{ mm}$$

$$\cos\gamma = \frac{e}{BE} = \frac{365.3}{391.2} = 0.934 \qquad \gamma = 20.97°$$

$$\sin\beta = \frac{b}{AB} = \frac{140}{300} = 0.467 \qquad \beta = 27.82°$$

$$\alpha = 27.82° - 20.97° = 6.85°$$

$$\sin \beta' = \frac{b'}{A_1B_1} = \frac{240 - 40}{300} = 0.667 \qquad \beta' = 41.81°$$

$$\therefore \quad \gamma' = \beta' - \alpha = 41.81° - 6.85° = 34.96° \qquad \text{(Refer Fig. 16.5b)}$$

$$e' = B_1E_1 \cos \gamma' = BE \cos \gamma' = 391.2 \cos 34.96° = 320.6 \text{ mm}$$

$$r' = B_1E_1 \sin \gamma' + 40 = 391.2 \sin 34.96° + 40 = 264.2 \text{ mm}$$

$$b' = 200 \text{ mm}$$

$$a' = A_1B_1 \cos \beta' = 300 \cos 41.81° = 223.6 \text{ mm}$$

$$\sin \theta' = \frac{240}{300} = 0.8 \qquad \theta' = 53.13°$$

$$c' = a' \tan \theta' = 223.6 \tan 53.13° = 298.1\text{mm}$$

Taking moments about I,

$$mr'\,\omega^2 e' = mg\,(c' + r - r') + \frac{Mg}{2}(c' + b')$$

$$7.5 \times 0.2642 \times \omega^2 \times 0.3206 = 7.5 \times 9.81\,(0.2981 + 0.24 + 0.2642) + \frac{80 \times 9.81}{2}(0.2981 + 0.2)$$

$$\omega^2 = 339.4$$

$$\omega = \frac{2\,\pi\,N}{60} = 18.4$$

$$N = \underline{175.9 \text{ rpm}}$$

16.6 HARTNELL GOVERNOR

In this type of governor, the balls are controlled by a spring as shown in Fig. 16.6a. Initially, the spring is fitted in compression so that a force is applied to the sleeve. Two bell-crank levers, each carrying a mass at one end and a roller at the other, are pivoted to a pair of arms which rotate with the spindle. The rollers fit into a groove in the sleeve.

As the speed increases and the balls move away from the spindle axis, the bell-crank levers move on the pivot and lift the sleeve against the spring force. If the speed decreases, the sleeve moves downwards. The movement of the sleeve is communicated to the throttle of the engine. The spring force can be adjusted with the help of a screw cap.

Figue 16.6b shows the forces acting on the bell-crank lever in two positions (assuming that the sleeve moves up so that f is taken positive).

Let F = centrifugal force = $mr\omega^2$

F_S = spring froce

Taking moments about the fulcrum A,

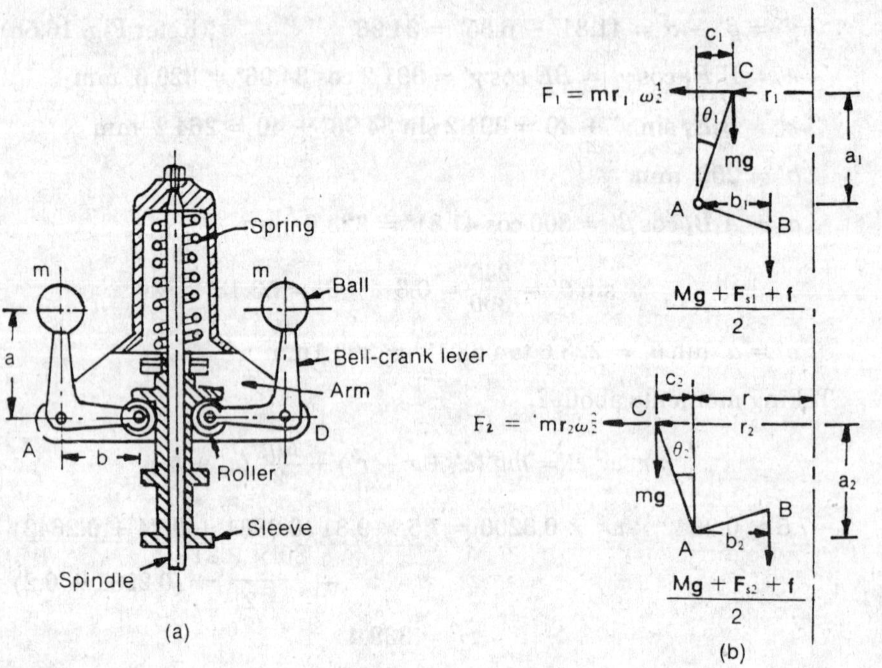

Fig. 16.6

$$F_1 a_1 = \frac{1}{2}(Mg + F_{S1} + f)\, b_1 + mg\, c_1 \tag{16.6}$$

$$F_2 a_2 = \frac{1}{2}(Mg + F_{S2} + f)\, b_2 - mg\, c_2 \tag{16.6a}$$

In the working range of the governor, θ is ussally small and so the obliquity effects of the arms of the bell-crank levers may be neglected. In that case,

$$a_1 = a_2 = a, \quad b_1 = b_2 = b, \quad c_1 = c_2 = 0$$

$$F_1 a = \frac{1}{2}(Mg + F_{S1} + f)\, b \tag{i}$$

and
$$F_2 a = \frac{1}{2}(Mg + F_{S2} + f)\, b \tag{ii}$$

Subtracting (i) from (ii)

$$(F_2 - F_1)\, a = \frac{1}{2}(F_{S2} - F_{S1})\, b$$

or
$$F_{S2} - F_{S1} = \frac{2a}{b}(F_2 - F_1)$$

Let s = stiffness of the spring

h_1 = movement of the sleeve

$$F_{S2} - F_{S1} = h_1 s = \frac{2a}{b}(F_2 - F_1)$$

or
$$(\theta \times b)\, s = \frac{2a}{b}(F_2 - F_1)$$

or
$$\frac{r_2 - r_1}{a} \times b\, s = \frac{2a}{b}(F_2 - F_1)$$

or
$$s = 2 \times \frac{a^2}{b^2}\left(\frac{F_2 - F_1}{r_2 - r_1}\right) \tag{16.7}$$

Example 16.5 The arms of a Hartnell governor are of equal length. When the sleeve is in the mid-position, the masses rotate in a circle of diameter 150 mm (the arms are vertical in the mid-position). Neglecting friction, the equilibrium speed for this position is 360 rpm. Maximum variation of speed, taking friction into account, is to be ± 6% of the mid-position speed for a maximum sleeve movement of 30 mm. The sleeve mass is 5 kg and the friction at the sleeve 35 N.

Assuming that the power of the governor is sufficient to overcome the friction by 1% change of speed on each side of the mid-position, find (neglecting obliquity effect of arms),

(i) the mass of each rotating ball
(ii) the spring stiffness
(iii) the initial compression of the spring.

Solution:

$$\omega = \frac{2\pi \times 360}{60} = 37.7 \text{ rad/s}$$

(i) Considering the friction at the mid-position,

$$mr\,\omega_1^2\, a = \frac{1}{2}(Mg + F_S + f)\, b$$

$$m \times \left(\frac{0.150}{2}\right) \times (37.7 \times 1.01)^2 = \frac{1}{2}(5 \times 9.81 + F_S + 35) \qquad (a = b) \quad \text{(i)}$$

and
$$mr\,\omega_2^2\, a = \frac{1}{2}(Mg + F_S - f)\, b$$

$$m \times \left(\frac{0.150}{2}\right) \times (37.7 \times 0.99)^2 = \frac{1}{2}[5 \times 9.81 + F_S - 35] \quad \text{(ii)}$$

Subtracting (ii) from (i)

$$m \times 0.075 \times (37.7)^2\,[(1.01)^2 - (0.99)^2] = \frac{1}{2} \times (35 + 35)$$

or $$\underline{(m = 8.21 \text{ kg})}$$

(ii) In the extreme positions,

$$mr_2\,\omega_2^2\,a = \frac{1}{2}\,(Mg + F_{S2} + f\,)\,b$$

$$8.21\times\left(0.075 + \frac{0.03}{2}\right)(37.7 \times 1.06)^2 = \frac{1}{2}\,(5 \times 9.81 + F_{S2} + 35) \qquad (a = b)$$

$$F_{S2} = 2275.8 \text{ N}$$

$$mr_1\,\omega_1^2\,a = \frac{1}{2}\,(Mg + F_{S1} - f)\,b$$

$$8.21\times\left(0.075 - \frac{0.03}{2}\right)(37.7 \times 0.94)^2 = \frac{1}{2}\,(5 \times 9.81 + F_{S1} - 35)$$

$$F_{S1} = 1223.2\text{N}$$

$$h_1\,s = F_{S2} - F_{S1}$$

$$0.03 \times s = 2275.8 - 1223.2$$

$$s = 35088 \text{ N/m} \quad \text{or} \quad \underline{35.088 \text{ N/mm}}$$

(iii) Initial compression $= \dfrac{F_{S1}}{s} = \dfrac{1223.2}{35.088} = \underline{34.86 \text{ mm}}$

Example 16.6 In a spring-loaded Hartnell type of governor, the mass of each ball is 4 kg and the lift of the sleeve is 40 mm. The governor begins to float at 200 rpm when the radius of the ball path is 90 mm. The mean working speed of the governor is 16 times the range of speed when friction is neglected. The lengths of the ball and roller arms of the bell-crank lever are 100 mm and 80 mm respectively. The pivot centre and the axis of governor are 115 mm apart. Determine the initial compression of the spring, taking into account the obliquity of arms.

Assuming the friction at the sleeve to be equivalent to a force of 15 N, determine the total alteration in speed before the sleeve begins to move from the mid-position.

Solution:

$m = 4$ kg $\qquad N_1 = 200$ rpm

$h_1 = 40$ mm $\qquad r_1 = 90$ mm

$a = 100$ mm $\qquad r = 115$ mm

$b = 80$ mm

$$\text{Mean speed, } N = \frac{N_1 + N_2}{2}$$

As $$N = 16\,(N_2 - N_1)$$

$$\therefore \qquad \frac{N_1 + N_2}{2} = 16\,(N_2 - N_1)$$

$$\text{or} \qquad \frac{200 + N_2}{2} = 16\,(N_2 - 200)$$

$$N_2 = 212.9 \text{ rpm}$$

Angle turned by bell-crank lever between two extreme positions

$$= \frac{\text{lift }(h_1)}{b} = \frac{c_1 + c_2}{a}$$

$$\text{or} \qquad c_1 + c_2 = h_1 \frac{a}{b} = 40 \times \frac{100}{80} = 50 \text{ mm}$$

$$\text{But} \qquad c_1 = r - r_1 = 115 - 90 = 25 \text{ mm}$$

$$c_2 = 50 - 25 = 25 \text{ mm}$$

$$r_2 = r + c_2 = 115 + 25 = 140 \text{ mm}$$

$$b_1 = b_2 = \sqrt{b^2 - (h/2)^2} = \sqrt{(80)^2 - (20)^2} = 77.46 \text{ mm}$$

$$a_1 = a_2 = \sqrt{(100)^2 - (25)^2} = 96.82 \text{ mm}$$

$$\omega_1 = \frac{2\,\pi \times 200}{600} = 20.94 \text{ rad/s}$$

$$\omega_2 = \frac{2\,\pi \times 212.9}{60} = 22.29 \text{ rad/s}$$

In the extreme positions,

$$m r_1\, \omega_1^2\, a_1 = \frac{1}{2} F_{S1}\, b_1 + mg\, c_1 \qquad (M = 0, f = 0)$$

$$4 \times 0.09 \times (20.94)^2 \times 0.096\,82 = \frac{1}{2} F_{S1} \times 0.077\,46 + 4 \times 9.81 \times 0.025$$

$$F_{S1} = 392.5 \text{ N}$$

$$m r_2\, \omega_2^2\, a_2 = \frac{1}{2} F_{S2}\, b_2 - mg\, c_2$$

$$4 \times 0.14 \times (22.29)^2 \times 0.096\,82 = \frac{1}{2} F_{S2} \times 0.077\,46 - 4 \times 9.81 \times 0.025$$

$$F_{S2} = 698 \text{ N}$$

$$h_1\, s = F_{S2} - F_{S1}$$

$$40 \times s = 698 - 392.5$$

$$s = 7.64 \text{ N/mm}$$

$$\text{Initial compression} = \frac{F_{S1}}{s} = \frac{392.5}{7.64} = \underline{51.37 \text{ mm}}$$

F_S at mid-position $= F_{S1} + 20\,s = 392.5 + 7.64 \times 20 = 545.3$ N

$$\text{Mean speed} = \frac{N_1 + N_2}{2} = \frac{212.9 - 200}{2} = 206.45 \text{ rpm}$$

At the mid-position, taking friction into account,

$$mr\omega^2\,a = \frac{1}{2}(F_S + f)\,b$$

$$4 \times 0.115 \times \omega^2 \times 0.1 = \frac{1}{2}(545.3 + 15) \times 0.08$$

$$\omega_1^2 = 487.2$$

$$\omega_1 = \frac{2\,\pi\,N_1}{60} = 22.07$$

$$N_1 = 210.8 \text{ rpm}$$

Also

$$mr\,\omega_2^2\,a = \frac{1}{2}(F_S - f)\,b$$

$$4 \times 0.115 \times \omega_2^2 \times 0.1 = \frac{1}{2}(545.3 - 15) \times 0.08$$

$$\omega_2^2 = 461.13$$

$$\omega_2 = \frac{2\,\pi\,N}{60} = 21.47$$

$$N_2 = 205.1 \text{ rpm}$$

Alteration in speed = 210.8 – 205.1 = 5.7 rpm

16.7 WILSON-HARTNELL GOVERNOR (RADIAL-SPRING GOVERNOR)

A Wilson-Hartnell governor is a spring-loaded type of governor. In this, two bell-crank levers are pivoted at the ends of two arms which rotate with the spindle (Fig. 16.7a). The vertical arms of the bell-crank levers support the two balls at their ends while the horizontal arms carry two rollers at their ends. The two balls are connected by two main springs arranged symmetrically on either side of the sleeve. While rotating, when the ball radius increases with the increase in speed, the springs exert an inward pull F_S on the balls and the rollers press against the sleeve which is raised, closing the throttle valve.

Usually, the main springs are not adjustable and, for this reason, an adjustable auxiliary spring is provided. It is attached to one end of a lever, the other end of which fits into a groove in the sleeve. The lever is pivoted at a fulcrum B. The auxiliary spring tends to keep the sleeve down so that it assists each main spring, i.e. main and the auxiliary springs are in tension simultaneously.

Let s = stiffness of each of the main springs

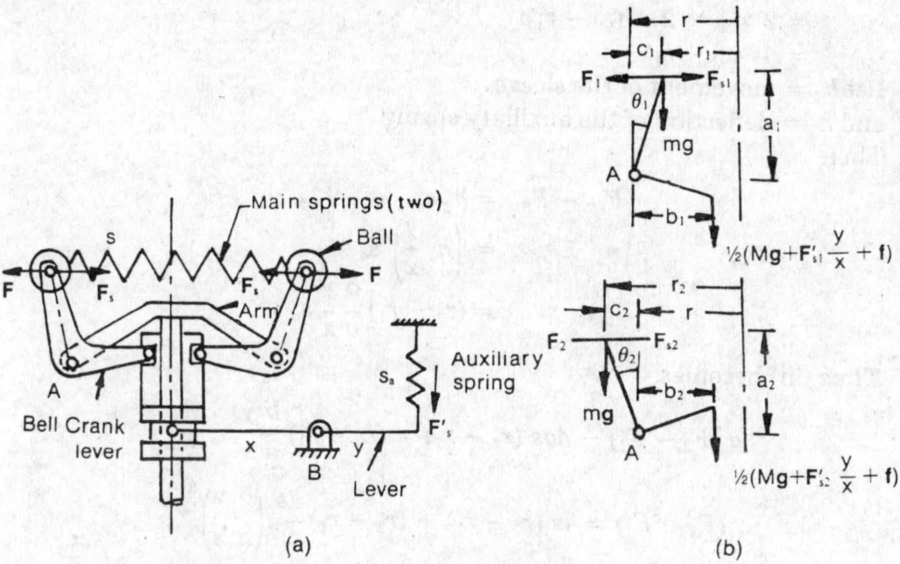

Fig. 16.7

s_a = stiffness of the auxiliary spring

F_S' = force applied by the auxiliary spring

Assuming that the sleeve moves up, take moments about the fulcrum A in two positions (Fig. 16.7b),

$$F_1\,a_1 - F_{S1}\,a_1 = \frac{1}{2}\left(Mg + F_{S1}'\frac{y}{x} + f\right) b_1 + mg\,c_1 \tag{16.8}$$

$$F_2\,a_2 - F_{S2}\,a_2 = \frac{1}{2}\left(Mg + F_{S2}'\frac{y}{x} + f\right) b_2 - mg\,c_2 \tag{16.8a}$$

If obliquity effects are neglected,

$$a_1 = a_2 = a, \quad b_1 = b_2 = b \quad \text{and} \quad c_1 = c_2 = 0$$

$$(F_1 - F_{S1})\,a = \frac{1}{2}\left(Mg + F_{S1}'\frac{y}{x} + f\right) b \tag{i}$$

$$(F_2 - F_{S2})\,a = \frac{1}{2}\left(Mg + F_{S2}'\frac{y}{x} + f\right) b \tag{ii}$$

Subtracting (i) form (ii),

$$a\,(F_2 - F_1) - a\,(F_{S2} - F_{S1}) = (F_{S2}' - F_{S1}')\,\frac{yb}{2x} \tag{iii}$$

The main spring consists of two springs. Therefore, the force exerted is given by,

$F_{S2} - F_{S1} = 2 \times$ Force exerted by each spring

$= 2 \times$ Stiffness of each spring $\times$ Elongation of each spring
$= 2 \times s \times 2 \times (r_2 - r_1)$
$= 4s\,(r_2 - r_1)$

Let h_1 = movement of the sleeve
and h_2 = deflection of the auxiliary spring
Then

$$F'_{S2} - F'_{S1} = h_2\, s_a$$
$$= \left(h_1 \frac{y}{x}\right) s_a$$
$$= (r_2 - r_1)\frac{b}{a}\frac{y}{x} s_a$$

Thus (iii) becomes,

$$a\,(F_2 - F_1) - 4as\,(r_2 - r_1) = (r_2 - r_1)\frac{b}{a}\frac{y}{x} s_a \frac{yb}{2x}$$

or
$$(F_2 - F_1) = 4s\,(r_2 - r_1) + (r_2 - r_1)\frac{s_a}{2}\left(\frac{b}{a}\frac{y}{x}\right)^2$$

or
$$\frac{F_2 - F_1}{r_2 - r_1} = 4s + \frac{s_a}{2}\left(\frac{b}{a}\frac{y}{x}\right)^2 \tag{16.9}$$

To find the stiffness of the main springs while using this equation, the stiffness of the auxiliary spring may be fixed first.

Example 16.7 In a Wilson-Hartnell type of governor, the mass of each ball is 5 kg. The lengths of the ball arm and the sleeve arm of each bell-crank lever are 100 mm and 80 mm respectively. The stiffness of each of the two springs attached directly to the balls is 0.4 N/mm. The lever for the auxiliary spring is pivoted at its midpoint. When the radius of rotation is 100 mm, the equilibrium speed is 200 rpm. If the sleeve is lifted by 8 mm for an increase of speed of 6%, find the required stiffness of the auxiliary spring.

Solution:

$m = 5$ kg $\qquad s = 0.4$ N/mm $= 400$ N/m
$r_1 = 100$ mm $\qquad a = 100$ mm
$N_1 = 200$ rpm $\qquad b = 80$ mm
$y/x = 1$

We have,

$$\frac{F_2 - F_1}{r_2 - r_1} = 4s + \frac{s_a}{2}\left(\frac{b}{a} \times \frac{y}{x}\right)^2$$

When $r_1 = 100$ mm, $N_1 = 200$ rpm.

$$\therefore \qquad \omega_1 = \frac{200 \times 2\pi}{60} = 20.94 \text{ rad/s}$$

$$F_1 = mr_1\,\omega_1^2 = 5 \times 0.1 \times (20.94)^2 = 219.2 \text{ N}$$

For 6% rise of speed,

$$\omega_2 = 20.94 \times 1.06 = 22.2 \text{ rad/s}$$

For sleeve rise of 8 mm,

$$\text{Increase in ball radius} = 8 \times \frac{100}{80} = 10 \text{ mm}$$

$$r_2 = 100 + 10 = 110 \text{ mm}$$

$$F_2 = mr_2\,\omega_2^2 = 5 \times 0.11 \times (22.2)^2 = 271.1 \text{ N}$$

$$\frac{271.1 - 219.2}{0.11 - 0.1} = 4 \times 400 + \frac{s_a}{2}\left(\frac{0.08}{0.1} \times 1\right)^2 \quad \text{[Refer Eq. (16.9)]}$$

$$s_a = 11\,219 \text{ N/m or } \underline{11.219 \text{ N/mm}}$$

16.8 SPRING-CONTROLLED GRAVITY GOVERNOR

In a spring-controlled gravity governor, two bell-crank levers are pivoted on the moving sleeve (Fig. 16.8a). The rollers at the ends of the horizontal arms of the levers press against a cap fixed to the governor shaft. Thus the motion of the pivots will be vertically upwards whereas the rollers will be able to move horizontally over the cap. As the speed increases, the balls move away, the

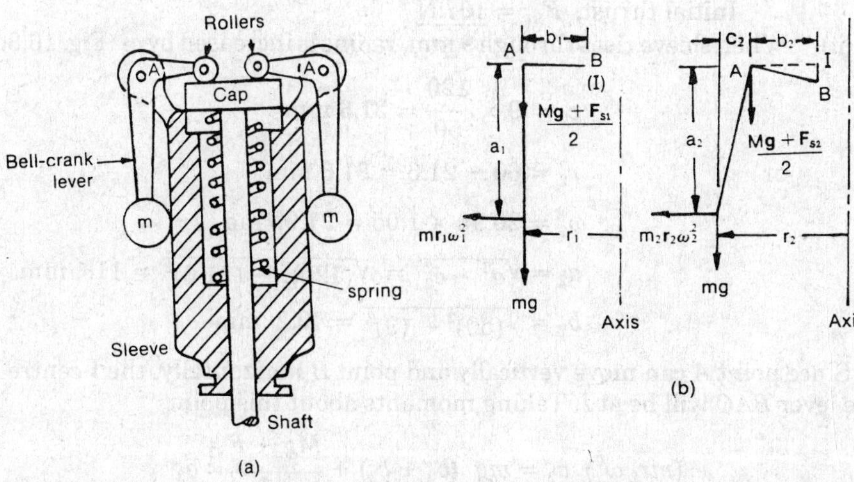

Fig. 16.8

pivots are raised and the spring is compressed between the sleeve and the cap.

Example 16.8 **In a spring-controlled gravity governor, the mass of each ball is 1.6 kg. Distance of fulcrum from the axis of rotation is 60 mm. The bell-crank lever has a vertical arm 120 mm long and a horizontal arm 50 mm long. The mass of the sleeve is 6.5 kg. The sleeve begins to rise at 200 rpm and the rise of sleeve for 5% increase is 9 mm. Determine the initial thrust in the spring and its stiffness.**

Solution:

$$m = 1.6 \text{ kg} \qquad N_1 = 200 \text{ rpm}$$
$$M = 6.5 \text{ kg} \qquad a = a_1 = 120 \text{ mm}$$
$$r_1 = 60 \text{ mm} \qquad b = b_1 = 50 \text{ mm}$$

$$\omega_1 = \frac{2\pi \times 200}{60} = 20.94 \text{ rad/s}$$

(i) For initial (neutral) position, taking moments about B, the I-centre (Fig. 16.8b),

$$mr_1 \omega_1^2 a_1 = mg\, b_1 + \frac{Mg + F_{S1}}{2} b_1$$

where F_{S1} is the spring load on the sleeve. The total sleeve load $(Mg + F_{S1})$ acts on the levers through the fulcrums $A - A$.
Thus,

$$[1.60 \times 0.06 \times (20.94)^2] \times 0.12 = \left[1.6 \times 9.81 + \frac{6.5 \times 9.81 + F_{S1}}{2}\right] \times 0.05$$

$\therefore$ Initial thrust, $F_{S1} = \underline{107 \text{ N}}$

(ii) When sleeve rises through 9 mm, radius is increased by c_2 (Fig. 16.8c),

$$c_2 = 9 \times \frac{120}{50} = 21.6 \text{ mm}$$

or $$r_2 = 60 + 21.6 = 81.6 \text{ mm}$$

$$\omega_2 = 20.94 \times 1.05 = 21.99 \text{ rad/s}$$

$$a_2 = \sqrt{a^2 - c_2^2} = \sqrt{(120)^2 - (21.6)^2} = 118 \text{ mm}$$

$$b_2 = \sqrt{(50)^2 - (9)^2} = 49.2 \text{ mm}$$

Since point A can move vertically and point B horizontally, the I-centre of the lever BAC will be at I. Taking moments about this point,

$$(mr_2 \omega_2^2)\, a_2 = mg\, (b_2 + c_2) + \frac{Mg + F_{S2}}{2} b_2$$

$$1.6 \times 0.0816 \times (21.99)^2 \times 0.118 = 1.6 \times 9.81\,(0.0492 + 0.0216) + \frac{6.5 \times 9.81 + F_{S2}}{2} \times 0.0492$$

$$F_{S2} = 193.8 \text{ N}$$

$$\text{Stiffness of spring} = \frac{F_{S2} - F_{S1}}{h_2} = \frac{193.8 - 107}{9} = \underline{9.6 \text{ N/mm}}$$

16.9 INERTIA GOVERNOR

As described earlier, an inertia governor is based on the principle of inertia of matter and is operated by the acceleration or deceleration of the rotating masses in addition to centrifugal forces.

In this type of governor, a mass m, having its centre at G, is fixed to an arm QG which is pivoted to a rotating disc on the engine shaft at Q. The points Q, G and the centre of rotation O are not to be collinear (Fig. 16.9). The arm QG is connected to an eccentric that operates the fuel supply valve. Whenever the arm moves relatively to the disc, it shifts the position of the eccentric which changes the fuel supply.

Let r = radial distance OG

ω = angular velocity of the disc

v = tangential velocity of G ($= \omega r$)

Centrifugal force of the rotating mass, $F = mr\,\omega^2$ (radially outwards)

If the engine shaft is accelerated due to increase in speed, the ball mass does not get accelerated at the same amount on account of its inertia, the inertia force being equal to

$$F_i = m f = m \frac{dv}{dt}$$

Moment of F about $Q = mr\,\omega^2 x$ (counter-clockwise)

Moment of F_i about $Q = m \dfrac{dv}{dt} y$ (counter-clockwise)

Thus it is seen that the moments due to the two forces add together to make the governor action rapid. Note that, as the mass moves outwards, the arm rotates in a direction opposite to that of the rotation of the shaft. In case the arm is arranged on the disc in the manner shown in Fig. 16.9b, the two moments due to F and F_i act in the opposite directions to make the governor action sluggish. This arrangement is, therefore, avoided.

It is also possible to use two ball masses fixed to the arm as shown in Fig. 16.9c. The arm is pivoted at its midpoint C. A change in the angular speed of the disc makes the ball masses to have an angular movement about C. If I_c is the moment of inertia of the arm and the masses about an axis through C, then

Torque on the arm $= I_c\, d\omega/dt$

Note that in an inertia governor, when the acceleration (or deceleration)

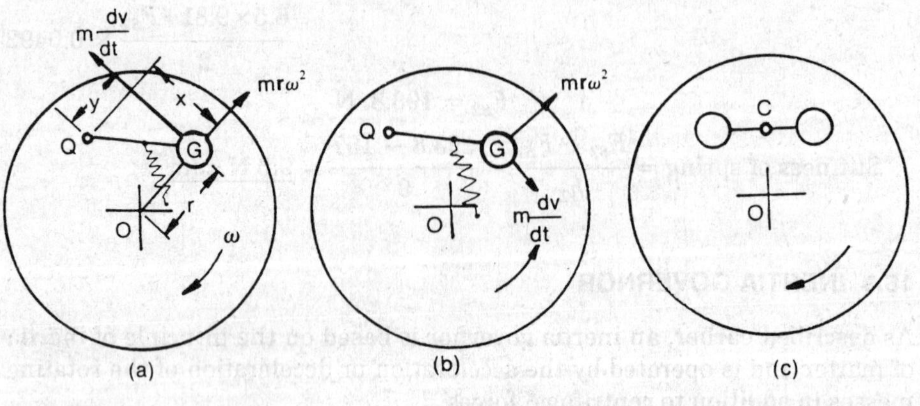

Fig. 16.9

is very small or the change in velocity is very slow, the additional inertia force is practically zero and an inertia governor in effect, becomes a centrifugal governor.

16.10 SENSITIVENESS OF A GOVERNOR

A governor is said to be sensitive when it readily responds to a small change of speed. The movement of the sleeve for a fractional change of speed is the measure of sensitivity.

As a governor is used to limit the change of speed of the engine between zero to full-load conditions, the sensitiveness of a governor is also defined as

$$\text{Sensitiveness} = \frac{\text{Mean speed}}{\text{Range of speed}}$$

$$= \frac{N}{N_2 - N_1}$$

$$= \frac{1}{2}\frac{N_2 + N_1}{N_2 - N_1} \qquad (16.10)$$

where N = mean speed

N_1 = minimum speed corresponding to full load conditions

N_2 = maximum speed corresponding to no-load conditions

16.11 HUNTING

Sensitiveness of a governor is a desirable quality. However, if a governor is too sensitive, it may fluctuate continuously, because when the load on the engine falls, the sleeve rises rapidly to a maximum position. This shuts off the fuel supply to the extent to affect a sudden fall in the speed. As the speed falls to below the mean value, the sleeve again moves rapidly and falls to a minimum position to increase the fuel supply. The speed subsequently rises and becomes more than the average with the result that the sleeve again falls to reduce the fuel supply. This process continues and is known as hunting.

16.12 ISOCHRONISM

A governor with a range of speed zero is known as an isochronous governor.

Sensitiveness $= \dfrac{\text{Mean speed}}{\text{Range of speed}} = \infty$ i.e. such a governor is infinitely sensitive. This means that for all positions of the sleeve or the balls, the governor has the same speed. Any change of speed results in moving the balls and the sleeve to their extreme positions. However, an isochronous governor is not practical due to friction at the sleeve.

For a Porter governor, with all arms equal in length and intersecting on the axis (neglecting friction),

$$h_1 = \frac{g}{\omega_1^2}\left(1 + \frac{M}{m}\right) \quad \text{and} \quad h_2 = \frac{g}{\omega_2^2}\left(1 + \frac{M}{m}\right)$$

For isochronism, $\omega_1 = \omega_2$ and thus $h_1 = h_2$. However, from the configuration of a Porter governor, it can be judged that it is impossible to have two positions of the balls at the same speed. Thus a pendulum type of governor cannot possibly be isochronous.

In the case of a Hartnell governor (neglecting friction),

At ω_1,

$$m r_1 \omega_1^2 a = \frac{1}{2}(Mg + F_{S1})\, b$$

At ω_2,

$$m r_2 \omega_2^2 a = \frac{1}{2}(Mg + F_{S2})\, b$$

For isochronism, $\omega_1 = \omega_2$.

$$\therefore \quad \frac{Mg + F_{S1}}{Mg + F_{S2}} = \frac{r_1}{r_2} \tag{16.11}$$

Which is the required condition of isochronism.

16.13 STABILITY

A governor is said to be stable if it brings the speed of the engine to the required value and there is not much hunting. The ball masses occupy a definite position for each speed of the engine within the working range.

Obviously, the stability and the sensitivity are two opposite characteristics.

16.14 EFFORT OF A GOVERNOR

The effort of the governor is the mean force acting on the sleeve to raise or lower it for a given change of speed. At constant speed, the governor is in equilibrium and the resultant force acting on the sleeve is zero. But when the speed of the governor increases or decreases, a force is required to be exerted on the sleeve which tends to move it. When the sleeve occupies a new steady position, the resultant force acting on it again becomes zero.

If the force acting at the sleeve changes gradually from zero (when the governor is in the equilibrium position) to a value E for an increased speed of the governor, the mean force or the effort is $E/2$.

For a Porter governor, the height is given by

$$h = \frac{g}{\omega^2} + \frac{Mg\,(1+k)}{2m\,\omega^2} = \frac{2mg + Mg\,(1+k)}{2m\,\omega^2} \qquad \text{(i)}$$

Let ω be increased to c times ω, and E be the force applied on the sleeve to prevent it from moving. Thus, the force on the sleeve is increased to $(Mg + E)$. Then

$$h = \frac{2mg + (Mg+E)\,(1+k)}{2m\,(c\,\omega)^2} \qquad \text{(ii)}$$

Dividing (ii) by (i),

$$\frac{2mg + (Mg+E)\,(1+k)}{2mg + Mg\,(1+k)} = \frac{c^2}{1}$$

or

$$\frac{[2mg + (Mg+E)\,(1+k)] - [2mg + Mg\,(1+k)]}{2mg + Mg\,(1+k)} = \frac{c^2-1}{1}$$

or

$$\frac{E\,(1+k)}{2mg + Mg\,(1+k)} = c^2 - 1$$

or

$$E = \frac{c^2-1}{(1+k)}\left[2mg + Mg\,(1+k)\right]$$

Effort

$$\frac{E}{2} = \frac{c^2-1}{1+k}\left[mg + \frac{Mg}{2}(1+k)\right] \qquad (16.12)$$

If $k = 1$,

$$\text{Effort, } \frac{E}{2} = \frac{c^2 - 1}{2}(m + M)\,g \tag{16.13}$$

For a Watt governor, $M = 0$,

$$\text{Effort, } \frac{E}{2} = \frac{c^2 - 1}{2} \times mg \tag{16.14}$$

Thus, the effort of a Watt governor is less than that of a Porter governor.

Note: Sometimes effort is defined as the force required to be applied for 1% change in speed, i.e.

$$\text{Effort} = \frac{(1.01)^2 - 1}{2}(m + M)\,g$$

$$= 0.01(m + M)\,g$$

In a Hartnell governor,

$$mr\,\omega^2\,a = \frac{1}{2}(Mg + F_S)\,b \tag{iii}$$

Let E be the force applied on the sleeve to prevent its movement when the speed changes from ω to $c\,\omega$.

$$mr\,(c\,\omega)^2\,a = \frac{1}{2}(Mg + E + F_S)\,b \tag{iv}$$

Dividing (iii) by (iv),

$$\frac{1}{c^2} = \frac{Mg + F_S}{Mg + E + F_S}$$

or $$\frac{Mg + E + F_S}{Mg + F_S} = c^2$$

or $$\frac{E}{Mg + F_S} = (c^2 - 1)$$

or $$\text{Effort, } \frac{E}{2} = \frac{(c^2 - 1)}{2}(Mg + F_S). \tag{16.15}$$

16.15 POWER OF A GOVERNOR

The power of a governor is the work done at the sleeve for a given percentage change of speed, i.e. it is the product of the effort and the displacement of the sleeve.

For a Porter governor, having all equal arms which intersect on the axis or pivoted at points equidistant from the spindle axis,

$$\text{Power} = \frac{E}{2} \times (2 \times \text{height of governor})$$

If height of the governor changes from h to h_1 when the speed changes from ω to $c\,\omega$.

$$h = \frac{2m + Mg\,(1 + k)}{2m\,\omega^2} \quad \text{and} \quad h_1 = \frac{2m + Mg\,(1 + k)}{2m\,(c\,\omega)^2}$$

or

$$\frac{h_1}{h} = \frac{1}{c^2}$$

$$\therefore \text{Displacement of sleeve} = 2\,(h - h_1)$$

$$= 2h\left(1 - \frac{h_1}{h}\right)$$

$$= 2h\left(1 - \frac{1}{c^2}\right)$$

$$= 2h\left(\frac{c^2 - 1}{c^2}\right)$$

$$\therefore \quad \text{Power} = \frac{c^2 - 1}{2}\,(m + M)\,g \times 2h\left(\frac{c^2 - 1}{c^2}\right)$$

$$= (m + M)\,g\,h\left(\frac{c^2 - 1}{c}\right)^2 \qquad (16.16)$$

16.16 CONTROLLING FORCE

When the balls of a governor rotate in their circular path, the centrifugal force on each ball tends to move it outwards. This is resisted by an equal and opposite force acting radially inwards and is known as the controlling force.

The controlling force is supplied by the weight of the rotating mass in a Watt governor, the weight of the mass and that of the sleeve in a Porter governor and by the compressed spring in the case of a Hartnell governor.

A graph showing the variation of the controlling force with the radius of rotation is called the controlling curve or diagram. This curve is useful in finding out the stability of a governor as discussed below.

$$\text{Controlling force} = \tan\theta\left[mg + \frac{Mg \pm f}{2}\,(1 + k)\right] \quad \text{for a Porter governor}$$

$$= \frac{1}{2}(Mg + F_S \pm f)\,\frac{b}{a} \quad \text{for a Hartnell governor}$$

From the above relations, the values of the controlling force may be calculated for different radii of the ball. Figure 16.10 a shows AB as the controlling force curve (neglecting friction) plotted against the ball radius. Alternatively, as the controlling force is equal and opposite to the centrifugal force, it may be computed from the relation $F = mr\omega^2$ for different radii and the corresponding speeds. This relation also indicates that, for a particular speed, the controlling force is proportional to the radius. Thus a number of

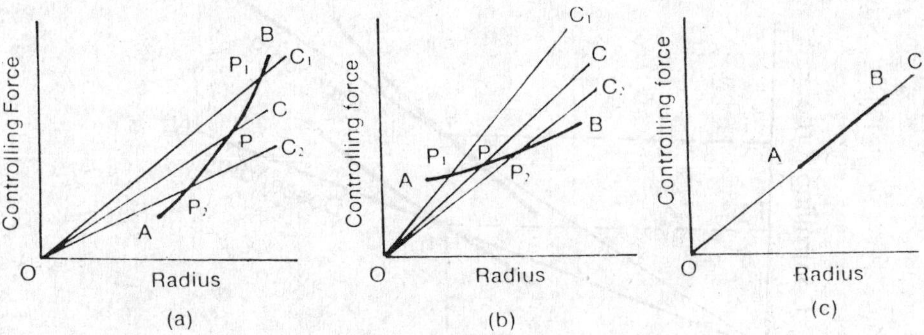

Fig. 16.10

lines, such as OC, OC_1, OC_2, etc. may be drawn on the same diagram providing the values of the controlling force for different radii at particular speeds. The intersection of the speed curves with the controlling force curve provides the speeds of the governor corresponding to the radii.

Suppose that the point P represents the mean speed of the governor. r is the corresponding radius of the balls. Now, if the speed increases to P_1 the radius of the balls increases to r_1, thus moving the sleeve up and closing the throttle valve to the required extent. Similarly, if the speed decreases to point P_2, the radius of the balls decreases to r_2 resulting in the lowering of the sleeve and opening the throttle valve further. This would increase the speed. This type of governor is said to be stable.

Now consider a controlling force curve of the type shown in Fig. 16.10 b. In this case, the point P again represents the mean speed of the governor. If the speed increases to P_1, the radius of the balls decreases to r_1. This means that the sleeve is lowered and the throttle valve is further opened to increase the fuel supply and consequently the speed. Similarly, on decreasing the speed, the sleeve is moved up, closing the valve and thus further reducing the speed. Such a governor is therefore unstable.

Thus, for a governor to be stable the slope of the controlling force curve must be greater than that of the speed curve.

Figure 16.10 c shows a controlling force curve AB which sometimes may be obtained in some spring-loaded governors by suitable adjustments. It can be observed that, at the speed represented by the line OC, the balls can take up any radius. Under such conditions, the governor is said to be isochronous.

If friction is taken into account, two more curves of the controlling force are obtained as shown in Fig. 16.11. Thus, in all, three curves of the controlling force are obtained as follows:

(i) For steady run (neglecting friction)
(ii) While the sleeve moves up (f positive)
(iii) While the sleeve moves down (f negative)

The vertical intercept gh signifies that between the speeds corresponding to gh, the radius of the balls does not change while the direction of movement

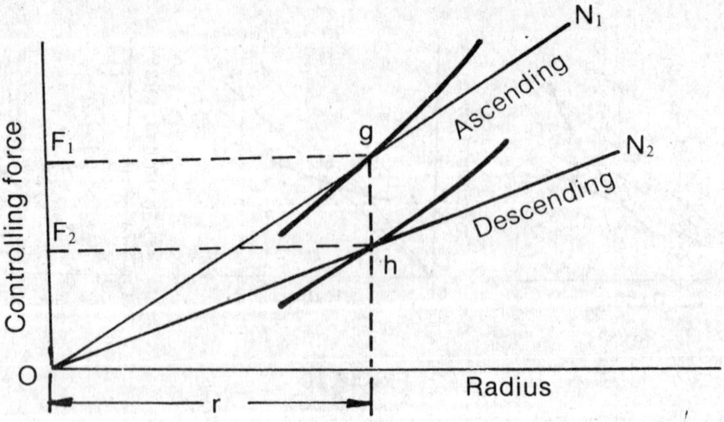

Fig. 16.11

of the sleeve does. In other words, between speeds N_1 and N_2, the governor is insensitive. At all radii of the balls within the range, there are two speeds for no change of the radius.

Coefficient of Insensitiveness: $\dfrac{N_1 - N_2}{N}$ is known as the coefficient of insensitiveness where N is the corresponding speed neglecting friction.

Example 16.9 In a spring-controlled governor, the controlling force curve is a straight line. The balls are 400 mm apart when the controlling force is 1500 N and 240 mm when it is 800 N. The mass of each ball is 10 kg. Determine the speed at which the governor runs when the balls are 300 mm apart. By how much should the initial tension be increased to make the governor isochronous? Also find the isochronous speed.

Solution: (i) The controlling force curve of a spring-controlling governor is a straight line and thus can be expressed as

(i) $F = ar + b$

where r = 200 mm and F = 1500 N

$$1500 = 0.2\,a + b \qquad \text{(i)}$$

When r = 120 mm, F = 800 N

$$800 = 0.12\,a + b \qquad \text{(ii)}$$

From (i) and (ii),

$$0.08a = 700$$

or $a = 8750$ and $b = -250$

$$F = mr\omega^2 = ar + b$$

When $r = \dfrac{300}{2} = 150$ mm,

$$10 \times 0.15 \times \omega^2 = 8750 \times 0.15 + (-250)$$

$$\therefore \qquad \omega^2 = 708.3$$

$$\omega = \frac{2\,\pi\,N}{60} = 26.6$$

$$N = \underline{254.2 \text{ rpm}}$$

(ii) To make the governor isochronous, the controlling force line must pass through the origin, i.e. b is to be zero. This is possible only if the initial tension is increased by 250 N (Refer to Fig. 16.12).

(iii) $F = mr\omega^2 = ar + b$

$$10 \times r \times \omega^2 = 8750\,r + 0$$

$$\therefore \qquad \omega^2 = 875$$

$$\omega = \frac{2\,\pi\,N}{60} = 29.58$$

$$N = \underline{282.5 \text{ rpm}}$$

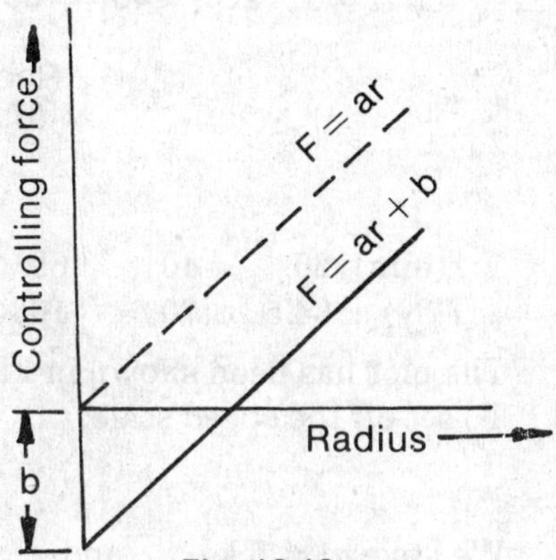

Fig. 16.12

Example 16.10 In a Porter governor, each arm is 200 mm long and is pivoted at the axis of rotation. The mass of each ball is 5 kg and the load on the sleeve is 30 kg. The extreme radii of rotation are 80 mm and 140 mm. Plot a graph of the controlling force vs. radius of rotation and set off a speed scale along the ordinate corresponding to a radius of 160 mm.

Solution: Controlling force of a Porter governor (Refer Fig. 16.27),

$$F = \frac{r}{h}\left[mg + \frac{Mg \pm f}{2}(1 + k)\right]$$

In this case, $k = 1$ and $f = 0$

$$\therefore \qquad F = \frac{r}{h}(m + M)\,g = \frac{(m + M)\,g}{l^2 - r^2}\,r$$

We have $m = 5$ kg, $M = 30$ kg, $l = OA = 200$ mm

$$\therefore \qquad F = \frac{(5 + 30) \times 9.81}{\sqrt{(200)^2 - r^2}}\,r = 343.35 \times \frac{r}{\sqrt{(200)^2 - r^2}}$$

Prepare the following table for different values of r and the corresponding force.

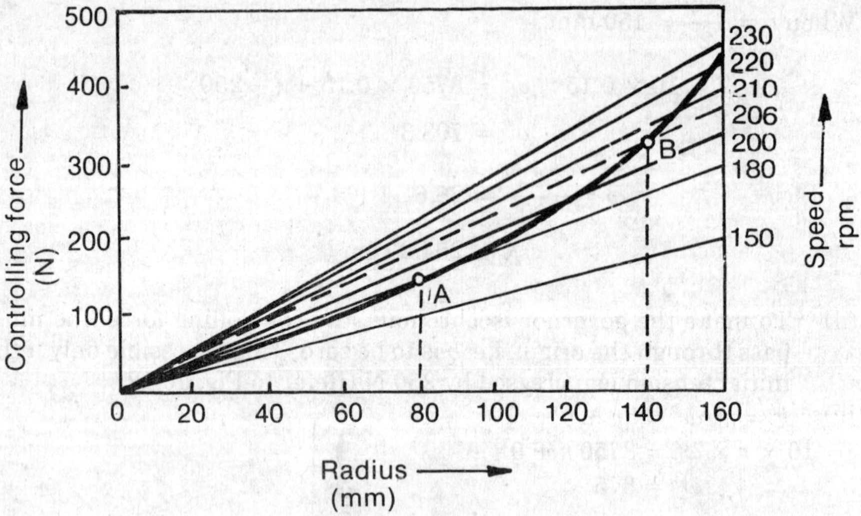

Fig. 16.13

r(mm)	20	40	60	80	100	120	140	160
F(N)	34.5	70	108	150	198	258	336	458

The plot has been shown in Fig. 16.13.

To set off the speed scale,

$$F = mr\omega^2$$

We have $m = 5$ kg and $r = 160$ mm

$$F = 5 \times 0.160 \times \left(\frac{2\,\pi\,N}{60}\right)^2 = 0.008\,77\ N^2$$

The following table can be prepared:

N(rpm)	100	150	160	170	180	190	200	210	220	230
F(N)	87.7	197.3	224	253	284	317	351	387	425	464

The speed scale can now be marked on the graph as shown in Fig. 16.13.

To obtain the range of equilibrium speeds, draw vertical lines through $r = 80$ mm and 140 mm meeting the controlling force curve at A and B respectively. Draw straight lines from the origin and through points A and B. It can be observed that the points A and B correspond to speeds 150 and 190 rpm respectively.

The range of speed is from 180 to 206 rpm

PROBLEMS

16.1 Figure 16.1 shows three forms of a watt governor. In Fig. 16.1a, length OA = 640 mm, in Fig. 16.1b, EA = 480 mm, EF = 160 mm and in Fig. 16.1c, EA =

800 mm, EF = 160 mm. The angle θ = 30° in each case. Show that for the given configurations, the speed of rotation is the same. What will be the percentage change in speed for a 50 mm rise in the level of the balls?

(4.8%; 9.8%; 1.8%)

16.2 Each arm of a Porter governor is 300 mm long and is pivoted on the axis of rotation. Each ball has a mass of 6 kg and the sleeve weighs 18 kg. The radius of rotation of the ball is 200 mm when the governor begins to lift and 250 mm when the speed is maximum. Determine the maximum and the minimum speed and the range of speed of the governor.

(146.8 rpm; 126.4 rpm; 20.4 rpm)

16.3 Each arm of a Porter governor is 400 mm long. The upper arms are pivoted on the axis of the sleeve and the lower arms are attached to the sleeve at a distance of 40 mm from the axis. Each ball has a mass of 6 kg and the weight on the sleeve is 50 kg. Find the range of speed of the governor if the extreme radii of rotation of the balls are 260 mm and 300 mm.

(10.6 rpm)

16.4 The mass of each ball of a Proell governor is 3 kg and the weight on the sleeve is 20 kg. Each arm is 220 mm long and the pivots of the upper and the lower arms are 20 mm from the axis. For the midposition of the sleeve, the extension link of the lower arm is vertical, and the radius of the path of rotation of the masses is 160 mm. The height of the governor is 180 mm. The governor speed is 150 rpm when in midposition. Determine the length of the extension link and the tension in the upper arm.

(125 mm; 137.9 N)

16.5 The following data relate to a Hartnell governor (Fig. 16.6a):
m = 1.5 kg; a = 100 mm; b = 40 mm; r_1 = 70 mm; r_2 = 110 mm; N_1 = 260 rpm; and N_2 = 275 rpm.
The axis of rotation is 80 mm from the fulcrum. Calculate the rate of the spring and the equilibrium speed when the radius of rotation of the balls is 80 mm.

(18.44 N/mm; 265.3 rpm)

16.6 The following data refer to a Wilson-Hartnell governor:
Mass of each ball = 2.5 kg
Minimum speed = 210 rpm
Maximum speed = 220 rpm
Minimum radius = 120 mm
Maximum radius = 160 mm
Length of ball arm of each bell-crank lever = 180 mm
Length of sleeve arm of each bell-crank lever = 120 mm
Combined stiffness of the two ball springs = 300 N/m
Determine the equivalent stiffness of the auxiliary spring referred to the sleeve.
[Equivalent stiffness of the auxiliary spring referred to the sleeve = $S_a\ (y/x)^2$]

(5.423 kN/m)

16.7 The mass of each ball of a spring controlled gravity governor is 1.4 kg. The bell-crank lever has its vertical arm 90 mm long and the horizontal arm 40 mm long. The distance of the fulcrum from the axis of rotation is 45 mm. The sleeve has a mass of 7.5 kg. The sleeve begins to rise at 220 rpm and the rise of sleeve for 6% rise is 8 mm. Find the initial thrust in the spring and its stiffness.

(49.45 N; 9.21 N/mm)

16.8 A Porter governor has equal arms. Each arm is 240 mm long and is pivoted on the axis of rotation. Each ball has a mass of 5 kg and the load on the sleeve is

18 kg. The ball radius is 150 mm when the sleeve begins to rise and 200 mm at the maximum speed. Find the range of speed. Also determine the coefficient of insensitivenes if the friction at the sleeve is equivalent to a force of 10 N.

(27.5 rpm; 0.044)

17

GYROSCOPE

17.1 ANGULAR VELOCITY

The angular velocity of a rotating body is specified by

- the magnitude of velocity
- the direction of the axis of rotor
- sense of rotation of the rotor, i.e. clockwise or counter-clockwise.

Angular velocity is represented by a vector in the following manner:

(i) Magnitude of the velocity is represented by the length of the vector.

(ii) Direction of axis of the rotor is represented by drawing the vector parallel to the axis of the rotor or normal to the plane of the angular velocity.

(iii) Sense of rotation of the rotor is denoted by taking the direction of the vector in a set rule. The general rule is that of a right-handed screw, i.e. if a screw is rotated in the clockwise direction, it goes away from the viewer and vice-versa.

For example, Fig. 17.1a shows a rotor which rotates in the clockwise direction when viewed from end A. Its angular motion has been shown vectorially in Fig. 17.1b. The vector has been taken to a suitable scale parallel to the axis of the rotor. The sense of direction of the vector is from **a** to **b**

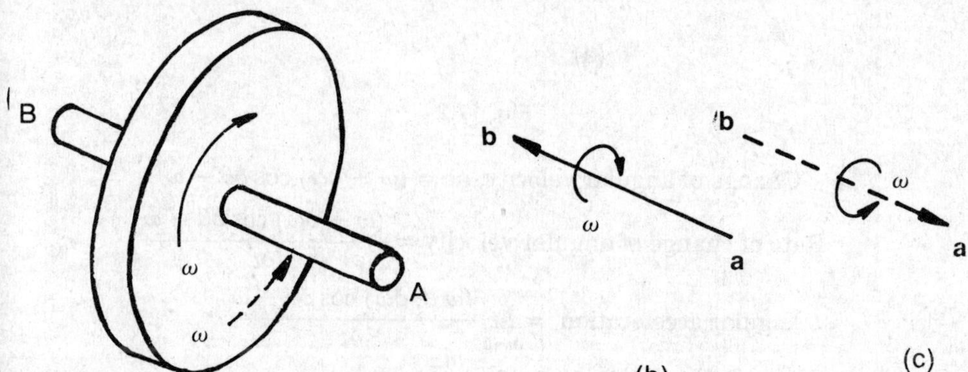

Fig. 17.1

according to the screw rule. However, if the direction of rotation of the rotor is reversed, it would be from **b** to **a** (Fig. 17.1c).

17.2 ANGULAR ACCELERATION

Let a rotor spin (rotate) about the horizontal axis *ox* at a speed of ω rad/s in the direction as shown in Fig. 17.2a. Let **oa** represent its angular velocity (Fig. 17.2 b).

Now, if the magnitude of the angular velocity changes to $(\omega + \delta\omega)$ and the direction of the axis of spin to Ox' (in time δt), the vector **ob** would represent its angular velocity in the new position. Join **ab** which represents the change in the angular velocity of the rotor. The vector **ab** can be resolved into two components,

(i) **ac** representing angular velocity change in a plane normal to **ac** or x-axis and

(ii) **cb** representing angular velocity change in a plane normal to **cb** or y-axis.

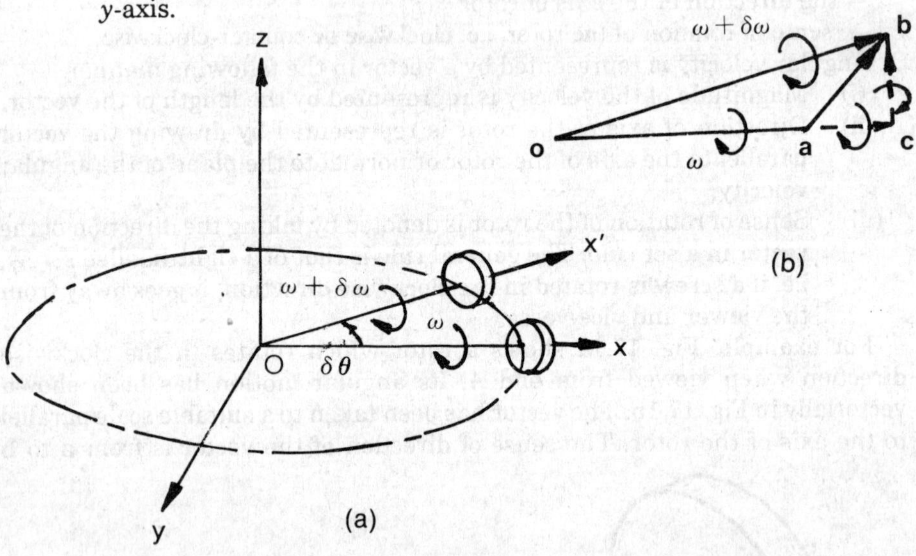

Fig. 17.2

$$\text{Change of angular velocity, } \mathbf{ac} = (\omega + \delta\omega)\cos\delta\theta - \omega$$

$$\text{Rate of change of angular velocity} = \frac{(\omega + \delta\omega)\cos\delta\theta - \omega}{\delta t}$$

$$\therefore \text{ Angular acceleration } = \underset{\delta t \to 0}{Lt} \frac{(\omega + \delta\omega)\cos\delta\theta - \omega}{\delta t}$$

As $\delta t \to 0$, $\delta\theta \to 0$ and $\cos\delta\theta \to 1$

$$\therefore \text{ Angular acceleration} = \underset{\delta t \to 0}{Lt} \frac{\omega + \delta\omega - \omega}{\delta t} = \frac{d\omega}{dt}$$

Change of angular velocity, **cb** = $(\omega + \delta\omega) \sin \delta\theta$

$$\text{Rate of change of angular velocity} = \frac{(\omega + \delta\omega) \sin \delta\theta}{\delta t}$$

$$\therefore \text{ Angular acceleration} = \underset{\delta t \to 0}{Lt} \frac{(\omega + \delta\omega) \sin \delta\theta}{\delta t}$$

As $\delta t \to 0$, $\delta\theta \to 0$ and $\sin \delta\theta \to \delta\theta$

$$\therefore \text{ Angular acceleration} = \underset{\delta t - 0}{Lt} \frac{(\omega + \delta\omega)\, \delta\theta}{\delta t} = \omega \frac{d\theta}{dt}$$

$$\text{Total angular acceleration, } \alpha = \frac{d\omega}{dt} + \omega \frac{d\theta}{dt} \tag{17.1}$$

This shows that the total angular acceleration of the rotor is the sum of,

(i) $d\omega/dt$, representing change in the magnitude of the angular velocity of the rotor.

(ii) $\omega d\theta/dt$, representing change in the direction of the axis of spin. The direction of **cb** is from **c** to **b** in the vector diagram (being a component of **ab**), the acceleration acts clockwise in the vertical plane *xz* (when viewed from front along the *y*-axis).

17.3 GYROSCOPIC TORQUE (COUPLE)

Let *I* be the moment of inertia of a rotor and ω its angular velocity about a horizontal axis of spin *Ox* in the direction as shown in Fig. 17.3 a. Let this axis of spin turn through a small angle $\delta\theta$ in the horizontal plane (*xy*) to the position *Ox'* in time δt.

Figure 17.3 b shows the vector diagram. **oa** represents the angular velocity vector when the axis is *Ox* and **ob** when the axis is changed to *Ox'*. Then **ab** represents the change in the angular velocity due to change in direction of the axis of spin of the rotor. This change in the angular velocity is clockwise when viewed from **a** towards **b** and is in the vertical plane *xz*. This change results in an angular acceleration, the sense and direction of which are the same as that of the change in the angular velocity.

$$\text{Change in angular velocity, } \mathbf{ab} = \omega \times \delta\theta$$

$$\text{Angular acceleration, } \alpha = \omega \frac{\delta\theta}{\delta t}$$

In the limit, when $\delta t \to 0$, $\alpha = \omega \dfrac{d\theta}{dt}$

Usually, $d\theta/dt$, the angular velocity of the axis of spin is called the angular velocity of precession and is denoted by ω_p.

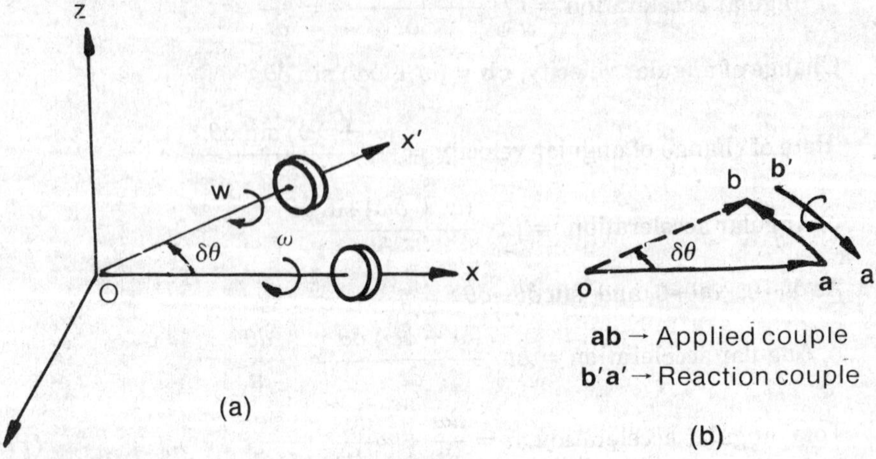

Fig. 17.3

$\therefore$ Angular acceleration $= \omega\, \omega_p$

Torque required to produce this acceleration is known as gyroscopic torque and is a couple which must be applied to the axis of spin to cause it to rotate with angular velocity ω_p about the axis of precession Oz.

$$\text{Acceleration torque, } T = I\,\alpha$$

$$= I\,\omega\,\omega_p \qquad (17.2)$$

For the configuration of Fig. 17.3 a,
Ox is known as the axis of spin
Oz is known as the axis of precession
Oy is known as the axis of gyroscopic couple
yz is the plane of spin (parallel to plane of rotor)
xy is the plane of precession
xz is the plane of gyroscopic couple

The torque obtained above is that which is required to cause the axis of spin to precess in the horizontal plane and is known as the active gyroscopic torque or the applied torque. A reactive gyroscopic torque or reaction torque is also applied to the axis which tends to rotate the axis of spin in the opposite direction, i.e. in the counter-clockwise direction in the above case. Just as the centrifugal force on a rotating body tends to move the body outwards while a centripetal acceleration (and thus centripetal force) acts on it inwards, in the same way, the effects of active and reactive gyroscopic torques can be understood.

The effect of the gyroscopic couple on a rotating body is known as the gyroscope effect on the body. A gyroscope is a spinning body which is free to move in other directions under the action of external forces.

Example 17.1 A uniform disc having a mass of 8 kg and a radius of gyration of 150 mm is mounted on one end of a horizontal arm of length 200 mm. The other end of the arm can rotate freely in a universal bearing. The disc is given a clockwise spin of 250 rpm as seen from the disc end of the arm. Determine the motion of the disc if the arm remains horizontal.

Solution:

$m = 8$ kg $\qquad l = 0.2$ m

$k = 0.15$ m $\qquad N = 240$ rpm

$$I = mk^2 = 8 \times (0.15)^2 = 0.18 \text{ kg m}^2$$

$$\omega = \frac{2\,\pi \times 240}{60} = 25.13 \text{ rad/s}$$

$$C = I\,\omega\,\omega_p$$

$$mg\,l = I\,\omega\,\omega_p$$

$$8 \times 9.81 \times 0.2 = 0.18 \times 25.13 \times \omega_p$$

$$\omega_p = 3.47 \text{ rad/s}$$

As the disc rotates, the weight of the disc acts downwards and thus a couple about the *y*-axis, in the clockwise direction is applied on the disc (Fig.17.4). The reaction couple is thus counter-clockwise which tends to keep the arm horizontal.

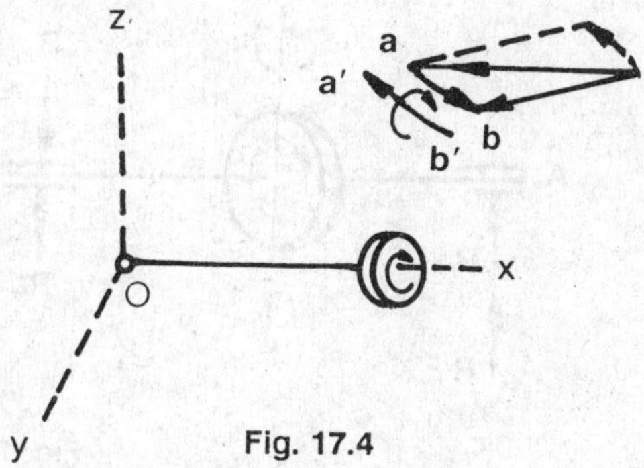

Fig. 17.4

Assuming that the axis of spin *Ox* precesses about the *z*-axis in the counter-clockwise direction, the vector **oa** would rotate to the position **ob** in a short period. Then **ab** is the applied couple and **b′ a′** is the reaction couple which is clockwise and tends to give the arm a clockwise rotation about the *y*-axis which is not true. Thus the axis of spin *Ox* must precess about the *z*-axis in the clockwise direction with an angular velocity of 3.47 rad/s.

Example 17.2 A disc with radius of gyration 60 mm and a mass of 4 kg is mounted centrally on a horizontal axle of 80 mm length between the bearings. It spins about the axle at 800 rpm counter-clockwise when viewed from the right-hand side bearing. The axle precesses about a vertical axis at 50 rpm in the clockwise direction when viewed from above. Determine the resultant reaction at each bearing due to the mass and the gyroscopic effect.

Solution:

$m = 4$ kg $\qquad N = 800$ rpm

$k = 0.06$ m $\qquad N_p = 50$ rpm

$$I = mk^2 = 4 \times (0.06)^2$$

$$= 0.0144 \text{ kg . m}^2$$

$$l = 80 \text{ mm} = 0.08 \text{ m}$$

$$\omega = \frac{2\pi \times 800}{60} = 83.78 \text{ rad/s}$$

$$\omega_p = \frac{2\pi \times 50}{60} = 5.24 \text{ rad/s}$$

$$\therefore \quad C = I\,\omega\,\omega_p$$

$$= 0.0144 \times 83.78 \times 5.24 = 6.32 \text{ N.m}$$

The applied (active) and reaction couples are shown in Fig. 17.5. The reaction couple is clockwise when viewed from front and tends to raise the bearing A and lower the bearing B.

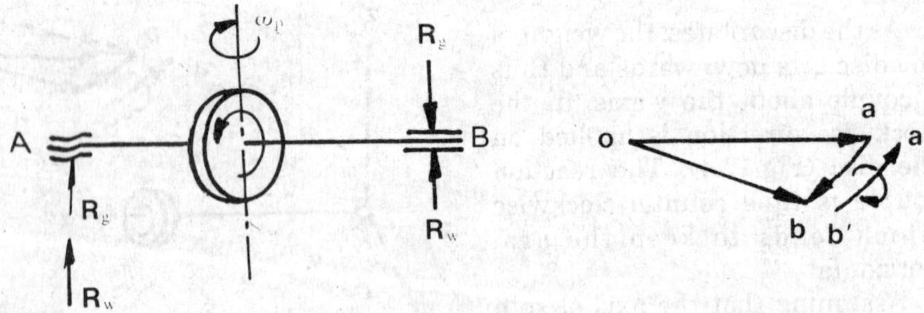

Fig. 17.5

Force at bearing A due to gyro. couple $= \dfrac{C}{l} = \dfrac{6.32}{0.08} = 79$ N (upwards)

Force at bearing B due to gyro. couple = 79 N (downwards)

Force at each bearing due to weight of the disc $= \dfrac{4 \times 9.81}{2} = 19.6$ N

or Reaction at each bearing due to weight = 19.6 N (upwards)

∴ Reaction at bearing A = 79 + 19.6 = 98.6 N (upwards)

Reaction at bearing B = 79 – 19.6 = 59.4 N (downwards)

Example 17.3 An aeroplane flying at 240 km/hr turns towards left and completes a quarter circle of 60 m radius. The mass of the rotary engine and the propeller of the plane amounts to 450 kg with a radius of gyration of 320 mm. The engine speed is 2000 rpm clockwise when viewed from the rear. Determine the gyroscopic couple on the aircraft and state its effect.

In what way is the effect changed when

(i) the aeroplane turns towards right

(ii) the engine rotates clockwise when viewed from the front (nose end) and the aeroplane turns (a) left (b) right?

Solution:

$$m = 450 \text{ kg} \qquad \omega = \frac{2\pi \times 2000}{60} = 209.4 \text{ rad/s}$$

$$k = 0.32 \text{ m} \qquad v = \frac{240 \times 10^3}{3600} = 66.67 \text{ m/s}$$

$$I = mk^2 = 450 \times (0.32)^2 = 46.08 \text{ kg.m}^2$$

$$\omega_p = \frac{v}{r} = \frac{66.67}{60} = 1.11 \text{ rad/s}$$

$$C = I\,\omega\,\omega_p = 46.08 \times 209.4 \times 1.11$$

$$= 10\,713 \text{ N.m} = 10.713 \text{ kN.m}$$

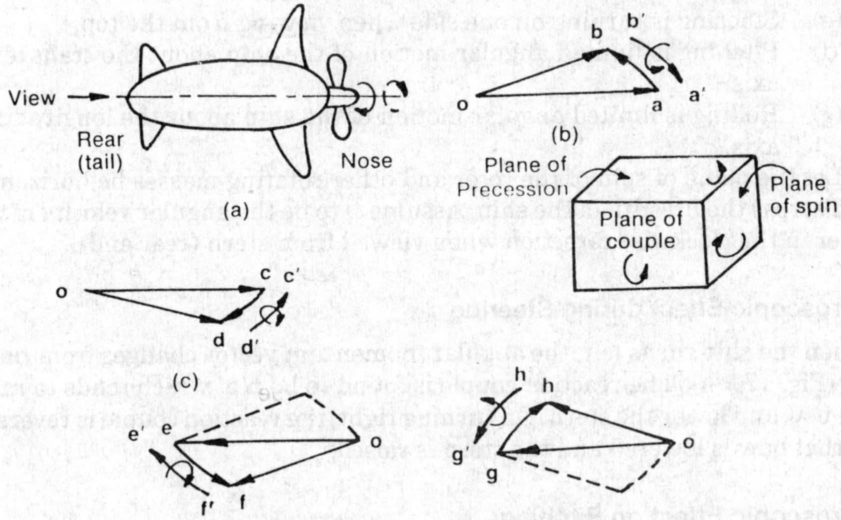

Fig. 17.6

Figure 17.6 a shows the aeroplane in space. When it turns left,
oa is the angular momentum vector before turning (Fig. 17.6 b);
ob is the angular momentum vector after turning;
ab is the applied or active gyroscopic couple;
b′ a′ is the reaction couple and is perpendicular to **oa** in the limit. This couple acts in the vertical plane and tends to raise the nose and depress the tail of the areoplane.

(i) If the aeroplane takes a right turn, the reaction couple is **d′c′** (Fig. 17.6 c), which shows that the nose is depressed and the tail is raised.

(ii) (a) When the engine rotates or spins clockwise on viewing from the nose end, **ef** is applied couple as the aeroplano turns left (Fig. 17.6 d). **f′e′** is the reaction couple indicating that the nose is depressed and the tail is raised.

(b) When the aeroplane takes a right turn, the tail is depressed and the nose is raised (Fig. 17.6 e).

It can be concluded from the above cases that if the direction of either the spin of the rotor or of the precession is changed, the gyroscopic effect is reversed, but if both are changed, the effect remains the same.

17.4 GYROSCOPIC EFFECT ON NAVAL SHIPS

Some of the terms used in connection with the motion of naval ships or sea-vessels are given below (Fig. 17.7 a):

(a) Bow is the fore or the front end.
(b) Stern or aft is the rear end.
(c) Starboard is the right-hand side when looking from the stern.
(d) Port is the left-hand side when looking from the stern.
(e) Steering is turning on one side when viewing from the top.
(f) Pitching is limited angular motion of the ship about the transverse axis.
(g) Rolling is limited angular motion of the ship about the longitudinal axis.

Let the plane of spin of the rotor and other rotating masses be horizontal and across the breadth of the ship. Assume ω to be the angular velocity of the rotor in the clockwise direction when viewed from stern (rear end).

Gyroscopic Effect during Steering

When the ship turns left, the angular momentum vector changes from **oa** to **ob** (Fig. 17.7 b). The reaction couple is found to be **b′a′** which tends to raise the bow and lower the stern. On turning right, the reaction couple is reversed so that bow is lowered and the stern is raised.

Gyroscopic Effect on Pitching

Pitching of the ship is usually considered to take place with simple harmonic motion. A simple harmonic motion is represented by, $x = X \sin \omega_0 t$.

Such a motion is obtained by the projection of a rotating vector X on a diameter while rotating around a circle with a constant angular velocity ω_0 and where x is the displacement from the mean position in time t. In the same way, angular displacement θ of the axis of spin from its mean position is given by

$$\theta = \varphi \sin \omega_0 t$$

where φ = Amplitude (angular) of swing or the maximum angle turned from the mean position in radians

$$\omega_0 = \text{Angular velocity of SHM} = \frac{2\pi}{\text{Time period}}$$

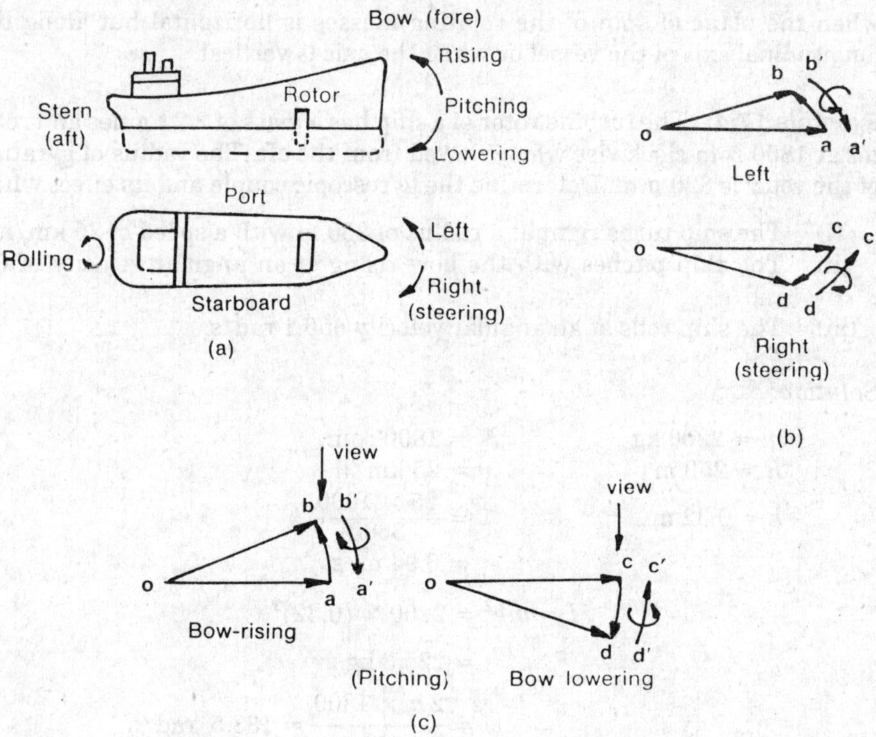

Fig. 17.7

Angular velocity of precession, $\frac{d\theta}{dt} = \varphi\, \omega_0 \cos \omega_0\, t$

This is maximum when $\cos \omega_0 t = 1$.

Therefore, maximum angular velocity of precession, $\omega_p = \varphi\, \omega_0$

$$\text{Gyroscopic couple} = I\,\omega\,\omega_p = I\,\omega \left(\varphi \times \frac{2\,\pi}{\text{Time period}}\right) \tag{17.4}$$

When the bow is rising, the reaction couple is clockwise on viewing from top and thus the ship would move towards right or starboard side. Similarly, when the bow is lowered, the ship turns towards left or port side (Fig. 17.7 c).

$$\text{Angular acceleration} = -\,\varphi\,\omega_0^2 \sin \omega_0 t \tag{17.5}$$

$$\text{Maximum angular acceleration} = \varphi\,\omega_0^2 \tag{17.6}$$

Gyroscopic Effect on Rolling

As the axes of the rolling of the ship and of the rotor are parallel, there is no precession of the axis of spin and thus there is no gyroscopic effect.

In the same way, the effects on steering, pitching or rolling can be observed when the plane of spin of the rotating masses is horizontal but along the longitudinal axis of the vessel or when the axis is vertical.

Example 17.4 The turbine rotor of a ship has a mass of 2.2 tonnes and rotates at 1800 rpm clockwise when viewed from the aft. The radius of gyration of the rotor is 320 mm. Determine the gyroscopic couple and its effect when

(i) The ship turns right at a radius of 250 m with a speed of 25 km/h.
(ii) The ship pitches with the bow rising at an angular velocity of 0.8 rad/s.
(iii) The ship rolls at an angular velocity of 0.1 rad/s.

Solution:

$m = 2200$ kg $\quad N = 1800$ rpm

$R = 250$ m $\quad v = 25$ km/h

$k = 0.32$ m $\quad = \dfrac{25 \times 1000}{3600}$

$= 6.94$ m/s

$$I = mk^2 = 2200 \times (0.32)^2$$

$$= 225.3 \text{ kg.m}^2$$

$$\omega = \frac{2\,\pi \times 1800}{60} = 188.5 \text{ rad/s}$$

$$\omega_p = \frac{v}{R} = \frac{6.94}{250} = 0.0278 \text{ rad/s}$$

(i) $$C = I\,\omega\,\omega_p$$

$$= 225.3 \times 188.5 \times 0.0278$$

$$= 1180 \text{ N.m}$$

The effect is to lower the bow (fore) and raise the stern (aft) when the ship turns right (Fig. 17.7b).

(ii) $$\omega_p = 0.8 \text{ rad/s}$$

$$C = I\,\omega\,\omega_p = 225.3 \times 188.5 \times 0.8$$

$$= 33\,972 \text{ N.m}$$

The effect of the reaction couple when the bow is rising, is to turn the ship towards right or towards starboard.

(iii) $$\omega_p = 0.1 \text{ rad/s}$$

$$C = 225.3 \times 188.5 \times 0.1 = 4246.5 \text{ N.m}$$

As the axis of spin is always parallel to the axis of precession for all positions, there is no gyroscopic effect on the ship.

Example 17.5 The rotor of the turbine of a ship has a mass of 2500 kg and rotates at a speed of 3200 rpm counter-clockwise when viewed from stern. The rotor has radius of gyration of 0.4 m. Determine the gyroscopic couple and its effect when

(i) the ship steers to the left in a curve of 80 m radius at a speed of 15 knots (1 knot = 1860 m/h)

(ii) the ship pitches 5 degrees above and 5 degrees below the normal position and the bow is descending with its maximum velocity. The pitching motion is simple harmonic with a periodic time of 40 seconds.

(iii) the ship rolls and at the instant, its angular velocity is 0.04 rad/s clockwise when viewed from stern.

Also find the maximum angular acceleration during pitching.

Solution:

$m = 2500$ kg $\qquad N = 3200$ rpm

$k = 0.4$ m $\qquad v = \dfrac{15 \times 1860}{3600} = 7.75$ m/s

$I = mk^2 = 2500 \times (0.4)^2 \qquad \omega = \dfrac{2\pi \times 3200}{60} = 335$ rad/s

$= 400$ kg.m^2

(i) $R = 80$ m

$$\omega_p = \frac{v}{R} = \frac{7.75}{80} = 0.097 \text{ rad/s}$$

$$C = 400 \times 335 \times 0.097$$

$$= 12\,981 \text{ N.m}$$

The effect is to lower the bow and raise the stern (Figs. 17.8 a and b).

(ii) $\varphi = 5° = 5 \times \dfrac{\pi}{180} = 0.0873$ rad

$T = 40$ s

$$\therefore\ \omega_0 = \frac{2\pi}{40} = 0.157 \text{ rad/s}$$

$$\omega_p = \varphi\,\omega_0 = 0.0873 \times 0.157 = 0.0137 \text{ rad/s}$$

$$C = I\,\omega\,\omega_p = 400 \times 335 \times 0.0137 = 1837.5 \text{ N.m}$$

As the bow descends during pitching, the ship would turn towards right or starboard side (Figs. 17.8 a and c).

(iii) $\omega_p = 0.04$ rad/s

$C = 400 \times 335 \times 0.04 = 5360$ N.m

No gyroscopic effect is there as discussed earlier.

Maximum Angular Acceleration during Pitching

$$\alpha_{max} = \varphi\,\omega_o^2 = 0.0873 \times (0.157)^2 = 0.002\,15 \text{ rad/s}^2$$

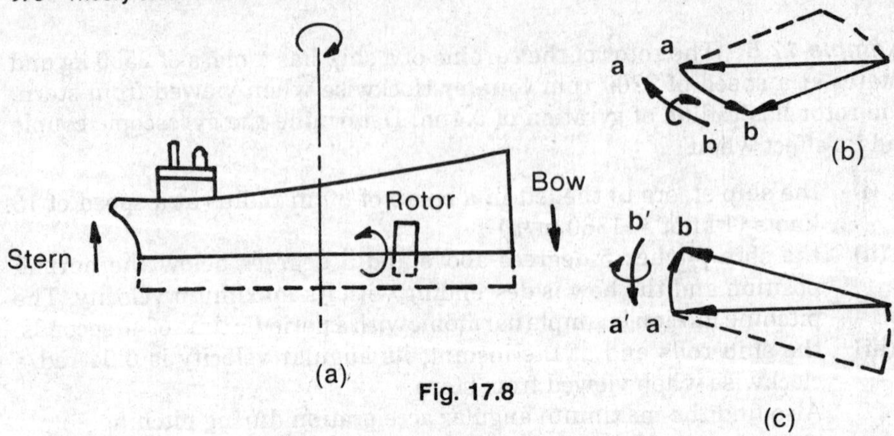

Fig. 17.8

17.5 STABILITY OF AN AUTOMOBILE

In case of a four-wheeled vechicle, it is essential that no wheel is lifted off the ground while the vehicle takes a turn. The condition is fulfilled as long as the vertical reaction of the ground on any of the wheels is positive (or upwards).

Fig. 17.9 shows a four-wheeled vehicle having a mass m. Assuming that the weight is equally divided among the four wheels,

$$\text{Weight on each wheel} = \frac{W}{4} = \frac{mg}{4} \text{ (downwards)}$$

$$\text{Reaction of ground on each wheel, } R_w = \frac{W}{4} = \frac{mg}{4} \text{ (upwards)}$$

Effect of Gyroscopic couple

Gyroscopic couple due to four wheels, $C_w = 4\, I_w\, \omega_w\, \omega_p = 4\, I_w \dfrac{v^2}{rR}$

where I_w = mass moment of inertia of each wheel

ω_w = angular velocity of wheels $= \dfrac{v}{r}$

ω_p = angular velocity of precession $= \dfrac{v}{R}$

v = linear velocity of the vehicle

R = radius of curvature

gyroscopic couple due to engine rotating parts,

$$C_e = I_e\, \omega_e\, \omega_p = I_e\, G\, \omega_w\, \omega_p$$

where G is the gear ratio $= \dfrac{\omega_e}{\omega_w}$

Total gyroscopic couple, $C_G = C_w + C_e$

Positive sign is used when the engine parts rotate in the same direction as the wheels and the negative sign when they rotate in the opposite.

Assuming that C_G is positive and the vehicle takes a left turn, the reaction

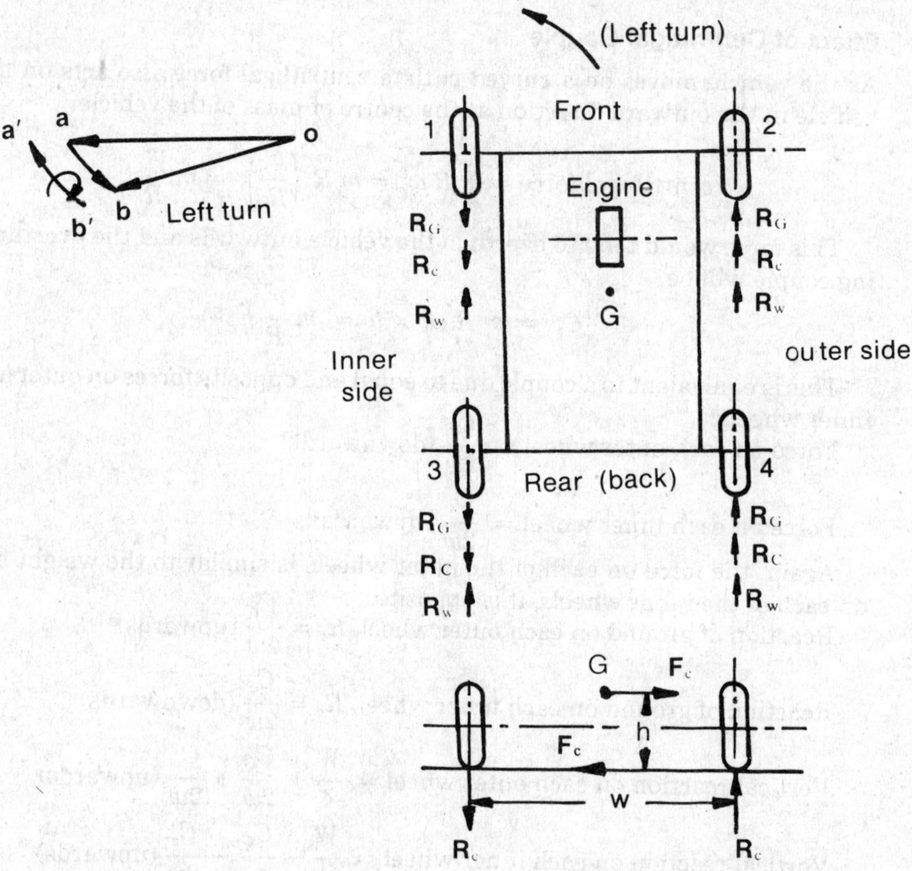

Fig. 17.9

gyroscopic couple on it is clockwise when viewed from the rear of the vehicle. The reaction couple is provided by equal and opposite forces on the outer and the inner wheels of the vehicle.

Forces on the two outer wheels $= \dfrac{C_G}{w}$ (downwards)

Forces on the two inner wheels $= \dfrac{C_G}{w}$ (upwards)

Forces on each of the outer wheels $= \dfrac{C_G}{2w}$ (downwards)

Forces on each of the inner wheels $= \dfrac{C_G}{2w}$ (upwards)

Thus the force on each of the outer wheels is similar to the weight. On the inner wheels it is in the opposite direction. Thus,

Reaction of ground on each outer wheel, $R_G = \dfrac{C_G}{2w}$ (upwards)

Reaction of ground on each inner wheel, $R_G = \dfrac{C_G}{2w}$ (downwards)

Effect of Centrifugal Couple

As the vehicle moves on a curved path, a centrifugal force also acts on the vehicle in the outward direction at the centre of mass of the vehicle.

$$\text{centrifugal force} = m\,R\,\omega_p^2 = m\,R\left(\frac{v}{R}\right)^2 = m\frac{v^2}{R}.$$

This force would tend to overturn the vehicle outwards and the overturning couple will be

$$C_c = m\,R\,\omega_p^2 \times h = m\frac{v^2}{R}h$$

This is equivalent to a couple due to equal and opposite forces on outer and inner wheels.

Force on each outer wheel $= \dfrac{C_c}{2w}$ (downwards)

Force on each inner wheel $= \dfrac{C_c}{2w}$ (upwards)

Again, the force on each of the outer wheels is similar to the weight and on each of the inner wheels, it is opposite.

Reaction of ground on each outer wheel, $R_c = \dfrac{C_c}{2w}$ (upwards)

Reaction of ground on each inner wheel, $R_c = \dfrac{C_c}{2w}$ (downwards)

Vertical reaction on each outer wheel $= \dfrac{W}{4} + \dfrac{C_G}{2w} + \dfrac{C_c}{2w}$ (upwards)

Vertical reaction on each inner wheel $= \dfrac{W}{4} - \dfrac{C_G}{2w} - \dfrac{C_c}{2w}$ (upwards)

It can be observed that there are chances that the reaction of the ground on the inner wheels may not be upwards and thus the wheels are lifted from the ground. For positive reaction, the condition will be

$$\frac{W}{4} - \frac{C_G}{2w} - \frac{C_c}{2w} \geq 0$$

or

$$\frac{W}{4} \geq \frac{C_G + C_c}{2w}$$

or

$$R_w \geq R_G + R_c \tag{17.7}$$

Example 17.6 Each wheel of a four-wheeled, rear engine automobile has a moment of intertia of 2.4 kg.m^2 and an effective diameter of 660 mm. The rotating parts of the engine have a moment of inertia of 1.2 kg.m^2. The gear ratio of engine to the back wheel is 3 to 1. The engine axis is parallel to the rear axle and the crankshaft rotates in the same sense as the road wheels. The mass of the vehicle is 2200 kg and the centre of the mass is 550 mm above the road level. The track width of the vehicle is 1.5 m. Determine the limiting speed of the vehicle around a curve with 80 m radius so that all the four wheels maintain contact with the road surface.

Solution:

$I_w = 2.4$ kg.m^2 $\quad m = 2200$ kg

$r = 0.33$ m $\quad h = 0.55$ m

$I_e = 1.2$ kg.m^2 $\quad w = 1.5$ m

$G = \dfrac{\omega_e}{\omega_w} = 3$ $\quad R = 80$ m

(i) *Reaction due to weight*

$$R_w = \frac{mg}{4} = \frac{2200 \times 9.81}{4} = 5395.5 \text{ N (upwards)}$$

(ii) *Reaction due to gyroscopic couple*

$$C_w = 4\, I_w \frac{v^2}{rR} = 4 \times 2.4 \times \frac{v^2}{0.33 \times 80} = 0.364\, v^2$$

$$C_e = I_e\, G\, \omega_w\, \omega_p = 1.2 \times 3 \times \frac{v^2}{0.33 \times 80} = 0.136\, v^2$$

$$\therefore \quad C_G = C_w + C_e = 0.364\, v^2 + 0.136\, v^2 = 0.5\, v^2$$

Reaction on each outer wheel, $R_{Go} = \dfrac{C_G}{2w} = \dfrac{0.5\, v^2}{2 \times 1.5} = 0.167\, v^2$ (upwards)

Reaction on each inner wheel, $R_{Gi} = 0.167\, v^2$ (downwards)

(iii) *Reaction due to centrifugal couple*

$$C_c = \frac{mv^2}{R} h = \frac{2200 \times v^2}{80} \times 0.55 = 15.125\, v^2$$

Reaction on each outer wheel, $R_{co} = \dfrac{C_c}{2w} = \dfrac{15.125\, v^2}{2 \times 1.5} = 5.042\, v^2$ (upwards)

Reaction on each inner wheel, $R_{ci} = 5.042\, v^2$ (downwards)

For maximum safe speed, the condition is

$$R_w = R_{Gi} + R_{ci}$$

$$5395.5 = (0.167 + 5.042)\, v^2$$

$$v^2 = 1035.8$$

$$v = 32.18 \text{ m/s}$$

or $$v = \frac{32.18 \times 3600}{1000} = 115.9 \text{ km/h}$$

Example 17.7 A four-wheeled trolley car has a total mass of 3000 kg. Each axle with its two wheels and gears has a total moment of inertia of 32 kg.m^2.

Each wheel is of 450 mm radius. The centre distance between two wheels on an axle is 1.4 m. Each axle is driven by a motor with a speed ratio of 1:3. Each motor along with its gear has a moment of inertia of 16 kg.m^2 and rotates in the opposite direction to that of the axle. The centre of mass of the car is 1 m above the rails. Calculate the limiting speed of the car when it has to travel around a curve of 250 m radius without the wheels leaving the rails.

Solution:

$$I_w = \frac{32}{2} = 16 \text{ kg.m}^2 \qquad m = 3000 \text{ kg}$$

$$r = 0.45 \text{ m} \qquad h = 1 \text{ m}$$

$$I_m = 16 \text{ kg.m}^2 \qquad w = 1.4 \text{ m}$$

$$R = 250 \text{ m}$$

(i) *Reaction due to weight:*

$$R_w = \frac{mg}{4} = \frac{3000 \times 9.81}{4} = 7357.5 \text{ N (upwards)}$$

(ii) *Reaction due to gyroscopic couple:*

$$C_w = 4\, I_w \frac{v^2}{r.R} = 4 \times 16 \times \frac{v^2}{0.45 \times 250} = 0.569\, v^2$$

$$C_m = 2\, I_m\, G\, \omega_w\, \omega_p \qquad \text{(as there are two motors)}$$

$$= 2 \times 16 \times 3 \times \frac{v^2}{0.45 \times 250}$$

$$= 0.853\, v^2$$

$$C_G = C_w - C_m \qquad \text{(Motors rotate in opposite direction)}$$

$$= 0.569\, v^2 - 0.853\, v^2 = -\, 0.284\, v^2$$

Reaction on each outer wheel,

$$R_{co} = \frac{C_G}{2w} = \frac{0.284\, v^2}{2 \times 1.4} = 0.1014\, v^2 \text{ (downwards)}$$

Reaction on each inner wheel, $R_{Go} = 0.1014\, v^2$ (upwards)

(iii) *Reaction due to centrifugal couple:*

$$C_c = \frac{mv^2}{R} h = 3000 \times \frac{v^2}{250} \times 1 = 12\, v^2$$

$$R_{co} = \frac{C_c}{2w} = \frac{12\, v^2}{2 \times 1.4} = 4.286\, v^2 \qquad \text{(upwards)}$$

$$R_{ci} = \frac{C_c}{2w} = 4.286\, v^2 \text{ (downwards)}$$

Total reaction on outer wheels $= 7357.5 - 0.1014\, v^2 + 4.286\, v^2$

$= 7357.5 + 4.1846\, v^2$.

Total reaction on inner wheels $= 7357.5 + 0.1014\, v^2 - 4.286\, v^2$

$$= 7357.5 - 4.1846\, v^2$$

Thus the reaction on the outer wheels is always positive (upwards). There are chances that the inner wheels leave the rails.

For maximum speed, $7357.5 - 4.1846\, v^2 = 0$

or $$v^2 = 1758.2$$

$$v = 41.93 \text{ m/s}$$

or $$v = \frac{41.93 \times 3600}{1000} = 151 \text{ km/h}$$

Example 17.8 A 2.2 tonne racing car has a wheel base of 2.4 m and a track of 1.4 m. The centre of mass of the car lies at 0.6 m above the ground and 1.4 m from the rear axle. Equivalent mass of engine parts is 140 kg with radius of gyration of 150 mm. The back axle ratio is 5. The engine shaft and flywheel rotate clockwise when viewed from front. Each wheel has a diameter of 0.8 m and a moment of inertia of 0.7 kg.m^2.

Determine the load distribution on the wheels when the car is rounding a curve of 100 m radius at a speed of 72 km/hr to the (i) left (ii) right.

Solution:

$M = 2200$ kg $\quad m = 140$ kg

$w = 1.4$ m $\quad k = 0.15$ m

$b = 2.4$ m $\quad I_w = 0.7$ kg.m^2

$R = 100$ m $\quad v = \dfrac{72 \times 1000}{3600} = 20$ m/s

(i) *Car turning left* (Refer Fig. 17.10)

(a) *Reaction due to weight:*

$$\text{Total weight} = 2200 \times 9.81 = 21\,582 \text{ N}$$

$$R_{w1,2} = \left(21\,582 \times \frac{1.4}{2.4}\right) \times \frac{1}{2} = 6295 \text{ N (upwards)}$$

$$R_{w3,4} = \left(21\,582 \times \frac{1}{2.4}\right) \times \frac{1}{2} = 4496 \text{ N (upwards)}$$

(b) *Reaction due to gyroscopic couples:*

$$C_w = 4\, I_w \frac{v^2}{rR} = 4 \times 0.7 \times \frac{(20)^2}{0.4 \times 100}\ 28 \text{ N.m}$$

For outer wheels, $R'_{G2,4} = \dfrac{C_w}{2w} = \dfrac{28}{2 \times 1.4} = 10$ N (upwards)

For inner wheels, $R'_{G1,3} = 10$ N (downwards)

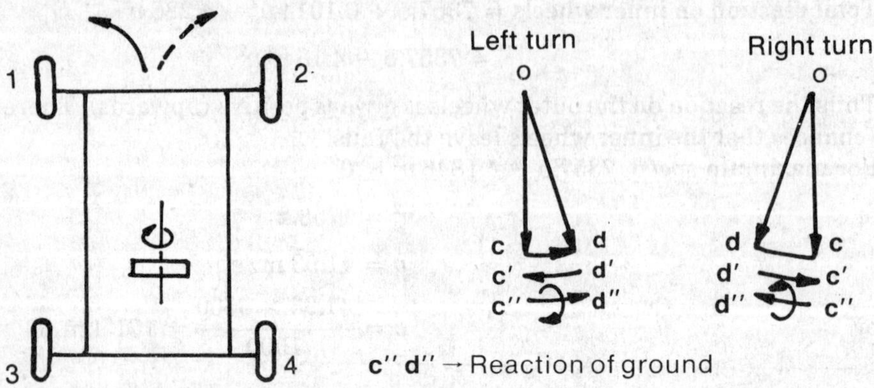

Fig. 17.10

$$I_e = mk^2 = 140 \times (0.15)^2 = 3.15 \text{ kg.m}^2$$

$$C_e = I_e\, G\, \omega_w\, \omega_p = 3.15 \times 5 \times \frac{(20)^2}{0.4 \times 100} = 157.5 \text{ N}$$

For front wheels, $R''_{G1,2} = \dfrac{C_e}{2b} = \dfrac{157.5}{2 \times 2.4} = 32.8$ N (upwards)

For rear wheels, $R''_{G3,4} = 32.8$ N (downwards)

(c) *Reaction due to centrifugal couple:*

$$C_c = M\frac{v^2}{R}h = 2200 \times \frac{(20)^2}{100} \times 0.6 = 5280 \text{ N.m}$$

For outer wheels, $R_{c2,4} = \dfrac{C_c}{2w} = \dfrac{5280}{2 \times 1.4} = 1886$ N (upwards)

For inner wheels, $R_{c1,3} = 1886$ N (downwards)

Therefore, reactions on wheels:

$$R = R_w + R'_G + R''_G + R_c$$

$$R_1 = 6295 - 10 + 32.8 - 1886 = 4431.8 \text{ N}$$

$$R_2 = 6295 + 10 + 32.8 + 1886 = 8223.8 \text{ N}$$

$$R_3 = 4496 - 10 - 32.8 - 1886 = 2567.2 \text{ N}$$

$$R_4 = 4496 + 10 - 32.8 + 1886 = 6359.2 \text{ N}$$

(ii) *Car turning right:*

All the reactions due to gyroscopic couples and centrifugal couple change signs. Therefore,

$$R_1 = 6295 + 10 - 32.8 + 1886 = 8158.2 \text{ N}$$

$$R_2 = 6295 - 10 - 32.8 - 1886 = 4366.2 \text{ N}$$

$$R_3 = 4496 + 10 + 32.8 + 1886 = 6424.8 \text{ N}$$

$$R_4 = 4496 - 10 + 32.8 - 1886 = 2632.8 \text{ N}$$

Example 17.9 The total mass of a four-wheeled trolley car is 1800 kg. The car runs on rails of 1.6 m gauge and rounds a curve of 24 m radius at 36 km/hr. The track is banked at 10°. The external diameter of the wheels is 600 mm and each pair with axle has a mass of 180 kg with radius of gyration of 240 mm. The height of the centre of mass of the car above the wheel base is 950 mm. Determine the pressure on each rail allowing for centrifugal force and gyroscopic couple actions.

Solution:

$M = 1800$ kg $\qquad \theta = 10°$
$w = 1.6$ m $\qquad h = 0.95$ m
$R = 24$ m $\qquad r = 0.3$ m
$m = 180$ kg $\qquad k = 0.24$ m

$$v = 36 \text{ kg/hr}$$

$$= \frac{36 \times 1000}{3600} = 10 \text{ m/s}$$

First, considering the effect of dead weight ($W = Mg$) of the car and that of the centrifugal force on it, determine the reactions R_A and R_B at the wheels A and B (Fig. 17.11).

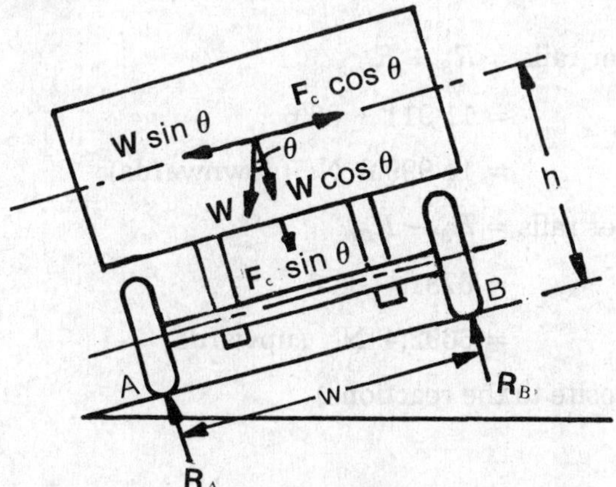

Fig. 17.11

Resolving the forces perpendicular to the track,

$$R_A + R_B = Mg \cos\theta + \frac{M v^2}{R} \sin\theta$$

$$= 1800 \times 9.81 \times \cos 10° + \frac{1800 \times (10)^2}{24} \sin 10°$$

$$= 18\,692 \text{ N}$$

Taking moments about B,

$$R_A \times w = Mg\cos\theta\,\frac{w}{2} + F_c\sin\theta\,\frac{w}{2} + Mg\sin\theta \times h - F_c\cos\theta \times h$$

$$R_A = \left(Mg\cos\theta + \frac{Mv^2}{R}\sin\theta\right)\times\frac{1}{2} + \left(Mg\sin\theta - \frac{Mv^2}{R}\cos\theta\right)\times\frac{h}{w}$$

$$R_A = \left(1800\times 9.81\times\cos 10^\circ + \frac{1800\times(10)^2}{24}\times\sin 10^\circ\right)\times\frac{1}{2}$$

$$+\left(1800\times 9.81\times\sin 10^\circ - \frac{1800\times(10)^2}{24}\times\cos 10^\circ\right)\times\frac{0.95}{1.6}$$

$$= 9346 - 2565 = 6781\text{ N}$$

$$R_B = 18692 - 6781 = 11911\text{ N}$$

Reaction due to gyroscopic couple

$$C_w = 2\,I_w\,\omega_w\cos\theta\times\omega_p = 2(mk^2)\,\frac{v^2}{rR}\times\cos\theta$$

$$= 2\times 180\times(0.24)^2\times\frac{(10)^2}{0.3\times 24}\times\cos 10^\circ$$

$$= 283.6\text{ N.m}$$

Reaction on each outer wheel, $R_{Go} = \dfrac{C_w}{2w} = \dfrac{283.6}{2\times 1.6} = 88.6$ N (upwards)

Reaction on each inner wheel, $R_{Gi} = 88.6$ N (downwards)

Therefore,

Pressure on outer rails $= R_B + R_{Go}$

$= 11\,911 + 88.6$

$= 11\,999.6$ N (downwards)

Pressure on inner rails $= R_A - R_{Gi}$

$= 6781 - 88.6$

$= 6692.4$ N (upwards)

(pressure is opposite to the reactions)

17.6 STABILITY OF A TWO-WHEEL VEHICLE

The case of a two-wheel vehicle can be taken in the same way as that of an automobile. However, it is easier to tilt such a vehicle inwards to neutralise the overturning effect and the vehicle can stay in equilibrium while taking a turn.

Let a vehicle take a left turn as shown in Fig. 17.12 a. The vehicle is inclined to the vertical (inwards) for equilibrium. The angle of inclination of the vehicle to the vertical is known as the angle of heel. Let

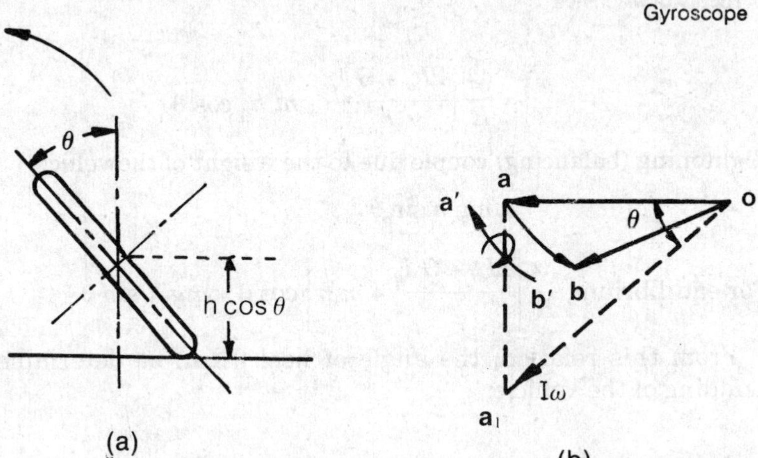

Fig. 17.12

v = linear velocity of vehicle on the track
r = radius of the wheel
R = radius of the track
I_w = moment of inertia of each wheel
I_e = moment of inertia of rotating parts of the engine
m = total mass of the vehicle and the rider
ω_w = angular velocity of the wheels
ω_e = angular velocity of rotating parts of the engine
G = gear ratio
h = height of centre of mass of the vehicle and the rider
θ = inclination of vehicle to the vertical (angle of heel)

As the axis of spin is not horizontal but inclined to the vertical at an angle θ and the axis of precession is vertical, it is necessary to take the horizontal component of the spin vector.

Spin vector (horizontal) $= I\omega \cos\theta$ (See Fig. 17.12b)

$$= (2I_w \omega_w + I_e \omega_e)\cos\theta$$

and Gyroscopic couple $= (2I_w \omega_w + I_e G \omega_w)\cos\theta\, \omega_p$

$$= (2I_w + G I_e)\frac{v}{r}\frac{v}{R}\cos\theta$$

$$= (2I_w + G I_e)\frac{v^2}{rR}\cos\theta$$

The reaction couple **b′ a′** is clockwise when viewed from the rear (back) of the vehicle and tends to overturn it in the outward direction.

Overturning couple due to centrifugal force $= \left(m\frac{v^2}{R}\right) h\cos\theta$

$\therefore$ Total overturning couple $= (2I_w + G I_e)\dfrac{v^2}{rR}\cos\theta + m\dfrac{v^2}{R} h\cos\theta$

$$= \frac{v^2}{R}\left(\frac{2I_w + G\,I_e}{r} + m\,h\right)\cos\theta$$

Rightening (balancing) couple due to the weight of the vehicle

$$= mg\,h\sin\theta$$

For equilibrium, $\frac{v^2}{R}\left[\frac{2I_w + G\,I_e}{r} + m\,h\right]\cos\theta = mg\,h\sin\theta$ (17.8)

From this relation, the angle of heel θ can be determined to avoid skidding of the vehicle.

Example 17.10 Each wheel of a motorcycle is of 600 mm diameter and has a moment of inertia of 1.2 kg.m^2. The total mass of the motorcycle and the rider is 180 kg and the combined centre of mass is 580 mm above the ground level when the motor cycle is upright. The moment of inertia of the rotating parts of the engine is 0.2 kg.m^2. The engine speed is 5 times the speed of the wheels and is in the same sense. Determine the angle of heel necessary when the motorcycle takes a turn of 35 m radius at a speed of 54 km/h.

Solution:

$m = 180$ kg $\quad I_w = 1.2$ kg m^2

$r = 0.3$ m $\quad I_e = 0.2$ kg m^2

$R = 35$ m $\quad v = \frac{54 \times 1000}{3600} = 15$ m/s

$G = \frac{\omega_e}{\omega_w} = 5 \quad h = 0.58$ m

Gyroscopic couple, $C_G = (2\,I_w + G\,I_e)\,\frac{v^2}{rR}\cos\theta$

$$= (2 \times 1.2 + 5 \times 0.2) \times \frac{(15)^2}{0.3 \times 35} \times \cos\theta$$

$$= 72.86\cos\theta$$

Centrifugal couple, $C_c = \mathrm{m}\,\frac{v^2}{R}\,h\cos\theta$

$$= 180 \times \frac{(15)^2}{35} \times 0.58 \times \cos\theta$$

$$= 671.14\cos\theta$$

Total overturning couple $= (72.86 + 671.14)\cos\theta = 744\cos\theta$

$$\text{Rightening couple} = mg\,h \sin\theta$$

$$= 180 \times 9.81 \times 0.58 \sin\theta$$

$$= 1024 \sin\theta$$

$$\therefore \quad = 1024 \sin\theta = 744 \cos\theta$$

or $$\tan\theta = \frac{744}{1024} = 0.727$$

$$\theta = 36°$$

PROBLEMS

17.1 A flywheel having a mass of 20 kg and a radius of gyration of 300 mm is given a spin of 500 rpm about its axis which is horizontal. The flywheel is suspended at a point 250 mm from the plane of rotation of the flywheel. Find the rate of precession of the wheel.

(0.52 rad/s)

17.2 A disc supported between two bearings on a shaft of negligible weight has a mass of 80 kg and a radius of gyration of 300 mm. The distances of the disc from the bearings are 300 mm to the right from the left hand bearing and 450 mm to the left from the right hand bearing. The bearings are supported by thin vertical cords. When the disc rotates at 100 rad/s in the clockwise direction looking from the left-hand bearing, the cord supporting the left hand side bearing breaks. Find the angular velocity of precession at the instant the cord is cut and discuss the motion of the disc.

(0.327 rad/s)

17.3 The moment of inertia of an aeroplane air screw is 20 kg.m^2 and the speed of rotation 1000 rpm clockwise when seen from the front. The speed of the flight is 200 km per hour. Find the gyroscopic reaction of the air screw on the aeroplane when it makes a left handed turn on a path of 150 m radius.

(775.5 N.m)

17.4 The rotor of a marine turbine has a moment of inertia of 750 kg.m^2 and rotates at 3000 rpm clockwise when viewed from aft. If the ship pitches with angular simple harmonic motion having a periodic time of 16 seconds and an amplitude of 0.1 radian, find
(i) the maximum angular velocity of the rotor axis
(ii) the maximum value of the gyroscopic couple
(iii) the gyroscopic effect as the bow dips

(0.0393 rad/s; 9261 N.m; Bow swings to port (left) as it dips)

17.5 The turbine rotor of a sea vessel having a mass of 950 kg rotates at 1200 rpm clockwise while looking from the stern. The vessel pitches with an angular velocity of 1.2 rad/s. What will be the gyroscopic couple transmitted to the hull when the bow rises? The radius of gyration of rotor is 300 mm.

(12.89 k N.m)

17.6 A ship is propelled by a turbine rotor having a mass of 6 tonnes and speed of 2400 rpm. The direction of rotation of the rotor is clockwise when viewed from the stern. The radius of gyration of the rotor is 450 mm. Determine the gyroscopic effect when

(i) the ship steers to the left in a curve of 60 m radius at a speed of 18 knots (1 knot = 1860 m/hr)

(ii) the ship pitches 7.5 degrees above and 7.5 degrees below the normal position and the bow is descending with its maximum velocity. The pitching motion is simple harmonic with a periodic time of 18 seconds.

(iii) the ship rolls and at the instant, its angular velocity is 0.035 rad/s counter-clockwise when viewed from the stern.

Also find the maximum angular acceleration during pitching.

(47.33 kN.m, bow is raised; 13.96 kN.m, ship turns towards port side; 10.69 kN.m, no gyroscopic effect; 0.016 rad/s²)

17.7 A rear engine automobile is travelling along a curve track of 120 m radius. Each of the four wheels has a moment of inertia of 2.2 kg.m^2 and an effective diameter of 600 mm. The rotating parts of the engine have a moment of inertia of 1.25 kg.m^2. The gear ratio of the engine to the back wheel is 3.2. The engine axis is parallel to the rear axle and the crankshaft rotates in the same sense as the road wheels. The mass of the vehicle is 2050 kg and the centre of mass is 520 mm above the road level. The width of the track is 1.6 m. What will be the limiting speed of the vehicle if all the four wheels maintain contact with the road surface?

(150 km/h)

17.8 The moment of inertia of a pair of locomotive driving wheels with the axle is 200 kg.m^2. The distance between the wheel centres is 1.6 m and the diameter of the wheel treads is 1.8 m. Due to defective ballasting, one wheel falls by 5 mm and rises again in a total time of 0.12 seconds while the locomotive travels on a level track at 100 km/hr. Assuming that the displacement of the wheel takes place with simple harmonic motion, determine the gyroscopic couple produced and the reaction between the wheel and rail due to this couple.

(505 N.m; 315.6 N)

17.9 Each road wheel of a motor cycle is of 600 mm diameter and has a moment of inertia of 1.1 kg.m^2. The motor cycle and the rider together weigh 220 kg and the combined centre of mass is 620 mm above the ground level when the motor cycle is upright. The moment of inertia of the rotating parts of the engine is 0.18 kg.m^2. The engine rotates at 4.5 times the speed of road wheels in the same sense. Find the angle of heel necessary when the motor cycle is taking a turn of 35 m radius at a speed of 72 km/hr.

(51°24′)

18

VIBRATIONS

18.1 INTRODUCTION

A body is said to vibrate if it has a to and fro motion. A pendulum swinging on either side of a mean position does so under the action of gravity. When the pendulum swings through the midposition, its centre of mass is at the lowest point and it possesses only kinetic energy. At each extremity of its swing, it has only potential energy. In the absence of any friction, the motion continues indefinitely. It can be shown that if the swings on either side of the mean position are very small, it approximates to simple harmonic motion.

Usually, vibrations are due to elastic forces. Whenever a body is displaced from its equilibrium position, work is done on the elastic constraints of the body and is stored as strain energy. Now, if the body is released, the internal forces cause the body to move towards its equilibrium position. If the motion is frictionless, the strain energy stored in the body is converted into kinetic energy during the period the body reaches the equilibrium position at which it has maximum kinetic energy. The body passes through the mean position, the kinetic energy is utilised to overcome the elastic forces and is stored in the form of strain energy, and so on.

18.2 DEFINITIONS

(i) Free (Natural) Vibrations Elastic vibrations in which there are no friction and external forces after the initial release of the body are known as free or natural vibrations.

(ii) Damped Vibrations When the energy of a vibrating system is gradually dissipated by friction and other resistances, the vibrations are said to be damped. The vibrations gradually cease and the system rests in its equilibrium position.

(iii) Forced Vibrations When a repeated force continuously acts on a system, the vibrations are said to be forced. The frequency of the vibrations is that of the applied force and is independent of their own natural frequency of vibrations.

(iv) Period It is the time taken by a motion to repeat itself, and is measured in seconds.

(v) Cycle It is the motion completed during one time period.

(vi) Frequency Frequency is the number of cycles of motion completed in one second. It is expressed in hertz (Hz) and is equal to one cycle per second.

(vii) Resonance When the frequency of the external force is the same as that of the natural frequency of the system, a state of resonance is said to have been reached. Resonance results in large amplitudes of vibrations and this may be dangerous.

18.3 TYPES OF VIBRATIONS

Consider a vibrating body, e.g. a rod, shaft or spring. Figure 18.1 shows a massless shaft, one end of which is fixed and the other end carrying a heavy disc. The system can execute the following types of vibrations.

(i) Longitudinal Vibrations If the shaft is elongated and shortened so that the mass moves up and down resulting in tensile and compressive stresses in the shaft, the vibrations are said to be longitudinal. The different particles of the body move parallel to the axis of the body (Fig.18.1a).

(ii) Transverse Vibrations When the shaft is bent alternately (Fig.18.1b) and tensile and compressive stresses due to bending result, the vibrations are

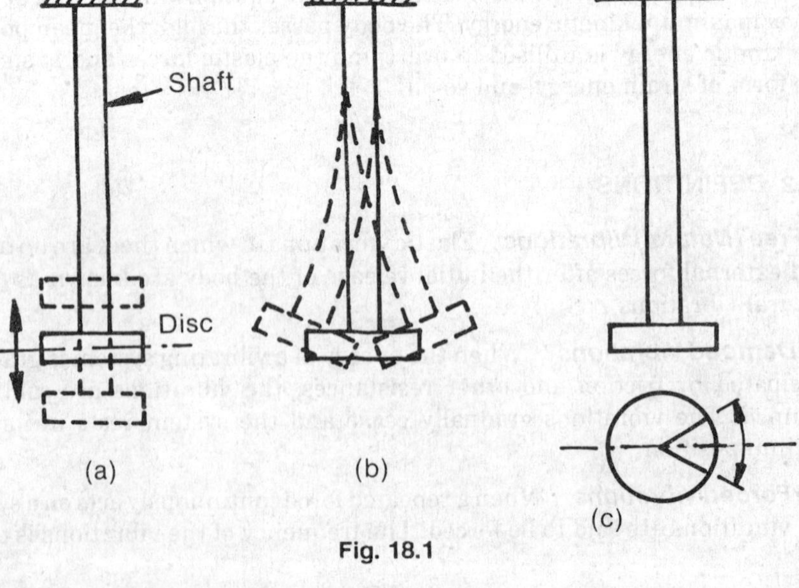

Fig. 18.1

said to be transverse. The particles of the body move approximately perpendicular to its axis.

(iii) Torsional Vibrations When the shaft is twisted and untwisted alternately and torsional shear stresses are induced, the vibrations are known as torsional vibrations. The particles of the body move in a circle about the axis of the shaft (Fig. 18.1 c).

In the following sections, different types of vibrations have been discussed separately.

SECTION-I (LONGITUDINAL VIBRATIONS)

18.4 FREE LONGITUDINAL VIBRATIONS

The natural frequency of a vibrating system may be found by any of the following methods.

(i) Equilibrium Method It is based on the principle that whenever a vibratory system is in equilibrium, the algebraic sum of forces and moments acting on it is zero. This is in accordance with D'Alembert's principle that the sum of the inertia forces and the external forces on a body in equilibrium must be zero.

Figure 18.2 a shows a helical spring suspended vertically from a rigid support with its free end at A-A.

If a mass m is suspended from the free end, the spring is stretched by a distance Δ and B-B becomes the equilibrium position (Fig. 18.2 b). Thus Δ is the static deflection of the spring under the weight of the mass m.

Let s = stiffness of the spring (force required per unit deflection)

In the static equilibrium position,

Upward force = Downward force

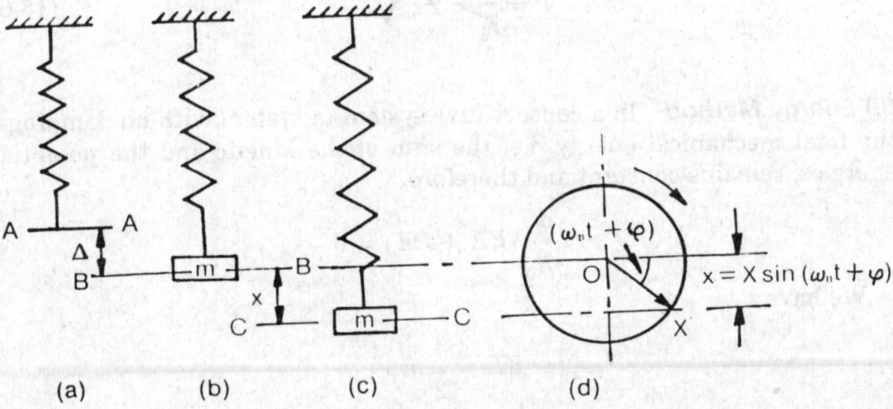

Fig. 18.2

$$s \times \Delta = mg \tag{18.1}$$

Now, if the mass m is pulled farther down through a distance (Fig.18.2c), the forces acting on the mass will be

Inertia force $= m\ddot{x}$ (upwards)
Spring force (restoring force) $= sx$ (upwards)

As the sum of the inertia force and the external force on the body in any direction is to be zero (D'Alembert's principle),

$$m\ddot{x} + sx = 0 \tag{18.2}$$

If the mass is released, it will start oscillating above and below the equilibrium position. The oscillations will continue for ever if there is no frictional resistance to the motion.

The above equation can be written as

$$\ddot{x} + \left(\frac{s}{m}\right) x = 0 \tag{18.2a}$$

This is the equation of a simple harmonic motion and is analogous to

$$\ddot{x} + \omega_n^2 x = 0 \tag{18.3}$$

where ω_n is the circular (natural) frequency of SHM.

Circular frequency of the vibrating system,

$$\omega_n = \sqrt{\frac{s}{m}} \tag{18.4}$$

and linear frequency of the vibrating system,

$$f_n = \frac{1}{2\pi}\sqrt{\frac{s}{m}} \tag{18.5}$$

Time period,

$$T = \frac{1}{f_n} = 2\pi\sqrt{\frac{m}{s}} \tag{18.6}$$

(ii) Energy Method In a conservative system (a system with no damping), the total mechanical energy, i.e. the sum of the kinetic and the potential energies, remains constant and therefore,

$$\frac{d}{dt}(KE + PE) = 0$$

We have

$$KE = \frac{1}{2} m \dot{x}^2$$

and

$$PE = \text{Mean force} \times \text{Displacement}$$

$$= \frac{0 + sx}{2} \times x$$

$$= \frac{sx^2}{2}$$

$$\frac{d}{dt}\left(\frac{1}{2} m\dot{x}^2 + \frac{sx^2}{2}\right) = 0$$

or

$$\frac{1}{2} m \times 2\dot{x}\ddot{x} + \frac{1}{2} s \times 2x\dot{x} = 0$$

or

$$m\ddot{x} + sx = 0$$

or

$$\omega_n = \sqrt{\frac{s}{m}}$$

(iii) Rayleigh's Method In this method, the maximum kinetic energy at the mean position (where potential energy is zero) is made equal to the maximum potential energy (or strain energy) at the extreme position (where the kinetic energy is zero).

Let the motion be simple harmonic.

Therefore, $x = X \sin \omega_n t$

where X = maximum displacement from the mean position to the extreme position.

$\therefore$

$$\dot{x} = \omega_n X \cos \omega_n t$$

or

$$\dot{x}_{\max} = \omega_n X$$

$$KE \text{ at mean position} = PE \text{ at extreme position}$$

$$\frac{1}{2} m (\omega_n X)^2 = \frac{1}{2} sX^2$$

or

$$m\,\omega_n^2 = s$$

or

$$\omega_n = \sqrt{\frac{s}{m}}$$

18.5 DISPLACEMENT, VELOCITY AND ACCELERATION

The displacement of the mass m from the mean position at any instant can be found from the solution of Eq. (18.3). The solution will be of the form

$$x = A \sin \omega_n t + B \cos \omega_n t \qquad (18.7)$$

Where A and B are constants and are to be found from the initial conditions.

The equation for the displacement can also be written in the following form by assuming $A = X\cos\varphi$ and $B = X\sin\varphi$,

$$x = X(\sin\omega_n t\cos\varphi + \cos\omega_n t\sin\varphi)$$

$$= X\sin(\omega_n t + \varphi) \qquad (18.8)$$

where X and φ are other constants and have to be found from initial conditions (refer Fig. 18.2 d).

Also velocity,

$$v = \dot{x} = X\omega_n\cos(\omega_n t + \varphi)$$

$$= X\omega_n\sin\left[\frac{\pi}{2} + (\omega_n t + \varphi)\right]$$

and acceleration,

$$f = \ddot{x} = -X\omega_n^2\sin(\omega_n t + \varphi)$$

$$= X\omega_n^2\sin[\pi + (\omega_n t + \varphi)]$$

These relationships indicate that the velocity vector leads the displacement vector by $\pi/2$ and the acceleration vector leads the displacement vector by π.

18.6 INERTIA EFFECT OF THE MASS OF SPRING

So far, the mass of the spring and thus the effect of inertia have been neglected. The same may be taken into account as follows:

Let m' = mass of the spring wire per unit length

v = velocity of the free end of the spring at the instant under consideration

l = total length of the spring wire

Consider an element of length δy at a length y measured round the coils from the fixed end.

$$KE \text{ of the element} = \frac{1}{2} \times \text{Mass of element} \times (\text{Velocity of element})^2$$

$$= \frac{1}{2}(m'\,\delta y) \times \left(\frac{y}{l}v\right)^2$$

$$KE \text{ of the spring} = \int_0^l \frac{1}{2}m'\,v^2\left(\frac{y}{l}\right)^2 dy$$

$$= \frac{1}{2}\frac{m'\,v^2}{l^2}\int_0^l y^2\cdot dy = \frac{1}{2}\frac{m'\,v^2}{l^2}\frac{l^3}{3}$$

$$= \frac{1}{3}\frac{1}{2}(m'\,l)v^2$$

$$= \frac{1}{3} \times \left[\frac{1}{2} \times \text{Mass of spring} \times (\text{Velocity of free end})^2\right]$$

$$= \frac{1}{3} \times KE \text{ of a mass equal to that of the spring moving with the same velocity as the free end.}$$

This shows that the inertia effect of the spring is equal to that of a mass one third of the mass of the spring, concentrated at its free end.

Thus

Equivalent mass at the free end $= m + \dfrac{m_1}{3}$

where

m_1 = mass of the spring

$$f_n = \frac{1}{2}\sqrt{\frac{s}{m + \dfrac{m_1}{3}}} \qquad (18.9)$$

It can also be noted that the net force on the spring at any instant tending to restore the vibrating mass to the equilibrium position is sx which is proportional to the displacement of the mass. This is true for any vibration due to the elastic forces. Thus in a vibrating system in which the restoring force is proportional to the displacement from the equilibrium position, the frequency of the system will always be given by

$$f_n = \frac{1}{2\pi}\sqrt{\frac{s}{m}} = \frac{1}{2\pi}\sqrt{\frac{mg/\Delta}{m}} = \frac{1}{2\pi}\sqrt{\frac{g}{\Delta}} \qquad (18.10)$$

where Δ is the static deflection under the suspended mass m.

For example, consider a rod of length l suspended vertically. A mass m is suspended at the free end (Fig. 18.1 a).

Then

Static deflection, $\Delta = \dfrac{mgl}{AE}$

where

A = cross-sectional area of the rod

l = length of the rod

E = Young's modulus of the rod material.

Frequency,

$$f_n = \frac{1}{2\pi}\sqrt{\frac{gAE}{mgl}} = \frac{1}{2\pi}\sqrt{\frac{AE}{ml}}$$

However, if the mass of the suspended rod is also to be considered,

$$f_n = \frac{1}{2\pi}\sqrt{\frac{AE}{\left(m + \frac{1}{3}m_1\right) l}}$$

where m_1 = mass of the rod

Example 18.1 Determine the equivalent spring stiffness and the natural frequency of the following vibrating systems when (refer to Figs. 18.3 a to e);

(a) the mass is suspended to a spring
(b) the mass is suspended at the bottom of two springs in series
(c) the mass is fixed in between two springs
(d) the mass is fixed to the midpoint of a spring
(e) the mass is fixed to a point on a bar joining free ends of two springs.

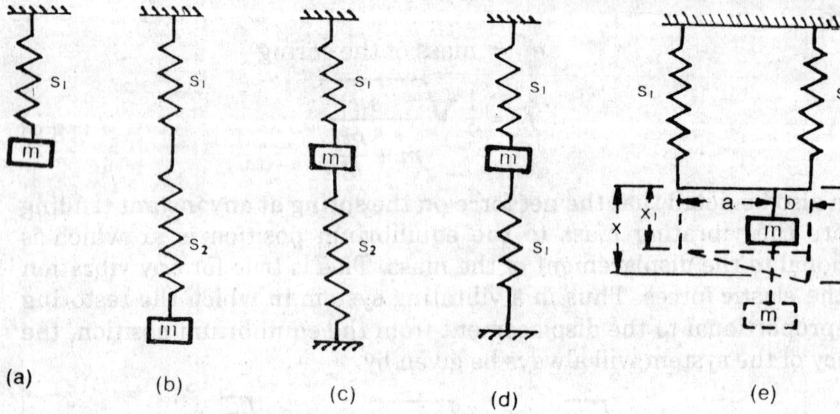

Fig. 18.3

Take

$s_1 = 5$ N/mm, $\quad s_2 = 8$ N/mm
$m = 10$ kg, $\quad a = 20$ mm and $b = 12$ mm

Solution: (a) As there is only one spring, the equivalent spring stiffness is the same, i.e. $s = s_1 = 5$ N/mm

$$f_n = \frac{1}{2n}\sqrt{\frac{s_1}{m}} = \frac{1}{2\pi}\sqrt{\frac{5 \times 10^3}{10}} = 3.56 \text{ Hz}$$

(b) Spring force will be the same in the two springs but static deflections will be different.

Let s = Equivalent spring stiffness of the two springs.

Deflection of mass m = Deflection of spring 1 + Deflection of spring 2

$$\frac{mg}{s} = \frac{mg}{s_1} + \frac{mg}{s_2}$$

or

$$\frac{1}{s} = \frac{1}{s_1} + \frac{1}{s_2}$$

or

$$s = \frac{s_1 s_2}{s_1 + s_2}$$

$$f_n = \frac{1}{2\pi}\sqrt{\frac{s}{m}} = \frac{1}{2\pi}\sqrt{\frac{s_1\, s_2}{(s_1 + s_2)\, m}}$$

$$= \frac{1}{2\pi}\sqrt{\frac{5 \times 10^3) \times (8 \times 10^3)}{(5 + 8) \times 10^3 \times 10}} = 2.79 \text{ Hz}$$

(c) The spring forces will be different but the deflections will be the same of the two springs and the mass m.

Let Δ = deflection of each spring and of mass m.

Net spring force = Spring force in 1 + Spring force in 2

$$s\Delta = s_1\,\Delta + s_2\,\Delta$$

or

$$s = s_1 + s_2 = 5 + 8 = 13 \text{ N/mm}$$

$$f_n = \frac{1}{2\pi}\sqrt{\frac{s}{m}} = \frac{1}{2\pi}\sqrt{\frac{13 \times 10^3}{10}} = 5.74 \text{ Hz}$$

(d) The spring stiffness of a coiled spring is inversely proportional to the number of coils in the spring. As the mass is fixed at the midpoint, the number of coils becomes half on each side.

Stiffness of spring on each side $= \dfrac{s_1}{1/2} = 2s_1$

Now the system is similar to case (iii).

Equivalent spring stiffness,

$$s = 2s_1 + 2s_1 = 4s_1$$

$$f_n = \frac{1}{2\pi}\sqrt{\frac{4s_1}{m}} = \frac{1}{2\pi}\sqrt{\frac{4 \times 5 \times 10^3}{10}} = \underline{7.12 \text{ Hz}}$$

(e) Spring forces as well as the static deflections of the two springs will be different.

Spring force in 1 $= mg\dfrac{b}{a + b}$

Spring force in 2 $= mg\dfrac{a}{a + b}$

Deflection of spring 1, $\Delta_1 = mg\dfrac{b}{a + b}\dfrac{1}{s_1}$

Deflection of spring 2, $\Delta_2 = mg\dfrac{a}{a + b}\dfrac{1}{s_2}$

Assuming that the deflection of 2 is more than that of 1

Deflection of mass m $\Delta = \Delta_2 - (\Delta_2 - \Delta_1)\dfrac{b}{a + b}$

$$= mg\left[\frac{a}{a + b}\frac{1}{s_2} - \left(\frac{a}{a + b}\frac{1}{s_2} - \frac{b}{a + b}\frac{1}{s_1}\right)\frac{b}{a + b}\right]$$

$$= \frac{mg}{a+b}\left[\frac{a}{s_2} - \frac{ab}{(a+b)\,s_2} + \frac{b^2}{(a+b)s_1}\right]$$

$$= \frac{mg}{a+b}\left[\frac{a^2 s_1 + abs_1 - abs_1 + b^2 s_2}{(a+b)s_1 s_2}\right]$$

$$= \frac{mg}{(a+b)^2}\left[\frac{a^2}{s_2} + \frac{b^2}{s_1}\right]$$

Total spirng force = Force in spring 1 + Force in spring 2

$$s\Delta = s_1\,\Delta_1 + s_2\,\Delta_2$$

$$s\,\frac{mg}{(a+b)^2}\left[\frac{a^2}{s_2} + \frac{b^2}{s_2}\right] = s_1\,mg\,\frac{b}{a+b} \times \frac{1}{s_1} + s_2\,mg\,\frac{a}{a+b} \times \frac{1}{s_2}$$

$$s\,\frac{1}{a+b}\left(\frac{a^2}{s_2} + \frac{b^2}{s_1}\right) = b + a$$

or
$$s = \frac{(a+b)^2}{\left(\dfrac{a^2}{s_2} + \dfrac{b^2}{s_1}\right)}$$

$$f_n = \frac{1}{2\pi}\sqrt{\frac{(a+b)^2}{\left(\dfrac{a^2}{s_2} + \dfrac{b^2}{s_1}\right) m}} = \sqrt{\frac{(0.02 + 0.012)^2}{\dfrac{(0.02)^2}{8 \times 10^3} + \dfrac{(0.012)^2}{5 \times 10^3}} \times \frac{1}{10}} = \underline{36\text{ Hz}}$$

Alternatively,

$$f_n = \frac{1}{2\pi}\sqrt{\frac{g}{\Delta}} = \frac{1}{2\pi}\sqrt{\frac{g}{\dfrac{mg}{(a+b)^2}\left(\dfrac{a^2}{s_2} + \dfrac{b^2}{s_1}\right)}} = \frac{1}{2\pi}\sqrt{\frac{(a+b)^2}{\left(\dfrac{a^2}{s_2} + \dfrac{b^2}{s_1}\right) m}}$$

i.e. the same expression.

Example 18.2 Determine the frequency (circular) of vibration of the systems shown in Figs. 18.4 a and b. Neglect the mass of the pulleys.

Solution: (a) Force in each spring = 2 W

Deflection of mass m, Δ = 2 (deflection of spring 1 + deflection of spring 2)

$$= 2\left(\frac{2W}{s_1} + \frac{2W}{s_2}\right)$$

$$= 4\,mg\left(\frac{s_1 + s_2}{s_1\,s_2}\right)$$

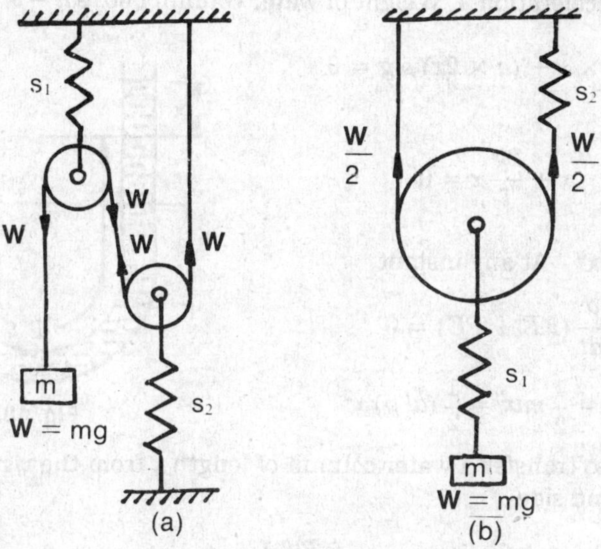

Fig. 18.4

$$\omega_n = \frac{g}{\Delta} = \sqrt{\frac{g(s_1\, s_2)}{4mg\,(s_1 + s_2)}} = \sqrt{\frac{s_1\, s_2}{4\,(s_1 + s_2)\, m}}$$

(b) Force in spring 1 = W

Force in spring 2 = $W/2$

Deflection of mass = Deflection of spring 1 + Deflection of spring 2

$$= \frac{W}{s_1} + \frac{1}{2}\frac{W/2}{s_2}$$

$$= mg\left(\frac{1}{s_1} + \frac{1}{4s_2}\right)$$

$$= mg\left(\frac{4s_2 + s_1}{4s_1\, s_2}\right)$$

$$\omega_n = \sqrt{\frac{g}{\Delta}} = \sqrt{\frac{4\, s_1\, s_2}{(s_1 + 4s_2)\, m}}$$

Example 18.3 Determine the equation of vibration of the water column in a U-tube shown in Fig. 18.5.

Solution: *(a) Newton's Method*:

Let a = area of cross-section of the tube

ρ = mass density of water

l = total length of water column

Inertia force + External force = 0

Mass × Acceleration + Weight of water column above $h - h = 0$

$$(al\rho) \times \ddot{x} + (a \times 2x)\,\rho\, g = 0$$

or

$$\ddot{x} + \frac{2g}{l}x = 0$$

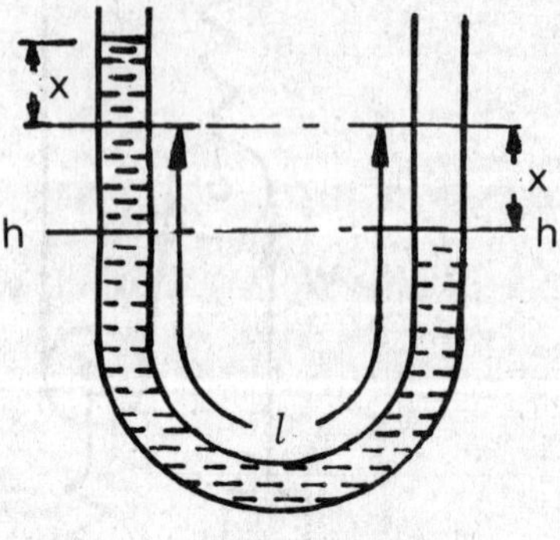

Fig. 18.5

Energy Method At any instant,

$$\frac{d}{dt}(KE + PE) = 0$$

$$KE = \frac{1}{2}mv^2 = \frac{1}{2}(al\,\rho)\,\dot{x}^2$$

PE = Work to transfer a water column of length x from the right-hand side to the left-hand side.

$$= mg\,x$$

$$= (a\,x\,\rho)\,g\,x$$

$$= a\,\rho\,g\,x^2$$

$$\frac{d}{dt}\left(\frac{1}{2}al\,\rho\,\dot{x}^2 + a\,\rho\,g\,x^2\right) = 0$$

$$\frac{1}{2}a\,l\,\rho \times 2\,\dot{x}\,\ddot{x} + a\,\rho\,g \times 2\,x\,\dot{x} = 0$$

$$\ddot{x} + \frac{2g}{l}x = 0$$

$$\omega_n = \sqrt{\frac{2g}{l}}$$

Example 18.4 Determine the natural frequency of a vibrating system shown in Fig.18.6.

Solution: Force in spring 1, $F_1 = W$

Force in spring 2, $F_2 = W\dfrac{l_1}{l_2}$ $(F_1 \times l_1 = F_2 \times l_2)$

Deflection of mass = Deflection of spring 1 + $\dfrac{l_1}{l_2}$ (Deflection of spring 2)

$$\Delta = \frac{W}{s_1} + \frac{l_1}{l_2} \times \frac{(W\,l_1/l_2)}{s_2}$$

$$= W\left[\frac{1}{s_1} + \frac{(l_1/l_2)^2}{s_2}\right]$$

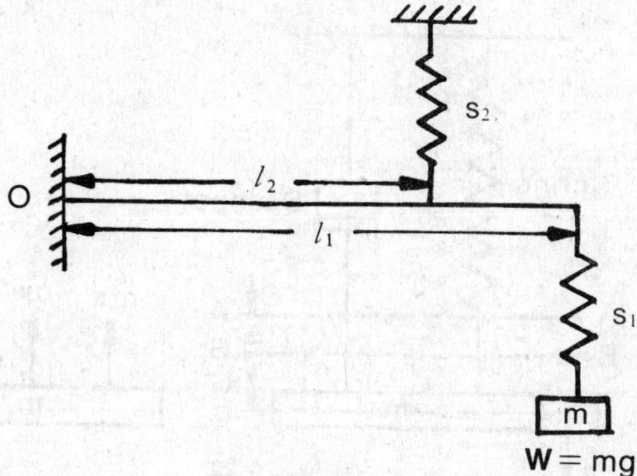

Fig. 18.6

$$= mg\left[\frac{s_2 + s_1\,(l_1/l_2)^2}{s_1\,s_2}\right]$$

$$\omega_n = \sqrt{\frac{g}{\Delta}} = \sqrt{\frac{s_1\,s_2}{[s_1\,(l_1/l_2)^2 + s_2]\,m}}$$

$$= \sqrt{\frac{s_1\,s_2\,(l_2/l_1)^2}{[s_1 + s_2\,(l_2/l_1)^2]\,m}}$$

18.7 DAMPED VIBRATIONS

When an elastic body is set in vibratory motion, the vibrations die out after some time due to the internal molecular friction of the mass of the body and the friction of the medium in which it vibrates.

The diminishing of vibrations with time is called damping. External damping can be increased by using dashpots or dampers. A dashpot has a piston which moves in a cylinder filled with some fluid. Shock absorbers, fitted in the suspension system of a motor vehicle, reduce the movement of the springs when there are sudden shocks, thus damping out the bouncing which could have occurred otherwise.

As before, consider a helical spring suspended from a fixed support (Fig. 18.7). A-A is the level of the free end before the mass m is suspended. B-B is the level of static equilibrium under the weight of the mass. The mass is attached to a dashpot to retard its movement.

It is usual to assume that the damping force is proportional to the velocity of vibration at lower values of speed and proportional to the square of the velocity at higher speeds. Only the former case will be considered in this chapter.

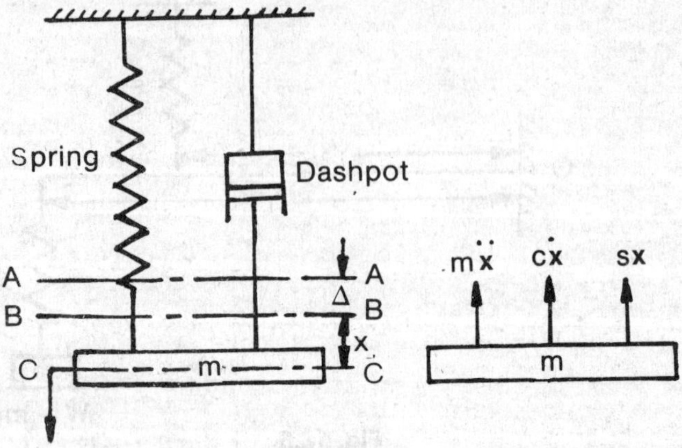

Fig. 18.7

Consider the forces on the mass m when it is displaced through a distance below the equilibrium position during vibratory motion.

Let s = stiffness of the spring

c = damping coefficient (damping force per unit velocity)

ω_n = frequency of natural undamped vibrations

x = displacement of mass from mean position at time t

$v = \dot{x}$ = velocity of the mass at time t

$f = \ddot{x}$ = acceleration of the mass at time t.

When the mass moves downwards, the friction force of the dashpot acts in the upward direction.

Now, the forces acting on the mass are

– Inertia force $= m\ddot{x}$ (upwards)

– Damping force $= c\dot{x}$ (upwards)

– Spring force (restoring force) $= sx$ (upwards)

As the sum of the inertia force and the external forces on a body in any direction is to be zero,

$$m\ddot{x} + c\dot{x} + sx = 0$$

or

$$\ddot{x} + \frac{c}{m}\dot{x} + \frac{s}{m}x = 0 \tag{18.11}$$

It is a differential equation of the second order. Its solution will be of the form

$$x = A\,e^{\alpha_1 t} + B\,e^{\alpha_2 t} \tag{18.12}$$

Where A and B are some constants. α_1 and α_2 are the roots of the auxiliary equation

$$\alpha^2 + \frac{c}{m}\alpha + \frac{s}{m} = 0 \tag{18.13}$$

i.e.

$$\alpha_{1,2} = -\frac{c}{2m} \pm \sqrt{\left(\frac{c}{2m}\right)^2 - \left(\frac{s}{m}\right)} \tag{18.14}$$

The ratio of $\left(\frac{c}{2m}\right)^2$ to $\left(\frac{s}{m}\right)$ represents the degree of dampness provided in the system and its square-root is known as *damping factor* ζ, i.e.

$$\zeta = \sqrt{\frac{(c/2m)^2}{s/m}} = \frac{c}{2\sqrt{sm}}$$

or

Damping coefficient,

$$c = 2\,\zeta\,\sqrt{sm} = 2\,\zeta\,m\,\omega_n = 2\,\zeta\,\frac{s}{\omega_n} \tag{18.15}$$

When $\zeta = 1$, the damping is known as critical. The corresponding value of damping coefficient c is denoted by c_c.

Thus under critical damping conditions,

$$c_c = 2\,\sqrt{sm} = 2m\,\omega_n = 2s/\omega_n \tag{18.16}$$

and

$$\zeta = \frac{c}{c_c} = \frac{\text{Actual damping coefficient}}{\text{Critical damping coefficient}} \tag{18.17}$$

Thus when

$\zeta = 1$, the damping is critical

$\zeta > 1$, the system is over-damped

$\zeta < 1$, the system is under-damped.

Equation (18.11) can also be written as

$$\ddot{x} + 2\,\zeta\,\omega_n\,\dot{x} + \omega_n^2\,x = 0 \tag{18.18}$$

and

$$\alpha_{1,2} = -\,\zeta\,\omega_n \pm \sqrt{\zeta^2\,\omega_n^2 - \omega_n^2}$$

$$= (-\,\zeta \pm \sqrt{\zeta^2 - 1}\,)\,\omega_n$$

The exact solution of Eq. (18.18) will depend upon whether the roots $\alpha_{1,2}$ are real or imaginary.

(i) $\zeta > 1$, i.e. the system is over-damped.

The roots of the auxiliary equation are real.

$$\alpha_{1,2} = (-\,\zeta \pm \sqrt{\zeta^2 - 1}\,)\,\omega_n$$

Therefore, the solution is

$$x = A\,e^{\alpha_1 t} + B\,e^{\alpha_2 t}$$

$$= A\,e^{(-\zeta + \sqrt{\zeta^2 - 1})\,\omega_n t} + B\,e^{(-\zeta - \sqrt{\zeta^2 - 1})\,\omega_n t} \tag{18.19}$$

Constants A and B can be determined from the initial conditions. This is the equation of an aperiodic motion, i.e. the system cannot vibrate due to over-damping. The magnitude of the resultant displacement approaches zero with time.

(ii) $\zeta < 1$, i.e. the system is underdamped.
The roots of the auxiliary equation are imaginary.

$$\alpha_{1,2} = (-\zeta \pm i\sqrt{1-\zeta^2})\,\omega_n$$

$$\therefore \quad x = A\,e^{(-\zeta + i\sqrt{1-\zeta^2})\,\omega_n t} + B\,e^{(-\zeta - i\sqrt{1-\zeta^2})\,\omega_n t}$$

$$= e^{-\zeta\omega_n t}\left[A\,e^{(i\sqrt{1-\zeta^2}\,\omega_n t)} + B\,e^{(-i\sqrt{1-\zeta^2}\,\omega_n t)}\right]$$

Put

$$\sqrt{1-\zeta^2}\,\omega_n = \omega_d$$

$$\text{Then } x = e^{-\zeta\omega_n t}\left[A\,e^{i\omega_d t} + B\,e^{-i\omega_d t}\right]$$

$$= e^{-\zeta\omega_n t}\left[A\,(\cos\omega_d t + i\sin\omega_d t) + B\,(\cos\omega_d t - i\sin\omega_d t)\right]$$

$$= e^{-\zeta\omega_n t}\left[(A+B)\cos\omega_d t + i\,(A-B)\sin\omega_d t\right]$$

$$= e^{-\zeta\omega_n t}\left[C\cos\omega_d t + D\sin\omega_d t\right] \quad (18.20)$$

where

$$C = A + B \quad \text{and} \quad D = i\,(A - B)$$

Constants C and D can be found from initial conditions. Alternatively, put

$$A + B = X\sin\varphi \quad \text{and} \quad i\,(A - B) = X\cos\varphi.$$

Thus

$$x = e^{-\zeta\omega_n t}\,(X\sin\varphi\cos\omega_d t + X\cos\varphi\sin\omega_d t)$$

$$= X\,e^{-\zeta\omega_n t}\sin(\omega_d t + \varphi) \quad (18.21)$$

Constants X and φ are to be determined from the initial conditions. This equation indicates that the system oscillates with frequency ω_d $(= \sqrt{1-\zeta^2}\,\omega_n)$. As ζ is less than 1, ω_d is always less than ω_n.

The solution consists of three terms:

- X, which is constant
- $e^{-\zeta\omega_n t}$, which decreases with time and finally $e^{-\infty} = 0$
- $\sin(\omega_d t + \varphi)$ which represents a repetition of motion.

Thus, the resultant motion is oscillatory with decreasing amplitudes having a frequency of ω_d. Ultimately, the motion dies down with time.

Also,

$$\text{Linear frequency, } f_d = \frac{\omega_d}{2\pi}$$

and

$$\text{Time period, } T_d = \frac{2\pi}{\omega_d}$$

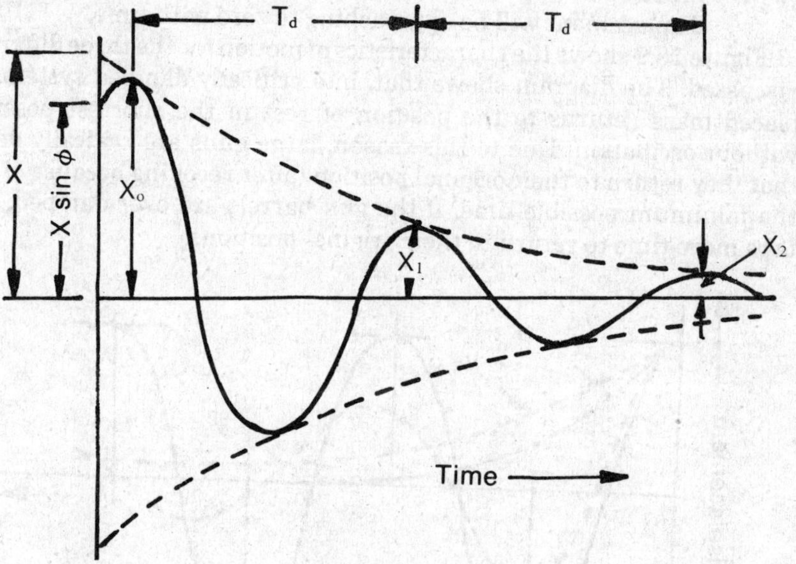

Fig. 18.8

let X_0 = displacement at the start of motion when $t = 0$

X_1 = displacement at the end of first oscillation when $t = T_d$

$$= X e^{-\zeta \omega_n T_d} \sin (\omega_d T_d + \varphi)$$

$$= X e^{-\zeta \omega_n T_d} \sin (\omega_d \frac{2\pi}{\omega_d} + \varphi)$$

$$= X e^{-\zeta \omega_n T_d} \sin \varphi$$

X_2 = displacement at the end of second oscillation

$$= X e^{-\zeta \omega_n \times 2 T_d} \sin \varphi$$

Similarly,

$$X_3 = X e^{-\zeta \omega_n \times 3 T_d} \sin \varphi$$

....

$$X_n = X e^{-\zeta \omega_n \times n T_d} \sin \varphi$$

$$X_{n+1} = X e^{-\zeta \omega_n (n+1) T_d} \sin \varphi$$

Then

$$\frac{X_n}{X_{n+1}} = e^{\zeta \omega_n T_d} = \frac{X_0}{X_1} = \frac{X_1}{X_2} = \frac{X_2}{X_3} = \ldots \quad (18.22)$$

which shows that the ratio of two successive oscillations is constant (Fig. 18.8).

(iii) $\zeta = 1$, i.e. the damping is critical.

The roots of the auxiliary equation are equal, each being equal to $-\omega_n$ and the solution is

$$x = (A + Bt)\, e^{-\omega_n t} \quad (18.23)$$

Since $e^{-\omega_n t}$ approaches zero as $t \to \infty$, the motion is aperiodic. The displacement will be approaching to zero with time.

Figure 18.9 shows the characteristics of motion for the three different cases discussed. The diagram shows that, in a critically damped system, the displaced mass returns to the position of rest in the shortest possible time without oscillation. Due to this reason, large guns are critically damped so that they return to their original position (after recoiling because of firing) in the minimum possible time. If the gun barrels are over-damped, they will take more time to return to their original positions.

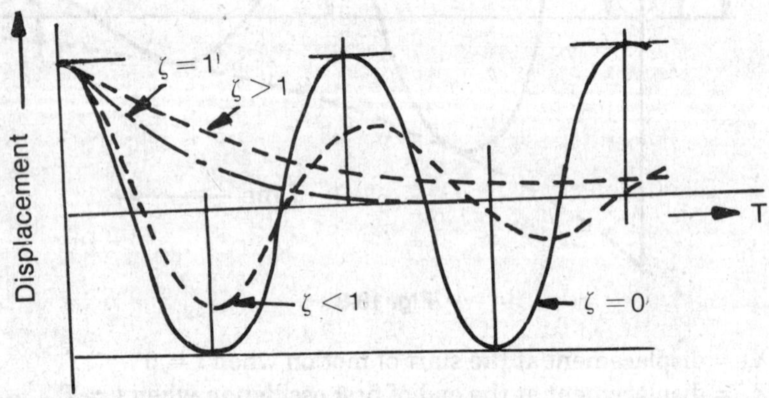

Fig. 18.9

It can be noted that:

(i) An undamped system ($\zeta = 0$) vibrates at its natural frequency which depends upon the static deflection under the weight of its mass ($\omega_n = \sqrt{g/\Delta}$).

(ii) When the system is underdamped ($\zeta < 1$), the frequency of the system decreases to ω_d ($= \sqrt{1 - \zeta^2}\, \omega_n$) and the time period increases to $T_d = 2\pi/\omega_d$. The amplitudes of the vibrations decrease with time, the ratio of successive amplitudes being constant. The vibrations die down with time.

(iii) At critical damping, $\zeta = 1$, $\omega_d = 0$ and $T_d = \infty$. The system does not vibrate and the mass m moves back slowly to the equilibrium position.

(iv) For an overdamped system, $\zeta > 1$, the system behaves in the same manner as for critical damping.

(v) ζ is the ratio of the existing damping in a system to that required for critical damping, i.e. $\zeta = c/c_c$.

18.8 LOGARITHMIC DECREMENT

In Section 18.7, it was observed that the ratio of two successive oscillations is constant in an underdamped system. Natural logarithm of this ratio is called logarithmic decrement and is denoted by

$$\delta = \ln\left(\frac{X_n}{X_{n+1}}\right) = \ln e^{(\zeta\,\omega_n\,T_d)} = \zeta\,\omega_n\,T_d$$

or

$$\delta = \zeta\,\omega_n \frac{2\pi}{\omega_d} = \zeta\,\omega_n \frac{2\pi}{\sqrt{1-\zeta^2}\,\omega_n} = \frac{2\pi\,\zeta}{\sqrt{1-\zeta^2}} \tag{18.24}$$

Example 18.5 **A vibrating system consists of a mass of 50 kg, a spring of stiffness 30 kN/m and a damper. The damping provided is only 20% of the critical value. Determine:**

(i) the damping factor
(ii) the critical damping coefficient
(iii) the natural frequency of damped vibrations
(iv) the logarithmic decrement
(v) the ratio of two consecutive amplitudes.

Solution:

$$m = 50 \text{ kg} \qquad s = 30\,000 \text{ N/m} \qquad c = 0.2c_c$$

(i) $\zeta = \dfrac{c}{c_c} = \underline{0.2}$

(ii) $c_c = 2\sqrt{sm} = 2\sqrt{30\,000 \times 50} = 2450 \text{ N/m/s} = 2.45 \underline{\text{ N/mm/s}}$

(iii) $\omega_d = \sqrt{1-\zeta^2}\,\omega_n$

where

$$\omega_n = \sqrt{\frac{s}{m}} = \sqrt{\frac{30\,000}{50}} = 24.5 \text{ rad/s}$$

$$\omega_d = \sqrt{1-(0.2)^2} \times 24.5 = \underline{24 \text{ rad/s}}$$

(iv) $\delta = \dfrac{2\pi\,\zeta}{\sqrt{1-\zeta^2}} = \dfrac{2\,\pi \times 0.2}{\sqrt{1-(0.2)^2}} = \underline{1.28}$

(v) $\dfrac{X_n}{X_{n+1}} = e^{\delta} = e^{1.28} = \underline{3.6}$

Example 18.6 **Determine the time in which the mass in a damped vibrating system would settle down to 1/50th of its initial deflection for the following data:**

$m = 200$ kg $\quad \zeta = 0.22 \quad s = 40$ N/mm

Also find the number of oscillations completed to reach this value of deflection.

Solution: We know

$$\frac{X_0}{X_N} = e^{\zeta\,\omega_n\,NT_d}$$

where

$$\omega_n = \sqrt{\frac{s}{m}} = \sqrt{\frac{40 \times 10^3}{200}} = 14.14 \text{ rad/s}$$

$\therefore$ $50 = e^{0.22 \times 14.14\, NT_d}$

or

Total time $NT_d = \underline{1.26 \text{ s}}$

$$T_d = \frac{2\pi}{\sqrt{1-\zeta^2}\,\omega_n} = \frac{2\pi}{(\sqrt{1-(0.22)^2}) \times 14.14} = 0.455 \text{ s}$$

Number of oscillations completed $= \dfrac{1.26}{0.455} = \underline{2.76}$

Example 18.7 In a single-degree damped vibrating system, a suspended mass of 8 kg makes 30 oscillations in 18 seconds. The amplitude decreases to 0.25 of the initial value after 5 oscillations. Determine:

(i) the stiffness of the spring
(ii) the logarithmic decrement
(iii) the damping factor, and
(iv) the damping coefficient.

Solution

$m = 8$ kg $\quad N = 30 \quad t = 18$ s

$f_n = \dfrac{30}{18} = 1.67$ Hz

$\omega_n = 2\pi f_n = 2\pi \times 1.67 = 10.47$ rad/s

(i) $\omega_n = \sqrt{\dfrac{s}{m}}$

$10.47 = \sqrt{\dfrac{s}{8}}$

$\therefore$ $s = 877$ N/m or $\underline{0.877 \text{ N/mm}}$

(ii) $$\frac{X_0}{X_5} = \frac{X_0}{X_1} \times \frac{X_1}{X_2} \times \frac{X_2}{X_3} \times \frac{X_3}{X_4} \times \frac{X_4}{X_5}$$

$$= \left(\frac{X_0}{X_1}\right)^5 \qquad \left(\frac{X_0}{X_1} = \frac{X_1}{X_2} = \frac{X_2}{X_3} = \frac{X_3}{X_4} = \frac{X_4}{X_5}\right)$$

$$\therefore \quad \left(\frac{X_0}{X_1}\right) = \left(\frac{X_0}{X_5}\right)^{1/5} = \left(\frac{1}{0.25}\right)^{1/5} = 1.32$$

$$\delta = \ln\left(\frac{X_0}{X_5}\right) = \ln 1.32 = 0.278$$

(iii) $$\frac{2\pi\zeta}{\sqrt{1-\zeta^2}} = 0.278$$

or

$$\sqrt{1-\zeta^2} = 22.6\ \zeta$$

$$1-\zeta^2 = 510.82\,\zeta^2$$

$$\zeta^2 = 0.00195$$

$$\zeta = \underline{0.0442}$$

(iv) $$c = 2\,m\,\omega_n\,\zeta$$

$$= 2 \times 8 \times 10.47 \times 0.0442$$

$$= \underline{7.4\ \text{N/m/s}}$$

Example 18.8 A machine mounted on springs and fitted with a dashpot has a mass of 60 kg. There are three springs, each of stiffness 12 N/mm. The amplitude of vibrations reduces from 45 to 8 mm in two complete oscillations. Assuming that the damping force varies as the velocity, determine:

(i) the damping coefficient,
(ii) the ratio of frequencies of damped and undamped vibrations, and
(iii) the periodic time of damped vibrations.

Solution: $m = 60$ kg

Stiffness of each spring = 12 N/mm

Combined stiffness, $s = 12 \times 3 = 36$ N/mm $= 36 \times 10^3$ N/m

$$\omega_n = \sqrt{\frac{s}{m}} = \sqrt{\frac{36 \times 10^3}{60}} = 24.49 \text{ rad/s}$$

(i) $$\frac{X_0}{X_2} = \frac{X_0}{X_1} \times \frac{X_1}{X_2}$$

$$= \left(\frac{X_0}{X_1}\right)^2 \qquad \left(\frac{X_1}{X_2} = \frac{X_2}{X_1}\right)$$

or

$$\frac{X_0}{X_1} = \left(\frac{X_0}{X_2}\right)^{1/2} = \left(\frac{45}{8}\right)^{1/2} = 2.37$$

$$\frac{2\,\pi\,\zeta}{\sqrt{1-\zeta^2}} = \ln 2.37 = 0.864$$

$$1-\zeta^2 = 52.88\,\zeta^2$$

$$\zeta^2 = 0.0185$$

$$\zeta = 0.136$$

$$c = 2m\,\omega_n\,\zeta = 2 \times 60 \times 24.49 \times 0.136 = 400 \text{ N/m/s}$$
$$= \underline{0.4 \text{ N/mm/s}}$$

(ii) $$\frac{\text{Damped frequency}}{\text{Undamped frequency}} = \frac{\omega_d}{\omega_n} = \frac{\sqrt{1-\zeta^2}\,\omega_n}{\omega_n} = \sqrt{1-\zeta^2}$$
$$= \sqrt{1-(0.136)^2} = \underline{0.99}$$

(iii) $$T_d = \frac{2\pi}{\omega_d} = \frac{2\pi}{\sqrt{1-\zeta^2}\,\omega_n} = \frac{2\pi}{(\sqrt{1-(0.136)^2}) \times 24.49} = \underline{0.259 \text{ s}}$$

Example 18.9 A machine weighs 18 kg and is supported on springs and dashpots. The total stiffness of the springs is 12 N/mm and damping is 0.2 N/mm/s. The system is initially at rest and a velocity of 120 mm/s is imparted to the mass. Determine:

(i) the displacement and velocity of mass as a function of time,
(ii) the displacement and velocity after 0.4 s.

Solution:

$m = 18$ kg $\quad v = 0.12$ m/s
$s = 12$ N/mm $= 12\,000$ N/m
$c = 0.2$ N/mm/s $= 200$ N/m/s

$$\omega_n = \sqrt{\frac{s}{m}} = \sqrt{\frac{12\,000}{18}} = 25.82 \text{ rad/s}$$
$$c = 2\,m\,\omega_n\,\zeta$$
$$200 = 2 \times 18 \times 25.82 \times \zeta$$
$$\zeta = 0.215$$
$$\omega_d = \sqrt{1-\zeta^2}\,\omega_n = \sqrt{1-(0.215)^2} \times 25.82 = 25.2 \text{ rad/s}$$

(i) $x = X\,e^{-\zeta\omega_n t} \sin(\omega_d t + \varphi)$ [Eq. (18.21)]

$x = 0 \quad$ at $t = 0$

$\therefore \quad X \sin\varphi = 0$

or

$\sin\varphi = 0 \quad$ (X cannot be zero)

or

$\varphi = 0$

$\therefore \quad x = X\,e^{-\zeta\omega_n t} \sin(\omega_d t)$

$\dot{x} = X\,e^{-\zeta\omega_n t}\,\omega_d \cos(\omega_d t) + X \sin\omega_d t\,(-\zeta\,\omega_n)\,e^{-\zeta\omega_n t}$

$\dot{x} = 0.12 \quad$ at $t = 0$

$\therefore \quad 0.12 = X\,\omega_d = 25.2\,X$

or $X = 0.00\,476 \text{ m} = 476 \text{ mm}$

Displacement, $x = 4.76\ e^{-0.215 \times 25.82\,t} \sin(25.2t)$

or $x = 4.76\, e^{-5.55\,t} \sin 25.2\, t$

Velocity,

$$\dot{x} = X e^{-\zeta \omega_n t} [\omega_d \cos \omega_d t - \zeta\, \omega_n \sin \omega_d t]$$

$$= 4.76\, e^{-5.55\,t} [25.2 \cos 25.2\, t - 5.55 \sin 25.2\, t]$$

$$= e^{-5.55\,t} [120 \cos 25.2\, t - 26.4 \sin 25.2\, t]$$

(ii) $x = 4.76\, e^{-5.55 \times 0.4} \sin(25.2 \times 0.4)]$

$= \underline{0.0905 \text{ mm}}$

$\dot{x} = e^{-5.55 \times 0.4} [120 \cos(25.2 \times 0.4) - 26.4 \sin(25.2 \times 0.4)]$

$= \underline{12.3 \text{ mm}}$

Example 18.10 A gun is so designed that, on firing, the barrel recoils against a spring. A dashpot, at the end of the recoil, allows the barrel to come back to its initial position within the minimum time without any oscillation. A gun barrel has a mass of 500 kg and a recoil spring of 300 N/mm. The barrel recoils 1 m on firing. Determine

(i) the initial recoil velocity of the gun barrel, and
(ii) the critical damping coefficient of the dashpot engaged at the end of the recoil stroke.

Solution:

$m = 500$ kg $\quad x = 300$ N/mm $\quad x = 1$ m

(i) The dashpot does not operate during the recoil.
KE of the barrel = work done on the spring

$$\frac{1}{2} m v^2 = \frac{1}{2} c x^2$$

$$\frac{1}{2} \times 500 \times v^2 = \frac{1}{2} \times (300 \times 10^3) \times (1)^2$$

$$v = \underline{24.5 \text{ m/s}}$$

(ii) $c_c = 2\, m\, \omega_n$
But

$$\omega_n = \sqrt{\frac{s}{m}} = \sqrt{\frac{300 \times 10^3}{500}} = 24.5 \text{ m/s}$$

$\therefore \quad c_c = 2 \times 500 \times 24.5 = 24\,500 \text{ N/m/s} \text{ or } \underline{24.5 \text{ N/mm/s}}$

18.9 FORCED-DAMPED VIBRATIONS

A mass m is attached to a helical spring and is suspended from a fixed support as before. Damping is also provided in the system with a dashpot (Fig.18.10).

Before the mass is set in motion, let B-B be the static equilibrium position under the weight of the mass. Now, if the mass is subjected to an oscillating force $F = F_0 \sin \omega t$, the forces acting on the mass at any instant will be

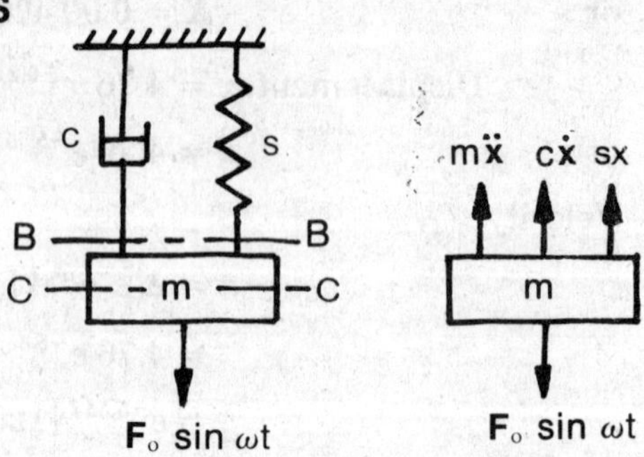

Fig. 18.10

- Impressed oscillating force $F = F_0 \sin \omega t$ (downwards)
- Inertia force $= m\ddot{x}$ (upwards)
- Damping force $= c\dot{x}$ (upwards)
- Spring force (restoring force) $= sx$ (upwards)

Thus the equation of motion will be

$$m\ddot{x} + c\dot{x} + sx - F_0 \sin \omega t = 0$$

or

$$m\ddot{x} + c\dot{x} + sx = F_0 \sin \omega t \qquad (18.25)$$

Complete solution of this equation consists of two parts, the complementary function (CF) and the particular integral (PI).

$$CF = X e^{-\zeta \omega_n t} \sin(\omega_d t + \varphi_1) \qquad \text{[refer to Eq. (18.21)]}$$

To obtain the PI, let

$$\frac{c}{m} = a \ , \ \frac{s}{m} = b \text{ and } \frac{F_0}{m} = d.$$

Then, using the operator D, the equation becomes

$$(D^2 + aD + b)\, x = d \sin \omega t$$

$$PI = \frac{d \sin \omega t}{D^2 + aD + b}$$

$$= \frac{d \sin \omega t}{-\omega^2 + aD + b}$$

$$= \frac{1}{(b - \omega^2) + aD} \times \frac{(b - \omega^2) - aD}{(b - \omega^2) - aD} d \sin \omega t$$

$$= d\left[\frac{\sin\omega t\,(b-\omega^2) - aD\sin\omega t}{(b-\omega^2)^2 - a^2D^2}\right]$$

$$= d\left[\frac{\sin\omega t\,(b-\omega^2) - a\,\omega\cos\omega t}{(b-\omega^2)^2 + (a\omega)^2}\right]$$

Take $(b-\omega^2) = R\cos\varphi$ and $a\,\omega = R\sin\varphi$

Constants R and φ are given by

$$R = \sqrt{(b-\omega^2)^2 + (a\,\omega)^2} \quad \text{and} \quad \varphi = \tan^{-1}\frac{a}{b-\omega^2}$$

$$PI = \frac{d\,R\,(\sin\omega t\cos\varphi - \cos\omega t\sin\varphi)}{(b-\omega^2)^2 + (a\omega)^2}$$

$$= \frac{d\sqrt{(b-\omega^2)^2 + (a\,\omega)^2}}{(b-\omega^2) + (a\,\omega)^2}\sin(\omega t - \varphi)$$

$$= \frac{d}{\sqrt{(b-\omega^2)^2 + (a\omega)^2}}\sin(\omega t - \varphi)$$

$$= \frac{F_0\,m}{\sqrt{\left(\frac{s}{m} - \omega^2\right)^2 + \left(\frac{c}{m}\omega\right)^2}}\sin(\omega t - \varphi)$$

$$= \frac{F_0}{\sqrt{(s - m\,\omega^2)^2 + (c\,\omega)^2}}\sin(\omega t - \varphi)$$

$$x = CF + PI$$

$$= X e^{-\zeta\omega_n t}\sin(\omega_d t + \varphi_1) + \frac{F_0}{\sqrt{(s - m\,\omega^2)^2 + (c\,\omega)^2}}\sin(\omega t - \varphi) \quad (18.26)$$

The damped-free vibrations represented by the first part (CF) becomes negligible with time as $e^{-\infty} = 0$. The steady-state response of the system is then given by the second part PI.

The amplitude of the steady-state response is given by

$$A = \frac{F_0}{\sqrt{(s - m\,\omega^2)^2 + (c\,\omega)^2}} \quad (18.27)$$

$$= \frac{F_0/s}{\sqrt{\left(1 - \frac{m\,\omega^2}{s}\right)^2 + \left(\frac{c}{s}\omega\right)^2}}$$

$$= \frac{F_0/s}{\sqrt{\left[1 - \left(\frac{\omega}{\omega_n}\right)^2\right]^2 + \left(2\zeta\frac{\omega}{\omega_n}\right)^2}} \quad (18.27a)$$

The frequency of the steady-state forced vibrations is the same as that of the impressed vibrations. φ is the phase lag for the displacement relative to the velocity vector.

$$\tan\varphi = \frac{a}{b-\omega^2} = \frac{\frac{c}{m}\omega}{\frac{s}{m}-\omega^2} = \frac{c\,\omega}{s-m\,\omega^2} = \frac{2\zeta\frac{\omega}{\omega_n}}{1-\left(\frac{\omega}{\omega_n}\right)^2} \quad (18.28)$$

The particular solution of the equation of motion can also be obtained graphically as follows:

Assuming that the displacement of the vibrating mass under the action of the applied simple harmonic force $F_0 \sin \omega t$ is also simple harmonic and lags by an amount φ. Then

$$x = A\sin(\omega t - \varphi)$$

and

$$\dot{x} = \omega A\cos(\omega t - \varphi) = \omega A \sin\left[\frac{\pi}{2} + (\omega t - \varphi)\right]$$

$$\ddot{x} = -\omega^2 A\sin(\omega t - \varphi)$$

where A is the amplitude of vibrations.

Substituting these values in the equation

$$m\ddot{x} + c\dot{x} + sx = F_0 \sin\omega t$$

$$-m\,\omega^2 A\sin(\omega t - \varphi) + c\,\omega A\sin\left[\frac{\pi}{2} + (\omega t - \varphi)\right] + sA\sin(\omega t - \varphi) - F_0\sin\omega t = 0$$

$$F_0\sin\omega t + m\,\omega^2 A\sin(\omega t - \varphi) - c\,\omega A\sin\left[\frac{\pi}{2} + (\omega t - \varphi)\right] - sA\sin(\omega t - \varphi) = 0$$

The forces and the vector sum of the same have been shown in Fig. 18.11. In triangle *abc*.

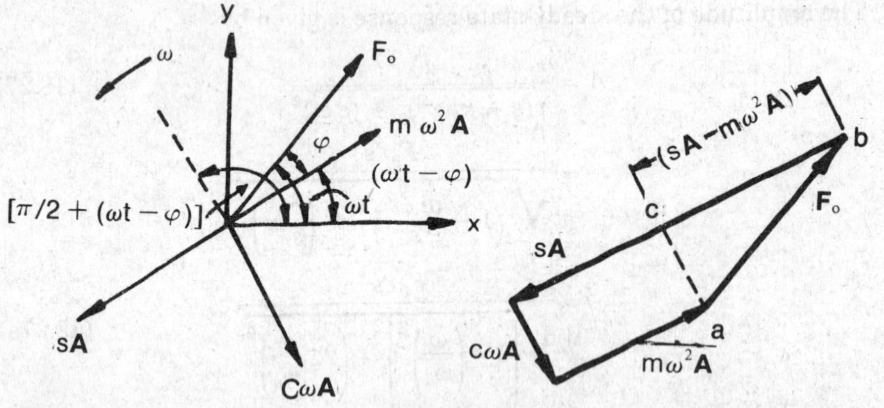

Fig. 18.11

$$\sqrt{(sA - m\,\omega^2 A)^2 + (c\,\omega\,A)^2} = F_0$$

or

$$A\sqrt{(s - m\,\omega^2) + (c\,\omega)^2} = F_0$$

or

$$A = \frac{F_0}{\sqrt{(s - m\,\omega^2)^2 + (c\,\omega)^2}}$$

and

$$\tan\varphi = \frac{c\,\omega}{s - m\,\omega^2}$$

The vectors as shown in the diagram are fixed relative to one another and rotate with angular velocity ω.

18.10 MAGNIFICATION FACTOR

The ratio of the amplitude of the steady-state response to the static deflection under the action of force F_0 is known as *magnification factor* (*MF*).

$$MF = \frac{F_0/\sqrt{(s - \omega^2)^2 + (c\omega)^2}}{F_0/s}$$

$$= \frac{s}{\sqrt{(s - m\,\omega^2)^2 + (c\,\omega)^2}}$$

$$= \frac{1}{\sqrt{\left(1 - \frac{m}{s}\,\omega^2\right)^2 + \left(\frac{c}{s}\,\omega\right)^2}}$$

$$= \frac{1}{\sqrt{\left[1 - \left(\frac{\omega}{\omega_n}\right)^2\right]^2 + \left(2\,\zeta\,\frac{\omega}{\omega_n}\right)^2}} \qquad (18.29)$$

Thus, the magnification factor depends upon:
(a) the ratio of frequencies, $\frac{\omega}{\omega_n}$, and
(b) the damping factor.

The plot of magnification factor against the ratio of frequencies (ω/ω_n) for different values of ζ is shown in Fig. 18.12. The curves show that as the damping increases or ζ increases, the maximum value of the magnification factor decreases and vice-versa. When there is no damping ($\zeta = 0$), it reaches infinity at $\omega/\omega_n = 1$, i.e. when the frequency of the forced vibrations is equal to the frequency of the free vibration. This condition is known as resonance.

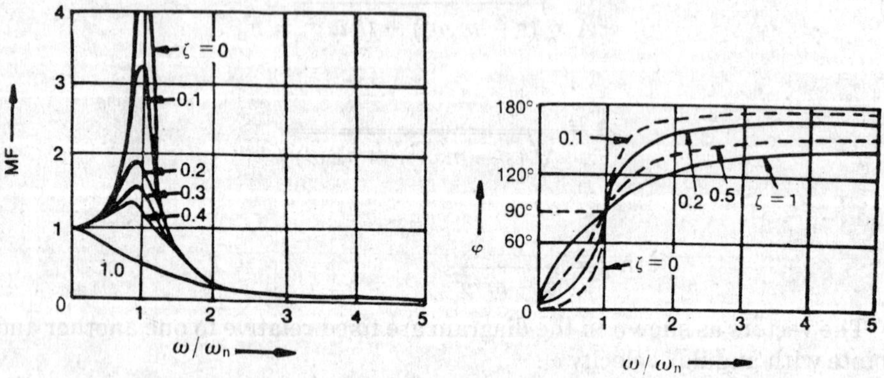

Fig. 18.12

In practice, the magnification factor cannot reach infinity owing to friction which tends to dampen the vibration. However, the amplitude can reach very high values.

Also observe that irrespective of the amount of damping, the maximum amplitudes of vibration occur before the ratio ω/ω_n reaches unity or when the frequency of the forced vibration is less than that of the undamped vibrations.

Example 18.11 A machine part having a mass of 2.5 kg vibrates in a viscous medium. A harmonic exciting force of 30 N acts on the part and causes a resonant amplitude of 14 mm with a period of 0.22 second. Find the damping coefficient.

If the frequency of the exciting force is changed to 4 Hz, determine the increase in the amplitude of the forced vibrations upon the removal of the damper.

Solution:

$$m = 2.5 \text{ kg} \qquad F_0 = 30 \text{ N}$$

$$A = 14 \text{ mm} \qquad T = 0.22 \text{ s}$$

$$\omega = \frac{2\pi}{T} = \frac{2\pi}{0.22} = 28.56 \text{ rad/s}$$

(i) At resonance, $\omega = \omega_n$

or

$$\omega_n = \sqrt{\frac{s}{m}} = 28.56 \text{ rad/s}$$

or

$$\sqrt{\frac{s}{2.5}} = 28.56$$

or

$$s = 2039 \text{ N/m} \quad \text{or} \quad 2.039 \text{ N/mm}$$

$$A = \frac{F_0/s}{\sqrt{\left[1 - \left(\frac{\omega}{\omega_n}\right)^2\right]^2 + \left(\frac{2\zeta\omega}{\omega_n}\right)^2}}$$

or

$$A = \frac{F_0/s}{2\zeta} \qquad \left(\frac{\omega}{\omega_n} = 1\right)$$

or

$$0.014 = \frac{30/2039}{2\zeta}$$

or

$$\zeta = 0.526$$

$$c = 2\, m\, \omega_n\, \zeta = 2 \times 2.5 \times 28.56 \times 0.526$$

$$= 75.04 \text{ N/m/s}$$

$$= \underline{0.07\,504 \text{ N/mm/s}}$$

(ii) $\omega = f_n \times 2\pi = 4 \times 2\pi = 25.13$ rad/s

With damper:

$$A = \frac{30/2039}{\sqrt{\left[1 - \left(\frac{25.13}{28.56}\right)^2\right]^2 + \left[2 \times 0.526 \times \frac{25.13}{28.56}\right]^2}}$$

$$= \frac{30/2039}{\sqrt{(0.2258)^2 + (0.9248)^2}}$$

$$= 0.0155 \text{ m}$$

Without damper: $\zeta = 0$

$$A = \frac{30/2039}{0.2258} = 0.0652 \text{ m}$$

$\therefore$ Increase in magnitude = 0.0652 – 0.0155 = 0.0497 m or $\underline{49.7 \text{ mm}}$

Example 18.12 A single-cylinder vertical diesel engine has a mass of 400 kg and is mounted on a steel chassis frame. The static deflection owing to the weight of the chassis is 2.4 mm. The reciprocating masses of the engine amounts to 18 kg and the stroke of the engine is 160 mm. A dashpot with a damping coefficient of 2 N/mm/s is also used to dampen the vibrations. In the steady-state of the vibrations, determine:

(i) the amplitude of the vibrations if the driving shaft rotates at 500 rpm
(ii) the speed of the driving shaft when the resonance occurs.

Solution:

$m = 400$ kg $\qquad N = 500$ rpm
$c = 2000$ N/m/s $\qquad \Delta = 2.4$ mm
$r = 80$ mm $\qquad = 0.0024$ m

$$\omega = \frac{2\pi \times 500}{60} = 52.36 \text{ rad/s}$$

Now $\quad s \times \Delta = mg$

$\therefore \quad s \times 0.0024 = 400 \times 9.81$

$s = 1.635 \times 10^6$ N/m

Centrifugal force due to reciprocating parts (or the static force),

$$F_0 = mr\,\omega^2 = 18 \times 0.08 \times (52.36)^2 = 3948 \text{ N}$$

(i) $$A = \frac{F_0}{\sqrt{(s - m\,\omega^2)^2 + (c\,\omega)^2}} \qquad \text{[Eq. (18.27)]}$$

$$= \frac{3948}{\sqrt{[1.635 \times 10^6 - 400(52.36)^2]^2 + (2000 \times 52.36)^2}}$$

$= 0.0072$ m $\quad$ or $\underline{7.2 \text{ mm}}$

(ii) Resonant speed:

$$\omega = \omega_n = \sqrt{\frac{s}{m}} = \sqrt{\frac{1.635 \times 10^6}{400}} = 63.93 \text{ rad/s}$$

or $$\frac{2\pi N}{60} = 63.93$$

$N = \underline{610.5 \text{ rpm}}$

Example 18.13 A body having a mass of 15 kg is suspended from a spring which deflects 12 mm under weight of the mass. Determine the frequency of the free vibrations. What is the viscous damping force needed to make the motion aperiodic at a speed of 1 mm/s?

If, when damped to this extent, a disturbing force having a maximum value of 100 N and vibrating at 6 Hz is made to act on the body, determine the amplitude of the ultimate motion.

Solution:

$m = 15$ kg $\qquad \Delta = 12$ mm
$F_0 = 100$ N $\qquad f = 6$ Hz

$$f_n = \frac{1}{2\pi}\sqrt{\frac{g}{\Delta}} = \frac{1}{2\pi}\sqrt{\frac{9.81}{0.012}} = 4.55 \text{ Hz}$$

The motion becomes aperiodic when the damped frequency is zero or when it is critically damped ($\zeta = 1$) and

$$\omega = \omega_n = \sqrt{\frac{g}{\Delta}} = \sqrt{\frac{9.81}{0.012}} = \underline{28.59 \text{ rad/s}}$$

$$c = c_c = 2\, m\, \omega_n = 2 \times 15 \times 28.59 = 857 \text{ N/m/s}$$

$$= 0.857 \text{ N/mm/s}$$

Thus, the force needed is 0.857 N at a speed of 1 mm/s.

$$A = \frac{F_0}{\sqrt{(s - m\,\omega^2)^2 + (c\,\omega)^2}}$$

But $\omega = 2\pi \times f = 2\pi \times 6 = 37.7 \text{ rad/s}$

and s can be found from

$$f_n = \frac{1}{2\,\pi}\sqrt{s/m}$$

or $$4.55 = \frac{1}{2\pi}\sqrt{s/15}$$

or $$s = 12{,}260 \text{ N/m}$$

$\therefore$ $$A = \frac{100}{\sqrt{[12{,}260 - 15 \times (37.7)^2]^2 + (857 \times 37.7)^2}}$$

$$= 0.00298 \text{ m}$$

$$= \underline{2.98 \text{ mm}}$$

18.11 VIBRATION ISOLATION AND TRANSMISSIBILITY

Vibrations are produced in machines having unbalanced masses. These vibrations will be transmitted to the foundation upon which the machines are installed. This is usuallay undesirable. To diminish the transmitted forces, machines are usually mounted on springs or dampers, or on some other vibration isolating material. Then the vibratory forces can reach the foundation only through these springs, dampers, or the isolating material used.

Transmissibility is defined as the ratio of the force transmitted (to the foundation) to the force applied. It is a measure of the effectiveness of the vibration isolating material.

As the transmitted force is the vector sum of the spring force (sA) and the damping force ($c\omega A$) which are at perpendicular to each other (Fig.18.11).

$$F_t = \sqrt{(sA)^2 + (c\,\omega\,A)^2}$$
$$= A\sqrt{s^2 + (c\,\omega)^2}$$
$$= \frac{F_0}{\sqrt{(s - m\,\omega^2)^2 + (c\,\omega)^2}}\sqrt{s^2 + (c\,\omega)^2}$$
$$= \frac{F_0\sqrt{1 + \left(\frac{c}{s}\,\omega\right)^2}}{\sqrt{\left(1 - \frac{m}{s}\,\omega^2\right)^2 + \left(\frac{c}{s}\,\omega\right)^2}}$$
$$= \frac{F_0\sqrt{1 + (2\,\zeta\,\omega/\omega_n)^2}}{\sqrt{[-(\omega/\omega_n)^2]^2 + (2\,\zeta\,\omega/\omega_n)^2}}$$

Transmissibility,

$$\varepsilon = \frac{F_t}{F_0} = \frac{\sqrt{1 + (2\,\zeta\,\omega/\omega_n)^2}}{\sqrt{[1 - (\omega/\omega_n)^2]^2 + (2\,\zeta\,\omega/\omega_n)^2}} \qquad (18.30)$$

At resonance,

$$\frac{\omega}{\omega_n} = 1,$$

∴
$$\varepsilon = \frac{\sqrt{1 + (2\,\zeta)^2}}{2\zeta} \qquad (18.31)$$

When no damper is used, $\zeta = 0$
and

$$\varepsilon = \frac{1}{\pm\,[1 - (\omega/\omega_n)^2]} \qquad (18.32)$$

Transmissibility has been plotted against ω/ω_n for different values of ζ in Fig. 18.13. Note that

(i) When $\omega/\omega_n < \sqrt{2}$, ε is more than 1, i.e. the transmitted force is always more than the exciting force.
(ii) When $\omega/\omega_n > \sqrt{2}$, ε is less than 1 i.e. the transmitted force is always less than the exciting force.
(iii) When $\omega/\omega_n = \sqrt{2}$, ε is 1, i.e. the transmitted force is equal to the exciting force.
(iv) When $\omega/\omega_n = 1$, the transmitted force is infinity. If damping is used, the magnitude of the transmitted force can be reduced.
(v) When $\omega/\omega_n > \sqrt{2}$, ε increases as the damping is increased.

Thus in a system where ω/ω_n can vary from zero to higher values, dampers should not be used. Instead, stops may be provided to limit the resonance amplitude (at resonance, the amplification factor is infinity).

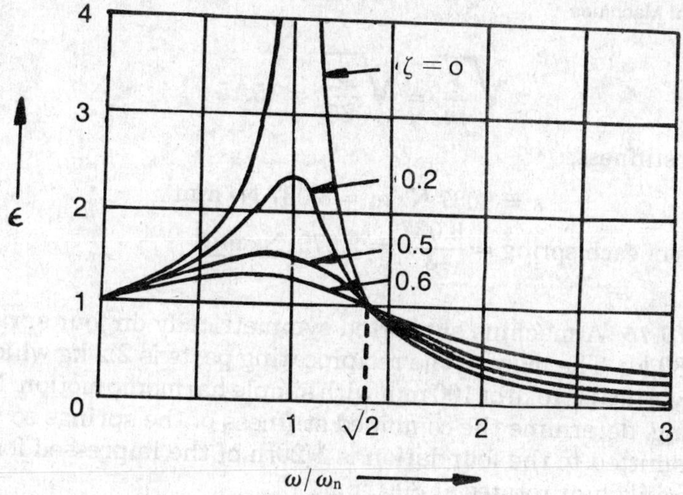

Fig. 18.13

Example 18.14 A refrigerator unit having a mass of 35 kg is to be supported on three springs, each having a spring stiffness s. The unit operates at 480 rpm. Find the value of stiffness s if only 10% of the shaking force is allowed to be transmitted to the supporting structure.

Solution: As no damper is used,

$$\varepsilon = \frac{1}{\pm\left[1 - \left(\frac{\omega}{\omega_n}\right)^2\right]}$$

$$\omega = \frac{2\pi \times 480}{60} = 16\pi \quad \text{and} \quad \varepsilon = 0.1$$

$$\therefore \quad 0.1 = \frac{1}{\pm\left[1 - \left(\frac{16\pi}{\omega_n}\right)^2\right]}$$

$$\pm\left[0.1 - 0.1\left(\frac{16\pi}{\omega_n}\right)^2\right] = 1$$

If the positive sign is taken, $\frac{16\pi}{\omega_n} = \sqrt{-9}$ which is not possible.

Therefore taking the negative sign, $\frac{16\pi}{\omega_n} = \sqrt{11}$

or

$$\omega_n = 15.15 \text{ rad/s}$$

or
$$\sqrt{\frac{s}{m}} = \sqrt{\frac{s}{35}} = 15.15$$

Equivalent stiffness,
$$s = 8037 \text{ N/m} = 8.037 \text{ N/mm}$$

Stiffness of each spring $= \frac{8.037}{3} = \underline{2.679 \text{ N/mm}}$

Example 18.15 A machine supported symmetrically on four springs has a mass of 80 kg. The mass of the reciprocating parts is 2.2 kg which move through a vertical stroke of 100 mm with simple harmonic motion. Neglecting damping, determine the combined stiffness of the springs so that the force transmitted to the foundation is 1/20th of the impressed force. The machine crankshaft rotates at 800 rpm.

If, under actual working conditions, the damping reduces the amplitudes of successive vibrations by 30%, find:

(i) the force transmitted to the foundation at 800 rpm,
(ii) the force transmitted to the foundation at resonance, and
(iii) the amplitude of the vibrations at resonance.

Solution:

$M = 80$ kg $\qquad \varepsilon = \frac{1}{20} = 0.05$

$m = 2.2$ kg $\qquad N = 800$ rpm

$$r = \frac{100}{2} = 50 \text{ mm}$$

$$\omega = \frac{2\pi N}{60} = \frac{2\pi\, 800}{60} = 83.78 \text{ rad/s}$$

In the absence of damping,
$$\varepsilon = \frac{1}{\left(\frac{\omega}{\omega_n}\right)^2 - 1}$$

or
$$0.05 = \frac{1}{\left(\frac{83.78}{\omega_n}\right)^2 - 1}$$

$$\omega_n = 18.28 \text{ rad/s}$$

or
$$\sqrt{\frac{s}{M}} = \sqrt{\frac{s}{80}} = 18.28$$

∴ Combined stiffness, $\quad s = 26\,739$ N/m

$$= 26.739 \text{ N/mm}$$

(i) $$\frac{2\pi\zeta}{\sqrt{1-\zeta^2}} = \ln\left(\frac{x_1}{x_2}\right) = \ln\left(\frac{1}{1-0.3}\right)$$

$$\frac{\zeta^2}{1-\zeta^2} = 0.00\,323$$

$$\zeta = 0.0567$$

$$\varepsilon = \frac{\sqrt{1+(2\,\zeta\,\omega/\omega_n)^2}}{\sqrt{[1-(\omega/\omega_n)^2]^2+(2\,\zeta\,\omega/\omega_n)^2}}$$

$$= \frac{\sqrt{1+\left(2\times 0.0567\times\dfrac{83.78}{18.28}\right)^2}}{\sqrt{\left[1-\left(\dfrac{83.78}{18.28}\right)^2\right]^2+\left(2\times 0.0567\times\dfrac{83.78}{18.28}\right)^2}}$$

$$= 0.0563$$

The maximum unbalanced force on the machine due to the reciprocating parts,

$$F = mr\omega^2 = 2.2\times 0.05\times(83.78)^2 = 772.1\text{ N}$$

$$\varepsilon = \frac{F_t}{F}$$

or

$$0.0563 = \frac{F_t}{772.1}$$

or

$$F_t = 43.47\text{ N}$$

(ii) At resonance, $\dfrac{\omega}{\omega_n} = 1$

$$\varepsilon = \frac{\sqrt{(1+(2\,\zeta)^2}}{2\,\zeta} = \frac{\sqrt{1+(2\times 0.0567)^2}}{2\times 0.0567} = 8.875$$

Maximum unbalanced force on the machine due to reciprocating parts at resonance, i.e. when $\omega = \omega_n$,

$$F = 2.2\times 0.05\times(18.28)^2 = 36.76\text{ N}$$

$$F_t = \varepsilon\times F = 8.875\times 36.76 = 326.25\text{ N}$$

(iii) $$\text{Amplitude} = \frac{\text{Force transmitted at resonance}}{\text{Stiffness}}$$

$$= \frac{326.25}{26.739}$$

$$= \underline{12.2\text{ mm}}$$

SECTION II (TRANSVERSE VIBRATIONS)

Natural frequency of shafts and beams under different types of loads and end conditions have been explained in the following sections.

18.12 SINGLE CONCENTRATED LOAD

In case of shafts and beams of negligible mass carrying a concentrated mass, the force is proportional to the deflection of the mass from the equilibrium position and the relation derived for natural frequency of longitudinal vibrations holds good, i.e.

$$f_n = \frac{1}{2\pi}\sqrt{g/\Delta}$$

where $\Delta = \dfrac{mgl^3}{3EI}$ for cantilevers, supporting a concentrated mass at the free end.

$= \dfrac{mga^2b^2}{3EIl}$ for simply supported beams

$= \dfrac{mga^3b^3}{3EIl^3}$ for beams fixed at both ends

These cases have been shown in Fig. 18.14.

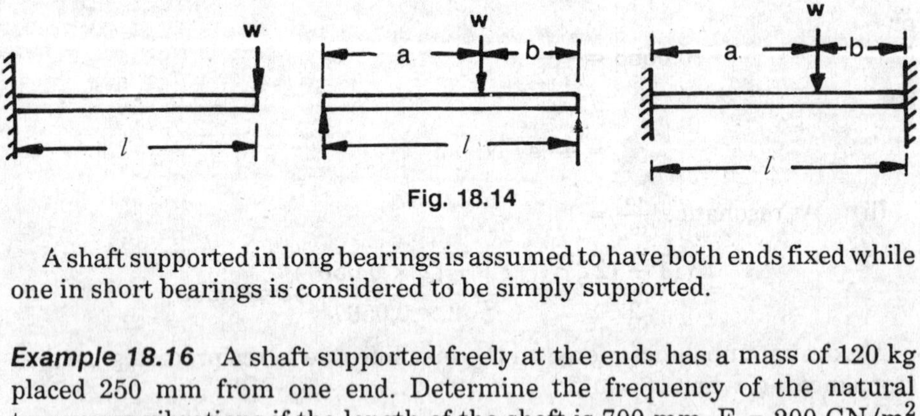

Fig. 18.14

A shaft supported in long bearings is assumed to have both ends fixed while one in short bearings is considered to be simply supported.

Example 18.16 A shaft supported freely at the ends has a mass of 120 kg placed 250 mm from one end. Determine the frequency of the natural transverse vibrations if the length of the shaft is 700 mm, E = 200 GN/m^2 and shaft diameter is 40 mm.

Solution

m = 120 kg $E = 200 \times 10^9$ N/m^2

l = 0.7 m d = 0.04 m

a = 0.25 m b = 0.7 – 0.25 = 0.45 m

$$I = \frac{\pi}{64} \times d^4 = \frac{\pi}{64} \times (0.04)^4 = 0.1256 \times 10^{-6} \text{ m}^4$$

Static deflection at the load point,

$$\Delta = \frac{mg\,a^2b^2}{3EIl}$$

$$= \frac{120 \times 9.81 \times (0.25)^2 \times (0.45)^2}{3 \times 200 \times 10^9 \times 0.1256 \times 10^{-6} \times 0.7}$$

$$= 0.282 \times 10^{-3} \text{ m}$$

$$f_n = \frac{1}{2\pi}\sqrt{\frac{g}{\Delta}} = \frac{1}{2\pi}\sqrt{\frac{9.81}{0.282 \times 10^{-3}}} = \underline{29.68 \text{ Hz}}$$

18.13 UNIFORMLY LOADED SHAFT

Figure 18.15 shows a shaft supported at its ends and carrying a uniform mass.

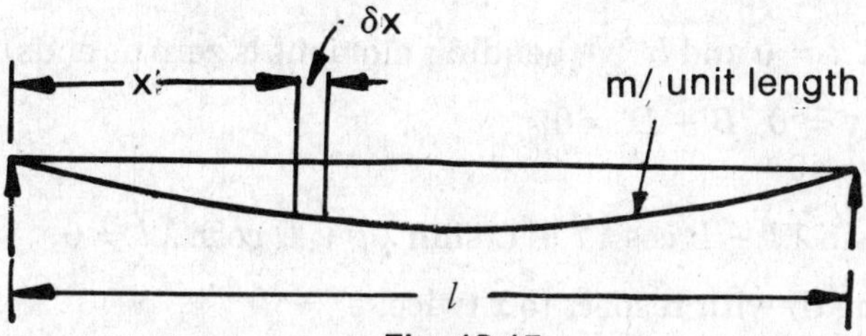

Fig. 18.15

Let m = distributed mass per unit length
l = length of the shaft

The shaft makes transverse vibrations due to elastic forces. At any instant, let it be deflected by an amount y at a distance x from the end A. The vibrations being free and due to elastic forces, will be simple harmonic motion type.

From the theory of bending of shafts,

$$EI\frac{d^4y}{dx^4} = \text{dynamic load per unit length}$$

$$= \text{centrifugal force per unit length.}$$

$$= my\,\omega^2$$

or
$$\frac{d^4y}{dx^4} - \frac{my\,\omega^2}{EI} = 0 \tag{18.33}$$

or
$$\frac{d^4y}{dx^4} - \lambda^4 y = 0$$

where
$$\lambda^4 = \frac{m\,\omega^2}{EI}$$

The auxiliary equation is

$$(D^4 - \lambda^4)\, y = 0$$

This gives

$$D = \pm \lambda \text{ and } \pm i\lambda$$

The solution will be of the form

$$y = A \sin \lambda x + B \cos \lambda x + C \sinh \lambda x + D \cosh \lambda x \qquad \text{(i)}$$

This is the general expression for the deflection in case of uniformly loaded shafts. Constants *A*, *B*, *C* and *D* have to be found from the end conditions.

Simply Supported Shaft

The boundary conditions are:

(a) $y = 0$ at $x = 0$ and l

(b) $\dfrac{d^2y}{dx^2} = 0$ at $x = 0$ and l (bending moment is zero at ends)

When $x = 0,\ y = 0;\ B + D = 0$ (ii)

When $x = l,\ y = 0$

$$A \sin \lambda l + B \cos \lambda l + C \sinh \lambda l + D \cosh \lambda l = 0 \qquad \text{(iii)}$$

Differentiating (i) with respect to *x* twice,

$$\frac{dy}{dx} = \lambda\, (A \cos \lambda x - B \sin \lambda x + C \cos \lambda x + D \sinh \lambda x)$$

$$\frac{d^2y}{dx^2} = \lambda^2(-A \sin \lambda x - B \cos \lambda x + C \sinh \lambda x + D \cosh \lambda x)$$

When $x = 0$,

$$\frac{d^2y}{dx^2} = 0$$

$$\therefore \qquad \lambda^2\, (-B + D) = 0 \qquad \text{(iv)}$$

When $x = l$,

$$\frac{d^2y}{dx^2} = 0$$

$$\lambda^2\, (-A \sin \lambda l - B \cos \lambda l + C \sinh \lambda l + D \cosh \lambda l) = 0 \qquad \text{(v)}$$

From (ii) and (iv),

$B = 0$ and $D = 0$

Thus (iii) and (v) can be written as

$$A \sin \lambda l + C \sinh \lambda l = 0$$

and

$$-A \sin \lambda l + C \sin h \lambda l = 0$$

Adding these, we get

$$C \sinh \lambda l = 0$$

Subtracting,

$$A \sinh \lambda l = 0$$

$\sinh \lambda l$ cannot be zero, because if $\lambda = 0,\ \lambda^4 = 0.$

or

$$\frac{m\,\omega^2}{EI} = 0$$

or

$$\frac{m}{EI}\,(2\,\pi\,f_n)^2 = 0$$

or

$$f_n = 0$$

which means that the system does not vibrate.

∴ $$C = 0$$

When $A \sinh \lambda l = 0$, A cannot be zero as B, C and D are already zero and if A is also zero, no vibration would be existing.

∴ $$\sin \lambda l = 0$$

or

$$\lambda l = 0,\ \pi,\ 2\pi,\ 3\pi, \ldots$$

But λl cannot be equal to zero; if so, there will be no vibration.

∴

$$\lambda = \frac{\pi}{l}, \frac{2\pi}{l}, \frac{3\pi}{l}, \ldots$$

or

$$\left[\frac{m\,\omega^2}{EI}\right]^{1/4} = \frac{\pi}{l}, \frac{2\pi}{l}, \frac{3\pi}{l}, \ldots$$

or

$$\omega^{1/2} = \frac{\pi}{l}\left(\frac{EI}{m}\right)^{1/4}, \frac{2\pi}{l}\left(\frac{EI}{m}\right)^{1/4}, \frac{3\pi}{l}\left(\frac{EI}{m}\right)^{1/4}, \ldots$$

$$\omega = (2\,\pi\,f_n) = \frac{\pi^2}{l^2}\sqrt{\frac{EI}{m}}, \frac{4\pi^2}{l^2}\sqrt{\frac{EI}{m}}, \frac{9\pi^2}{l^2}\sqrt{\frac{EI}{m}}, \ldots$$

$$f_n - \frac{\pi}{2}\sqrt{\frac{EI}{ml^4}}, \frac{4\pi}{2}\sqrt{\frac{EI}{ml^4}}, \frac{9\pi}{2}\sqrt{\frac{EI}{ml^4}}, \ldots$$

A simply supported shaft carrying a uniformly distributed mass has maximum deflection at the mid-span.

$$\Delta = \frac{5mgl^4}{384EI}$$

or

$$\frac{EI}{ml^4} = \frac{5g}{384\Delta}$$

Then, taking the smallest value of f_n,

$$f_n = \frac{\pi}{2}\sqrt{\frac{5g}{348\Delta}} \tag{18.34}$$

This is the lowest frequency of transverse vibrations and is called the fundamental frequency. It has a node at each end.

The next higher frequency is four times the fundamental frequency. It has three nodes, two at the ends and one at the centre.

The next higher frequency is nine times the fundamental frequency. It has four nodes dividing the shaft into three equal parts, and so on (Fig. 18.16).

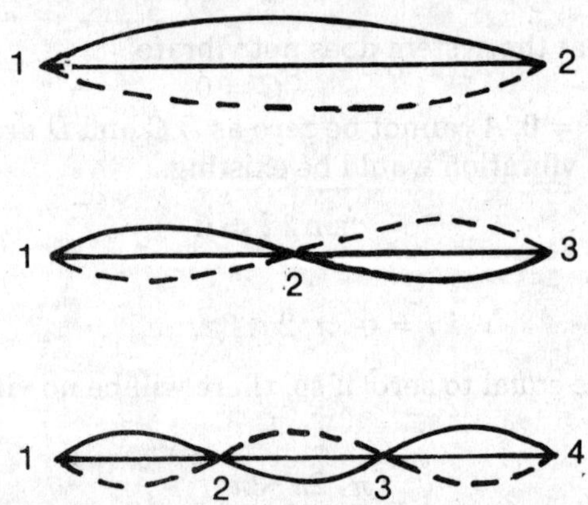

Fig. 18.16

Thus a simply supported shaft will have an infinite number of frequencies under uniformly distributed load.

Similarly, the cases of cantilevers and shafts fixed at both ends can be considered. The end conditions will be as follows.

(i) Cantilevers

$$y = 0 \text{ at } x = 0 \qquad \text{(zero deflection)}$$

$$\frac{dy}{dx} = 0 \text{ at } x = 0 \qquad \text{(zero slope)}$$

$$\frac{d^2y}{dx^2} = 0 \text{ at } x = l \qquad \text{(zero bending moment)}$$

$$\frac{d^3y}{dx^3} = 0 \text{ at } x = l \qquad \text{(zero shear force)}$$

and

$$\Delta = \frac{mgl^3}{8EI}$$

(ii) Both Ends Fixed

$$y = 0 \text{ at } x = 0 \text{ and } l$$

$$\frac{dy}{dx} = 0 \text{ at } x = 0 \text{ and } l$$

and

$$\Delta = \frac{mgl^4}{384\,EI}$$

18.14 SHAFT CARRYING SEVERAL LOADS

There are two methods to find the natural frequency of the system:

(i) Dunkerley's method which is semi-empirical. This gives approximate results but is simple.

(ii) The energy method which gives accurate results but involves heavy calculations if there are many loads.

(i) Dunkerley's Method Let W_1, W_2, W_3, ... be the concentrated loads on the shaft due to masses m_1, m_2, m_3, ... and Δ_1, Δ_2, Δ_3, ... the static deflections of this shaft under each load when that load acts alone on the shaft. Let the shaft carry a uniformly distributed mass of m per unit length over its whole span and the static deflection at mid-span due to the load of this mass be Δ_s. Also, let

f_n = frequency of transverse vibration of the whole system

f_{ns} = frequency with the distributed load acting alone

f_{n1}, f_{n2}, f_{n3}, ... = frequencies of transverse vibrations when each of W_1, W_2, W_3, ... acts alone.

Then, according to Dunkerley's empirical formula,

$$\frac{1}{f_n^2} = \frac{1}{f_{n1}^2} + \frac{1}{f_{n2}^2} + \frac{1}{f_{n3}^2} + \ldots\ldots + \frac{1}{f_{ns}^2} \qquad (18.35)$$

where

$$f_{n1} = \frac{1}{2\pi}\sqrt{\frac{g}{\Delta_1}} = \frac{\sqrt{9.81}}{2\pi}\frac{1}{\sqrt{\Delta_1}} = \frac{0.4985}{\sqrt{\Delta_1}}$$

Similarly,

$$f_{n2} = \frac{0.4985}{\sqrt{\Delta_2}};\ f_{n3} = \frac{0.4985}{\sqrt{\Delta_3}}, \text{ and so on.}$$

$$f_{ns} = \frac{\pi}{2}\sqrt{\frac{5g}{384\,\Delta_s}} = \frac{\pi}{2}\sqrt{\frac{5\times 9.81}{384}}\times\frac{1}{\sqrt{\Delta_s}} = \frac{0.5614}{\sqrt{\Delta_s}}$$

$$\frac{1}{f_n^2} = \frac{1}{(0.4985)^2}(\Delta_1+\Delta_2+\Delta_3+\ldots) + \frac{1}{(0.5614)^2}\Delta_s$$

$$= \frac{1}{(0.4985)^2}\left(\Delta_1+\Delta_2+\Delta_3+\ldots+\frac{\Delta_s}{1.27}\right)$$

$$f_n = \frac{0.4985}{\sqrt{\Delta_1+\Delta_2+\Delta_3+\ldots+\dfrac{\Delta_s}{1.27}}} \tag{18.36}$$

(ii) Energy Method Consider a shaft with negligible mass, carrying point loads $W_1, W_2, W_3, \ldots$ due to masses $m_1, m_2, m_3 \ldots$ as shown in Fig. 18.17. Let $y_1, y_2, y_3, \ldots$ be total deflections under these loads.

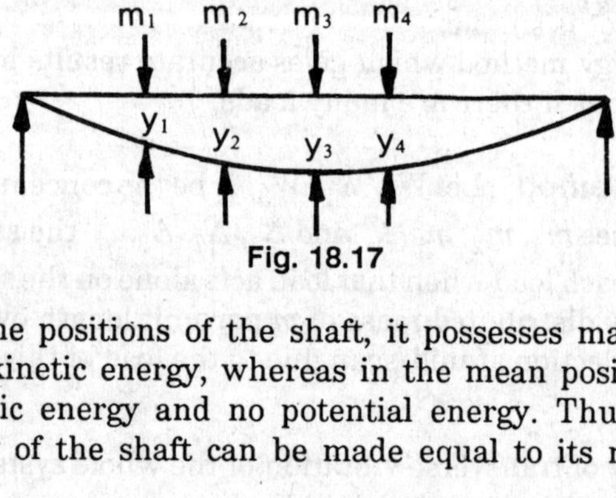

Fig. 18.17

In the extreme positions of the shaft, it possesses maximum potential energy and no kinetic energy, whereas in the mean position, it possesses maximum kinetic energy and no potential energy. Thus, the maximum potential energy of the shaft can be made equal to its maximum kinetic energy.

$$\text{Maximum } PE = \frac{1}{2}W_1 y_1 + \frac{1}{2}W_2 y_2 + \frac{1}{2}W_3 y_3 + \ldots$$

$$= \frac{g}{2}(m_1 y_1 + m_2 y_2 + m_3 y_3 + \ldots)$$

$$= \frac{g}{2}\Sigma\, my$$

$$\text{Maximum } KE = \frac{1}{2} m_1 v_1^2 + \frac{1}{2} m_2 v_2^2 + \frac{1}{2} m_3 v_3^3 + \ldots$$

$$= \frac{1}{2} m_1 (\omega y_1)^2 + \frac{1}{2} m_2 (\omega y_2)^2 + \frac{1}{2} m_3 (\omega y_3)^2 + \ldots$$

$$= \frac{\omega^2}{2} (m_1 y_1^2 + m_2 y_2^2 + m_3 y_3^3 + \ldots$$

$$= \frac{\omega^2}{2} \Sigma my^2$$

where ω is the circular frequency of vibration. Equating maximum PE and maximum KE,

$$\frac{g}{2} \Sigma my = \frac{\omega^2}{2} \Sigma my^2$$

$$\omega = \sqrt{\frac{g \Sigma my}{\Sigma my^2}}$$

$$f_n = \frac{\omega}{2\pi} = \frac{1}{2\pi} \sqrt{\frac{g \Sigma my}{\Sigma my^2}} \tag{18.37}$$

Example 18.17 A shaft 40 mm diameter and 2.5 m long has a mass of 15 kg per meter length. It is simply supported at the ends and carries three masses 90 kg, 140 kg and 60 kg at 0.8 m, 1.5 m and 2 m respectively from the left support. Taking $E = 200$ GN/m^2, find the frequency of the transverse vibrations.

Solution:

$d = 40$ mm $= 0.04$ m $\quad l = 2.5$ m

$$I = \frac{\pi}{64} \times d^4 = \frac{\pi}{64} \times (0.04)^4 = 0.1257 \times 10^{-6}\ m^4$$

We have,

$$f_n = \frac{0.4985}{\sqrt{\Delta_1 + \Delta_2 + \Delta_3 + \dfrac{\Delta_s}{1.27}}}$$

$$\Delta_1 = \frac{mg\, a^2 b^2}{3EIl}$$

Here $m = 90$ kg, $a = 0.8$ m and $b = 1.7$ m.

$$\therefore \quad \Delta_1 = \frac{90 \times 9.81 \times (0.8)^2 \times (1.7)^2}{3 \times 200 \times 10^9 \times 0.1257 \times 10^{-6} \times 2.5} = 0.00\,866 \text{ m}$$

For Δ_2, $\quad m = 140$ kg, $\quad a = 1.5$ m, $\quad b = 1$ m

$$\therefore \quad \Delta_2 = \frac{140 \times 0.81 \times (1.5)^? \times (1)^?}{3 \times 200 \times 10^9 \times 0.1257 \times 10^{-6} \times 2.5} = 0.1639 \text{ m}$$

For Δ_3, m = 60 kg, a = 2 m, b = 0.5 m

$$\therefore \quad \Delta_3 = \frac{60 \times 9.81 \times (2)^2 \times (0.5)^2}{3 \times 200 \times 10^9 \times 0.1257 \times 10^{-6} \times 2.5} = 0.00\,312 \text{ m}$$

$$\Delta_s = \frac{5mgl^4}{384EI} = \frac{5 \times 15 \times 9.81 \times (2.5)^4}{384 \times 200 \times 10^9 \times 0.1257 \times 10^{-6}} = 0.00\,298 \text{ m}$$

$$f_n = \frac{0.4985}{\sqrt{0.00\,866 + 0.01\,639 + 0.00\,312 + \dfrac{0.00\,298}{1.27}}} = 2.85 \text{ Hz}$$

18.15 WHIRLING OF SHAFTS

When a rotor is mounted on a shaft, its centre of mass does not usually coincide with the centre line of the shaft. Therefore, when the shaft rotates, it is subjected to a centrifugal force which makes the shaft bend in the direction of eccentricity of the centre of mass. This further increases the eccentricity, and hence the centrigugal force. In this way, the effect is cumulative and ultimately the shaft may even fail. The bending of the shaft depends upon the eccentricity of the centre of mass of the rotor as well as upon the speed at which the shaft rotates.

Critical or *whirling* or *whipping speed* is the speed at which the shaft tends to vibrate violently in the transverse direction.

It has been observed that if the critical speed is instantly run through, the shaft again becomes almost straight. But at some other speed, the same phenomenon recurs, the only difference being that the shaft now bends in two bows and so on.

Figure 18.18 shows a rotor having a mass m attached to a shaft.

Let s = stiffness of shaft

e = initial eccentricity of centre of mass of rotor

m = mass of rotor

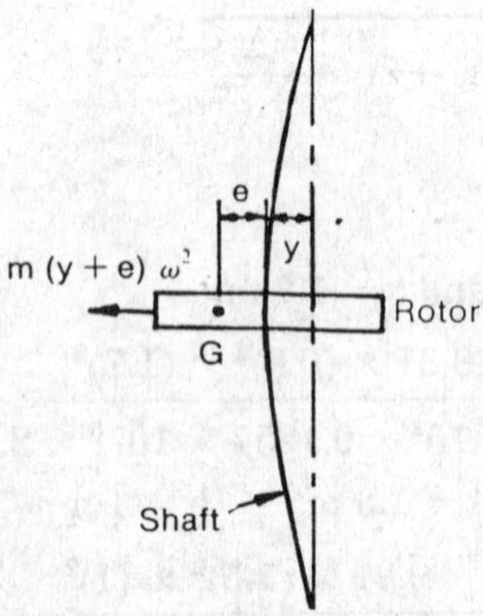

Fig. 18.18

y = additional deflection of rotor due to centrifugal force
ω = angular velocity of shaft.

Then

Centrifugal force $= m(y + e)\,\omega^2$
Force resisting the deflection $= sy$
For equilibrium,

$$sy = m\,(y + e)\,\omega^2 = my\,\omega^2 + me\,\omega^2$$

or $$y\,(s - m\,\omega^2) = me\,\omega^2$$

$$y = \frac{me\,\omega^2}{s - m\,\omega^2}$$

$$= \frac{e}{\dfrac{s}{m\,\omega^2} - 1}$$

$$= \frac{e}{\left(\dfrac{\omega_n}{\omega}\right)^2 - 1}$$

Thus, when $\omega = \omega_n$, the deflection y is infinitely large (resonance occurs) and the speed ω is the critical speed, i.e.

$$\omega_c = \omega_n = \sqrt{\frac{s}{m}} = \sqrt{\frac{g}{\Delta}} \tag{18.38}$$

If the speed of the shaft is increased rapidly beyond the critical speed, $\omega > \omega_n$ or $(\omega_n/\omega)^2 < 1$ or y is negative. This means that the shaft deflects in the opposite direction. As the speed continues to increase, y approaches the value $-e$ or the centre of mass of the rotor approaches the centre line of rotation. This principle is used in running high-speed turbines by speeding up the rotor rapidly to beyond the critical speed. When y approaches the value of $-e$, the rotor runs steadily.

Example 18.18 Determine the critical speed of the shaft of Example 18.17 loaded in the same way.

Solution: The critical speed of the shaft in rps is equal to the natural frequency of transverse vibrations in Hz.

f_n = 2.85 Hz
N_c = 2.85 rps = (2.85 × 60) rpm = 171 rpm

Example 18.19 A rotor has a mass of 12 kg and is mounted midway on a 24 mm diameter horizontal shaft supported at the ends by two bearings. The bearings are 1 m apart. The shaft rotates at 2400 rpm. If the centre of mass of the rotor is 0.11 mm away from the geometric centre of the rotor due to a certain manufacturing defect, find the amplitude of the steady-state vibration and the dynamic force transmitted to the bearing. E = 200 GN/m^2.

Solution: Assuming the bearings to be short so that the shaft can be assumed to be simply supported,

$m = 12$ kg $\qquad l = 1$ m
$d = 0.024$ m $\qquad N = 2400$ rpm
$e = 0.11$ mm $\qquad E = 200 \times 10^9$ N/m^2

$$I = \frac{\pi}{64} \times d^4 = \frac{\pi}{64} \times (0.024)^4 = 16.3 \times 10^{-9} \text{ m}^4$$

$$\Delta = \frac{mgl^3}{48EI} = \frac{12 \times 9.81 \times (1)^3}{48 \times 200 \times 10^9 \times 16.3 \times 10^{-9}} = 0.000\,752 \text{ m}$$

$$\omega_n = \sqrt{\frac{g}{\Delta}} = \sqrt{\frac{9.81}{0.000\,752}} = 114.2 \text{ rad/s}$$

$$\omega = \frac{2\pi N}{60} = \frac{2\pi \times 2400}{60} = 251.3 \text{ rad/s}$$

Amplitude,

$$y = \frac{e}{\left(\frac{\omega_n}{\omega}\right)^2 - 1} = \frac{0.11}{\left(\frac{114.2}{251.3}\right)^2 - 1} = -0.139 \text{ mm}$$

$$= -0.000\,139 \text{ m}$$

The negative sign indicates that the displacement is out of phase with the centrifugal force.

Dynamic force on the bearings $= sy$

$= m\,\omega_n^2\,y$

$= 12 \times (114.2)^2 \times 0.000139$

$= \underline{21.7 \text{ N}}$

The total load on each bearing can also be found.

$$\text{Total load on each bearing} = \frac{mg}{2} + \frac{sy}{2} = \frac{12 \times 9.81}{2} + \frac{21.7}{2} = 69.7 \text{ N}$$

Example 18.20 The following data relate to a shaft held in long bearings.

Length of shaft	= 1.2 m
Diameter of shaft	= 14 mm
Mass of a rotor at midpoint	= 16 kg
Eccentricity of centre of mass of rotor from centre of rotor	= 0.4 mm
Modulus of elasticity of shaft material	= 200 GN/m^2
Permissible stress in shaft material	= 70×10^6 N/m^2

Determine the critical speed of the shaft and the range of speed over which it is unsafe to run the shaft. Assume the shaft to be massless.

Solution:

$m = 16$ kg $\qquad e = 0.0004$ m

$l = 1.2$ m $\qquad E = 200 \times 10^9$ N/m^2

$d = 0.014$ m $\qquad f = 70 \times 10^6$ N/m^2

(i) As the shaft is held in long bearings, it may be assumed to be fixed at the ends.

$$\Delta = \frac{mgl^3}{192EI} = \frac{16 \times 9.81 \times (1.2)^3}{192 \times 200 \times 10^9 \times \frac{\pi}{64} \times (0.014)^4} = 0.00\,375 \text{ m}$$

$$f_n = \frac{1}{2}\sqrt{\frac{g}{\Delta}} = \frac{1}{2\pi}\sqrt{\frac{9.81}{0.00\,375}} = 8.145 \text{ Hz}$$

The critical speed of the shaft in rps is equal to the natural frequency of transverse vibrations in Hz, i.e.

$$N_c = 8.145 \text{ rps} = \underline{489 \text{ rpm}}$$

(ii) When the shaft rotates, additional dynamic load on the shaft can be obtained from the relation

$$\frac{M}{I} = \frac{f}{y}$$

$$\frac{\frac{W_1 l}{8}}{\frac{\pi}{64} \times d^4} = \frac{f}{d/2} \quad \text{or} \quad \frac{\frac{W_1 \times 1.2}{8}}{\frac{\pi}{64} \times (0.014)^4} = \frac{70 \times 10^6}{\frac{0.014}{2}}$$

$$W_1 = 125.7 \text{ N}$$

Additional deflection due to this load $= \dfrac{W_1}{W} \times \Delta$

$$= \frac{W_1}{mg} \times \Delta$$

$$= \frac{125.7}{16 \times 9.81} \times 0.00375$$

$$= 0.003 \text{ m}$$

Also,

Additional deflection, $\qquad y = \dfrac{\pm e}{\left(\dfrac{\omega_c}{\omega}\right)^2 - 1}$

or

$$0.003 = \frac{\pm 0.0004}{\left(\frac{N_c}{N}\right)^2 - 1}$$

$$\left(\frac{489}{N}\right)^2 - 1 = \pm 0.1333$$

$$N = \underline{459} \quad \text{and} \quad \underline{525}$$

Thus the range of unsafe speed is from 459 rpm to 525 rpm.

SECTION III (TORSIONAL VIBRATION)

18.16 FREE TORSIONAL VIBRATIONS (SINGLE ROTOR)

Consider a uniform shaft of length l rigidly fixed at its upper end and carrying a disc of moment of inertia I at its lower end (Fig.18.19). The shaft is assumed to be massless. If the disc is given a twist about its vertical axis and then released, it will start oscillating about the axis and will perform torsional vibrations.

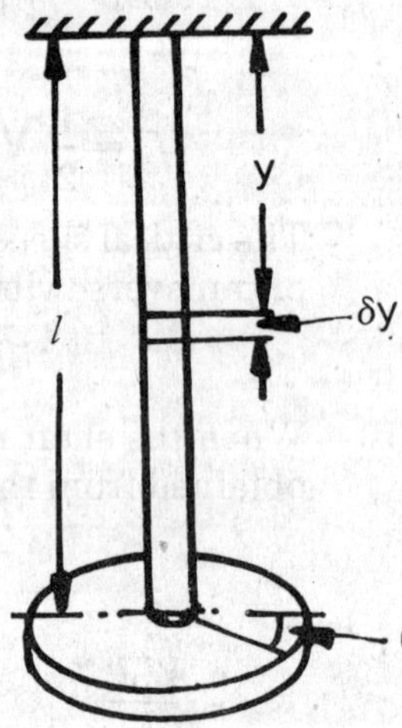

Fig. 18.19

Let θ = angular displacement of the disc from its equilibrium position at any instant

q = torsional stiffness of the shaft

= torque required to twist the shaft per radian within elastic limits $\left(= \dfrac{GJ}{l}\right)$

where

G = modulus of rigidity of the shaft material

J = polar moment of inertia of the shaft cross-section

At any instant, the torques acting on the disc are

– Inertia torque $= -I\theta$

– Restoring torque (spring (torque) $= -q\theta$

Negative signs have been used as both of these torques act opposite to the angular displacement.

For equilibrium, the sum of all torques acting on the disc must be zero. Therefore,

$$I\,\theta + q\,\theta = 0$$

or

$$\ddot{\theta} + \frac{q}{I}\theta = 0$$

This is the equation of simple harmonic motion.

$$\omega_n = \sqrt{\frac{q}{I}} \qquad (18.39)$$

$$f_n = \frac{1}{2\pi}\sqrt{\frac{q}{I}} \qquad (18.39a)$$

18.17 INERTIA EFFECT OF MASS OF SHAFT

Let I_1 = moment of inertia of shaft

ω = angular velocity of free end

Consider an element of length δy at a distance y from the fixed end. Then

$$KE \text{ of element} = \frac{1}{2} \times (MOI \text{ of element}) \times (\text{angular velocity})^2$$

$$= \frac{1}{2} \times \left(I_1 \frac{\delta y}{1}\right) \left(\frac{\omega y}{l}\right)^2$$

$$KE \text{ of shaft} = \int_0^l \frac{1}{2} \times \frac{I_1}{l} \left(\frac{y}{l} \omega\right)^2 dy$$

$$= \frac{I_1 \omega^2}{2l^2} \int_0^l y^2 \, dy$$

$$= \frac{I_1 \omega^2}{2l^3} \times \frac{l^3}{3}$$

$$= \frac{1}{3} \frac{1}{2} I_1 \omega^2$$

$$= \frac{1}{3} \times \left[\frac{1}{2} (MOI \text{ of shaft}) \times (\text{angular velocity of free end})^2\right]$$

$$= \frac{1}{3} \times KE \text{ of a disc of } MOI \text{ equal to that of the shaft}$$

attached to the free end of the shaft.

Thus to consider the inertia of the shaft, the moment of inertia of the disc is increased by an amount equal to one-third of that of the shaft.
Then

$$f_n = \frac{1}{2\pi} \sqrt{\frac{q}{I + \frac{I_1}{3}}} \tag{18.40}$$

18.18 MULTIFILAR SYSTEMS

Multifilar systems are used to determine the moment of inertia of irregular bodies such as unsymmetrical castings, spoked flywheels, connecting rods, etc. for which it is quite difficult to find their moment of inertia from their dimensions.

(i) Bifilar Suspension Figure 18.20 represents a disc of mass m suspended from a rigid support with the help of two cords.

Let l = length of each cord

a and b = distance of centre of mass of the disc from the points of suspension of cords 1 and 2 respectively.

If the disc is now turned through a small angle θ about a vertical axis through the centre of mass, the cords will inclined to the vertical. On release, the disc will oscillate about the vertical axis and execute a torsional vibration.

Let θ = angular displacement of disc

φ_1, φ_2 = inclination of cords to the vertical

F_1, F_2 = tensions in cords 1 and 2 respectively.

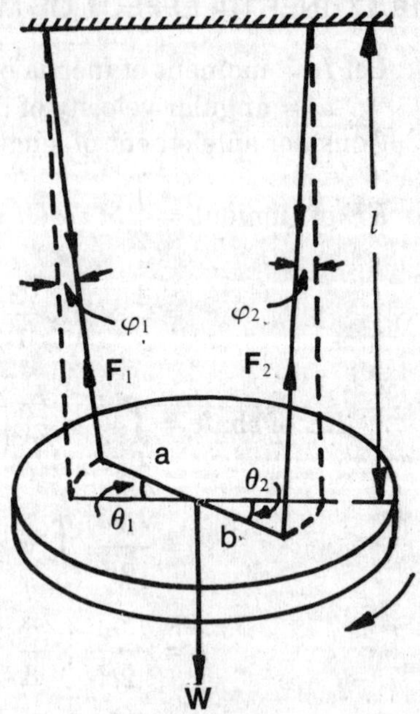

Fig. 18.20

In the oscillating position, the effects of vertical accelerations can be neglected as the angles φ_1 and φ_2 are small and the torque produced can be considered to be only due to horizontal torques.

Restoring torque = (horizontal force on cord 1 × a) + (horizontal force on cord 2 × b)

$$= -[F_1 a \sin \varphi_1 + F_2 b \sin \varphi_2]$$

$$= -[F_1 a \varphi_1 + F_2 b \varphi_2] \quad \text{(as } \varphi_1 \text{ and } \varphi_2 \text{ are small)}$$

$$= -\left[\frac{Wb}{a+b} a \varphi_1 + \frac{Wa}{a+b} b \varphi_2\right] \quad (W = \text{weight of disc, } MG)$$

$$= -\frac{Wab}{a+b}(\varphi_1 + \varphi_2)$$

$$= -\frac{Wab}{a+b}\left(\frac{a\theta}{l} + \frac{b\theta}{l}\right) \quad (\varphi_1 l = a\theta \text{ and } \varphi_2 l = b\theta)$$

$$= -\frac{Wab}{(a+b)} \times \frac{\theta(a+b)}{l}$$

$$= -\frac{mgab}{l}\theta$$

Inertia torque $= -I\ddot{\theta} = -mk^2\ddot{\theta}$

where k is the radius of gyration of the disc about the vertical axis through the centre of mass.

For equilibrium,

Inertia torque + Restoring torque = 0

$$mk^2\ddot{\theta} + \frac{mgab}{l}\theta = 0$$

or
$$\ddot{\theta} + \frac{g\,ab}{lk^2}\,\theta = 0$$

$\therefore$
$$f_n = \frac{1}{2\pi}\sqrt{\frac{g\,ab}{lk^2}}$$

or
$$k = \frac{1}{2\pi f_n}\sqrt{\frac{gab}{l}} \qquad (18.41)$$

Thus radius of gyration can be found out by finding the natural frequency of vibration of the body.

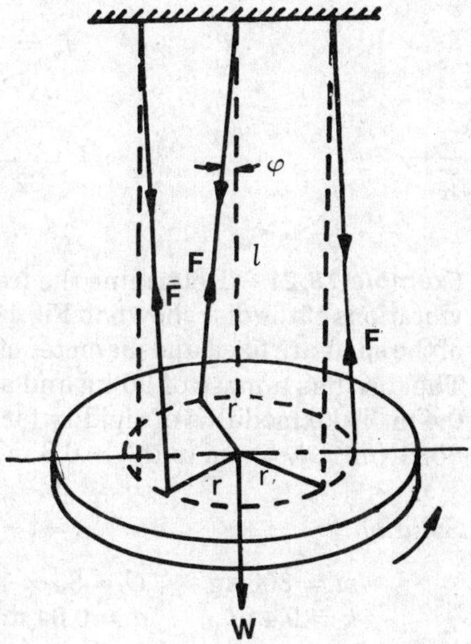

Fig. 18.21

(ii) Trifilar Suspension Consider a disc of mass m (weight W), suspended by three vertical cords, each of length l, from a fixed support as shown in Fig. 18.21. Each cord is symmetrically attached to the disc at the same distance r from the centre of mass of the disc.

If the disc is now turned through a small angle about its vertical axis, the cords become inclined. On being released the disc will perform oscillations about the vertical axis. At any instant

Let θ = angular displacement of the disc
φ = inclination of the cords to the vertical
F = tension in each cord = $W/3$
Inertia torque = $-\,I\ddot{\theta} = -\,mk^2\,\ddot{\theta}$

Restoring torque = $-3\times$(Horizontal component of force in each string $\times\, r$)

$$= -\,3 \times Fr\sin\varphi$$

$$= -\,3\,Fr\,\varphi \qquad (\text{ as } \varphi \text{ is small})$$

$$= -\,3F\,r\,\frac{\theta\,r}{l} \qquad (\because \varphi l = \theta r)$$

$$= -\,\frac{3W}{3} \times \frac{r^2}{l}\,\theta$$

$$= -\,\frac{mg\,r^2}{l}\,\theta$$

$\therefore$
$$mk^2\,\ddot{\theta} + \frac{mg\,r^2}{l}\,\theta = 0$$

$$\ddot{\theta} + \frac{gr^2}{lk^2}\,\theta - 0$$

$$f_n = \frac{1}{2\pi}\sqrt{\frac{gr^2}{lk^2}}$$

$$k = \frac{1}{2\,\pi\, f_n}\sqrt{\frac{gr^2}{l}} \qquad (18.42)$$

Example 18.21 Determine the frequency of torsional vibrations of the disc shown in Fig. 18.22 if both the ends of the shaft are fixed and diameter of the shaft is 40 mm. The disc has a mass of 96 kg and a radius of gyration of 0.4 m. Take modulus of rigidity for the shaft material as 85 GN/m^2. l_1 = 1 m and l_2 = 0.8 m.

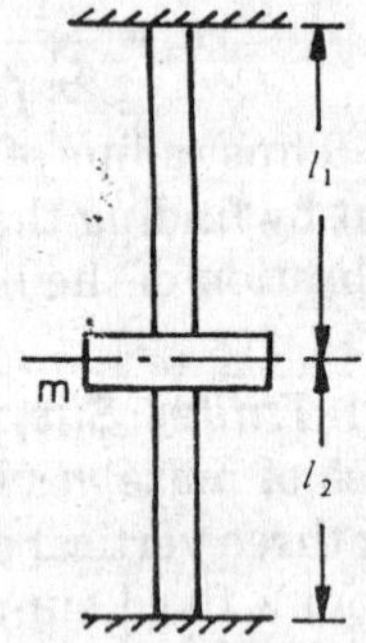

Fig. 18.22

Solution:

m = 600 kg $\qquad G = 85 \times 10^9$ N/m^2

k = 0.4 m $\qquad d$ = 0.04 m

$$I = mk^2 = 96 \times (0.4)^2$$

$$= 15.36 \text{ kg.m}^2$$

$$J = \frac{\pi}{32}d^4 = \frac{\pi}{32} \times (0.04)^4$$

$$= 0.251 \times 10^{-6} \text{ m}^4$$

Total torsional stiffness of shaft, $q = q_1 + q_2$

$$= \frac{GJ}{l_1} + \frac{GJ}{l_2}$$

$$= 85 \times 10^9 \times 0.251 \times 10^{-6}\left(\frac{1}{1} + \frac{1}{0.8}\right)$$

$$= 48\,004 \text{ N.m}$$

$$f_n = \frac{1}{2\,\pi}\sqrt{\frac{q}{I}}$$

$$= \frac{1}{2\,\pi}\sqrt{\frac{48\,004}{15.36}}$$

$$= \underline{8.9 \text{ Hz}}$$

Example 18.22 Determine the natural frequency of a simple pendulum (Fig. 18.23), taking the mass of the rod into consideration.

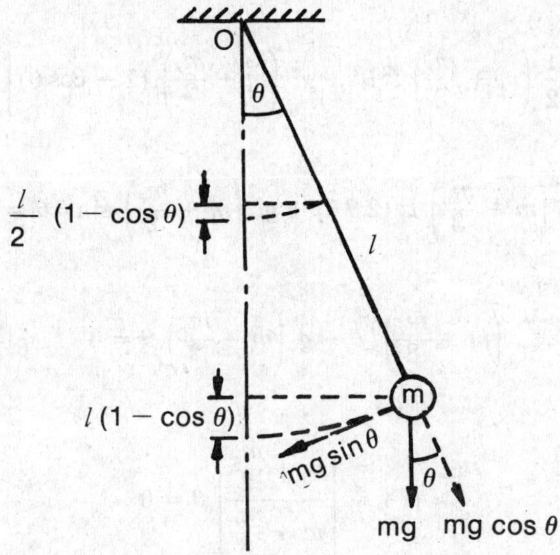

Fig. 18.23

Solution:

Equilibrium Method
Taking moments about θ,

$$I_0\theta + mg\,(l\sin\theta) + m_r\,g\left(\frac{l}{2}\sin\theta\right) = 0$$

$$(ml^2 + \frac{m_r}{3}l^2)\,\ddot{\theta} + mgl\,\theta + m_r\,gl\,\frac{\theta}{2} = 0 \qquad (\theta \text{ is small})$$

$$\left(m + \frac{m_r}{3}\right)l^2\,\ddot{\theta} + gl\left(m + \frac{m_r}{2}\right)\theta = 0$$

$$\theta + \frac{g}{l}\left(\frac{m + \dfrac{m_r}{2}}{m + \dfrac{m_r}{3}}\right)\theta = 0$$

$$f_n = \frac{1}{2\pi}\sqrt{\frac{g}{l}\,\frac{m + (m_r/2)}{m + (m_r/3)}}$$

Energy Method At any instant,

$$\frac{d}{dt}(KE + PE) = 0$$

or

$$\frac{d}{dt}\left[\frac{1}{2}I_0\,\theta^2 + \left\{mgl(1 - \cos\theta) + m_r\,g\,\frac{l}{2}(1 - \cos\theta)\right\}\right] = 0$$

or

$$\frac{d}{dt}\left[\frac{1}{2}\left(m+\frac{m_r}{3}\right) l^2 \dot{\theta}^2 + gl\left(m+\frac{m_r}{2}\right)(1-\cos\theta)\right] = 0$$

or

$$\frac{1}{2}\left(m+\frac{m_r}{3}\right) l^2 (2\,\dot{\theta}\,\ddot{\theta}) + gl\left(m+\frac{m_r}{2}\right)\sin\theta\,\dot{\theta} = 0$$

or

$$\left(m+\frac{m_r}{3}\right) l\,\ddot{\theta} + g\left(m+\frac{m_r}{2}\right)\theta = 0\ldots \qquad (\theta \text{ is small})$$

or

$$\ddot{\theta} + \frac{g}{l}\left(\frac{m+\frac{m_r}{2}}{m+\frac{m_r}{3}}\right)\theta = 0$$

$$f_n = \frac{1}{2\,\pi}\sqrt{\frac{g}{l}\,\frac{m+\frac{m_r}{2}}{m+\frac{m_r}{3}}}$$

If the mass of the rod is neglected,

$$f_n = \frac{1}{2\,\pi}\sqrt{g/l}$$

Example 18.23 Find the natural frequency of the oscillation in the cases shown in Figs. 18.24 a and b. The roller rolls on the surface without slipping.

Solution:

(a) Equilibrium (Newton's) Method Taking moments about the instantaneous centre A, considering small oscillations of the disc,

$$I_a\,\ddot{\theta} + (sx)\,r = 0$$

or

$$(I_0 + mr^2)\,\ddot{\theta} + (s\,\theta\,r)\,r = 0$$

or

$$\left(\frac{1}{2}mr^2 + mr^2\right)\ddot{\theta} + sr^2\,\theta = 0$$

or

$$\ddot{\theta} + \frac{sr^2}{\frac{3}{2}mr^2}\theta = 0$$

or

$$\ddot{\theta} + \frac{2s}{3m}\theta = 0$$

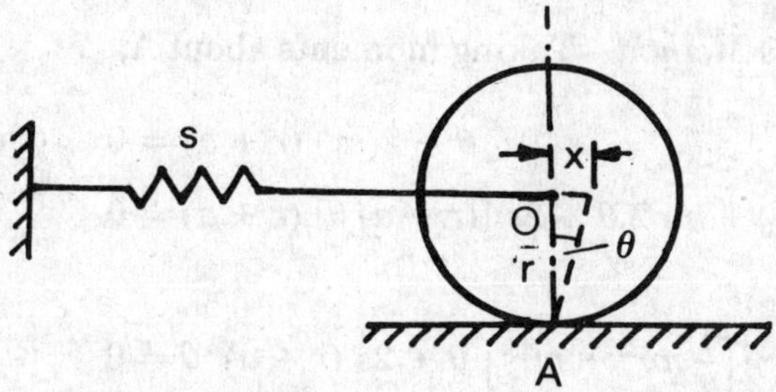

(a)

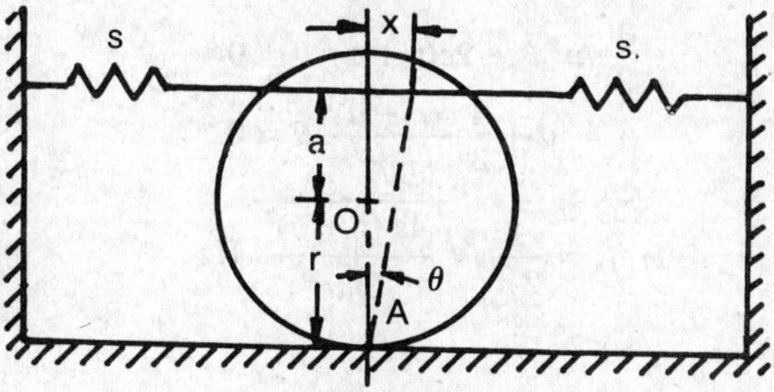

(b)

Fig. 18.24

$$f_n = \frac{1}{2\pi}\sqrt{\frac{2s}{3m}} \text{ Hz}$$

Energy Method

$$\frac{d}{dt}(KE + PE) = 0$$

$$\frac{d}{dt}\left[\frac{1}{2} I_a \dot{\theta}^2 + \frac{1}{2} sx^2\right] = 0$$

$$\frac{d}{dt}\left[\frac{1}{2}\left(\frac{3}{2} mr^2\right)\dot{\theta}^2 + \frac{1}{2} s(\theta r)^2\right] = 0$$

$$\frac{d}{dt}\left[\frac{3}{4} m r^2 \dot{\theta}^2 + \frac{1}{2} sr^2 \theta^2\right] = 0$$

$$\frac{3}{4} mr^2 \times 2\dot{\theta}\ddot{\theta} + \frac{1}{2} sr^2 \times 2\theta\dot{\theta} = 0$$

$$\ddot{\theta} + \frac{2s}{3m}\theta = 0$$

i.e. the same equation as before.

(b) Newton's Method Taking moments about A,

$$I_a \ddot{\theta} + 2(sx)\,(r + a) = 0 \qquad \text{(there are two springs)}$$

$$(I_0 + mr^2)\,\ddot{\theta} + 2s\,[(r + a)\,\theta]\,(r + a) = 0$$

or

$$\left(\frac{1}{2}mr^2 + mr^2\right)\ddot{\theta} + 2s\,(r + a)^2\,\theta = 0$$

or

$$\frac{3}{2}mr^2\,\ddot{\theta} + 2s(r + a)^2\,\theta = 0$$

$$\ddot{\theta} + \frac{4}{3}\frac{s(r + a)^2}{mr^2}\theta = 0$$

$$f_n = \frac{1}{2\pi}\sqrt{\frac{4s\,(r + a)^2}{3mr^2}}\ \text{Hz}$$

Energy Method

$$\frac{d}{dt}(KE + PE) = 0$$

$$\frac{d}{dt}\left[\frac{1}{2}\left(\frac{3}{2}mr^2\right)\dot{\theta}^2 + 2 \times \frac{1}{2}sx^2\right] = 0$$

$$\frac{d}{dt}\left[\frac{3}{4}mr^2\dot{\theta}^2 + s\{(r + a)\theta\}^2\right] = 0$$

$$\frac{3}{4}mr^2 \times 2\,\dot{\theta}\,\ddot{\theta} + s \times 2\,(r + a)\,\theta\,(r + a) = 0$$

$$\ddot{\theta} + \frac{4}{3}\frac{s(r + a)^2}{mr^2}\theta = 0$$

i.e. the same equation.

Example 18.24 Find the equation of motion and the natural frequency of the systems shown in Figs. 18.25 a and b.

Solution:

(a) Equilibrium Method

Let I_0 = combined *MOI* of disc about *O*.

$$x = r\,\theta$$
$$\dot{x} = r\,\dot{\theta}$$
$$\ddot{x} = r\,\ddot{\theta}$$
$$x_1 = R\,\theta$$

Taking moments about *O*,

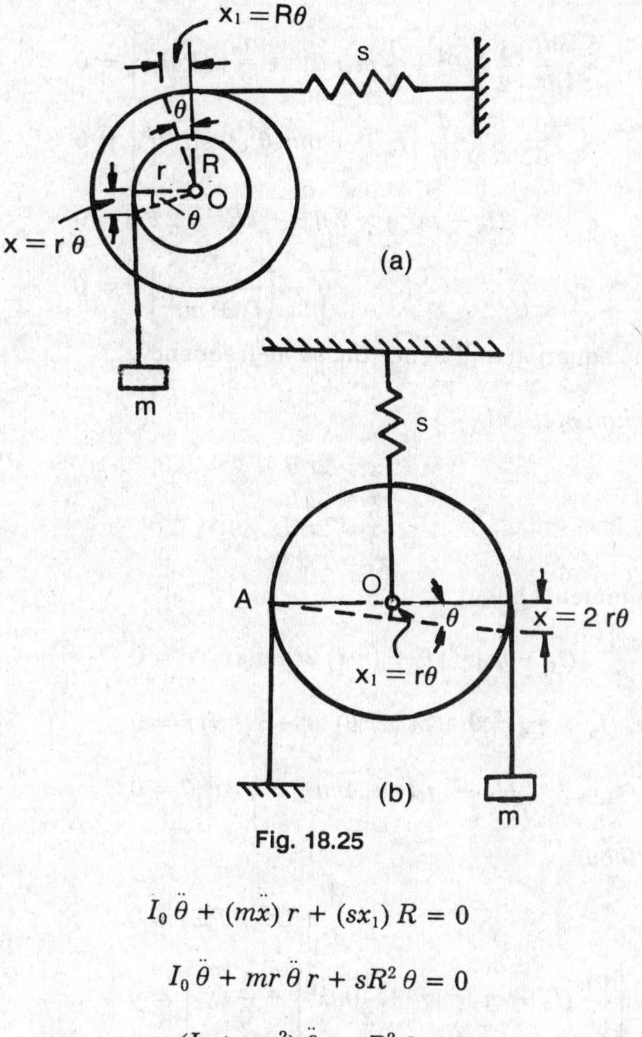

Fig. 18.25

$$I_0 \ddot{\theta} + (m\ddot{x})\, r + (s x_1)\, R = 0$$

$$I_0 \ddot{\theta} + mr\,\ddot{\theta}\, r + sR^2\, \theta = 0$$

$$(I_0 + mr^2)\,\ddot{\theta} + sR^2\,\theta = 0$$

$$\ddot{\theta} + \left(\frac{sR^2}{I_0 + mr^2}\right)\theta = 0$$

$$f_n = \frac{1}{2\pi}\sqrt{\frac{sR^2}{I_0 + mr^2}}\ \text{Hz}$$

Energy Method

$$\frac{d}{dt}\,(KE + PE) = 0$$

$$\frac{d}{dt}\left[\left(\frac{1}{2}\, I_0\, \dot{\theta}^2 + \frac{1}{2} m\dot{x}^2\right) + \left(\frac{1}{2}\, s x_1^2\right)\right] = 0$$

$$\frac{d}{dt}\left[\frac{1}{2} I_0 \dot{\theta}^2 + \frac{1}{2} m(r\dot{\theta})^2 + \frac{1}{2} s\,(R\,\theta)^2\right] = 0$$

$$\frac{1}{2}\frac{d}{dt}\left[I_0 \dot{\theta}^2 + mr^2 \dot{\theta}^2 + sR^2 \theta^2\right] = 0$$

$$(I_0 + mr^2) \times 2\,\dot{\theta}\,\ddot{\theta} + sR^2 \times 2\,\theta\,\dot{\theta} = 0$$

$$\ddot{\theta} + \left(\frac{sR^2}{I_0 + mr^2}\right)\theta = 0$$

i.e. the same equation and hence the same frequency.

(b) Equilibrium Method

$$x = 2r\,\theta$$
$$\dot{x} = 2r\,\dot{\theta}$$
$$\ddot{x} = 2r\,\ddot{\theta}$$
$$x_1 = r\,\theta$$

Taking moments about A,

$$(I_0 + m_0\, r^2)\,\ddot{\theta} + (m\ddot{x})\,2r + (sx_1)\,r = 0$$

$$(I_0 + m_0\, r^2)\ddot{\theta} + m\,(2r\,\ddot{\theta})\,2r + \varepsilon(r\,\theta)\,r = 0$$

$$(I_0 + m_0 r^2 + 4mr^2)\,\ddot{\theta} + sr^2\,\theta = 0$$

Energy Method

$$\frac{d}{dt}(KE + PE) = 0$$

$$\frac{d}{dt}\left[\left\{\frac{1}{2}\,(I_0 + m_0\, r^2)\dot{\theta}^2 + \frac{1}{2}\, m\dot{x}^2\right\} + \frac{1}{2}\, sx_1^2\right] = 0$$

$$\frac{d}{dt}\left[\frac{1}{2}\,(I_0 + m_0 r^2)\,\dot{\theta}^2 + \frac{1}{2}\, m(2r\dot{\theta})^2 + \frac{1}{2}\, s\,(r\,\theta)^2\right] = 0$$

$$\frac{d}{dt}\left[\frac{1}{2}(I_0 + m_0 r)^2\,\dot{\theta}^2 + 2mr^2\,\dot{\theta}^2 + \frac{sr^2}{2}\,\theta^2\right] = 0$$

$$\frac{1}{2}\,(I_0 + m_0\, r^2)\,2\,\dot{\theta}\,\ddot{\theta} + 2mr^2 \times 2\,\dot{\theta}\,\ddot{\theta} + \frac{sr^2}{2} \times 2\,\theta\,\dot{\theta} = 0$$

$$(I_0 + m_0\, r^2 + 4mr^2)\,\ddot{\theta} + sr^2\,\theta = 0 \quad \text{(the same equation)}$$

or

$$\ddot{\theta} + \frac{sr^2}{I_0 + m_0\, r^2 + 4mr^2}\,\theta = 0$$

$$f_n = \frac{1}{2\pi}\sqrt{\frac{s\,r^2}{I_0 + m_0 r^2 + 4\,mr^2}}$$

18.19 FREE TORSIONAL VIBRATIONS (TWO-ROTOR SYSTEM)

If a shaft held in bearings carries a rotor at each end, it can vibrate torsionally such that the two rotors move in the opposite directions. Thus some length of the shaft is twisted in one direction while the rest is twisted in the other. The section which does not undergo any twist is called the nodal section. The shaft behaves as if clamped at the nodal section and the two sections vibrate as two separate shafts with equal frequencies (Fig. 18.26).

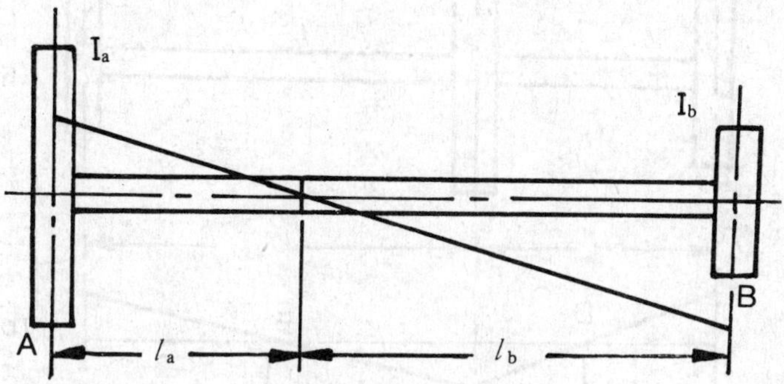

Fig. 18.26

Let

l_a and l_b = lengths of two portions of the shaft

I_a and I_b = *MOI* of rotors *A* and *B* respectively

q_a and q_b = torsional stiffness of lengths l_a and l_b of the shaft respectively

f_{na} and f_{nb} = natural frequencies of torsional vibrations of rotors *A* and *B* respectively.

Then

$$f_{na} = f_{nb}$$

$\therefore$

$$\frac{1}{2\pi}\sqrt{\frac{q_a}{I_a}} = \frac{1}{2\pi}\sqrt{\frac{q_b}{I_b}}$$

or

$$\frac{q_a}{I_a} = \frac{q_b}{I_b}$$

or

$$\frac{GJ}{l_a I_a} = \frac{GJ}{l_a I_a}$$

or

$$\frac{l_a}{l_b} = \frac{I_b}{I_a} \quad \text{or} \quad I_a\, l_a = I_b\, l_b \qquad (18.43)$$

Thus the node divides the length of the shaft in the inverse ratio of moment of inertias of the two rotors.

Also

$$\frac{\text{Amplitude of rotor } A}{\text{Amplitude of rotor } B} = \frac{A_a}{A_b} = \frac{l_a}{l_b}$$

18.20 FREE TORSIONAL VIBRATIONS (THREE-ROTOR SYSTEM)

Consider a three-rotor system (Fig. 18.27 a) in which the two rotors A and B

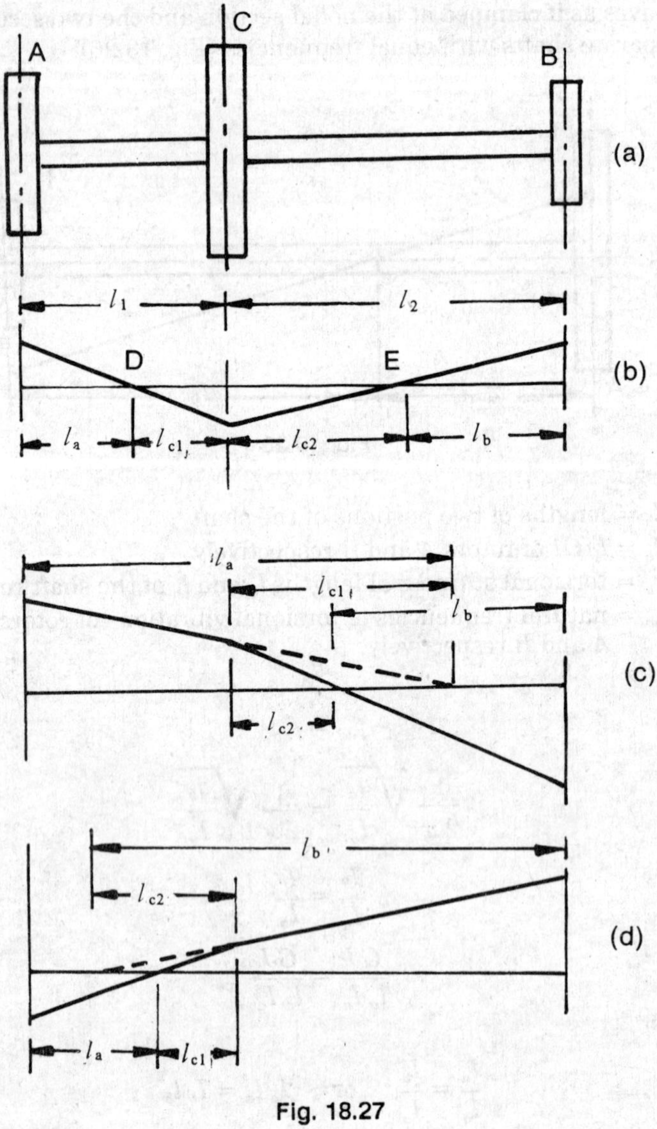

Fig. 18.27

are fixed to the ends of the shaft, and the rotor C is in between. Let the rotors A and B rotate in the same direction and C in the opposite direction and the node points lie at D and E as shown in Fig. 18.27 b.

Let I_a, I_b and I_c be the mass moments of inertia of rotors A, B and C respectively. l_1, l_2, l_a and l_c are the distances as indicated in the diagram. Now,

$$f_{na} = f_{nb} = f_{nc}$$

or

$$\frac{1}{2\pi}\sqrt{\frac{q_a}{I_a}} = \frac{1}{2\pi}\sqrt{\frac{q_b}{I_b}} = \frac{1}{2\pi}\sqrt{\frac{q_c}{I_c}}$$

or

$$\frac{q_a}{I_a} = \frac{q_b}{I_b} = \frac{q_c}{I_c}$$

The torque required to produce unit twist of C is the sum of torques required to produce a unit twist in each of the lengths l_{c1} and l_{c2}.

$$\therefore \quad \frac{GJ}{l_a I_a} = \left(\frac{GJ}{l_{c1}} + \frac{GJ}{l_{c2}}\right)\frac{1}{I_c} = \frac{GJ}{l_b I_b}$$

or

$$\frac{1}{l_a I_a} = \left(\frac{1}{l_1 - l_a} + \frac{1}{l_2 - l_b}\right)\frac{1}{I_c} - \frac{1}{l_b I_b} \qquad (18.44)$$

As $l_a I_a = l_b I_b$, length l_a can be expressed in terms of l_b and a quadratic equation in l_b can be obtained.

– One set of values given by the quadratic equation gives the position of two nodes and the frequency thus obtained is known as two-node frequency.

– In the other set of values, one gives the position of a single node and the other is beyond the physical limits of the equation. The frequency so obtained is known as single-node frequency or fundamental frequency.

If A and C rotate in the same direction and B rotates in the opposite direction, a single node is obtained between B and C (Fig. 18.27 c). (l_a does not give the actual node point.)

Similarly, if B and C rotate in the same direction and A in the opposite direction, single node is obtained between A and C (Fig. 18.27 d). (l_b does not give the actual node point.)

Let A_a, A_b and A_c be the amplitudes of rotors A, B and C respectively. Then

$$\frac{A_a}{A_c} = \frac{l_a}{l_{c1}} \quad \text{and} \quad \frac{A_b}{A_c} = \frac{l_b}{l_{c2}}$$

In general, possible number of node points and frequencies is one less than the number of rotors in torsional vibrating system.

18.21 TORSIONALLY EQUIVALENT SHAFT

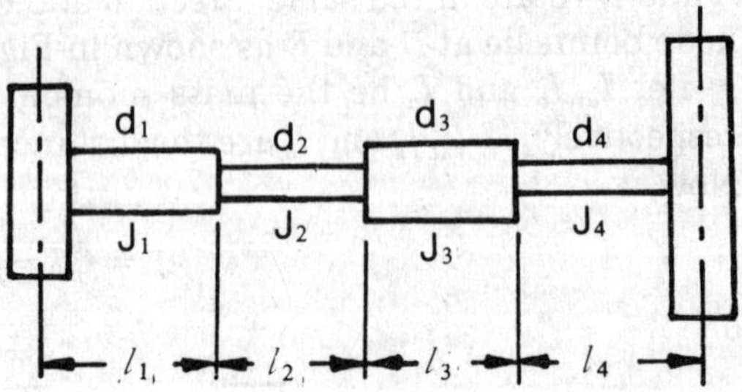

Fig. 18.28

Sometimes, rotors are fixed to a shaft of various diameters at different sections. The most convenient manner of finding the frequency of such a system is by replacing the shaft with a torsionally equavalent shaft having a suitable diameter. A torsionally equivalent shaft is one which has the same torsional stiffness as that of the stepped shaft so that it twists to the same extent under a given torque as the stepped shaft would.

Let θ be the total twist in the shaft under an applied torque T, and $\theta_1, \theta_2, \theta_3$ and θ_4 be the twists in sections l_1, l_2, l_3 and l_4 respectively (Fig.18.28).

$$\theta = \theta_1 + \theta_2, + \theta_3 + \theta_4$$

$$= \frac{Tl_1}{GJ_1} + \frac{Tl_2}{GJ_2} + \frac{Tl_3}{GJ_3} + \frac{Tl_4}{GJ_4}$$

Let d and l be the diameter and the length respectively of the torsionally equivalent shaft.

Then

$$\frac{Tl}{GJ} = \frac{Tl_1}{GJ_1} + \frac{Tl_2}{GJ_2} + \frac{Tl_3}{GJ_3} + \frac{Tl_4}{GJ_4}$$

or

$$\frac{l}{J} = \frac{l_1}{J_1} + \frac{l_2}{J_2} + \frac{l_3}{J_3} + \frac{l_4}{J_4}$$

or

$$\frac{l}{\frac{\pi}{32} d^4} = \frac{l_1}{\frac{\pi}{32} d_1^4} + \frac{l_2}{\frac{\pi}{32} d_2^4} + \frac{l_3}{\frac{\pi}{32} d_3^4} + \frac{l_4}{\frac{\pi}{32} d_4^4}$$

or

$$\frac{l}{d^4} = \frac{l_1}{d_1^4} + \frac{l_2}{d_2^4} + \frac{l_3}{d_3^4} + \frac{l_4}{d_4^4}$$

or

$$l = l_1 \left(\frac{d}{d_1}\right)^4 + l_2 \left(\frac{d}{d_2}\right)^4 + l_3 \left(\frac{d}{d_3}\right)^4 + l_4 \left(\frac{d}{d_4}\right)^4$$

The diameter d is usually chosen as one of the existing diameters of the stepped shaft.

Example 18.25 The shaft shown in Fig.18.29 carries two masses. The mass *A* is 300 kg with radius of gyration of 0.75 m and the mass B is 500 kg with radius of gyration of 0.9 m. Determine the frequency of the torsional vibrations. It is desired to have the node at the mid-section of the shaft of 120 mm diameter by changing the diameter of the section having a 90 mm diameter. What will be the new diameter?

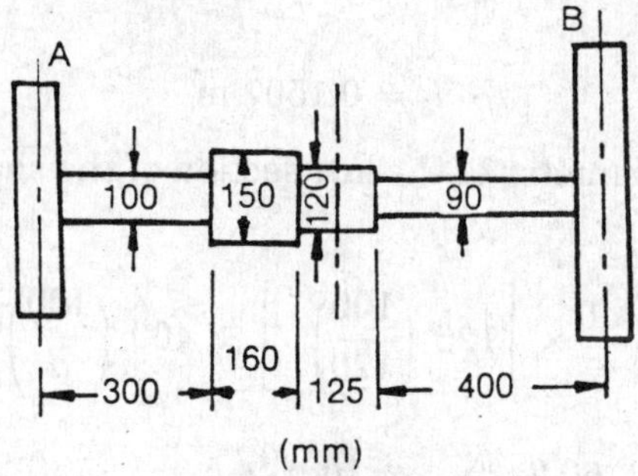

Fig. 18.29

$$G = 84 \times 10^9 \text{ N/m}^2.$$

Solution: (a) Reducing the shaft to torsionally equivalent shaft of 100 mm diameter.

$$l = 300\left(\frac{100}{100}\right)^4 + 160\left(\frac{100}{150}\right)^4 + 125\left(\frac{100}{120}\right)^4 + 400\left(\frac{100}{90}\right)^4$$

$$= 300 + 31.6 + 60.2 + 609.7$$

$$= 1001.5 \text{ mm or } 1.0015 \text{ m}$$

To locate the node point,

$$I_a\, l_a = I_b\, l_b$$

or

$$m_a\, k_a^2\, l_a = m_b\, k_b^2\, l_b$$

or

$$300 \times (0.75)^2\, l_a = 500 \times (0.9)^2\, (1.0015 - l_a)$$

or

$$l_a = 0.707 \text{ m}$$

$$f_n = \frac{1}{2\pi}\sqrt{\frac{GJ}{I_a\, l_a}} = \frac{1}{2\pi}\sqrt{\frac{84 \times 10^9 \times \frac{\pi}{32}(0.1)^4}{300 \times (0.75)^2 \times 0.707}} = \underline{13.2 \text{ Hz}}$$

(b) If the node point is to be at the mid section of the shaft with diameter 120 mm.

$$l_a = 300 + 31.6 + \frac{60.2}{2} = 361.7 \text{ mm} \quad \text{or} \quad 0.3617 \text{ m}$$

$$m_a k_a^2 l_a = m_b k_b^2 l_b$$

$$300 \times (0.75)^2 \times 0.3617 = 500 \times (0.9)^2 \times l_b$$

or

$$l_b = 0.1507 \text{ m}$$

Let d be the new diameter of the last section of the shaft. Then

$$l_b = \frac{1}{2} \times \left[125 \left(\frac{100}{120}\right)^4\right] + 400 \left(\frac{100}{d}\right)^4$$

or

$$150.7 = \frac{60.2}{2} + 400 \left(\frac{100}{d}\right)^4$$

$$d = 135 \text{ mm}$$

Example 18.26 A torsional system is shown in Fig. 18.30 a. Find the frequencies of torsional vibrations and the positions of the nodes. Also find the amplitudes of vibrations.

$$G = 84 \times 10^9 \text{ N/m}^2.$$

Solution: Let the engine, propeller and the flywheel are represented by the rotors A, B and C respectively. Reducing the system to torsionally equivalent shaft of 40 mm diameter,

$$l_1 = 2 \left(\frac{40}{45}\right)^4 = 1.247 \text{ m}$$

$$I_a l_a = I_b l_b$$

$$30 l_a = 50 l_b \quad \text{or} \quad l_b = 0.6 l_a$$

Also,

$$\frac{1}{I_a l_a} = \frac{1}{I_c} \left(\frac{1}{l_1 - l_a} + \frac{1}{l_2 - l_b}\right)$$

$$\frac{1}{30\, l_a} = \frac{1}{90} \left(\frac{1}{1.247 - l_a} + \frac{1}{4 - 0.6 l_a}\right)$$

On solving,
$l_a = 4.821$ m and 0.913 m
$l_b = 2.893$ m and 0.548 m
For two-node frequency (Fig.18.30c),
$l_a = 0.913$ m and $l_b = 0.548$ m

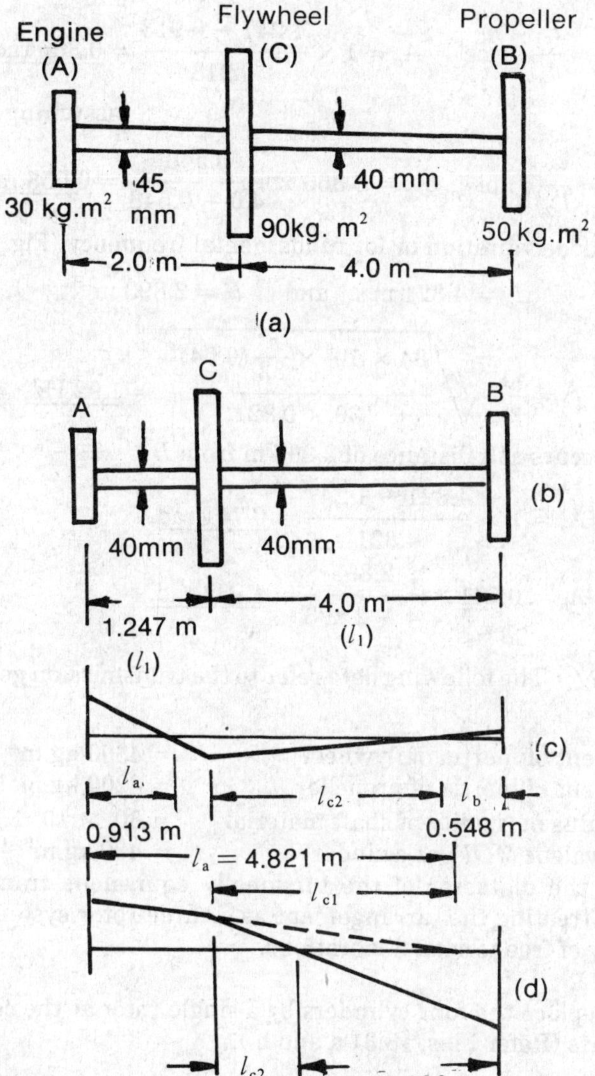

Fig. 18.30

However,

actual, $l_a = 0.913 \times \left(\frac{45}{40}\right)^4 = 1.462$ m

$$f_n = \frac{1}{2\pi}\sqrt{\frac{GJ}{I_a l_a}} = \frac{1}{2\pi}\sqrt{\frac{84 \times 10^9 \times \frac{\pi}{32}(0.04)^4}{30 \times 0.913}} = \underline{4.41 \text{ Hz}}$$

Amplitudes:

$$\frac{A_c}{A_a} = \frac{l_{c1}}{l_a} \quad \text{or} \quad A_c = 1 \times \frac{1.247 - 0.913}{0.913} = \underline{0.366 \text{ rad}}$$

(assuming $A_a = 1$ rad.)

$$\frac{A_b}{A_c} = \frac{l_b}{l_{c2}} \quad \text{or} \quad A_b = 0.366 \times \frac{0.548}{4.0 - 0.548} = \underline{0.058 \text{ rad}}$$

For single-node vibration or for fundamental frequency (Fig. 18.30 d)

$$l_a = 4.821 \text{ m} \quad \text{and} \quad l_b = 2.893 \text{ m}$$

$$f_n = \frac{1}{2\pi}\sqrt{\frac{84 \times 10^9 \times \frac{\pi}{32}(0.04)^4}{30 \times 0.821}} = \underline{1.92 \text{ Hz}}$$

The node occurs at a distance of 2.893 m from B.

$$A_c = 1 \times \frac{4.821 - 1.247}{4.821} = \underline{0.741 \text{ rad.}}$$

$$A_b = 0.741 \times \frac{2.893}{4 - 2.893} = \underline{1.937 \text{ rad.}}$$

Example 18.27 The following data refer to the transmission gear of a motor ship:

Moment of inertia of flywheel	$= 4800$ kg.m^2
Moment of inertia of propeller	$= 3200$ kg.m^2
Modulus of rigidity of shaft material	$= 80 \times 10^9$ N/m^2
Equivalent *MOI* per cylinder	$= 400$ kg.m^2

Assuming the diameter of the torsionally equivalent crankshaft to be 320 mm and treating the arrangement as a three-rotor system, determine the frequency of free torsional vibrations.

Solution: Replace the four cylinders by a single rotor at the centre of their combined mass (Refer Figs. 18.31 a and b).

The length of the torsionally equivalent shaft between the flywheel and the propeller,

$$l_2 = 4\left(\frac{320}{300}\right)^4 + 6\left(\frac{320}{280}\right)^4 + 5\left(\frac{320}{300}\right)^4$$

$$= 5.178 + 10.236 + 6.473$$

$$= 21.89 \text{ m}$$

$$I_a l_a = I_b l_b$$

$$(400 \times 4)\, l_a = 3200\, l_b$$

$$l_a = 2 l_b$$

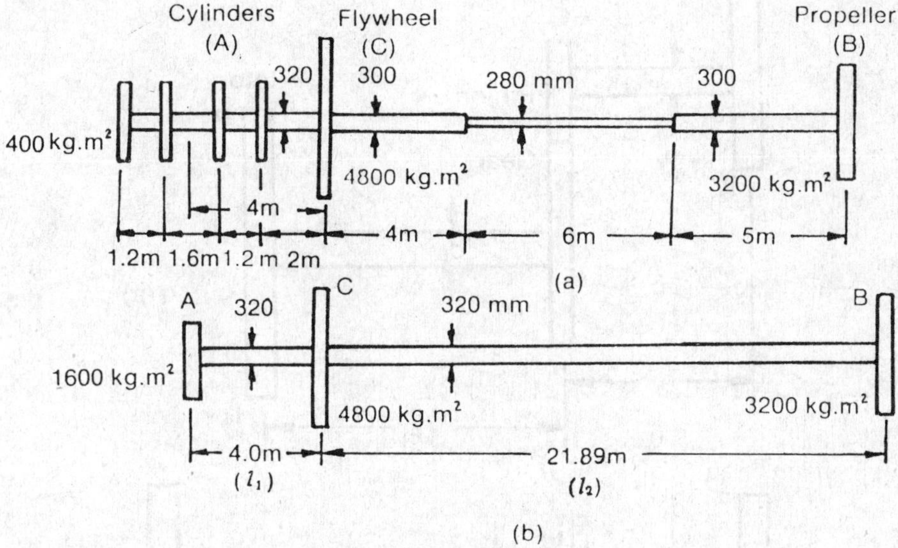

Fig. 18.31

$$\frac{1}{I_b\, l_b} = \frac{1}{I_c}\left[\frac{1}{l_1 - l_a} + \frac{1}{l_2 - l_b}\right]$$

$$\frac{1}{3200\, l_b} = \frac{1}{4800}\left[\frac{1}{4 - 2l_b} + \frac{1}{21.89 - l_b}\right]$$

On solving,

$l_b = 14.76$ m and 1.48 m

$l_a = 29.52$ m and 2.96 m

$$f_{n1} = \frac{1}{2\pi}\sqrt{\frac{GJ}{I_b\, l_b}} = \frac{1}{2\pi}\sqrt{\frac{80\times 10^9 \times \frac{\pi}{32}(0.32)^4}{3200 \times 14.76}} = \underline{6.65 \text{ Hz}}$$

(Single-node)

$$f_{n2} = 6.65 \times \sqrt{\frac{14.76}{1.48}} = \underline{21 \text{ Hz}}$$

(two-node)

18.22 FREE TORSIONAL VIBRATIONS OF GEARED SYSTEM

Figure 18.32a shows a geared system. Shaft 1 carries a rotor *A* on one end and a pinion on the other (at *C*). Shaft 2 carries a gear meshing with the pinion at one end and a rotor *B* on the other. The system may be replaced by an equivalent shaft system. Assume that:

the inertia of the gears (at *C*) and shafts is negligible.
– the load is within elastic limits of gear teeth, i.e. they are rigid.
– no backlash or slip occurs in the gear drive.

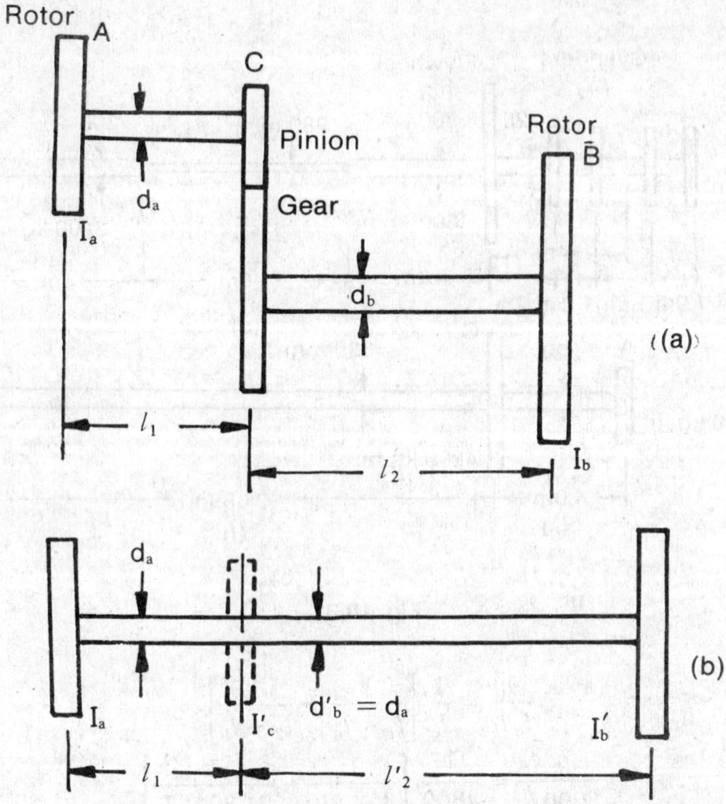

Fig. 18.32

With these assumptions, the system reduces to a two-rotor system as shown in Fig. 18.32 b.

The system 'b' will be equivalent to the system 'a' if

(i) the kinetic energy of system 'b' is equal to that of 'a' and

(ii) the strain energy of system 'b' is equal to that of 'a'.

Equating the kinetic energies

KE of original system = KE of section l_1 + KE of section l_2

KE of equivalent system = KE of section l_1 + KE of section l_2'

KE of section l_2' = KE of section l_2

$$\frac{1}{2} I'_b\, \omega'^2_b = \frac{1}{2} I_b\, \omega_b^2$$

$$\frac{1}{2} I'_b\, \omega_a^2 = \frac{1}{2} I_b\, \omega_b^2 \qquad (\omega'_b = \omega_a)$$

or

$$I'_b = I_b \left(\frac{\omega_b}{\omega_a}\right)^2 = \frac{I_b}{G_r^2} \qquad \left(G_r = \frac{\omega_a}{\omega_b}\right)$$

Equating the strain energies
SE of section $l_2' = SE$ of section l_2

$$\frac{1}{2} T_b' \theta_b' = \frac{1}{2} T_b \theta_b$$

$$\frac{J_b' G}{l_2'} \theta_b' \theta_b' = \frac{J_b G}{l_2} \theta_b \theta_b \qquad \left(\theta = \frac{Tl}{JG}\right)$$

$$l_2' = \left(\frac{\theta_b'}{\theta_b}\right)^2 l_2 \left(\frac{J_b'}{J_b}\right)$$

$$= \left(\frac{\omega_b'}{\omega_b}\right)^2 l_2 \left(\frac{d_b'}{d_b}\right)^4 \qquad (\theta = \omega\, t)$$

$$= \left(\frac{\omega_a}{\omega_b}\right)^2 l_2 \left(\frac{d_b'}{d_b}\right)^4 \qquad (\omega'_l = \omega_a)$$

$$= G_r^2\, l_2 \left(\frac{d_b'}{d_b}\right)^4 \qquad (18.45)$$

Assuming the diameter of the equivalent shaft to be equal to that of shaft 1,

$$l'_2 = G_r^2\, l_2 \left(\frac{d_a}{d_b}\right)^4$$

Length of equivalent shaft

$$= l_1 + G_r^2\, l_2 \left(\frac{d_a}{d_b}\right)^4 \qquad (18.46)$$

In case the inertia of the gearing is not negligible, an additional rotor has to be considered at a distance l_1 from the rotor A. The mass moment of inertia of the rotor is given by

$$I_c' = I_{ca} + \frac{I_{cb}}{G_r^2} \qquad (18.47)$$

Where
I_{ca} = *MOI* of gear
I_{cb} = *MOI* of pinion.
In this way, the system will act as a three-rotor system.

Example 18.28 A reciprocating *IC* engine is coupled to a centrifugal pump through a pair of gears. The shaft from the flywheel of the engine to the gear wheel has a 48 mm diameter and is 800 mm long. The shaft from the pinion to the pump has a 32 mm diameter and is 280 mm long. Pump speed is four times the engine speed. Moments of inertia of flywheel, gear-wheel, pinion and pump impeller are 1000 kg.m^2, 14 kg.m^2, 5 kg.m^2 and 18 kg.m^2 respectively. Find the natural frequency of the torsional oscillation of the system. G = 80 GN/m^2.

Solution: Let rotors A and B in Fig.18.32 represent the engine flywheel and the centrifugal pump respectively. As the pump speed is four times the engine speed (the pinion is supported on the pump shaft and the gear on the engine shaft), therefore,

$$G_r = \frac{1}{4} = 0.25$$

Also,

$$d_a = 48 \text{ mm} \qquad l_1 = 800 \text{ mm} \qquad I_a = 1000 \text{ kg.m}^2$$
$$d_b = 32 \text{ mm} \qquad l_2 = 280 \text{ mm} \qquad I_a = 18 \text{ kg.m}^2$$
$$I_{ca} = 14 \text{ kg.m}^2 \qquad I_{cb} = 5 \text{ kg.m}^2 \qquad G = 80 \text{ GN/m}^2$$

$$I'_b = \frac{I_b}{G_r^2} = \frac{18}{(0.25)^2} = 288 \text{ kg.m}^2$$

$$I'_c = I_{ca} + \frac{I_{cb}}{G_r^2} = 14 + \frac{5}{(0.25)^2} = 94 \text{ kg.m}^2$$

Thus the system, becomes a three-rotor one.

$$l'_2 = (0.25)^2 \times 280 \times \left(\frac{48}{32}\right)^4 = 88.6 \text{ mm}$$
$$= 0.0886 \text{ m}$$

Now,

$$I_a\, l_a = I'_b\, l'_b$$

or

$$1000\, l_a = 288\, l'_b$$

or

$$l'_b = 3.47\, l_a$$

Also

$$\frac{1}{I_a\, l_a} = \frac{1}{I'_c}\left[\frac{1}{l_1 - l_a} + \frac{1}{l'_2 - l'_b}\right]$$

$$\frac{1}{1000\, l_a} = \frac{1}{94}\left[\frac{1}{0.8 - l_a} + \frac{1}{0.0886 - 3.47\, l_a}\right]$$

$$l_a = 0.2354 \text{ m} \quad \text{or} \quad 0.0059 \text{ m}$$

Single-node frequency,

$$f_{n1} = \frac{1}{2\pi}\sqrt{\frac{GJ}{I_a\, l_a}} = \frac{1}{2\pi}\sqrt{\frac{80 \times 10^9 \times \frac{\pi}{32}(0.048)^4}{1000 \times 0.2354}} = \underline{2.12 \text{ Hz}}$$

Two-node frequency,

$$f_{n2} = 2.12 \times \sqrt{\frac{0.2354}{0.0059}} = \underline{13.4 \text{ Hz}}$$

PROBLEMS

18.1 The string shown in Fig. 18.33 is under tension T. Determine the natural frequency of vibration in a plane perpendicular to the string assuming that for small displacements T remains constant. Also prove that the period of vibration is maximum when $a = l/2$.

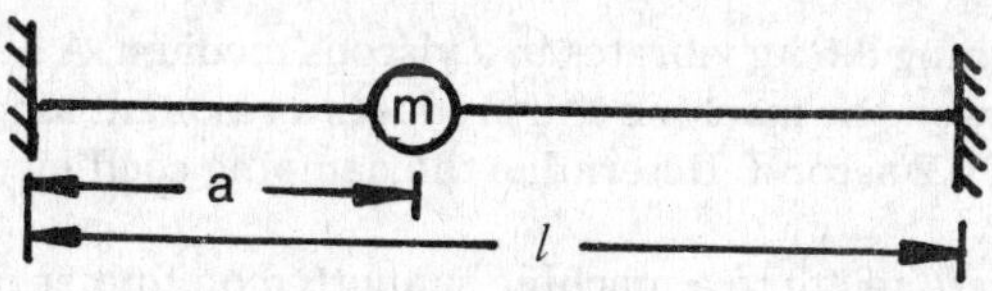

Fig. 18.33

$$\left(f_n = \frac{1}{2\pi}\sqrt{\frac{Tl}{m(l-a)\,a}} \right)$$

18.2 In a spring mass vibrating system, the natural frequency of vibration is reduced to half the value when a second spring is added to the first spring in series. Determine the stiffness of the second spring in terms of that of the first spring.

($s_2 = s_1/3$)

18.3 In a spring-mass vibrating system, the natural frequency of vibration is 3.56 Hz. When the amount of the suspended mass is increased by 5 kg, the natural frequency is lowered to 2.9 Hz. Determine the original unknown mass and the spring constant.

(10 kg; 5 N/mm)

18.4 A vibrating system consists of a mass of 20 kg, a spring of stiffness 20 kN/m and a damper. The damping provided is only 30% of the critical value. Determine the natural frequency of the damped vibration and the ratio of two consecutive amplitudes.

(30.2 rad/s ; 7.21)

18.5 The following data relate to a damped vibrating system:
$m = 140$ kg $s = 50$ N/mm and $\zeta = 0.25$
Determine the time in which the mass would settle down to 1/80th of its initial deflection. Also, what will be the number of oscillations complete to reach this value?

(0.927 s ; 2.7)

18.6 In a single-degree damped vibrating system, the suspended mass of 4 kg makes 24 oscillations in 20 seconds. The amplitude decreases to 0.3 of the initial value after 4 oscillations. Find the stiffness of the spring, the logarithmic decrement, the damping factor and damping coefficient.

(227 N/m; 0.3; 0.0478; 2.88 N/m/s)

18.7 A gun barrel weighs 300 kg and has a recoil spring of stiffness 250 N/mm. The barrel recoils 0.8 m on firing. Determine:

(i) the critical recoil velocity of the gun
(ii) the critical damping coefficient of the dashpot engaged at the end of the recoil stroke.

(23.1 m/s; 13.86 N/mm/s)

18.8 A spring mass system is excited by a force $F \sin \omega t$. On measuring, the amplitude of vibration is found to be 12 mm at resonance. However, at a frequency 0.8 times the resonant frequency, the amplitude reduces to 8 mm. Determine the damping ratio of the system.

(0.2)

18.9 An aircraft radio transmitter weighs 26 kg and is mounted on five springs which deflect 8 mm when the transmitter is placed upon them. Neglecting damping, find the percentage of engine vibration received by the transmitter for engine speeds of 1500 rpm and 2500 rpm.

(5.22%; 1.82%)

18.10 A machine weighing 3.5 kg vibrates in a viscous medium. A harmonic exciting force of 40 N acts on the machine and produces a resonant amplitude of 18 mm with a period of 0.2 second. Determine the damping coefficient.

(70.77 N/m/s)

18.11 The following data relate to a machine supported on four springs:
Mass of machine = 120 kg, stroke = 90 mm, mass of reciprocating parts = 2.5 kg and speed = 750 rpm.
Springs are symmetrically placed with respect to the centre of mass of the machine. Neglecting any damping, find the combined stiffness of the springs so that force transmitted to the foundation is 1/22 of the impressed force.
If under actual working conditions, the damping reduces the amplitude of the successive vibrations by 25% determine the forces transmitted to the foundation at 750 rpm and at resonance. Also find the amplitude of the vibrations at resonance.

(32.18 N/mm ; 34.42 N ; 305.62 N ; 9.5 mm)

18.12 A steel bar 22 mm wide and 45 mm deep is freely supported at two points 800 mm apart and carries a load of 180 kg midway between them. Determine the natural frequency of the transverse vibration, neglecting the weight of the bar. Also find the frequency of vibration if an additional load of 180 kg is distributed uniformly along the length of the shaft. Take E = 250 GN/m^2.

(23.5 Hz; 19.2 Hz)

18.13 A hollow shaft 1.8 m long is supported in flexible bearings at the ends. It carries two wheels each of 60 kg. mass, one at the centre of the shaft and the other at 450 mm from the centre. The external and internal diameters of the shaft are 80 mm and 50 mm respectively. Determine the lowest whirling speed of the shaft. The density of the shaft material is 7500 kg/m^3 and the modulus of elasticity 210 GN/m^2.

(169.4 rpm)

18.14 An equilateral triangular plate of 800 mm side is suspended by three strings. Each string is 2 m long and is connected at each corner of the plate. Find the frequency of oscillation and the time period.

(0.704 Hz; 1.42 s)

18.15 The following particulars relate to a three-cylinder oil engine coupled to a generator (Fig. 18.34).
Moment of inertia of engine flywheel = 600 kg.m^2
Moment of inertia of generator rotor = 120 kg.m^2

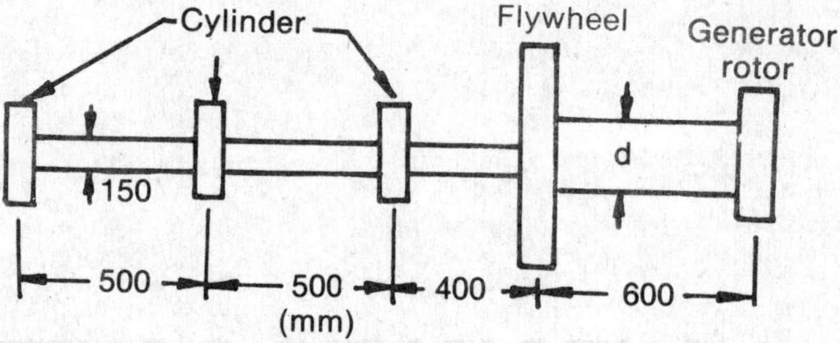

Fig. 18.34

Equivalent moment of inertia for each cylinder = 25 kg.m^2
Equivalent diameter of the crankshaft = 150 mm
Two-node frequency of torsional vibration = 50 Hz
Treating the arrangement as a three-rotor system and the modulus of rigidity for shaft material as 84 kN/mm^2, determine the shaft diameter d.

(162 mm)

18.16 A shaft 1.2 m long has diameter of 45 mm for half the length and 60 mm for the remaining length. One end of the shaft is fixed and the other carries a rotor of mass 200 kg with a radius of gyration of 45 mm. Find the frequency of free torsional vibration neglecting the inertia of the shaft. Take G = 84 GN/m^2.

(3.88 Hz)

18.17 Referring to Fig. 18.28, the moments of inertia of the left and right side rotors are 75 kg.m^2 and 50 kg.m^2 respectively. The lengths l_1, l_2, l_3 and l_4 are 300 mm, 400 mm, 100 mm and 260 mm and the diameters d_1, d_2, d_3 and d_4 are 150 mm, 100 mm, 190 mm and 130 mm respectively. Determine the frequency of natural torsional oscillation of the system. G = 85 GN/m^2.

(35.5 Hz)

18.18 A centrifugal pump rotating at 400 rpm is driven by an electric motor at 1200 rpm through a single stage reduction gearing. The moments of inertia of the pump impeller and the motor are 1500 kg.m^2 and 450 kg.m^2 respectively. The lengths of the pump shaft and the motor shaft are 500 and 200 mm, and their diameters are 100 and 50 mm respectively. Neglecting the inertia of the gears, find the frequency of torsional oscillations of the system. G = 85 GN/m^2.

(3.78 Hz)

19

AUTOMATIC CONTROL

19.1 INTRODUCTION

Physical systems or mechanisms are required to be adjusted or controlled so that they perform their specific duties. This is done either manually or automatically. Automatic control is desired in order to save the human operator from drudgery. It is also more efficient.

Self-actuated or automatic control of mechanisms is not a new thing. A centrifugally actuated ball governor which controls the throttle valve to maintain a constant speed of an engine is an example of an automatically controlled mechanism. In process industries where production is the outcome of continuous flow through consecutive treatments, automatic control ensures the uniformly and quality of products and reduces the time of machine watching and thus the wage bills. Examples are paper, chemical or foodstuff industries.

In control systems, the result of the act of adjustment, i.e. closing a valve, moving a lever, pressing a button, etc. is known as *command* and the subsequent result or behaviour of the system as *response. Process control* is the automatic control of variables, e.g., length, temperature, radius etc. that occur in industrial flow production.

19.2 OPEN AND CLOSED-LOOP CONTROL (UNMONITORED AND MONITORED CONTROL)

Control in which the input command is not influenced by the behaviour of the system response (output) is called an open-loop or unmonitored control. Examples are:

(i) Traffic control on the roads by lights where the timing mechanism is preset irrespective of the intensity of traffic.

(ii) Switching off the street lights of a town at a predetermined time by a time-switch irrespective of the fact that the sun rises at a different time each day.

(iii) Switching off an electric heater by a time-switch irrespective of whether the dish has been prepared or not.

A closed-loop control system is one in which the actual value of a controlled quality is measured and compared continuously with the desired value. The quantity measured may be load or input. In the traffic control system mentioned above, if the flow of traffic is measured either by direct human observation or by counting impulses due to the vehicles passing over a pressure pad, and then changing the time setting accordingly, it becomes a closed-loop control.

Another example of closed-loop control can be that of a water heater. The desired temperature may be maintained by an operator with the help of a thermometer and an electric rod fitted with an on/off switch. He will observe the thermometer continuously and will switch on the heater whenever the temperature of water rises above the desired temperature and switch it off when it falls below the same. Instead of the operator, a thermostat can also serve the purpose of switching on/off the heater thus making the system fully automatic.

A differential device used to measure the actual controlled quantity and to compare it continuously with the desired value is known as an *error detector* or *deviation sensor*.

Measuring the output for comparison with the input is known as *feedback*.

19.3 AMPLIFICATION

Usually, an error detector itself has insufficient power output to actuate the correcting mechanism directly. The error signal has to be amplified by using a gear-box, lever system or a hydraulic/pneumatic relay.

19.4 ACTUATOR (SERVOMOTOR)

An actuator is an external source of power connected to the input of the controlled machine and serves to reduce the error. Thus it is a device which produces a limited angular or linear motion and may be mechanical, hydraulic or electric. A servomotor is usually hydraulic or electric and has a continuous output.

19.5 TRANSDUCER

A transducer is a converting device that converts the measurement of the quantity to be controlled into different units, e.g. angular velocity into potential difference, pressure change into angular rotation as in a bourdon tube, temperature change into emf as in a thermocouple, linear strain into resistance change as in strain gauge, and so on. It is also usual to convert the error signal into an electric quantity due to the relative ease with which minute electric signals can be detected, transmitted, amplified and manipulated by using solid state electronic means. However, when reliability under adverse operating conditions (high temperature, radiation, dustful atmosphere, corrosiveness, etc.) is desired, pneumatic means should be preferred.

19.6 BLOCK DIAGRAMS

A block diagram is a symbolic outline of a system in which various components or operations are represented by rectangles in an ordered sequence. The rectangles are connected by arrows showing the flow of the working medium or of information.

Figure 19.1 shows the block diagram of an ordinary carburettor, and Fig. 19.2 a rocket launching system and its block diagram. Figure 19.3 shows the controlling system of a steam turbine and its block diagram.

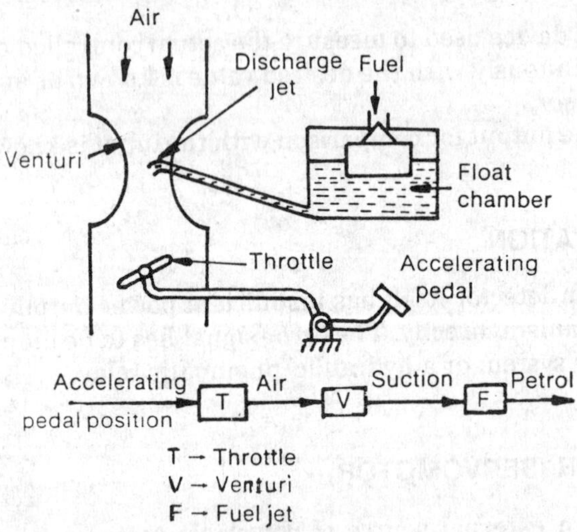

Fig. 19.1

19.7 LAG IN RESPONSE

In any system, usually, there is a delay in response (output) due to some inherent cause and it becomes difficult to measure the input and output simultaneously. In a shaft transmitting torque, there is an angular lag of one end of the shaft behind the other. Inertia delays a motor attaining its required velocity after the application of a torque. In a steam turbine, if the load is suddenly reduced, there will be some lag in the closing of the steam valve by the hydraulic relay as the first movement of the piston valve will not be sufficient to open the ports. This lag increases the probability of unstable operation.

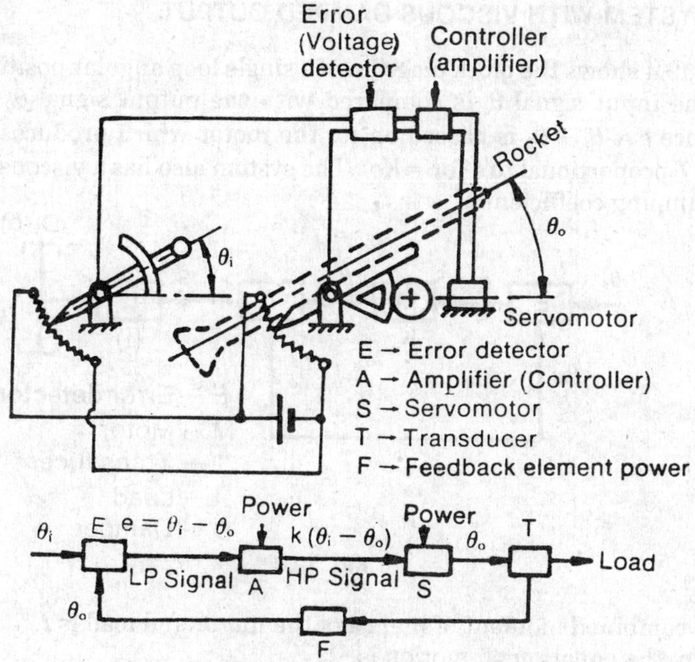

Fig. 19.2

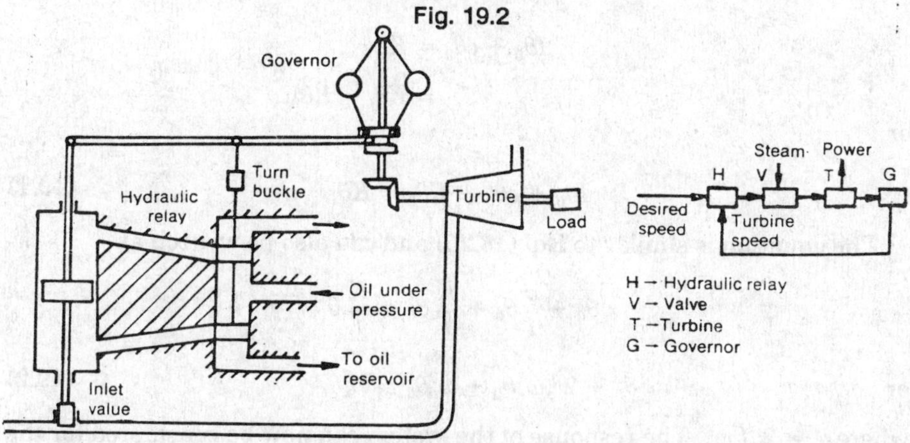

Fig. 19.3

19.8 DAMPING

When a torque is applied in a system in a direction opposite to its motion, it is known as damping. In case of Coulomb damping, the opposition is constant and thus there will be a constant difference (error) between the input and the output under steady conditions. In the viscous damping provided by a dash-pot, the opposition is proportional to the relative velocity. As the relative velocity is zero in the steady state, the damping is also zero.

19.9 SYSTEM WITH VISCOUS DAMPED OUTPUT

Figure 19.4 shows the block diagram of a single loop angular position system. First the input signal θ_i is compared with the output signal θ_0. Then the difference $e = \theta_i - \theta_0$ is passed on to the motor which produces an output torque T proportional to e (or $= Ke$). The system also has a viscous resistance with damping coefficient c.

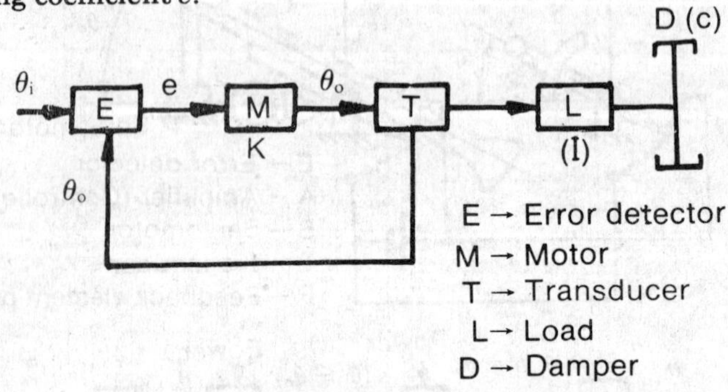

Fig. 19.4

The combined moment of inertia of the motor and load is I. Then, the equation of motion is

$$I\ddot{\theta}_0 + c\dot{\theta}_0 = Ke$$

$$= K\theta_i - K\theta_0$$

or

$$I\ddot{\theta}_0 + c\dot{\theta}_0 + K\theta_0 = K\theta_i \tag{19.1}$$

The equation is similar to Eq. (18.25) and can also be written as

$$\ddot{\theta}_0 + \frac{c}{I}\dot{\theta}_0 + \frac{K}{I}\theta_0 = \frac{K}{I}\theta_i$$

or

$$\ddot{\theta}_0 + 2\,\zeta\omega_n\dot{\theta}_0 + \omega_n^2\theta_0 = \omega_n^2\theta_i \tag{19.2}$$

where $c = 2\zeta I\,\omega_n$. The response of the system can now be considered for the following types of input:

(i) Unit step displacement,
(ii) Step velocity (ramp displacement), and
(iii) Harmonic.

19.10 UNIT STEP DISPLACEMENT INPUT

Unit step displacement input is one in which $\theta_i = 1$ for $t > 0$, i.e. the input signal is made at an angle θ_i from the zero position and is held in the new position.

Therefore, the equation becomes,

$$\ddot{\theta}_0 + 2\zeta\,\omega_n\,\dot{\theta}_0 + \omega_n^2\,\theta_0 = \omega_n^2 \tag{19.3}$$

The full response of the system will be the sum of transient and steady-state values, i.e. the sum of complementary and particular solutions.

The complementary function is the solution of the equation,

$$\ddot{\theta}_0 + 2\,\zeta\,\omega_n\,\dot{\theta}_0 + \omega_n^2\,\theta_0 = 0$$

This is similar to Eq. (18.18), and the solution is;

$$CF = Xe^{-\zeta\,\omega_n^{\,t}}\sin(\omega_d\,t + \varphi) \qquad \text{[Refer Eq. (18.21)]}$$

The particular solution is given by,

$$PS = \frac{\omega_n^2}{D^2 + 2\zeta\,\omega_n D + \omega_n^2}$$

In the steady-state, frequency of oscillation of the output becomes equal to that of the input.

$$\dot{\theta}_0 = \dot{\theta}_i = 0 \qquad \left(\because \theta_i = 1 = \text{constant}\right.$$

and

$$\ddot{\theta}_0 = 0 \qquad \left.\therefore\ \dot{\theta}_i = 0\right)$$

$$\text{PS} = \frac{\omega_n^2}{\omega_n^2} = 1$$

Thus, the complete solution is

$$\theta_0 = 1 + Xe^{-\zeta\,\omega_n t}\sin\,(\omega_d\,t + \varphi) \tag{19.4}$$

The constants X and φ can be found from initial conditions.

When $t = 0$, $\theta_0 = 0$, and

$0 = 1 + X\sin\varphi$ or $\sin\varphi = -\dfrac{1}{X}$

When $t = 0$, $\dot{\theta}_0 = 0$, and

$$0 = X\,[e^{-\zeta\,\omega_n t}\,\omega_d\cos\,(\omega_d\,t + \varphi) + \sin(\omega_d\,t + \varphi)\,(-\zeta\omega_n)e^{-\zeta\omega_n t}]$$

or

$$0 = \omega_d\cos\varphi - \zeta\,\omega_n\sin\varphi$$

or

$$\tan\varphi = \frac{\omega_d}{\zeta\,\omega_n}$$

Thus, in the right-angle triangle ABC (Fig. 19.5),

$$AB = \omega_d,\ BC = \zeta\,\omega_n,$$

$$AC = \sqrt{(\zeta\omega_n)^2 + \omega_d^2}$$

$$= \sqrt{\zeta^2\,\omega_n^2 + (1-\zeta^2)\,\omega_n^2}$$

$$= \omega_n$$

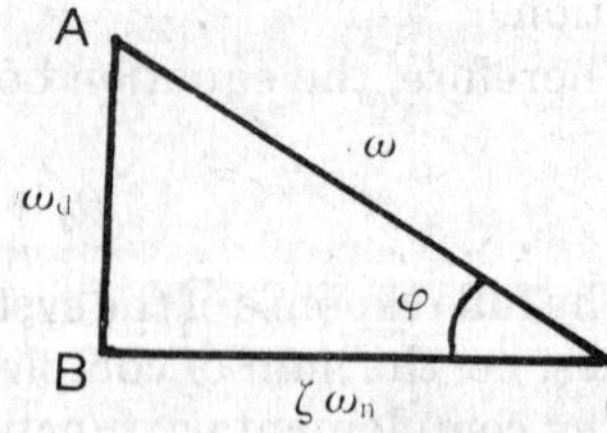

Fig. 19.5

$$X = -\frac{1}{\sin\varphi} = -\frac{1}{\omega_d/\omega_n} = -\frac{\omega_n}{\omega_d}$$

$$\therefore\ \theta_0 = 1 - \frac{\omega_n}{\omega_d}\,e^{-\zeta\,\omega_n t}\,\sin\left[\omega_d t + \tan^{-1}\frac{\omega_d}{\zeta\,\omega_n}\right]$$

or

$$\theta_0 = 1 - \frac{1}{\sqrt{1-\zeta^2}}\,e^{-\zeta\,\omega_n t}\,\sin\left[\sqrt{1-\zeta^2}\,\omega_n t + \tan^{-1}\frac{\sqrt{1-\zeta^2}}{\zeta}\right] \qquad (19.5)$$

The plots of θ_0 vs. t have been shown in Fig. 19.6 for various values of damping coefficient ζ. It can be observed that;

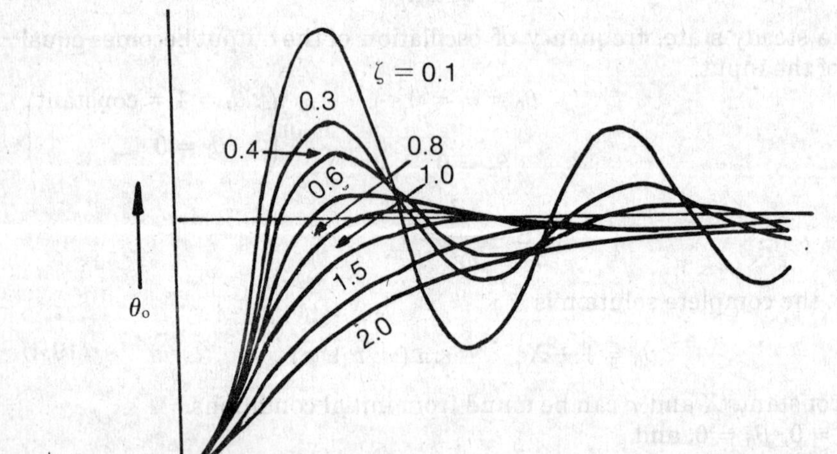

Fig. 19.6

(i) The amount of damping influences the magnitude or amplitude of overshoot.

(ii) If damping (ζ) is increased, the frequency of oscillation is reduced.

(iii) The quickest response with no overshoot is at critical damping ($\zeta = 1$).

(iv) For $\zeta = 0.6$ (approximately), the response is even faster with very little overshoot.

(v) There is no steady-state lag (error).

(vi) To reach the steady state as quickly as possible, ω_n should be high or K as large as possible and I as small as possible.

19.11 STEP VELOCITY (RAMP DISPLACEMENT) INPUT

When the system is given a uniform angular velocity rather than an angular dispalcement, it is known as step function input. Thus we have

$\theta_i = \omega_i t$ or $\dot{\theta}_i = \omega_i$

The equation of motion becomes

$$\ddot{\theta}_0 + 2\zeta\,\omega_n\,\dot{\theta}_0 + \omega_n^2\,\theta_0 = \omega_n^2 \omega_i t \tag{19.6}$$

The transient solution is the same as for the unit step displacement input. The steady-state or particular solution is

$$\theta_{ss} = \frac{\omega_n^2\,\omega_i t}{D^2 + 2\zeta\omega_n D + \omega_n^2}$$

In the steady-state,

$$\dot{\theta}_0 = \dot{\theta}_i = \omega_i$$

and

$$\ddot{\theta}_0 = 0$$

$$\theta_{ss} = \frac{\omega_n^2\,\omega_i t}{2\zeta\,\omega_n D + \omega_n^2}$$

$$= \frac{\omega_i t}{1 + \dfrac{2\zeta}{\omega_n} D}$$

$$= \left(1 + \frac{2\zeta}{\omega_n} D\right)^{-1} \omega_i\,t$$

$$= \left[1 \frac{2\zeta}{\omega_n} D + \frac{(-1)(-2)}{2} \left(\frac{2\zeta}{\omega_n}\right)^2 D^2 + \ldots\right] \omega_i\,t$$

$$= \omega_i t - \frac{2\zeta}{\omega_n}\,\omega_i$$

$$= \theta_i - \frac{2\zeta}{\omega_n}\,\omega_i \tag{19.7}$$

Thus the steady-state speeds are equal $(=\omega_n)$ or there is no error in velocity, but the output θ_0 lags the input by $(2\zeta/\omega_n)\,\omega_i$.

This lag increases with the increase in damping and the velocity.

The plot of θ_0 vs. t has been shown in Fig. 19.7.

Steady-state error,

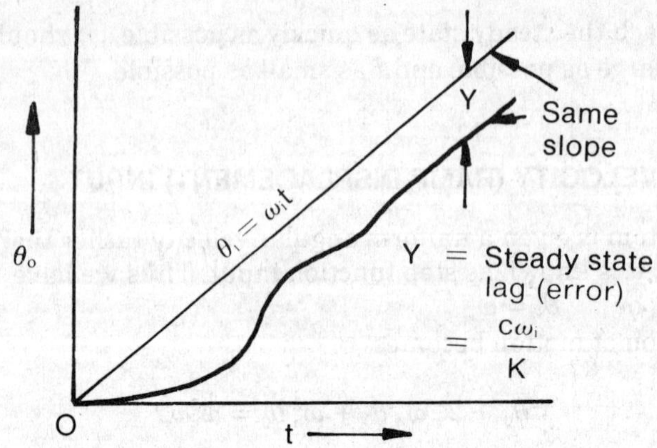

Fig. 19.7

$$\theta_i - \theta_0 = \frac{2\zeta\omega_i}{\omega_n} = \frac{2c}{c_c}\frac{\omega_i}{\sqrt{K/I}} = \frac{2c\omega_i}{2\sqrt{IK}\sqrt{K/I}} = \frac{c\omega_i}{K}$$

19.12 HARMONIC INPUT

For a sinusoidal or harmonic input,

$\theta_i = \theta_0 \sin \omega t$

The equation of motion becomes

$$\ddot{\theta} + 2\zeta\omega_n \dot{\theta} + \omega_n^2\theta = \omega_n^2 \theta_0 \sin \omega t \qquad (19.8)$$

The equation is exactly similar to Eq. (18.25) and therefore has a similar solution. The response is harmonic with a steady-state lag.

19.13 PERFORMANCE IMPROVEMENT

From the discussion of the previous section, it follows that the main defects in the response of a single-loop control system are the large overshoot in the transient motion, dynamic lags in the steady-state motion and the effects due to load disturbances.

The quantities which can be varied in the system shown in Fig. 19.4 are:

– Inertia of load and motor
– Damping
– Gain or proportional constant K.

As the inertia of the load and motor is usually kept small, the variation can be made in the damping and proportional constant.

If damping is increased, the overshoot in transient motion is reduced but steady-state lag is increased. If gain is increased, the lag is reduced but the transient error is increased. Thus it is clear that some sort of compensation is necessary for keeping the proportional gain high. The usual methods are by using the derivative or integral of the error with the existing proportional control.

19.14 STEP VELOCITY INPUT WITH ERROR RATE DAMPING (DERIVATIVE CONTROL)

Let the controller provide a proportional plus a derivative action of the error. Therefore, the equation will be

$$I\ddot{\theta}_0 + c\dot{\theta}_0 = Ke + K_1\dot{e} \qquad (19.9)$$

$$= (K\theta_i - K\theta_0) + (K_1\dot{\theta}_i - K_i\dot{\theta}_0)$$

or

$$I\ddot{\theta}_0 + (c + K_1)\dot{\theta}_0 + K\theta_0 = K\theta_i + K_1\dot{\theta}_i \qquad (19.10)$$

or

$$\ddot{\theta}_0 + \left(\frac{c + K_1}{I}\right)\dot{\theta}_0 + \left(\frac{K}{I}\right)\theta_0 = \left(\frac{K}{I}\right)\theta_i + \left(\frac{K_1}{I}\right)\dot{\theta}_i$$

Comparing it with the equation

$$\ddot{\theta}_0 + 2\zeta\omega_n\dot{\theta}_0 + \omega_n^2\theta_0 = \omega_n^2\omega_i t$$

ω_n is unchanged and is equal to $\sqrt{\dfrac{K}{I}}$.

$$2\zeta\omega_n = \frac{c + K_l}{I}$$

or

$$\zeta = \frac{c + K_1}{2I\omega_n} = \frac{c + K_1}{2I\sqrt{K/I}} = \frac{c + K_1}{2\sqrt{KI}} = \frac{c + K_1}{c_c}$$

Also, in the steady-state,

$$\dot{\theta}_i = \dot{\theta}_0 = \omega_i \quad \text{and} \quad \ddot{\theta}_0 = 0$$

$$\therefore \quad (\theta_i - \theta_0)\frac{K}{I} = \omega_i\left(\frac{c + K_1 - K_1}{I}\right) = \frac{c\omega_i}{I}$$

And steady-state error

$$\theta_i - \theta_0 = \frac{c\omega_i}{K} \qquad (19.11)$$

Thus, there is no change in the steady-state error (lag) by using a derivative control. It is the same as was found to be without using the derivative control (Sec. 19.11). However, now the damping ratio is increased from c/c_c to $(c + K_1)/c_c$. This improves the transient response (overshoot) without the use of excesive damping torque on the load. In other words, for the same value of overshoot, the value of c can be reduced, and this can reduce the steady-state lag correspondingly.

19.15 STEP VELOCITY INPUT WITH INTEGRAL CONTROL

The equation of motion will be

$$I\ddot{\theta}_0 + c\dot{\theta}_0 = Ke + K_1 \int_0^t e\,dt$$

$$= K\theta_i - K_1\theta_0 + K_i \int_0^t \theta_i dt - K_1 \int_0^t \theta_0\,dt$$

Differentiating to eliminate the integral sign,

$$I\dddot{\theta}_0 + c\ddot{\theta}_0 + K\dot{\theta}_0 + K_1\theta_0 = K\dot{\theta}_i + K_1\theta_i \qquad (19.12)$$

In the steady-state,

$$\dot{\theta}_0 = \dot{\theta}_i = \omega_i \quad \text{and} \quad \ddot{\theta}_0 = \theta\ldots_0 = 0$$

$$\therefore \qquad K\omega_i + K_1\,\theta_0 = K\theta_i + K_1\omega_i$$

or

$$\theta_0 = \theta_i$$

Therefore, the positional lag or steady-state error is zero. The transient solution depends upon the value of K_1. If K_1 is small, the time taken by the transients to die out may be too large, though peak values will not be different from the ones without integral control. If K_1 large, so may be the overshoots, and the system may become unstable.

19.16 TRANSFER FUNCTION

In control systems, the relationship between the input and the output (response) is given by a differential equation of motion. If the differential equation is expressed in symbolic form by substituting D for d/dt or by the Laplace transformation, the transfer function is the operational relationship of the output and the input.

Rewriting Eq. (19.2) using the operator D,

$$(D^2 + 2\,\zeta\,\omega_n\,D + \omega_n^2)\,\theta_0 = \omega_n^2\,\theta_i$$

Transfer function,

$$\frac{\theta_0}{\theta_i} = \frac{\omega_n^2}{D^2 + 2\,\zeta\omega_n + \omega_n^2}$$

Consider a Hartnell governor as shown in Fig. 16.6 a. Assume that the load on the sleeve, the weight of the balls and the friction force are negligible as compared to the inertia forces. The sleeve of the governor is also provided with some viscous damping having a damping coefficient c.

Further assume that in the equilibrium position, when the arms holding the balls are vertical, the compression of the spring is x_0 at the equilibrium speed ω_0. Then

$$mr\,\omega_0^2\,a = \frac{1}{2}\,s\,x_0\,b$$

Now, if due to a small change ω in speed, the displacement of the sleeve is x, then the forces acting on the bell-crank lever will be (refer Fig. 19.8),

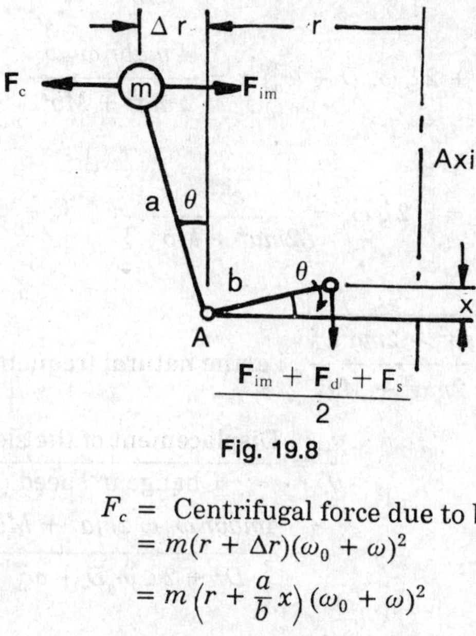

Fig. 19.8

$$\begin{aligned} F_c &= \text{Centrifugal force due to ball mass} \\ &= m(r + \Delta r)(\omega_0 + \omega)^2 \\ &= m\left(r + \frac{a}{b}x\right)(\omega_0 + \omega)^2 \end{aligned}$$

F_{im} = inertia force of the balls $= m\left(\frac{a}{b}\ddot{x}\right)$

F_{iM} = inertia force of the sleeve mass $= M\ddot{x}$

F_d = Damping force $= c\dot{x}$

F_s = spring force $= s(x_0 + x)$

Taking moments about the fulcrum A (taking only one half of the governor into consideration).

$$m\left(r + \frac{a}{b}x\right)(\omega_0 + \omega)^2\,a = m\frac{a}{b}\,\ddot{x}a + \frac{1}{2}\,M\ddot{x}b + \frac{1}{2}\,c\dot{x}\,b + \frac{1}{2}\,s(x_0 + x)b$$

Neglecting second order small terms,

$$mr\omega_0^2 a + m\frac{a}{b}x\omega_0^2 a + 2mr\,\omega\,\omega_0\,a = m\frac{a^2}{b}\ddot{x} + \frac{1}{2}M\ddot{x}b + \frac{1}{2}c\dot{x}\,b + \frac{1}{2}s(x_0+x)b$$

But
$$mr\,\omega_0^2\,a = \frac{1}{2}s\,x_0\,b$$

$$\therefore \quad \left(\frac{ma^2}{b} + \frac{Mb}{2}\right)\ddot{x} + \frac{1}{2}cb\dot{x} + \left(\frac{1}{2}sb - \frac{ma^2}{b}\omega_0^2\right)x = 2mr\,\omega_0\,\omega\,a$$

Multiplying throughout by $2b$ and using operator D,

$$2ma^2 + Mb^2)\,D^2\,x + cb^2DX + (sb^2 - 2ma^2\omega_0^2)x = 2m\,a.b.r.\omega_0\omega$$

or

$$\left(D^2 + \frac{cb^2}{2ma^2 + Mb^2}D + \frac{sb^2 - 2ma^2\omega_0^2}{2ma^2 + Mb^2}\right)x = \frac{4mabr\,\omega_0\,\omega}{2ma^2 + Mb^2}$$

or

$$\left(D^2 + 2\,\zeta\,\omega_n\,D + \omega_n^2\right)x = \frac{4\,mabr\,\omega_0\,\omega}{2\,ma^2 + Mb^2}$$

where

$$2\,\zeta\,\omega_n = \frac{cb^2}{2ma^2 + Mb^2} \qquad (\zeta = \text{damping factor})$$

and

$$\omega_n = \sqrt{\frac{sb^2 - 2ma^2\omega_0^2}{2ma^2 + Mb^2}}, \text{ i.e. the natural frequency}$$

$$\text{Transfer function,} \quad \frac{\theta_0}{\theta_i} = \frac{\text{Displacement of the sleeve } (x)}{\text{Change in speed } (\omega)}$$

$$= \frac{4mabr\,\omega_0\,\omega\;2ma^2 + Mb^2}{D^2 + 2\,\zeta\,\omega_n D + \omega_n^2}$$

19.17 TRANSFER FUNCTION RELATIONSHIPS

A control system can have several loops and components, each having its characteristic transfer function.

(i) Open-loop Transfer Function An open-loop or forward-loop control system has several components having individual transfer functions such as $F_1(D)$, $F_2(D)$, $F_3(D)$, etc. as shown in Fig. 19.9.

$$\text{TF} = \frac{\theta_0}{\theta_i} = \frac{\theta_0}{\theta_3}\frac{\theta_3}{\theta_2}\frac{\theta_2}{\theta_1}\frac{\theta_1}{\theta_i} = F_4(D)\,F_3(D)\,F_2(D)\,F_1(D) = KG(D)$$

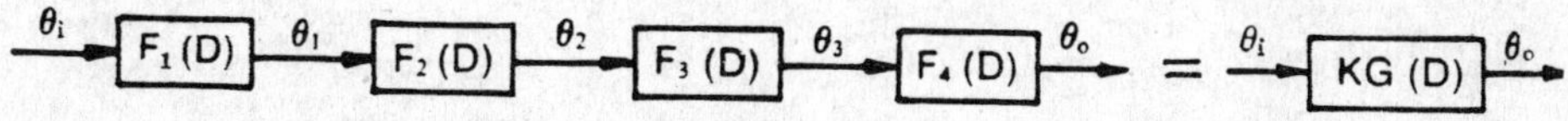

Fig. 19.9

(ii) Closed-loop Transfer Function A closed-loop or feedback loop is shown in Fig. 19.10.

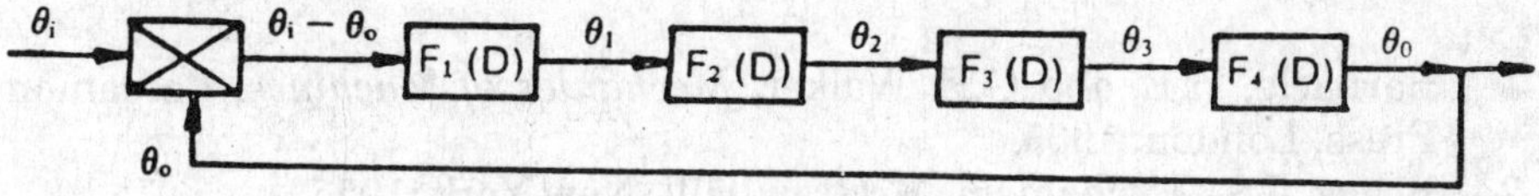

Fig. 19.10

$$\frac{\theta_0}{\theta_i - \theta_0} = KG(D)$$

or $$\theta_0 = KG(D)\theta_i - KG(D)\theta_0$$

or $$[1 + KG(D)]\theta_0 = KG(D)\theta_i$$

$$\therefore \quad \text{TF} = \frac{\theta_0}{\theta_i} = \frac{KG(D)}{1 + KG(D)} = \frac{\text{Open-loop TF}}{1 + \text{Open-loop TF}}$$

BIBLIOGRAPHY

1. Barnacle, H.E. and G.E. Walker, *Mechanics of Machines*, Pergamon Press, London, 1965.
2. Beggs, J.S., *Mechanism*, McGraw-Hill, New York, 1955.
3. Black, P., *Mechanics of Machines*, Pergamon Press, London, 1967.
4. Church, A.H., *Mechanical Vibrations*, John Wiley and Sons, New York, 1963.
5. Doughtie, V.L. and W.L. James, *Elements of Mechanism*, John Wiley and Sons, New York, 1954.
6. Doughtie, V.L. and A. Vallance, *Design of Machine Members*, McGraw-Hill, New York, 1964.
7. Dyson, F., *Principles of Mechanisms*, Oxford University Press, Oxford, 1964.
8. Erdman, A.G. and G.N. Sandoor, *Mechanism Design*, Prentice-Hall, 1988.
9. Fairs, V.M. and R.M. Keown, *Mechanism*, McGraw-Hill, New York, 1960.
10. Green, W.G., *Theory of Machines*, Blackie and Son (India) Ltd., 1963.
11. Ham, C.W., E.J. Crane, and W.L. Rogers, *Mechanics of Machinery*, McGraw-Hill, New York, 1958.
12. Hannah, J. and R.C. Stephens, *Mechanics of Machines*, Edward Arnold Ltd., London, 1964.
13. Harland Billings, J., *Applied Kinematics*, Affiliated East-West Press, New Delhi, 1969.
14. Harrisberger, L., *Mechanization of Motion*, John Wiley and Sons, New York, 1961.
15. Howard, P.J., *Theory of Machines*, Macdonald, London, 1966.
16. Lent, D., *Analysis and Design of Mechanisms*, Prentice-Hall, 1961.
17. Levitskii, N.I., *Analysis and Synthesis of Mechanisms*, Amerind Publishing Co., New Delhi, 1975.
18. Low, B.B., *Theory of Machines*, Longmans, Green and Co., 1964.
19. Marbie, H.H. and F.W. Ocvirk, *Dynamics of Machinery*, John Wiley and Sons, 1963.
20. Morrison, J.L.M. and B. Crossland, *Mechanics of Machines*, ELBS and Longmans, Green and Co., 1966.
21. Pujara, K., *Vibrations and Noise*, Dhanpat Rai and Sons, New Delhi, 1977.

22. Ruzinov, L.D., *Design of Mechanisms by Geometric Transformations*, London Iliffe Books Ltd., 1968.
23. Ryder, G.R., *Theory of Machines through Worked Examples*, B.I. Publications, 1961.
24. Shigley, J.E., *Mechanical Engineering Design*, McGraw-Hill, 1986.
25. Soni, A.H., *Mechanism Synthesis and Analysis*, McGraw-Hill, 1974.
26. Thomson, W.T., *Theory of Vibrations with Applications*, Prentice-Hall of India, New Delhi, 1975.
27. Wilson, C.E., J.P. Sadler, and W.J. Michels, *Kinematics and Dynamics of Machinery*, Harper and Row Publishers, New York, 1983.
28. Zimmerman, J.R., *Elementary Kinematics of Mechanisms*, John Wiley and Sons, 1962.

INDEX

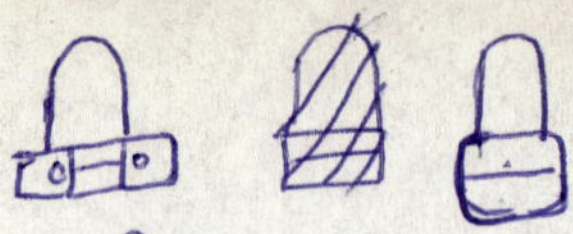

518 - $\frac{5200,~~00~~}{518}$ = 10 + 150 tonnes

1

3